THE HALF-MULTIPLIER OPERATOR,
NESTED ARRAYS,
MATRIX PROGRAMMING,
AND
THE UNIFIELD EQUATION

THE HALF-MULTIPLIER OPERATOR,

$$[A]\mathtt{Ti}j \; o \; [B]jk = [C]JK$$

NESTED ARRAYS,

$$\begin{pmatrix} A_{11} & A_{21} & A_{31} \\ A_{12} & A_{22} & A_{32} \\ A_{13} & A_{23} & A_{33} \\ A_{14} & A_{24} & A_{34} \end{pmatrix} \cdot o \cdot \begin{pmatrix} B_{11} & B_{21} & B_{31} \\ B_{12} & B_{22} & B_{32} \\ B_{13} & B_{23} & B_{33} \\ B_{14} & B_{24} & B_{34} \end{pmatrix} := \left[\left[\begin{pmatrix} A_{11}B_{11} & A_{11}B_{21} & A_{11}B_{31} \\ A_{12}B_{12} & A_{12}B_{22} & A_{12}B_{32} \\ A_{13}B_{13} & A_{13}B_{23} & A_{13}B_{33} \\ A_{14}B_{14} & A_{14}B_{24} & A_{14}B_{34} \end{pmatrix} \right]\left(\begin{pmatrix} A_{21}B_{11} & A_{21}B_{21} & A_{21}B_{31} \\ A_{22}B_{12} & A_{22}B_{22} & A_{22}B_{32} \\ A_{23}B_{13} & A_{23}B_{23} & A_{23}B_{33} \\ A_{24}B_{14} & A_{24}B_{24} & A_{24}B_{34} \end{pmatrix} \right)\left(\begin{pmatrix} A_{31}B_{11} & A_{31}B_{21} & A_{31}B_{31} \\ A_{32}B_{12} & A_{32}B_{22} & A_{32}B_{32} \\ A_{33}B_{13} & A_{33}B_{23} & A_{33}B_{33} \\ A_{34}B_{14} & A_{34}B_{24} & A_{34}B_{34} \end{pmatrix} \right)\right]^{\blacksquare}$$

MATRIX PROGRAMMING,

$$\mathrm{SStrx2LQC} := \left(\frac{1}{\mathrm{Ntrx2}}\right) \cdot \left[\mathrm{diag}\left(\left[\left((\mathrm{ONE7\,ALGSUM\,DB2})\cdot\left(\mathrm{diag}(\mathrm{ONE7\,ALGSUM\,DB2})^T\right)\mathrm{DB2}\right]^T\right)\cdot\mathrm{diag}\left[\left(\mathrm{SUMLQC}^T\right)^{-1}\right]\right] - \mathrm{diag}\left(\mathrm{SStrLQC}^T\right)^{\blacksquare}$$

AND

THE UNIFIELD EQUATION

$$c\left([A]^T_{ij} \; o \; [B]_{jk}\right) = [C]_{jk}$$

CLINTON L. HOLT

To order additional copies of this book, contact:
Xlibris Corporation
1-888-795-4274
www.Xlibris.com
Orders@Xlibris.com
67426

CLASSICAL PHYSICS

$C([A]^T_{ij} \circ [B]_{jk}) = [C]_{jk}$

Let $c = m$, and $[B]_{jk} = [a]_{jk}$ (acceleration) and $[C]_{jk} = F_{jk}$, then
$[C]_{jk} = m([A]^T_{ij} \circ [B]_{jk})$ but $[A]_{ij} = [I]_{jj}$ and $[B]_{jk} = [a]_{jk}$ and
$[C]_{jk} = F_{jk}$, so
$F_{jk} = m([I]_{jj} \circ [a]_{jk})$ for m = to a constant.
$F_{jk} = m([a]_{jk})$
If $[A]_{ij} \neq [I]_{jj}$, then we have $\mathbf{F_{jk} = m([A]^T_{ij} \circ [a]_{jk})}$

EINSTEIN'S FIELD EQUATION FOR GRAVITATION

$C([A]^T_{ij} \circ [B]_{jk}) = [C]_{jk}$

Let $[C]_{jk} = [G]_{jk}$, $[B]_{jk} = [T]_{ij}$ and $[A]_{ij} = [I]_{jj}$ and $c = 8\pi\rho$.

Then we have:

$8\pi\rho([A]^T_{ij} \circ [T]_{jk}) = [G]_{jk}$ but $[A]_{ij} = [I]_{jj}$ so we have
$\mathbf{8\pi\rho[T]_{jk} = [G]_{jk}}$ which is Einstein's First Field Equation. This is
the equation as we now understand it, but the true equation
is $\mathbf{8\pi\rho([A]^T_{ij} \circ [T]_{jk}) = [G]_{jk}}$.

GRAVITATIONAL WAVE EQUATION:

$$[G]_{jk} = \frac{8\pi\rho \ \Psi^T[T]_{jk}\Psi}{\Psi^T \ \Psi} \quad \text{or} \quad [G]_{jk} = \frac{8\pi\rho \ \Psi^T([A]^T_{ij} \circ [T]_{jk})\Psi}{\Psi^T \ \Psi}$$

STATISTICS

There was, until now, no field equation describing the field
of statistics. Going back to the Unified Field Equation:

$C([A]^T_{ij} \circ [B]_{jk}) = [C]_{jk}$ and for the simple case, letting $[B]_{jk} = [I]_{jk}$ and $c = 1/N$, the equation becomes:

$1/N([A]^T_{ij} \circ [B]_{jk}) = [C]_{jk} = 1/N([A]^T_{ij} \circ [I]_{jk}) = \mathbf{1/N([A]^T_{ij}) = [C]_{ij}}$

Since $[A][I]$ is a straight multiplication problem, we do
not need to transpose it. But we do need to sum the columns of
$[A]$, so the basic statistical equation becomes:

$1/N([1]_{1,I}[A]_{ij}) = [C]_{ij}$

But to make this statistical, we must square the above expression so that it becomes:

$$1/N([1]_{1,i}[A]_{ij})^2 = [C]^2_{ij}$$ **Where** $[C]^2_{ij} = CC^T$. (this gives us a one by one matrix as a solution).

Suppose $[B]_{jk}$ is not = to $[I]_{jk}$. Then we have our basic complex statistical field equation, the simplest form of which is the Analysis of Variance.

$$1/N([1]_{1,i}[A]_{ij}[B]_{jk})^2 = [C]^2_{1,k} \text{ where } [C]^2 = CC^{Tequation}$$

But we also need to subtract the correction factor(s), so the total statistical field equation becomes:

$$1/N([1]_{1,i}[A]_{ij}[B]_{jk})^2 - \text{correction factor(s)} = C^2_{1,k} - \text{correction}$$
factor(s). There are actually two basic equations, they are
$$1/N([1]_{1,i}[A]_{ij}[B]_{jk})^2 = [C]^2_{1k} \text{ Between Subjects equation}$$
$$1/N([A]_{ij}[B]_{jk}\{1\}_{k,i})^2 = [C]^2_{k1} \text{ Within Subject}$$

There are only 3, perhaps 4 operators from which statistics (perhaps all of statistics) may be computed:

$$^{1/2}\Sigma([A]^T_{ij} \circ [B]_{jk} = i \ C_{jk} \text{ matrices (Half-Multiplier mode)}$$
$$(\text{where } i = \#\text{rows in un-transposed matrix})$$
$$^c\Sigma \ 1/N([A]^T_{ij} \circ [B]_{jk}) = 1/N[C]_{ik} \text{ regular matrix multiplication.}$$
$$^R\Sigma \ 1/N([A]^T_{ij} \circ [B]_{jk}) = 1/N([B]_{jk}[1]_{k,1}\circ[A]^T_{ij})$$
$$^M\Sigma \ 1/N([A]^T_{ij} \circ [B]_{jk}) = 1/N([A]^T_{ij}[1]_{i,1}\circ[B]_{jk})$$

QUANTUM STATISTICS:

$$[A]^T_{MN}[A]_{MN} = [A]_{NM}[A]_{MN} = [A]^2_{NN}$$

Then we do the following:

$$\frac{1/N[DB1]_{iN}[A]^2_{NN}[DB1]^T_{iN}}{[DB1]_{iN}[DB1]^T_{iN}} = \frac{1/N[DB1]_{iN}[A]^2_{NN}[DB1]_{Ni}}{[DB1]_{iN}[DB1]_{Ni}} = \frac{1/N[DB1]_{iN}[A]^2_{NN}[DB1]_{Ni}}{[DB1]^2_{ii}} =$$

$$\frac{1/N[C]^2_{ii}}{[DB1]^2_{ii}}$$

QUANTUM MECHANICS FROM EINSTEIN'S EQUATION:

$$[G]_{jk} = \frac{8\pi\rho \ \Psi^T[T]_{jk}\Psi}{\Psi^T \ \Psi}$$

$$E_{jk} \ \frac{c \ \Psi^T_{jk}[H]_{jk}\Psi_{jk}}{\Psi^T \ \Psi} \quad or \quad [E]_{jk} = \frac{c\Psi^T([A]^T_{ij} \ o \ [H]_{jk})\ \Psi}{\Psi^T \ \Psi}$$

INVENTORY/ACCOUNTING SYSTEM:

$[A]^T_{ij} \ o \ [B]_{jk} = [C]_{jk} =$
$[INV]_{ij}[DB]_{jk} = [SOL]_{ik}$ This is the sum total of everything bought, sold, manufactured, etc.

$^i[A]^T_{ij} \ o \ [B]_{jk} = \ ^i[C]_{jk} =$
$^i[INV]^T_{ij} \ o \ [DB]_{jk} = \ ^i[AP]_{jk}$ This keeps and individual item accounting of everything bought, sold, manufactured, etc. The superscript i just tells us which row in the un-transposed matrix we have hollow-dotted onto the [DB] matrix.

$^j[B]_{jk} \ o \ [A]^T_{ij} = \ ^j[C]_{jk} =$
$^j[DB]_{jk} \ o \ [INV]^T_{ij} = \ ^j[IP]_{jk}$ This takes an individual item from the database matrix (a column) and multiplies it across the inventory(or accounting page if we wish) giving us a slice of the total pie, so to speak. It tells us how much of that item was used, by whom or what machine or smokestack, and shows how it is distributed throughout the total inventory.

$[A]^T_{ij} \ o \ [B]_{jk} = i, \ [C]_{jk}$ sub-matrices.
$[INV]^T_{ij} \ o \ [DB]_{jk} = i, \ [SOL]_{jk}$ sub-matrices

This is the pure half-multiplier operation giving all the accountpage matrices, i of them, in a single operation.

TABLE OF CONTENTS

OPEN LETTER   Pg. 15

INTRODUCTION   Pg. 19

THE THEORY OF INFORMATION AND THE UNIFIED
    FIELD EQUATION   Pg. 31

PROOF FOR THE HALF-MULTIPLIER OPERATOR   Pg. 31

PROOF FOR REGULAR MATRIX MULTIPLICATION   Pg. 32

ROW PRODUCT OF A MATRIX   Pg. 48

MATRIC PRODUCT OF A MATRIX   Pg. 58

TRANSPOSE COMMUTIVITY OF THE HALF-MULTIPLIER OPERATOR   Pg. 66

NESTED ARRAYS   Pg. 68

ONTO MULTIPLICATION OF MATRICES   Pg. 74

THE THEORY OF INFORMATION AND THE UNIFIED
    FIELD EQUATION   Pg. 91

PHILOSOPHY   Pg. 98

GENERALIZED INVENTORY/ACCOUNTING SYSTEM   Pg. 105

CHEMICAL USAGE INVENTORY   Pg. 120

MULTIPLE CHEMICAL USAGE INVENTORY   Pg. 126

CALCULATION OF A WATER BILL   Pg. 142

TO BILL THE AREA, STATE, USA, WORLD   Pg. 150

TWO FACTOR MIXED DESIGN: REPEATED MEASURES ON ONE
    FACTOR FOR STATISTICS AND ACCOUNTING/INVENTORY
    SYSTEMS (EXAMPLE)   Pg. 160

STATISTICAL ANALYSIS OF PROBLEM   Pg. 183

PHILOSOPHICAL MEANING WHEN $\Sigma [A]_{ij}^{T} \circ [B]_{jk} = \infty$ Pg. 200

EXAMPLE: COMPANY THAT MAKES SAUSAGE AND BOLOGNA   Pg. 202

ARGONIA, KS STORE RECIPIES   Pg. 231

LOVE BOX CO. INVENTORY   Pg. 242

MATRIX SOLUTION FOR GAUSSIAN REDUCTION   Pg. 250

SIMPLIFYING   Pg. 268

INVERSES   Pg. 270

SYMMETRY   Pg. 273

NESTED ARRAYS AND GAUSSIAN REDUCTION   Pg. 275

GENERAL MATRIX SOLUTION SET FOR ½ H ↓ AND ↑   Pg. 282

ELEMENTARY MOLECULAR ORBITAL METHODS   Pg. 283

ETHYLENE   Pg. 288

ACCOUNTING FOR OVERLAP INTERGAL   Pg. 290

BOND ORDER, MOBILE BOND ORDER, TOTAL BOND ORDER
    AND FREE VALENCE INDEX   Pg. 292

BUTADIENE   Pg. 293

MO'S AND NESTED ARRAYS   Pg. 307
CYCLOBUTADIENE   Pg. 310
OZONE   Pg. 313
UTILIZING SYMMETRY IN BUTADIENE   Pg. 319
COEFFICIENTS BY PROPERTIES OF NON-BONDING
    MOLECULAR ORBITALS (NBMO's)   Pg. 323
HETEROCYCLIC COMPOUNDS   Pg. 327
PERTURBATION THEORY (SIMPLE)   Pg. 332
PART 2. PERTURBATION THEORY (COMPLETE)   Pg. 338
WAVE FUNCTION OF ACROLEIN FROM BUTADIENE   Pg. 341
COMPUTING ΔE   Pg. 351
SECOND ORDER ENERGY CORRECTION   Pg. 364
HIGHER ORDER ENERGY CORRECTIONS   Pg. 366
MULTIPLE PERTURBATIONS   Pg. 371
BIBLIOGRAPHY   Pg. 378
STATISTICS   Pg. 379
THE MEAN   Pg. 382
STANDARD DEVIATION   Pg. 383
t-TEST FOR A DIFFERENCE BETWEEN A SAMPLE MEAN &
    POPULATION MEAN   Pg. 387
t-TEST FOR THE DIFFERENCE BETWEEN TWO
    INDEPENDENT MEANS   Pg. 390
t-TEST FOR RELATED MEASURES   Pg. 398
SANDLERS A TEST   Pg. 406
THE ANALYSIS OF VARIANCE   Pg. 408
BASIC TRANSLATIONS FOR STATISTICS   Pg. 414
COMPLETELY RANDOMIZED DESIGN   Pg. 415
FACTORIAL DESIGN: TWO FACTORS   Pg. 424
FACTORIAL DESIGN: THREE FACTORS (COMPLETE ANALYSIS
    OF STATISTICAL METHOD WITH NESTED ARRAYS)   Pg. 440
TREATMENT BY LEVELS DESIGN (WITH NESTED ARRAYS)   Pg. 478
TREATMENT BY SUBJECTS: REPEATED MEASURES DESIGN   Pg. 491
TREATMENT BY TREATMENT BY SUBJECTS, OR
    REPEATED MEASURES: TWO FACTOR DESIGN   Pg. 506
TWO FACTOR MIXED DESIGN: REPEATED MEASURES ON
    ONE FACTOR   Pg. 526
THREE FACTOR MIXED DESIGN: REPEATED MEASURES ON
    ONE FACTOR   Pg. 538
THREE FACTOR MIXED DESIGN: REPEATED MEASURES ON
    TWO FACTORS   Pg. 554
LATIN SQUARE DESIGN: SIMPLE   Pg. 574
LATIN SQUARE DESIGN: COMPLEX   Pg. 587

USE OF ORTHOGONAL COMPONENTS IN TESTS FOR TREND Pg. 604
 EXAMPLE 2 Pg. 611
 EXAMPLE 3 Pg. 623
TEST FOR DIFFERENCE BETWEEN VARIANCES OF TWO
 INDEPENDENT SAMPLES Pg. 643
TEST FOR DIFFERENCE BETWEEN VARIANCES OF TWO
 RELATED SAMPLES Pg. 646
PEARSON PRODUCT-MOMENT CORRELATION Pg. 649
t-TEST FOR DIFFERENCES AMONG SEVERAL MEANS Pg. 652
PEARSON PRODUCT-MOMENT CORRELATION Pg. 656
SIMPLE ANALYSIS OF COVARIANCE:
 ONE TREATMENT VARIABLE Pg. 658
FACTORIAL ANALYSIS OF COVARIANCE:
 TWO TREATMENT VARIABLES Pg. 672
SIMPLE CHI-SQUARE AND THE **PHI** COEFFICIENT Pg. 701
COMPLEX CHI-SQUARE AND THE CONTINGENCY COEFFICIENT Pg. 707
THE QUANTUM STATISTICS Pg. 717
LINEAR REGRESSION Pg. 721
STRAIGHT LINE Pg. 723
SEMI-LOG Pg. 724
LOG-LOG Pg. 727
PARABOLIC FIT Pg. 729
CRYPTOGRAPHY Pg. 731
PHYSICS 101 Pg. 745
COMPUTATION OF SEVERAL SETS OF DATA ALL AT ONCE Pg. 759
A o B o C Pg. 767
Y-NAUGHT RYTE Pg. 769

OPEN LETTER

Several years ago, quite by accident, I discovered an interesting property of matrices that has been overlooked in the mainstream of Mathematics. If we have two conformable matrices, transpose the pre-multiplier, take the columns of the transposed pre-multiplier, in order, and multiply them straight across the post-multiplier matrix (as we do Gausian Reduction), we obtain a set of half-multiplied matrices, or Nested Arrays. If the pre-multiplier was a row matrix, this half-multiplication gives us a new matrix, if the pre-multiplier was a matrix, we obtain a nested array. I call this the Half-Multiplier because, unlike in regular matrix multiplication where we multiply and sum, with this operator we do the multiplication but do not necessarily sum the products. If we transpose the resulting nested array and sum the columns, we acheive the same results as in regular matrix multiplication (see proof). If we sum the rows instead of the columns, we acheive the Cross Product of a matrix (submitted for your verification, as of yet undefined in modern mathematics). If we add the individual matrices (sub-matrices) of the nested array together, we achieve the Matric Product of a matrix. Also, under Half-Multiplication, matrices become commutative and/or transpose cummutative. This is important because what was believed to be a non-commutative operator is now commutative. This operator seems to generate and govern Nested Arrays.

With this operator, just by adding a constant term, there seems to be generated a generalized field equation (the Mother Equation) from which local field equations may be formulated. Ie. we can derive the field equations for Classical Physics, Quantum Mechanics, Astro-physics, Accounting/Inventory systems and with a little "twist" can derive a field equation for statistics which can be derived for all the above field equations. These equations all have an added term not found in the known field equations which, if the added term is the identity matrix, the equations all reduce to the field equations in their familiar forms with their connection to statistics.

Because this operator is new, no known math packages can handle it. But fortunately it has equivalent operations in regular matrix math which can be substituted. I.e. Multiplying by a diagonal matrix gives the same solution as half-multiplying.

Also, since no programs can handle nested arrays, we overcome this difficulty by writing the nested array as a single 'partitioned' array and multiply as needed to achieve the same results as multiplying by a nested array.

One final note, a number can be considered to be a tensor of rank zero. Reversing this, if a tensor of any rank can be considered to be a single number, then the math in this book reduces to simple arithmetic. The new "number" has all the properties of single numbers except we do all operations all at the same time. This goes contrary to the modern assumption that there is only one way to multiply a matrix and that there is no element by element multiplication of matrices. I have offered a proof for this as well as offer a proof for Gaussian Reduction in this book.

I have found a 7 step method of writing Analysis of Variance into a single equation that works for all AOV if the mathematical processes are the same, two Mathematical sentences if we have both Between and Within Subjects computations, but the two can still be summed into a single sentence (expression). The book needs to be shortened, so I have cut down on the easy Statistics and concentrated on the harder, more advanced problems. All can be written as a single equation in their final form.

A note about the proofs, just to keep them short and simple, I have shown an example for a small matrix, then for the proof, I proved the theorems for M,N and M+1, N+1 matrices both at the same time. I could prove them both separately, but this book is too long as it is.

When I took some classes in physics in college, I was impressed by some of the derivations because the mathematics was pretty. In most cases the math was boring or utilitarian, but in some cases it was just intellectually beautiful. In the case of the Half-Multiplier Operator, the math is quite pretty. It is also seductive. Just when I thought I was through with a problem, the operator would show me a simpler, easier way to solve the problem. But just the fact that we can solve all parts of a problem in one operation instead of many separate, similar calculations is only one aspect of this equations beauty (see Sect. 2.3 and 2.4 in Statistics). We can solve everything all at the same time if the math is set up correctly. Also, even though I derived Classical Physics, Einstein's First Field

Equation for Gravity and Quantum Mechanics, most of this book covers the solution for the field equations for accounting/ inventory systems and the field equations for statistics. The main reason is that I have forgotten all my physics, (never used it) although there is some at the end of this book and some quantum mechanics is covered in the section on quantum chemistry. Anyway, you folks have already covered these equations extremely well but no one has ever done statistics and inventories the way utilized by this operator. I think that even the calculus will fall easily under this operator. Another part of the beauty of these equations is that the mathematics, once completed, is automatically it's own computer program.

For the separate book on statistics, you will need a copy of THE COMPUTATIONAL HANDBOOK OF STATISTICS, 2nd Edition by James L Bruning and B. L. Kintz, 2nd Edition, Scott Foresman and Co., Glenview, Ill. to be able to understand my approach (unless you're a professional statistician) as I explain nothing of what I do, I have left the explanations to them. This is an example of the beauty of this operator. I have never taken a statistics class above the most elementary, yet with the Statistical Field Equation, even I was able to re-write the entire field of Statistics very easily and simply. In fact, I was writing the programs after just looking at the way the problem was done without having to solve the problem in the first place or see the mathematical equations. The most advanced techniques are in the 9th grade level (Logs and simple algebra). Most of the math is simple addition, subtraction, multiplication and division.

About 200 years ago, some mathematician said there was only one way to multiply a matrix and that notion has been accepted as gospel ever since. No one has ever looked to see if this idea was true or not. In this book you will see the first effort to explore matrices as they are meant to be used.

Well, I'll let you get on with the reading of this book.

Sincerely,

Clinton L. Holt
8-8-2009

INTRODUCTION

A little over a year ago, I occupied my time by trying to write a simple program so that I could perform elementary Molecular Orbital computations on my HP-48SX calculator. While I was working with Gaussian Reduction to solve for the roots of the secular equations of simple organic molecules, I made a singular discovery. If I took every step I computed in a Gaussian reduction sequence and wrote each solution step in a series of columns and made these columns into a matrix, then transposed and multiplied this new matrix to the secular determinant, this matrix multiplication automatically carried out the Gaussian reduction of computing zeros in each position under the principal diagonal. By all accounts, what I had done was impossible as defined by the mathematics of today. I mean, I had been told since I was a senior in high school and by my professors all through college that we can multiply matrices only one way, that all other multiplications any other way have no meaning. I had also been told that there never has been a proof for Gaussian reduction. It is used to solve for systems of linear equations because it works, but has no other mathematical validity or use. What I had done by accident was to discover a proof for Gaussian Reduction. What I also discovered, but did not know until later, was that this Half-Multiplier Operator may be the connection between our regular mathematics that we all know (addition, subtraction, multiplication and division) and tensors and matrices. What this means is that if 1, 2, 3, . . . ∞ are numbers with all the properties mathematicians have discovered about numbers up to and including logarithms, calculus, trigonometry, differential equations, etc., that an m x n array of numbers can be considered to be a single number, not just an array of separate numbers. In other words, a two by two matrix is not a matrix containing four separate numbers 1, 2, 3 and 4, but can be considered as a single number all by itself. To add, subtract, multiply and divide this number, all the other numbers we wish to operate on must be of the same size. To add 50 to 1, 50 can be considered as a one by one matrix and 1 is a one by one matrix, and we get $[50] + [1] = [51]$. We cannot add 1 to a number of a different size, i.e. $[1] + [50\ 2]$ cannot be summed, but $[1\ -2] + [50\ 2] = [51\ 0]$

which can be summed. Let's take two matrices with elements 1, 2, 3 and 4 and 5, 6, 7 and 8 respectively and see how their properties mimic our regular concept of mathematics. First we shall add the two numbers together and see that we add all four numbers at the same time to their counterparts in the second matrix in one operation.

$$\begin{pmatrix} 1 & 2 \\ 3 & 4 \end{pmatrix} + \begin{pmatrix} 5 & 6 \\ 7 & 8 \end{pmatrix} = \begin{pmatrix} 6 & 8 \\ 10 & 12 \end{pmatrix}$$

Here we add 1+5 = 6, 2+6=8, 3+7=10 and 4+8=12 all at the same time. They are commutative under addition, i.e. if we reverse the order of addition, we get the same answer:

$$\begin{pmatrix} 5 & 6 \\ 7 & 8 \end{pmatrix} + \begin{pmatrix} 1 & 2 \\ 3 & 4 \end{pmatrix} = \begin{pmatrix} 6 & 8 \\ 10 & 12 \end{pmatrix}$$

Let's now subtract the two numbers, and we will do it both ways:

$$\begin{pmatrix} 5 & 6 \\ 7 & 8 \end{pmatrix} - \begin{pmatrix} 1 & 2 \\ 3 & 4 \end{pmatrix} = \begin{pmatrix} 4 & 4 \\ 4 & 4 \end{pmatrix}$$

$$\begin{pmatrix} 1 & 2 \\ 3 & 4 \end{pmatrix} - \begin{pmatrix} 5 & 6 \\ 7 & 8 \end{pmatrix} = \begin{pmatrix} -4 & -4 \\ -4 & -4 \end{pmatrix}$$

Note that switching the order of subtraction changes the signs, so these two numbers are not in general commutative, but the absolute value of the solution would be commutative.

If we multiply the two matrices in the manner that is accepted by mathematics, we would proceed in the following manner:

$$\begin{pmatrix} 1 & 2 \\ 3 & 4 \end{pmatrix} \cdot \begin{pmatrix} 5 & 6 \\ 7 & 8 \end{pmatrix} = \begin{pmatrix} 19 & 22 \\ 43 & 50 \end{pmatrix}$$

$$\begin{pmatrix} 5 & 6 \\ 7 & 8 \end{pmatrix} \cdot \begin{pmatrix} 1 & 2 \\ 3 & 4 \end{pmatrix} = \begin{pmatrix} 23 & 34 \\ 31 & 46 \end{pmatrix}$$

To accomplish this, we proceed as follows: The top row in the first Matrix is multiplied to the first column in the second Matrix and the two products are added, i.e.

1x5 + 2x7 = 5+14 = 19. The first row is now multiplied to the second column in the second Matrix and the sum of the products is places in the second top-hand position. 1x6 + 2x8 = 6+16=22. We have run out of columns in the second matrix to multiply [1 2] by, so now we go to the second row in the first matrix and multiply it to the first column in the second Matrix: 3x5 + 4x7 = 15 + 28 = 43. This number is put in the first position in the second row. To finish up, we finally multiply the second row in the first Matrix to the second column in the second Matrix and sum their products. This solution is put in the second position in the bottom row of the solution Matrix. 3x6 + 4x8 = 18 + 32 = 50. This doesn't look like our regular multiplication that we are familiar with. If we multiply two numbers we should use the single numbers we are familiar with. So now we get to the next step in our new definition of number. Mathematicians say this operation is illegal, that it doesn't work and the results have no meaning, but we will look at it anyway.

I will define the operator ⊗ as meaning we will multiply the first number in the top row of the first matrix by the first number in the top row of the second matrix, the second number in the top row of the first Matrix by the second number in the top row of the second matrix, the first number in the second row of the first matrix by the first number in the bottom row of the second matrix and the second number in the bottom row of the first matrix by the second number in the bottom row of the second matrix, i.e.

$$\begin{vmatrix} 1 & 2 \\ 3 & 4 \end{vmatrix} \otimes \begin{vmatrix} 5 & 6 \\ 7 & 8 \end{vmatrix} = \begin{vmatrix} 5 & 12 \\ 21 & 32 \end{vmatrix}$$

This actually makes more sense since it is logical that 5x1=5, 6x2=12, 7x3=21 and 8x4=32. Although this form of multiplication cannot work and is defined as impossible, it's use is found throughout statistics. We can write the sum of squares more simply and elegantly as

$$[1]_{1i}\left([A]_{ij}\otimes[A]_{ij}\right)[1]_{j1}$$

To obtain the Sum of Squares for the second Matrix above, we would compute it as follows:

$$\text{ONE2} := (1 \quad 1) \quad \text{TWO1} := \begin{pmatrix} 1 \\ 1 \end{pmatrix}$$

$$\text{ONE2}\overrightarrow{\begin{pmatrix} 5 & 6 \\ 7 & 8 \end{pmatrix}^2}\cdot\text{TWO1} = 174$$

$$(1 \quad 1)\cdot\overrightarrow{\begin{pmatrix} 5 & 6 \\ 7 & 8 \end{pmatrix}^2}\cdot\begin{pmatrix} 1 \\ 1 \end{pmatrix} = 174$$

Where $\overrightarrow{\begin{pmatrix} 5 & 6 \\ 7 & 8 \end{pmatrix}^2} = \begin{pmatrix} 25 & 36 \\ 49 & 64 \end{pmatrix}$

The arrow above the squared matrix is MathCad's of saying we square each individual element in the Matrix rather than squaring the matrix itself using regular matrix multiplication. So the grand sum is 25+36+49+64 = 174. (I use MathCad 13 as my computer's math program.) This sum of squares is the correction factor in statistics. Note also that the $\otimes$ operation is commutative, just like in our regular mathematics i.e.

$$\overrightarrow{\left[\begin{pmatrix} 1 & 2 \\ 3 & 4 \end{pmatrix}\cdot\begin{pmatrix} 5 & 6 \\ 7 & 8 \end{pmatrix}\right]} = \begin{pmatrix} 5 & 12 \\ 21 & 32 \end{pmatrix} \text{ and } \overrightarrow{\left[\begin{pmatrix} 5 & 6 \\ 7 & 8 \end{pmatrix}\cdot\begin{pmatrix} 1 & 2 \\ 3 & 4 \end{pmatrix}\right]} = \begin{pmatrix} 5 & 12 \\ 21 & 32 \end{pmatrix}$$

Finally we get to division, there is no way to divide a number by a Matrix, so we have to multiply the Matrix by an inverse to accomplish division, but on the one-to-one system as shown just above, we can divide Matrices. Let's divide the second Matrix by the first.

$$\overrightarrow{\left[\begin{pmatrix} 1 & \frac{1}{2} \\ \frac{1}{3} & 1 \end{pmatrix}\cdot\begin{pmatrix} 5 & 6 \\ 7 & 8 \end{pmatrix}\right]} = \begin{pmatrix} 5 & 3 \\ 2.333 & 2 \end{pmatrix}$$

or

$$\overrightarrow{\begin{pmatrix} 1 & 2 \\ 3 & 4 \end{pmatrix}^{-1}} = \begin{pmatrix} 1 & 0.5 \\ 0.333 & 0.25 \end{pmatrix}$$

and

$$\overrightarrow{\left[\overrightarrow{\begin{pmatrix} 1 & 2 \\ 3 & 4 \end{pmatrix}^{-1}} \begin{pmatrix} 5 & 6 \\ 7 & 8 \end{pmatrix} \right]} = \begin{pmatrix} 5 & 3 \\ 2.333 & 2 \end{pmatrix}$$

The rightmost term under the arrow takes the inverse of each element in the 1, 2, 3, 4 Matrix and multiplies it one-to-one to the 5,6,7,8 Matrix (under the second, longer arrow). All computer programs are written using the inverse of Matrices, so we must use their terminology to obtain the results we desire.

Let's look at how the half-multiplier operation works using the two 4x4 Matrices above. To half multiply the two Matrices, we must first transpose the Matrix containing the elements 1, 2, 3 and 4 and hollow dot multiply into the second matrix, i.e.

$$\begin{pmatrix} 1 & 2 \\ 3 & 4 \end{pmatrix}^{T} = \begin{pmatrix} 1 & 3 \\ 2 & 4 \end{pmatrix}$$

$$\begin{pmatrix} 1 & 3 \\ 2 & 4 \end{pmatrix} \circ \begin{pmatrix} 5 & 6 \\ 7 & 8 \end{pmatrix} = \left[\begin{pmatrix} 1 \\ 2 \end{pmatrix} \circ \begin{pmatrix} 5 & 6 \\ 7 & 8 \end{pmatrix} \quad \begin{pmatrix} 3 \\ 4 \end{pmatrix} \circ \begin{pmatrix} 5 & 6 \\ 7 & 8 \end{pmatrix} \right]$$

$$\left[\begin{pmatrix} 1 \\ 2 \end{pmatrix} \circ \begin{pmatrix} 5 & 6 \\ 7 & 8 \end{pmatrix} \begin{pmatrix} 3 \\ 4 \end{pmatrix} \circ \begin{pmatrix} 5 & 6 \\ 7 & 8 \end{pmatrix} \right] = \left[\begin{pmatrix} 1.5 & 1.6 \\ 2.7 & 2.8 \end{pmatrix} \begin{pmatrix} 3.5 & 3.6 \\ 4.7 & 4.8 \end{pmatrix} \right] = \left[\begin{pmatrix} 5 & 6 \\ 14 & 16 \end{pmatrix} \begin{pmatrix} 15 & 18 \\ 28 & 32 \end{pmatrix} \right]$$

This is the multiplied matrix in half-multiplier mode. There are three ways to manipulate this data, actually there are four, we can leave the matrices as they are, or we can align the two matrices on top of each other and sum their rows: (this is actually a nested array).

$$(1 \ 1) \left[\begin{pmatrix} 5 & 6 \\ 14 & 16 \end{pmatrix} \right] = \left[(19 \ 22) \right]$$

$$(1 \ 1) \left[\begin{pmatrix} 15 & 18 \\ 28 & 32 \end{pmatrix} \right] = \left[(43 \ 50) \right]$$

Deleting the inner brackets we get

$$\begin{pmatrix} 19 & 22 \\ 43 & 50 \end{pmatrix}.$$

which is the solution for regular matrix multiplication.

Instead of adding the rows, let's add the columns of each of the 2 sub-matrices:

Deleting the inner brackets and combining the two column matrices into a 2x2 matrix we get:

$$\left[\begin{pmatrix} 5 & 6 \\ 14 & 16 \end{pmatrix} \cdot \begin{pmatrix} 1 \\ 1 \end{pmatrix} = \begin{pmatrix} 11 \\ 30 \end{pmatrix} \quad \begin{pmatrix} 15 & 18 \\ 28 & 32 \end{pmatrix} \cdot \begin{pmatrix} 1 \\ 1 \end{pmatrix} = \begin{pmatrix} 33 \\ 60 \end{pmatrix} \right] = \begin{pmatrix} 11 & 33 \\ 30 & 60 \end{pmatrix}$$

To solve this mathematically according to the proof, we sum the columns in the 5, 6, 7, 8 matrix and half-multiply into the 1, 2, 3, 4 matrix, i.e.

$$\begin{pmatrix} 1 & 3 \\ 2 & 4 \end{pmatrix} \circ \begin{pmatrix} 5+6 \\ 7+8 \end{pmatrix} = \begin{pmatrix} 11 \\ 15 \end{pmatrix} \circ \begin{pmatrix} 1 & 3 \\ 2 & 4 \end{pmatrix} = \begin{pmatrix} 11 & 33 \\ 30 & 60 \end{pmatrix}$$

This is may be the Cross Product of a matrix.

The next thing we can do is add the two sub-matrices together, i.e.

$$\left| \begin{pmatrix} 5 & 6 \\ 14 & 16 \end{pmatrix} + \begin{pmatrix} 15 & 18 \\ 28 & 32 \end{pmatrix} \right| = \begin{pmatrix} 20 & 24 \\ 42 & 48 \end{pmatrix}$$

To solve this mathematically according to the proof, we sum the columns in the 1, 2, 3, 4 matrix and half-multiply into the 5, 6, 7, 8 matrix, i.e.

$$\begin{pmatrix} 1+3 \\ 2+4 \end{pmatrix} \circ \begin{pmatrix} 5 & 6 \\ 7 & 8 \end{pmatrix} = \begin{pmatrix} 4 \\ 6 \end{pmatrix} \circ \begin{pmatrix} 5 & 6 \\ 7 & 8 \end{pmatrix} = \begin{pmatrix} 4\cdot5 & 4\cdot6 \\ 6\cdot7 & 6\cdot8 \end{pmatrix} = \begin{pmatrix} 20 & 24 \\ 42 & 48 \end{pmatrix}$$

This is defined as the Matric Product of a matrix.

Three of the above four computations, although derived by using Gaussian Reduction, are illegal, or at least unknown to modern mathematics. Examples of their use will follow in this book.

Now that our concept of number has been more completely defined, I wish that other mathematicians and physicists would check on this work and see how it applies or does not apply to other problems in mathematics, especially Statistics and calculus. I have derived Newton's Equation and Einstein's First Field Equation of Gravity, so this equation should hold through all of Physics.

Before we get into the book proper, I will present a further tutorial on how this math is used.

TUTORIAL

The first thing we will do is sum the columns of the following matrix. The row matrix ONE is the operator which will sum the columns.

$$\text{ONE} := (1 \ 1 \ 1 \ 1 \ 1 \ 1) \qquad M := \begin{bmatrix} 1 & 7 & 13 & 19 & 25 & 31 \\ 2 & 8 & 14 & 20 & 26 & 32 \\ 3 & 9 & 15 & 21 & 27 & 33 \\ 4 & 10 & 16 & 22 & 28 & 34 \\ 5 & 11 & 17 & 23 & 29 & 35 \\ 6 & 12 & 18 & 24 & 30 & 36 \end{bmatrix}$$

The problem looks like:

$$(1 \ 1 \ 1 \ 1 \ 1 \ 1) \cdot \begin{bmatrix} 1 & 7 & 13 & 19 & 25 & 31 \\ 2 & 8 & 14 & 20 & 26 & 32 \\ 3 & 9 & 15 & 21 & 27 & 33 \\ 4 & 10 & 16 & 22 & 28 & 34 \\ 5 & 11 & 17 & 23 & 29 & 35 \\ 6 & 12 & 18 & 24 & 30 & 36 \end{bmatrix}$$

$$\text{ONE} \cdot M = (21 \ \ 57 \ \ 93 \ \ 129 \ \ 165 \ \ 201)$$

To sum the rows we post multiply by a column Matrix composed of ones equal to the number of columns in the pre-multiplier Matrix. ie:

$$\text{ONET} := \begin{bmatrix} 1 \\ 1 \\ 1 \\ 1 \\ 1 \\ 1 \end{bmatrix} \qquad M = \begin{bmatrix} 1 & 7 & 13 & 19 & 25 & 31 \\ 2 & 8 & 14 & 20 & 26 & 32 \\ 3 & 9 & 15 & 21 & 27 & 33 \\ 4 & 10 & 16 & 22 & 28 & 34 \\ 5 & 11 & 17 & 23 & 29 & 35 \\ 6 & 12 & 18 & 24 & 30 & 36 \end{bmatrix}$$

The problem for summing the rows looks like:

$$M = \begin{bmatrix} 1 & 7 & 13 & 19 & 25 & 31 \\ 2 & 8 & 14 & 20 & 26 & 32 \\ 3 & 9 & 15 & 21 & 27 & 33 \\ 4 & 10 & 16 & 22 & 28 & 34 \\ 5 & 11 & 17 & 23 & 29 & 35 \\ 6 & 12 & 18 & 24 & 30 & 36 \end{bmatrix} \begin{bmatrix} 1 \\ 1 \\ 1 \\ 1 \\ 1 \\ 1 \end{bmatrix} \qquad M \cdot ONET = \begin{bmatrix} 96 \\ 102 \\ 108 \\ 114 \\ 120 \\ 126 \end{bmatrix}$$

To multiply each row in a Matrix by a constant, proceed as follows:

$$\begin{matrix} 3\text{ X} \\ 4\text{ X} \\ 5\text{ X} \\ 6\text{ X} \\ 7\text{ X} \\ 8\text{ X} \end{matrix} \begin{vmatrix} 1 & 7 & 13 & 19 & 25 & 31 \\ 2 & 8 & 14 & 20 & 26 & 32 \\ 3 & 9 & 15 & 21 & 27 & 33 \\ 4 & 10 & 16 & 22 & 28 & 34 \\ 5 & 11 & 17 & 23 & 29 & 35 \\ 6 & 12 & 18 & 24 & 30 & 36 \end{vmatrix} =$$

$$N := \begin{bmatrix} 3 \\ 4 \\ 5 \\ 6 \\ 7 \\ 8 \end{bmatrix}$$

Modern computer programs allow us to multiply in a one-to-one correspondence with two matrices, but none allow us to multiply in the manner of Gauss. This method has not been invented yet. Fortunately, there is an operation that can get us the same answer, but at the cost of more multiplication steps and computer memory. But we must do the math this way until programmers write more efficient programs. To do this, we must diagonalize the pre-multiplier matrix, in this example N:

$$O := diag(N)$$

$$O = \begin{bmatrix} 3 & 0 & 0 & 0 & 0 & 0 \\ 0 & 4 & 0 & 0 & 0 & 0 \\ 0 & 0 & 5 & 0 & 0 & 0 \\ 0 & 0 & 0 & 6 & 0 & 0 \\ 0 & 0 & 0 & 0 & 7 & 0 \\ 0 & 0 & 0 & 0 & 0 & 8 \end{bmatrix} \qquad M = \begin{bmatrix} 1 & 7 & 13 & 19 & 25 & 31 \\ 2 & 8 & 14 & 20 & 26 & 32 \\ 3 & 9 & 15 & 21 & 27 & 33 \\ 4 & 10 & 16 & 22 & 28 & 34 \\ 5 & 11 & 17 & 23 & 29 & 35 \\ 6 & 12 & 18 & 24 & 30 & 36 \end{bmatrix}$$

Then

$$O \cdot M = \begin{bmatrix} 3 & 21 & 39 & 57 & 75 & 93 \\ 8 & 32 & 56 & 80 & 104 & 128 \\ 15 & 45 & 75 & 105 & 135 & 165 \\ 24 & 60 & 96 & 132 & 168 & 204 \\ 35 & 77 & 119 & 161 & 203 & 245 \\ 48 & 96 & 144 & 192 & 240 & 288 \end{bmatrix}$$

To multiply the columns of the Matrix by the same constants, we post-multiply (instead of pre-multiplying) by the diagonal Matrix. i.e.:

$$M = \begin{bmatrix} 1 & 7 & 13 & 19 & 25 & 31 \\ 2 & 8 & 14 & 20 & 26 & 32 \\ 3 & 9 & 15 & 21 & 27 & 33 \\ 4 & 10 & 16 & 22 & 28 & 34 \\ 5 & 11 & 17 & 23 & 29 & 35 \\ 6 & 12 & 18 & 24 & 30 & 36 \end{bmatrix} \begin{bmatrix} 3 & 0 & 0 & 0 & 0 & 0 \\ 0 & 4 & 0 & 0 & 0 & 0 \\ 0 & 0 & 5 & 0 & 0 & 0 \\ 0 & 0 & 0 & 6 & 0 & 0 \\ 0 & 0 & 0 & 0 & 7 & 0 \\ 0 & 0 & 0 & 0 & 0 & 8 \end{bmatrix}$$

$$M \cdot O = \begin{bmatrix} 3 & 28 & 65 & 114 & 175 & 248 \\ 6 & 32 & 70 & 120 & 182 & 256 \\ 9 & 36 & 75 & 126 & 189 & 264 \\ 12 & 40 & 80 & 132 & 196 & 272 \\ 15 & 44 & 85 & 138 & 203 & 280 \\ 18 & 48 & 90 & 144 & 210 & 288 \end{bmatrix}$$

ELEMENTARY STATISTICAL CALCULATIONS:

Suppose we want to add the first two numbers, the third and fourth numbers and the fifth and sixth numbers together and keep the solutions separate:

$$\text{ONE·M} = (21 \quad 57 \quad 93 \quad 129 \quad 165 \quad 201)$$

We can proceed as follows:

$$(21 \ 57 \ 93 \ 129 \ 165 \ 201)$$

$$21 + 57 = 78$$
$$93 + 129 = 222$$
$$165 + 201 = 366$$

Or we can set up a database matrix that will do the same multiplications all at the same time, i.e.

$$\text{DB1} := \begin{bmatrix} 1 & 0 & 0 \\ 1 & 0 & 0 \\ 0 & 1 & 0 \\ 0 & 1 & 0 \\ 0 & 0 & 1 \\ 0 & 0 & 1 \end{bmatrix} \qquad (21 \ 57 \ 93 \ 129 \ 165 \ 201) \cdot \begin{bmatrix} 1 & 0 & 0 \\ 1 & 0 & 0 \\ 0 & 1 & 0 \\ 0 & 1 & 0 \\ 0 & 0 & 1 \\ 0 & 0 & 1 \end{bmatrix} = (78 \quad 222 \quad 366)$$

or

$$\text{ONE·M·DB1} = (78 \quad 222 \quad 366)$$

Suppose we want to add the first three numbers and the second three numbers and keep the solutions separate:

$$\text{ONE·M} = (21 \quad 57 \quad 93 \quad 129 \quad 165 \quad 201)$$

$$DB2 := \begin{bmatrix} 1 & 0 \\ 1 & 0 \\ 1 & 0 \\ 0 & 1 \\ 0 & 1 \\ 0 & 1 \end{bmatrix} \qquad (21\ 57\ 93\ 129\ 165\ 201) \cdot \begin{bmatrix} 1 & 0 \\ 1 & 0 \\ 1 & 0 \\ 0 & 1 \\ 0 & 1 \\ 0 & 1 \end{bmatrix} = (\ 171 \quad 495\ )$$

Then

$$ONE \cdot M \cdot DB2 = (\ 171 \quad 495\ )$$

Suppose we want to add the first and third numbers, the second and fourth numbers and the fifth and sixth numbers together:

$$DB3 := \begin{bmatrix} 1 & 0 & 0 \\ 0 & 1 & 0 \\ 1 & 0 & 0 \\ 0 & 1 & 0 \\ 0 & 0 & 1 \\ 0 & 0 & 1 \end{bmatrix} \qquad (21\ 57\ 93\ 129\ 165\ 201) \cdot \begin{bmatrix} 1 & 0 & 0 \\ 0 & 1 & 0 \\ 1 & 0 & 0 \\ 0 & 1 & 0 \\ 0 & 0 & 1 \\ 0 & 0 & 1 \end{bmatrix} = (\ 114 \quad 186 \quad 366\ )$$

THEN

$$ONE \cdot M \cdot DB3 = (\ 114 \quad 186 \quad 366\ )$$

THE THEORY OF INFORMATION
AND
THE UNIFIED FIELD EQUATION

PROOF FOR THE HALF-MULTIPLIER OPERATOR
PROOF OF $[A]^T$ O $[B]=$ DIAGONAL(S)$[A]^T[B]$

I must do this proof first, since I will need it to prove the rest of the math to follow. Suppose we have the matrices:

$$[A]^T = \begin{vmatrix} A_{11} \\ \cdot \\ A_{1J} \\ \cdot \\ A_{1N} \\ \cdot \\ A_{1,N+1} \end{vmatrix} \quad [B]= \begin{vmatrix} B_{11} \cdot \cdot \cdot B_{1K} \cdot \cdot \cdot B_{1M} \cdot \cdot \cdot B_{1,M+1} \\ \cdot \quad\quad \cdot \quad\quad \cdot \quad\quad \cdot \\ B_{J1} \cdot \cdot \cdot B_{JK} \cdot \cdot \cdot B_{JM} \cdot \cdot \cdot B_{J,M+1} \\ \cdot \quad\quad \cdot \quad\quad \cdot \quad\quad \cdot \\ B_{N1} \cdot \cdot \cdot B_{NK} \cdot \cdot \cdot B_{NM} \cdot \cdot \cdot B_{N,M+1} \\ \cdot \quad\quad \cdot \quad\quad \cdot \quad\quad \cdot \\ B_{N+1,1} \cdot \cdot B_{N+1,K} \cdot \cdot B_{N+1,M} \cdot \cdot B_{N+1,M+1} \end{vmatrix}$$

Changing the pre-multiplier into a diagonal matrix and multiplying:

$$\begin{vmatrix} A_{11} \cdot \cdot \cdot 0 \cdot \cdot \cdot 0 \cdot \cdot \cdot 0 \\ \cdot \cdot \cdot \cdot \cdot \cdot \cdot \cdot \cdot \cdot \cdot \\ 0 \cdot \cdot \cdot A_{1J} \cdot \cdot \cdot 0 \cdot \cdot \cdot 0 \\ \cdot \cdot \cdot \cdot \cdot \cdot \cdot \cdot \cdot \cdot \cdot \\ 0 \cdot \cdot \cdot 0 \cdot \cdot \cdot A_{1M} \cdot \cdot 0 \\ \cdot \cdot \cdot \cdot \cdot \cdot \cdot \cdot \cdot \cdot \cdot \\ 0 \cdot \cdot \cdot 0 \cdot \cdot \cdot 0 \cdot \cdot A_{1,M+1} \end{vmatrix} \begin{vmatrix} B_{11} \cdot \cdot \cdot B_{1K} \cdot \cdot \cdot B_{1M} \cdot \cdot \cdot B_{1,M+1} \\ \cdot \quad\quad \cdot \quad\quad \cdot \quad\quad \cdot \\ B_{J1} \cdot \cdot \cdot B_{JK} \cdot \cdot \cdot B_{JM} \cdot \cdot \cdot B_{J,M+1} \\ \cdot \quad\quad \cdot \quad\quad \cdot \quad\quad \cdot \\ B_{N1} \cdot \cdot \cdot B_{NK} \cdot \cdot \cdot B_{NM} \cdot \cdot \cdot B_{N,M+1} \\ \cdot \quad\quad \cdot \quad\quad \cdot \quad\quad \cdot \\ B_{N+1,1} \cdot \cdot B_{N+1,K} \cdot \cdot B_{N+1,M} \cdot \cdot B_{N+1,M+1} \end{vmatrix}$$

We get the solution:

$$\begin{vmatrix} A_{11}B_{11} \cdot \cdot \cdot + \cdot A_{11}B_{1K} \cdot + \cdot \cdot \cdot A_{11}B_{1M} \cdot + \cdot \cdot \cdot \cdot A_{11}B_{1,M+1} \\ \cdot \quad\quad\quad \cdot \quad\quad\quad \cdot \quad\quad\quad \cdot \\ A_{1J}B_{J1} \cdot \cdot \cdot + \cdot A_{1J}B_{JK} \cdot + \cdot \cdot \cdot A_{1J}B_{JM} \cdot + \cdot \cdot \cdot \cdot A_{1J}B_{J,M+1} \\ \cdot \quad\quad\quad \cdot \quad\quad\quad \cdot \quad\quad\quad \cdot \\ A_{1M}B_{N1} \cdot \cdot \cdot + \cdot A_{1M}B_{NK} \cdot + \cdot \cdot \cdot A_{1M}B_{NM} \cdot + \cdot \cdot \cdot \cdot A_{1M}B_{N,M+1} \\ \cdot \quad\quad\quad \cdot \quad\quad\quad \cdot \quad\quad\quad \cdot \\ A_{1,M+1}B_{N+1,1} \cdot + \cdot A_{1,M+1}B_{N+1,K} \cdot + \cdot A_{1,M+1}B_{N+1,M} \cdot + \cdot A_{1,M+1}B_{N+1,M+1}) \end{vmatrix}$$

This proof is in thousands of textbooks from High school to college to PhD so I won't prove it here. Taking the pre-multiplier and multiplying each element across it's corresponding row we get:

$$
\begin{vmatrix} A_{11} \\ \cdot \\ \cdot \\ A_{1J} \\ \cdot \\ A_{1M} \end{vmatrix} \circ \begin{vmatrix} B_{11} \cdot \cdot \cdot B_{1K} \cdot \cdot \cdot B_{1M} \\ \cdot \qquad \cdot \qquad \cdot \\ B_{J1} \cdot \cdot \cdot B_{JK} \cdot \cdot \cdot B_{JM} \\ \cdot \qquad \cdot \qquad \cdot \\ B_{N1} \cdot \cdot \cdot B_{NK} \cdot \cdot \cdot B_{NM} \cdot \end{vmatrix} \qquad =
$$

$$
\begin{vmatrix} A_{11}(B_{11} \cdot \cdot \cdot B_{1K} \cdot \cdot \cdot B_{1M}) \\ \cdot \qquad \cdot \qquad \cdot \\ A_{1J}(B_{J1} \cdot \cdot \cdot B_{JK} \cdot \cdot \cdot B_{JM}) \\ \cdot \qquad \cdot \qquad \cdot \\ A_{1M}(B_{N1} \cdot \cdot \cdot B_{NK} \cdot \cdot \cdot B_{NM}) \end{vmatrix} = \begin{vmatrix} A_{11}B_{11} \cdot + \cdot A_{11}B_{1K} \cdot + \cdot A_{11}B_{1M} \\ \cdot \qquad \cdot \qquad \cdot \\ A_{1J}B_{J1} \cdot + \cdot A_{1J}B_{JK} \cdot + \cdot A_{1J}B_{JM} \\ \cdot \qquad \cdot \qquad \cdot \\ A_{1M}B_{N1} \cdot + \cdot A_{1M}B_{NK} \cdot + \cdot A_{1M}B_{NM} \end{vmatrix}
$$

This is for all mxn Matrices, now to prove it works for all matrices m+1, n+1:

$$
\begin{vmatrix} A_{11} \\ \cdot \\ A_{1J} \\ \cdot \\ A_{1M} \\ \cdot \\ A_{1,M+1} \end{vmatrix} \circ \begin{vmatrix} B_{11} \cdot \cdot \cdot B_{1K} \cdot \cdot \cdot B_{1M} \cdot \cdot \cdot B_{1,M+1} \\ \cdot \qquad \cdot \qquad \cdot \qquad \cdot \\ B_{J1} \cdot \cdot \cdot B_{JK} \cdot \cdot \cdot B_{JM} \cdot \cdot \cdot B_{J,M+1} \\ \cdot \qquad \cdot \qquad \cdot \qquad \cdot \\ B_{N1} \cdot \cdot \cdot B_{NK} \cdot \cdot \cdot B_{NM} \cdot \cdot \cdot B_{N,M+1} \\ \cdot \qquad \cdot \qquad \cdot \qquad \cdot \\ B_{N+1,1} \cdot \cdot B_{N+1,K} \cdot B_{N+1,M} \cdot \cdot B_{N+1,M+1} \end{vmatrix} =
$$

$$
\begin{vmatrix} A_{11}(B_{11} \cdot \cdot \cdot B_{1K} \cdot \cdot \cdot B_{1M} \cdot \cdot \cdot B_{1,M+1}) \\ \cdot \qquad \cdot \qquad \cdot \qquad \cdot \\ A_{1J}(B_{J1} \cdot \cdot \cdot B_{JK} \cdot \cdot \cdot B_{JM} \cdot \cdot \cdot B_{J,M+1}) \\ \cdot \qquad \cdot \qquad \cdot \qquad \cdot \\ A_{1M}(B_{N1} \cdot \cdot \cdot B_{NK} \cdot \cdot \cdot B_{NM} \cdot \cdot \cdot B_{N,M+1}) \\ \cdot \qquad \cdot \qquad \cdot \qquad \cdot \\ A_{1,M+1}(B_{N+1,1} \cdot \cdot B_{N+1,K} \cdot \cdot B_{N+1,M} \cdot \cdot B_{N+1,M+1}) \end{vmatrix} =
$$

$$
\begin{vmatrix} A_{11}B_{11} \cdot + \cdot A_{11}B_{1K} \cdot + \cdot A_{11}B_{1M} \cdot + \cdot A_{11}B_{1,M+1} \\ \cdot \qquad \cdot \qquad \cdot \qquad \cdot \\ A_{1J}B_{J1} \cdot + \cdot A_{1J}B_{JK} \cdot + \cdot A_{1J}B_{JM} \cdot + \cdot A_{1J}B_{J,M+1} \\ \cdot \qquad \cdot \qquad \cdot \qquad \cdot \\ A_{1M}B_{N1} \cdot + \cdot A_{1M}B_{NK} \cdot + \cdot A_{1M}B_{NM} \cdot + \cdot A_{1M}B_{N,M+1} \\ \cdot \qquad \cdot \qquad \cdot \qquad \cdot \\ A_{1,M+1}B_{N+1,1} \cdot + \cdot A_{1,M+1}B_{N+1,K} \cdot + \cdot A_{1,M+1}B_{N+1,M} \cdot + \cdot A_{1,M+1}B_{N+1,M+1} \end{vmatrix}
$$

QED

PROOF #1: Regular matrix multiplication:

$$[A]_{ij}[B]_{jk} = [C]_{ik} \cdot [A]^{T}_{ij} \circ [B]_{jk} = C_{jk}$$

First I will do a micro-proof for simplicity, then a regular proof.

$$
\begin{vmatrix} A_{11} & A_{21} & A_{31} \\ A_{12} & A_{22} & A_{32} \\ A_{13} & A_{23} & A_{33} \\ A_{14} & A_{24} & A_{34} \end{vmatrix} \circ \begin{vmatrix} B_{11} & B_{12} & B_{13} \\ B_{21} & B_{22} & B_{23} \\ B_{31} & B_{32} & B_{33} \\ B_{41} & B_{42} & B_{43} \end{vmatrix} \qquad =
$$

Separating each column in the $[A]^{T}$ matrix into a column matrix, we cross multiply and get the following Nested Array:

$$\left|\begin{matrix} A_{11} \\ A_{12} \\ A_{13} \\ A_{14} \end{matrix}\right| \circ \left|\begin{matrix} B_{11} & B_{12} & B_{13} \\ B_{21} & B_{22} & B_{23} \\ B_{31} & B_{32} & B_{33} \\ B_{41} & B_{42} & B_{43} \end{matrix}\right| \quad \left|\begin{matrix} A_{21} \\ A_{22} \\ A_{23} \\ A_{24} \end{matrix}\right| \circ \left|\begin{matrix} B_{11} & B_{12} & B_{13} \\ B_{21} & B_{22} & B_{23} \\ B_{31} & B_{32} & B_{33} \\ B_{41} & B_{42} & B_{43} \end{matrix}\right| \quad \left|\begin{matrix} A_{31} \\ A_{32} \\ A_{33} \\ A_{34} \end{matrix}\right| \circ \left|\begin{matrix} B_{11} & B_{12} & B_{13} \\ B_{21} & B_{22} & B_{23} \\ B_{31} & B_{32} & B_{33} \\ B_{41} & B_{42} & B_{43} \end{matrix}\right|$$

We take the first column in $[A]^{\mathrm{T}}$ and multiply it straight across the $[B]_{jk}$ matrix. This operation is equivalent to diagonalizing the first column of $[A]_{ij}$ and multiplying across $[B]$. i.e.

$$\begin{bmatrix} A_{11} & 0 & 0 & 0 \\ 0 & A_{12} & 0 & 0 \\ 0 & 0 & A_{13} & 0 \\ 0 & 0 & 0 & A_{14} \end{bmatrix} \cdot \begin{bmatrix} B_{11} & B_{12} & B_{13} \\ B_{21} & B_{22} & B_{23} \\ B_{31} & B_{32} & B_{33} \\ B_{41} & B_{42} & B_{43} \end{bmatrix} \rightarrow \begin{bmatrix} A_{11}B_{11} & A_{11}B_{12} & A_{11}B_{13} \\ A_{12}B_{21} & A_{12}B_{22} & A_{12}B_{23} \\ A_{13}B_{31} & A_{13}B_{32} & A_{13}B_{33} \\ A_{14}B_{41} & A_{14}B_{42} & A_{14}B_{43} \end{bmatrix}$$

$$\begin{bmatrix} A_{21} & 0 & 0 & 0 \\ 0 & A_{22} & 0 & 0 \\ 0 & 0 & A_{23} & 0 \\ 0 & 0 & 0 & A_{24} \end{bmatrix} \cdot \begin{bmatrix} B_{11} & B_{12} & B_{13} \\ B_{21} & B_{22} & B_{23} \\ B_{31} & B_{32} & B_{33} \\ B_{41} & B_{42} & B_{43} \end{bmatrix} \rightarrow \begin{bmatrix} A_{21}B_{11} & A_{21}B_{12} & A_{21}B_{13} \\ A_{22}B_{21} & A_{22}B_{22} & A_{22}B_{23} \\ A_{23}B_{31} & A_{23}B_{32} & A_{23}B_{33} \\ A_{24}B_{41} & A_{24}B_{42} & A_{24}B_{43} \end{bmatrix}$$

$$\begin{bmatrix} A_{31} & 0 & 0 & 0 \\ 0 & A_{32} & 0 & 0 \\ 0 & 0 & A_{33} & 0 \\ 0 & 0 & 0 & A_{34} \end{bmatrix} \cdot \begin{bmatrix} B_{11} & B_{12} & B_{13} \\ B_{21} & B_{22} & B_{23} \\ B_{31} & B_{32} & B_{33} \\ B_{41} & B_{42} & B_{43} \end{bmatrix} \rightarrow \begin{bmatrix} A_{31}B_{11} & A_{31}B_{12} & A_{31}B_{13} \\ A_{32}B_{21} & A_{32}B_{22} & A_{32}B_{23} \\ A_{33}B_{31} & A_{33}B_{32} & A_{33}B_{33} \\ A_{34}B_{41} & A_{34}B_{42} & A_{34}B_{43} \end{bmatrix}$$

This is equal to:

$$\begin{bmatrix} \begin{bmatrix} A_{11}B_{11} & A_{11}B_{12} & A_{11}B_{13} \\ A_{12}B_{21} & A_{12}B_{22} & A_{12}B_{23} \\ A_{13}B_{31} & A_{13}B_{32} & A_{13}B_{33} \\ A_{14}B_{41} & A_{14}B_{42} & A_{14}B_{43} \end{bmatrix} & \begin{bmatrix} A_{21}B_{11} & A_{21}B_{12} & A_{21}B_{13} \\ A_{22}B_{21} & A_{22}B_{22} & A_{22}B_{23} \\ A_{23}B_{31} & A_{23}B_{32} & A_{23}B_{33} \\ A_{24}B_{41} & A_{24}B_{42} & A_{24}B_{43} \end{bmatrix} & \begin{bmatrix} A_{31}B_{11} & A_{31}B_{12} & A_{31}B_{13} \\ A_{32}B_{21} & A_{32}B_{22} & A_{32}B_{23} \\ A_{33}B_{31} & A_{33}B_{32} & A_{33}B_{33} \\ A_{34}B_{41} & A_{34}B_{42} & A_{34}B_{43} \end{bmatrix} \end{bmatrix}$$

But this is a nested array, and I will show later that when we transpose a nested array, only the matrices are transposed and not their elements. Since in Half-Multiplying we originally transposed the Spreadsheet Matrix, we must now re-transpose the nested array back in it's un-transposed state. Then we will remove the inner brackets of the Nested Array using Partitioning and put the sub-Matrices into a single array.

$$\begin{bmatrix} A_{11}B_{11} & A_{11}B_{12} & A_{11}B_{13} \\ A_{12}B_{21} & A_{12}B_{22} & A_{12}B_{23} \\ A_{13}B_{31} & A_{13}B_{32} & A_{13}B_{33} \\ A_{14}B_{41} & A_{14}B_{42} & A_{14}B_{43} \\ A_{21}B_{11} & A_{21}B_{12} & A_{21}B_{13} \\ A_{22}B_{21} & A_{22}B_{22} & A_{22}B_{23} \\ A_{23}B_{31} & A_{23}B_{32} & A_{23}B_{33} \\ A_{24}B_{41} & A_{24}B_{42} & A_{24}B_{43} \\ A_{31}B_{11} & A_{31}B_{12} & A_{31}B_{13} \\ A_{32}B_{21} & A_{32}B_{22} & A_{32}B_{23} \\ A_{33}B_{31} & A_{33}B_{32} & A_{33}B_{33} \\ A_{34}B_{41} & A_{34}B_{42} & A_{34}B_{43} \end{bmatrix}$$

Now we will sum the columns for each individual matrix: (we must remember that we are working with nested arrays, but computers are not programmed to handle these yet, so we must set up the pre-multiplier so that it operates on the individual Sub-Matrices)

$$\begin{pmatrix} 1 & 1 & 1 & 1 & 0 & 0 & 0 & 0 & 0 & 0 & 0 & 0 \\ 0 & 0 & 0 & 0 & 1 & 1 & 1 & 1 & 0 & 0 & 0 & 0 \\ 0 & 0 & 0 & 0 & 0 & 0 & 0 & 0 & 1 & 1 & 1 & 1 \end{pmatrix} \cdot \begin{bmatrix} A_{1,1} \cdot B_{1,1} & A_{1,1} \cdot B_{1,2} & A_{1,1} \cdot B_{1,3} \\ A_{1,2} \cdot B_{2,1} & A_{1,2} \cdot B_{2,2} & A_{1,2} \cdot B_{2,3} \\ A_{1,3} \cdot B_{3,1} & A_{1,3} \cdot B_{3,2} & A_{1,3} \cdot B_{3,3} \\ A_{1,4} \cdot B_{4,1} & A_{1,4} \cdot B_{4,2} & A_{1,4} \cdot B_{4,3} \\ A_{2,1} \cdot B_{1,1} & A_{2,1} \cdot B_{1,2} & A_{2,1} \cdot B_{1,3} \\ A_{2,2} \cdot B_{2,1} & A_{2,2} \cdot B_{2,2} & A_{2,2} \cdot B_{2,3} \\ A_{2,3} \cdot B_{3,1} & A_{2,3} \cdot B_{3,2} & A_{2,3} \cdot B_{3,3} \\ A_{2,4} \cdot B_{4,1} & A_{2,4} \cdot B_{4,2} & A_{2,4} \cdot B_{4,3} \\ A_{3,1} \cdot B_{1,1} & A_{3,1} \cdot B_{1,2} & A_{3,1} \cdot B_{1,3} \\ A_{3,2} \cdot B_{2,1} & A_{3,2} \cdot B_{2,2} & A_{3,2} \cdot B_{2,3} \\ A_{3,3} \cdot B_{3,1} & A_{3,3} \cdot B_{3,2} & A_{3,3} \cdot B_{3,3} \\ A_{3,4} \cdot B_{4,1} & A_{3,4} \cdot B_{4,2} & A_{3,4} \cdot B_{4,3} \end{bmatrix}$$

The answer to this is quite long on mathcad, but since we are working with nested arrays, we may also look at the multiplication in this manner:

$$(1\ 1\ 1\ 1)\cdot\begin{bmatrix} A_{11}\cdot B_{11} & A_{11}\cdot B_{12} & A_{11}\cdot B_{13} \\ A_{12}\cdot B_{21} & A_{12}\cdot B_{22} & A_{12}\cdot B_{23} \\ A_{13}\cdot B_{31} & A_{13}\cdot B_{32} & A_{13}\cdot B_{33} \\ A_{14}\cdot B_{41} & A_{14}\cdot B_{42} & A_{14}\cdot B_{43} \end{bmatrix} \rightarrow \left(A_{11}\cdot B_{11}+A_{12}\cdot B_{21}+A_{13}\cdot B_{31}+A_{14}\cdot B_{41} \quad A_{11}\cdot B_{12}+A_{12}\cdot B_{22}+A_{13}\cdot B_{32}+A_{14}\cdot B_{42} \quad A_{11}\cdot B_{13}+A_{12}\cdot B_{23}+A_{13}\cdot B_{33}+A_{14}\cdot B_{43} \right)$$

$$(1\ 1\ 1\ 1)\cdot\begin{bmatrix} A_{21}\cdot B_{11} & A_{21}\cdot B_{12} & A_{21}\cdot B_{13} \\ A_{22}\cdot B_{21} & A_{22}\cdot B_{22} & A_{22}\cdot B_{23} \\ A_{23}\cdot B_{31} & A_{23}\cdot B_{32} & A_{23}\cdot B_{33} \\ A_{24}\cdot B_{41} & A_{24}\cdot B_{42} & A_{24}\cdot B_{43} \end{bmatrix} \rightarrow \left(A_{21}\cdot B_{11}+A_{22}\cdot B_{21}+A_{23}\cdot B_{31}+A_{24}\cdot B_{41} \quad A_{21}\cdot B_{12}+A_{22}\cdot B_{22}+A_{23}\cdot B_{32}+A_{24}\cdot B_{42} \quad A_{21}\cdot B_{13}+A_{22}\cdot B_{23}+A_{23}\cdot B_{33}+A_{24}\cdot B_{43} \right)$$

$$(1\ 1\ 1\ 1)\cdot\begin{bmatrix} A_{31}\cdot B_{11} & A_{31}\cdot B_{12} & A_{31}\cdot B_{13} \\ A_{32}\cdot B_{21} & A_{32}\cdot B_{22} & A_{32}\cdot B_{23} \\ A_{33}\cdot B_{31} & A_{33}\cdot B_{32} & A_{33}\cdot B_{33} \\ A_{34}\cdot B_{41} & A_{34}\cdot B_{42} & A_{34}\cdot B_{43} \end{bmatrix} \rightarrow \left(A_{31}\cdot B_{11}+A_{32}\cdot B_{21}+A_{33}\cdot B_{31}+A_{34}\cdot B_{41} \quad A_{31}\cdot B_{12}+A_{32}\cdot B_{22}+A_{33}\cdot B_{32}+A_{34}\cdot B_{42} \quad A_{31}\cdot B_{13}+A_{32}\cdot B_{23}+A_{33}\cdot B_{33}+A_{34}\cdot B_{43} \right)$$

$$\begin{bmatrix} \left(A_{11}\cdot B_{11}+A_{12}\cdot B_{21}+A_{13}\cdot B_{31}+A_{14}\cdot B_{41} \quad A_{11}\cdot B_{12}+A_{12}\cdot B_{22}+A_{13}\cdot B_{32}+A_{14}\cdot B_{42} \quad A_{11}\cdot B_{13}+A_{12}\cdot B_{23}+A_{13}\cdot B_{33}+A_{14}\cdot B_{43}\right) \\ \left(A_{21}\cdot B_{11}+A_{22}\cdot B_{21}+A_{23}\cdot B_{31}+A_{24}\cdot B_{41} \quad A_{21}\cdot B_{12}+A_{22}\cdot B_{22}+A_{23}\cdot B_{32}+A_{24}\cdot B_{42} \quad A_{21}\cdot B_{13}+A_{22}\cdot B_{23}+A_{23}\cdot B_{33}+A_{24}\cdot B_{43}\right) \\ \left(A_{31}\cdot B_{11}+A_{32}\cdot B_{21}+A_{33}\cdot B_{31}+A_{34}\cdot B_{41} \quad A_{31}\cdot B_{12}+A_{32}\cdot B_{22}+A_{33}\cdot B_{32}+A_{34}\cdot B_{42} \quad A_{31}\cdot B_{13}+A_{32}\cdot B_{23}+A_{33}\cdot B_{33}+A_{34}\cdot B_{43}\right) \end{bmatrix}$$

But this is equal to:

$$\begin{bmatrix} A_{11} & A_{12} & A_{13} & A_{14} \\ A_{21} & A_{22} & A_{23} & A_{24} \\ A_{31} & A_{32} & A_{33} & A_{34} \end{bmatrix}\cdot\begin{bmatrix} B_{11} & B_{12} & B_{13} \\ B_{21} & B_{22} & B_{23} \\ B_{31} & B_{32} & B_{33} \\ B_{41} & B_{42} & B_{43} \end{bmatrix} \rightarrow \begin{bmatrix} A_{11}\cdot B_{11}+A_{12}\cdot B_{21}+A_{13}\cdot B_{31}+A_{14}\cdot B_{41} \quad A_{11}\cdot B_{12}+A_{12}\cdot B_{22}+A_{13}\cdot B_{32}+A_{14}\cdot B_{42} \quad A_{11}\cdot B_{13}+A_{12}\cdot B_{23}+A_{13}\cdot B_{33}+A_{14}\cdot B_{43} \\ A_{21}\cdot B_{11}+A_{22}\cdot B_{21}+A_{23}\cdot B_{31}+A_{24}\cdot B_{41} \quad A_{21}\cdot B_{12}+A_{22}\cdot B_{22}+A_{23}\cdot B_{32}+A_{24}\cdot B_{42} \quad A_{21}\cdot B_{13}+A_{22}\cdot B_{23}+A_{23}\cdot B_{33}+A_{24}\cdot B_{43} \\ A_{31}\cdot B_{11}+A_{32}\cdot B_{21}+A_{33}\cdot B_{31}+A_{34}\cdot B_{41} \quad A_{31}\cdot B_{12}+A_{32}\cdot B_{22}+A_{33}\cdot B_{32}+A_{34}\cdot B_{42} \quad A_{31}\cdot B_{13}+A_{32}\cdot B_{23}+A_{33}\cdot B_{33}+A_{34}\cdot B_{43} \end{bmatrix}$$

micro-QED

Generalized Proof

$$\begin{bmatrix} A_{11} & A_{1J} & A_{1N} & A_{1,N+1} \\ A_{I1} & A_{IJ} & A_{IN} & A_{I,N+1} \\ A_{M1} & A_{MJ} & A_{MN} & A_{M,N+1} \\ A_{M+1,1} & A_{M+1,J} & A_{M+1,N} & A_{M+1,N+1} \end{bmatrix}\begin{bmatrix} B_{11} & B_{1K} & B_{1M} & B_{1,M+1} \\ B_{J1} & B_{JK} & B_{JM} & B_{J,M+1} \\ B_{N1} & B_{NK} & B_{NM} & B_{N,M+1} \\ B_{N+1,1} & B_{N+1,K} & B_{N+1,M} & B_{N+1,M+1} \end{bmatrix}$$

I am going to have to type this in by hand, Word 7 doesn't have the memory to handle the math here. The Solution for a regular M x N Matrix Multiplication is given by:

$$\begin{vmatrix} A_{11}B_{11}. \ .+ \ .A_{13}B_{J1}+. \ . \ . \ A_{1N}B_{N1} \\ \cdot \\ A_{11}B_{11}. \ .+ \ .A_{13}B_{21}+. \ . \ . \ A_{1N}B_{N1} \\ \cdot \\ A_{M1}B_{11}. \ .+ \ .A_{M3}B_{21}+. \ . \ . \ A_{MN}B_{N1} \end{vmatrix} \quad \begin{matrix} A_{11}B_{1K}+. \ . \ . \ .A_{13}B_{JK}. \ .+ \ . \ A_{1N}B_{NK}. \\ \cdot \\ A_{11}B_{1K}+. \ . \ . \ .A_{13}B_{JK}. \ + \ A_{1N}B_{NK} \\ \cdot \\ A_{M1}B_{1K}+. \ . \ . \ .A_{M3}B_{JK}. \ + \ A_{1N}B_{NK} \end{matrix}$$

$$\begin{matrix} A_{11}B_{1M} \ . \ + \ .A_{13}B_{JM} \ . \ .+. \ A_{1N}B_{NM} \\ \cdot \\ A_{11}B_{1M} \ . \ +. \ A_{13}B_{JM} \ . \ .+..A_{1N}B_{NM} \\ \cdot \\ A_{M1}B_{1M} \ . \ + \ .A_{M3}B_{JM}+. \ .+. \ A_{MN}B_{NM} \end{matrix} \quad \begin{vmatrix} A_{11}B_{1M+1} \ .+.A_{13}B_{JM} \ . \ . \ +.A_{1N}B_{NM} \\ \cdot \\ A_{11}B_{1M+1}. \ .+.A_{13}B_{JM} \ . \ . \ +.A_{1N}B_{NM} \\ \cdot \\ A_{M1}B_{1M}. \ .+.A_{MJ}B_{JM}. \ . \ .+.A_{MN}B_{NM} \end{vmatrix}$$

Now we shall achieve the same answer by the properties of the Half-Multiplier Operator. We shall first accomplish this by multiplying by the diagonal form for all matrices size MxN as this is the most familiar to mathematicians.

Diagonalizing the first transposed column (remember, for this to work, A_{iN} must be transposed and we multiply across the B_{nm} matrix.

$$\begin{vmatrix} A_{11} \\ \cdot \\ & \cdot \\ & & A_{1J} \\ & & & \cdot \\ & & & & \cdot \\ & & & & & A_{1N} \end{vmatrix} \ \begin{vmatrix} B_{11}. \ . \ .B_{1K}. \ . \ .B_{1M} \\ \cdot \quad\quad \cdot \quad\quad \cdot \\ \cdot \quad\quad \cdot \quad\quad \cdot \\ B_{J1}. \ . \ .B_{JK}. \ . \ .B_{JM} \\ \cdot \quad\quad \cdot \quad\quad \cdot \\ \cdot \quad\quad \cdot \quad\quad \cdot \\ B_{N1}. \ . \ .B_{NK}. \ . \ .B_{NM} \end{vmatrix} \quad =$$

$$\begin{vmatrix} A_{11}(B_{11}. \ . \ .B_{1K}. \ . \ .B_{1M}) \\ \cdot \quad\quad\quad \cdot \quad\quad \cdot \\ \cdot \quad\quad\quad \cdot \quad\quad \cdot \\ A_{1J} (B_{J1}. \ . \ .B_{JK}. \ . \ .B_{JM}) \\ \cdot \quad\quad\quad \cdot \quad\quad \cdot \\ \cdot \quad\quad\quad \cdot \quad\quad \cdot \\ A_{1N} (B_{N1}. \ . \ .B_{NK}. \ . \ .B_{NM}) \end{vmatrix} \quad =$$

$$\begin{vmatrix} A_{11}B_{11}. \ . \ .A_{11}B_{1K}. \ . \ .A_{11}B_{1M} \\ \cdot \quad\quad\quad \cdot \quad\quad \cdot \\ \cdot \quad\quad\quad \cdot \quad\quad \cdot \\ A_{1J} B_{J1}. \ . \ .A_{1J}B_{JK}. \ . \ .A_{1J}B_{JM} \\ \cdot \quad\quad\quad \cdot \quad\quad \cdot \\ \cdot \quad\quad\quad \cdot \quad\quad \cdot \\ A_{1N} B_{N1}. \ . \ .A_{1N}B_{NK}. \ . \ .A_{1N}B_{NM} \end{vmatrix}$$

Now for the i th column:

$$
\begin{vmatrix}
A_{11} & & & \\
& \cdot & & \\
& & A_{1\,J} & \\
& & & \cdot \\
& & & A_{1\,N}
\end{vmatrix}
\begin{vmatrix}
B_{11}\cdot & \cdot B_{1K}\cdot & \cdot B_{1M} \\
& \cdot & \cdot \\
B_{J1}\cdot & \cdot B_{JK}\cdot & \cdot B_{JM} \\
& \cdot & \cdot \\
B_{N1}\cdot & \cdot B_{NK}\cdot & \cdot B_{NM}
\end{vmatrix}
=
$$

$$
\begin{vmatrix}
A_{11}(B_{11}\cdot & \cdot B_{1K}\cdot & \cdot B_{1M}) \\
& \cdot & \cdot \\
A_{1\,J}(B_{J1}\cdot & \cdot B_{JK}\cdot & \cdot B_{JM}) \\
& \cdot & \cdot \\
A_{1\,N}(B_{N1}\cdot & \cdot B_{NK}\cdot & \cdot B_{NM})
\end{vmatrix}
=
$$

$$
\begin{vmatrix}
A_{11}B_{11}\cdot & \cdot A_{11}B_{1K}\cdot & \cdot A_{11}B_{1M} \\
& \cdot & \cdot \\
A_{1J}B_{J1}\cdot & \cdot A_{1J}B_{JK}\cdot & \cdot A_{1J}B_{JM} \\
& \cdot & \cdot \\
A_{1N}B_{N1}\cdot & \cdot A_{1N}B_{NK}\cdot & \cdot A_{1N}B_{NM}
\end{vmatrix}
$$

And now for the Mth column:

$$
\begin{vmatrix}
A_{M1} & & & \\
& \cdot & & \\
& & A_{MJ} & \\
& & & \cdot \\
& & & A_{MN}
\end{vmatrix}
\begin{vmatrix}
B_{11}\cdot & \cdot B_{1K}\cdot & \cdot B_{1M} \\
& \cdot & \cdot \\
B_{J1}\cdot & \cdot B_{JK}\cdot & \cdot B_{JM} \\
& \cdot & \cdot \\
B_{N1}\cdot & \cdot B_{NK}\cdot & \cdot B_{NM}
\end{vmatrix}
=
\begin{vmatrix}
A_{M1}(B_{11}\cdot & \cdot B_{1K}\cdot & \cdot B_{1M}) \\
& \cdot & \cdot \\
A_{MJ}(B_{J1}\cdot & \cdot B_{JK}\cdot & \cdot B_{JM}) \\
& \cdot & \cdot \\
A_{MN}(B_{N1}\cdot & \cdot B_{NK}\cdot & \cdot B_{NM})
\end{vmatrix}
=
$$

$$
\begin{vmatrix}
A_{M1}B_{11}\cdot & \cdot A_{M1}B_{1K}\cdot & \cdot A_{M1}B_{1M} \\
& \cdot & \cdot \\
A_{MJ}B_{J1}\cdot & \cdot A_{MJ}B_{JK}\cdot & \cdot A_{MJ}B_{JM} \\
& \cdot & \cdot \\
A_{MN}B_{N1}\cdot & \cdot A_{MN}B_{NK}\cdot & \cdot A_{MN}B_{NM}
\end{vmatrix}
$$

Now let's make these into a Nested array and sum each column
of the individual sub-matrix.

$$\begin{vmatrix} \begin{vmatrix} A_{11}B_{11}. & . & .A_{11}B_{1K}. & . & .A_{11}B_{1M} \\ . & & . & & . \\ A_{1J}B_{J1}. & . & .A_{1J}B_{JK}. & . & .A_{1J}B_{JM} \\ . & & . & & . \\ A_{1N}B_{N1}. & . & .A_{1N}B_{NK}. & . & .A_{1N}B_{NM} \end{vmatrix} \\ \begin{vmatrix} A_{11}B_{11}. & . & .A_{11}B_{1K}. & . & .A_{11}B_{1M} \\ . & & . & & . \\ A_{1J}B_{J1}. & . & .A_{1J}B_{JK}. & . & .A_{1J}B_{JM} \\ . & & . & & . \\ A_{1N}B_{N1}. & . & .A_{1N}B_{NK}. & . & .A_{1N}B_{NM} \end{vmatrix} \\ \begin{vmatrix} A_{M1}B_{11}. & . & .A_{M1}B_{1K}. & . & .A_{M1}B_{1M} \\ . & & . & & . \\ A_{MJ}B_{J1}. & . & .A_{MJ}B_{JK}. & . & .A_{MJ}B_{JM} \\ . & & . & & . \\ A_{MN}B_{N1}. & . & .A_{MN}B_{NK}. & . & .A_{MN}B_{NM} \end{vmatrix} \end{vmatrix}$$

$$\begin{vmatrix} A_{11}B_{11}..+.. A_{1J}B_{J1}..+.. A_{1N}B_{N1} & A_{11}B_{1K}..+..A_{1J}B_{JK}..+.. A_{1N}B_{NK} & A_{11}B_{1M}..+.. A_{1J}B_{JM}..+..A_{1N}B_{NM} \\ A_{11}B_{11}..+.. A_{1J}B_{J1}..+.. A_{1N}B_{N1} & A_{11}B_{1K}..+..A_{1J}B_{JK}..+.. A_{1N}B_{NK} & A_{11}B_{1M}..+..A_{1J}B_{JM}..+.. A_{1N}B_{NM} \\ A_{M1}B_{11}..+.. A_{MJ}B_{J1}..+..A_{MN}B_{N1} & A_{M1}B_{1K}..+.. A_{MJ}B_{JK}..+.. A_{MN}B_{NK} & A_{M1}B_{1M}..+.. A_{MJ}B_{JM}..+..A_{MN}B_{NM} \end{vmatrix}$$

Now let's do the same for all M+1, N+1 Matrices:

$$\begin{vmatrix} A_{11} & & & \\ . & & & \\ & A_{11} & & \\ . & & & \\ & & A_{1N} & \\ & & & A1,_{M+1} \end{vmatrix} \begin{vmatrix} B_{11}. & . & .B_{1K}. & . & .B_{1M}. & . & .B_{1,M+1} \\ . & & . & & . & & . \\ B_{J1}. & . & .B_{JK}. & . & .B_{JM}. & . & .B_{J,M+1} \\ . & & . & & . & & . \\ B_{N1}. & . & .B_{NK}. & . & .B_{NM}. & . & .B_{N,M+1} \\ . & & . & & . & & . \\ B_{N+1,1}. & .B_{N+1,K}. & .B_{N+1,M}. & .B_{N+1,M+1} \end{vmatrix} =$$

$$\begin{vmatrix} A_{11} (B_{11}. & . & .B_{1K}. & . & .B_{1M}. & . & .B_{1,M+1}) \\ . & . & . & & . & & . \\ A_{1J} (B_{J1}. & . & .B_{JK}. & . & .B_{JM}. & . & .B_{J,M+1}) \\ . & . & . & & . & & . \\ A_{1N} (B_{N1}. & . & .B_{NK}. & . & .B_{NM}. & . & .B_{N,M+1}) \\ . & . & . & & . & & . \\ A_{1,N+1} (B_{N+1,1}. & .B_{N+1,K}. & .B_{N+1,M}. & .B_{N+1,M+1}) \end{vmatrix} = \begin{vmatrix} A_{11}B_{11}. & . & A_{11}B_{1K}. & . & .A_{11}B_{1M}. & . & .A_{11}B_{1,M+1} \\ . & & . & & . & & . \\ A_{1J}B_{J1}. & . & A_{1J}B_{JK}. & . & A_{1J}B_{JM}. & . & .A_{1J}B_{J,M+1} \\ . & & . & & . & & . \\ A_{1N}B_{N1}. & . & A_{1N}B_{NK}. & . & A_{1N}B_{NM}. & . & A_{1N}B_{N,M+1} \\ . & & . & & . & & . \\ A_{1,N+1}B_{N+1,1}. & .A_{1,N+1}B_{N+1,K}. & .A_{1,N+1}B_{N+1,M}. & .A_{1,N+1}B_{N+1,M+1} \end{vmatrix}$$

The summed columns are equal to:

$$\begin{aligned} A_{11}B_{11}. \ .+ \ .A_{1J}B_{J1}+. \ . \ . \ A_{1N}B_{N1} + \ \ \ A_{1,N+1}B_{N+1,1} \ ] & \ [A_{11}B_{1K}+. \ . \ . \ .A_{1J}B_{JK}. \ .+ \ . \ A_{1N}B_{NK} + A_{1,N+1}B_{N+1,K}] \\ A_{11}B_{1M} \ . \ + \ .A_{1J}B_{JM}+. \ .+. \ A_{1N}B_{NM} + . \ A_{1,N+1}B_{N+1,M}] & \ [A_{11}B_{1,M+1}. \ .+.A_{1J}B_{J,M+1}. \ .+.A_{1N}B_{N,M+1} + \ \ \ A_{1,N+1}B_{N+1,M+1}] \end{aligned}$$

Now we multiply by the Jth column:

$$
\begin{vmatrix}
A_{i1} & & & \\
\ddots & & & \\
& A_{iJ} & & \\
& & A_{iN} & \\
& & & A_{1,N+1}
\end{vmatrix}
\begin{vmatrix}
B_{11} & \cdots & B_{1K} & \cdots & B_{1M} & \cdots & B_{1,M+1} \\
\vdots & & \vdots & & \vdots & & \vdots \\
B_{J1} & \cdots & B_{JK} & \cdots & B_{JM} & \cdots & B_{J,M+1} \\
\vdots & & \vdots & & \vdots & & \vdots \\
B_{N1} & \cdots & B_{NK} & \cdots & B_{NM} & \cdots & B_{N,M+1} \\
\vdots & & \vdots & & \vdots & & \vdots \\
B_{N+1,1} & \cdots & B_{N+1,K} & \cdots & B_{N+1,M} & \cdots & B_{N+1,M+1}
\end{vmatrix} =
$$

$$
\begin{vmatrix}
A_{i1}\,(B_{11} \cdots B_{1K} \cdots B_{1M} \cdots B_{1,M+1}) \\
\vdots \\
A_{iJ}\,(B_{J1} \cdots B_{JK} \cdots B_{JM} \cdots B_{J,M+1}) \\
\vdots \\
A_{iN}\,(B_{N1} \cdots B_{NK} \cdots B_{NM} \cdots B_{N,M+1}) \\
\vdots \\
A_{1,N+1}\,(B_{N+1,1} \cdots B_{N+1,K} \cdots B_{N+1,M} \cdots B_{N+1,M+1})
\end{vmatrix}
=
\begin{vmatrix}
A_{i1}B_{11} \cdots A_{i1}B_{1K} \cdots A_{i1}B_{1M} \cdots A_{i1}B_{1,M+1} \\
\vdots \\
A_{iJ}B_{J1} \cdots A_{iJ}B_{JK} \cdots A_{iJ}B_{JM} \cdots A_{iJ}B_{J,M+1} \\
\vdots \\
A_{iN}B_{N1} \cdots A_{iN}B_{NK} \cdots A_{iN}B_{NM} \cdots A_{iN}B_{N,M+1} \\
\vdots \\
A_{1,N+1}B_{N+1,1} \cdots A_{1,N+1}B_{N+1,K} \cdots A_{1,N+1}B_{N+1,M} \cdots A_{1,N+1}B_{N+1,M+1}
\end{vmatrix}
$$

Summing the columns we get:

$$[A_{i1}B_{11} \cdots + A_{iJ}B_{J1} + \cdots A_{iN}B_{N1} + A_{i,N+1}B_{N+1,1}] \quad [A_{i1}B_{1K} + \cdots A_{iJ}B_{JK} \cdots + A_{iN}B_{NK} + A_{i,N+1}B_{N+1,K}]$$

$$[A_{i1}B_{1M} \cdots + A_{iJ}B_{JM} \cdots + \cdots A_{iN}B_{NM} + A_{i,N+1}B_{N+1,M}] \quad [A_{i1}B_{1M+1} \cdots + A_{iJ}B_{JM+1} \cdots + \cdots + A_{iN}B_{NM+1} + A_{i,N+1}B_{N+1,M+1}]$$

Now we multiply by the Mth column:

$$
\begin{vmatrix}
A_{M1} & & & \\
\ddots & & & \\
& A_{MJ} & & \\
& & A_{MN} & \\
& & & A_{M,N+1}
\end{vmatrix}
\begin{vmatrix}
B_{11} & \cdots & B_{1K} & \cdots & B_{1M} & \cdots & B_{1,M+1} \\
\vdots & & \vdots & & \vdots & & \vdots \\
B_{J1} & \cdots & B_{JK} & \cdots & B_{JM} & \cdots & B_{J,M+1} \\
\vdots & & \vdots & & \vdots & & \vdots \\
B_{N1} & \cdots & B_{NK} & \cdots & B_{NM} & \cdots & B_{N,M+1} \\
\vdots & & \vdots & & \vdots & & \vdots \\
B_{N+1,1} & \cdots & B_{N+1,K} & \cdots & B_{N+1,M} & \cdots & B_{N+1,M+1}
\end{vmatrix} =
$$

$$
\begin{vmatrix}
A_{M1}\,(B_{11} \cdots B_{1K} \cdots B_{1M} \cdots B_{1,M+1}) \\
\vdots \\
A_{MJ}\,(B_{J1} \cdots B_{JK} \cdots B_{JM} \cdots B_{J,M+1}) \\
\vdots \\
A_{MN}\,(B_{N1} \cdots B_{NK} \cdots B_{NM} \cdots B_{N,M+1}) \\
\vdots \\
A_{M,N+1}\,(B_{N+1,1} \cdots B_{N+1,K} \cdots B_{N+1,M} \cdots B_{N+1,M+1})
\end{vmatrix}
=
\begin{vmatrix}
A_{M1}B_{11} \cdots A_{M1}B_{1K} \cdots A_{M1}B_{1M} \cdots A_{M1}B_{1,M+1} \\
\vdots \\
A_{MJ}B_{J1} \cdots A_{MJ}B_{JK} \cdots A_{MJ}B_{JM} \cdots A_{MJ}B_{J,M+1} \\
\vdots \\
A_{MN}B_{N1} \cdots A_{MN}B_{NK} \cdots A_{MN}B_{NM} \cdots A_{MN}B_{N,M+1} \\
\vdots \\
A_{M,N+1}B_{N+1,1} \cdots A_{M,N+1}B_{N+1,K} \cdots A_{M,N+1}B_{N+1,M} \cdots A_{M,N+1}B_{N+1,M+1}
\end{vmatrix}
$$

Summing the columns we get:

$$[A_{M1}B_{11} \cdots + \cdots A_{MJ}B_{J1} + \cdots A_{MN}B_{N1} \cdots + \cdots A_{M,N+1}B_{N+1,1}] \quad [A_{M1}B_{1K} \cdots + \cdots A_{MJ}B_{JK} \cdots + \cdots A_{MN}B_{NK} \cdots + \cdots A_{M,N+1}B_{N+1,K}]$$

$$[A_{M1}B_{1M} \cdots + \cdots A_{MJ}B_{JM} + \cdots + \cdots A_{MN}B_{NM} \cdots + \cdots A_{M,N+1}B_{N+1,M}] \quad [A_{M1}B_{1,M+1} \cdots + \cdots A_{MJ}B_{J,M+1} \cdots + \cdots + A_{MN}B_{N,M+1} + \cdots A_{M,N+1}B_{N+1,M+1}]$$

And finally we multiply by the M+1 th column:

$$\begin{vmatrix} A_{M+1,1} \\ \quad \cdot \\ \quad\quad A_{M+1,J} \\ \quad\quad\quad \cdot \\ \quad\quad\quad\quad A_{M+1,N} \\ \quad\quad\quad\quad\quad \cdot \\ \quad\quad\quad\quad\quad\quad A_{M+1,N+1} \end{vmatrix} \begin{Vmatrix} B_{11}\cdot\ \cdot\cdot B_{1K}\cdot\ \cdot\cdot B_{1M}\cdot\ \cdot\cdot B_{1,M+1} \\ \cdot\quad\quad\cdot\quad\quad\cdot\quad\quad\cdot \\ B_{J1}\cdot\ \cdot\cdot B_{JK}\cdot\ \cdot\cdot B_{JM}\cdot\ \cdot\cdot B_{J,M+1} \\ \cdot\quad\quad\cdot\quad\quad\cdot\quad\quad\cdot \\ B_{N1}\cdot\ \cdot\cdot B_{NK}\cdot\ \cdot\cdot B_{NM}\cdot\ \cdot\cdot B_{N,M+1} \\ \cdot\quad\quad\cdot\quad\quad\cdot\quad\quad\cdot \\ B_{N+1,1}\quad\quad\ \cdot\cdot B_{N+1,K}\cdot\ \cdot\cdot B_{N+1,M}\cdot\cdot B_{N+1,M+1} \end{Vmatrix} \quad =$$

$$\begin{vmatrix} A_{M+1,1}\ (B_{11}\cdot\ \cdot\cdot B_{1K}\cdot\ \cdot\cdot B_{1M}\cdot\ \cdot\cdot B_{1,M+1}) \\ \cdot\quad\quad\quad\quad\cdot\quad\quad\cdot\quad\quad\cdot \\ A_{M+1,J}\ (B_{J1}\cdot\ \cdot\cdot B_{JK}\cdot\ \cdot\cdot B_{JM}\cdot\ \cdot\cdot B_{J,M+1}) \\ \cdot\quad\quad\quad\quad\cdot\quad\quad\cdot\quad\quad\cdot \\ A_{M+1,N}\ (B_{N1}\cdot\ \cdot\cdot B_{NK}\cdot\ \cdot\cdot B_{NM}\cdot\ \cdot\cdot B_{N,M+1}) \\ \cdot\quad\quad\quad\quad\cdot\quad\quad\cdot\quad\quad\cdot \\ A_{M+1,N+1}\ (B_{N+1,1}\cdot\ \cdot B_{N+1,K}\cdot\ \cdot B_{N+1,M}\cdot\ \cdot B_{N+1,M+1}) \end{vmatrix} =$$

$$\begin{vmatrix} A_{M+1,1}B_{11}\cdot\ A_{M+1,1}B_{1K}\cdot\ \cdot\ A_{M+1,1}B_{1M}\cdot\ \cdot\ A_{M+1,1}B_{1,M+1} \\ \cdot\quad\quad\quad\cdot\quad\quad\quad\cdot\quad\quad\quad\cdot \\ A_{M+1,J}B_{J1}\cdot A_{M+1,J}\ B_{JK}\cdot\quad A_{M+1,J}\ B_{JM}\cdot\ \cdot\ A_{M+1,J}\ B_{J,M+1} \\ \cdot\quad\quad\quad\cdot\quad\quad\quad\cdot\quad\quad\quad\cdot \\ A_{M+1,N}B_{N1}\cdot A_{M+1,N}B_{NK}\cdot\ \cdot\ A_{M+1,N}B_{NM}\cdot\ \cdot\ A_{M+1,N}\ B_{N,M+1} \\ \cdot\quad\quad\quad\cdot\quad\quad\quad\cdot\quad\quad\quad\cdot \\ A_{M+1,N+1}B_{N+1,1}\cdot\ A_{M+1,N+1}B_{N+1,K}\cdot\ A_{M+1,N+1}B_{N+1,M}\cdot\ A_{M+1,N+1}B_{N+1,M+1} \end{vmatrix}$$

$[A_{M+1,1}B_{11}\cdot+\ \cdot A_{M+1,J}B_{J1}+\cdot\ \cdot\ \cdot\ A_{M+1,N}B_{N,1}\ +\ A_{M+1,N+1}B_{N+1,1}]$ $[A_{M+1,1}B_{1K}+\cdot\cdot A_{M+1,J}B_{JK}\cdot +\ \cdot\ A_{M+1,N}B_{NK}\ +\ A_{M+1,N+1}B_{N+1,K}]$

$A_{M+1,1}B_{1M}\ +\ \cdot A_{M+1,J}B_{JM}\cdot\ +\cdot\ A_{M+1,N}B_{NM}\ +\ A_{M+1,N+1}B_{N+1,M}]$ $[A_{M+1,1}B_{1,M+1}\cdot+\cdot A_{M+1,J}B_{J,M+1}\cdot+\cdot\cdot A_{M+1,N}B_{N,M+1}\ +\ A_{M+1,N+1}B_{N+1,M+1}]$

Transposing them (Stacking the nested arrays on top of each other we get):

Note: When we first Half-Multiply, we transpose the Pre-multiplier matrix. In order to get our regular matrix multiplication results we have to re-transpose the nested arrays, then sum their columns. (We could sum the columns then transpose also.)

$$\left|\begin{array}{l}
A_{11}\,B_{11} \cdot \cdot \cdot A_{11}B_{1K} \cdot \cdot \cdot A_{11}B_{1M} \cdot \cdot \cdot A_{11}B_{1,M+1} \\
\qquad \cdot \qquad\qquad \cdot \qquad\qquad \cdot \qquad\qquad \cdot \\
A_{1J}\,B_{J1} \cdot \cdot \cdot A_{1J}\,B_{JK} \cdot \cdot \cdot A_{1J}\,B_{JM} \cdot \cdot \cdot A_{1J}\,B_{J,M+1} \\
\qquad \cdot \qquad\qquad \cdot \qquad\qquad \cdot \qquad\qquad \cdot \\
A_{1N}\,B_{N1} \cdot \cdot \cdot A_{1N}\,B_{NK} \cdot \cdot \cdot A_{1N}\,B_{NM} \cdot \cdot \cdot A_{1N}\,B_{N,M+1} \\
\qquad \cdot \qquad\qquad \cdot \qquad\qquad \cdot \qquad\qquad \cdot \\
A_{1,N+1}B_{N+1,1} \cdot \cdot A_{1,N+1}B_{N+1,K} \cdot A_{1,N+1}B_{N+1,M} \cdot A_{1,N+1}B_{N+1,M+1}
\end{array}\right.$$

$$\left.\begin{array}{l}
A_{i1}B_{11} \cdot \cdot \cdot A_{i1}B_{1K} \cdot \cdot \cdot A_{i1}B_{1M} \cdot \cdot \cdot A_{i1}B_{1,M+1} \\
\qquad \cdot \qquad\qquad \cdot \qquad\qquad \cdot \qquad\qquad \cdot \\
A_{iJ}\,B_{J1} \cdot \cdot \cdot A_{iJ}B_{JK} \cdot \cdot \cdot A_{iJ}B_{JM} \cdot \cdot \cdot A_{iJ}B_{J,M+1} \\
\qquad \cdot \qquad\qquad \cdot \qquad\qquad \cdot \qquad\qquad \cdot \\
A_{iN}\,B_{N1} \cdot \cdot \cdot A_{iN}\,B_{NK} \cdot \cdot \cdot A_{iN}\,B_{NM} \cdot \cdot \cdot A_{iN}\,B_{N,M+1} \\
\qquad \cdot \qquad\qquad \cdot \qquad\qquad \cdot \qquad\qquad \cdot \\
A_{i,N+1}\,B_{N+1,1} \cdot \cdot A_{i,N+1}\,B_{N+1,K} \cdot A_{i,N+1}\,B_{N+1,M} \cdot A_{i,N+1}\,B_{N+1,M+1}
\end{array}\right.$$

$$\left|\begin{array}{l}
A_{M1}B_{11} \cdot \cdot \cdot A_{M1}B_{1K} \cdot \cdot \cdot A_{M1}B_{1M} \cdot \cdot \cdot A_{M1}B_{1,M+1} \\
\qquad \cdot \qquad\qquad \cdot \qquad\qquad \cdot \qquad\qquad \cdot \\
A_{MJ}B_{J1} \cdot \cdot \cdot A_{MJ}B_{JK} \cdot \cdot \cdot A_{MJ}B_{JM} \cdot \cdot \cdot A_{MJ}B_{J,M+1} \\
\qquad \cdot \qquad\qquad \cdot \qquad\qquad \cdot \qquad\qquad \cdot \\
A_{MN}B_{N1} \cdot \cdot \cdot A_{MN}B_{NK} \cdot \cdot A_{MN}B_{NM} \cdot \cdot \cdot A_{MN}B_{N,M+1} \\
\qquad \cdot \qquad\qquad \cdot \qquad\qquad \cdot \qquad\qquad \cdot \\
A_{M,N+1}B_{N+1,1} \cdot \cdot A_{M,N+1}B_{N+1,K} \cdot A_{M,N+1}B_{N+1,M} \cdot A_{M,N+1}B_{N+1,M+1}
\end{array}\right.$$

$$\left|\begin{array}{l}
A_{M+1,1}B_{11} \cdot \cdot \cdot A_{M+1,1}B_{1K} \cdot \cdot A_{M+1,1}B_{1M} \cdot \cdot A_{M+1,1}B_{1,M+1} \\
\qquad \cdot \qquad\qquad \cdot \qquad\qquad \cdot \qquad\qquad \cdot \\
A_{M+1,J}B_{J1} \cdot A_{M+1,J}\,B_{JK} \cdot \quad A_{M+1,J}\,B_{JM} \cdot \cdot A_{M+1,J}\,B_{J,M+1} \\
\qquad \cdot \qquad\qquad \cdot \qquad\qquad \cdot \qquad\qquad \cdot \\
A_{M+1,N}B_{N1} \cdot A_{M+1,N}B_{NK} \cdot \cdot A_{M+1,N}B_{NM} \cdot \cdot A_{M+1,N}\,B_{N,M+1} \\
\qquad \cdot \qquad\qquad \cdot \qquad\qquad \cdot \qquad\qquad \cdot \\
A_{M+1,N+1}B_{N+1,1} \cdot A_{M+1,N+1}B_{N+1,K} \cdot A_{M+1,N+1}B_{N+1,M} \cdot A_{M+1,N+1}B_{N+1,M+1}
\end{array}\right.$$

Summing the columns in each sub-matrix we get:

$$[A_{11}B_{11} \cdot \cdot + \cdot A_{1J}B_{J1}+ \cdot \cdot \cdot A_{1N}B_{N1} + \quad A_{1,N+1}B_{N+1,1} \;] \quad [A_{11}B_{1K}+ \cdot \cdot \cdot A_{1J}B_{JK} \cdot \cdot + \cdot A_{1N}B_{NK} + A_{1,N+1}B_{N+1,K}]$$
$$A_{11}B_{1M} \cdot + \cdot A_{1J}B_{JM}+ \cdot + \cdot A_{1N}B_{NM} + \cdot A_{1,N+1}B_{N+1,M}] \quad [A_{11}B_{1,M+1} \cdot + \cdot A_{1J}B_{J,M+1} \cdot + \cdot A_{1N}B_{N,M+1} + \quad A_{1,N+1}B_{N+1,M+1}]$$

$$[A_{11}B_{11} \cdot \cdot + \cdot A_{1J}B_{J1}+ \cdot \cdot \cdot A_{1N}B_{N1} + \quad A_{i,N+1}B_{N+1,1}] \quad [\; A_{11}B_{1K}+ \cdot \cdot A_{1J}B_{JK} \cdot + \cdot A_{1N}B_{NK} + A_{i,N+1}B_{N+1,K} \;]$$
$$[A_{11}B_{1M} \cdot + \cdot A_{1J}B_{JM} \cdot \cdot + \cdot A_{1N}B_{NM} + \cdot A_{i,N+1}B_{N+1,M}] \quad [A_{11}B_{1,M+1} \cdot + \cdot A_{i3}B_{JM+1} \cdot + \cdot A_{1N}B_{NM+1} + A_{i,N+1}B_{N+1,M+1}]$$

$$[A_{M1}B_{11} \cdot \cdot + \cdot \cdot A_{MJ}B_{J1}+ \cdot \cdot \cdot A_{MN}B_{N1} \cdot \cdot + \cdot \cdot A_{M,N+1}B_{N+1,1} \;] \quad [A_{M1}B_{1K} \cdot \cdot + \cdot \quad A_{MJ}B_{JK} \cdot \cdot + \cdot \quad A_{MN}B_{NK} \cdot \cdot + \cdot \cdot A_{M,N+1}B_{N+1,K}]$$
$$[A_{M1}B_{1M} \cdot \cdot + \cdot \cdot A_{MJ}B_{JM}+ \cdot \cdot + \cdot \quad A_{MN}B_{NM} \cdot + \cdot \cdot A_{M,N+1}B_{N+1,M}] \quad [A_{M1}B_{1,M+1} \cdot \cdot + \cdot \cdot A_{MJ}B_{J,M+1} \cdot + \cdot \cdot A_{MN}B_{N,M+1} + \cdot \cdot A_{M,N+1}B_{N+1,M+1}]$$

$$[A_{M+1,1}B_{11} \cdot + \cdot A_{M+1,J}B_{J1}+ \cdot \cdot \cdot A_{M+1,N}B_{N,1} + A_{M+1,N+1}B_{N+1,1}] \quad [A_{M+1,1}B_{1K}+ \cdot A_{M+1,J}B_{JK} \cdot + \cdot A_{M+1,N}B_{NK} + A_{M+1,N+1}B_{N+1,K}]$$
$$A_{M+1,1}B_{1M} + \cdot A_{M+1,J}B_{JM} \cdot + \cdot A_{M+1,N}B_{NM} + A_{M+1,,N+1}B_{N+1,M}] \quad [A_{M+1,1}B_{1,M+1} \cdot + \cdot A_{M+1,J}B_{J,M+1} \cdot + \cdot A_{M+1,N}B_{N,M+1} + A_{M+1,N+1}B_{N+1,M+1}]$$

Separating them from each other we get:

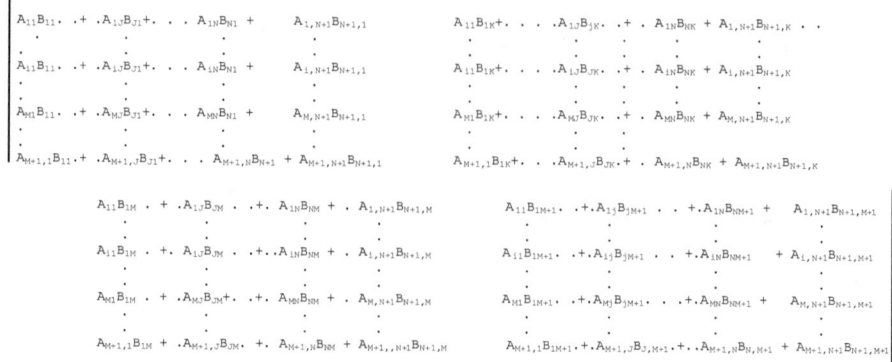

Which is the solution for regular Matrix Multiplication.

<u>QED</u>

Now we will achieve the same results using the Half-Multiplier Operator (we proved the top portion with diagonals, here we will use the Half-Multiplier Operator:

First we will prove for all matrices sized M x N:

$$
\begin{vmatrix} A_{11} \\ \cdot \\ \cdot \\ A_{1J} \\ \cdot \\ A_{1N} \end{vmatrix}
\circ
\begin{vmatrix} B_{11} \cdot & \cdot & \cdot B_{1K} \cdot & \cdot & \cdot B_{1M} \\ \cdot & & \cdot & & \cdot \\ \cdot & & \cdot & & \cdot \\ B_{J1} \cdot & \cdot & \cdot B_{JK} \cdot & \cdot & \cdot B_{JM} \\ \cdot & & \cdot & & \cdot \\ B_{N1} \cdot & \cdot & \cdot B_{NK} \cdot & \cdot & \cdot B_{NM} \end{vmatrix} =
$$

$$
\begin{vmatrix} A_{11}(B_{11} \cdot & \cdot & \cdot B_{1K} \cdot & \cdot & \cdot B_{1M}) \\ \cdot & & \cdot & & \cdot \\ \cdot & & \cdot & & \cdot \\ A_{1J}(B_{J1} \cdot & \cdot & \cdot B_{JK} \cdot & \cdot & \cdot B_{JM}) \\ \cdot & & \cdot & & \cdot \\ A_{1N}(B_{N1} \cdot & \cdot & \cdot B_{NK} \cdot & \cdot & \cdot B_{NM}) \end{vmatrix} =
$$

$$
\begin{vmatrix} A_{11}B_{11} \cdot & \cdot & \cdot A_{11}B_{1K} \cdot & \cdot & \cdot A_{11}B_{1M} \\ \cdot & & \cdot & & \cdot \\ \cdot & & \cdot & & \cdot \\ A_{1J}B_{J1} \cdot & \cdot & \cdot A_{1J}B_{JK} \cdot & \cdot & \cdot A_{1J}B_{JM} \\ \cdot & & \cdot & & \cdot \\ A_{1N}B_{N1} \cdot & \cdot & \cdot A_{1N}B_{NK} \cdot & \cdot & \cdot A_{1N}B_{NM} \end{vmatrix}
$$

Now for the ith column:

$$
\begin{vmatrix} A_{i1} \\ \cdot \\ \cdot \\ A_{iJ} \\ \cdot \\ \cdot \\ A_{iN} \end{vmatrix} \circ
\begin{vmatrix} B_{11} \cdot \cdot \cdot B_{1K} \cdot \cdot \cdot B_{1M} \\ \cdot \quad\quad \cdot \quad\quad \cdot \\ \cdot \quad\quad \cdot \quad\quad \cdot \\ B_{J1} \cdot \cdot \cdot B_{JK} \cdot \cdot \cdot B_{JM} \\ \cdot \quad\quad \cdot \quad\quad \cdot \\ \cdot \quad\quad \cdot \quad\quad \cdot \\ B_{N1} \cdot \cdot \cdot B_{NK} \cdot \cdot \cdot B_{NM} \end{vmatrix} =
\begin{vmatrix} A_{i1}(B_{11} \cdot \cdot \cdot B_{1K} \cdot \cdot \cdot B_{1M}) \\ \cdot \quad\quad \cdot \quad\quad \cdot \\ \cdot \quad\quad \cdot \quad\quad \cdot \\ A_{iJ}(B_{J1} \cdot \cdot \cdot B_{JK} \cdot \cdot \cdot B_{JM}) \\ \cdot \quad\quad \cdot \quad\quad \cdot \\ \cdot \quad\quad \cdot \quad\quad \cdot \\ A_{iN}(B_{N1} \cdot \cdot \cdot B_{NK} \cdot \cdot \cdot B_{NM}) \end{vmatrix} =
$$

$$
\begin{vmatrix} A_{i1}B_{11} \cdot \cdot \cdot A_{i1}B_{1K} \cdot \cdot \cdot A_{i1}B_{1M} \\ \cdot \quad\quad \cdot \quad\quad \cdot \\ \cdot \quad\quad \cdot \quad\quad \cdot \\ A_{iJ}B_{J1} \cdot \cdot \cdot A_{iJ}B_{JK} \cdot \cdot \cdot A_{iJ}B_{JM} \\ \cdot \quad\quad \cdot \quad\quad \cdot \\ \cdot \quad\quad \cdot \quad\quad \cdot \\ A_{iN}B_{N1} \cdot \cdot \cdot A_{iN}B_{NK} \cdot \cdot \cdot A_{iN}B_{NM} \end{vmatrix}
$$

And now for the Mth column:

$$
\begin{vmatrix} A_{M1} \\ \cdot \\ \cdot \\ A_{MJ} \\ \cdot \\ \cdot \\ A_{MN} \end{vmatrix} \circ
\begin{vmatrix} B_{11} \cdot \cdot \cdot B_{1K} \cdot \cdot \cdot B_{1M} \\ \cdot \quad\quad \cdot \quad\quad \cdot \\ \cdot \quad\quad \cdot \quad\quad \cdot \\ B_{J1} \cdot \cdot \cdot B_{JK} \cdot \cdot \cdot B_{JM} \\ \cdot \quad\quad \cdot \quad\quad \cdot \\ \cdot \quad\quad \cdot \quad\quad \cdot \\ B_{N1} \cdot \cdot \cdot B_{NK} \cdot \cdot \cdot B_{NM} \end{vmatrix} =
\begin{vmatrix} A_{M1}(B_{11} \cdot \cdot \cdot B_{1K} \cdot \cdot \cdot B_{1M}) \\ \cdot \quad\quad \cdot \quad\quad \cdot \\ \cdot \quad\quad \cdot \quad\quad \cdot \\ A_{MJ}(B_{J1} \cdot \cdot \cdot B_{JK} \cdot \cdot \cdot B_{JM}) \\ \cdot \quad\quad \cdot \quad\quad \cdot \\ \cdot \quad\quad \cdot \quad\quad \cdot \\ A_{MN}(B_{N1} \cdot \cdot \cdot B_{NK} \cdot \cdot \cdot B_{NM}) \end{vmatrix} =
$$

$$
\begin{vmatrix} A_{M1}B_{11} \cdot \cdot \cdot A_{M1}B_{1K} \cdot \cdot \cdot A_{M1}B_{1M} \\ \cdot \quad\quad \cdot \quad\quad \cdot \\ \cdot \quad\quad \cdot \quad\quad \cdot \\ A_{MJ}B_{J1} \cdot \cdot \cdot A_{MJ}B_{JK} \cdot \cdot \cdot A_{MJ}B_{JM} \\ \cdot \quad\quad \cdot \quad\quad \cdot \\ \cdot \quad\quad \cdot \quad\quad \cdot \\ A_{MN}B_{N1} \cdot \cdot \cdot A_{MN}B_{NK} \cdot \cdot \cdot A_{MN}B_{NM} \end{vmatrix}
$$

Now let's make these into a Nested Array and sum each column
of the individual Sub-matrices:

$$
\begin{vmatrix}
\begin{vmatrix}
A_{11}B_{11} . & . . A_{11}B_{1K} . & . . A_{11}B_{1M} \\
\cdot & \cdot & \cdot \\
\cdot & \cdot & \cdot \\
A_{1J}B_{J1} . & . . A_{1J}B_{JK} . & . . A_{1J}B_{JM} \\
\cdot & \cdot & \cdot \\
\cdot & \cdot & \cdot \\
A_{1N}B_{N1} . & . . A_{1N}B_{NK} . & . . A_{1N}B_{NM}
\end{vmatrix} \\[2em]
\begin{vmatrix}
A_{i1}B_{11} . & . . A_{i1}B_{1K} . & . . A_{i1}B_{1M} \\
\cdot & \cdot & \cdot \\
\cdot & \cdot & \cdot \\
A_{iJ}B_{J1} . & . . A_{iJ}B_{JK} . & . . A_{iJ}B_{JM} \\
\cdot & \cdot & \cdot \\
A_{iN}B_{N1} . & . . A_{iN}B_{NK} . & . . A_{iN}B_{NM}
\end{vmatrix} \\[2em]
\begin{vmatrix}
A_{M1}B_{11} . & . . A_{M1}B_{1K} . & . . A_{M1}B_{1M} \\
\cdot & \cdot & \cdot \\
\cdot & \cdot & \cdot \\
A_{MJ}B_{J1} . & . . A_{MJ}B_{JK} . & . . A_{MJ}B_{JM} \\
\cdot & \cdot & \cdot \\
\cdot & \cdot & \cdot \\
A_{MN}B_{N1} . & . . A_{MN}B_{NK} . & . . A_{MN}B_{NM}
\end{vmatrix}
\end{vmatrix}
$$

Summing the columns we get:

$$
\begin{vmatrix}
A_{11}B_{11}..+.. A_{1J}B_{J1}..+.. A_{1N}B_{N1} & A_{11}B_{1K}..+..A_{1J}B_{JK}..+.. A_{1N}B_{NK} & A_{11}B_{1M}..+.. A_{1J}B_{JM}..+..A_{1N}B_{NM} \\
A_{i1}B_{11}..+.. A_{1J}B_{J1}..+.. A_{1N}B_{N1} & A_{i1}B_{1K}..+..A_{iJ}B_{JK}..+.. A_{iN}B_{NK} & A_{i1}B_{1M}..+..A_{iJ}B_{JM}..+.. A_{iN}B_{NM} \\
A_{M1}B_{11}..+.. A_{MJ}B_{J1}..+..A_{MN}B_{N1} & A_{M1}B_{1K}..+.. A_{MJ}B_{JK}..+.. A_{MN}B_{NK} & A_{M1}B_{1M}..+.. A_{MJ}B_{JM}..+..A_{MN}B_{NM}
\end{vmatrix}
$$

Which is true for all M x N Matrices, now to prove it is true for all Matrices M+1, N+1

$$
\begin{vmatrix}
A_{11} \\
\cdot \\
A_{1j} \\
\cdot \\
A_{1N} \\
\cdot \\
A_{1,N+1}
\end{vmatrix}
\circ
\begin{vmatrix}
B_{11} . & . . B_{1K} . & . . B_{1M} . & . . B_{1,M+1} \\
\cdot & \cdot & \cdot & \cdot \\
\cdot & \cdot & \cdot & \cdot \\
B_{J1} . & . . B_{JK} . & . . B_{JM} . & . . B_{J,M+1} \\
\cdot & \cdot & \cdot & \cdot \\
\cdot & \cdot & \cdot & \cdot \\
B_{N1} . & . . B_{NK} . & . . B_{NM} . & . . B_{N,M+1} \\
\cdot & \cdot & \cdot & \cdot \\
B_{N+1,1} . & . B_{N+1,K} . & . B_{N+1,M} . & . B_{N+1,M+1}
\end{vmatrix}
=
$$

$$
\begin{vmatrix}
A_{11} (B_{11} . & . . B_{1K} . & . . B_{1M} . & . . B_{1,M+1}) \\
\cdot & \cdot & \cdot & \cdot \\
A_{iJ} (B_{J1} . & . . B_{JK} . & . . B_{JM} . & . . B_{J,M+1}) \\
\cdot & \cdot & \cdot & \cdot \\
A_{1N} (B_{N1} . & . . B_{NK} . & . . B_{NM} . & . . B_{N,M+1}) \\
\cdot & \cdot & \cdot & \cdot \\
A_{1,N+1} (B_{N+1,1} . & . B_{N+1,K} . & . B_{N+1,M} . & . B_{N+1,M+1})
\end{vmatrix}
=
\begin{vmatrix}
A_{11}B_{11} . & . . A_{11}B_{1K} . & . . A_{11}B_{1M} . & . . A_{11}B_{1,M+1} \\
\cdot & \cdot & \cdot & \cdot \\
A_{1J}B_{J1} . & . . A_{1J}B_{JK} . & . . A_{1J}B_{JM} . & . . A_{1J}B_{J,M+1} \\
\cdot & \cdot & \cdot & \cdot \\
A_{1N}B_{N1} . & . . A_{1N}B_{NK} . & . . A_{1N}B_{NM} . & . . A_{1N}B_{N,M+1} \\
\cdot & \cdot & \cdot & \cdot \\
A_{1,N+1}B_{N+1,1} . & . A_{1,N+1}B_{N+1,K} . & . A_{1,N+1}B_{N+1,M} . & . A_{1,N+1}B_{N+1,M+1}
\end{vmatrix}
$$

The summed columns are equal to:

$$[A_{11}B_{11}. .+ .A_{1J}B_{J1}+. . . A_{1N}B_{N1} + \quad A_{1,N+1}B_{N+1,1}] \quad [A_{11}B_{1K}+. . . .A_{1J}B_{JK}. .+ . A_{1N}B_{NK} + A_{1,N+1}B_{N+1,K}]$$
$$A_{11}B_{1M} . + .A_{1J}B_{JM}+. .+. A_{1N}B_{NM} + . A_{1,N+1}B_{N+1,M}] \quad [A_{11}B_{1,M+1}. .+.A_{1J}B_{J,M+1}. .+.A_{1N}B_{N,M+1} + \quad A_{1,N+1}B_{N+1,M+1}]$$

$$\begin{vmatrix} A_{11} \\ \cdot \\ A_{1J} \\ \cdot \\ A_{1N} \\ \cdot \\ A_{1,N+1} \end{vmatrix} \begin{vmatrix} B_{11}. . .B_{1K}. . .B_{1M}. . .B_{1,M+1} \\ \cdot \quad \cdot \quad \cdot \quad \cdot \\ B_{J1}. . .B_{JK}. . .B_{JM}. . .B_{J,M+1} \\ \cdot \quad \cdot \quad \cdot \quad \cdot \\ B_{N1}. . .B_{NK}. . .B_{NM}. . .B_{N,M+1} \\ \cdot \quad \cdot \quad \cdot \quad \cdot \\ B_{N+1,1} . .B_{N+1,K} . .B_{N+1,M} . .B_{N+1,M+1} \end{vmatrix} =$$

$$\begin{vmatrix} A_{11} (B_{11}. . .B_{1K}. . .B_{1M}. . .B_{1,M+1}) \\ \cdot \quad \cdot \quad \cdot \quad \cdot \\ A_{1J} (B_{J1}. . .B_{JK}. . .B_{JM}. . .B_{J,M+1}) \\ \cdot \quad \cdot \quad \cdot \quad \cdot \\ A_{1N} (B_{N1}. . .B_{NK}. . .B_{NM}. . .B_{N,M+1}) \\ \cdot \quad \cdot \quad \cdot \quad \cdot \\ A_{1,N+1} (B_{N+1,1}. .B_{N+1,K}. .B_{N+1,M}. .B_{N+1,M+1}) \end{vmatrix} = \begin{vmatrix} A_{11}B_{11}. . . A_{11}B_{1K}. . . A_{11}B_{1M}. . . A_{11}B_{1,M+1} \\ \cdot \quad \cdot \quad \cdot \quad \cdot \\ A_{1J}B_{J1}. . . A_{1J}B_{JK}. . . A_{1J}B_{JM}. . . A_{1J}B_{J,M+1} \\ \cdot \quad \cdot \quad \cdot \quad \cdot \\ A_{1N}B_{N1}. . . A_{1N}B_{NK}. . A_{1N}B_{NM}. . . A_{1N}B_{N,M+1} \\ \cdot \quad \cdot \quad \cdot \quad \cdot \\ A_{1,N+1}B_{N+1,1}. . A_{1,N+1}B_{N+1,K}. . A_{1,N+1}B_{N+1,M}. . A_{1,N+1}B_{N+1,M+1} \end{vmatrix}$$

Summing the columns we get:

$$[A_{11}B_{11}. .+ .A_{1J}B_{J1}+. . . A_{1N}B_{N1} + \quad A_{1,N+1}B_{N+1,1}] \quad [A_{11}B_{1K}+. . .A_{1J}B_{JK}. .+ . A_{1N}B_{NK} + A_{1,N+1}B_{N+1,K}]$$
$$[A_{11}B_{1M} . +. A_{1J}B_{JM} . .+..A_{1N}B_{NM} + . A_{1,N+1}B_{N+1,M}] \quad [A_{11}B_{1,M+1}. .+.A_{1J}B_{J,M+1} . +.A_{1N}B_{N,M+1} + A_{1,N+1}B_{N+1,M+1}]$$

$$\begin{vmatrix} A_{M1} \\ A_{MJ} \\ \cdot \\ A_{MN} \\ \cdot \\ A_{M,N+1} \end{vmatrix} \circ \begin{vmatrix} B_{11}. . .B_{1K}. . .B_{1M}. . .B_{1,M+1} \\ \cdot \quad \cdot \quad \cdot \quad \cdot \\ B_{J1}. . .B_{JK}. . .B_{JM}. . .B_{J,M+1} \\ \cdot \quad \cdot \quad \cdot \quad \cdot \\ B_{N1}. . .B_{NK}. . .B_{NM}. . .B_{N,M+1} \\ \cdot \quad \cdot \quad \cdot \quad \cdot \\ B_{N+1,1} . .B_{N+1,K} . .B_{N+1,M} . .B_{N+1,M+1} \end{vmatrix} =$$

$$\begin{vmatrix} A_{M1} (B_{11}. . .B_{1K}. . .B_{1M}. . .B_{1,M+1}) \\ \cdot \quad \cdot \quad \cdot \quad \cdot \\ A_{MJ} (B_{J1}. . .B_{JK}. . .B_{JM}. . .B_{J,M+1}) \\ \cdot \quad \cdot \quad \cdot \quad \cdot \\ A_{MN} (B_{N1}. . .B_{NK}. . .B_{NM}. . .B_{N,M+1}) \\ \cdot \quad \cdot \quad \cdot \quad \cdot \\ A_{M,N+1} (B_{N+1,1}. .B_{N+1,K}. .B_{N+1,M}. .B_{N+1,M+1}) \end{vmatrix} = \begin{vmatrix} A_{M1}B_{11}. . . A_{M1}B_{1K}. . . A_{M1}B_{1M}. . . A_{M1}B_{1,M+1} \\ \cdot \quad \cdot \quad \cdot \quad \cdot \\ A_{MJ}B_{J1}. . . A_{MJ}B_{JK}. . . A_{MJ}B_{JM}. . . A_{MJ}B_{J,M+1} \\ \cdot \quad \cdot \quad \cdot \quad \cdot \\ A_{MN}B_{N1}. . . A_{MN}B_{NK}. . . A_{MN}B_{NM}. . . A_{MN}B_{N,M+1} \\ \cdot \quad \cdot \quad \cdot \quad \cdot \\ A_{M,N+1}B_{N+1,1}. . A_{M,N+1}B_{N+1,K}. . A_{M,N+1}B_{N+1,M}. . A_{M,N+1}B_{N+1,M+1} \end{vmatrix}$$

Summing the columns we get:

$$[A_{M1}B_{11}. .+..A_{MJ}B_{J1}+. . . A_{MN}B_{N1}..+..A_{M,N+1}B_{N+1,1}] \quad [A_{M1}B_{1K}..+. .A_{MJ}B_{JK}. .+ . . A_{MN}B_{NK}..+.. A_{M,N+1}B_{N+1,K}]$$
$$[A_{M1}B_{1M}..+..A_{MJ}B_{JM}+. .+. A_{MN}B_{NM}..+..A_{M,N+1}B_{N+1,M}] \quad [A_{M1}B_{1,M+1}. .+..A_{MJ}B_{J,M+1}..+..A_{MN}B_{N,M+1} +..A_{M,N+1}B_{N+1,M+1}]$$

For the M+1th column:

$$
\begin{vmatrix} A_{M+1,1} \\ \\ A_{M+1,J} \\ \\ A_{M+1,N} \\ \\ A_{M+1,N+1} \end{vmatrix}
\begin{vmatrix} B_{11} \cdots B_{1K} \cdots B_{1M} \cdots B_{1,M+1} \\ \\ B_{J1} \cdots B_{JK} \cdots B_{JM} \cdots B_{J,M+1} \\ \\ B_{N1} \cdots B_{NK} \cdots B_{NM} \cdots B_{N,M+1} \\ \\ B_{N+1,1} \cdots B_{N+1,K} \cdots B_{N+1,M} \cdots B_{N+1,M+1} \end{vmatrix} =
$$

$$
\begin{vmatrix} A_{M+1,1} \ (B_{11} \cdots B_{1K} \cdots B_{1M} \cdots B_{1,M+1}) \\ \\ A_{M+1,J} \ (B_{J1} \cdots B_{JK} \cdots B_{JM} \cdots B_{J,M+1}) \\ \\ A_{M+1,N} \ (B_{N1} \cdots B_{NK} \cdots B_{NM} \cdots B_{N,M+1}) \\ \\ A_{M+1,N+1} \ (B_{N+1,1} \cdots B_{N+1,K} \cdots B_{N+1,M} \cdots B_{N+1,M+1}) \end{vmatrix}
=
\begin{vmatrix} A_{M+1,1}B_{11} \ A_{M+1,1}B_{1K} \cdots A_{M+1,1}B_{1M} \cdots A_{M+1,1}B_{1,M+1} \\ \\ A_{M+1,J}B_{J1} \ A_{M+1,J} B_{JK} \ A_{M+1,J} B_{JM} \cdots A_{M+1,J} B_{J,M+1} \\ \\ A_{M+1,N}B_{N1} \ A_{M+1,N}B_{NK} \cdots A_{M+1,N}B_{NM} \cdots A_{M+1,N} B_{N,M+1} \\ \\ A_{M+1,N+1}B_{N+1,1} \ A_{M+1,N+1}B_{N+1,K} \cdots A_{M+1,N+1}B_{N+1,M} \ A_{M+1,N+1}B_{N+1,M+1} \end{vmatrix}
$$

$[A_{M+1,1}B_{11} + . A_{M+1,J}B_{J1} + \ldots A_{M+1,N}B_{N,1} + A_{M+1,N+1}B_{N+1,1}]$ $[A_{M+1,1}B_{1K} + .. A_{M+1,J}B_{JK} + . A_{M+1,N}B_{NK} + A_{M+1,N+1}B_{N+1,K}]$
$A_{M+1,1}B_{1M} + . A_{M+1,J}B_{JM} + . A_{M+1,N}B_{NM} + A_{M+1,N+1}B_{N+1,M}]$ $[A_{M+1,1}B_{1,M+1} + . A_{M+1,J}B_{J,M+1} + .. A_{M+1,N}B_{N,M+1} + A_{M+1,N+1}B_{N+1,M+1}]$

Transposing the Nested Array (Stacking the sub-matrices on top of each other) we get

$$
\begin{vmatrix}
A_{11} B_{11} \cdots A_{11}B_{1K} \cdots A_{11}B_{1M} \cdots A_{11}B_{1,M+1} \\
\\
A_{1J} B_{J1} \cdots A_{1J} B_{JK} \cdots A_{1J} B_{JM} \cdots A_{1J} B_{J,M+1} \\
\\
A_{1N} B_{N1} \cdots A_{1N} B_{NK} \cdots A_{1N} B_{NM} \cdots A_{1N} B_{N,M+1} \\
\\
A_{1,N+1}B_{N+1,1} \cdots A_{1,N+1}B_{N+1,K} \cdots A_{1,N+1}B_{N+1,M} \cdots A_{1,N+1}B_{N+1,M+1} \\
\\
A_{11}B_{11} \cdots A_{11}B_{1K} \cdots A_{11}B_{1M} \cdots A_{11}B_{1,M+1} \\
\\
A_{1J} B_{J1} \cdots A_{1J}B_{JK} \cdots A_{1J}B_{JM} \cdots A_{1J}B_{J,M+1} \\
\\
A_{1N} B_{N1} \cdots A_{1N} B_{NK} \cdots A_{1N} B_{NM} \cdots A_{1N} B_{N,M+1} \\
\\
A_{1,N+1} B_{N+1,1} \cdots A_{1,N+1} B_{N+1,K} \cdots A_{1,N+1} B_{N+1,M} \cdots A_{1,N+1} B_{N+1,M+1} \\
\\
A_{M1}B_{11} \cdots A_{M1}B_{1K} \cdots A_{M1}B_{1M} \cdots A_{M1}B_{1,M+1} \\
\\
A_{MJ}B_{J1} \cdots A_{MJ}B_{JK} \cdots A_{MJ}B_{JM} \cdots A_{MJ}B_{J,M+1} \\
\\
A_{MN}B_{N1} \cdots A_{MN}B_{NK} \cdots A_{MN}B_{NM} \cdots A_{MN}B_{N,M+1} \\
\\
A_{M,N+1}B_{N+1,1} \cdots A_{M,N+1}B_{N+1,K} \cdots A_{M,N+1}B_{N+1,M} \cdots A_{M,N+1}B_{N+1,M+1} \\
\\
A_{M+1,1}B_{11} \ A_{M+1,1}B_{1K} \cdots A_{M+1,1}B_{1M} \cdots A_{M+1,1}B_{1,M+1} \\
\\
A_{M+1,J}B_{J1} \ A_{M+1,J} B_{JK} \ A_{M+1,J} B_{JM} \cdots A_{M+1,J} B_{J,M+1} \\
\\
A_{M+1,N}B_{N1} \ A_{M+1,N}B_{NK} \cdots A_{M+1,N}B_{NM} \cdots A_{M+1,N} B_{N,M+1} \\
\\
A_{M+1,N+1}B_{N+1,1} \ A_{M+1,N+1}B_{N+1,K} \ A_{M+1,N+1}B_{N+1,M} \ A_{M+1,N+1}B_{N+1,M+1}
\end{vmatrix}
$$

Summing the columns in each Sub-matrix we get:

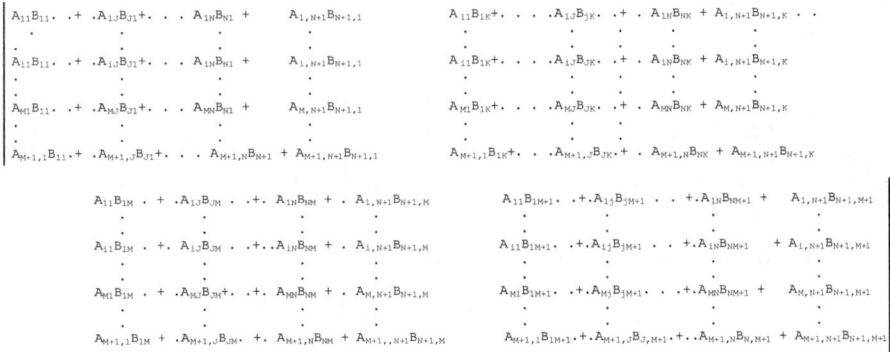

Separating each new column we get:

$$A_{11}B_{11}\cdot .+ .A_{1J}B_{J1}+\cdot .\cdot A_{1N}B_{N1} + \quad A_{1,N+1}B_{N+1,1} \qquad A_{11}B_{1K}+\cdot .\cdot .A_{1J}B_{JK}\cdot .+ . A_{1N}B_{NK} + A_{1,N+1}B_{N+1,K}\cdot .\cdot$$

$$A_{11}B_{11}\cdot .+ .A_{1J}B_{J1}+\cdot .\cdot A_{1N}B_{N1} + \quad A_{1,N+1}B_{N+1,1} \qquad A_{11}B_{1K}+\cdot .\cdot .A_{1J}B_{JK}\cdot .+ . A_{1N}B_{NK} + A_{1,N+1}B_{N+1,K}$$

$$A_{M1}B_{11}\cdot .+ .A_{MJ}B_{J1}+\cdot .\cdot A_{MN}B_{N1} + \quad A_{M,N+1}B_{N+1,1} \qquad A_{M1}B_{1K}+\cdot .\cdot .A_{MJ}B_{JK}\cdot .+ . A_{MN}B_{NK} + A_{M,N+1}B_{N+1,K}$$

$$A_{M+1,1}B_{11}\cdot+ .A_{M+1,J}B_{J1}+\cdot .\cdot .A_{M+1,N}B_{N+1} + A_{M+1,N+1}B_{N+1,1} \qquad A_{M+1,1}B_{1K}+\cdot .\cdot .A_{M+1,J}B_{JK}\cdot + . A_{M+1,N}B_{NK} + A_{M+1,N+1}B_{N+1,K}$$

$$A_{11}B_{1M} \cdot + .A_{1J}B_{JM} \cdot .\cdot+. A_{1N}B_{NM} + . A_{1,N+1}B_{N+1,M} \qquad A_{11}B_{1M+1}\cdot .+.A_{1J}B_{JM+1} . . .+.A_{1N}B_{NM+1} + \quad A_{1,N+1}B_{N+1,M+1}$$

$$A_{11}B_{1M} \cdot +. A_{1J}B_{JM} \cdot .+..A_{1N}B_{NM} + . A_{1,N+1}B_{N+1,M} \qquad A_{11}B_{1M+1}\cdot .+.A_{1J}B_{JM+1} . . .+.A_{1N}B_{NM+1} + A_{1,N+1}B_{N+1,M+1}$$

$$A_{M1}B_{1M} \cdot + .A_{MJ}B_{JM}+. .+. A_{MN}B_{NM} + . A_{M,N+1}B_{N+1,M} \qquad A_{M1}B_{1M+1}\cdot .+.A_{MJ}B_{JM+1} . .\cdot .+.A_{MN}B_{NM+1} + \quad A_{M,N+1}B_{N+1,M+1}$$

$$A_{M+1,1}B_{1M} + .A_{M+1,J}B_{JM}\cdot +. A_{M+1,N}B_{NM} + A_{M+1,,N+1}B_{N+1,M} \qquad A_{M+1,1}B_{1M+1}\cdot+.A_{M+1,J}B_{J,M+1}\cdot+..A_{M+1,N}B_{N,M+1} + A_{M+1,N+1}B_{N+1,M+1}$$

QED

    This is the same answer we get as when we multiply the matrices the regular way.

We have now proved that the Half-Multiplier Operator when half multiplied into sub-matrices, transposing to a column matrix (we have to re-transpose, actually, because we transposed the A_{ij} Matrix to permit the half-multiplication), summing the columns gives the same solution as regular Matrix Multiplication $A_{ij}B_{jk} = C_{ik}$.

THE ROW PRODUCT OF A MATRIX
OR
THE CROSS PRODUCT OF A MATRIX

When we took the half-multiplied sub-matrices and summed the columns, the solution was the same as matrix multiplication. Let's look at what happens if we take the half-multiplied sub-matrices and sum their rows instead of their columns.

$$^R\textstyle\sum [A]^T_{ij} \circ [B]_{JK} = ([B]_{JK}[1]_{k1}) \circ [A]^T_{iJ}$$

$$\begin{vmatrix} A_{11} & A_{21} & A_{31} \\ A_{12} & A_{22} & A_{32} \\ A_{13} & A_{23} & A_{33} \\ A_{14} & A_{24} & A_{34} \end{vmatrix} \circ \begin{vmatrix} B_{11} & B_{12} & B_{13} \\ B_{21} & B_{22} & B_{23} \\ B_{31} & B_{32} & B_{33} \\ B_{41} & B_{42} & B_{43} \end{vmatrix} =$$

$$\begin{vmatrix} A_{11} \\ A_{12} \\ A_{13} \\ A_{14} \end{vmatrix} \circ \begin{vmatrix} B_{11} & B_{12} & B_{13} \\ B_{21} & B_{22} & B_{23} \\ B_{31} & B_{32} & B_{33} \\ B_{41} & B_{42} & B_{43} \end{vmatrix} \begin{vmatrix} A_{21} \\ A_{22} \\ A_{23} \\ A_{24} \end{vmatrix} \circ \begin{vmatrix} B_{11} & B_{12} & B_{13} \\ B_{21} & B_{22} & B_{23} \\ B_{31} & B_{32} & B_{33} \\ B_{41} & B_{42} & B_{43} \end{vmatrix} \begin{vmatrix} A_{31} \\ A_{32} \\ A_{33} \\ A_{34} \end{vmatrix} \circ \begin{vmatrix} B_{11} & B_{12} & B_{13} \\ B_{21} & B_{22} & B_{23} \\ B_{31} & B_{32} & B_{33} \\ B_{41} & B_{42} & B_{43} \end{vmatrix}$$

$$\begin{bmatrix} A_{11} & 0 & 0 & 0 \\ 0 & A_{12} & 0 & 0 \\ 0 & 0 & A_{13} & 0 \\ 0 & 0 & 0 & A_{14} \end{bmatrix} \cdot \begin{bmatrix} B_{11} & B_{12} & B_{13} \\ B_{21} & B_{22} & B_{23} \\ B_{31} & B_{32} & B_{33} \\ B_{41} & B_{42} & B_{43} \end{bmatrix} \rightarrow \begin{bmatrix} A_{11} \cdot B_{11} & A_{11} \cdot B_{12} & A_{11} \cdot B_{13} \\ A_{12} \cdot B_{21} & A_{12} \cdot B_{22} & A_{12} \cdot B_{23} \\ A_{13} \cdot B_{31} & A_{13} \cdot B_{32} & A_{13} \cdot B_{33} \\ A_{14} \cdot B_{41} & A_{14} \cdot B_{42} & A_{14} \cdot B_{43} \end{bmatrix}$$

$$\begin{bmatrix} A_{21} & 0 & 0 & 0 \\ 0 & A_{22} & 0 & 0 \\ 0 & 0 & A_{23} & 0 \\ 0 & 0 & 0 & A_{24} \end{bmatrix} \cdot \begin{bmatrix} B_{11} & B_{12} & B_{13} \\ B_{21} & B_{22} & B_{23} \\ B_{31} & B_{32} & B_{33} \\ B_{41} & B_{42} & B_{43} \end{bmatrix} \rightarrow \begin{bmatrix} A_{21} \cdot B_{11} & A_{21} \cdot B_{12} & A_{21} \cdot B_{13} \\ A_{22} \cdot B_{21} & A_{22} \cdot B_{22} & A_{22} \cdot B_{23} \\ A_{23} \cdot B_{31} & A_{23} \cdot B_{32} & A_{23} \cdot B_{33} \\ A_{24} \cdot B_{41} & A_{24} \cdot B_{42} & A_{24} \cdot B_{43} \end{bmatrix}$$

$$
\begin{bmatrix} A_{31} & 0 & 0 & 0 \\ 0 & A_{32} & 0 & 0 \\ 0 & 0 & A_{33} & 0 \\ 0 & 0 & 0 & A_{34} \end{bmatrix} \cdot \begin{bmatrix} B_{11} & B_{12} & B_{13} \\ B_{21} & B_{22} & B_{23} \\ B_{31} & B_{32} & B_{33} \\ B_{41} & B_{42} & B_{43} \end{bmatrix} \rightarrow \begin{bmatrix} A_{31}{\cdot}B_{11} & A_{31}{\cdot}B_{12} & A_{31}{\cdot}B_{13} \\ A_{32}{\cdot}B_{21} & A_{32}{\cdot}B_{22} & A_{32}{\cdot}B_{23} \\ A_{33}{\cdot}B_{31} & A_{33}{\cdot}B_{32} & A_{33}{\cdot}B_{33} \\ A_{34}{\cdot}B_{41} & A_{34}{\cdot}B_{42} & A_{34}{\cdot}B_{43} \end{bmatrix}
$$

This Nested Array is equal to:

$$
\begin{bmatrix} \begin{bmatrix} A_{11}{\cdot}B_{11} & A_{11}{\cdot}B_{12} & A_{11}{\cdot}B_{13} \\ A_{12}{\cdot}B_{21} & A_{12}{\cdot}B_{22} & A_{12}{\cdot}B_{23} \\ A_{13}{\cdot}B_{31} & A_{13}{\cdot}B_{32} & A_{13}{\cdot}B_{33} \\ A_{14}{\cdot}B_{41} & A_{14}{\cdot}B_{42} & A_{14}{\cdot}B_{43} \end{bmatrix} \cdot \begin{bmatrix} A_{21}{\cdot}B_{11} & A_{21}{\cdot}B_{12} & A_{21}{\cdot}B_{13} \\ A_{22}{\cdot}B_{21} & A_{22}{\cdot}B_{22} & A_{22}{\cdot}B_{23} \\ A_{23}{\cdot}B_{31} & A_{23}{\cdot}B_{32} & A_{23}{\cdot}B_{33} \\ A_{24}{\cdot}B_{41} & A_{24}{\cdot}B_{42} & A_{24}{\cdot}B_{43} \end{bmatrix} \cdot \begin{bmatrix} A_{31}{\cdot}B_{11} & A_{31}{\cdot}B_{12} & A_{31}{\cdot}B_{13} \\ A_{32}{\cdot}B_{21} & A_{32}{\cdot}B_{22} & A_{32}{\cdot}B_{23} \\ A_{33}{\cdot}B_{31} & A_{33}{\cdot}B_{32} & A_{33}{\cdot}B_{33} \\ A_{34}{\cdot}B_{41} & A_{34}{\cdot}B_{42} & A_{34}{\cdot}B_{43} \end{bmatrix} \end{bmatrix}
$$

Here we are back to a Nested Array. We do not need to transpose this array, since after the row sums are done, the matrix is already in the form that is needed for the proper solution. I think that this the Cross Product of a matrix, (which is up until now undefined) and wait for better mathematicians than I to confirm or deny this hypothesis. Let's go ahead and sum the columns and see what we get.

$$
\begin{vmatrix} A_{11}(B_{11}{+}B_{12}{+}B_{13}) & A_{21}(B_{11}{+}B_{12}{+}B_{13}) & A_{31}(B_{11}{+}B_{12}{+}B_{13}) \\ A_{12}(B_{21}{+}B_{22}{+}B_{23}) & A_{22}(B_{21}{+}B_{22}{+}B_{23}) & A_{32}(B_{21}{+}B_{22}{+}B_{23}) \\ A_{13}(B_{31}{+}B_{32}{+}B_{33}) & A_{23}(B_{31}{+}B_{32}{+}B_{33}) & A_{33}(B_{31}{+}B_{32}{+}B_{33}) \\ A_{14}(B_{41}{+}B_{42}{+}B_{43}) & A_{24}(B_{41}{+}B_{42}{+}B_{43}) & A_{34}(B_{41}{+}B_{42}{+}B_{43}) \end{vmatrix} =
$$

$$
\begin{vmatrix} A_{11}{\sum}B_{1} & A_{21}{\sum}B_{1} & A_{31}{\sum}B_{1} \\ A_{12}{\sum}B_{2} & A_{22}{\sum}B_{2} & A_{32}{\sum}B_{2} \\ A_{13}{\sum}B_{3} & A_{23}{\sum}B_{3} & A_{33}{\sum}B_{3} \\ A_{14}{\sum}B_{4} & A_{24}{\sum}B_{4} & A_{34}{\sum}B_{4} \end{vmatrix}
$$

But this is equal to:

$$
\begin{vmatrix} \Sigma B_1 \\ \Sigma B_2 \\ \Sigma B_3 \\ \Sigma B_4 \end{vmatrix} \circ \begin{vmatrix} A_{11} & A_{21} & A_{31} \\ A_{12} & A_{22} & A_{32} \\ A_{13} & A_{23} & A_{33} \\ A_{14} & A_{24} & A_{34} \end{vmatrix}
$$

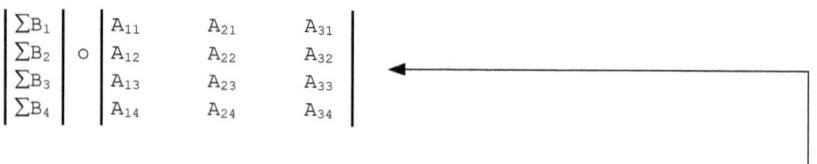

So this is equivalent to the Itempage matrix in the inventory/ accounting system. The neat thing about this operator is that it has it's counterparts in regular mathematics. That is, the above expression can be calculated by the matric equation:

$$
([B]_{JK} [1]_{k1}) \circ [A]^{T}_{iJ}
$$

$$
\begin{vmatrix} B_{11} & B_{12} & B_{13} \\ B_{21} & B_{22} & B_{23} \\ B_{31} & B_{32} & B_{33} \\ B_{41} & B_{42} & B_{43} \end{vmatrix} \begin{vmatrix} 1 \\ 1 \\ 1 \\ 1 \end{vmatrix} \circ \begin{vmatrix} A_{11} & A_{21} & A_{31} \\ A_{12} & A_{22} & A_{32} \\ A_{13} & A_{23} & A_{33} \\ A_{14} & A_{24} & A_{34} \end{vmatrix} =
$$

This is an important property. It says that if you transpose [A] then [B] becomes commutative with matrix [A]. Let's look at a simple example for illustration. Bill has just come into some money. He wants to buy a boat for $10,000, two cars, one for $20,000 and the other for $15,000, a TV/VCR for $500 and some new clothes, two suits for $250 each. Lets make a matrix out of what Bill wants to buy, and how much of each.

The one boat, one car, one TV/VCR and two suits become: [1 1 1 1 2]

And the one for price is:
$$
\begin{vmatrix} 10,000 \\ 20,000 \\ 15,000 \\ 500 \\ 250 \end{vmatrix}
$$

To find the total, we multiply the two together in the regular way:

$$
[1 \; 1 \; 1 \; 1 \; 2] \begin{vmatrix} 10,000 \\ 20,000 \\ 15,000 \\ 500 \\ 250 \end{vmatrix} = \$46,000
$$

Now let's transpose [A] and see what happens:

$$\begin{vmatrix} 1 \\ 1 \\ 1 \\ 1 \\ 2 \end{vmatrix} \circ \begin{vmatrix} 10,000 \\ 20,000 \\ 15,000 \\ 500 \\ 250 \end{vmatrix} = \begin{vmatrix} 10,000 \\ 20,000 \\ 15,000 \\ 500 \\ 250 \end{vmatrix} \circ \begin{vmatrix} 1 \\ 1 \\ 1 \\ 1 \\ 2 \end{vmatrix}$$

One boat equals $10,000 no matter how you look at it, the first car is still $20,000 multiplied both ways, etc.

This works for [A] and [B] if [A] is a row or column matrix, whether [A] is a N x M matrix or just a row or column matrix. But if [B] and [A] both become a matrix of more than just a row or column, then multiplying $[B] \circ [A]^T$ gives the transpose of [A][B]. (This is proved somewhat a little later on in this section of my paper). i.e. Suppose:

$$[B] = \begin{vmatrix} D & E & F \\ G & H & I \\ J & K & L \end{vmatrix} \text{ and } [A] = [A \ B \ C]$$

$$[A \ B \ C] \begin{vmatrix} D & E & F \\ G & H & I \\ J & K & L \end{vmatrix} = [AD+BG+CJ \quad AE+BH+CK \quad AF+BI+CL]$$

$$[B] = \begin{vmatrix} D & E & F \\ G & H & I \\ J & K & L \end{vmatrix}$$

$$^R\sum[B] \circ [A] = \begin{vmatrix} D & E & F \\ G & H & I \\ J & K & L \end{vmatrix} \begin{vmatrix} 1 \\ 1 \\ 1 \end{vmatrix} \circ \begin{vmatrix} A \\ B \\ C \end{vmatrix} = \begin{vmatrix} A(D+E+F) \\ B(G+H+I) \\ C(J+K+L) \end{vmatrix} = \begin{vmatrix} \begin{vmatrix} AD \\ BG \\ CJ \end{vmatrix} + \begin{vmatrix} AE \\ BH \\ CK \end{vmatrix} + \begin{vmatrix} AF \\ BI \\ CL \end{vmatrix} \end{vmatrix}$$

Sum the columns and we have the same answer as in [A] [B] above. Remove the inner brackets from the nested array (ie, partition the Matrix) above and the solutions are equal. But this is also a valid solution all in it's own. Even each individual column is a valid solution (or valid sub-solution) also. A good example of this you will see in the section on Itempage accounting.

An interesting effect occurs when

$$B_{11} + B_{12} + B_{13} = 1$$
$$B_{21} + B_{22} + B_{23} = 1$$

$$B_{31} + B_{32} + B_{33} = 1$$
$$B_{41} + B_{42} + B_{43} = 1$$

The equation reduces to:

$$^{R}\Sigma [A]_{iJ}[B]_{JK} = ([B]_{JK}[1]_{K1}) \text{ o } [A]^{T}_{iJ} = [A]^{T}_{iJ}$$

I am going to call the above inventory where $[B]_{JK}[1]_{K1} = [1]_{J1}$ a Closed System (or closed inventory) as opposed to an Open Inventory where $[B]_{JK}[1]_{K1} = [IP]_{J1}$.

NOW LET'S PROVE THIS FOR ALL N X M MATRICES:

$$\begin{vmatrix} B_{11} & . . & .B_{1K} & . . & .B_{1M} \\ . & & . & & . \\ . & & . & & . \\ B_{J1} & . . & .B_{JK} & . . & .B_{JM} \\ +. & & . & & . \\ . & & . & & . \\ B_{N1} & . . & .B_{NK} & . . & .B_{NM} \end{vmatrix}_{NM} \begin{vmatrix} 1 \\ . \\ . \\ 1 \\ . \\ . \\ 1 \end{vmatrix}_{N,1} =$$

$$\begin{vmatrix} B_{11} & . .+. & .B_{1K} & .+. & .B_{1M} \\ . & . . . & . . & . & . . \\ . & . . . & . . & . & . . \\ B_{J1} & . .+. & .B_{JK} & .+. & .B_{JM}. \\ . & . . . & . . & . & . . \\ . & . . . & . . & . & . . \\ B_{N1} & . .+. & .B_{NK} & .+. & .B_{NM}. \end{vmatrix}_{NM}$$

Then $([B]_{JK}[1]_{K1}) \text{ o } [A]^{T}_{iJ} =$ (this will give us i sub-matrices)

$$\begin{vmatrix} B_{11} & . .+. & .B_{1K} & .+. & .B_{1M}. \\ . & . . . & . . & . & . . \\ . & . . . & . . & . & . . \\ B_{J1} & . .+. & .B_{JK} & .+. & .B_{JM}. \\ . & . . . & . . & . & . . \\ . & . . . & . . & . & . . \\ B_{N1} & . .+. & .B_{NK} & .+. & .B_{NM}. \end{vmatrix} \text{ o } \begin{vmatrix} A_{11} & . . & .A_{i1} & . . & .A_{M1} \\ & & & & \\ A_{1J} & . . & .A_{iJ} & . . & .A_{MJ}. \\ & & & & \\ A_{1N} & . . & .A_{iN} & . . & .A_{MN} \end{vmatrix}$$

$$[A]_{MN} = \begin{vmatrix} A_{11} & . . & .A_{1J} & . . & .A_{1N} \\ . & & . & & . \\ A_{i1} & . . & .A_{iJ} & . . & .A_{iN} \\ . & & . & & . \\ A_{M1} & . . & .A_{MJ} & . . & .A_{MN} \end{vmatrix}$$

LET ΣB1 $= B_{11}. \ .+. \ .B_{1K}. \ .+. \ .B_{1M}.$
LET ΣB2 $= B_{J1}. \ .+. \ .B_{JK}. \ .+. \ .B_{JM}.$

LET $\sum B3 = B_{N1}. \;.+. \;.B_{NK}. \;.+. \;.B_{NM}.$

And $^R\sum[B] \circ [A]^T = (\sum B1 \circ [A]^T) + (\sum B2 \circ [A]^T) + (\sum B3 \circ [A]^T)$

$(\sum B1 \circ [A]^T) =$

$$
\begin{vmatrix}
B_{11}. & . & . & .0. & . & . & .0 \\
. & . & . & . & . & . & . & . \\
. & . & . & . & . & . & . & . \\
0. & . & . & .B_{1K}. & . & . & 0 \\
. & . & . & . & . & . & . & . \\
. & . & . & . & . & . & . & . \\
0. & . & . & .0. & . & . & .B_{1M}.
\end{vmatrix}
\circ
\begin{vmatrix}
A_{11} & . & . & .A_{i1} & . & . & .A_{M1} \\
A_{1J} & . & . & .A_{iJ} & . & . & .A_{MJ}. \\
A_{1N} & . & . & .A_{iN} & . & . & .A_{MN}
\end{vmatrix}
$$

$$
\begin{vmatrix}
B_{11}. & . & . & .0. & . & . & .0 \\
. & . & . & . & . & . & . & . \\
0. & . & . & .B_{1J}. & . & . & 0 \\
. & . & . & . & . & . & . & . \\
. & . & . & . & . & . & . & . \\
0. & . & . & .0. & . & . & .B_{N1}
\end{vmatrix}
\circ
\begin{vmatrix}
A_{11} & . & . & .A_{i1} & . & . & .A_{M1} \\
A_{1J} & . & . & .A_{iJ} & . & . & .A_{MJ}. \\
A_{1N} & . & . & .A_{iN} & . & . & .A_{MN}
\end{vmatrix}
$$

$$
\begin{vmatrix}
B_{11}(A_{11} & . & . & .A_{i1} & . & . & .A_{M1}) \\
. & . & . & . \\
B_{J1}(A_{1J} & . & . & .A_{iJ} & . & . & .A_{MJ}) \\
. & . & . & . \\
B_{N1}(A_{1N} & . & . & .A_{iN} & . & . & .A_{MN})
\end{vmatrix}
=
\begin{vmatrix}
B_{11}A_{11} & .. & B_{11}A_{i1} & .. & B_{11}A_{M1} \\
. & . & . \\
B_{J1}A_{1J} & .. & B_{J1}A_{iJ} & .. & B_{J1}A_{MJ} \\
. & . & . \\
B_{N1}A_{1N} & .. & B_{N1}A_{iN} & .. & B_{N1}A_{MN}
\end{vmatrix}
$$

$(\sum B2 \circ [A]^T) =$ (MAKE COLUMN 2 OF [A] INTO A DIAGONAL MATRIX AND MULTIPLY x [A], WE GET:

$$
\begin{vmatrix}
B_{1K}(A_{11} & . & . & .A_{i1} & . & . & .A_{M1}) \\
. & . & . & . \\
B_{JK}(A_{1J} & . & . & .A_{iJ} & . & . & .A_{MJ}) \\
. & . & . & . \\
B_{MK}(A_{1N} & . & . & .A_{iN} & . & . & .A_{MN})
\end{vmatrix}
=
\begin{vmatrix}
B_{1K}A_{11} & .. & B_{1K}A_{i1} & .. & B_{1K}A_{M1} \\
. & . & . \\
B_{JK}A_{1J} & .. & B_{JK}A_{iJ} & .. & B_{JK}A_{MJ} \\
. & . & . \\
B_{NK}A_{1N} & .. & B_{NK}A_{iN} & .. & B_{NK}A_{MN}
\end{vmatrix}
$$

$(\sum B3 \; o \; [A]^{\mathsf{T}}) =$ (MAKE COLUMN 3 OF [B] INTO A DIAGONAL MATRIX AND MULTIPLY x [A], WE GET:

$$
\begin{vmatrix}
B_{1M}(A_{11} & \cdots & A_{i1} & \cdots & A_{M1}) \\
\cdot & \cdot & \cdot & & \cdot \\
\cdot & \cdot & \cdot & & \cdot \\
B_{JM}(A_{1J} & \cdots & A_{iJ} & \cdots & A_{MJ}) \\
\cdot & \cdot & \cdot & & \cdot \\
\cdot & \cdot & \cdot & & \cdot \\
B_{NM}(A_{1N} & \cdots & A_{iN} & \cdots & A_{MN})
\end{vmatrix}
=
\begin{vmatrix}
B_{1M}A_{11} & \cdots & B_{1M}A_{i1} & \cdots & B_{1M}A_{M1} \\
\cdot & & \cdot & & \cdot \\
\cdot & & \cdot & & \cdot \\
B_{JM}A_{1J} & \cdots & B_{JM}A_{iJ} & \cdots & B_{JM}A_{MJ} \\
\cdot & & \cdot & & \cdot \\
\cdot & & \cdot & & \cdot \\
B_{NM}A_{1N} & \cdots & B_{NM}A_{iN} & \cdots & B_{NM}A_{MN}
\end{vmatrix}
$$

Now that we've computed the three sub-matrices, we must add them together. I don't have a lot of room on this computer to add the whole thing and keep the computations short and clear. Let's first add the first three sums that occupy A_{11}:

$$B_{11}A_{11} + B_{1K}A_{11} + B_{1M}A_{11} = A_{11}(B_{11} + B_{1K} + B_{1M}).$$

These check, so let's add them all together. Do you see anything that happens when we add the Sub-matrices? That is the subject of my next proof.

$$
\begin{vmatrix}
A_{11}B_{11} . + . A_{11}B_{1K} . + . A_{11}B_{1M} . \\
A_{1J}B_{J1} . + . A_{1J}B_{JK} . + . A_{1J}B_{JM} . \\
A_{1N}B_{N1} . + . A_{1N}B_{NK} . + . A_{1N}B_{NM}
\end{vmatrix}
\quad
\begin{matrix}
A_{11}B_{11} . + . A_{11}B_{1K} . + . A_{11}B_{1M} \\
A_{1J}B_{J1} . + . A_{1J}B_{JK} . + . A_{1J}B_{JM} \\
A_{1N}B_{N1} . + . A_{1N}B_{NK} . + . A_{1N}B_{NM}
\end{matrix}
\quad
\begin{vmatrix}
A_{M1}B_{11} . + . \; A_{M1}B_{1K} . + . \; A_{M1}B_{1M} \\
A_{MJ}B_{J1} . + . \; A_{MJ}B_{JK} . + . \; A_{MJ}B_{JM} \\
A_{MN}B_{N1} . + . \; A_{MN}B_{NK} . + . \; A_{MN}B_{NM}
\end{vmatrix}
$$

Which is equal to:

$$
\begin{vmatrix}
A_{11}(B_{11} . + . B_{1K} . + . B_{1M}) \\
A_{1J}(B_{J1} . + . . B_{JK} . . + . . B_{JM}) \\
A_{1N}(B_{N1} . . + . . B_{NK} . . + . . B_{NM})
\end{vmatrix}
\quad
\begin{matrix}
A_{11}(B_{11} . + . B_{1K} . + . B_{1M}) \\
A_{1J}(B_{J1} . + . . B_{JK} . . + . . B_{JM}) \\
A_{1N}(B_{N1} . . + . . B_{NK} . . + . . B_{NM})
\end{matrix}
\quad
\begin{vmatrix}
A_{M1}(B_{11} . + . B_{1K} . + . B_{1M}) \\
A_{MJ}(B_{J1} . . + . . B_{JK} . . + . . B_{JM}) \\
A_{MN}(B_{N1} . . + . . B_{NK} . . + . . B_{NM})
\end{vmatrix}
$$

QED

This is the proof for all N x M Matrices. Now to prove it works for all Matrices N+1,M+1.

Now let's prove this:

$$\begin{vmatrix} B_{11} & . & . & B_{1K} & . & . & B_{1M} & . & . & B_{1,M+1} \\ & . & & . & & . & . & & . \\ & . & & . & & . & . & & . \\ & . & & . & & . & . & & . \\ B_{J1} & . & . & B_{JK} & . & . & B_{JM} & . & . & B_{J,M1} \\ + & . & & . & & . & . & & . \\ & . & & . & & . & . & & . \\ B_{N1} & . & . & B_{NK} & . & . & B_{NM} & . & . & B_{N,M+1} \\ & . & & . & & . & . & & . \\ & . & & . & & . & . & & . \\ B_{N+1,1} & . & . & B_{N+1,K} & . & B_{N+1,M} & . & . & B_{N+1,M+1} \end{vmatrix} \begin{vmatrix} 1 \\ . \\ . \\ . \\ 1 \\ . \\ . \\ 1 \\ . \\ . \\ 1 \end{vmatrix} = $$

$$\begin{vmatrix} B_{11} & . & +. & . B_{1K} & . & +. & . B_{1M} & . & . & +. & . B_{1,M+1} \\ . & . & . & . & . & . & . & . & . & . & . \\ . & . & . & . & . & . & . & . & . & . & . \\ B_{J1} & . & +. & . B_{JK} & . & +. & . B_{JM} & . & . & +. & . B_{J,M1} \\ . & . & . & . & . & . & . & . & . & . & . \\ . & . & . & . & . & . & . & . & . & . & . \\ B_{N1} & . & +. & . B_{NK} & . & +. & . B_{NM} & . & . & +. & . B_{N,M+1} \\ . & . & . & . & . & . & . & . & . & . & . \\ . & . & . & . & . & . & . & . & . & . & . \\ B_{N+1,1} & . +. & . B_{N+1,K} & . +. & . B_{N+1,M} & . +. & . B_{N+1,M+1} \end{vmatrix}$$

Then $([\mathbf{B}]_{JK}[\mathbf{1}]_{K1}) \circ [\mathbf{A}]^{T}_{iJ} =$

$$\begin{vmatrix} B_{11} . +. . B_{1K} . +. . B_{1M} . . +. . B_{1,M+1} \\ . . . . \\ B_{J1} . +. . B_{JK} . +. . B_{JM} . . +. . B_{J,M1} \\ . . . . \\ B_{N1} . +. . B_{NK} . +. . B_{NM} . . +. . B_{N,M+1} \\ . . . . \\ B_{N+1,1} . +. . B_{N+1,K} . +. . B_{N+1,M} +. . B_{N+1,M+1} \end{vmatrix} \circ \begin{vmatrix} A_{11} & . & . & . A_{11} & . & . & . A_{M1} & . & . & . A_{M+1,1} \\ A_{1J} & . & . & . A_{1J} & . & . & . A_{MJ} & . & . & . A_{M+1,J} \\ A_{1N} & . & . & . A_{1N} & . & . & A_{MN} & . & . & A_{M+1,N} \\ A_{1,N+1} & . & . & A_{1,N+1} & . & . & A_{M,N+1} & . & . & A_{M+1,N+1} \end{vmatrix}$$

$$[\mathbf{A}]_{MN} = \begin{vmatrix} A_{11} & . & . & . A_{1J} & . & . & . A_{1N} & . & . & . A_{1,N+1} \\ A_{11} & . & . & . A_{1J} & . & . & . A_{1N} & . & . & . A_{1,N+1} \\ A_{M1} & . & . & . A_{MJ} & . & . & . A_{MN} & . & . & . A_{M,N+1} \\ A_{M+1,1} & . & . & A_{M+1,J} & . & . & A_{M+1,N} & . & . & A_{M+1,N+1} \end{vmatrix}$$

LET $\sum B1 = B_{11} . +. . B_{1K} . +. . B_{1M} . . +. . B_{1,M+1}$

LET $\sum B2 = B_{J1} . +. . B_{JK} . +. . B_{JM} . . +. . B_{J,M1}$

LET $\sum B3 = B_{N1} . +. . B_{NK} . +. . B_{NM} . . +. . B_{N,M+1}$

LET $\sum B4 = B_{N+1,1} . +. . B_{N+1,K} . +. . B_{N+1,M} . +. . B_{N+1,M+1}$

And $^{R}\sum[B] \circ [A]^{T} = (\sum B1 \circ [A]^{T}) + (\sum B2 \circ [A]^{T}) + (\sum B3 \circ [A]^{T}) + (\sum B4 \circ [A]^{T})$ $(\sum B1 \circ [A]^{T}) =$

$$
\begin{vmatrix}
B_{11} & \cdots & 0 & \cdots & 0 & \cdots & 0 \\
0 & \cdots & B_{1K} & \cdots & 0 & \cdots & 0 \\
0 & \cdots & 0 & \cdots & B_{1M} & \cdots & 0 \\
0 & \cdots & 0 & \cdots & 0 & \cdots & B_{1,M+1}
\end{vmatrix}_{N+1,M+1}
\quad \circ \quad
\begin{vmatrix}
A_{11} & \cdots & A_{i1} & \cdots & A_{M1} & \cdots & A_{M+1,1} \\
A_{1J} & \cdots & A_{iJ} & \cdots & A_{MJ} & \cdots & A_{M+1,J} \\
A_{1N} & \cdots & A_{iN} & \cdots & A_{MN} & \cdots & A_{M+1,N} \\
A_{1,N+1} & \cdots & A_{i,N+1} & \cdots & A_{M,N+1} & \cdots & A_{M+1,N+1}
\end{vmatrix}_{N+1,M+1}
$$

$$
\begin{vmatrix}
B_{11} & \cdots & 0 & \cdots & 0 & \cdots & 0 \\
0 & \cdots & B_{1J} & \cdots & 0 & \cdots & 0 \\
0 & \cdots & 0 & \cdots & B_{N1} & \cdots & 0 \\
0 & \cdots & 0 & \cdots & 0 & \cdots & B_{N+1,1}
\end{vmatrix}_{N+1,M+1}
\quad \circ \quad
\begin{vmatrix}
A_{11} & \cdots & A_{i1} & \cdots & A_{M1} & \cdots & A_{M+1,1} \\
A_{1J} & \cdots & A_{iJ} & \cdots & A_{MJ} & \cdots & A_{M+1,J} \\
A_{1N} & \cdots & A_{iN} & \cdots & A_{MN} & \cdots & A_{M+1,N} \\
A_{1,N+1} & \cdots & A_{i,N+1} & \cdots & A_{M,N+1} & \cdots & A_{M+1,N+1}
\end{vmatrix}_{N+1,M+1}
$$

$$
\begin{vmatrix}
B_{11}(A_{11} & \cdots & A_{i1} & \cdots & A_{M1} & \cdots & A_{M+1,1}) \\
B_{J1}(A_{1J} & \cdots & A_{iJ} & \cdots & A_{MJ} & \cdots & A_{M+1,J}) \\
B_{N1}(A_{1N} & \cdots & A_{iN} & \cdots & A_{MN} & \cdots & A_{M+1,N}) \\
B_{N+1,1}(A_{1,N+1} & \cdots & A_{i,N+1} & \cdots & A_{M,N+1} & \cdots & A_{M+1,N+1})
\end{vmatrix}
=
\begin{vmatrix}
B_{11}A_{11} & \cdots & B_{11}A_{i1} & \cdots & B_{11}A_{M1} & \cdots & B_{11}A_{M+1,1} \\
B_{J1}A_{1J} & \cdots & B_{J1}A_{iJ} & \cdots & B_{J1}A_{MJ} & \cdots & B_{J1}A_{M+1,J} \\
B_{N1}A_{1N} & \cdots & B_{N1}A_{iN} & \cdots & B_{N1}A_{MN} & \cdots & B_{N1}A_{M+1,N} \\
B_{N+1,1}A_{1,N+1} & \cdots & B_{N+1,1}A_{i,N+1} & \cdots & B_{N+1,1}A_{M,N+1} & \cdots & B_{N+1,1}A_{M+1,N+1}
\end{vmatrix}
$$

$(\Sigma B2 \circ [A]^{T}) =$ (MAKE COLUMN 2 OF [A] INTO A DIAGONAL MATRIX AND MULTIPLY x [A], WE GET:

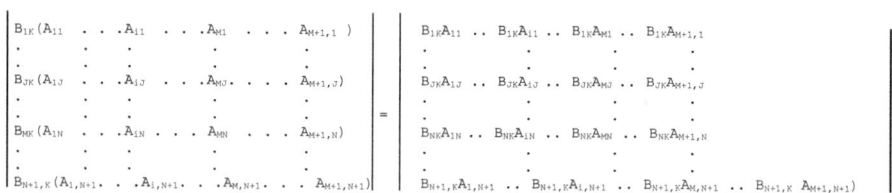

$(\Sigma B3 \circ [A]^{T}) =$ (MAKE COLUMN 3 OF [B] INTO A DIAGONAL MATRIX AND MULTIPLY x [A], WE GET:

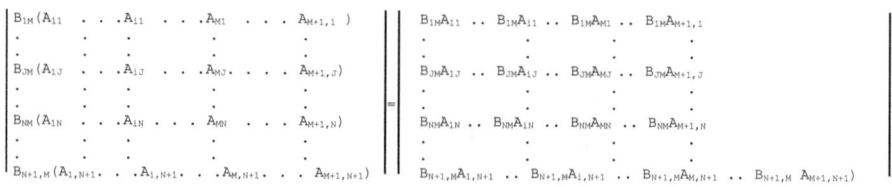

$(\Sigma B4 \circ [A]^{T}) =$ (MAKE COLUMN 4 OF [B] INTO A DIAGONAL MATRIX AND MULTIPLY x [A], WE GET

$$
\begin{Vmatrix}
B_{1,M+1}(A_{11} & \cdots & A_{11} & \cdots & A_{M1} & \cdots & A_{M+1,1}) \\
\vdots & & \vdots & & \vdots & & \vdots \\
B_{J,M+1}(A_{1J} & \cdots & A_{1J} & \cdots & A_{MJ} & \cdots & A_{M+1,J}) \\
\vdots & & \vdots & & \vdots & & \vdots \\
B_{N,M+1}(A_{1N} & \cdots & A_{1N} & \cdots & A_{MN} & \cdots & A_{M+1,N}) \\
\vdots & & \vdots & & \vdots & & \vdots \\
B_{N+1,M+1}(A_{1,N+1} & \cdots & A_{1,N+1} & \cdots & A_{M,N+1} & \cdots & A_{M+1,N+1})
\end{Vmatrix}
=
\begin{Vmatrix}
B_{1,M+1}A_{11} & \cdots & B_{1,M+1}A_{11} & \cdots & B_{1,M+1}A_{M1} & \cdots & B_{1,M+1}A_{M+1,1} \\
\vdots & & \vdots & & \vdots & & \vdots \\
B_{J,M+1}A_{1J} & \cdots & B_{J,M+1}A_{1J} & \cdots & B_{J,M+1}A_{MJ} & \cdots & B_{J,M+1}A_{M+1,J} \\
\vdots & & \vdots & & \vdots & & \vdots \\
B_{N,M+1}A_{1N} & \cdots & B_{N,M+1}A_{1N} & \cdots & B_{N,M+1}A_{MN} & \cdots & B_{N,M+1}A_{M+1,N} \\
\vdots & & \vdots & & \vdots & & \vdots \\
B_{N+1,M+1}A_{1,N+1} & \cdots & B_{N+1,M+1}A_{1,N+1} & \cdots & B_{N+1,M+1}A_{M,N+1} & \cdots & B_{N+1,M+1}A_{M+1,N+1}
\end{Vmatrix}
$$

Now that we've computed the four sub-matrices, we must add them together. Let's first add the first four sums that occupy A_{11}:

$$B_{11}A_{11} + B_{1K}A_{11} + B_{1M}A_{11} +, B_{1,M+1}A_{11} = A_{11}(B_{11} + B_{1K} + B_{1M} + B_{1,M+1}).$$

These check, so let's add them all together.

$$
\begin{Vmatrix}
A_{11}B_{11} + . A_{11}B_{1K} + . A_{11}B_{1M} + . A_{11}B_{1,M+} \\
A_{1J}B_{J1} + . A_{1J}B_{JK} + . A_{1J}B_{JM} + . A_{1J}B_{J,M1} \\
A_{1N}B_{N1} + . A_{1N}B_{NK} + . A_{1N}B_{NM} + . A_{1N}B_{N,M+1} \\
A_{1,N+1}B_{N+1,1} + . A_{1,N+1}B_{N+1,K} + . A_{1,N+1}B_{N+1,M} + . A_{1,N+1}B_{N+1,M+1} \\
\\
A_{M1}B_{11} + . \quad A_{M1}B_{1K} + . \quad A_{M1}B_{1M} + . \quad A_{M1}B_{1,M+} \\
A_{MJ}B_{J1} + . \quad A_{MJ}B_{JK} + . \quad A_{MJ}B_{JM} + . \quad A_{MJ}B_{J,M1} \\
A_{M N}B_{N1} + . \quad A_{M N}B_{NK} + . \quad A_{M N}B_{NM} + . \quad A_{M N}B_{N,M+1} \\
A_{M,N+1}B_{N+1,1} + . A_{M,N+1}B_{N+1,K} + . A_{M,N+1}B_{N+1,M} + . A_{M,N+1}B_{N+1,M+1}
\end{Vmatrix}
\begin{Vmatrix}
A_{11}B_{11} + . A_{11}B_{1K} + . A_{11}B_{1M} + . A_{11}B_{1,M+} \\
A_{1J}B_{J1} + . A_{1J}B_{JK} + . A_{1J}B_{JM} + . A_{1J}B_{J,M1} \\
A_{1N}B_{N1} + . A_{1N}B_{NK} + . A_{1N}B_{NM} + . A_{1N}B_{N,M+1} \\
A_{1,N+1}B_{N+1,1} + . A_{1,N+1}B_{N+1,K} + . A_{1,N+1}B_{N+1,M} + . A_{1,N+1}B_{N+1,M+1} \\
\\
A_{M+1,1}B_{11} + . A_{M+1,1}B_{1K} + . A_{M+1,1}B_{1M} + . A_{M+1,1}B_{1,M+} \\
A_{M+1,J}B_{J1} + . A_{M+1,J}B_{JK} + . A_{M+1,J}B_{JM} + . A_{M+1,J}B_{J,M1} \\
A_{M+1,N}B_{N1} + . A_{M+1,N}B_{NK} + . A_{M+1,N}B_{NM} + . A_{M+1,N}B_{N,M+1} \\
A_{M+1,N+1}B_{N+1,1} + . A_{M+1,N+1}B_{N+1,K} + . A_{M+1,N+1}B_{N+1,M} + . A_{M+1,N+1}B_{N+1,M+1}
\end{Vmatrix}
$$

Which is equal to:

$$
\begin{Vmatrix}
A_{11}(B_{11} + . B_{1K} + . B_{1M} + . B_{1,M+}) \\
A_{1J}(B_{J1} . . + . . B_{JK} . . + . . B_{JM} . . + . . B_{J,M1}) \\
A_{1N}(B_{N1} . . + . . B_{NK} . . + . . B_{NM} . . + . B_{N,M+1}) \\
A_{1,N+1}(B_{N+1,1} + . . B_{N+1,K} + . . B_{N+1,M} + . . B_{N+1,M+1}) \\
\\
A_{M1}(B_{11} + . B_{1K} + . B_{1M} + . B_{1,M+})) \\
A_{MJ}(B_{J1} . . + . . B_{JK} . . + . . B_{JM} . + . . B_{J,M1}) \\
A_{M N}(B_{N1} . . + . . B_{NK} . . + . . B_{NM} . . + . . B_{N,M+1}) \\
A_{M,N+1}(B_{N+1,1} + . . B_{N+1,K} + . . B_{N+1,M} + . . B_{N+1,M+1})
\end{Vmatrix}
\begin{Vmatrix}
A_{i1}(B_{11} + . B_{1K} + . B_{1M} + . B_{1,M+}) \\
A_{1J}(B_{J1} . . + . . B_{JK} . . + . . B_{JM} . . + . . B_{J,M1}) \\
A_{1N}(B_{N1} . . + . . B_{NK} . . + . . B_{NM} . . + . B_{N,M+1}) \\
A_{i,N+1}(B_{N+1,1} + . . B_{N+1,K} + . . B_{N+1,M} + . . B_{N+1,M+1}) \\
\\
A_{M+1,1}(B_{11} + . B_{1K} + . B_{1M} + . B_{1,M+})) \\
A_{M+1,J}(B_{J1} . . + . . B_{JK} . . + . . B_{JM} . + . . B_{J,M1}) \\
A_{M+1,N}(B_{N1} . . + . . B_{NK} . . + . . B_{NM} . . + . . B_{N,M+1}) \\
A_{M+1,N+1}(B_{N+1,1} + . . B_{N+1,K} + . . B_{N+1,M} + . . B_{N+1,M+1})
\end{Vmatrix}
$$

QED

SUMMARY OF 3 EQUATIONS DESCRIBED

$${}^{R}\sum [A]^{T}_{iJ} \circ [B]_{JK} = ([B]_{JK}[1]_{K1}) \circ [A]^{T}_{iJ} \text{ FOR OPEN SYSTEMS}$$

$${}^{R}\sum [A]_{iJ}[B]_{JK} = ([B]_{JK}[1]_{K1}) \circ [A]^{T}_{iJ} = [I]_{jj}[A]^{T}_{iJ} = [1]_{J1} \circ [A]^{T}_{iJ} = [A]^{T}_{iJ} \text{ FOR CLOSED SYSTEMS}$$

AND

$${}^{R}\sum [B] \circ [A]^{T} = (\sum B1 \circ [A]^{T}) + (\sum B2 \circ [A]^{T}) + (\sum B3 \circ [A]^{T}) + (\sum B4 \circ [A]^{T} = [IP1] + [IP2] + [IP3] + [IP4]$$

FOR EACH ITEM IN THE DATABASE MATRIX [B].

THE MATRIC PRODUCT OF A MATRIX

Note: We will do this part quickly as it is very similar to the first two proofs, except in this example we sum the 3 Sub-matrices together.)

$$^M\sum [A]^T_{iJ} \circ [B]_{JK} = ([A]^T_{iJ}[1]_{J1}) \circ [B]_{JK}$$

$$\begin{vmatrix} A_{11} & A_{21} & A_{31} \\ A_{12} & A_{22} & A_{32} \\ A_{13} & A_{23} & A_{33} \\ A_{14} & A_{24} & A_{34} \end{vmatrix} \circ \begin{vmatrix} B_{11} & B_{12} & B_{13} \\ B_{21} & B_{22} & B_{23} \\ B_{31} & B_{32} & B_{33} \\ B_{41} & B_{42} & B_{43} \end{vmatrix} =$$

$$\begin{vmatrix} A_{11} \\ A_{12} \\ A_{13} \\ A_{14} \end{vmatrix} \circ \begin{vmatrix} B_{11} & B_{12} & B_{13} \\ B_{21} & B_{22} & B_{23} \\ B_{31} & B_{32} & B_{33} \\ B_{41} & B_{42} & B_{43} \end{vmatrix} \quad \begin{vmatrix} A_{21} \\ A_{22} \\ A_{23} \\ A_{24} \end{vmatrix} \circ \begin{vmatrix} B_{11} & B_{12} & B_{13} \\ B_{21} & B_{22} & B_{23} \\ B_{31} & B_{32} & B_{33} \\ B_{41} & B_{42} & B_{43} \end{vmatrix} \quad \begin{vmatrix} A_{31} \\ A_{32} \\ A_{33} \\ A_{34} \end{vmatrix} \circ \begin{vmatrix} B_{11} & B_{12} & B_{13} \\ B_{21} & B_{22} & B_{23} \\ B_{31} & B_{32} & B_{33} \\ B_{41} & B_{42} & B_{43} \end{vmatrix}$$

$$\begin{bmatrix} A_{11} & 0 & 0 & 0 \\ 0 & A_{12} & 0 & 0 \\ 0 & 0 & A_{13} & 0 \\ 0 & 0 & 0 & A_{14} \end{bmatrix} \begin{bmatrix} B_{11} & B_{12} & B_{13} \\ B_{21} & B_{22} & B_{23} \\ B_{31} & B_{32} & B_{33} \\ B_{41} & B_{42} & B_{43} \end{bmatrix} \rightarrow \begin{bmatrix} A_{11}{\cdot}B_{11} & A_{11}{\cdot}B_{12} & A_{11}{\cdot}B_{13} \\ A_{12}{\cdot}B_{21} & A_{12}{\cdot}B_{22} & A_{12}{\cdot}B_{23} \\ A_{13}{\cdot}B_{31} & A_{13}{\cdot}B_{32} & A_{13}{\cdot}B_{33} \\ A_{14}{\cdot}B_{41} & A_{14}{\cdot}B_{42} & A_{14}{\cdot}B_{43} \end{bmatrix}$$

$$\begin{bmatrix} A_{21} & 0 & 0 & 0 \\ 0 & A_{22} & 0 & 0 \\ 0 & 0 & A_{23} & 0 \\ 0 & 0 & 0 & A_{24} \end{bmatrix} \begin{bmatrix} B_{11} & B_{12} & B_{13} \\ B_{21} & B_{22} & B_{23} \\ B_{31} & B_{32} & B_{33} \\ B_{41} & B_{42} & B_{43} \end{bmatrix} \rightarrow \begin{bmatrix} A_{21}{\cdot}B_{11} & A_{21}{\cdot}B_{12} & A_{21}{\cdot}B_{13} \\ A_{22}{\cdot}B_{21} & A_{22}{\cdot}B_{22} & A_{22}{\cdot}B_{23} \\ A_{23}{\cdot}B_{31} & A_{23}{\cdot}B_{32} & A_{23}{\cdot}B_{33} \\ A_{24}{\cdot}B_{41} & A_{24}{\cdot}B_{42} & A_{24}{\cdot}B_{43} \end{bmatrix}$$

$$\begin{bmatrix} A_{31} & 0 & 0 & 0 \\ 0 & A_{32} & 0 & 0 \\ 0 & 0 & A_{33} & 0 \\ 0 & 0 & 0 & A_{34} \end{bmatrix} \begin{bmatrix} B_{11} & B_{12} & B_{13} \\ B_{21} & B_{22} & B_{23} \\ B_{31} & B_{32} & B_{33} \\ B_{41} & B_{42} & B_{43} \end{bmatrix} \rightarrow \begin{bmatrix} A_{31}{\cdot}B_{11} & A_{31}{\cdot}B_{12} & A_{31}{\cdot}B_{13} \\ A_{32}{\cdot}B_{21} & A_{32}{\cdot}B_{22} & A_{32}{\cdot}B_{23} \\ A_{33}{\cdot}B_{31} & A_{33}{\cdot}B_{32} & A_{33}{\cdot}B_{33} \\ A_{34}{\cdot}B_{41} & A_{34}{\cdot}B_{42} & A_{34}{\cdot}B_{43} \end{bmatrix}$$

This is equal to:

$$\begin{bmatrix} \begin{bmatrix} A_{11}{\cdot}B_{11} & A_{11}{\cdot}B_{12} & A_{11}{\cdot}B_{13} \\ A_{12}{\cdot}B_{21} & A_{12}{\cdot}B_{22} & A_{12}{\cdot}B_{23} \\ A_{13}{\cdot}B_{31} & A_{13}{\cdot}B_{32} & A_{13}{\cdot}B_{33} \\ A_{14}{\cdot}B_{41} & A_{14}{\cdot}B_{42} & A_{14}{\cdot}B_{43} \end{bmatrix} \begin{bmatrix} A_{21}{\cdot}B_{11} & A_{21}{\cdot}B_{12} & A_{21}{\cdot}B_{13} \\ A_{22}{\cdot}B_{21} & A_{22}{\cdot}B_{22} & A_{22}{\cdot}B_{23} \\ A_{23}{\cdot}B_{31} & A_{23}{\cdot}B_{32} & A_{23}{\cdot}B_{33} \\ A_{24}{\cdot}B_{41} & A_{24}{\cdot}B_{42} & A_{24}{\cdot}B_{43} \end{bmatrix} \begin{bmatrix} A_{31}{\cdot}B_{11} & A_{31}{\cdot}B_{12} & A_{31}{\cdot}B_{13} \\ A_{32}{\cdot}B_{21} & A_{32}{\cdot}B_{22} & A_{32}{\cdot}B_{23} \\ A_{33}{\cdot}B_{31} & A_{33}{\cdot}B_{32} & A_{33}{\cdot}B_{33} \\ A_{34}{\cdot}B_{41} & A_{34}{\cdot}B_{42} & A_{34}{\cdot}B_{43} \end{bmatrix} \end{bmatrix}$$

Here we are back to a Nested Array. We do not need to transpose this array, since after the row sums are done, the matrix is already in the form that is needed for the proper solution. To conserve on space, I am going to omit the first step here and just go ahead and collect terms, the complex stuff will be gotten to later. Let's go ahead and sum the matrices and see what we get:

$$
\left|
\begin{array}{lll}
B_{11}(A_{11}+A_{21}+A_{31}) & B_{12}(A_{11}+A_{21}+A_{31}) & B_{13}(A_{11}+A_{21}+A_{31}) \\
B_{21}(A_{12}+A_{22}+A_{32}) & B_{22}(A_{12}+A_{22}+A_{32}) & B_{23}(A_{12}+A_{22}+A_{32}) \\
B_{31}(A_{13}+A_{23}+A_{33}) & B_{32}(A_{13}+A_{23}+A_{33}) & B_{33}(A_{13}+A_{23}+A_{33}) \\
B_{41}(A_{14}+A_{24}+A_{34}) & B_{42}(A_{14}+A_{24}+A_{34}) & B_{43}(A_{14}+A_{24}+A_{34})
\end{array}
\right| = {}^{M}\Sigma[A]^{T} \circ [B] =
$$

$$
\left|
\begin{array}{lll}
B_{11}\Sigma A_{1} & B_{12}\Sigma A_{1} & B_{13}\Sigma A_{1} \\
B_{21}\Sigma A_{2} & B_{22}\Sigma A_{2} & B_{23}\Sigma A_{2} \\
B_{31}\Sigma A_{3} & B_{32}\Sigma A_{3} & B_{33}\Sigma A_{3} \\
B_{41}\Sigma A_{4} & B_{42}\Sigma A_{4} & B_{43}\Sigma A_{4}
\end{array}
\right|
$$

But this is equal to:

$$
\left|
\begin{array}{l}
\Sigma A_{1} \\
\Sigma A_{2} \\
\Sigma A_{3} \\
\Sigma A_{4}
\end{array}
\right| \circ
\left|
\begin{array}{lll}
B_{11} & B_{21} & B_{31} \\
B_{12} & B_{22} & B_{32} \\
B_{13} & B_{23} & B_{33} \\
B_{14} & B_{24} & B_{34}
\end{array}
\right|
$$

So this is equivalent to the Accountpage matrix in the inventory/accounting system. The neat thing about this operator is that it also has it's counterparts in regular mathematics. That is, the above expression can be calculated by the matric equation:

$$
\mathbf{diag}([A]^{T}_{iJ}[1]_{J1})[B]_{JK}
$$

This is also an important property. It says that if you transpose [A] and sum the rows of [A] and o multiply across [B], the solution is the same as if we took the half-multiplied matrices and added them all together. [A] is now commutative with [B].

For an open system, suppose:

$$
\begin{array}{l}
(A_{11}+A_{21}+A_{31})= 1 \\
(A_{12}+A_{22}+A_{32})= 1 \\
(A_{13}+A_{23}+A_{33})= 1 \\
(A_{14}+A_{24}+A_{34})= 1
\end{array}
\quad , \quad \text{Then } {}^{M}\Sigma[A]^{T}_{iJ} \circ [B]_{JK}= [I]_{JJ} \circ [B]_{JK} = [B]_{JK}
$$

PROOF FOR ALL M x N MATRICES

$$
\begin{vmatrix}
A_{11} & \cdots & A_{i1} & \cdots & A_{M1} \\
\vdots & & \vdots & & \vdots \\
A_{1J} & \cdots & A_{iJ} & \cdots & A_{MJ} \\
\vdots & & \vdots & & \vdots \\
A_{1N} & \cdots & A_{iN} & \cdots & A_{MN}
\end{vmatrix}
\begin{vmatrix} 1 \\ \cdot \\ 1 \\ \cdot \\ 1 \end{vmatrix} =
$$

$$
\begin{vmatrix}
A_{11} & .+. & A_{i1} & .+. & A_{M1} \\
\vdots & & \vdots & & \vdots \\
A_{1J} & .+. & A_{iJ} & .+. & A_{MJ} \\
\vdots & & \vdots & & \vdots \\
A_{1N} & .+. & A_{iN} & .+. & A_{MN}
\end{vmatrix}
$$

Then $[A]^T_{iJ}[1]_{i1} \circ [B]_{JK} =$

$$
\begin{vmatrix}
A_{11} & \cdots & A_{i1} & \cdots & A_{M1} \\
\vdots & & \vdots & & \vdots \\
A_{1J} & \cdots & A_{iJ} & \cdots & A_{MJ} \\
\vdots & & \vdots & & \vdots \\
A_{1N} & \cdots & A_{iN} & \cdots & A_{MN}
\end{vmatrix}
\begin{vmatrix} 1 \\ \cdot \\ 1 \\ \cdot \\ 1 \end{vmatrix} \circ
\begin{vmatrix}
B_{11} & \cdots & B_{1K} & \cdots & B_{1M} \\
\vdots & & \vdots & & \vdots \\
B_{J1} & \cdots & B_{JK} & \cdots & B_{JM} \\
\vdots & & \vdots & & \vdots \\
B_{N1} & \cdots & B_{NK} & \cdots & B_{NM}
\end{vmatrix}
$$

LET $\sum A1^T = A_{11} \ .+. \ A_{i1} \ .+. \ A_{M1}$

LET $\sum A2^T = A_{1J} \ .+. \ A_{iJ} \ .+. \ A_{MJ}$

LET $\sum A3^T = A_{1N} \ .+. \ A_{iN} \ .+. \ A_{MN}$

$$
\begin{vmatrix}
\sum A1^T & \cdots & 0 & \cdots & 0 \\
\vdots & & \vdots & & \vdots \\
0 & \cdots & \sum A2^T & \cdots & 0 \\
\vdots & & \vdots & & \vdots \\
0 & \cdots & 0 & \cdots & \sum A3^T
\end{vmatrix}
\; \; \;
\begin{vmatrix}
B_{11} & \cdots & B_{1K} & \cdots & B_{1M} \\
& & \vdots & & \vdots \\
B_{J1} & \cdots & B_{JK} & \cdots & B_{JM} \\
& & \vdots & & \vdots \\
B_{N1} & \cdots & B_{NK} & \cdots & B_{NM}
\end{vmatrix} =
$$

$$
\begin{vmatrix}
B_{11}\sum A1^T & \cdots & B_{1K}\sum A1^T & \cdots & B_{1M}\sum A1^T \\
\vdots & & \vdots & & \vdots \\
B_{J1}\sum A2^T & \cdots & B_{JK}\sum A2^T & \cdots & B_{JM}\sum A2^T \\
\vdots & & \vdots & & \vdots \\
B_{N1}\sum A3^T & \cdots & B_{NK}\sum A3^T & \cdots & B_{NM}\sum A3^T
\end{vmatrix} =
$$

This matrix is also equal to the sum of each individual half-multiplied Sub-matrices.

$${}^{M}\Sigma[A]^{T} \circ [B] = (\Sigma[A1]^{T} \circ [B]) + (\Sigma[A2]^{T} \circ [B]) + (\Sigma[A3]^{T} \circ [B])$$

$(\Sigma[A1]^{T} \circ [B])$

$$
\begin{vmatrix}
A_{11} \cdot & \cdot & \cdot & 0 \cdot & \cdot & \cdot & 0 \\
\cdot & \cdot & \cdot & \cdot & \cdot & \cdot & \cdot \\
\cdot & \cdot & \cdot & \cdot & \cdot & \cdot & \cdot \\
0 \cdot & \cdot & \cdot & A_{1J} \cdot & \cdot & \cdot & 0 \\
\cdot & \cdot & \cdot & \cdot & \cdot & \cdot & \cdot \\
\cdot & \cdot & \cdot & \cdot & \cdot & \cdot & \cdot \\
0 \cdot & \cdot & \cdot & 0 \cdot & \cdot & \cdot & A_{1M} \cdot
\end{vmatrix}
\begin{vmatrix}
B_{11} \cdot \cdot \cdot B_{1K} \cdot \cdot B_{1M} \cdot \cdot \cdot B_{1,M+1} \\
\cdot \quad\quad \cdot \quad\quad \cdot \quad\quad \cdot \\
B_{J1} \cdot \cdot \cdot B_{JK} \cdot \cdot B_{JM} \cdot \cdot \cdot B_{J,M+1} \\
\cdot \quad\quad \cdot \quad\quad \cdot \quad\quad \cdot \\
\cdot \quad\quad \cdot \quad\quad \cdot \quad\quad \cdot \\
B_{N1} \cdot \cdot \cdot B_{NK} \cdot \cdot B_{NM} \cdot \cdot \cdot B_{N,M+1}
\end{vmatrix}
=
$$

$$
\begin{vmatrix}
A_{11}(B_{11} \cdot \cdot \cdot B_{1K} \cdot \cdot \cdot B_{1M}) \\
\cdot \quad\quad \cdot \quad\quad \cdot \\
A_{1J}(B_{J1} \cdot \cdot \cdot B_{JK} \cdot \cdot \cdot B_{JM}) \\
\cdot \quad\quad \cdot \quad\quad \cdot \\
A_{1M}(B_{N1} \cdot \cdot \cdot B_{NK} \cdot \cdot \cdot B_{NM})
\end{vmatrix}
=
\begin{vmatrix}
A_{11}B_{11} \cdot + \cdot A_{11}B_{1K} \cdot + \cdot A_{11}B_{1M} \\
\cdot \quad\quad \cdot \quad\quad \cdot \\
A_{1J}B_{J1} \cdot + \cdot A_{1J}B_{JK} \cdot + \cdot A_{1J}B_{JM} \\
\cdot \quad\quad \cdot \quad\quad \cdot \\
A_{1M}B_{N1} \cdot + \cdot A_{1M}B_{NK} \cdot + \cdot A_{1M}B_{NM}
\end{vmatrix}
$$

$(\Sigma[A2]^{T} \circ [B]) =$ (MAKE COLUMN 2 OF [A] INTO A DIAGONAL MATRIX
AND MULTIPLY x [B], WE GET:

$$
\begin{vmatrix}
A_{i1}(B_{11} \cdot \cdot \cdot B_{1K} \cdot \cdot \cdot B_{1M}) \\
\cdot \quad\quad \cdot \quad\quad \cdot \\
A_{iJ}(B_{J1} \cdot \cdot \cdot B_{JK} \cdot \cdot \cdot B_{JM}) \\
\cdot \quad\quad \cdot \quad\quad \cdot \\
A_{iN}(B_{N1} \cdot \cdot \cdot B_{NK} \cdot \cdot \cdot B_{NM})
\end{vmatrix}
=
\begin{vmatrix}
A_{i1}B_{11} \cdot + \cdot A_{i1}B_{1K} \cdot + \cdot A_{i1}B_{1M} \\
\cdot \quad\quad \cdot \quad\quad \cdot \\
A_{iJ}B_{J1} \cdot + \cdot A_{iJ}B_{JK} \cdot + \cdot A_{iJ}B_{JM} \\
\cdot \quad\quad \cdot \quad\quad \cdot \\
A_{iN}B_{N1} \cdot + \cdot A_{iN}B_{NK} \cdot + \cdot A_{iN}B_{NM}
\end{vmatrix}
$$

$(\Sigma[A3]^{T} \circ [B]) =$ (MAKE COLUMN 3 OF [A] INTO A DIAGONAL MATRIX
AND MULTIPLY x [B], WE GET:

$$
\begin{vmatrix}
A_{M1}(B_{11} \cdot \cdot \cdot B_{1K} \cdot \cdot \cdot B_{1M}) \\
\cdot \quad\quad \cdot \quad\quad \cdot \\
A_{MJ}(B_{J1} \cdot \cdot \cdot B_{JK} \cdot \cdot \cdot B_{JM}) \\
\cdot \quad\quad \cdot \quad\quad \cdot \\
A_{MN}(B_{N1} \cdot \cdot \cdot B_{NK} \cdot \cdot \cdot B_{NM})
\end{vmatrix}
=
\begin{vmatrix}
A_{M1}B_{11} \cdot + \cdot A_{M1}B_{1K} \cdot + \cdot A_{M1}B_{1M} \\
\cdot \quad\quad \cdot \quad\quad \cdot \\
A_{MJ}B_{J1} \cdot + \cdot A_{MJ}B_{JK} \cdot + \cdot A_{MJ}B_{JM} \\
\cdot \quad\quad \cdot \quad\quad \cdot \\
A_{MN}B_{N1} \cdot + \cdot A_{MN}B_{Ni} \cdot + \cdot A_{MN}B_{NM}
\end{vmatrix}
$$

Now that we've computed the four Sub-matrices, we must add
them together. Let's first add the first four sums that occupy A_{11}:

$$A_{11}B_{11} + B_{11}A_{i1} + A_{M1}B_{11} = B_{11}(A_{11} + A_{i1} + A_{M1}). = B_{11}\Sigma A1^{T}$$

These check, so let's add them all together.

$$
\begin{vmatrix}
B_{11}\Sigma A1^T & \cdots & B_{1K}\Sigma A1^T & \cdots & B_{1M}\Sigma A1^T \\
\cdot & \cdot & \cdot & & \cdot \\
B_{J1}\Sigma A2^T & \cdots & B_{JK}\Sigma A2^T & \cdots & B_{JM}\Sigma A2^T \\
\cdot & \cdot & \cdot & & \cdot \\
\cdot & \cdot & \cdot & & \cdot \\
B_{N1}\Sigma A3^T & \cdots & B_{NK}\Sigma A3^T & \cdots & B_{NM}\Sigma A3^T
\end{vmatrix} =
$$

$$
\begin{vmatrix}
B_{11}(A_{11}.+.A_{i1}.+.A_{M1}) & B_{1K}(A_{11}.+.A_{i1}.+.A_{M1}) & B_{1M}(A_{11}.+.A_{i1}.+.A_{M1}) \\
B_{J1}(A_{1J}.+.A_{iJ}.+.A_{MJ}) & B_{JK}(A_{1J}.+.A_{iJ}.+.A_{MJ}) & B_{JM}(A_{1J}.+.A_{iJ}.+.A_{MJ}) \\
B_{N1}(A_{1N}.+.A_{iN}.+.A_{MN}) & B_{NK}(A_{1N}.+.A_{iN}.+.A_{MN}) & B_{NM}(A_{1N}.+.A_{iN}.+.A_{MN})
\end{vmatrix}
$$

QED

This proves the operator works for all N x M matrices. Now to prove, by induction, that it works for all M+1, N+1 matrices. We will solve for a Nx1,Mx1 Matrix:

$$
\begin{vmatrix}
A_{11} & \cdots & A_{i1} & \cdots & A_{M1} & \cdots & A_{M+1,1} \\
\cdot & & \cdot & & \cdot & & \cdot \\
A_{1J} & \cdots & A_{iJ} & \cdots & A_{MJ} & \cdots & A_{M+1,J} \\
\cdot & & \cdot & & \cdot & & \cdot \\
A_{1N} & \cdots & A_{iN} & \cdots & A_{MN} & \cdots & A_{M+1,N} \\
\cdot & & \cdot & & \cdot & & \cdot \\
A_{1,N+1} & & A_{i,N+1} & \cdots & A_{M,N+1} & \cdots & A_{M+1,N+1}
\end{vmatrix}
\begin{vmatrix}
1 \\
\cdot \\
1 \\
\cdot \\
1 \\
\cdot \\
1
\end{vmatrix} =
$$

$$
\begin{vmatrix}
A_{11} & .+. & A_{i1} & .+. & A_{M1} & .+. & A_{M+1,1} \\
\cdot & & \cdot & & \cdot & & \cdot \\
A_{1J} & .+. & A_{iJ} & .+. & A_{MJ} & .+. & A_{M+1,J} \\
\cdot & & \cdot & & \cdot & & \cdot \\
A_{1N} & .+. & A_{iN} & .+. & A_{MN} & .+. & A_{M+1,N} \\
\cdot & & \cdot & & \cdot & & \cdot \\
A_{1,N+1} & & A_{i,N+1}.+. & & A_{M,N+1} & .+. & A_{M+1,N+1}
\end{vmatrix}
$$

Then $([A]^T_{iJ}[1]_{i1}) \circ [B]_{JK} =$

$$
\begin{vmatrix}
A_{11} & . & . & A_{i1} & . & . & . & A_{M1} & . & . & .A_{M+1,1} \\
. & . & . & . & . & . & . & . & . & . & . \\
A_{1J} & . & . & . & A_{iJ} & . & . & A_{MJ} & . & . & .A_{M+1,J}. \\
. & . & . & . & . & . & . & . & . & . & . \\
A_{1N} & . & . & A_{iN} & . & . & . & A_{MN} & . & . & .A_{M+1,N} \\
. & . & . & . & . & . & . & . & . & . & . \\
A_{1,N+1} & . & . & A_{i,N+1} & . & . & . & A_{M,N+1} & . & .A_{M+1,N+1}
\end{vmatrix}
\begin{vmatrix} 1 \\ . \\ . \\ 1 \\ . \\ . \\ 1 \\ . \\ . \\ 1 \end{vmatrix}
\circ
\begin{vmatrix}
B_{11} & . & . & B_{1K} & . & . & . & B_{1M} & . & . & . & B_{1,M+1} \\
. & & & . & & & . & & & . \\
. & & & . & & & . & & & . \\
B_{J1} & . & . & B_{JK} & . & . & . & B_{JM} & . & . & . & B_{J,M+1} \\
. & & & . & & & . & & & . \\
B_{N1} & . & . & B_{NK} & . & . & . & B_{NM} & . & . & . & B_{N,M+1} \\
. & & & . & & & . & & & . \\
B_{N+1,1} & . & . & B_{N+1,K} & . & . & B_{N+1,M} & . & . & B_{N+1,M+1}
\end{vmatrix}_{N+1,M+1}
$$

LET $\sum A1^T = A_{11}. \; .+. \; . \; A_{i1}. \; .+. \; . \; A_{M1}. \; .+. \; .A_{M+1,1}$

LET $\sum A2^T = A_{1J}. \; .+. \; . \; A_{iJ}. \; .+. \; . \; A_{MJ}. \; . \; +. \; A_{M+1,J}.$

LET $\sum A3^T = A_{1N}. \; .+. \; . \; A_{iN}. \; .+. \; . \; A_{MN}. \; . \; +. \; A_{M+1,N}$

LET $\sum A4^T = A_{1,N+1}. \; + A_{i,N+1}. \; +. \; . \; A_{M,N+1}. \; +. \; .A_{M+1,N+1}$

$$
\begin{vmatrix}
\sum A1^T & . & . & . & 0 & . & . & . & 0 & . & .0 \\
. & . & . & . & . & . & . & . & . & . & . \\
0 & . & . & . & \sum A2^T & . & . & . & 0 & . & . & 0. \\
. & . & . & . & . & . & . & . & . & . & . \\
0 & . & . & . & 0 & . & . & . & \sum A3^T & . & . & . & 0 \\
. & . & . & . & . & . & . & . & . & . & . \\
0 & . & . & . & 0 & . & . & & 0. & . & \sum A4^T
\end{vmatrix}
\begin{vmatrix}
B_{11} & . & . & . & B_{1K} & . & . & . & B_{1M} & . & . & B_{1,M+1} \\
. & & & & . & & & & . & & & . \\
B_{J1} & . & . & . & B_{JK} & . & . & . & B_{JM} & . & . & .B_{J,M+1} \\
. & & & & . & & & & . & & & . \\
B_{N1} & . & . & . & B_{NK} & . & . & . & B_{NM} & . & . & B_{N,M+1} \\
. & & & & . & & & & . & & & . \\
B_{N+1,1} & . & . & B_{N+1,K} & . & . & B_{N+1,M} & . & .B_{N+1,M+1}
\end{vmatrix} =
$$

$$
\begin{vmatrix}
B_{11}\sum A1^T & . & . & . & B_{1K}\sum A1^T & . & . & . & B_{1M}\sum A1^T & . & . & . & B_{1,M+1}\sum A1^T \\
. & & & & . & & & & . \\
. & & & & . & & & & . \\
B_{J1}\sum A2^T & . & . & . & B_{JK}\sum A2^T & . & . & . & B_{JM}\sum A2^T & . & . & . & .B_{J,M+1}\sum A2^T \\
. & & & & . & & & & . \\
B_{N1}\sum A3^T & . & . & . & B_{NK}\sum A3^T & . & . & . & B_{NM}\sum A3^T & . & . & . & B_{N,M+1}\sum A3^T \\
. & & & & . & & & & . \\
B_{N+1,1}\sum A4^T & . & . & B_{N+1,K}\sum A4^T & . & .B_{N+1,M}\sum A4^T & . & .B_{N+1,M+1}\sum A4^T
\end{vmatrix}
$$

with $=$ at center right.

This matrix is also equal to the sum of each individual Half-Multiplied Sub-matrix.

$^M\sum [A]^T \circ [B] = (\sum [A1]^T \circ [B]) + (\sum [A2]^T \circ [B]) + (w\sum [A3]^T \circ [B]) + (\sum [A4]^T \circ [B])$

$(\sum [A1]^T \circ [B])$

$$
\begin{vmatrix}
A_{11} & . & . & 0 & . & . & 0 & . & . & 0 \\
 & . & & . & & & . & & & . \\
0 & . & . & A_{1J} & . & . & 0 & . & . & 0 \\
 & . & & . & & & . & & & . \\
0 & . & . & 0 & . & . & A_{1M} & . & . & 0 \\
 & . & & . & & & . & & & . \\
0 & . & . & 0 & . & . & 0 & . & A_{1,M+1}
\end{vmatrix}
\begin{vmatrix}
B_{11} & . & . & B_{1K} & . & . & B_{1M} & . & . & B_{1,M+1} \\
 & . & & . & & & . & & & . \\
B_{J1} & . & . & B_{JK} & . & . & B_{JM} & . & . & B_{J,M+1} \\
 & . & & . & & & . & & & . \\
B_{N1} & . & . & B_{NK} & . & . & B_{NM} & . & . & B_{N,M+1} \\
 & . & & . & & & . & & & . \\
B_{N+1,1} & . & . & B_{N+1,K} & . & . & B_{N+1,M} & . & . & B_{N+1,M+1}
\end{vmatrix} =
$$

$$
\begin{vmatrix}
A_{11}(B_{11} & . & . & B_{1K} & . & . & B_{1M} & . & . & B_{1,M+1}) \\
 . & & & . & & & . & & & . \\
A_{1J}(B_{J1} & . & . & B_{JK} & . & . & B_{JM} & . & . & B_{J,M+1}) \\
 . & & & . & & & . & & & . \\
A_{1M}(B_{N1} & . & . & B_{NK} & . & . & B_{NM} & . & . & B_{N,M+1}) \\
 . & & & . & & & . & & & . \\
A_{1,M+1}(B_{N+1,1} & . & . & B_{N+1,K} & . & . & B_{N+1,M} & . & . & B_{N+1,M+1})
\end{vmatrix} =
\begin{vmatrix}
A_{11}B_{11} & .+. & A_{11}B_{1K} & .+. & A_{11}B_{1M} & .+. & A_{11}B_{1,M+1} \\
 . & & . & & . & & . \\
A_{1J}B_{J1} & .+. & A_{1J}B_{JK} & .+. & A_{1J}B_{JM} & .+. & A_{1J}B_{J,M+1} \\
 . & & . & & . & & . \\
A_{1M}B_{N1} & .+. & A_{1M}B_{NK} & .+. & A_{1M}B_{NM} & .+. & A_{1M}B_{N,M+1} \\
 . & & . & & . & & . \\
A_{1,M+1}B_{N+1,1} & .+. & A_{1,M+1}B_{N+1,K} & .+. & A_{1,M+1}B_{N+1,M} & .+. & A_{1,M+1}B_{N+1,M+1}
\end{vmatrix}
$$

$(\sum[A2]^{T} \circ [B]) =$ (MAKE COLUMN 2 OF [A] INTO A DIAGONAL MATRIX AND MULTIPLY x [B], WE GET:

$$
\begin{vmatrix}
A_{11}(B_{11} & . & . & B_{1K} & . & . & B_{1M} & . & . & B_{1,M+1}) \\
 . & & & . & & & . & & & . \\
A_{1J}(B_{J1} & . & . & B_{JK} & . & . & B_{JM} & . & . & B_{J,M+1}) \\
 . & & & . & & & . & & & . \\
A_{1N}(B_{N1} & . & . & B_{NK} & . & . & B_{NM} & . & . & B_{N,M+1}) \\
 . & & & . & & & . & & & . \\
A_{1,N+1}(B_{N+1,1} & . & . & B_{N+1,K} & . & . & B_{N+1,M} & . & . & B_{N+1,M+1})
\end{vmatrix} =
\begin{vmatrix}
A_{11}B_{11} & .+. & A_{11}B_{1K} & .+. & A_{11}B_{1M} & .+. & A_{11}B_{1,M+1} \\
 . & & . & & . & & . \\
A_{1J}B_{J1} & .+. & A_{1J}B_{JK} & .+. & A_{1J}B_{JM} & .+. & A_{1J}B_{J,M+1} \\
 . & & . & & . & & . \\
A_{1N}B_{N1} & .+. & A_{1N}B_{NK} & .+. & A_{1N}B_{NM} & .+. & A_{1N}B_{N,M+1} \\
 . & & . & & . & & . \\
A_{1,N+1}B_{N+1,1} & .+. & A_{1,N+1}B_{N+1,K} & .+. & A_{1,N+1}B_{N+1,M} & .+. & A_{1,N+1}B_{N+1,M+1}
\end{vmatrix}
$$

$(\sum[A3]^{T} \circ [B]) =$ (MAKE COLUMN 3 OF [A] INTO A DIAGONAL MATRIX AND MULTIPLY x [B], WE GET:

$$
\begin{vmatrix}
A_{M1}(B_{11} & . & . & B_{1K} & . & . & B_{1M} & . & . & B_{1,M+1}) \\
 . & & & . & & & . & & & . \\
A_{MJ}(B_{J1} & . & . & B_{JK} & . & . & B_{JM} & . & . & B_{J,M+1}) \\
 . & & & . & & & . & & & . \\
A_{MN}(B_{N1} & . & . & B_{NK} & . & . & B_{NM} & . & . & B_{N,M+1}) \\
 . & & & . & & & . & & & . \\
A_{M,N+1}(B_{N+1,1} & . & . & B_{N+1,i} & . & . & B_{N+1,M} & . & . & B_{N+1,M+1})
\end{vmatrix} =
\begin{vmatrix}
A_{M1}B_{11} & .+. & A_{M1}B_{1K} & .+. & A_{M1}B_{1M} & .+. & A_{M1}B_{1,M+1} \\
 . & & . & & . & & . \\
A_{MJ}B_{J1} & .+. & A_{MJ}B_{JK} & .+. & A_{MJ}B_{JM} & .+. & A_{MJ}B_{J,M+1} \\
 . & & . & & . & & . \\
A_{MN}B_{N1} & .+. & A_{MN}B_{N1} & .+. & A_{MN}B_{NM} & .+. & A_{MN}B_{N,M+1} \\
 . & & . & & . & & . \\
A_{M,N+1}B_{N+1,1} & .+. & A_{M,N+1}B_{N+1,K} & .+. & A_{M,N+1}B_{N+1,M} & .+. & A_{M,N+1}B_{N+1,M+1}
\end{vmatrix}
$$

$(\sum[A4]^{T} \circ [B]) =$ (MAKE COLUMN 4 OF [A] INTO A DIAGONAL MATRIX AND MULTIPLY x [B], WE GET

$$
\begin{vmatrix}
A_{M+1,1}(B_{11} & . & . & B_{1K} & . & . & B_{1M} & . & . & B_{1,M+1}) \\
 . & & & . & & & . & & & . \\
A_{M+1,J}(B_{J1} & . & . & B_{JK} & . & . & B_{JM} & . & . & B_{J,M+1}) \\
 . & & & . & & & . & & & . \\
A_{M+1,N}(B_{N1} & . & . & B_{NK} & . & . & B_{NM} & . & . & B_{N,M+1}) \\
 . & & & . & & & . & & & . \\
A_{M+1,N+1}(B_{N+1,1} & . & . & B_{N+1,K} & . & . & B_{N+1,M} & . & . & B_{N+1,M+1})
\end{vmatrix} =
\begin{vmatrix}
A_{M+1,1}B_{11} & .+. & A_{M+1,1}B_{1K} & .+. & A_{M+1,1}B_{1M} & .+. & A_{M+1,1}B_{1,M+1} \\
 . & & . & & . & & . \\
A_{M+1,J}B_{J1} & .+. & A_{M+1,J}B_{JK} & .+. & A_{M+1,J}B_{JM} & .+. & A_{M+1,J}B_{J,M+1} \\
 . & & . & & . & & . \\
A_{M+1,N}B_{N1} & .+. & A_{M+1,N}B_{NK} & .+. & A_{M+1,N}B_{NM} & .+. & A_{M+1,N}B_{N,M+1} \\
 . & & . & & . & & . \\
A_{M+1,N+1}B_{N+1,1} & .+. & A_{M+1,N+1}B_{N+1,K} & .+. & A_{M+1,N+1}B_{N+1,M} & .+. & A_{M+1,N+1}B_{N+1,M+1}
\end{vmatrix}
$$

Now that we've computed the four Sub-matrices, we must add them together. Let's first add the first four sums that occupy A_{11}:

$$A_{11}B_{11} + B_{11}A_{i1} + A_{M1}B_{11} +, A_{M+1,1} B_{11} = B_{11}(A_{11} + A_{i1} + A_{M1} + A_{M+1,1}). = B_{11}\Sigma A1^T$$

These check, so let's add them all together.

$$\begin{vmatrix} B_{11}\Sigma A1^T \ldots & B_{1K}\Sigma A1^T \ldots & B_{1M}\Sigma A1^T \ldots & B_{1,M+1}\Sigma A1^T \\ \cdot & \cdot & \cdot & \\ B_{J1}\Sigma A2^T \ldots & B_{JK}\Sigma A2^T \ldots & B_{JM}\Sigma A2^T \ldots & .B_{J,M+1}\Sigma A2^T \\ \cdot & \cdot & \cdot & = \\ B_{N1}\Sigma A3^T \ldots & B_{NK}\Sigma A3^T \ldots & B_{NM}\Sigma A3^T \ldots & B_{N,M+1}\Sigma A3^T \\ \cdot & \cdot & \cdot & \\ B_{N+1,1}\Sigma A4^T .. & B_{N+1,K}\Sigma A4^T .. & .B_{N+1,M}\Sigma A4^T .. & .B_{N+1,M+1}\Sigma A4^T \end{vmatrix}$$

$$\begin{vmatrix} B_{11}(A_{11}.+.A_{i1}.+.A_{M1}.+.A_{M+1,1}) & B_{1K}(A_{11}.+.A_{i1}.+.A_{M1}.+.A_{M+1,1}) \\ B_{J1}(A_{1J}.+.A_{iJ}.+.A_{MJ}.+.A_{M+1,J}.) & B_{JK}(A_{1J}.+.A_{iJ}.+.A_{MJ}.+.A_{M+1,J}.) \\ B_{N1}(A_{1N}.+.A_{iN}.+.A_{MN}.+.A_{M+1,N}) & B_{NK}(A_{1N}.+.A_{iN}.+.A_{MN}.+.A_{M+1,N}) \\ B_{N+1,1}(A_{1,N+1}.+.A_{i,N+1}.+.A_{M,N+1}.+.A_{M+1,N+1}) & B_{N+1,K}(A_{1,N+1}.+.A_{i,N+1}.+.A_{M,N+1}.+.A_{M+1,N+1}) \end{vmatrix}$$

$$\begin{vmatrix} B_{1M}(A_{11}.+.A_{i1}.+.A_{M1}.+.A_{M+1,1}) & B_{1,M+1}(A_{11}.+.A_{i1}.+.A_{M1}.+.A_{M+1,1}) \\ B_{JM}(A_{1J}.+.A_{iJ}.+.A_{MJ}.+.A_{M+1,J}.) & B_{J,M+1}(A_{1J}.+.A_{iJ}.+.A_{MJ}.+.A_{M+1,J}.) \\ B_{NM}(A_{1N}.+.A_{iN}.+.A_{MN}.+.A_{M+1,N}) & B_{N,M+1}(A_{1N}.+.A_{iN}.+.A_{MN}.+.A_{M+1,N}) \\ B_{N+1,M}(A_{1,N+1}.+.A_{i,N+1}.+.A_{M,N+1}.+.A_{M+1,N+1}) & B_{N+1,M+1}(A_{1,N+1}.+.A_{i,N+1}.+.A_{M,N+1}.+.A_{M+1,N+1}) \end{vmatrix}$$

QED

We have now proved by induction that this operator works for Matrices of any size or dimension.

SUMMARY OF 3 EQUATIONS DESCRIBED

$${}^M\Sigma[A]^T{}_{iJ} \circ [B]_{JK} = ([A]^T{}_{iJ}[1]_{i1}) \circ [B]_{JK} \text{ FOR OPEN SYSTEMS}$$

$${}^M\Sigma[A]^T{}_{iJ} \circ [B]_{JK} = ([A]^T{}_{iJ}[1]_{i1}) \circ [B]_{JK} = [I]_{JJ}[B]_{JK} = [1]_{J1} \circ [B]$$
$$_{JK} = [B]_{JK} \text{ FOR CLOSED SYSTEMS}$$

AND
$${}^M\Sigma[A]^T\circ[B] = (\Sigma[A1]^T \circ [B]) + (\Sigma[A2]^T \circ [B]) + (\Sigma[A3]^T \circ$$
$$[B]) + (\Sigma[A4]^T \circ [B])[AP1]+[AP2]+[AP3]+[AP4]$$

FOR EACH ITEM IN THE SPREADSHEET MATRIX [A].

THE TRANSPOSE COMMUTIVITY OF
THE HALF-MULTIPLIER OPERATOR

The proofs are long and confusing. I'm not even sure what I'm proving since we have no solutions in mathematics to compare them too, so I just tried to prove that the whole matrix is a sum of it's sub-matrices. In view of this, I think I'll just do a micro-proof here. If we need to, we can substitute I, J, N and M for matrix [A] and J, K, M and N for matrix [B].

$$\overleftarrow{\sum [A]^T_{iJ} \circ [B]_{JK}} =$$

$$\begin{vmatrix} B_{11} & B_{12} & B_{13} \\ B_{21} & B_{22} & B_{23} \\ B_{31} & B_{32} & B_{33} \\ B_{41} & B_{42} & B_{43} \end{vmatrix} \circ \begin{vmatrix} A_{11} & A_{21} & A_{31} \\ A_{12} & A_{22} & A_{32} \\ A_{13} & A_{23} & A_{33} \\ A_{14} & A_{24} & A_{34} \end{vmatrix} =$$

$$\begin{vmatrix} B_{11}A_{11} & B_{11}A_{21} & B_{11}A_{31} \\ B_{21}A_{12} & B_{21}A_{22} & B_{21}A_{32} \\ B_{31}A_{13} & B_{31}A_{23} & B_{31}A_{33} \\ B_{41}A_{14} & B_{41}A_{24} & B_{41}A_{34} \end{vmatrix} \begin{vmatrix} B_{12}A_{11} & B_{12}A_{21} & B_{12}A_{31} \\ B_{22}A_{12} & B_{22}A_{22} & B_{22}A_{32} \\ B_{32}A_{13} & B_{32}A_{23} & B_{32}A_{33} \\ B_{42}A_{14} & B_{42}A_{24} & B_{42}A_{34} \end{vmatrix} \begin{vmatrix} B_{13}A_{11} & B_{13}A_{21} & B_{13}A_{31} \\ B_{23}A_{12} & B_{23}A_{22} & B_{23}A_{32} \\ B_{33}A_{13} & B_{33}A_{23} & B_{33}A_{33} \\ B_{43}A_{14} & B_{43}A_{24} & B_{43}A_{34} \end{vmatrix}$$

Now we will transpose the nested array to put the array into a form we can mathematically use. Unlike the proof for the Half-multiplier Operator at the beginning of this section, when I transpose this time, we will have a single stacked matrix to contend with. We will sum using a Database Matrix. i.e. (I have separated the 3 sub-matrices just to make it easier to see what I'm doing, we do not need to do it when we are programming this on a computer. The [DB] matrix takes care of this).

$$\begin{vmatrix} 1 & 1 & 1 & 1 & 0 & 0 & 0 & 0 & 0 & 0 & 0 & 0 \\ 0 & 0 & 0 & 0 & 1 & 1 & 1 & 1 & 0 & 0 & 0 & 0 \\ 0 & 0 & 0 & 0 & 0 & 0 & 0 & 0 & 1 & 1 & 1 & 1 \end{vmatrix}$$

$$\begin{vmatrix} B_{11}A_{11} & B_{11}A_{21} & B_{11}A_{31} \\ B_{21}A_{12} & B_{21}A_{22} & B_{21}A_{32} \\ B_{31}A_{13} & B_{31}A_{23} & B_{31}A_{33} \\ B_{41}A_{14} & B_{41}A_{24} & B_{41}A_{34} \\ \\ B_{12}A_{11} & B_{12}A_{21} & B_{12}A_{31} \\ B_{22}A_{12} & B_{22}A_{22} & B_{22}A_{32} \\ B_{32}A_{13} & B_{32}A_{23} & B_{32}A_{33} \\ B_{42}A_{14} & B_{42}A_{24} & B_{42}A_{34} \\ \\ B_{13}A_{11} & B_{13}A_{21} & B_{13}A_{31} \\ B_{23}A_{12} & B_{23}A_{22} & B_{23}A_{32} \\ B_{33}A_{13} & B_{33}A_{23} & B_{33}A_{33} \\ B_{43}A_{14} & B_{43}A_{24} & B_{43}A_{34} \end{vmatrix} =$$

$$\left|\begin{array}{ccc} B_{11}A_{11}+B_{21}A_{12}+B_{31}A_{13}+B_{41}A_{14} & B_{11}A_{21}+B_{21}A_{22}+B_{31}A_{23}+B_{41}A_{24} & B_{11}A_{31}+B_{21}A_{32}+B_{31}A_{33}+B_{41}A_{34} \\ B_{12}A_{11}+B_{22}A_{12}+B_{32}A_{13}+B_{42}A_{14} & B_{12}A_{21}+B_{22}A_{22}+B_{32}A_{23}+B_{42}A_{24} & B_{12}A_{31}+B_{22}A_{32}+B_{32}A_{33}+B_{42}A_{34} \\ B_{13}A_{11}+B_{23}A_{12}+B_{33}A_{13}+B_{43}A_{14} & B_{13}A_{21}+B_{23}A_{22}+B_{33}A_{23}+B_{43}A_{24} & B_{13}A_{31}+B_{23}A_{32}+B_{33}A_{33}+B_{43}A_{34} \end{array}\right|$$

There is no need to re-transpose back into the nested array because this is the final form we want. When we re-transpose, we can take two steps of action. The outer brackets can be discarded and using partitioning, we are left with the Sub-matrices as a solution or as working operators. Or, we may leave it in the Matrix-Matrix form, manipulate the inner brackets to choose the size (columns only) of the solution matrices, then remove the outer bracket for the engineered solution. Let's multiply [A]x[B] in the normal way:

$$\begin{bmatrix} A_{11} & A_{12} & A_{13} & A_{14} \\ A_{21} & A_{22} & A_{23} & A_{24} \\ A_{31} & A_{32} & A_{33} & A_{34} \end{bmatrix} \cdot \begin{bmatrix} B_{11} & B_{12} & B_{13} \\ B_{21} & B_{22} & B_{23} \\ B_{31} & B_{32} & B_{33} \\ B_{41} & B_{42} & B_{43} \end{bmatrix}$$

Multiplying these together we get:

$$\begin{pmatrix} A_{11}{\cdot}B_{11}+A_{12}{\cdot}B_{21}+A_{13}{\cdot}B_{31}+A_{14}{\cdot}B_{41} & A_{11}{\cdot}B_{12}+A_{12}{\cdot}B_{22}+A_{13}{\cdot}B_{32}+A_{14}{\cdot}B_{42} & A_{11}{\cdot}B_{13}+A_{12}{\cdot}B_{23}+A_{13}{\cdot}B_{33}+A_{14}{\cdot}B_{43} \\ A_{21}{\cdot}B_{11}+A_{22}{\cdot}B_{21}+A_{23}{\cdot}B_{31}+A_{24}{\cdot}B_{41} & A_{21}{\cdot}B_{12}+A_{22}{\cdot}B_{22}+A_{23}{\cdot}B_{32}+A_{24}{\cdot}B_{42} & A_{21}{\cdot}B_{13}+A_{22}{\cdot}B_{23}+A_{23}{\cdot}B_{33}+A_{24}{\cdot}B_{43} \\ A_{31}{\cdot}B_{11}+A_{32}{\cdot}B_{21}+A_{33}{\cdot}B_{31}+A_{34}{\cdot}B_{41} & A_{31}{\cdot}B_{12}+A_{32}{\cdot}B_{22}+A_{33}{\cdot}B_{32}+A_{34}{\cdot}B_{42} & A_{31}{\cdot}B_{13}+A_{32}{\cdot}B_{23}+A_{33}{\cdot}B_{33}+A_{34}{\cdot}B_{43} \end{pmatrix}$$

Which is the transpose of the value computed above. It is also equal to $[B]^{T}[A]^{T}$ i.e.

$$\begin{bmatrix} B_{11} & B_{12} & B_{13} \\ B_{21} & B_{22} & B_{23} \\ B_{31} & B_{32} & B_{33} \\ B_{41} & B_{42} & B_{43} \end{bmatrix}^{T} \cdot \begin{bmatrix} A_{11} & A_{12} & A_{13} & A_{14} \\ A_{21} & A_{22} & A_{23} & A_{24} \\ A_{31} & A_{32} & A_{33} & A_{34} \end{bmatrix}^{T} =$$

$$\begin{bmatrix} A_{11}{\cdot}B_{11}+A_{12}{\cdot}B_{21}+A_{13}{\cdot}B_{31}+A_{14}{\cdot}B_{41} & A_{21}{\cdot}B_{11}+A_{22}{\cdot}B_{21}+A_{23}{\cdot}B_{31}+A_{24}{\cdot}B_{41} & A_{31}{\cdot}B_{11}+A_{32}{\cdot}B_{21}+A_{33}{\cdot}B_{31}+A_{34}{\cdot}B_{41} \\ A_{11}{\cdot}B_{12}+A_{12}{\cdot}B_{22}+A_{13}{\cdot}B_{32}+A_{14}{\cdot}B_{42} & A_{21}{\cdot}B_{12}+A_{22}{\cdot}B_{22}+A_{23}{\cdot}B_{32}+A_{24}{\cdot}B_{42} & A_{31}{\cdot}B_{12}+A_{32}{\cdot}B_{22}+A_{33}{\cdot}B_{32}+A_{34}{\cdot}B_{42} \\ A_{11}{\cdot}B_{13}+A_{12}{\cdot}B_{23}+A_{13}{\cdot}B_{33}+A_{14}{\cdot}B_{43} & A_{21}{\cdot}B_{13}+A_{22}{\cdot}B_{23}+A_{23}{\cdot}B_{33}+A_{24}{\cdot}B_{43} & A_{31}{\cdot}B_{13}+A_{32}{\cdot}B_{23}+A_{33}{\cdot}B_{33}+A_{34}{\cdot}B_{43} \end{bmatrix}$$

micro-**QED**

NESTED ARRAYS

The process of half-multiplying two matrices seems to naturally produce a nested array. These nested arrays have interesting properties, but properties I am unable to prove. They are free wheeling and we can do almost anything we want with them, as long as we do not break the rules of mathematics. When we half-multiply and obtain the i sub-matrices, if we transpose the nested array, the elements do not change their order, just the matrices. Also, the inner sets of brackets around the Sub-matrices may disappear (by partitioning) and we can treat this transposed nested array as a regular matrix for purposes of multiplication, addition and subtraction. We can then re-transpose the final matrix if we wish, or leave it in the same transposed form. Another interesting and perhaps a very important property is that after Half-Multiplication, we can adjust the brackets around the Sub-matrices to any size we wish, though normally they would all be of the same dimensions, this need not always be the case (see the examples on Statistics, Section 2.3 and 2.4). Let's look at a simple example where we take two sets of data, X and Y. We will wish to multiply each X by it's corresponding Y value.

First though, I must show how to transpose a nested array.

TRANSPOSING NESTED ARRAYS

I do not know how to prove this yet, the transpose of a Nested Array comes from the work done in Statistics and Quantum Chemistry. Suppose we have 4 arrays:

$$A := \begin{bmatrix} 1 & 4 & 7 \\ 2 & 5 & 8 \\ 3 & 6 & 9 \end{bmatrix} \qquad B := \begin{bmatrix} 2 & 8 & 5 \\ 4 & 1 & 9 \\ 6 & 3 & 7 \end{bmatrix} \qquad C := \begin{bmatrix} 1 & 7 & 4 \\ 3 & 9 & 6 \\ 5 & 2 & 8 \end{bmatrix} \qquad D := \begin{bmatrix} 1 & 2 & 3 \\ 4 & 5 & 6 \\ 7 & 8 & 9 \end{bmatrix}$$

The Nested Array becomes:

$$
\left[\begin{bmatrix} 1 & 4 & 7 \\ 2 & 5 & 8 \\ 3 & 6 & 9 \end{bmatrix} \cdot \begin{bmatrix} 2 & 8 & 5 \\ 4 & 1 & 9 \\ 6 & 3 & 7 \end{bmatrix} \cdot \begin{bmatrix} 1 & 7 & 4 \\ 3 & 9 & 6 \\ 5 & 2 & 8 \end{bmatrix} \cdot \begin{bmatrix} 1 & 2 & 3 \\ 4 & 5 & 6 \\ 7 & 8 & 9 \end{bmatrix}\right]
$$

Or we can write it in terms of the sub-matrices:

(A B C D)

When we transpose this we get:

$$
(A \; B \; C \; D)^T \; = \; \begin{bmatrix} A \\ B \\ C \\ D \end{bmatrix}
$$

Let

$$
NA := \begin{bmatrix} A \\ B \\ C \\ D \end{bmatrix}
$$

Which becomes:

$$
NA = \begin{bmatrix} \begin{bmatrix} 1 & 4 & 7 \\ 2 & 5 & 8 \\ 3 & 6 & 9 \end{bmatrix} \\ \begin{bmatrix} 2 & 8 & 5 \\ 4 & 1 & 9 \\ 6 & 3 & 7 \end{bmatrix} \\ \begin{bmatrix} 1 & 7 & 4 \\ 3 & 9 & 6 \\ 5 & 2 & 8 \end{bmatrix} \\ \begin{bmatrix} 1 & 2 & 3 \\ 4 & 5 & 6 \\ 7 & 8 & 9 \end{bmatrix} \end{bmatrix} \; \blacksquare
$$

The arrays transpose, but the individual elements inside the arrays do not.

Now let's look at a simple example where we take two sets of data, X and Y. We will wish to multiply each X by it's corresponding Y value.

$$\begin{bmatrix} X1 & Y1 & X5 & Y5 & X9 & Y9 \\ X2 & Y2 & X6 & Y6 & X10 & Y10 \\ X3 & Y3 & X7 & Y7 & X11 & Y11 \\ X4 & Y4 & X8 & Y8 & X12 & Y12 \end{bmatrix}$$

Half-Multiply by $[1]_{4,1}$ to create the Nested Array

$$\begin{bmatrix} 1 \\ 1 \\ 1 \\ 1 \end{bmatrix} \circ \begin{bmatrix} X1 & Y1 & X5 & Y5 & X9 & Y9 \\ X2 & Y2 & X6 & Y6 & X10 & Y10 \\ X3 & Y3 & X7 & Y7 & X11 & Y11 \\ X4 & Y4 & X8 & Y8 & X12 & Y12 \end{bmatrix} =$$

$$\begin{bmatrix} \begin{bmatrix} X1 & Y1 & X5 & Y5 & X9 & Y9 \\ X2 & Y2 & X6 & Y6 & X10 & Y10 \\ X3 & Y3 & X7 & Y7 & X11 & Y11 \\ X4 & Y4 & X8 & Y8 & X12 & Y12 \end{bmatrix} \end{bmatrix}$$

Now we remove the inner bracket and create three 2x4 Sub-matrices (we do this by hand):

$$\begin{bmatrix} \begin{bmatrix} X1 & Y1 \\ X2 & Y2 \\ X3 & Y3 \\ X4 & Y4 \end{bmatrix} \cdot \begin{bmatrix} X5 & Y5 \\ X6 & Y6 \\ X7 & Y7 \\ X8 & Y8 \end{bmatrix} \cdot \begin{bmatrix} X9 & Y9 \\ X10 & Y10 \\ X11 & Y11 \\ X12 & Y12 \end{bmatrix} \end{bmatrix}$$

Now we transpose this Nested Array (and partitioning):

$$\begin{bmatrix} \begin{bmatrix} X1 & Y1 \\ X2 & Y2 \\ X3 & Y3 \\ X4 & Y4 \end{bmatrix} \cdot \begin{bmatrix} X5 & Y5 \\ X6 & Y6 \\ X7 & Y7 \\ X8 & Y8 \end{bmatrix} \cdot \begin{bmatrix} X9 & Y9 \\ X10 & Y10 \\ X11 & Y11 \\ X12 & Y12 \end{bmatrix} \end{bmatrix}^T =$$

$$\begin{bmatrix} X1 & Y1 \\ X2 & Y2 \\ X3 & Y3 \\ X4 & Y4 \\ X5 & Y5 \\ X6 & Y6 \\ X7 & Y7 \\ X8 & Y8 \\ X9 & Y9 \\ X10 & Y10 \\ X11 & Y11 \\ X12 & Y12 \end{bmatrix}$$

Actually, I am not sure how the 'disappearance' of the inner brackets occur, can we just arbitrarily remove them, or do they actually disappear as we transpose, or are we just partitioning the Matrix to get it in the form we desire, or are Nested Arrays so free wheeling we can treat them as we wish and still have them in correct form?

Note the inner brackets are gone, the elements themselves are not transposed. To multiply by adding, we need to take the log of the elements of the array (treating the Matrix as a number):

$$\begin{bmatrix} LOGX1 & LOGY1 \\ LOGX2 & LOGY2 \\ LOGX3 & LOGY3 \\ LOGX4 & LOGY4 \\ LOGX5 & LOGY5 \\ LOGX6 & LOGY6 \\ LOGX7 & LOGY7 \\ LOGX8 & LOGY8 \\ LOGX9 & LOGY9 \\ LOGX10 & LOGY10 \\ LOGX11 & LOGY11 \\ LOGX12 & LOGY12 \end{bmatrix}$$

To take the log in MathCad, we must name the above matrix. I can't quite do it here because if you name a symbolic array, all the elements turn black. But if these were numbers and I called the transposed array A, then

LOGA = log($\vec{A}$)vectorize (vectorize is the button on the matrix palette with the X $\vec{\cdot}$ Y with an arrow over it. Now let's multiply the X's and Y's together, take the anti-log and obtain our solution.

$$TWO1 := \begin{pmatrix} 1 \\ 1 \end{pmatrix}$$

$$\begin{bmatrix} LOGX1 & LOGY1 \\ LOGX2 & LOGY2 \\ LOGX3 & LOGY3 \\ LOGX4 & LOGY4 \\ LOGX5 & LOGY5 \\ LOGX6 & LOGY6 \\ LOGX7 & LOGY7 \\ LOGX8 & LOGY8 \\ LOGX9 & LOGY9 \\ LOGX10 & LOGY10 \\ LOGX11 & LOGY11 \\ LOGX12 & LOGY12 \end{bmatrix} \cdot \begin{pmatrix} 1 \\ 1 \end{pmatrix} \rightarrow \begin{bmatrix} LOGX1+LOGY1 \\ LOGX2+LOGY2 \\ LOGX3+LOGY3 \\ LOGX4+LOGY4 \\ LOGX5+LOGY5 \\ LOGX6+LOGY6 \\ LOGX7+LOGY7 \\ LOGX8+LOGY8 \\ LOGX9+LOGY9 \\ LOGX10+LOGY10 \\ LOGX11+LOGY11 \\ LOGX12+LOGY12 \end{bmatrix}$$

To convert this to regular math, we do the following: XY= $10^{LOGA \times TWO1}$ vectorize, where TWO1 represents a matrix with two rows and one column. Generally, the way I use these variables throughout this paper, every ONE2 (one row, 2 columns) are filled with ones only. Thus NINE1 has 9 rows and one column all filled with ones. Suppose one of the X or Y values is zero, how do we take the log of zero which equals minus infinity? All we need is a number that is very close to zero, not zero itself. For purposes of this book, I define zero by default as 10EEX-100. This gives a log of -100, close enough to zero for the problems presented here and for most problems involving the universe. So I can show the program, I will call A=[1 1]. This program works if all the elements are numbers rather than symbols.

$A := (1 \quad 1)$

$LOGA := \overrightarrow{log(A)}$

$A1 := LOGA \cdot TWO1$

$$XY := 10^{\overrightarrow{A1}}$$

The solution we get back is:

$$
\begin{bmatrix}
X1Y1 \\
X2Y2 \\
X3Y3 \\
X4Y4 \\
X5Y5 \\
X6Y6 \\
X7Y7 \\
X8Y8 \\
X9Y9 \\
X10Y10 \\
X11Y11 \\
X12Y12
\end{bmatrix}
$$

Now we re-transpose the matrix:

$$
\begin{bmatrix}
X1Y1 & X5Y5 & X9Y9 \\
X2Y2 & X6Y6 & X10Y10 \\
X3Y3 & X7Y7 & X11Y11 \\
X4Y4 & X8Y8 & X12Y12
\end{bmatrix}
$$

And this is our statistical matrix with the corresponding X's and Y's multiplied.

ONTO MULTIPLICATION

Let's now check on a micro-proof concerning onto multiplication of matrices (element by corresponding element, rather than the sum of row x column). This is illegal according to modern mathematics. I claim 3 x 2 is always equal to 6, no matter how you multiply the two numbers together. This is part of the connection between our numbers we are familiar with and matrices. Whatever we can do with single numbers, we can do thousands of times in one operation with matrices. Let's take two general matrices:

$$\begin{pmatrix} J & M & P \\ K & N & Q \\ L & O & R \end{pmatrix} \qquad \begin{pmatrix} A & D & G \\ B & E & H \\ C & F & I \end{pmatrix}$$

and multiply them such that we get JxA, MxD, PxG, etc. We first need to take the logs of the two matrices, add them to multiply the numbers, and take the anti-log to return the numbers in familiar form:

$$\log \begin{pmatrix} A & D & G \\ B & E & H \\ C & F & I \end{pmatrix} \rightarrow \begin{bmatrix} \dfrac{\ln(A)}{\ln(10)} & \dfrac{\ln(D)}{\ln(10)} & \dfrac{\ln(G)}{\ln(10)} \\[2mm] \dfrac{\ln(B)}{\ln(10)} & \dfrac{\ln(E)}{\ln(10)} & \dfrac{\ln(H)}{\ln(10)} \\[2mm] \dfrac{\ln(C)}{\ln(10)} & \dfrac{\ln(F)}{\ln(10)} & \dfrac{\ln(I)}{\ln(10)} \end{bmatrix}$$

$$\log \begin{pmatrix} J & M & P \\ K & N & Q \\ L & O & R \end{pmatrix} \rightarrow \begin{bmatrix} \dfrac{\ln(J)}{\ln(10)} & \dfrac{\ln(M)}{\ln(10)} & \dfrac{\ln(P)}{\ln(10)} \\[2mm] \dfrac{\ln(K)}{\ln(10)} & \dfrac{\ln(N)}{\ln(10)} & \dfrac{\ln(Q)}{\ln(10)} \\[2mm] \dfrac{\ln(L)}{\ln(10)} & \dfrac{\ln(O)}{\ln(10)} & \dfrac{\ln(R)}{\ln(10)} \end{bmatrix}$$

$$\log \begin{pmatrix} A & D & G \\ B & E & H \\ C & F & I \end{pmatrix} + \log \begin{pmatrix} J & M & P \\ K & N & Q \\ L & O & R \end{pmatrix} \rightarrow \begin{bmatrix} \dfrac{\ln(A)}{\ln(10)}+\dfrac{\ln(J)}{\ln(10)} & \dfrac{\ln(D)}{\ln(10)}+\dfrac{\ln(M)}{\ln(10)} & \dfrac{\ln(G)}{\ln(10)}+\dfrac{\ln(P)}{\ln(10)} \\[2mm] \dfrac{\ln(B)}{\ln(10)}+\dfrac{\ln(K)}{\ln(10)} & \dfrac{\ln(E)}{\ln(10)}+\dfrac{\ln(N)}{\ln(10)} & \dfrac{\ln(H)}{\ln(10)}+\dfrac{\ln(Q)}{\ln(10)} \\[2mm] \dfrac{\ln(C)}{\ln(10)}+\dfrac{\ln(L)}{\ln(10)} & \dfrac{\ln(F)}{\ln(10)}+\dfrac{\ln(O)}{\ln(10)} & \dfrac{\ln(I)}{\ln(10)}+\dfrac{\ln(R)}{\ln(10)} \end{bmatrix}$$

$$10^{\begin{bmatrix} \dfrac{\ln(A)}{\ln(10)}+\dfrac{\ln(J)}{\ln(10)} & \dfrac{\ln(D)}{\ln(10)}+\dfrac{\ln(M)}{\ln(10)} & \dfrac{\ln(G)}{\ln(10)}+\dfrac{\ln(P)}{\ln(10)} \\[2mm] \dfrac{\ln(B)}{\ln(10)}+\dfrac{\ln(K)}{\ln(10)} & \dfrac{\ln(E)}{\ln(10)}+\dfrac{\ln(N)}{\ln(10)} & \dfrac{\ln(H)}{\ln(10)}+\dfrac{\ln(Q)}{\ln(10)} \\[2mm] \dfrac{\ln(C)}{\ln(10)}+\dfrac{\ln(L)}{\ln(10)} & \dfrac{\ln(F)}{\ln(10)}+\dfrac{\ln(O)}{\ln(10)} & \dfrac{\ln(I)}{\ln(10)}+\dfrac{\ln(R)}{\ln(10)} \end{bmatrix}}$$

(Here we take the anti-log 10^x. We must do this by hand, MathCad can't take the anti-log of a symbolic.)

$$= \begin{pmatrix} AJ & DM & GP \\ BK & EN & HQ \\ CL & FO & IR \end{pmatrix}$$

Or putting in numbers:

$$M1 := \begin{bmatrix} 1 & 4 & 3 & 2 \\ 2 & 3 & 4 & 1 \\ 3 & 2 & 1 & 4 \\ 4 & 1 & 2 & 3 \end{bmatrix} \qquad M2 := \begin{bmatrix} 4 & 1 & 2 & 3 \\ 3 & 2 & 4 & 1 \\ 2 & 3 & 1 & 4 \\ 1 & 4 & 3 & 2 \end{bmatrix}$$

$$\overrightarrow{LOGM1M2} := \overrightarrow{\log(M1)} + \overrightarrow{\log(M2)}$$

$$M1M2 := 10^{\overrightarrow{LOGM1M2}}$$

$$M1M2 = \begin{bmatrix} 4 & 4 & 6 & 6 \\ 6 & 6 & 16 & 1 \\ 6 & 6 & 1 & 16 \\ 4 & 4 & 6 & 6 \end{bmatrix}$$

Suppose one or more of the elements in M1 and M2 are zero's? We proceed as follows, letting zero, by default, equal 10EEX-100:

$$M3 := \begin{bmatrix} 1 & 4 & 3 & 2 \\ 2 & 10^{-100} & 4 & 1 \\ 3 & 2 & 1 & 4 \\ 4 & 1 & 2 & 3 \end{bmatrix} \qquad M4 := \begin{bmatrix} 4 & 1 & 2 & 3 \\ 3 & 2 & 4 & 1 \\ 2 & 3 & 1 & 4 \\ 1 & 10^{-100} & 3 & 2 \end{bmatrix}$$

$$\overrightarrow{LOGM3M4} := \overrightarrow{\log(M3)} + \overrightarrow{\log(M4)}$$

$$M3M4 := 10^{\overrightarrow{LOGM3M4}}$$

$$M3M4 = \begin{bmatrix} 4 & 4 & 6 & 6 \\ 6 & 0 & 16 & 1 \\ 6 & 6 & 1 & 16 \\ 4 & 0 & 6 & 6 \end{bmatrix}$$

MathCad cannot remove a diagonal from a matrix, it can take a column matrix and make a diagonal, but not vice versa. In many applications, especially since the Half-Multiplier's matrix equivalent is multiplication by a diagonal matrix, we need to be able to remove the diagonal from a matrix and make a separate matrix out of it. The following is how I do it with MathCad. The HP-48G will remove the diagonal from a matrix as a column matrix and allows us to re-make it into a diagonal matrix.

MATHCAD +6:

Suppose we have a matrix and we wish to square every element in the matrix and add them to get the grand sum of squares. We take the matrix, transpose it and multiply it by itself. i.e. $[A]^2 = [A]^T[A]$. The sum of the squares in each column lie in the diagonal. We need to separate the diagonal from the rest of the matrix, and then add the elements together to get a single sum. We proceed as follows:

$$A := \begin{bmatrix} 1 & 5 & 9 & 13 \\ 2 & 6 & 10 & 14 \\ 3 & 7 & 11 & 15 \\ 4 & 8 & 12 & 16 \end{bmatrix} \qquad A^T = \begin{bmatrix} 1 & 2 & 3 & 4 \\ 5 & 6 & 7 & 8 \\ 9 & 10 & 11 & 12 \\ 13 & 14 & 15 & 16 \end{bmatrix}$$

$$ASQ := A^T \cdot A$$

$$ASQ = \begin{bmatrix} 30 & 70 & 110 & 150 \\ 70 & 174 & 278 & 382 \\ 110 & 278 & 446 & 614 \\ 150 & 382 & 614 & 846 \end{bmatrix}$$

CHECK:

$$16 + 9 + 4 + 1 = 30$$
$$25 + 36 + 49 + 64 = 174$$
$$81 + 100 + 121 + 144 = 446$$
$$13^2 + 14^2 + 15^2 + 16^2 = 846$$

We need to remove the diagonal, MathCad will not do this for us. This is the best way: Write a template 4x4 identity matrix, but write it as the log (this helps take care of the problem of the log of zero). Take the log of ASQ, add the two and take the anti-log. This will return the diagonal of ASQ.

$$\text{LOGASQ} := \overrightarrow{\log(\text{ASQ})}$$ Here I take the log of each individual element in $[A]^2$.

$$\text{LOGI4} = \begin{bmatrix} 0 & -100 & -100 & -100 \\ -100 & 0 & -100 & -100 \\ -100 & -100 & 0 & -100 \\ -100 & -100 & -100 & 0 \end{bmatrix}$$ This is the log of the $[I]_{4,4}$ matrix.

$$\text{LOGDIAGASQ} := \text{LOGI4} + \text{LOGASQ}$$

$$\text{DIAGASQ} := 10^{\overrightarrow{\text{LOGDIAGASQ}}}$$ Here I take the anti-log of each element to obtain the final solution.

$$\text{DIAGASQ} = \begin{bmatrix} 30 & 0 & 0 & 0 \\ 0 & 174 & 0 & 0 \\ 0 & 0 & 446 & 0 \\ 0 & 0 & 0 & 846 \end{bmatrix}$$

Now we need to sum the values. Define:

$$\text{ONE4} := (1 \quad 1 \quad 1 \quad 1)$$

$$\text{GRANDSUMSQUARESOFA} := \text{ONE4} \, \text{DIAGASQ} \cdot \text{ONE4}^T$$

$$(1 \quad 1 \quad 1 \quad 1) \cdot \begin{bmatrix} 30 & 0 & 0 & 0 \\ 0 & 174 & 0 & 0 \\ 0 & 0 & 446 & 0 \\ 0 & 0 & 0 & 846 \end{bmatrix} \cdot \begin{bmatrix} 1 \\ 1 \\ 1 \\ 1 \end{bmatrix} = 1496 \ \blacksquare$$

$$\text{GRANDSUMSQUARESOFA} = 1496$$

HP-48G PROGRAM: ➭ MATRIX, ENTER VALUES, ↑↑, αA STO (this stores the matrix in A)

RCL A
↑
αα TRN
↑
SWAP
x
MTH,MATR,NXT,→DIAG [30 174 446 846]
4
↑
DIAG→
RCL ONE4   MUST ENTER ONE4 AS A COLUMN MATRIX FIRST, THEN
SWAP   TRANSPOSE TO GET ROW MATRIX, OR MATH WILL NOT WORK.
x
RCL ONE4
αα TRN
↑
x   1496

We can also accomplish the same thing by multiplying element by element then compute the grand sum of the matrix.

$$\text{AVECTSQ} := \begin{bmatrix} 1 & 5 & 9 & 13 \\ 2 & 6 & 10 & 14 \\ 3 & 7 & 11 & 15 \\ 4 & 8 & 12 & 16 \end{bmatrix} \cdot \begin{bmatrix} 1 & 5 & 9 & 13 \\ 2 & 6 & 10 & 14 \\ 3 & 7 & 11 & 15 \\ 4 & 8 & 12 & 16 \end{bmatrix} \qquad \text{ONE4} := (1\ 1\ 1\ 1)$$

$$\text{AVECTSQ} = \begin{bmatrix} 1 & 25 & 81 & 169 \\ 4 & 36 & 100 & 196 \\ 9 & 49 & 121 & 225 \\ 16 & 64 & 144 & 256 \end{bmatrix}$$

$$\text{GRANDSUMASQ} := \text{ONE4} \cdot \text{AVECTSQ} \cdot \text{ONE4}^{T}$$

$$\text{GRANDSUMASQ} = 1496$$

I have done the math this way with Statistics, bur did not really see how simple it was until now (8-12-97).

Also, when we transpose a Nested Array, we may put it in the form of a diagonal matrixinstead of a column or row matrix.

I'll not get into that here, but will field some examples in Quantum Chemistry, Gaussian Reduction and Statistics.

Below are some simple computations of Nested arrays from a request from Dr. Monroy in Juarez, Mexico.

NESTED ARRAYS WITH MATRICES OF DIFFERENT SIZES

Dear Dr. Monroy:

These are some of the properties of nested arrays as apply to the operator, since your problem uses square matrices I used examples as square matrices although they may be M x N just as easily. Hope they might be of help to you. The three arrays are defined as C, D and E and their nested form is defined as A. Imagine they are stacked on top of each other (like crackers) in 3-D but the only way to display them is in 2-D form. MathCad cannot perform computations on Nested Arrays, so in order to utilize them we ignore the brackets and just remember that they are there.

$$C := \begin{pmatrix} 1 & 2 & 3 \\ 4 & 5 & 6 \\ 7 & 8 & 9 \end{pmatrix} \qquad D := \begin{pmatrix} 10 & 11 & 12 \\ 13 & 14 & 15 \\ 16 & 17 & 18 \end{pmatrix} \qquad E := \begin{pmatrix} 19 & 20 & 21 \\ 22 & 23 & 24 \\ 25 & 26 & 27 \end{pmatrix}$$

First we combine the separate arrays into one single array A. They are still nested, but since computers can't handle the math we must "remember" that there are brackets around them and work on them from this premise. This is the premise behind the mathematics in the section on Statistics later on in this book.

These are the Matrices we will use:

$$A := \begin{pmatrix} 1 & 2 & 3 & 10 & 11 & 12 & 19 & 20 & 21 \\ 4 & 5 & 6 & 13 & 14 & 15 & 22 & 23 & 24 \\ 7 & 8 & 9 & 16 & 17 & 18 & 25 & 26 & 27 \end{pmatrix} \qquad B := \begin{bmatrix} 1 & 0 & 0 \\ 0 & 1 & 0 \\ 0 & 0 & 1 \\ 1 & 0 & 0 \\ 0 & 1 & 0 \\ 0 & 0 & 1 \\ 1 & 0 & 0 \\ 0 & 1 & 0 \\ 0 & 0 & 1 \end{bmatrix}$$

TO ADD EACH THREE MATRICES SEPARATELY:

$$A \cdot B = \begin{pmatrix} 30 & 33 & 36 \\ 39 & 42 & 45 \\ 48 & 51 & 54 \end{pmatrix}$$

$$\text{CHECK: } C + D + E = \begin{pmatrix} 30 & 33 & 36 \\ 39 & 42 & 45 \\ 48 & 51 & 54 \end{pmatrix}$$

LET'S ADD MATRIX 1 + MATRIX THREE AND IGNORE MATRIX 2:

$$A = \begin{pmatrix} 1 & 2 & 3 & 10 & 11 & 12 & 19 & 20 & 21 \\ 4 & 5 & 6 & 13 & 14 & 15 & 22 & 23 & 24 \\ 7 & 8 & 9 & 16 & 17 & 18 & 25 & 26 & 27 \end{pmatrix} \qquad R := \begin{bmatrix} 1 & 0 & 0 \\ 0 & 1 & 0 \\ 0 & 0 & 1 \\ 0 & 0 & 0 \\ 0 & 0 & 0 \\ 0 & 0 & 0 \\ 1 & 0 & 0 \\ 0 & 1 & 0 \\ 0 & 0 & 1 \end{bmatrix}$$

$$A \cdot R = \begin{vmatrix} 20 & 22 & 24 \\ 26 & 28 & 30 \\ 32 & 34 & 36 \end{vmatrix} \qquad \text{CHECK:} \quad C + E = \begin{vmatrix} 20 & 22 & 24 \\ 26 & 28 & 30 \\ 32 & 34 & 36 \end{vmatrix}$$

TO ONTO MULTIPLY EACH SEPARATE MATRIX BY ANOTHER:

F1 := augment(E, D)
F := augment(F1, C)

$$F = \begin{vmatrix} 19 & 20 & 21 & 10 & 11 & 12 & 1 & 2 & 3 \\ 22 & 23 & 24 & 13 & 14 & 15 & 4 & 5 & 6 \\ 25 & 26 & 27 & 16 & 17 & 18 & 7 & 8 & 9 \end{vmatrix}$$

$$\overrightarrow{(A \cdot F)} = \begin{vmatrix} 19 & 40 & 63 & 100 & 121 & 144 & 19 & 40 & 63 \\ 88 & 115 & 144 & 169 & 196 & 225 & 88 & 115 & 144 \\ 175 & 208 & 243 & 256 & 289 & 324 & 175 & 208 & 243 \end{vmatrix}$$

TO MULTIPLY THE THREE MATRICES THE REGULAR WAY:

$$A = \begin{vmatrix} 1 & 2 & 3 & 10 & 11 & 12 & 19 & 20 & 21 \\ 4 & 5 & 6 & 13 & 14 & 15 & 22 & 23 & 24 \\ 7 & 8 & 9 & 16 & 17 & 18 & 25 & 26 & 27 \end{vmatrix}$$

$$G := \begin{bmatrix} 1 & 2 & 3 & 0 & 0 & 0 & 0 & 0 & 0 \\ 4 & 5 & 6 & 0 & 0 & 0 & 0 & 0 & 0 \\ 7 & 8 & 9 & 0 & 0 & 0 & 0 & 0 & 0 \\ 0 & 0 & 0 & 10 & 11 & 12 & 0 & 0 & 0 \\ 0 & 0 & 0 & 13 & 14 & 15 & 0 & 0 & 0 \\ 0 & 0 & 0 & 16 & 17 & 18 & 0 & 0 & 0 \\ 0 & 0 & 0 & 0 & 0 & 0 & 19 & 20 & 21 \\ 0 & 0 & 0 & 0 & 0 & 0 & 22 & 23 & 24 \\ 0 & 0 & 0 & 0 & 0 & 0 & 25 & 26 & 27 \end{bmatrix}$$

$$A \cdot G = \begin{pmatrix} 30 & 36 & 42 & 435 & 468 & 501 & 1326 & 1386 & 1446 \\ 66 & 81 & 96 & 552 & 594 & 636 & 1524 & 1593 & 1662 \\ 102 & 126 & 150 & 669 & 720 & 771 & 1722 & 1800 & 1878 \end{pmatrix}$$

CHECK:

$$C \cdot C = \begin{pmatrix} 30 & 36 & 42 \\ 66 & 81 & 96 \\ 102 & 126 & 150 \end{pmatrix} \quad D \cdot D = \begin{pmatrix} 435 & 468 & 501 \\ 552 & 594 & 636 \\ 669 & 720 & 771 \end{pmatrix} \quad E \cdot E = \begin{pmatrix} 1326 & 1386 & 1446 \\ 1524 & 1593 & 1662 \\ 1722 & 1800 & 1878 \end{pmatrix}$$

TO HALF MULTIPLY THE THREE MATRICES AND GET ALL THE SUB-MATRICES:

$$I := \begin{pmatrix} 1 & 0 & 0 \\ 0 & 4 & 0 \\ 0 & 0 & 7 \end{pmatrix} \quad J := \begin{pmatrix} 2 & 0 & 0 \\ 0 & 5 & 0 \\ 0 & 0 & 8 \end{pmatrix} \quad K := \begin{pmatrix} 3 & 0 & 0 \\ 0 & 6 & 0 \\ 0 & 0 & 9 \end{pmatrix}$$

$$L := \begin{pmatrix} 10 & 0 & 0 \\ 0 & 13 & 0 \\ 0 & 0 & 16 \end{pmatrix} \quad M := \begin{pmatrix} 11 & 0 & 0 \\ 0 & 14 & 0 \\ 0 & 0 & 15 \end{pmatrix} \quad N := \begin{pmatrix} 12 & 0 & 0 \\ 0 & 15 & 0 \\ 0 & 0 & 16 \end{pmatrix}$$

$$O := \begin{pmatrix} 19 & 0 & 0 \\ 0 & 22 & 0 \\ 0 & 0 & 25 \end{pmatrix} \quad P := \begin{pmatrix} 20 & 0 & 0 \\ 0 & 23 & 0 \\ 0 & 0 & 26 \end{pmatrix} \quad Q := \begin{pmatrix} 21 & 0 & 0 \\ 0 & 24 & 0 \\ 0 & 0 & 27 \end{pmatrix}$$

$$I \cdot C = \begin{pmatrix} 1 & 2 & 3 \\ 16 & 20 & 24 \\ 49 & 56 & 63 \end{pmatrix} \quad J \cdot D = \begin{pmatrix} 20 & 22 & 24 \\ 65 & 70 & 75 \\ 128 & 136 & 144 \end{pmatrix} \quad K \cdot E = \begin{pmatrix} 57 & 60 & 63 \\ 132 & 138 & 144 \\ 225 & 234 & 243 \end{pmatrix}$$

$$L \cdot C = \begin{pmatrix} 10 & 20 & 30 \\ 52 & 65 & 78 \\ 112 & 128 & 144 \end{pmatrix} \quad M \cdot D = \begin{pmatrix} 110 & 121 & 132 \\ 182 & 196 & 210 \\ 240 & 255 & 270 \end{pmatrix} \quad N \cdot E = \begin{pmatrix} 228 & 240 & 252 \\ 330 & 345 & 360 \\ 400 & 416 & 432 \end{pmatrix}$$

$$O \cdot C = \begin{pmatrix} 19 & 38 & 57 \\ 88 & 110 & 132 \\ 175 & 200 & 225 \end{pmatrix} \quad P \cdot D = \begin{pmatrix} 200 & 220 & 240 \\ 299 & 322 & 345 \\ 416 & 442 & 468 \end{pmatrix} \quad Q \cdot E = \begin{pmatrix} 399 & 420 & 441 \\ 528 & 552 & 576 \\ 675 & 702 & 729 \end{pmatrix}$$

$$H := \begin{bmatrix} 1 & 0 & 0 & 2 & 0 & 0 & 3 & 0 & 0 \\ 0 & 4 & 0 & 0 & 5 & 0 & 0 & 6 & 0 \\ 0 & 0 & 7 & 0 & 0 & 8 & 0 & 0 & 9 \\ 10 & 0 & 0 & 11 & 0 & 0 & 12 & 0 & 0 \\ 0 & 13 & 0 & 0 & 14 & 0 & 0 & 15 & 0 \\ 0 & 0 & 16 & 0 & 0 & 17 & 0 & 0 & 18 \\ 19 & 0 & 0 & 20 & 0 & 0 & 21 & 0 & 0 \\ 0 & 22 & 0 & 0 & 23 & 0 & 0 & 24 & 0 \\ 0 & 0 & 25 & 0 & 0 & 26 & 0 & 0 & 27 \end{bmatrix} \quad G := \begin{bmatrix} 1 & 2 & 3 & 0 & 0 & 0 & 0 & 0 & 0 \\ 4 & 5 & 6 & 0 & 0 & 0 & 0 & 0 & 0 \\ 7 & 8 & 9 & 0 & 0 & 0 & 0 & 0 & 0 \\ 0 & 0 & 0 & 10 & 11 & 12 & 0 & 0 & 0 \\ 0 & 0 & 0 & 13 & 14 & 15 & 0 & 0 & 0 \\ 0 & 0 & 0 & 16 & 17 & 18 & 0 & 0 & 0 \\ 0 & 0 & 0 & 0 & 0 & 0 & 19 & 20 & 21 \\ 0 & 0 & 0 & 0 & 0 & 0 & 22 & 23 & 24 \\ 0 & 0 & 0 & 0 & 0 & 0 & 25 & 26 & 27 \end{bmatrix}$$

The final multiplication of the Nested Arrays becomes:

$$H \cdot G = \begin{bmatrix} 1 & 2 & 3 & 20 & 22 & 24 & 57 & 60 & 63 \\ 16 & 20 & 24 & 65 & 70 & 75 & 132 & 138 & 144 \\ 49 & 56 & 63 & 128 & 136 & 144 & 225 & 234 & 243 \\ 10 & 20 & 30 & 110 & 121 & 132 & 228 & 240 & 252 \\ 52 & 65 & 78 & 182 & 196 & 210 & 330 & 345 & 360 \\ 112 & 128 & 144 & 272 & 289 & 306 & 450 & 468 & 486 \\ 19 & 38 & 57 & 200 & 220 & 240 & 399 & 420 & 441 \\ 88 & 110 & 132 & 299 & 322 & 345 & 528 & 552 & 576 \\ 175 & 200 & 225 & 416 & 442 & 468 & 675 & 702 & 729 \end{bmatrix}$$

Suppose the Sub-Matrices in the Nested Array are not all the
same dimension? This is how we would Multiply and Half-Multiply
them together.

$$I := \begin{bmatrix} 1 & 2 & 3 \\ 4 & 5 & 6 \\ 7 & 8 & 9 \\ 10 & 11 & 12 \end{bmatrix} \qquad J := \begin{pmatrix} 10 & 11 & 12 & 13 \\ 14 & 15 & 16 & 17 \\ 18 & 19 & 20 & 21 \end{pmatrix} \qquad K := \begin{pmatrix} 22 & 23 & 24 & 25 & 26 \\ 27 & 28 & 29 & 30 & 31 \end{pmatrix}$$

Now we make the Nested Array. To make the Nested Array
conformable, we must add zero's where necessary to complete
the array and legalize it mathematically. The Nested Array
looks like:

$$
L = \left[\begin{pmatrix} 1 & 2 & 3 \\ 4 & 5 & 6 \\ 7 & 8 & 9 \\ 10 & 11 & 12 \end{pmatrix} \begin{pmatrix} 10 & 11 & 12 & 13 \\ 14 & 15 & 16 & 17 \\ 18 & 19 & 20 & 21 \end{pmatrix} \begin{pmatrix} 22 & 23 & 24 & 25 & 26 \\ 27 & 28 & 29 & 30 & 31 \end{pmatrix} \right]
$$

But we have to partition it out to put it in the form we can use. The partitioned Nested array looks like

$$
L := \begin{bmatrix} 1 & 2 & 3 & 10 & 11 & 12 & 13 & 22 & 23 & 24 & 25 & 26 \\ 4 & 5 & 6 & 14 & 15 & 16 & 17 & 27 & 28 & 29 & 30 & 31 \\ 7 & 8 & 9 & 18 & 19 & 20 & 21 & 0 & 0 & 0 & 0 & 0 \\ 10 & 11 & 12 & 0 & 0 & 0 & 0 & 0 & 0 & 0 & 0 & 0 \end{bmatrix}
$$

Or diagonalizing we get the Nested Array M.

$$
M = \left[\begin{matrix} \begin{pmatrix} 1 & 2 & 3 \\ 4 & 5 & 6 \\ 7 & 8 & 9 \\ 10 & 11 & 12 \end{pmatrix} & 0 & 0 \\ 0 & \begin{pmatrix} 10 & 11 & 12 & 13 \\ 14 & 15 & 16 & 17 \\ 18 & 19 & 20 & 21 \end{pmatrix} & 0 \\ 0 & 0 & \begin{pmatrix} 22 & 23 & 24 & 25 & 26 \\ 27 & 28 & 29 & 30 & 31 \end{pmatrix} \end{matrix} \right]
$$

Partitioning the Nested Array we get the following Matrix.

$$
M = \begin{bmatrix}
1 & 2 & 3 & 0 & 0 & 0 & 0 & 0 & 0 & 0 & 0 & 0 \\
4 & 5 & 6 & 0 & 0 & 0 & 0 & 0 & 0 & 0 & 0 & 0 \\
7 & 8 & 9 & 0 & 0 & 0 & 0 & 0 & 0 & 0 & 0 & 0 \\
10 & 11 & 12 & 0 & 0 & 0 & 0 & 0 & 0 & 0 & 0 & 0 \\
0 & 0 & 0 & 10 & 11 & 12 & 13 & 0 & 0 & 0 & 0 & 0 \\
0 & 0 & 0 & 14 & 15 & 16 & 17 & 0 & 0 & 0 & 0 & 0 \\
0 & 0 & 0 & 18 & 19 & 20 & 21 & 0 & 0 & 0 & 0 & 0 \\
0 & 0 & 0 & 0 & 0 & 0 & 0 & 0 & 0 & 0 & 0 & 0 \\
0 & 0 & 0 & 0 & 0 & 0 & 0 & 22 & 23 & 24 & 25 & 26 \\
0 & 0 & 0 & 0 & 0 & 0 & 0 & 27 & 28 & 29 & 30 & 31 \\
0 & 0 & 0 & 0 & 0 & 0 & 0 & 0 & 0 & 0 & 0 & 0 \\
0 & 0 & 0 & 0 & 0 & 0 & 0 & 0 & 0 & 0 & 0 & 0
\end{bmatrix}
$$

Transposing M we get

$$
M^T = \begin{bmatrix}
1 & 4 & 7 & 10 & 0 & 0 & 0 & 0 & 0 & 0 & 0 & 0 \\
2 & 5 & 8 & 11 & 0 & 0 & 0 & 0 & 0 & 0 & 0 & 0 \\
3 & 6 & 9 & 12 & 0 & 0 & 0 & 0 & 0 & 0 & 0 & 0 \\
0 & 0 & 0 & 0 & 10 & 14 & 18 & 0 & 0 & 0 & 0 & 0 \\
0 & 0 & 0 & 0 & 11 & 15 & 19 & 0 & 0 & 0 & 0 & 0 \\
0 & 0 & 0 & 0 & 12 & 16 & 20 & 0 & 0 & 0 & 0 & 0 \\
0 & 0 & 0 & 0 & 13 & 17 & 21 & 0 & 0 & 0 & 0 & 0 \\
0 & 0 & 0 & 0 & 0 & 0 & 0 & 0 & 22 & 27 & 0 & 0 \\
0 & 0 & 0 & 0 & 0 & 0 & 0 & 0 & 23 & 28 & 0 & 0 \\
0 & 0 & 0 & 0 & 0 & 0 & 0 & 0 & 24 & 29 & 0 & 0 \\
0 & 0 & 0 & 0 & 0 & 0 & 0 & 0 & 25 & 30 & 0 & 0 \\
0 & 0 & 0 & 0 & 0 & 0 & 0 & 0 & 26 & 31 & 0 & 0
\end{bmatrix}
$$

L=12,4 M=12,12

WE CAN MULTIPLY THIS IN A CONDENSED MANNER ONLY ONE WAY, BECAUSE
$K^T \times K$ WOULD EQUAL A 5x5 MATRIX, WHICH IS OUT OF BOUNDS (INDICES
DON'T CONFORM).

$$
L \cdot M^T =
\begin{bmatrix}
14 & 32 & 50 & 68 & 534 & 718 & 902 & 0 & 2890 & 3490 & 0 & 0 \\
32 & 77 & 122 & 167 & 718 & 966 & 1214 & 0 & 3490 & 4215 & 0 & 0 \\
50 & 122 & 194 & 266 & 902 & 1214 & 1526 & 0 & 0 & 0 & 0 & 0 \\
68 & 167 & 266 & 365 & 0 & 0 & 0 & 0 & 0 & 0 & 0 & 0
\end{bmatrix}
$$

$$
I :=
\begin{bmatrix}
1 & 2 & 3 \\
4 & 5 & 6 \\
7 & 8 & 9 \\
10 & 11 & 12
\end{bmatrix}
\qquad
J :=
\begin{pmatrix}
10 & 11 & 12 & 13 \\
14 & 15 & 16 & 17 \\
18 & 19 & 20 & 21
\end{pmatrix}
\qquad
K :=
\begin{pmatrix}
22 & 23 & 24 & 25 & 26 \\
27 & 28 & 29 & 30 & 31
\end{pmatrix}
$$

$$
M \cdot L^T =
\begin{bmatrix}
14 & 32 & 50 & 68 \\
32 & 77 & 122 & 167 \\
50 & 122 & 194 & 266 \\
68 & 167 & 266 & 365 \\
534 & 718 & 902 & 0 \\
718 & 966 & 1214 & 0 \\
902 & 1214 & 1526 & 0 \\
0 & 0 & 0 & 0 \\
2890 & 3490 & 0 & 0 \\
3490 & 4215 & 0 & 0 \\
0 & 0 & 0 & 0 \\
0 & 0 & 0 & 0
\end{bmatrix}
$$

$$
I \cdot I^T =
\begin{bmatrix}
14 & 32 & 50 & 68 \\
32 & 77 & 122 & 167 \\
50 & 122 & 194 & 266 \\
68 & 167 & 266 & 365
\end{bmatrix}
\qquad
I^T \cdot I =
\begin{pmatrix}
166 & 188 & 210 \\
188 & 214 & 240 \\
210 & 240 & 270
\end{pmatrix}
$$

$$
J^T \cdot J =
\begin{bmatrix}
620 & 662 & 704 & 746 \\
662 & 707 & 752 & 797 \\
704 & 752 & 800 & 848 \\
746 & 797 & 848 & 899
\end{bmatrix}
\qquad
J \cdot J^T =
\begin{pmatrix}
534 & 718 & 902 \\
718 & 966 & 1214 \\
902 & 1214 & 1526
\end{pmatrix}
$$

$$K \cdot K^T = \begin{pmatrix} 2890 & 3490 \\ 3490 & 4215 \end{pmatrix} \qquad K^T \cdot K = \begin{bmatrix} 1213 & 1262 & 1311 & 1360 & 1409 \\ 1262 & 1313 & 1364 & 1415 & 1466 \\ 1311 & 1364 & 1417 & 1470 & 1523 \\ 1360 & 1415 & 1470 & 1525 & 1580 \\ 1409 & 1466 & 1523 & 1580 & 1637 \end{bmatrix}$$

BUT WE CAN MULTIPLY BOTH WAYS BY COMPUTING WITH THE DIAGONALIZED
NESTED ARRAYS IN THIS MANNER:

$$M^T \cdot M = \begin{bmatrix}
166 & 188 & 210 & 0 & 0 & 0 & 0 & 0 & 0 & 0 & 0 & 0 \\
188 & 214 & 240 & 0 & 0 & 0 & 0 & 0 & 0 & 0 & 0 & 0 \\
210 & 240 & 270 & 0 & 0 & 0 & 0 & 0 & 0 & 0 & 0 & 0 \\
0 & 0 & 0 & 620 & 662 & 704 & 746 & 0 & 0 & 0 & 0 & 0 \\
0 & 0 & 0 & 662 & 707 & 752 & 797 & 0 & 0 & 0 & 0 & 0 \\
0 & 0 & 0 & 704 & 752 & 800 & 848 & 0 & 0 & 0 & 0 & 0 \\
0 & 0 & 0 & 746 & 797 & 848 & 899 & 0 & 0 & 0 & 0 & 0 \\
0 & 0 & 0 & 0 & 0 & 0 & 0 & 1213 & 1262 & 1311 & 1360 & 1409 \\
0 & 0 & 0 & 0 & 0 & 0 & 0 & 1262 & 1313 & 1364 & 1415 & 1466 \\
0 & 0 & 0 & 0 & 0 & 0 & 0 & 1311 & 1364 & 1417 & 1470 & 1523 \\
0 & 0 & 0 & 0 & 0 & 0 & 0 & 1360 & 1415 & 1470 & 1525 & 1580 \\
0 & 0 & 0 & 0 & 0 & 0 & 0 & 1409 & 1466 & 1523 & 1580 & 1637
\end{bmatrix}$$

$$M \cdot M^T = \begin{bmatrix}
14 & 32 & 50 & 68 & 0 & 0 & 0 & 0 & 0 & 0 & 0 & 0 \\
32 & 77 & 122 & 167 & 0 & 0 & 0 & 0 & 0 & 0 & 0 & 0 \\
50 & 122 & 194 & 266 & 0 & 0 & 0 & 0 & 0 & 0 & 0 & 0 \\
68 & 167 & 266 & 365 & 0 & 0 & 0 & 0 & 0 & 0 & 0 & 0 \\
0 & 0 & 0 & 0 & 534 & 718 & 902 & 0 & 0 & 0 & 0 & 0 \\
0 & 0 & 0 & 0 & 718 & 966 & 1214 & 0 & 0 & 0 & 0 & 0 \\
0 & 0 & 0 & 0 & 902 & 1214 & 1526 & 0 & 0 & 0 & 0 & 0 \\
0 & 0 & 0 & 0 & 0 & 0 & 0 & 0 & 0 & 0 & 0 & 0 \\
0 & 0 & 0 & 0 & 0 & 0 & 0 & 0 & 2890 & 3490 & 0 & 0 \\
0 & 0 & 0 & 0 & 0 & 0 & 0 & 0 & 3490 & 4215 & 0 & 0 \\
0 & 0 & 0 & 0 & 0 & 0 & 0 & 0 & 0 & 0 & 0 & 0 \\
0 & 0 & 0 & 0 & 0 & 0 & 0 & 0 & 0 & 0 & 0 & 0
\end{bmatrix}$$

Which gives us the product multiplying both ways.

Now let's Half-Multiply the Nested Array. Remember, computers cannot o multiply yet, this is a new discovery, so we must multiply by it's equivalent diagonal matrix operator to get the solution.

$$N := \begin{bmatrix}
1 & 0 & 0 & 0 & 2 & 0 & 0 & 0 & 3 & 0 & 0 & 0 & 0 & 0 & 0 & 0 \\
0 & 4 & 0 & 0 & 0 & 5 & 0 & 0 & 0 & 6 & 0 & 0 & 0 & 0 & 0 & 0 \\
0 & 0 & 7 & 0 & 0 & 0 & 8 & 0 & 0 & 0 & 9 & 0 & 0 & 0 & 0 & 0 \\
0 & 0 & 0 & 10 & 0 & 0 & 0 & 11 & 0 & 0 & 0 & 12 & 0 & 0 & 0 & 0 \\
10 & 0 & 0 & 0 & 11 & 0 & 0 & 0 & 12 & 0 & 0 & 0 & 13 & 0 & 0 & 0 \\
0 & 14 & 0 & 0 & 0 & 15 & 0 & 0 & 0 & 16 & 0 & 0 & 0 & 17 & 0 & 0 \\
0 & 0 & 18 & 0 & 0 & 0 & 19 & 0 & 0 & 0 & 20 & 0 & 0 & 0 & 21 & 0 \\
0 & 0 & 0 & 0 & 0 & 0 & 0 & 0 & 0 & 0 & 0 & 0 & 0 & 0 & 0 & 0 \\
22 & 0 & 0 & 0 & 23 & 0 & 0 & 0 & 24 & 0 & 0 & 0 & 25 & 0 & 0 & 0 \\
0 & 27 & 0 & 0 & 0 & 28 & 0 & 0 & 0 & 29 & 0 & 0 & 0 & 30 & 0 & 0 \\
0 & 0 & 0 & 0 & 0 & 0 & 0 & 0 & 0 & 0 & 0 & 0 & 0 & 0 & 0 & 0 \\
0 & 0 & 0 & 0 & 0 & 0 & 0 & 0 & 0 & 0 & 0 & 0 & 0 & 0 & 0 & 0
\end{bmatrix}
\qquad
O := \begin{bmatrix}
0 & 0 & 0 & 0 \\
0 & 0 & 0 & 0 \\
0 & 0 & 0 & 0 \\
0 & 0 & 0 & 0 \\
0 & 0 & 0 & 0 \\
0 & 0 & 0 & 0 \\
0 & 0 & 0 & 0 \\
0 & 0 & 0 & 0 \\
26 & 0 & 0 & 0 \\
0 & 31 & 0 & 0 \\
0 & 0 & 0 & 0 \\
0 & 0 & 0 & 0
\end{bmatrix}$$

P := augment(N, O)

$$P = \begin{bmatrix}
1 & 0 & 0 & 0 & 2 & 0 & 0 & 0 & 3 & 0 & 0 & 0 & 0 & 0 & 0 & 0 & 0 & 0 & 0 & 0 \\
0 & 4 & 0 & 0 & 0 & 5 & 0 & 0 & 0 & 6 & 0 & 0 & 0 & 0 & 0 & 0 & 0 & 0 & 0 & 0 \\
0 & 0 & 7 & 0 & 0 & 0 & 8 & 0 & 0 & 0 & 9 & 0 & 0 & 0 & 0 & 0 & 0 & 0 & 0 & 0 \\
0 & 0 & 0 & 10 & 0 & 0 & 0 & 11 & 0 & 0 & 0 & 12 & 0 & 0 & 0 & 0 & 0 & 0 & 0 & 0 \\
10 & 0 & 0 & 0 & 11 & 0 & 0 & 0 & 12 & 0 & 0 & 0 & 13 & 0 & 0 & 0 & 0 & 0 & 0 & 0 \\
0 & 14 & 0 & 0 & 0 & 15 & 0 & 0 & 0 & 16 & 0 & 0 & 0 & 17 & 0 & 0 & 0 & 0 & 0 & 0 \\
0 & 0 & 18 & 0 & 0 & 0 & 19 & 0 & 0 & 0 & 20 & 0 & 0 & 0 & 21 & 0 & 0 & 0 & 0 & 0 \\
0 & 0 & 0 & 0 & 0 & 0 & 0 & 0 & 0 & 0 & 0 & 0 & 0 & 0 & 0 & 0 & 0 & 0 & 0 & 0 \\
22 & 0 & 0 & 0 & 23 & 0 & 0 & 0 & 24 & 0 & 0 & 0 & 25 & 0 & 0 & 0 & 26 & 0 & 0 & 0 \\
0 & 27 & 0 & 0 & 0 & 28 & 0 & 0 & 0 & 29 & 0 & 0 & 0 & 30 & 0 & 0 & 0 & 31 & 0 & 0 \\
0 & 0 & 0 & 0 & 0 & 0 & 0 & 0 & 0 & 0 & 0 & 0 & 0 & 0 & 0 & 0 & 0 & 0 & 0 & 0 \\
0 & 0 & 0 & 0 & 0 & 0 & 0 & 0 & 0 & 0 & 0 & 0 & 0 & 0 & 0 & 0 & 0 & 0 & 0 & 0
\end{bmatrix}$$

This multiplication gives us the transpose of the half-multiplication. Note the Nested Arrays are separated better than in the other multiplication.

$$M^{T} \cdot P =
\begin{bmatrix}
1 & 16 & 49 & 100 & 2 & 20 & 56 & 110 & 3 & 24 & 63 & 120 & 0 & 0 & 0 & 0 & 0 & 0 & 0 & 0 \\
2 & 20 & 56 & 110 & 4 & 25 & 64 & 121 & 6 & 30 & 72 & 132 & 0 & 0 & 0 & 0 & 0 & 0 & 0 & 0 \\
3 & 24 & 63 & 120 & 6 & 30 & 72 & 132 & 9 & 36 & 81 & 144 & 0 & 0 & 0 & 0 & 0 & 0 & 0 & 0 \\
100 & 196 & 324 & 0 & 110 & 210 & 342 & 0 & 120 & 224 & 360 & 0 & 130 & 238 & 378 & 0 & 0 & 0 & 0 & 0 \\
110 & 210 & 342 & 0 & 121 & 225 & 361 & 0 & 132 & 240 & 380 & 0 & 143 & 255 & 399 & 0 & 0 & 0 & 0 & 0 \\
120 & 224 & 360 & 0 & 132 & 240 & 380 & 0 & 144 & 256 & 400 & 0 & 156 & 272 & 420 & 0 & 0 & 0 & 0 & 0 \\
130 & 238 & 378 & 0 & 143 & 255 & 399 & 0 & 156 & 272 & 420 & 0 & 169 & 289 & 441 & 0 & 0 & 0 & 0 & 0 \\
484 & 729 & 0 & 0 & 506 & 756 & 0 & 0 & 528 & 783 & 0 & 0 & 550 & 810 & 0 & 0 & 572 & 837 & 0 & 0 \\
506 & 756 & 0 & 0 & 529 & 784 & 0 & 0 & 552 & 812 & 0 & 0 & 575 & 840 & 0 & 0 & 598 & 868 & 0 & 0 \\
528 & 783 & 0 & 0 & 552 & 812 & 0 & 0 & 576 & 841 & 0 & 0 & 600 & 870 & 0 & 0 & 624 & 899 & 0 & 0 \\
550 & 810 & 0 & 0 & 575 & 840 & 0 & 0 & 600 & 870 & 0 & 0 & 625 & 900 & 0 & 0 & 650 & 930 & 0 & 0 \\
572 & 837 & 0 & 0 & 598 & 868 & 0 & 0 & 624 & 899 & 0 & 0 & 650 & 930 & 0 & 0 & 676 & 961 & 0 & 0
\end{bmatrix}$$

$$P^{T} \cdot M =
\begin{bmatrix}
1 & 2 & 3 & 100 & 110 & 120 & 130 & 484 & 506 & 528 & 550 & 572 \\
16 & 20 & 24 & 196 & 210 & 224 & 238 & 729 & 756 & 783 & 810 & 837 \\
49 & 56 & 63 & 324 & 342 & 360 & 378 & 0 & 0 & 0 & 0 & 0 \\
100 & 110 & 120 & 0 & 0 & 0 & 0 & 0 & 0 & 0 & 0 & 0 \\
2 & 4 & 6 & 110 & 121 & 132 & 143 & 506 & 529 & 552 & 575 & 598 \\
20 & 25 & 30 & 210 & 225 & 240 & 255 & 756 & 784 & 812 & 840 & 868 \\
56 & 64 & 72 & 342 & 361 & 380 & 399 & 0 & 0 & 0 & 0 & 0 \\
110 & 121 & 132 & 0 & 0 & 0 & 0 & 0 & 0 & 0 & 0 & 0 \\
3 & 6 & 9 & 120 & 132 & 144 & 156 & 528 & 552 & 576 & 600 & 624 \\
24 & 30 & 36 & 224 & 240 & 256 & 272 & 783 & 812 & 841 & 870 & 899 \\
63 & 72 & 81 & 360 & 380 & 400 & 420 & 0 & 0 & 0 & 0 & 0 \\
120 & 132 & 144 & 0 & 0 & 0 & 0 & 0 & 0 & 0 & 0 & 0 \\
0 & 0 & 0 & 130 & 143 & 156 & 169 & 550 & 575 & 600 & 625 & 650 \\
0 & 0 & 0 & 238 & 255 & 272 & 289 & 810 & 840 & 870 & 900 & 930 \\
0 & 0 & 0 & 378 & 399 & 420 & 441 & 0 & 0 & 0 & 0 & 0 \\
0 & 0 & 0 & 0 & 0 & 0 & 0 & 0 & 0 & 0 & 0 & 0 \\
0 & 0 & 0 & 0 & 0 & 0 & 0 & 572 & 598 & 624 & 650 & 676 \\
0 & 0 & 0 & 0 & 0 & 0 & 0 & 837 & 868 & 899 & 930 & 961 \\
0 & 0 & 0 & 0 & 0 & 0 & 0 & 0 & 0 & 0 & 0 & 0 \\
0 & 0 & 0 & 0 & 0 & 0 & 0 & 0 & 0 & 0 & 0 & 0
\end{bmatrix}$$

CHECK:

$$
\begin{bmatrix} 1 & 0 & 0 & 0 \\ 0 & 4 & 0 & 0 \\ 0 & 0 & 7 & 0 \\ 0 & 0 & 0 & 10 \end{bmatrix} \cdot \begin{bmatrix} 1 & 2 & 3 \\ 4 & 5 & 6 \\ 7 & 8 & 9 \\ 10 & 11 & 12 \end{bmatrix} = \begin{bmatrix} 1 & 2 & 3 \\ 16 & 20 & 24 \\ 49 & 56 & 63 \\ 100 & 110 & 120 \end{bmatrix}
$$

$$
\begin{bmatrix} 2 & 0 & 0 & 0 \\ 0 & 5 & 0 & 0 \\ 0 & 0 & 8 & 0 \\ 0 & 0 & 0 & 11 \end{bmatrix} \cdot \begin{bmatrix} 1 & 2 & 3 \\ 4 & 5 & 6 \\ 7 & 8 & 9 \\ 10 & 11 & 12 \end{bmatrix} = \begin{bmatrix} 2 & 4 & 6 \\ 20 & 25 & 30 \\ 56 & 64 & 72 \\ 110 & 121 & 132 \end{bmatrix}
$$

$$
\begin{pmatrix} 11 & 0 & 0 \\ 0 & 15 & 0 \\ 0 & 0 & 19 \end{pmatrix} \cdot \begin{pmatrix} 10 & 11 & 12 & 13 \\ 14 & 15 & 16 & 17 \\ 18 & 19 & 20 & 21 \end{pmatrix} = \begin{pmatrix} 110 & 121 & 132 & 143 \\ 210 & 225 & 240 & 255 \\ 342 & 361 & 380 & 399 \end{pmatrix}
$$

$$
\begin{pmatrix} 10 & 0 & 0 \\ 0 & 14 & 0 \\ 0 & 0 & 18 \end{pmatrix} \cdot \begin{pmatrix} 10 & 11 & 12 & 13 \\ 14 & 15 & 16 & 17 \\ 18 & 19 & 20 & 21 \end{pmatrix} = \begin{pmatrix} 100 & 110 & 120 & 130 \\ 196 & 210 & 224 & 238 \\ 324 & 342 & 360 & 378 \end{pmatrix}
$$

$$
\begin{pmatrix} 22 & 0 \\ 0 & 27 \end{pmatrix} \cdot \begin{pmatrix} 22 & 23 & 24 & 25 & 26 \\ 27 & 28 & 29 & 30 & 31 \end{pmatrix} = \begin{pmatrix} 484 & 506 & 528 & 550 & 572 \\ 729 & 756 & 783 & 810 & 837 \end{pmatrix}
$$

$$
\begin{pmatrix} 26 & 0 \\ 0 & 31 \end{pmatrix} \cdot \begin{pmatrix} 22 & 23 & 24 & 25 & 26 \\ 27 & 28 & 29 & 30 & 31 \end{pmatrix} = \begin{pmatrix} 572 & 598 & 624 & 650 & 676 \\ 837 & 868 & 899 & 930 & 961 \end{pmatrix}
$$

The solutions check. QED

THE THEORY OF INFORMATION AND THE UNIFIED FIELD EQUATION

FOR EVERY NUMERICAL OPERATION THAT EXISTS, THERE EXISTS A MATRIX OPERATION THAT WILL REPEAT THAT OPERATION ixj TIMES FOR ADDITION AND SUBTRACTION, OR (ixj)(jxk) TIMES FOR MULTIPLICATION.

Suppose we have a thousand numbers that we have to add to another thousand numbers. The form is A+B=C. There exists a matrix $[A]_{ij} + [B]_{ij} = [C]_{ij}$ that will add A + B ixj times, all at the same time under one mathematical operation.

Suppose we have A x B ± C = 0. This has four matrix solutions, two additive and two multiplicative.

OPERATION TAKEN ONE AT A TIME	MATRIX EQUIVALENT	
A x B = C	$[A]_{ij}[B]_{jk} = [C]_{ik}$	MULTIPLICATIVE (indices change)
A x B = -C	$[A]_{ij}[B]_{jk} = -[C]_{ik}$	MULTIPLICATIVE (indices change)
A x B + C = 0	$[A]^T_{ij} \circ [B]_{jk} = -[C]_{jk}$	ADDITIVE (indices don't change)
A x B - C = 0	$[A]^T_{ij} \circ [B]_{jk} = [C]_{jk}$	ADDITIVE (indices don't change)

There are actually an infinite number of combinations. Suppose we have A/B = C, the matrix equivalents would be $[A]_{ij}[B]^{-1}_{jk} = [C]_{ik}$ and/or $[A]^T_{ij} \circ [B]^{-1}_{jk} = [C]_{jk}$; for A x B x C = D the matric equivalents are $[A]_{ij}[B]_{jk}[C]_{kL} = [D]_{iL}$ or $[A]^T_{ij} \circ [B]_{jk} [C]_{kL} = [D]_{iL}$ or $([A]^T_{ij} \circ [B]_{jk})^T \circ [C]_{kL} = [D]_{iL}$. Whatever the problem, there is a matrix solution that will solve everything all in one operation instead of perhaps thousands of separate operations conducted one at a time.

There is one equation above that seems to be really important, so I am going to concentrate mainly on this one. It is the equation $[A]^T_{ij} \circ [B]_{jk} = [C]_{jk}$. This looked so similar to Einstein's First Field Equation in his theory of gravitation that I just had to explore it further. It was this exploration that leads me to believe that this is the Unified Field Equation.

$[A]^T_{ij} \circ [B]_{jk} = [C]_{jk}$ describes a universal Inventory/Accounting system. If we multiply this equation through by a constant c, the equation is converted to a physics or Statistics Field Equation. All we need to do is define c, [B] and [A]. Let's look at some examples.

CLASSICAL PHYSICS

$C([A]^T_{ij} \circ [B]_{jk}) = [C]_{jk}$
Let $c = m$, and $[B]_{jk} = [a]_{jk}$ (acceleration) and $[C]_{jk} = F_{jk}$, then
$[C]_{jk} = m([A]^T_{ij} \circ [B]_{jk})$ but $[A]_{ij} = [I]_{jj}$ and $[B]_{jk} = [a]_{jk}$ and
$[C]_{jk} = F_{jk}$, so
$F_{jk} = m([I]_{jj} \circ [a]_{jk})$ for $m = $ to a constant.
$F_{jk} = m([a]_{jk})$

If $[A]_{ij} \neq [I]_{jj}$, then we have

$$F_{jk} = m([A]^T_{ij} \circ [a]_{jk})$$

NOTE: We may be able to express this backwards, I don't know.
I mean, if $[A]_{ij} = [a]_{ij}$, $[B]_{jk} = [I]_{jk}$ and $[C]_{ij} = F_{ij}$.

Then $F_{ij} = m([a]_{ij})$ and $\mathbf{F_{ij} = m([a]^T_{ij} \circ [B]_{jk})}$

Since there is a connection between Accounting/Inventories and Statistics. there ought to be a connection between Physics and Statistics...lets check out the Between Subjects association:

$$F_{jk} = m([A]^T_{ij} \circ [a]_{jk})$$

$$m([1]([A]^T_{ij} \circ [a]_{jk})^2 = [F]^2_{jk}$$

If $[A]_{ij} = [I]$ this equation reduces to:

$$m([1]_{ij}[a]_{jk})^2 = [F]^2_{jk}$$

But since this is Statistics, we might have to divide by N, then the equation connecting statistics to physics becomes:

$$m/N([1]_{ij}[a]_{jk})^2 = [F]^2_{jk}$$

But this is the equation of Statistics for regular matrix multiplication. We can find for Astro-physics in the same way.

EINSTEIN'S FIELD EQUATION FOR GRAVITATION

$C([A]^T_{ij} \circ [B]_{jk}) = [C]_{jk}$

Let $[C]_{jk} = [G]_{jk}$, $[B]_{jk} = [T]_{ij}$ and $[A]_{ij} = [I]_{jj}$ and $c = 8\pi\rho$.

Then we have:

$8\pi\rho([A]^T_{ij} \circ [T]_{jk}) = [G]_{jk}$ but $[A]_{ij} = [I]_{jj}$ so we have

$8\pi\rho[T]_{jk} = [G]_{jk}$ which is Einstein's First Field Equation. This is the equation as we now understand it, but the true equation might be **$8\pi\rho([A]^T_{ij} \circ [T]_{jk}) = [G]_{jk}$**.

    Let's see if we can derive a wave equation for gravity from this equation.
    Let's multiply Einstein's equation through by the wave function $\Psi^T\Psi$.

$\Psi^T 8\pi\rho[T]_{jk}\Psi = \Psi^T[G]_{jk}\Psi$ but G is an energy term and, $8\pi\rho$ is a constant. The equation becomes:

$8\pi\rho \ \Psi^T[T]_{jk}\Psi = [G]_{jk} \ \Psi^T \ \Psi$

and therefore

$$[G]_{jk} = \frac{8\pi\rho \ \Psi^T[T]_{jk}\Psi}{\Psi^T \ \Psi} \quad \text{or} \quad [G]_{jk} = \frac{8\pi\rho \ \Psi^T([A]^T_{ij} \circ [T]_{jk})\Psi}{\Psi^T \ \Psi}$$

G_{jk} is now solvable.

    Astrophysicists say Einstein's equation is not complete. Knowing nothing about Astrophysics, I assume the mass term is included in the stress-metric tensor T_{jk}. Since Einstein's equation resembles Newton's equation, I suggest that the mass term may need to be removed from T_{jk} and hollow dotted into it. i.e.

$[G]_{jk} = 8\pi\rho([m]^T_{ij} \circ [T]_{jk})$ or
$[G]_{jk} = 8\pi\rho [m]^T_{ij} \circ ([A]^T_{ij} \circ [T]_{jk})$

or maybe: $[G]_{jk} = 8\pi\rho m([A]^T_{ij} \circ [T]_{jk})$ (I'm not sure if any are correct.)

    Which has the form of Newton's Second Equation. This is all I can do with the math for this kind of mathematics. I never could get a handle on the Geometrodynamics equations of Space-time. I have no idea how to solve it or use it. Are there any Astrophysicists out there who can make any sense out of it?

Einstein's Equation and Statistics:

$$G_{jk} = 8\pi\rho \ ([A]^T_{ij} \ o \ [T]_{jk})$$

$$8\pi\rho \ ([1]([A]^T_{ij} \ o \ [T]_{jk})^2 = [G]^2_{jk}$$

If $[A]_{ij} = [I]$ this equation reduces to:

$$8\pi\rho \ ([1]_{1,j}[T]_{jk})^2 = [G]^2_{jk}$$

But since this is Statistics, we might have to divide by N, not sure. Then the equation connecting Statistics to Physics becomes:

$$8\pi\rho/N \ ([1]_{1,j}[a]_{jk})^2 = [G]^2_{jk}$$

But this is the equation of Statistics for regular matrix multiplication.

STATISTICS

There was, until now, no Field Equation describing the mathematics" of Statistics. Going back to the Unified Field Equation: $C([A]^T_{ij} \ o \ [B]_{jk}) = [C]_{jk}$ and for the simple case, letting $[B]_{jk} = [I]_{jk}$ and $c = 1/N$, the equation becomes: (this means there is no Database Matrix to multiply to, we manipulate only the data directly on matrix [A].)

$$1/N([A]^T_{ij} \ o \ [B]_{jk}) = [C]_{jk} = 1/N([A]^T_{ij} \ o \ [I]_{jk}) = \mathbf{1/N([A]^T_{ij}) = [C]_{ij}}$$

Since [A][I] is a straight multiplication problem, we do not need to transpose it. But we do need to sum the columns of [A], so the basic Statistical equation becomes:

$$1/N([1]_{1,I}[A]_{ij}) = [C]_{ij}$$

But to make this Statistical, we must square the above expression so that it becomes:

$1/N([1]_{1,i}[A]_{ij})^2 = [C]^2_{1j}$ **Where** $[C]^2_{1j} = CC^T$. (this gives us a one by one matrix as a solution).

Suppose $[B]_{jk}$ is not = to $[I]_{jk}$. Then we have our basic complex Statistical Field Equation, the simplest form of which is the Analysis of Variance. This is a straight multiplication so [A] does not need to be transposed. (We could transpose [A] first, and then sum the columns in a second operation). Therefore the standard Complex Statistical Equation becomes:

$1/N([1]_{1,i}[A]_{ij}[B]_{jk})^2 = [C]^2_{1,k}$ where $[C]^2 = CC^T$

But we also need to subtract the correction factor(s), so the total Statistical Field Equation becomes:

$1/N([1]_{1,i}[A]_{ij}[B]_{jk})^2$ - **correction factor(s)** $= C^2_{1,k}$ - **correction factor(s)**.

This is for Between Subjects Statistic.

The Within Subjects Statistics the Field Equation becomes:

$1/N([A]_{ij}[B]_{jk}[1]_{kl})^2 = [C]^2_{1,k}$ where $[C]^2 = CC^T$

In other words, Between Subjects Statistics has the form of 1xAxB and Within Subjects Statistics has the form AxBx1.

Note, there are no Σ's anywhere, since this approach to Statistics is derived from Accounting and Inventories rather than random variables.

So there are only 3, perhaps 4 operators from which Statistics (perhaps all of Statistics) may be computed:

$1/2\sum([A]^T_{ij} \circ [B]_{jk} = i\ C_{jk}$ matrices (Half-Multiplier mode) (where i = #rows in un-transposed matrix)

$^C\sum 1/N([A]^T_{ij} \circ [B]_{jk}) = 1/N[C]_{ik}$ regular matrix multiplication.

$^R\sum 1/N([A]^T_{ij} \circ [B]_{jk}) = 1/N([B]_{jk}[1]_{k,1} \circ [A]^T_{ij})$

$^M\sum 1/N([A]^T_{ij} \circ [B]_{jk}) = 1/N([A]^T_{ij}[1]_{i,1} \circ [B]_{jk})$

Note: These equations also hold for Physics, Astrophysics and Inventory/Accounting systems also.

The mathematics also suggests that we may be able to create a Quantum Statistics (or perhaps Quantum Accounting/Inventories?). Or more fundamentally, a Quantum Probability. What I am going to derive I will give an example of at the end of the section of Statistics. I do not know how to interpret the solution, that will be left to the professional Statisticians.

Suppose we have completed the experimental data on an experiment and put into the matrix form $[A]_{MN}$. Now $[A]_{MN}$ is not a square matrix, so we must make a square out of it before we can do any sort of Quantum Statistical analysis on it. Squaring the data changes the matrix $[A]$ from a Statistical Matrix to a Probability Matrix. To keep track of each of the sums of the squares for each column in the matrix, we must square it such that we end up with an NxN matrix.

$$[A]^{T}_{MN}[A]_{MN} = [A]_{NM}[A]_{MN} = [A]^{2}_{NN}$$

Then we do the following:

$$\frac{1/N[DB1]_{iN} \ [A]^{2}_{NN}[DB1]^{T}_{iN}}{[DB1]_{iN}[DB1]^{T}_{iN}} = \frac{1/N[DB1]_{iN} \ [A]^{2}_{NN}[DB1]_{Ni}}{[DB1]_{iN}[DB1]_{Ni}} = \frac{1/N[DB1]_{iN} \ [A]^{2}_{NN}[DB1]_{Ni}}{[DB1]^{2}_{ii}} =$$

$$\frac{1/N[C]^{2}_{ii}}{[DB1]^{2}_{ii}}$$

Now Statistics and Probability are directly associated with each other. How, for instance, can we compute 5! ? That is 5x4x3x2x1=120? Easy!

$$\text{Anti-log}\left(\text{Log}\ ([5\ 4\ 3\ 2\ 1]) \begin{vmatrix} 1 \\ 1 \\ 1 \\ 1 \\ 1 \end{vmatrix}\ \right) = 120$$

Of course, we don't have to just use [DB1], we can use any of the Database Matrices that are pertinent to the analysis under scrutiny.

DERIVATION OF QUANTUM MECHANICS FROM EINSTEIN'S EQUATION

The constant at this time is unknown and will revert to c. Substituting the new values into the Wave equation we get:

$$E_{jk} \quad \frac{c \; \Psi^T_{jk}[H]_{jk}\Psi_{jk}}{\Psi^T \; \Psi} \quad \text{or} \quad [E]_{jk} = \frac{c\Psi^T([A]^T_{ij} \; o \; [H]_{jk})\Psi}{\Psi^T \; \Psi}$$

Physics and Astrophysics field equations are already known, so it should be simple for people proficient in these fields to check the correctness of these derivations. The field equations for Statistics and Accounting/Inventory systems are brand new and have not been discovered before.

INVENTORY/ACCOUNTING SYSTEM:

$[A]^T_{ij} \; o \; [B]_{jk} = [C]_{jk} =$
$[INV]_{ij}[DB]_{jk} = [SOL]_{ik}$ This is the sum total of everything bought, sold, manufactured, etc.

$^i[A]^T_{ij} \; o \; [B]_{jk} = \;^i[C]_{jk} =$
$^i[INV]^T_{ij} \; o \; [DB]_{jk} = \;^i[AP]_{jk}$ This keeps and individual item accounting of everything bought, sold, manufactured, etc. The superscript i just tells us which row in the un-transposed matrix we have hollow-dotted onto the [DB] matrix.

$^j[B]_{jk} \; o \; [A]^T_{ij} = \;^j[C]_{jk} = \;^j[DB]_{jk} \; o \; [INV]^T_{ij} = \;^j[IP]_{jk}$

This takes an individual item from the database matrix (a column) and multiplies it across the inventory(or accounting page if we wish) giving us a slice of the total pie, so to speak. It tells us how much of that item was used, by whom or what machine or smokestack, and shows how it is distributed throughout the total inventory. This one is best used taken one item at a time. If we want two items at a time, we have to add them together and compute a total for both. It would be like saying 35 apples and oranges instead of 15 apples and 20 oranges.

$[A]^T_{ij} \; o \; [B]_{jk} = i, \; [C]_{jk}$ Sub-matrices.
$[INV]^T_{ij} \; o \; [DB]_{jk} = i, \; [SOL]_{jk}$ Sub-matrices
This is the pure Half-Multiplier operation giving all the accountpage matrices, i of them, in a single operation.

PHILOSOPHY

There are more things under these heavens and earth
than are dreamed of in your philosophy,
Horatio. W. Shakespeare

Truth lies not in the heights, but is at the bottom of all things.
SKS Book 1

This math can easily be taught, but its beauty cannot be
taught but must be caught, but once caught, it's simplicity
is staggering to the intellect.

SKS Book 1 paraphrased

In the history of mathematics, the origins of math follow two
seemingly separate paths. The first was when man kept track of
the rising of the sun to predict the seasons and later to predict
eclipses. Careful records were kept, even of the positions of
the planets. This system of accounting was discarded with the
advent of Newton's three equations of motion when classical
physics was developed. The second system of mathematics was
developed to keep track of what merchants and kings owned,
what was owed them, and how much they owed to others. This
system developed into modern accounting systems. Two seemingly
different mathematical systems. With the Half-Multiplier
Operator, the two divergent systems are again brought together
under the dominion of a single equation. Mathematics has now
returned to it's roots. The Half-Multiplier Operator seems to
say that the universe prefers addition and subtraction rather
than the differential equations and the calculus mathematicians
and physicists have developed to explain the nature of matter
and energy. This is the missing operator, the missing link in
math, so to speak, that has been sought by everyone but has
escaped notice until now. We have used the math all the time but
have never consolidated it into one comprehensive set of proofs
that I try to develop in this book, but these equations seem to
hold for the whole of mathematics. Mathematicians say that $3x2=6$
in most cases, but not in all cases. I prove here that $3x2=6$
in all cases, even though mathematicians some 200 to 300 years
ago said that what I am about to show you is impossible to do.

Because everyone believed this, they missed out on discovering
one of the most beautiful mathematical operator's of all. This
is basically all that this book is about, that the operation
of multiplication holds in all cases i.e., 5x2=10 no matter
how you multiply it. With this simple premise, we can do all
our mathematical operations at the same time instead of doing
just one calculation at a time.

There seems to be a direct relationship between ordinary
addition, subtraction, multiplication, division and matrices.
Suppose we want to add the numbers

$$
\begin{array}{r}
24 \\
-11 \\
9 \\
44 \\
\hline
-6
\end{array}
\qquad \text{We can re-write this as}
$$

$$
\begin{bmatrix} 1 \\ 1 \\ -1 \\ 1 \\ -1 \end{bmatrix} \circ \begin{bmatrix} 24 \\ 16 \\ 11 \\ 9 \\ 44 \end{bmatrix} \;=\; (1 \;\; 1 \;\; -1 \;\; 1 \;\; -1) \cdot \begin{bmatrix} 24 \\ 16 \\ 11 \\ 9 \\ 44 \end{bmatrix} = {}^-6
$$

Transposing and multiplying sums the numbers for us.

Suppose we wish to subtract 99, 22, 111 and 55 from 333. Ordinary
subtraction looks like:

$$
\begin{bmatrix} 1 \\ -1 \\ -1 \\ -1 \\ -1 \end{bmatrix} \circ \begin{bmatrix} 333 \\ 99 \\ 22 \\ 111 \\ 55 \end{bmatrix} = \begin{bmatrix} 1 \\ 1 \\ 1 \\ 1 \\ 1 \end{bmatrix} \circ \begin{bmatrix} 333 \\ -99 \\ -22 \\ -111 \\ -55 \end{bmatrix} = (1 \;\; -1 \;\; -1 \;\; -1 \;\; -1) \cdot \begin{bmatrix} 333 \\ 99 \\ 22 \\ 111 \\ 55 \end{bmatrix} = 46
$$

But in ordinary Matrix multiplication we must sum the solutions
into a single number, with the Half-Multiplier we are under no
such constraint.

So here we can convert ordinary addition and subtraction to
matrix multiplication. This simple conversion is the missing
link of mathematics, it is the matrix/tensor connection to our
real number system operations of addition and subtraction.
That is, if the pre-multiplier is a matrix all of one's,

we are adding/subtracting the numbers in the columns of the post-multiplier matrix.

The operations of multiplication and division we know about intuitively from it's use in Gaussian reduction. i.e. 3x6=18, by using the Half-Multiplier we get the equal, but trivial equation $[3]^T$ o $[6]$ = 18. Suppose we need to multiply 3x6, 4x5, 9x3, 22x11 and 7x11. I want to subtract the second product from the first, ignore the third product, double the fourth product and subtract and triple the fifth product and add. We have:

$$\begin{bmatrix} 1 \\ -1 \\ 0 \\ -2 \\ 3 \end{bmatrix} o \begin{bmatrix} 3 \\ 4 \\ 9 \\ 22 \\ 7 \end{bmatrix} o \begin{bmatrix} 6 \\ 5 \\ 3 \\ 11 \\ 11 \end{bmatrix} = \begin{bmatrix} \begin{bmatrix} 1 & 0 & 0 & 0 & 0 \\ 0 & -1 & 0 & 0 & 0 \\ 0 & 0 & 0 & 0 & 0 \\ 0 & 0 & 0 & -2 & 0 \\ 0 & 0 & 0 & 0 & 3 \end{bmatrix} \cdot \begin{bmatrix} 3 \\ 4 \\ 9 \\ 22 \\ 7 \end{bmatrix} \end{bmatrix}^T \cdot \begin{bmatrix} 6 \\ 5 \\ 3 \\ 11 \\ 11 \end{bmatrix} = {}^-255$$

Or it is equal to:

$$(3 \quad {}^-4 \quad 0 \quad {}^-44 \quad 21) \cdot \begin{bmatrix} 6 \\ 5 \\ 3 \\ 11 \\ 11 \end{bmatrix} = {}^-255$$

Or in terms of Mathcad programs:

$$\begin{bmatrix} \begin{bmatrix} 1 \\ -1 \\ 0 \\ -2 \\ 3 \end{bmatrix} \cdot \begin{bmatrix} 3 \\ 4 \\ 9 \\ 22 \\ 7 \end{bmatrix} \end{bmatrix}^T \cdot \begin{bmatrix} 6 \\ 5 \\ 3 \\ 11 \\ 11 \end{bmatrix} = {}^-255$$

The three hollow dot multiplication's on the left are illegal according to modern math, diagonalizing the first column and multiplying to the second column matrix is legal (I do it this way because it is the only way a computer or calculator can come up with the same solution, no one has ever thought about

multiplying matrices in this manner before). And yet, what the
hollow dot multiplication on the far left is saying is

```
 1x3x6    = 18
-1x4x5    = -20
 0x9x3    = 0
-2x22x11  = -484
 3x7x11   = 231
```

    Summing the separate answers we get 18 - 20 + 0 - 484 +
231 = -255.

    Division is carried out the same as multiplication except
we multiply by the inverse of the divisor.

This is all there is to this operator. In every case, there
is an operation we already know and use that can be used to
compute the solution. But look at the problem above. If we
totally half-multiply all the numbers, we go through 15 separate
multiplication's and 5 sums for a total of 20 operations. When
we have to diagonalize, the diagonal times the second column
needs 25 multiplication's and 25 sums. We then transpose this
first product and multiply again to the third column matrix,
needing 5 multiplication's and 5 sums. This is a total of
25+25+5+5 = 60 separate mathematical operations to compute
the same answer as 20 steps using the half-multiplier. That's
three times as many. If the problem concerned column matrices a
million rows long, diagonalizing would produce a square matrix
of dimensions $1x10^6$ x $1x10^6$ with $1x10^{12}$ elements, that would be a
trillion elements, all zero's except the diagonal. So we would
need to perform a trillion multiplication's and a trillion sums
instead of a million multiplication's and a million sums. Quite
a savings on computer memory. Actually, MathCad can perform
this half-multiplication using the vectorize operation, but it
cannot perform the following multiplication's. Suppose we have
the following 3x3 matrix, and we wish to multiply the first
row by 3, the second row by 4 and the third row by 5. This is
also illegal, but we do it all the time when we use Gaussian
reduction. i.e.

$$\begin{vmatrix} 3 \\ 4 \\ 5 \end{vmatrix} \circ \begin{vmatrix} 1 & 2 & 3 \\ 4 & 5 & 6 \\ 7 & 8 & 9 \end{vmatrix} = \begin{vmatrix} 3 & 0 & 0 \\ 0 & 4 & 0 \\ 0 & 0 & 5 \end{vmatrix} \begin{vmatrix} 1 & 2 & 3 \\ 4 & 5 & 6 \\ 7 & 8 & 9 \end{vmatrix} = \begin{vmatrix} 3 & 6 & 9 \\ 16 & 20 & 24 \\ 35 & 40 & 45 \end{vmatrix}$$

Suppose, on the other hand, we wish to multiply the first column in the 3x3 matrix by 3, the second column by 4 and the third column by 5. We proceed as follows:

$$\begin{vmatrix} 1 & 2 & 3 \\ 4 & 5 & 6 \\ 7 & 8 & 9 \end{vmatrix} \begin{vmatrix} 3 \\ 4 \\ 5 \end{vmatrix} = \begin{vmatrix} 1 & 2 & 3 \\ 4 & 5 & 6 \\ 7 & 8 & 9 \end{vmatrix} \begin{vmatrix} 3 & 0 & 0 \\ 0 & 4 & 0 \\ 0 & 0 & 5 \end{vmatrix} = \begin{vmatrix} 3 & 8 & 15 \\ 12 & 20 & 30 \\ 21 & 32 & 45 \end{vmatrix}$$

These are important, because suppose we wish to deal only with the second row in the matrix and get rid of the first and third rows. We proceed as follows:

$$\begin{vmatrix} 0 & 0 & 0 \\ 0 & 1 & 0 \\ 0 & 0 & 0 \end{vmatrix} \begin{vmatrix} 1 & 2 & 3 \\ 4 & 5 & 6 \\ 7 & 8 & 9 \end{vmatrix} = \begin{vmatrix} 0 & 0 & 0 \\ 4 & 5 & 6 \\ 0 & 0 & 0 \end{vmatrix}.$$

Or suppose we want only the first and third columns of the matrix, and ignore the middle column. We proceed as follows:

$$\begin{vmatrix} 1 & 2 & 3 \\ 4 & 5 & 6 \\ 7 & 8 & 9 \end{vmatrix} \begin{vmatrix} 1 & 0 & 0 \\ 0 & 0 & 0 \\ 0 & 0 & 1 \end{vmatrix} = \begin{vmatrix} 1 & 0 & 3 \\ 4 & 0 & 6 \\ 7 & 0 & 9 \end{vmatrix}$$

These are very important properties in accounting and inventories and Statistics, especially when we have a large amount of data, but wish to compare only two or so columns or rows with each other in Statistics, or to find the individual accountpage and itempage matrices in the generalized accounting/ inventory system. These you will see more of in the examples to follow.

But, you might say, by the definition I give, that

$$\begin{vmatrix} 1 & 2 & 3 \\ 4 & 5 & 6 \\ 7 & 8 & 9 \end{vmatrix} \circ \begin{vmatrix} 3 \\ 4 \\ 5 \end{vmatrix} = \begin{vmatrix} 3 \\ 16 \\ 35 \end{vmatrix} \begin{vmatrix} 6 \\ 20 \\ 40 \end{vmatrix} \begin{vmatrix} 9 \\ 24 \\ 45 \end{vmatrix} \quad \text{and is also equivalent to} \quad \begin{vmatrix} 3 & 6 & 9 \\ 16 & 20 & 24 \\ 35 & 40 & 45 \end{vmatrix}$$

Depending upon which way we multiply. Right to left for the Nested Array and left to right for the larger single 3x3 matrix.

Our choice, depending on the information we wish to obtain
from the system. The Nested Array is ready for a transpose and
further manipulation as a larger column matrix (in this case,
a 9x1 column matrix), or just to separate the answers into
separate arrays (for printing purposes say) or the matrix can
be used any way we can use a regular 3x3 matrix.

I will show later (I can't prove it) that when we are done
with the math, we can drop the inner or the outer brackets of
the Nested Array to obtain either three 3x1 matrices or one 3x3
matrix, whatever form we desire the solution to be. Matrices
are freewheeling and we can do almost anything with them that
we can do with single numbers, just as long as we keep to the
rules of regular mathematics.

There is one further thing to notice. When we transpose
the pre-multiplier matrix and half-multiply across the
post-multiplier matrix, the transposed matrix becomes like a
real set of numbers and is not like a matrix anymore. So just
as we can take the square root of a single number, or the log
of a single number, we can now take the square root of every
number individually or the log of each number individually, all
at the same time. If we then re-transpose the pre-multiplier
matrix, it is in tensor or matrix form and we can multiply the
matrices as they are now defined.

So now the concept of number has been re-defined. Just as
1, 2, 3, . . . are considered individual numbers, now

$$\begin{vmatrix} 1 & 2 & 3 \\ 4 & 5 & 6 \\ 7 & 8 & 9 \end{vmatrix}$$ can be considered as a single number with all the
properties of a single number. It can also be treated as a Matrix
as Matrices are defined. In other words, a Matrix has a dual
nature, it can either be treated as a number or a Matrix.

But just as $1 + 1 = 2$, the dimensions must be the same or
conform with the rules of matrix multiplication. i.e.

Suppose we wish to multiply the two following numbers as we
would two regular numbers:

$$\begin{vmatrix} 1 & 2 & 3 \\ 4 & 5 & 6 \\ 7 & 8 & 9 \end{vmatrix} \otimes \begin{vmatrix} 9 & 8 & 7 \\ 6 & 5 & 4 \\ 3 & 2 & 1 \end{vmatrix} = \begin{vmatrix} 9 & 16 & 21 \\ 24 & 25 & 24 \\ 21 & 16 & 9 \end{vmatrix}$$

Or we may add all the numbers at the same time, just as long
as the matrices are the same size:

$$
\begin{vmatrix} 1 & 2 & 3 \\ 4 & 5 & 6 \\ 7 & 8 & 9 \end{vmatrix} \quad + \quad \begin{vmatrix} 9 & 8 & 7 \\ 6 & 5 & 4 \\ 3 & 2 & 1 \end{vmatrix} \quad = \quad \begin{vmatrix} 10 & 10 & 10 \\ 10 & 10 & 10 \\ 10 & 10 & 10 \end{vmatrix}
$$

The former multiplication is very illegal in mathematics, but
can anyone out there prove to me that 8x2 ≠ 16? Or 9x1 ≠ 9? This
I would like to see. Not only is this multiplication legal, but
it works! We cannot multiply all numbers in all problems at the
same time without it. Statistics as a branch of mathematics
would fall apart without it. But because mathematicians say it
is illegal, Statisticians must solve statistical problems the
hard way, one multiplication and sum at a time, rather than all
at once in one simplified operation.

GENERALIZED INVENTORY/ACCOUNTING SYSTEM

Let $[A]_{ij}$ = Inventory matrix
Let $[B]_{jk}$ = Database matrix

Then

$$[A]_{ij} \ [B]_{jk} = C_{ik} = [SOL]_{ik}$$
$$[INV]_{ij}[DB]_{jk} = [SOL]_{ik}$$

where C_{ik} = solution matrix for the inventory.

And

$$[A]^T_{ij} \ o \ [B]_{jk} = D_{jk} = [AP]_{jk}$$
$$[INV]^T_{ij} \ o \ [DB]_{jk} = [AP]_{jk}$$

Where $D_{jk} = [AP]_{jk}$ = the accounting matrix.

And

$$[A]^T_{ij} \ o \ \overleftarrow{[B]}_{jk} = I_{jk} = [IP]_{jk}$$
$$[INV]^T_{ij} \ o \ \overleftarrow{[DB]}_{jk} = [IP]_{jk}$$

Where $I_{jk} = [IP]_{jk}$ = The Itempage matrix.

The connection between the Accounting/Inventory Field Equation and Statistics is given by:

$$[INV]_{ij}[DB]_{jk} = [SOL]_{ik}$$
$$1/N([1]_{1,i}[INV]_{ij}[DB_i]_{jk})^2 = [SOL]_{1,k}[SOL]_{k,1}^T - \text{CORRECTION FACTOR}$$

The best way to show how the inventory system works is to work an example. The first example will be a chemical usage inventory (CUI). For the following example:

1. the number of chemicals with unknown fractions = 9 = j.
2. the number of reportable chemical fractions = 15 = k.

3. the number of machines using smokestacks + one total inventory = 8 machines + 1 total inventory = 8 + 1 = 9 = i.

Therefore, the dimensional analysis of this example inventory =

1. For Inventory: $[A]_{ij}$ $[B]_{jk}$ = C_{ik} = $[A]_{9,9}$ $[B]_{9,15}$ = $C_{9,15}$

2. For Accounting: $[A]^T_{ij}$ o $[B]_{jk}$ = D_{jk} = $[A]^T_{9,9}$ o $[B]_{9,15}$ = $D_{9,15}$

$$\text{For Itempage:} [A]^{\overleftarrow{T}}_{ij} \text{ o } [B]_{jk} = I_{jk} = [A]^{\overleftarrow{T}}_{9,9} \text{ o } [B]_{9,15} = I_{9,15}$$

THE INVENTORY MATRIX

We will write the inventory matrix first. We enter the inventory weights for the total yearly inventory (in pounds) in the first row of the matrix, for the second through 9th rows we will write the pounds of chemicals used in machines 1 through 8. Note: the sum of the columns 2 - 9 should equal the amount in the first row. i.e.

	AP	CSS	DGP	FC-37	KFR-18	NW-3A	RW-41	TGC	ZON
TOTAL INVENTORY	5590	55366	725	8610	49547	2750	21649	11000	880
MACHINE #1	5590	41605	0	0	0	0	0	0	0
MACHINE #2	0	12863	0	0	0	0	0	0	0
MACHINE #3	0	898.1	0	0	0	0	0	0	0
MACHINE #4	0	0	0	0	0	0	0	0	0
MACHINE #5	0	0	362.5	638	0	476.7	10824.5	5500	826.2
MACHINE #6	0	0	362.5	4023	0	325.6	10824.5	5500	0
MACHINE #7	0	0	0	3949.42	4952.7	1947.83	0	0	53.8
MACHINE #8	0	0	0	0	44574.3	0	0	0	0

Note: For the column under CSS, summing rows 2 through 8, we get 41605+12863+898.1 = 55366.1 which is the total yearly use for all machines.

This finished matrix I will call $[INV]_{9,9}$. For HP-48G program, STO INV99.

THE DATABASE MATRIX

The matrix $[B]_{jk}$ is written directly from the MSDS sheets for each chemical listed. It is permanent and will not change unless new chemicals are added to the inventory, old chemicals are not used anymore, or the manufacturer changes the proportions of the ingredients. First we make a vertical list of the 9 chemicals we use. In a row across the top, make a column for each of the 15 chemical fractions. Using the MSDS sheets and doing one chemical at a time, go along the row to the column where the chemical fraction is listed and enter the percent of that chemical as a decimal. Where there are no chemicals present, enter a zero. Approximately 5% of the DB matrix will be chemical fractions, and 95% will be zero's. I will call this matrix $[DB]_{9,15}$.

	APE	DPE	NAP	ISOP	2-EH	o-DCB	2EE	EG	AMM	FORM	2BA	DIOX	ACRYL	DEG	ACET
AP	.02	.06	0	0	0	0	0	0	0	0	0	0	0	0	0
CSS	0	0	.12	.04	.10	.03	0	0	0	0	0	0	0	0	0
DGP	0	0	0	.02	0	0	0	0	0	0	0	0	0	0	0
FC-37	0	0	0	0	0	0	.0001	.001	0	0	0	0	0	0	0
KFR-18	0	0	0	0	0	0	0	0	.01	0	0	0	0	0	0
NW-3A	0	0	0	0	0	0	0	0	0	.006	0	0	0	0	0
RW-41	0	0	0	0	0	0	0	0	0	.001	.0002	.0002	.0001	0	0
TGC	0	0	0	0	0	0	0	0	0	.005	0	0	0	.02	0
ZON	0	0	0	0	0	0	0	.04	0	0	0	0	0	0	.075

Where:

APE: ALKYL PHENOLIC ETHER

DPE: DIPHENYL ETHER

NAP: NAPTHALENE

ISOP: ISOPROPANOL

2-EH: 2-ETHYLHEXANE

oDCB: ortho-DICHLOROBENZENE

2EE: 2 ETHOXYETHANOL

EG: ETHYLENE GLYCOL

AMM: AMMONIA

FORM: FORMALDEHYDE

2BA: 2-BUTYLALCOHOL

DIOX: DIOXANE

ACRYL: ACRYLALDEHYDE

DEG: DIETHYLENEGLYCOL

ACET: ACETONE

We will store this matrix in DB915.

PROGRAM FOR HP-48G

```
ENTER  INV99
STO    INV99
ENTER  DB915
STO    DB915
RCL    DB915
RCL    INV99
x
```

MATHCAD +6 PROGRAM

$$
INV_{9,9} := \begin{bmatrix}
5590 & 55366 & 725 & 8610 & 49547 & 2750 & 21649 & 11000 & 880 \\
5590 & 41605 & 0 & 0 & 0 & 0 & 0 & 0 & 0 \\
0 & 12863 & 0 & 0 & 0 & 0 & 0 & 0 & 0 \\
0 & 898.1 & 0 & 0 & 0 & 0 & 0 & 0 & 0 \\
0 & 0 & 0 & 0 & 0 & 0 & 0 & 0 & 0 \\
0 & 0 & 362.5 & 638 & 0 & 476.7 & 10824.5 & 5500 & 826.2 \\
0 & 0 & 362.5 & 4023 & 0 & 325.6 & 10824.5 & 5500 & 0 \\
0 & 0 & 0 & 3949.42 & 4952.7 & 1947.83 & 0 & 0 & 53.8 \\
0 & 0 & 0 & 0 & 44574.3 & 0 & 0 & 0 & 0
\end{bmatrix}
$$

Inventory Matrix in pounds

$$
DB_{9,15} := \begin{bmatrix}
0.02 & 0.06 & 0 & 0 & 0 & 0 & 0 & 0 & 0 & 0 & 0 & 0 & 0 & 0 & 0 \\
0 & 0 & 0.12 & 0.04 & 0.1 & 0.03 & 0 & 0 & 0 & 0 & 0 & 0 & 0 & 0 & 0 \\
0 & 0 & 0 & 0.02 & 0 & 0 & 0 & 0 & 0 & 0 & 0 & 0 & 0 & 0 & 0 \\
0 & 0 & 0 & 0 & 0 & 0 & .0001 & 0.001 & 0 & 0 & 0 & 0 & 0 & 0 & 0 \\
0 & 0 & 0 & 0 & 0 & 0 & 0 & 0 & 0.1 & 0 & 0 & 0 & 0 & 0 & 0 \\
0 & 0 & 0 & 0 & 0 & 0 & 0 & 0 & 0 & 0.006 & 0 & 0 & 0 & 0 & 0 \\
0 & 0 & 0 & 0 & 0 & 0 & 0 & 0 & 0 & 0.006 & .0002 & .0002 & .0001 & 0 & 0 \\
0 & 0 & 0 & 0 & 0 & 0 & 0 & 0 & 0 & 0.005 & 0 & 0 & 0 & 0.02 & 0 \\
0 & 0 & 0 & 0 & 0 & 0 & 0 & 0.04 & 0 & 0 & 0 & 0 & 0 & 0 & 0.075
\end{bmatrix}
$$

Database Matrix

Multiplying the two together we get the pounds of chemical fractions used in a year.

$$INV_{9,9} * DB_{9,15} =$$

$$\begin{bmatrix}
5590 & 55366 & 725 & 8610 & 49547 & 2750 & 21649 & 11000 & 880 \\
5590 & 41605 & 0 & 0 & 0 & 0 & 0 & 0 & 0 \\
0 & 12863 & 0 & 0 & 0 & 0 & 0 & 0 & 0 \\
0 & 898.1 & 0 & 0 & 0 & 0 & 0 & 0 & 0 \\
0 & 0 & 0 & 0 & 0 & 0 & 0 & 0 & 0 \\
0 & 0 & 362.5 & 638 & 0 & 476.7 & 10824.5 & 5500 & 826.2 \\
0 & 0 & 362.5 & 4023 & 0 & 325.6 & 10824.5 & 5500 & 0 \\
0 & 0 & 0 & 3949.42 & 4952.7 & 1947.83 & 0 & 0 & 53.8 \\
0 & 0 & 0 & 0 & 44574.3 & 0 & 0 & 0 & 0
\end{bmatrix} \cdot \begin{bmatrix}
0.02 & 0.06 & 0 & 0 & 0 & 0 & 0 & 0 & 0 & 0 & 0 & 0 & 0 & 0 & 0 \\
0 & 0 & 0.12 & 0.04 & 0.1 & 0.03 & 0 & 0 & 0 & 0 & 0 & 0 & 0 & 0 & 0 \\
0 & 0 & 0 & 0.02 & 0 & 0 & 0 & 0 & 0 & 0 & 0 & 0 & 0 & 0 & 0 \\
0 & 0 & 0 & 0 & 0 & 0 & .0001 & 0.001 & 0 & 0 & 0 & 0 & 0 & 0 & 0 \\
0 & 0 & 0 & 0 & 0 & 0 & 0 & 0 & 0.1 & 0 & 0 & 0 & 0 & 0 & 0 \\
0 & 0 & 0 & 0 & 0 & 0 & 0 & 0 & 0 & 0.006 & 0 & 0 & 0 & 0 & 0 \\
0 & 0 & 0 & 0 & 0 & 0 & 0 & 0 & 0 & 0.006 & .0002 & .0002 & .0001 & 0 & 0 \\
0 & 0 & 0 & 0 & 0 & 0 & 0 & 0 & 0 & 0.005 & 0 & 0 & 0 & 0.02 & 0 \\
0 & 0 & 0 & 0 & 0 & 0 & 0 & 0.04 & 0 & 0 & 0 & 0 & 0 & 0 & 0.075
\end{bmatrix}$$

	APE	DPE	NAP	ISOP	2-EH	o-DCB	2EE	EG	AMM	FORM	2BA	DIOX	ACRYL	DEG	ACET
	111.8	335.4	6643.92	2229.14	5536.6	1660.98	0.861	43.81	4954.7	201.394	4.3298	4.3298	2.1649	220	66
	111.8	335.4	4992.6	1664.2	4160.5	1248.15	0	0	0	0	0	0	0	0	0
	0	0	1543.56	514.52	1286.3	385.89	0	0	0	0	0	0	0	0	0
	0	0	107.772	35.924	89.81	26.943	0	0	0	0	0	0	0	0	0
$INV_{9,9} \cdot DB_{9,15} =$	0	0	0	0	0	0	0	0	0	0	0	0	0	0	0
	0	0	0	7.25	0	0	0.0638	33.686	0	95.3072	2.1649	2.1649	1.08245	110	61.965
	0	0	0	7.25	0	0	0.4023	4.023	0	94.4006	2.1649	2.1649	1.08245	110	0
	0	0	0	0	0	0	0.39494	6.10142	495.27	11.68698	0	0	0	0	4.035
	0	0	0	0	0	0	0	0	4457.43	0	0	0	0	0	0

$[SOL]_{9,15} = INV_{9,9}*DB_{9,15}$

INTERPRETATION: In the total inventory (row 1) we used 6643.92 lbs naphthalene in 1995. Of this 6643.92 lbs, 4492.6 lbs were used in machine 1, 1543.56 lbs were used in machine 2 and 107.772 lbs were used in machine 3 and none was used in any other machine. Or we used 43.81 lbs ethylene glycol total for 1995, of which 33.686 lbs were used in machine 5, 4.023 lbs were used in machine 6 and 6.1014 lbs was used in machine 7, etc.

Suppose we are not a Synthetic Minor or Title V user and only need the total inventory, but not any breakdown by machine. We just take the first row in the inventory matrix and multiply it to the $DB_{9,15}$ matrix. Rather than write out a new row matrix for the inventory, we will compute it from the $INV_{9,9}$ matrix. R1 takes only the the values in row 1 (total inventory) and R8 takes only the values in row 7, ignoring all other machines.

$$R1 := \begin{bmatrix}
1 & 0 & 0 & 0 & 0 & 0 & 0 & 0 & 0 \\
0 & 0 & 0 & 0 & 0 & 0 & 0 & 0 & 0 \\
0 & 0 & 0 & 0 & 0 & 0 & 0 & 0 & 0 \\
0 & 0 & 0 & 0 & 0 & 0 & 0 & 0 & 0 \\
0 & 0 & 0 & 0 & 0 & 0 & 0 & 0 & 0 \\
0 & 0 & 0 & 0 & 0 & 0 & 0 & 0 & 0 \\
0 & 0 & 0 & 0 & 0 & 0 & 0 & 0 & 0 \\
0 & 0 & 0 & 0 & 0 & 0 & 0 & 0 & 0 \\
0 & 0 & 0 & 0 & 0 & 0 & 0 & 0 & 0
\end{bmatrix} \qquad R8 := \begin{bmatrix}
0 & 0 & 0 & 0 & 0 & 0 & 0 & 0 & 0 \\
0 & 0 & 0 & 0 & 0 & 0 & 0 & 0 & 0 \\
0 & 0 & 0 & 0 & 0 & 0 & 0 & 0 & 0 \\
0 & 0 & 0 & 0 & 0 & 0 & 0 & 0 & 0 \\
0 & 0 & 0 & 0 & 0 & 0 & 0 & 0 & 0 \\
0 & 0 & 0 & 0 & 0 & 0 & 0 & 0 & 0 \\
0 & 0 & 0 & 0 & 0 & 0 & 0 & 0 & 0 \\
0 & 0 & 0 & 0 & 0 & 0 & 1 & 0 & 0 \\
0 & 0 & 0 & 0 & 0 & 0 & 0 & 0 & 0
\end{bmatrix}$$

$TOTALINV := R1 \cdot INV_{9,9} \cdot DB_{9,15}$

ONE9 $:=$ (1 1 1 1 1 1 1 1 1)

TOTALINV1 $:=$ ONE9·TOTALINV

TOTALINV1 $=$ (111.8   335.4   6643.92   2229.14   5536.6   1660.98   0.861   43.81   4954.7   201.394   4.3298   4.3298   2.1649   220   66)

Suppose we want the total inventory just for machine 7, then

FOR THE INVENTORY FOR MACHINE 7:

TOTALINV $:=$ R8·INV$_{9,9}$·DB$_{9,15}$

TOTALINV7 $:=$ ONE9·TOTALINV

TOTALINV7 $=$ (0   0   0   0   0   0   0.39494   6.10142   495.27   11.68698   0   0   0   0   4.035)

Where R1 is the reduction matrix that returns only row one of INV$_{9,9}$ upon multiplication, and R8 is the reduction matrix that returns only row 8 upon multiplication.

Suppose the EPA or the company managers wish to know only the throughput of machines 5,6 and 7, since they are the major users of the chemicals. Then

$$
R567 := \begin{bmatrix}
0 & 0 & 0 & 0 & 0 & 0 & 0 & 0 & 0 \\
0 & 0 & 0 & 0 & 0 & 0 & 0 & 0 & 0 \\
0 & 0 & 0 & 0 & 0 & 0 & 0 & 0 & 0 \\
0 & 0 & 0 & 0 & 0 & 0 & 0 & 0 & 0 \\
0 & 0 & 0 & 0 & 0 & 0 & 0 & 0 & 0 \\
0 & 0 & 0 & 0 & 0 & 1 & 0 & 0 & 0 \\
0 & 0 & 0 & 0 & 0 & 0 & 1 & 0 & 0 \\
0 & 0 & 0 & 0 & 0 & 0 & 0 & 1 & 0 \\
0 & 0 & 0 & 0 & 0 & 0 & 0 & 0 & 0
\end{bmatrix}
$$

TOTALINV $:=$ R567·INV$_{9,9}$·DB$_{9,15}$

TOTALINV567 $:=$ ONE9·TOTALINV

	APE	DPE	NAP	ISOP	2-EH	o-DCB	2EE	EG	AMM	FORM	2BA	DIOX	ACRYL	DEG	ACET
TOTALINV=	0	0	0	0	0	0	0	0	0	0	0	0	0	0	
	0	0	0	0	0	0	0	0	0	0	0	0	0	0	
	0	0	0	0	0	0	0	0	0	0	0	0	0	0	
	0	0	0	0	0	0	0	0	0	0	0	0	0	0	
	0	0	0	0	0	0	0	0	0	0	0	0	0	0	
	0	0	0	7.25	0	0	0.0638	33.686	0	95.3072	2.1649	2.1649	1.08245	110	61.965
	0	0	0	7.25	0	0	0.4023	4.023	0	94.4006	2.1649	2.1649	1.08245	110	0
	0	0	0	0	0	0	0.39494	6.10142	495.27	11.68698	0	0	0	0	4.035
	0	0	0	0	0	0	0	0	0	0	0	0	0	0	

TOTALINV567= (0   0   0   14.5   0   0   0.86104   43.81042   495.27   201.39478   4.3298   4.3298   2.1649   220   66)

ACCOUNTPAGE MATRIX FOR TOTAL INVENTORY: ROW 1

$$^1[INV]^T_{9,9} \; o \; [DB]_{9,15} = \; ^1[AP]_{9,15}$$

In the Accountpage Matrix, we can keep track of all the chemicals used in each machine, or keep track of the total inventory. The solution is the un-summed form of the inventory solution matrix. This solution remains a 9x15 matrix. If summed, it is the total inventory matrix for that particular machine. The 1 in the upper left hand corner of $^1[INV]$ and $^1[AP]$ means we are working with the first row transposed in the inventory matrix. For instance, for the total inventory, $^1[AP]_{9,15} =$

	APE	DPE	NAP	ISOP	2-EH	o-DCB	2EE	EG	AMM	FORM	2BA	DIOX	ACRYL	DEG	ACET
5590	.02	.06	0	0	0	0	0	0	0	0	0	0	0	0	0
55371	0	0	.12	.04	.10	.03	0	0	0	0	0	0	0	0	0
725	0	0	0	.02	0	0	0	0	0	0	0	0	0	0	0
8630	0	0	0	0	0	0	.0001	.001	0	0	0	0	0	0	0
49547	0	0	0	0	0	0	0	0	.01	0	0	0	0	0	0
2750	0	0	0	0	0	0	0	0	0	.006	0	0	0	0	0
21649	0	0	0	0	0	0	0	0	0	.001	.0002	.0002	.0001	0	0
11000	0	0	0	0	0	0	0	0	0	.005	0	0	0	.02	0
880	0	0	0	0	0	0	0	.04	0	0	0	0	0	0	.075

But since the mathematics doesn't exist in which this multiplication can take place, we must convert it to an equivalent form to multiply. i.e. We will take the column for the total inventory and make a diagonal matrix out of it and then multiply.

$$INV_{9,9}{}^{T} = \begin{bmatrix} 5590 & 5590 & 0 & 0 & 0 & 0 & 0 & 0 & 0 \\ 55366 & 41605 & 12863 & 898.1 & 0 & 0 & 0 & 0 & 0 \\ 725 & 0 & 0 & 0 & 0 & 362.5 & 362.5 & 0 & 0 \\ 8610 & 0 & 0 & 0 & 0 & 638 & 4023 & 3949.42 & 0 \\ 49547 & 0 & 0 & 0 & 0 & 0 & 0 & 4952.7 & 44574.3 \\ 2750 & 0 & 0 & 0 & 0 & 476.7 & 325.6 & 1947.83 & 0 \\ 21649 & 0 & 0 & 0 & 0 & 10824.5 & 10824.5 & 0 & 0 \\ 11000 & 0 & 0 & 0 & 0 & 5500 & 5500 & 0 & 0 \\ 880 & 0 & 0 & 0 & 0 & 826.2 & 0 & 53.8 & 0 \end{bmatrix}$$

$$INV_{9,9}{}^{T} \cdot R1 = \begin{bmatrix} 5590 & 0 & 0 & 0 & 0 & 0 & 0 & 0 & 0 \\ 55366 & 0 & 0 & 0 & 0 & 0 & 0 & 0 & 0 \\ 725 & 0 & 0 & 0 & 0 & 0 & 0 & 0 & 0 \\ 8610 & 0 & 0 & 0 & 0 & 0 & 0 & 0 & 0 \\ 49547 & 0 & 0 & 0 & 0 & 0 & 0 & 0 & 0 \\ 2750 & 0 & 0 & 0 & 0 & 0 & 0 & 0 & 0 \\ 21649 & 0 & 0 & 0 & 0 & 0 & 0 & 0 & 0 \\ 11000 & 0 & 0 & 0 & 0 & 0 & 0 & 0 & 0 \\ 880 & 0 & 0 & 0 & 0 & 0 & 0 & 0 & 0 \end{bmatrix}$$

$NINE1 = ONE9^{T}$ Here we transpose $[1]_{1,9}$ to $[1]_{9,1}$ so we can post-multiply to sum the rows

$AP = INV_{9,9}[R1]_{9,9}[1]_{9,1}$. This multiplication returns only column 1 of the transposed inventory matrix.

$DIAGAP = diag(AP)$ Here I diagonalize the above column matrix.

$AP1 = [DIAGAP]_{9,9}[DB]_{9,15}$ The accountpage matrix for the total inventory.

$$AP1 = \begin{bmatrix} 111.8 & 335.4 & 0 & 0 & 0 & 0 & 0 & 0 & 0 & 0 & 0 & 0 & 0 & 0 & 0 \\ 0 & 0 & 6643.92 & 2214.64 & 5536.6 & 1660.98 & 0 & 0 & 0 & 0 & 0 & 0 & 0 & 0 & 0 \\ 0 & 0 & 0 & 14.5 & 0 & 0 & 0 & 0 & 0 & 0 & 0 & 0 & 0 & 0 & 0 \\ 0 & 0 & 0 & 0 & 0 & 0 & 0.861 & 8.61 & 0 & 0 & 0 & 0 & 0 & 0 & 0 \\ 0 & 0 & 0 & 0 & 0 & 0 & 0 & 0 & 4954.7 & 0 & 0 & 0 & 0 & 0 & 0 \\ 0 & 0 & 0 & 0 & 0 & 0 & 0 & 0 & 0 & 16.5 & 0 & 0 & 0 & 0 & 0 \\ 0 & 0 & 0 & 0 & 0 & 0 & 0 & 0 & 0 & 129.894 & 4.3298 & 4.3298 & 2.1649 & 0 & 0 \\ 0 & 0 & 0 & 0 & 0 & 0 & 0 & 0 & 0 & 55 & 0 & 0 & 0 & 220 & 0 \\ 0 & 0 & 0 & 0 & 0 & 0 & 0 & 35.2 & 0 & 0 & 0 & 0 & 0 & 0 & 66 \end{bmatrix}$$

INTERPRETATION: In 1995, we used a total of 111.8 lbs of alkyl phenolic ether in the chemical AP. The chemical CSS (row 2), of which the company used 55,366 lbs in 1995, used 6643.92 lbs naphthalene, 2214.64 lbs of isopropanol, 5536.6 lbs of 2-ethylhexane and 1660.98 lbs of o-dichlorobenzene. For the chemical RW-41 (row 7), the company used 129.894 lbs of formaldehyde, 4.3298 lbs 2-butoxyalcohol, 4.3298 lbs dioxane and 2.1694 lbs acrylaldehyde, etc.

Now let's look at the total accountpage matrix for machine 7:

$$AP7 := INV_{9,9}{}^{T} \cdot R8 \cdot NINE1$$

$$DIAGAP7 := diag(AP7)$$

$$AP7 = \begin{bmatrix} 0 \\ 0 \\ 0 \\ 3949.42 \\ 4952.7 \\ 1947.83 \\ 0 \\ 0 \\ 53.8 \end{bmatrix} \qquad DIAGAP7 = \begin{bmatrix} 0 & 0 & 0 & 0 & 0 & 0 & 0 & 0 & 0 \\ 0 & 0 & 0 & 0 & 0 & 0 & 0 & 0 & 0 \\ 0 & 0 & 0 & 0 & 0 & 0 & 0 & 0 & 0 \\ 0 & 0 & 0 & 3949.42 & 0 & 0 & 0 & 0 & 0 \\ 0 & 0 & 0 & 0 & 4952.7 & 0 & 0 & 0 & 0 \\ 0 & 0 & 0 & 0 & 0 & 1947.83 & 0 & 0 & 0 \\ 0 & 0 & 0 & 0 & 0 & 0 & 0 & 0 & 0 \\ 0 & 0 & 0 & 0 & 0 & 0 & 0 & 0 & 0 \\ 0 & 0 & 0 & 0 & 0 & 0 & 0 & 0 & 53.8 \end{bmatrix} \blacksquare$$

$$AP8 := DIAGAP7 \cdot DB_{9,15}$$

$$AP8 = \begin{bmatrix} 0 & 0 & 0 & 0 & 0 & 0 & 0 & 0 & 0 & 0 & 0 & 0 & 0 & 0 & 0 \\ 0 & 0 & 0 & 0 & 0 & 0 & 0 & 0 & 0 & 0 & 0 & 0 & 0 & 0 & 0 \\ 0 & 0 & 0 & 0 & 0 & 0 & 0 & 0 & 0 & 0 & 0 & 0 & 0 & 0 & 0 \\ 0 & 0 & 0 & 0 & 0 & 0 & 0.39494 & 3.94942 & 0 & 0 & 0 & 0 & 0 & 0 & 0 \\ 0 & 0 & 0 & 0 & 0 & 0 & 0 & 0 & 495.27 & 0 & 0 & 0 & 0 & 0 & 0 \\ 0 & 0 & 0 & 0 & 0 & 0 & 0 & 0 & 0 & 11.68698 & 0 & 0 & 0 & 0 & 0 \\ 0 & 0 & 0 & 0 & 0 & 0 & 0 & 0 & 0 & 0 & 0 & 0 & 0 & 0 & 0 \\ 0 & 0 & 0 & 0 & 0 & 0 & 0 & 0 & 0 & 0 & 0 & 0 & 0 & 0 & 0 \\ 0 & 0 & 0 & 0 & 0 & 0 & 0 & 2.152 & 0 & 0 & 0 & 0 & 0 & 0 & 4.035 \end{bmatrix}$$

INTERPRETATION: In the chemical FC-37, there was used .39494 lbs 2-ethoxyethylene and 3.94942 lbs ethylene glycol. For the chemical ZON, there is 2.152 lbs ethylene glycol and 4.035 lbs acetone, etc.

There will be 9 of these accountpage matrices, one for the total inventory, and 8 more, 1 for each of the eight machines.

Suppose we are Title V or Synthetic Minor and again the EPA or the state is mainly interested in machines 5, 6 and 7 because the bulk of the organic volatiles are used in these machines and expelled in their stacks which the EPA monitors. Then we have:

$$\sum_{5}^{7} [INV][DB] =$$

$$\begin{vmatrix} 0 & 0 & 362.5 & 638 & 0 & 476.7 & 10824.5 & 5500 & 826 \\ 0 & 0 & 362.5 & 4023 & 0 & 325.6 & 10824.5 & 5500 & 0 \\ 0 & 0 & 0 & 3949.42 & 4952.7 & 1947.83 & 0 & 0 & 53.8 \end{vmatrix} [DB]_{9,15}$$

Or we can transpose and hollow dot the three of them separately to obtain three 9x15 sub-matrices that account for just these three particular stacks. Note also that we can take the three accountpage solutions and add each of the separate matrices together to get a single accountpage matrix that is the total accounting of the chemical usage of machines 5, 6 and 7.

$$R6 := \begin{bmatrix} 0&0&0&0&0&0&0&0&0 \\ 0&0&0&0&0&0&0&0&0 \\ 0&0&0&0&0&0&0&0&0 \\ 0&0&0&0&0&0&0&0&0 \\ 0&0&0&0&0&0&0&0&0 \\ 0&0&0&0&0&1&0&0&0 \\ 0&0&0&0&0&0&0&0&0 \\ 0&0&0&0&0&0&0&0&0 \\ 0&0&0&0&0&0&0&0&0 \end{bmatrix} \quad R7 := \begin{bmatrix} 0&0&0&0&0&0&0&0&0 \\ 0&0&0&0&0&0&0&0&0 \\ 0&0&0&0&0&0&0&0&0 \\ 0&0&0&0&0&0&0&0&0 \\ 0&0&0&0&0&0&0&0&0 \\ 0&0&0&0&0&0&0&0&0 \\ 0&0&0&0&0&0&1&0&0 \\ 0&0&0&0&0&0&0&0&0 \\ 0&0&0&0&0&0&0&0&0 \end{bmatrix} \quad R8 := \begin{bmatrix} 0&0&0&0&0&0&0&0&0 \\ 0&0&0&0&0&0&0&0&0 \\ 0&0&0&0&0&0&0&0&0 \\ 0&0&0&0&0&0&0&0&0 \\ 0&0&0&0&0&0&0&0&0 \\ 0&0&0&0&0&0&0&0&0 \\ 0&0&0&0&0&0&0&0&0 \\ 0&0&0&0&0&0&0&1&0 \\ 0&0&0&0&0&0&0&0&0 \end{bmatrix}$$

$AP5A := INV_{9,9}^{T} \cdot R6 \cdot NINE1 \qquad DIAGAP5A := diag(AP5A) \qquad (^{6}[INV]^{T}_{9,9} \ [R6]_{9,9}[1]_{9,1}) \circ [DB]_{9,15}$

$AP6A := INV_{9,9}^{T} \cdot R7 \cdot NINE1 \qquad DIAGAP6A := diag(AP6A) \qquad (^{7}[INV]^{T}_{9,9} \ [R7]_{9,9}[1]_{9,1}) \circ [DB]_{9,15}$

$AP7A := INV_{9,9}^{T} \cdot R8 \cdot NINE1 \qquad DIAGAP7A := diag(AP7A) \qquad (^{8}[INV]^{T}_{9,9} \ [R8]_{9,9}[1]_{9,1}) \circ [DB]_{9,15}$

$AP5 := DIAGAP5A \cdot DB_{9,15}$

$AP6 := DIAGAP6A \cdot DB_{9,15}$

AP7 := DIAGAP7A·DB$_{9,15}$

AP567 := AP5 + AP6 + AP7

Results for Machine 5:

$$
AP5 =
\begin{bmatrix}
0 & 0 & 0 & 0 & 0 & 0 & 0 & 0 & 0 & 0 & 0 & 0 & 0 & 0 & 0 \\
0 & 0 & 0 & 0 & 0 & 0 & 0 & 0 & 0 & 0 & 0 & 0 & 0 & 0 & 0 \\
0 & 0 & 0 & 7.25 & 0 & 0 & 0 & 0 & 0 & 0 & 0 & 0 & 0 & 0 & 0 \\
0 & 0 & 0 & 0 & 0 & 0 & 0.0638 & 0.638 & 0 & 0 & 0 & 0 & 0 & 0 & 0 \\
0 & 0 & 0 & 0 & 0 & 0 & 0 & 0 & 0 & 0 & 0 & 0 & 0 & 0 & 0 \\
0 & 0 & 0 & 0 & 0 & 0 & 0 & 0 & 0 & 2.8602 & 0 & 0 & 0 & 0 & 0 \\
0 & 0 & 0 & 0 & 0 & 0 & 0 & 0 & 0 & 64.947 & 2.1649 & 2.1649 & 1.08245 & 0 & 0 \\
0 & 0 & 0 & 0 & 0 & 0 & 0 & 0 & 0 & 27.5 & 0 & 0 & 0 & 110 & 0 \\
0 & 0 & 0 & 0 & 0 & 0 & 0 & 33.048 & 0 & 0 & 0 & 0 & 0 & 0 & 61.965
\end{bmatrix}
$$

Results for Machine 6:

$$
AP6 =
\begin{bmatrix}
0 & 0 & 0 & 0 & 0 & 0 & 0 & 0 & 0 & 0 & 0 & 0 & 0 & 0 & 0 \\
0 & 0 & 0 & 0 & 0 & 0 & 0 & 0 & 0 & 0 & 0 & 0 & 0 & 0 & 0 \\
0 & 0 & 0 & 7.25 & 0 & 0 & 0 & 0 & 0 & 0 & 0 & 0 & 0 & 0 & 0 \\
0 & 0 & 0 & 0 & 0 & 0 & 0.4023 & 4.023 & 0 & 0 & 0 & 0 & 0 & 0 & 0 \\
0 & 0 & 0 & 0 & 0 & 0 & 0 & 0 & 0 & 0 & 0 & 0 & 0 & 0 & 0 \\
0 & 0 & 0 & 0 & 0 & 0 & 0 & 0 & 0 & 1.9536 & 0 & 0 & 0 & 0 & 0 \\
0 & 0 & 0 & 0 & 0 & 0 & 0 & 0 & 0 & 64.947 & 2.1649 & 2.1649 & 1.08245 & 0 & 0 \\
0 & 0 & 0 & 0 & 0 & 0 & 0 & 0 & 0 & 27.5 & 0 & 0 & 0 & 110 & 0 \\
0 & 0 & 0 & 0 & 0 & 0 & 0 & 0 & 0 & 0 & 0 & 0 & 0 & 0 & 0
\end{bmatrix}
$$

Results for Machine 7:

$$
AP7 =
\begin{bmatrix}
0 & 0 & 0 & 0 & 0 & 0 & 0 & 0 & 0 & 0 & 0 & 0 & 0 & 0 & 0 \\
0 & 0 & 0 & 0 & 0 & 0 & 0 & 0 & 0 & 0 & 0 & 0 & 0 & 0 & 0 \\
0 & 0 & 0 & 0 & 0 & 0 & 0 & 0 & 0 & 0 & 0 & 0 & 0 & 0 & 0 \\
0 & 0 & 0 & 0 & 0 & 0 & 0.39494 & 3.94942 & 0 & 0 & 0 & 0 & 0 & 0 & 0 \\
0 & 0 & 0 & 0 & 0 & 0 & 0 & 0 & 495.27 & 0 & 0 & 0 & 0 & 0 & 0 \\
0 & 0 & 0 & 0 & 0 & 0 & 0 & 0 & 0 & 11.68698 & 0 & 0 & 0 & 0 & 0 \\
0 & 0 & 0 & 0 & 0 & 0 & 0 & 0 & 0 & 0 & 0 & 0 & 0 & 0 & 0 \\
0 & 0 & 0 & 0 & 0 & 0 & 0 & 0 & 0 & 0 & 0 & 0 & 0 & 0 & 0 \\
0 & 0 & 0 & 0 & 0 & 0 & 0 & 2.152 & 0 & 0 & 0 & 0 & 0 & 0 & 4.035
\end{bmatrix}
$$

AP567 := AP5 + AP6 + AP7

Results for Machines 5, 6 & 7 combined.

$AP567 =$

0	0	0	0	0	0	0	0	0	0	0	0	0	0	0
0	0	0	0	0	0	0	0	0	0	0	0	0	0	0
0	0	0	14.5	0	0	0	0	0	0	0	0	0	0	0
0	0	0	0	0	0	0.86104	8.61042	0	0	0	0	0	0	0
0	0	0	0	0	0	0	0	495.27	0	0	0	0	0	0
0	0	0	0	0	0	0	0	16.50078	0	0	0	0	0	0
0	0	0	0	0	0	0	0	129.894	4.3298	4.3298	2.1649	0	0	
0	0	0	0	0	0	0	0	55	0	0	0	220	0	
0	0	0	0	0	0	0	35.2	0	0	0	0	0	0	66

Or, we can transpose $INV_{9,9}$ for 5, 6 and 7, sum the rows, diagonalize the column and multiply by DB9,15. i.e. (Note: R567 = R678 above).

$INV567 := INV_{9,9}^T \cdot R567 \cdot NINE1$ $(^{567}[INV]^T_{9,9} \ [R567]_{9,9}[1]_{9,1}) \circ [DB]_{9,15}$

$DIAGINV567 := diag(INV567)$ (MathCad will not take matrices in parenthesis, so we must do the problem in steps).

$AP567 := DIAGINV567 \, DB_{9,15}$

$AP567 =$

0	0	0	0	0	0	0	0	0	0	0	0	0	0	0
0	0	0	0	0	0	0	0	0	0	0	0	0	0	0
0	0	0	14.5	0	0	0	0	0	0	0	0	0	0	0
0	0	0	0	0	0	0.86104	8.61042	0	0	0	0	0	0	0
0	0	0	0	0	0	0	0	495.27	0	0	0	0	0	0
0	0	0	0	0	0	0	0	16.50078	0	0	0	0	0	0
0	0	0	0	0	0	0	0	129.894	4.3298	4.3298	2.1649	0	0	
0	0	0	0	0	0	0	0	55	0	0	0	220	0	
0	0	0	0	0	0	0	35.2	0	0	0	0	0	0	66

Because this may look complicated, let's look at the last problem step by step.

$$
\text{NINE1} = \begin{bmatrix} 1 \\ 1 \\ 1 \\ 1 \\ 1 \\ 1 \\ 1 \\ 1 \\ 1 \end{bmatrix}
\qquad
\text{R567} = \begin{bmatrix}
0 & 0 & 0 & 0 & 0 & 0 & 0 & 0 & 0 \\
0 & 0 & 0 & 0 & 0 & 0 & 0 & 0 & 0 \\
0 & 0 & 0 & 0 & 0 & 0 & 0 & 0 & 0 \\
0 & 0 & 0 & 0 & 0 & 0 & 0 & 0 & 0 \\
0 & 0 & 0 & 0 & 0 & 0 & 0 & 0 & 0 \\
0 & 0 & 0 & 0 & 0 & 1 & 0 & 0 & 0 \\
0 & 0 & 0 & 0 & 0 & 0 & 1 & 0 & 0 \\
0 & 0 & 0 & 0 & 0 & 0 & 0 & 1 & 0 \\
0 & 0 & 0 & 0 & 0 & 0 & 0 & 0 & 0
\end{bmatrix}
$$

$$
\text{INV}_{9,9}^{\,T} = \begin{bmatrix}
5590 & 5590 & 0 & 0 & 0 & 0 & 0 & 0 & 0 \\
55366 & 41605 & 12863 & 898.1 & 0 & 0 & 0 & 0 & 0 \\
725 & 0 & 0 & 0 & 0 & 362.5 & 362.5 & 0 & 0 \\
8610 & 0 & 0 & 0 & 0 & 638 & 4023 & 3949.42 & 0 \\
49547 & 0 & 0 & 0 & 0 & 0 & 0 & 4952.7 & 44574.3 \\
2750 & 0 & 0 & 0 & 0 & 476.7 & 325.6 & 1947.83 & 0 \\
21649 & 0 & 0 & 0 & 0 & 10824.5 & 10824.5 & 0 & 0 \\
11000 & 0 & 0 & 0 & 0 & 5500 & 5500 & 0 & 0 \\
880 & 0 & 0 & 0 & 0 & 826.2 & 0 & 53.8 & 0
\end{bmatrix}
$$

Multiplying we get:

$$
\text{INV567} := \text{INV}_{9,9}^{\,T} \cdot \text{R567} \cdot \text{NINE1}
$$

$$
\begin{bmatrix}
5590 & 5590 & 0 & 0 & 0 & 0 & 0 & 0 & 0 \\
55366 & 41605 & 12863 & 898.1 & 0 & 0 & 0 & 0 & 0 \\
725 & 0 & 0 & 0 & 0 & 362.5 & 362.5 & 0 & 0 \\
8610 & 0 & 0 & 0 & 0 & 638 & 4023 & 3949.42 & 0 \\
49547 & 0 & 0 & 0 & 0 & 0 & 0 & 4952.7 & 44574.3 \\
2750 & 0 & 0 & 0 & 0 & 476.7 & 325.6 & 1947.83 & 0 \\
21649 & 0 & 0 & 0 & 0 & 10824.5 & 10824.5 & 0 & 0 \\
11000 & 0 & 0 & 0 & 0 & 5500 & 5500 & 0 & 0 \\
880 & 0 & 0 & 0 & 0 & 826.2 & 0 & 53.8 & 0
\end{bmatrix}
\cdot
\begin{bmatrix}
0 & 0 & 0 & 0 & 0 & 0 & 0 & 0 & 0 \\
0 & 0 & 0 & 0 & 0 & 0 & 0 & 0 & 0 \\
0 & 0 & 0 & 0 & 0 & 0 & 0 & 0 & 0 \\
0 & 0 & 0 & 0 & 0 & 0 & 0 & 0 & 0 \\
0 & 0 & 0 & 0 & 0 & 0 & 0 & 0 & 0 \\
0 & 0 & 0 & 0 & 0 & 1 & 0 & 0 & 0 \\
0 & 0 & 0 & 0 & 0 & 0 & 1 & 0 & 0 \\
0 & 0 & 0 & 0 & 0 & 0 & 0 & 1 & 0 \\
0 & 0 & 0 & 0 & 0 & 0 & 0 & 0 & 0
\end{bmatrix}
\cdot
\begin{bmatrix} 1 \\ 1 \\ 1 \\ 1 \\ 1 \\ 1 \\ 1 \\ 1 \\ 1 \end{bmatrix}
$$

Removing the column for Machine 1 we have:

$$
\text{INV567} =
\begin{bmatrix}
0 \\
0 \\
725 \\
8610.42 \\
4952.7 \\
2750.13 \\
21649 \\
11000 \\
880
\end{bmatrix}
$$

Diagonalizing this Column Matrix we get:

$$
\text{DIAGINV567} =
\begin{bmatrix}
0 & 0 & 0 & 0 & 0 & 0 & 0 & 0 & 0 \\
0 & 0 & 0 & 0 & 0 & 0 & 0 & 0 & 0 \\
0 & 0 & 725 & 0 & 0 & 0 & 0 & 0 & 0 \\
0 & 0 & 0 & 8610.42 & 0 & 0 & 0 & 0 & 0 \\
0 & 0 & 0 & 0 & 4952.7 & 0 & 0 & 0 & 0 \\
0 & 0 & 0 & 0 & 0 & 2750.13 & 0 & 0 & 0 \\
0 & 0 & 0 & 0 & 0 & 0 & 21649 & 0 & 0 \\
0 & 0 & 0 & 0 & 0 & 0 & 0 & 11000 & 0 \\
0 & 0 & 0 & 0 & 0 & 0 & 0 & 0 & 880
\end{bmatrix}
$$

And now we multiply (or Half-Multiply if computer could handle the math).

$$\text{AP567} = \text{DIAGINV567DB}_{9,15}$$

$$
\begin{bmatrix}
0 & 0 & 0 & 0 & 0 & 0 & 0 & 0 & 0 \\
0 & 0 & 0 & 0 & 0 & 0 & 0 & 0 & 0 \\
0 & 0 & 725 & 0 & 0 & 0 & 0 & 0 & 0 \\
0 & 0 & 0 & 8610.42 & 0 & 0 & 0 & 0 & 0 \\
0 & 0 & 0 & 0 & 4952.7 & 0 & 0 & 0 & 0 \\
0 & 0 & 0 & 0 & 0 & 2750.13 & 0 & 0 & 0 \\
0 & 0 & 0 & 0 & 0 & 0 & 21649 & 0 & 0 \\
0 & 0 & 0 & 0 & 0 & 0 & 0 & 11000 & 0 \\
0 & 0 & 0 & 0 & 0 & 0 & 0 & 0 & 880
\end{bmatrix}
\begin{bmatrix}
0.02 & 0.06 & 0 & 0 & 0 & 0 & 0 & 0 & 0 & 0 & 0 & 0 & 0 & 0 \\
0 & 0 & 0.12 & 0.04 & 0.1 & 0.03 & 0 & 0 & 0 & 0 & 0 & 0 & 0 & 0 \\
0 & 0 & 0 & 0.02 & 0 & 0 & 0 & 0 & 0 & 0 & 0 & 0 & 0 & 0 \\
0 & 0 & 0 & 0 & 0 & 0 & 0.001 & 0 & 0 & 0 & 0 & 0 & 0 & 0 \\
0 & 0 & 0 & 0 & 0 & 0 & 0 & 0 & 0.1 & 0 & 0 & 0 & 0 & 0 \\
0 & 0 & 0 & 0 & 0 & 0 & 0 & 0 & 0 & 0.006 & 0 & 0 & 0 & 0 \\
0 & 0 & 0 & 0 & 0 & 0 & 0 & 0 & 0 & 0.001 & 0 & 0 & 0 & 0 \\
0 & 0 & 0 & 0 & 0 & 0 & 0 & 0 & 0 & 0.005 & 0 & 0 & 0.02 & 0 \\
0 & 0 & 0 & 0 & 0 & 0 & 0 & 0.04 & 0 & 0 & 0 & 0 & 0 & 0.075
\end{bmatrix}
$$

Let's look at this more generally. In this example, I included the total inventory as row one in the inventory matrix. Then I broke this down to chemical usage per machine as an individual row each. Including the whole matrix except row 1, summing the columns equals the value of row 1. Therefore, we don't even need row 1 for our computations. I just put it in for illustration and convenience. If we don't include the total inventory in the inventory matrix, just the machine breakdowns, we get:

$$^1[AP]_{9,15} = \Sigma \; [AP]_{9,15} = \; ^2\Sigma \; [AP]_{9,15} \; +^3\Sigma \; [AP]_{9,15}+ \; \cdot \; \cdot \; \cdot \; ^9\Sigma \; [AP]_{9,15}$$

Or if only the fractional parts are included but not the total inventory:

$$^T[AP]_{9,15}= \sum_{1}^{9} [AP]_{9,15}$$

Let's do a few computations by hand and see if they check.

We had 5590 lbs AP total, of which 2% was alkylphenolic ether. So 5590 x .020 = 111.8 lbs APE. In machine 7, we used 3949.42 lbs FC-37 of which .01% is 2-ethoxyethylene. This computes as 3949.42 x .0001 = .3949 lbs 2-EE. Also, .1% of FC-37 is composed of ethylene glycol, so we have 3949.42 x .001 = 3.94942 lbs EG. To check on the total inventory, let's do the formaldehyde total, since there is more in this column than any other. Of 2750 lbs NW-3A, .6% is formaldehyde: 2750 x .006 = 16.5 lbs. Of 21,649 lbs RW-41, .6% is also formaldehyde, so we have 21,649 x .006 = 129.894bs formaldehyde. The chemical mix called TGC contains .5% formaldehyde, so we have 11,000 x .005 = 55 lbs formaldehyde. The total formaldehyde is then 16.5+129.849+55 = 201.394 lbs formaldehyde used in 1995, which matches the matrix calculation.

THE ITEMPAGE MATRIX

$$^c[DB]_{9,15} \circ [INV]^T_{9,9}$$

By all accounts, the solution to the itempage matrix is impossible, for we multiply the inventory matrix by the database matrix, multiplying from right to left, rather than left to right. (See proof in the theory section at the beginning of this paper.) What I'm doing is taking the individual chemical (item) and hollow dotting it into the transposed inventory matrix, receiving a sub-matrix that documents **only that chemicals** distribution through the inventory It's main power is realized when we hollow dot multiply by one column in the database at a time. The lowercase c to the upper left of $^c[DB]$ stands for the number of the column we are multiplying by. There will be $c = j = 15$ itempage matrices if we wish to compute them all. Again, I will work with formaldehyde, since there are 3 entries in it's column. In this case, $c = 10$ since formaldehyde is listed in the 10th column. We remove this column from $[DB]_{9,15}$ and hollow dot multiply into the $[INV]^T_{9,9}$ matrix. i.e. Note also when we are done, that the solution is defined in four variables, rather than three. That is, in **machine 5**, the **chemical** RW-41 contains a total of 129.894 **lbs** of **formaldehyde.** We pick up an extra descriptive variable using this reverse multiplication.

$$
\begin{bmatrix}
0 \\ 0 \\ 0 \\ 0 \\ 0 \\ .006 \\ .006 \\ .005 \\ 0
\end{bmatrix}
\circ
\begin{bmatrix}
5590 & 5590 & 0 & 0 & 0 & 0 & 0 & 0 & 0 \\
55366 & 41605 & 12863 & 898.1 & 0 & 0 & 0 & 0 & 0 \\
725 & 0 & 0 & 0 & 0 & 362.5 & 362.5 & 0 & 0 \\
8610 & 0 & 0 & 0 & 0 & 638 & 4023 & 3949.42 & 0 \\
49547 & 0 & 0 & 0 & 0 & 0 & 0 & 4952.7 & 44574.3 \\
2750 & 0 & 0 & 0 & 0 & 476.7 & 325.6 & 1947.83 & 0 \\
21649 & 0 & 0 & 0 & 0 & 10824.5 & 10824.5 & 0 & 0 \\
11000 & 0 & 0 & 0 & 0 & 5500 & 5500 & 0 & 0 \\
880 & 0 & 0 & 0 & 0 & 826.2 & 0 & 53.8 & 0
\end{bmatrix}
=
$$

But a computer cannot multiply across, so we must diagonalize the column matrix first.

$$INV_{9,9}{}^T = \begin{bmatrix} 5590 & 5590 & 0 & 0 & 0 & 0 & 0 & 0 & 0 \\ 55366 & 41605 & 12863 & 898.1 & 0 & 0 & 0 & 0 & 0 \\ 725 & 0 & 0 & 0 & 0 & 362.5 & 362.5 & 0 & 0 \\ 8610 & 0 & 0 & 0 & 0 & 638 & 4023 & 3949.42 & 0 \\ 49547 & 0 & 0 & 0 & 0 & 0 & 0 & 4952.7 & 44574.3 \\ 2750 & 0 & 0 & 0 & 0 & 476.7 & 325.6 & 1947.83 & 0 \\ 21649 & 0 & 0 & 0 & 0 & 10824.5 & 10824.5 & 0 & 0 \\ 11000 & 0 & 0 & 0 & 0 & 5500 & 5500 & 0 & 0 \\ 880 & 0 & 0 & 0 & 0 & 826.2 & 0 & 53.8 & 0 \end{bmatrix} \quad \text{and} \quad {}^{10}[DB]_{9,15} = \begin{bmatrix} 0 \\ 0 \\ 0 \\ 0 \\ 0 \\ .006 \\ .006 \\ .005 \\ 0 \end{bmatrix}$$

$${}^{10}[DB]_{9,15} = \text{FORM}$$

Here we will diagonalize the Formaldehyde Matrix.

$$\text{DIAGFORM} = \text{diag(FORM)}$$

And now we multiply them together.

$$IP_{10} = \text{DIAGFORM} \cdot INV_{9,9}{}^T$$

=

$$\begin{bmatrix} 0 & 0 & 0 & 0 & 0 & 0 & 0 & 0 & 0 \\ 0 & 0 & 0 & 0 & 0 & 0 & 0 & 0 & 0 \\ 0 & 0 & 0 & 0 & 0 & 0 & 0 & 0 & 0 \\ 0 & 0 & 0 & 0 & 0 & 0 & 0 & 0 & 0 \\ 0 & 0 & 0 & 0 & 0 & 0 & 0 & 0 & 0 \\ 0 & 0 & 0 & 0 & 0 & 0.006 & 0 & 0 & 0 \\ 0 & 0 & 0 & 0 & 0 & 0 & 0.006 & 0 & 0 \\ 0 & 0 & 0 & 0 & 0 & 0 & 0 & 0.005 & 0 \\ 0 & 0 & 0 & 0 & 0 & 0 & 0 & 0 & 0 \end{bmatrix} \cdot \begin{bmatrix} 5590 & 5590 & 0 & 0 & 0 & 0 & 0 & 0 & 0 \\ 55366 & 41605 & 12863 & 898.1 & 0 & 0 & 0 & 0 & 0 \\ 725 & 0 & 0 & 0 & 0 & 362.5 & 362.5 & 0 & 0 \\ 8610 & 0 & 0 & 0 & 0 & 638 & 4023 & 3949.42 & 0 \\ 49547 & 0 & 0 & 0 & 0 & 0 & 0 & 4952.7 & 44574.3 \\ 2750 & 0 & 0 & 0 & 0 & 476.7 & 325.6 & 1947.83 & 0 \\ 21649 & 0 & 0 & 0 & 0 & 10824.5 & 10824.5 & 0 & 0 \\ 11000 & 0 & 0 & 0 & 0 & 5500 & 5500 & 0 & 0 \\ 880 & 0 & 0 & 0 & 0 & 826.2 & 0 & 53.8 & 0 \end{bmatrix}$$

	Tot	#1	#2	#3	#4	#5	#6	#7	#8
	0	0	0	0	0	0	0	0	0
	0	0	0	0	0	0	0	0	0
	0	0	0	0	0	0	0	0	0
	0	0	0	0	0	0	0	0	0
$IP_{10} =$	0	0	0	0	0	0	0	0	0
	16.5	0	0	0	0	2.8602	1.9536	11.68698	0
	129.894	0	0	0	0	64.947	64.947	0	0
	55	0	0	0	0	27.5	27.5	0	0
	0	0	0	0	0	0	0	0	0

INTERPRETATION: Looking at the Itempage matrix for formaldehyde, we can instantly see that three chemicals contain formaldehyde and their use is distributed among machines 5, 6 and 7 in the poundage's listed. I really wish I could place the chemical names by the matrix, but MathCad won't let me and I do not wish to type out the matrix again, so we just have to remember that in order, the chemicals are:

AP

CSS

DGP

FC-37

KFR-18

NW-3A

RW-41

TGC

ZON

Let me theorize here for a moment. I don't really think that putting the ^{c}[DB] in front of the [INV]^T matrix is mathematically correct. I think it's fundamentally simpler than this. I mean it works, but if you transpose the [DB] matrix and multiply by [INV]^T there is no solution (the indices do not match), but according to the proof of the Half-Multiplier, if you can hollow dot multiply the matrices and then sum the columns, you can also transpose and multiply the regular way and get the same answer. i.e. [DB]^T[INV]^T=[SOL]^T; but [INV]^T[DB]^T≠[SOL]^T.

What this seems to suggest is that the hollow dot operation is transpose commutative. That is [SOL]^T = [IP].

$$DB_{9,15}{}^T \cdot INV_{9,9}{}^T =$$

111.8	111.8	0	0	0	0	0	0	0
335.4	335.4	0	0	0	0	0	0	0
6643.92	4992.6	1543.56	107.772	0	0	0	0	0
2229.14	1664.9	514.52	35.924	0	7.25	7.25	0	0
5536.6	4160.5	1286.3	89.81	0	0	0	0	0
1660.98	1248.15	385.89	26.943	0	0	0	0	0
0.861	0	0	0	0	0.0638	0.4023	0.39494	0
43.81	0	0	0	0	33.686	4.023	6.10142	0
4954.7	0	0	0	0	0	0	495.27	4457.43
201.394	0	0	0	0	95.3072	94.4006	11.68698	0
4.3298	0	0	0	0	2.1649	2.1649	0	0
4.3298	0	0	0	0	2.1649	2.1649	0	0
2.1649	0	0	0	0	1.08245	1.08245	0	0
220	0	0	0	0	110	110	0	0
66	0	0	0	0	61.965	0	4.035	0

Or

$$\left(\mathrm{INV}_{9,9} \cdot \mathrm{DB}_{9,15}\right)^T =$$

111.8	111.8	0	0	0	0	0	0	0
335.4	335.4	0	0	0	0	0	0	0
6643.92	4992.6	1543.56	107.772	0	0	0	0	0
2229.14	1664.2	514.52	35.924	0	7.25	7.25	0	0
5536.6	4160.5	1286.3	89.81	0	0	0	0	0
1660.98	1248.15	385.89	26.943	0	0	0	0	0
0.861	0	0	0	0	0.0638	0.4023	0.39494	0
43.81	0	0	0	0	33.686	4.023	6.10142	0
4954.7	0	0	0	0	0	0	495.27	4457.43
201.394	0	0	0	0	95.3072	94.4006	11.68698	0
4.3298	0	0	0	0	2.1649	2.1649	0	0
4.3298	0	0	0	0	2.1649	2.1649	0	0
2.1649	0	0	0	0	1.08245	1.08245	0	0
220	0	0	0	0	110	110	0	0
66	0	0	0	0	61.965	0	4.035	0

Let's see if this makes sense. Let's transpose the [DB] matrix, sum the new rows, and o multiply across the [INV]^T matrix and see if we get the same answer.

$$\mathrm{ONE15} = (\mathrm{FIFTEEN1})^T$$

$$\mathrm{FIFTEEN1} := \begin{bmatrix} 1 \\ 1 \\ 1 \\ 1 \\ 1 \\ 1 \\ 1 \\ 1 \\ 1 \\ 1 \\ 1 \\ 1 \\ 1 \\ 1 \\ 1 \end{bmatrix} \qquad \mathrm{DB}_{9,15} \cdot \mathrm{FIFTEEN1} = \begin{bmatrix} 0.08 \\ 0.29 \\ 0.02 \\ 0.0011 \\ 0.1 \\ 0.006 \\ 0.0065 \\ 0.025 \\ 0.115 \end{bmatrix}$$

$$\mathrm{SUMROWDB} := \mathrm{DB}_{9,15} \cdot \mathrm{FIFTEEN1} \qquad \text{Here I sum the rows of the database matrix}$$

DIAGSUMROWDB $=$ diag(SUMROWDB)   Here I diagonalize the above solution

IPtotal $=$ DIAGSUMROWDB·INV$_{9,9}{}^{T}$ This is the total itempage solution

$$
\text{IPtotal} =
\begin{bmatrix}
447.2 & 447.2 & 0 & 0 & 0 & 0 & 0 & 0 & 0 \\
16056.14 & 12065.45 & 3730.27 & 260.449 & 0 & 0 & 0 & 0 & 0 \\
14.5 & 0 & 0 & 0 & 0 & 7.25 & 7.25 & 0 & 0 \\
9.471 & 0 & 0 & 0 & 0 & 0.7018 & 4.4253 & 4.34436 & 0 \\
4954.7 & 0 & 0 & 0 & 0 & 0 & 0 & 495.27 & 4457.43 \\
16.5 & 0 & 0 & 0 & 0 & 2.8602 & 1.9536 & 11.68698 & 0 \\
140.7185 & 0 & 0 & 0 & 0 & 70.35925 & 70.35925 & 0 & 0 \\
275 & 0 & 0 & 0 & 0 & 137.5 & 137.5 & 0 & 0 \\
101.2 & 0 & 0 & 0 & 0 & 95.013 & 0 & 6.187 & 0
\end{bmatrix}
$$

INTERPRETATION: The chemical CSS (second row) has a total of 16,056.14 lbs of all organic's, of which 12,065.45 lbs were used in machine 1, 3730.27 lbs were used in machine 2 and 260.449 lbs were used in machine 3. In the first row, the chemical AP contains a total of 447.2 lbs organic's used in machine 1 only. Since this is a simple computation, let's check and see if it is right. The chemical AP contains two organic's, alkylphenolic ether (2%) and diphenyl ether (6%). We used a total of 5590 lbs in 1995, so we have 5590x.02 + 5590x.06 = 111.8 + 335.4 = 447.2 lbs APE and DPE combined.

If we compute all the itempage matrices and add these sub-matrices together, we will come up with the above matrix. The itempage matrix does not keep track of the individual components, we must itempage for each separate chemical to find this out. It just gives us the total of volatiles used. Note also, that the post-multiplier matrix operates on the pre-multiplier matrix, giving a solution of chemicals per machine rather than chemical fraction per bulk chemical.

We can also take $\text{Log}[\text{IP}_{TOT}]_{9,9} - \text{Log}[\text{INV}]_{9,9}$, then take the anti-log which will give us the total % volatiles per chemical per machine.

We can also multiply the $[\text{AP}_{TOT}]_{jk}$ matrix by $[1]_{1,j}$ to sum the columns. i.e. $[1]_{1,j} = [1]_{1,9}$; $[1]_{1,9}$ $[\text{AP}_{TOT}]_{9,15} = [111.8 \ 335.4$ $6642 \ 2229.3 \ 5537 \ 1661 \ .863 \ 43.83 \ 4955 \ 84.5 \ 4.33 \ 4.33 \ 2.16 \ 220$ $66] = [\text{SOL}]_{1,15}$.

Which, looking at the total inventory (row 1) in the
$[SOL]_{9,15}$ matrix are equal. Multiplying $[SOL]_{1,15}$ by $[1]_{15,1}$ we get
$[SOL]_{1,15}[1]_{15,1} = [LBS]_{1,1}$ (a single number) which gives the total
poundage for volatile organic's and/or criteria pollutants in
the inventory. i.e.

$$[SOL]_{1,15}[1]_{15,1} = 21,897.513 \text{ lbs total in the inventory.}$$

NOTE: I had the Statistics completed for the following section,
but my computer was stolen And I just decided to leave well
enough alone and not redeo the Statistical analysis.

MULTIPLE CHEMICAL USAGE INVENTORY

Here I'm going to make simple computations comparing inventories over two years, comparing chemical usage in 1995 and 1996. This could also be for day after day comparisons, weekly, monthly, semi-annually, etc.

CHEMICALS USED (IN POUNDS) →

$$
INV1995 = \begin{bmatrix}
5590 & 55371 & 725 & 8630 & 49547 & 2750 & 21649 & 11000 & 880 \\
5590 & 41605 & 0 & 0 & 0 & 0 & 0 & 0 & 0 \\
0 & 12863 & 0 & 0 & 0 & 0 & 0 & 0 & 0 \\
0 & 898.1 & 0 & 0 & 0 & 0 & 0 & 0 & 0 \\
0 & 0 & 0 & 0 & 0 & 0 & 0 & 0 & 0 \\
0 & 0 & 362.5 & 638 & 0 & 476.7 & 10824.5 & 5500 & 826.2 \\
0 & 0 & 362.5 & 4023 & 0 & 325.6 & 10824.5 & 5500 & 0 \\
0 & 0 & 0 & 3949.42 & 4952.7 & 1947.83 & 0 & 0 & 53.8 \\
0 & 0 & 0 & 0 & 44574.3 & 0 & 0 & 0 & 0
\end{bmatrix}
$$

(column headers: AP, CSS, DGP, FC-37, KFR-18, NW-3A, RW-41, TGC, ZON)

$$
INV1996 = \begin{bmatrix}
5600 & 56481 & 800 & 8530 & 50547 & 2780 & 22750 & 12000 & 980 \\
5600 & 42439 & 0 & 0 & 0 & 0 & 0 & 0 & 0 \\
0 & 13093.3 & 0 & 0 & 0 & 0 & 0 & 0 & 0 \\
0 & 916.1 & 0 & 0 & 0 & 0 & 0 & 0 & 0 \\
0 & 0 & 0 & 0 & 0 & 0 & 0 & 0 & 0 \\
0 & 0 & 400 & 630.6 & 0 & 481.9 & 11375 & 6000 & 919.9 \\
0 & 0 & 400 & 3976.4 & 0 & 329.2 & 11375 & 6000 & 0 \\
0 & 0 & 0 & 3903.66 & 5054.7 & 1969.2 & 0 & 0 & 60.14 \\
0 & 0 & 0 & 0 & 44592.3 & 0 & 0 & 0 & 0
\end{bmatrix}
$$

MathCad cannot create a matrix larger than 100 elements at a time. To create larger matrices, we must augment them (join two matrices side by side) or stack them, (join two matrices on top of each other). Examples are as follows.

$$\text{stack(INV1995 INV1996)} = \begin{bmatrix} 5590 & 55371 & 725 & 8630 & 49547 & 2750 & 21649 & 11000 & 880 \\ 5590 & 41605 & 0 & 0 & 0 & 0 & 0 & 0 & 0 \\ 0 & 12863 & 0 & 0 & 0 & 0 & 0 & 0 & 0 \\ 0 & 898.1 & 0 & 0 & 0 & 0 & 0 & 0 & 0 \\ 0 & 0 & 0 & 0 & 0 & 0 & 0 & 0 & 0 \\ 0 & 0 & 362.5 & 638 & 0 & 476.7 & 10824.5 & 5500 & 826.2 \\ 0 & 0 & 362.5 & 4023 & 0 & 325.6 & 10824.5 & 5500 & 0 \\ 0 & 0 & 0 & 3949.42 & 4952.7 & 1947.83 & 0 & 0 & 53.8 \\ 0 & 0 & 0 & 0 & 44574.3 & 0 & 0 & 0 & 0 \\ 5600 & 56481 & 800 & 8530 & 50547 & 2780 & 22750 & 12000 & 980 \\ 5600 & 42439 & 0 & 0 & 0 & 0 & 0 & 0 & 0 \\ 0 & 13093.3 & 0 & 0 & 0 & 0 & 0 & 0 & 0 \\ 0 & 916.1 & 0 & 0 & 0 & 0 & 0 & 0 & 0 \\ 0 & 0 & 0 & 0 & 0 & 0 & 0 & 0 & 0 \\ 0 & 0 & 400 & 630.6 & 0 & 481.9 & 11375 & 6000 & 919.9 \\ 0 & 0 & 400 & 3976.4 & 0 & 329.2 & 11375 & 6000 & 0 \\ 0 & 0 & 0 & 3903.66 & 5054.7 & 1969.2 & 0 & 0 & 60.14 \\ 0 & 0 & 0 & 0 & 44592.3 & 0 & 0 & 0 & 0 \end{bmatrix}$$

$$DB1 := \begin{bmatrix} .02 & .06 & 0 & 0 & 0 & 0 & 0 & 0 & 0 \\ 0 & 0 & .12 & .04 & .10 & .03 & 0 & 0 & 0 \\ 0 & 0 & 0 & .02 & 0 & 0 & 0 & 0 & 0 \\ 0 & 0 & 0 & 0 & 0 & 0 & .0001 & .001 & 0 \\ 0 & 0 & 0 & 0 & 0 & 0 & 0 & 0 & .10 \\ 0 & 0 & 0 & 0 & 0 & 0 & 0 & 0 & 0 \\ 0 & 0 & 0 & 0 & 0 & 0 & 0 & 0 & 0 \\ 0 & 0 & 0 & 0 & 0 & 0 & 0 & 0 & 0 \\ 0 & 0 & 0 & 0 & 0 & 0 & 0 & .04 & 0 \end{bmatrix} \quad DB2 := \begin{bmatrix} 0 & 0 & 0 & 0 & 0 & 0 \\ 0 & 0 & 0 & 0 & 0 & 0 \\ 0 & 0 & 0 & 0 & 0 & 0 \\ 0 & 0 & 0 & 0 & 0 & 0 \\ 0 & 0 & 0 & 0 & 0 & 0 \\ .006 & 0 & 0 & 0 & 0 & 0 \\ .0006 & .0002 & .0002 & .0001 & 0 & 0 \\ .005 & 0 & 0 & 0 & .02 & 0 \\ 0 & 0 & 0 & 0 & 0 & .075 \end{bmatrix}$$

augment(DB1, DB2) =

APE	DPE	NAPTH	ISOP	2-EH	oDCB	2EtOH	EtGly	AMM	FORM	2-BtOH	DIOX	ACRYL	DEG	ACET
0.02	0.06	0	0	0	0	0	0	0	0	0	0	0	0	0
0	0	0.12	0.04	0.1	0.03	0	0	0	0	0	0	0	0	0
0	0	0	0.02	0	0	0	0	0	0	0	0	0	0	0
0	0	0	0	0	0	0.0001	0.001	0	0	0	0	0	0	0
0	0	0	0	0	0	0	0	0.1	0	0	0	0	0	0
0	0	0	0	0	0	0	0	0	0.006	0	0	0	0	0
0	0	0	0	0	0	0	0	0	0.0006	0.0002	0.0002	0.0001	0	0
0	0	0	0	0	0	0	0	0	0.005	0	0	0	0.02	0
0	0	0	0	0	0	0	0.04	0	0	0	0	0	0	0.075

$$INV := \text{stack(INV1995 INV1996)}$$

$$DB := augment(DB1, DB2)$$

$$SOL := INV \cdot DB$$

$$
\text{INVDB} =
\begin{bmatrix}
111.8 & 335.4 & 6644.52 & 2229.34 & 5537.1 & 1661.13 & 0.863 & 43.83 & 4954.7 & 84.489 & 4.33 & 4.33 & 2.165 & 220 & 66 \\
111.8 & 335.4 & 4992.6 & 1664.2 & 4160.5 & 1248.15 & 0 & 0 & 0 & 0 & 0 & 0 & 0 & 0 & 0 \\
0 & 0 & 1543.56 & 514.52 & 1286.3 & 385.89 & 0 & 0 & 0 & 0 & 0 & 0 & 0 & 0 & 0 \\
0 & 0 & 107.772 & 35.924 & 89.81 & 26.943 & 0 & 0 & 0 & 0 & 0 & 0 & 0 & 0 & 0 \\
0 & 0 & 0 & 0 & 0 & 0 & 0 & 0 & 0 & 0 & 0 & 0 & 0 & 0 & 0 \\
0 & 0 & 0 & 7.25 & 0 & 0 & 0.064 & 33.686 & 0 & 36.855 & 2.165 & 2.165 & 1.082 & 110 & 61.965 \\
0 & 0 & 0 & 7.25 & 0 & 0 & 0.402 & 4.023 & 0 & 35.948 & 2.165 & 2.165 & 1.082 & 110 & 0 \\
0 & 0 & 0 & 0 & 0 & 0 & 0.395 & 6.101 & 495.27 & 11.687 & 0 & 0 & 0 & 0 & 4.035 \\
0 & 0 & 0 & 0 & 0 & 0 & 0 & 0 & 4457.43 & 0 & 0 & 0 & 0 & 0 & 0 \\
112 & 336 & 6777.72 & 2275.24 & 5648.1 & 1694.43 & 0.853 & 47.73 & 5054.7 & 90.33 & 4.55 & 4.55 & 2.275 & 240 & 73.5 \\
112 & 336 & 5092.68 & 1697.56 & 4243.9 & 1273.17 & 0 & 0 & 0 & 0 & 0 & 0 & 0 & 0 & 0 \\
0 & 0 & 1571.196 & 523.732 & 1309.33 & 392.799 & 0 & 0 & 0 & 0 & 0 & 0 & 0 & 0 & 0 \\
0 & 0 & 109.932 & 36.644 & 91.61 & 27.483 & 0 & 0 & 0 & 0 & 0 & 0 & 0 & 0 & 0 \\
0 & 0 & 0 & 0 & 0 & 0 & 0 & 0 & 0 & 0 & 0 & 0 & 0 & 0 & 0 \\
0 & 0 & 0 & 8 & 0 & 0 & 0.063 & 37.427 & 0 & 39.716 & 2.275 & 2.275 & 1.138 & 120 & 68.992 \\
0 & 0 & 0 & 8 & 0 & 0 & 0.398 & 3.976 & 0 & 38.8 & 2.275 & 2.275 & 1.138 & 120 & 0 \\
0 & 0 & 0 & 0 & 0 & 0 & 0.39 & 6.309 & 505.47 & 11.815 & 0 & 0 & 0 & 0 & 4.51 \\
0 & 0 & 0 & 0 & 0 & 0 & 0 & 0 & 4459.23 & 0 & 0 & 0 & 0 & 0 & 0
\end{bmatrix}
$$

INTERPRETATION: In 1995, there was a total of 111.8 lbs alkylphenolic ether used, 111.8 lbs in machine #1. There was a total of 6644.52 lbs naphthalene used, 4992.6 lbs in machine #1, 1543.56 lbs in machine #2 and 107.772 lbs in machine #3. In 1996, there was 6777.72 lbs used, 5092.68 lbs in machine #1, 1571.196 lbs in machine #2 and 109.932 lbs used in machine #3. In 1996, a total of 4.55 lbs dioxane was used, 2.275 lbs in machine #5 and 2.275 lbs in machine #6 etc.

Now, to compare how much more or less of chemicals was used in 1996 than 1995, we proceed as follows

$$
\text{DB3} :=
\begin{bmatrix}
-1 & 0 & 0 & 0 & 0 & 0 & 0 & 0 & 0 \\
0 & -1 & 0 & 0 & 0 & 0 & 0 & 0 & 0 \\
0 & 0 & -1 & 0 & 0 & 0 & 0 & 0 & 0 \\
0 & 0 & 0 & -1 & 0 & 0 & 0 & 0 & 0 \\
0 & 0 & 0 & 0 & -1 & 0 & 0 & 0 & 0 \\
0 & 0 & 0 & 0 & 0 & -1 & 0 & 0 & 0 \\
0 & 0 & 0 & 0 & 0 & 0 & -1 & 0 & 0 \\
0 & 0 & 0 & 0 & 0 & 0 & 0 & -1 & 0 \\
0 & 0 & 0 & 0 & 0 & 0 & 0 & 0 & -1
\end{bmatrix}
* \quad
\text{DB4} :=
\begin{bmatrix}
1 & 0 & 0 & 0 & 0 & 0 & 0 & 0 & 0 \\
0 & 1 & 0 & 0 & 0 & 0 & 0 & 0 & 0 \\
0 & 0 & 1 & 0 & 0 & 0 & 0 & 0 & 0 \\
0 & 0 & 0 & 1 & 0 & 0 & 0 & 0 & 0 \\
0 & 0 & 0 & 0 & 1 & 0 & 0 & 0 & 0 \\
0 & 0 & 0 & 0 & 0 & 1 & 0 & 0 & 0 \\
0 & 0 & 0 & 0 & 0 & 0 & 1 & 0 & 0 \\
0 & 0 & 0 & 0 & 0 & 0 & 0 & 1 & 0 \\
0 & 0 & 0 & 0 & 0 & 0 & 0 & 0 & 1
\end{bmatrix}
*
$$

Putting these together into a single matrix we get:

$$\text{augment}(DB3, DB4) = \begin{bmatrix} -1 & 0 & 0 & 0 & 0 & 0 & 0 & 0 & 0 & 1 & 0 & 0 & 0 & 0 & 0 & 0 & 0 & 0 \\ 0 & -1 & 0 & 0 & 0 & 0 & 0 & 0 & 0 & 0 & 1 & 0 & 0 & 0 & 0 & 0 & 0 & 0 \\ 0 & 0 & -1 & 0 & 0 & 0 & 0 & 0 & 0 & 0 & 0 & 1 & 0 & 0 & 0 & 0 & 0 & 0 \\ 0 & 0 & 0 & -1 & 0 & 0 & 0 & 0 & 0 & 0 & 0 & 0 & 1 & 0 & 0 & 0 & 0 & 0 \\ 0 & 0 & 0 & 0 & -1 & 0 & 0 & 0 & 0 & 0 & 0 & 0 & 0 & 1 & 0 & 0 & 0 & 0 \\ 0 & 0 & 0 & 0 & 0 & -1 & 0 & 0 & 0 & 0 & 0 & 0 & 0 & 0 & 1 & 0 & 0 & 0 \\ 0 & 0 & 0 & 0 & 0 & 0 & -1 & 0 & 0 & 0 & 0 & 0 & 0 & 0 & 0 & 1 & 0 & 0 \\ 0 & 0 & 0 & 0 & 0 & 0 & 0 & -1 & 0 & 0 & 0 & 0 & 0 & 0 & 0 & 0 & 1 & 0 \\ 0 & 0 & 0 & 0 & 0 & 0 & 0 & 0 & -1 & 0 & 0 & 0 & 0 & 0 & 0 & 0 & 0 & 1 \end{bmatrix} ,$$

And now we can compare the usage between the years 1995 and 1996.

DB9 := augment(DB3, DB4)

DB9 SOL =

	APE	DPE	NAP	ISOP	2-EH	o-DCB	2EE	EG	AMM	FORM	2BA	DIOX	ACRYL	DEG	ACET
	0.2	0.6	133.2	45.9	111	33.3	−0.01	3.9	100	5.841	0.22	0.22	0.11	20	7.5
	0.2	0.6	100.08	33.36	83.4	25.02	0	0	0	0	0	0	0	0	0
	0	0	27.636	9.212	23.03	6.909	0	0	0	0	0	0	0	0	0
	0	0	2.16	0.72	1.8	0.54	0	0	0	0	0	0	0	0	0
	0	0	0	0	0	0	0	0	0	0	0	0	0	0	0
	0	0	0	0.75	0	0	−0.001	3.741	0	2.861	0.11	0.11	0.055	10	7.027
	0	0	0	0.75	0	0	−0.005	−0.047	0	2.852	0.11	0.11	0.055	10	0
	0	0	0	0	0	0	−0.005	0.208	10.2	0.128	0	0	0	0	0.476
	0	0	0	0	0	0	0	0	1.8	0	0	0	0	0	0

INTERPRETATION: We used .20 lbs more alkylphenolic ether in 1996 than in 1995, 133.2 lbs more naphthalene in 1996 than 1995 and -.01 lbs less of 2-ethylhexane in 1996 than in 1995.

Since we are using a stacked matrix for the above calculation, we need to use the statistical database DB9. We can also use the un-stacked Inventory matrices to calculate the differences in chemical usage from 1995 to 1996. The stacked matrix is used in the HP-48G only, but we can use the un-stacked matrices as follows: (this method gives the same answer as above, but it is computed differently).

INVTOTAL $:=$ INV1996 $-$ INV1995

$$INV1996-INV1995 = \begin{bmatrix} 10 & 1110 & 75 & -100 & 1000 & 30 & 1101 & 1000 & 100 \\ 10 & 834 & 0 & 0 & 0 & 0 & 0 & 0 & 0 \\ 0 & 230.3 & 0 & 0 & 0 & 0 & 0 & 0 & 0 \\ 0 & 18 & 0 & 0 & 0 & 0 & 0 & 0 & 0 \\ 0 & 0 & 0 & 0 & 0 & 0 & 0 & 0 & 0 \\ 0 & 0 & 37.5 & -7.4 & 0 & 5.2 & 550.5 & 500 & 93.7 \\ 0 & 0 & 37.5 & -46.6 & 0 & 3.6 & 550.5 & 500 & 0 \\ 0 & 0 & 0 & -45.76 & 102 & 21.37 & 0 & 0 & 6.34 \\ 0 & 0 & 0 & 0 & 18 & 0 & 0 & 0 & 0 \end{bmatrix}$$

Therefore:

$$INVTOTAL \cdot DB = \begin{bmatrix} 0.2 & 0.6 & 133.2 & 45.9 & 111 & 33.3 & -0.01 & 3.9 & 100 & 5.841 & 0.22 & 0.22 & 0.11 & 20 & 7.5 \\ 0.2 & 0.6 & 100.08 & 33.36 & 83.4 & 25.02 & 0 & 0 & 0 & 0 & 0 & 0 & 0 & 0 & 0 \\ 0 & 0 & 27.636 & 9.212 & 23.03 & 6.909 & 0 & 0 & 0 & 0 & 0 & 0 & 0 & 0 & 0 \\ 0 & 0 & 2.16 & 0.72 & 1.8 & 0.54 & 0 & 0 & 0 & 0 & 0 & 0 & 0 & 0 & 0 \\ 0 & 0 & 0 & 0 & 0 & 0 & 0 & 0 & 0 & 0 & 0 & 0 & 0 & 0 & 0 \\ 0 & 0 & 0 & 0.75 & 0 & 0 & -0.001 & 3.741 & 0 & 2.861 & 0.11 & 0.11 & 0.055 & 10 & 7.027 \\ 0 & 0 & 0 & 0.75 & 0 & 0 & -0.005 & -0.047 & 0 & 2.852 & 0.11 & 0.11 & 0.055 & 10 & 0 \\ 0 & 0 & 0 & 0 & 0 & 0 & -0.005 & 0.208 & 10.2 & 0.128 & 0 & 0 & 0 & 0 & 0.476 \\ 0 & 0 & 0 & 0 & 0 & 0 & 0 & 0 & 1.8 & 0 & 0 & 0 & 0 & 0 & 0 \end{bmatrix}$$

Now we will look at some Accountpage Matrices. We can only extract a column at a time, not a row, so we have to transpose the Inventory Matrix to obtain whatever row we wish to account for. For this example, we shall extract the total inventory for 1995 and show its account page, and we shall extract the column for Machine 7 and show its account page. We proceed as follows:

$$\text{INV1995}^T = \begin{bmatrix} 5590 & 5590 & 0 & 0 & 0 & 0 & 0 & 0 & 0 \\ 55371 & 41605 & 12863 & 898.1 & 0 & 0 & 0 & 0 & 0 \\ 725 & 0 & 0 & 0 & 0 & 362.5 & 362.5 & 0 & 0 \\ 8630 & 0 & 0 & 0 & 0 & 638 & 4023 & 3949.42 & 0 \\ 49547 & 0 & 0 & 0 & 0 & 0 & 0 & 4952.7 & 44574.3 \\ 2750 & 0 & 0 & 0 & 0 & 476.7 & 325.6 & 1947.83 & 0 \\ 21649 & 0 & 0 & 0 & 0 & 10824.5 & 10824.5 & 0 & 0 \\ 11000 & 0 & 0 & 0 & 0 & 5500 & 5500 & 0 & 0 \\ 880 & 0 & 0 & 0 & 0 & 826.2 & 0 & 53.8 & 0 \end{bmatrix}$$

We will extract the column, diagonalize it and multiply it to the DB Matrix.

To extract column 0:

Press Matrix palette, type in INV1995, CLICK M TRANSPOSE, CLICK M <>, TYPE IN 0, PRESS =.

$$V := \left(\text{INV1995}^T\right)^{<0>}$$

$$\left(\text{INV1995}^T\right)^{<0>} = \begin{bmatrix} 5590 \\ 55371 \\ 725 \\ 8630 \\ 49547 \\ 2750 \\ 21649 \\ 11000 \\ 880 \end{bmatrix} \qquad \text{diag}(V) = \begin{bmatrix} 5590 & 0 & 0 & 0 & 0 & 0 & 0 & 0 & 0 \\ 0 & 55371 & 0 & 0 & 0 & 0 & 0 & 0 & 0 \\ 0 & 0 & 725 & 0 & 0 & 0 & 0 & 0 & 0 \\ 0 & 0 & 0 & 8630 & 0 & 0 & 0 & 0 & 0 \\ 0 & 0 & 0 & 0 & 49547 & 0 & 0 & 0 & 0 \\ 0 & 0 & 0 & 0 & 0 & 2750 & 0 & 0 & 0 \\ 0 & 0 & 0 & 0 & 0 & 0 & 21649 & 0 & 0 \\ 0 & 0 & 0 & 0 & 0 & 0 & 0 & 11000 & 0 \\ 0 & 0 & 0 & 0 & 0 & 0 & 0 & 0 & 880 \end{bmatrix}$$

$$\text{AP1} := \text{diag}(V) \cdot \text{DB}$$

CLINTON L. HOLT

$$AP1 = \begin{bmatrix}
\end{bmatrix}$$

	APE	DPE	NAPTH	ISOP	2-ETHEX	o-DCB	2-EtOH	EtGly	AMM	FORMAL	2-BtOH	DIOX	ACRYL	DEG	ACE
	111.8	335.4	0	0	0	0	0	0	0	0	0	0	0	0	0
	0	0	6644.52	2214.84	5537.1	1661.13	0	0	0	0	0	0	0	0	0
	0	0	0	14.5	0	0	0	0	0	0	0	0	0	0	0
	0	0	0	0	0	0	0.863	8.63	0	0	0	0	0	0	0
AP1 =	0	0	0	0	0	0	0	0	4954.7	0	0	0	0	0	0
	0	0	0	0	0	0	0	0	0	16.5	0	0	0	0	0
	0	0	0	0	0	0	0	0	0	12.989	4.33	4.33	2.165	0	0
	0	0	0	0	0	0	0	0	0	55	0	0	0	220	0
	0	0	0	0	0	0	0	35.2	0	0	0	0	0	0	66

ANALYSIS: 5500 LBS OF BURCO AP CONTAINS 111.8 LBS ALKYLPHENOLIC ETHER AND 335.4 LBS DIPHENYL ETHER, 55,371 LBS SUPERSOL CSS CONTAINS 6644.52 LBS NAPTHALENE, 2214.84 LBS ISOPROPANOL, 5537.1 LBS 2-ETHYLHEXANE, AND 1661.13 LBS O-DICHLOROBENZENE, 725 LBS DGP CONTAINS 14.5 LBS ISOPROPANOL, 8630 LBS FC-37 CONTAINS .863 LBS 2-ETHOXYETHOXY ALCOHOL AND 8.63 LBS ETHYLENE GLYCOL, 49,547 LBS KFR-18 CONTAINS 4954.7 LBS AMMONIA, 2750 LBS NW-3A CONTAINS 16.5 LBS FORMALDEHYDE, 21,649 LBS RW-41 CONTAINS 12.989 LBS FORMALDEHYDE, 4.33 LBS 2-BUTYL ALCOHOL, 4.33 LBS DIOXANE, AND 2.165 LBS ACRYLALDEHYDE. 11,000 LBS TGC CONTAINS 220 LBS DIETHYLENE GLYCOL AND 880 LBS ZONYL CONTAINS 66 LBS ACETONE.

Now lets look at the account page Matrix for Machine 6 (it uses the most chemicals):

$$W := \left(INV1995^T\right)^{<6>}$$

$$diag(W) = \begin{bmatrix}
0 & 0 & 0 & 0 & 0 & 0 & 0 & 0 & 0 \\
0 & 0 & 0 & 0 & 0 & 0 & 0 & 0 & 0 \\
0 & 0 & 362.5 & 0 & 0 & 0 & 0 & 0 & 0 \\
0 & 0 & 0 & 4023 & 0 & 0 & 0 & 0 & 0 \\
0 & 0 & 0 & 0 & 0 & 0 & 0 & 0 & 0 \\
0 & 0 & 0 & 0 & 0 & 325.6 & 0 & 0 & 0 \\
0 & 0 & 0 & 0 & 0 & 0 & 10824.5 & 0 & 0 \\
0 & 0 & 0 & 0 & 0 & 0 & 0 & 5500 & 0 \\
0 & 0 & 0 & 0 & 0 & 0 & 0 & 0 & 0
\end{bmatrix} \qquad \left(INV1995^T\right)^{<6>} = \begin{bmatrix}
0 \\
0 \\
362.5 \\
4023 \\
0 \\
325.6 \\
10824.5 \\
5500 \\
0
\end{bmatrix}$$

$$AP6 := diag(W) \cdot DB$$

$$AP6 = \begin{bmatrix} 0 & 0 & 0 & 0 & 0 & 0 & 0 & 0 & 0 & 0 & 0 & 0 & 0 & 0 & 0 \\ 0 & 0 & 0 & 0 & 0 & 0 & 0 & 0 & 0 & 0 & 0 & 0 & 0 & 0 & 0 \\ 0 & 0 & 0 & 7.25 & 0 & 0 & 0 & 0 & 0 & 0 & 0 & 0 & 0 & 0 & 0 \\ 0 & 0 & 0 & 0 & 0 & 0 & 0.402 & 4.023 & 0 & 0 & 0 & 0 & 0 & 0 & 0 \\ 0 & 0 & 0 & 0 & 0 & 0 & 0 & 0 & 0 & 0 & 0 & 0 & 0 & 0 & 0 \\ 0 & 0 & 0 & 0 & 0 & 0 & 0 & 0 & 0 & 1.954 & 0 & 0 & 0 & 0 & 0 \\ 0 & 0 & 0 & 0 & 0 & 0 & 0 & 0 & 0 & 6.495 & 2.165 & 2.165 & 1.082 & 0 & 0 \\ 0 & 0 & 0 & 0 & 0 & 0 & 0 & 0 & 0 & 27.5 & 0 & 0 & 0 & 110 & 0 \\ 0 & 0 & 0 & 0 & 0 & 0 & 0 & 0 & 0 & 0 & 0 & 0 & 0 & 0 & 0 \end{bmatrix}$$

ANALYSIS: MACHINE 6 USED 362.5 LBS DGP CONTAINING 7.25 LBS ISOPROPANOL. 4023 LBS FC-37 CONTAINING .402 LBS 2-ETHOXYETHOXY ALCOHOL, 325.6 LBS NW-3A 1.954 LBS FORMALDEHYDE

10,824.5 LBS RW-41 CONTAINING 6.495 LBS FORMALDEHYDE, 2.165 LBS 2-BUTYL ALCOHOL, 2.165 LBS DIOXANE AND 1.082 LBS ACRYLALDEHYDE 5500 LBS TGC CONTAINING 27.5 LBS FORMALDEHYDE AND 110 LBS DIETHYLENE GLYCOL

AND FINALLY, FOR THIS SET OF EXAMPLES, WE WILL CALCULATE THE ITEMPAGE MATRIX FOR FORMALDEHYDE AND ISOPROPANOL.

FIRST WE MUST TRANSPOSE THE 1995 INVENTORY MATRIX, THEN FROM THE FORMALDEHYDE COLUMN ON THE DATABASE MATRIX WE MUST REMOVE THE COLUMNS REPRESENTING FORMALDEHYDE AND ISOPROPANOL. i.e.:

$$FORM := (DB)^{<9>}$$

$$IP9 := diag(FORM) \cdot INV1995^T$$

$$INV1995^T = \begin{bmatrix} 5590 & 5590 & 0 & 0 & 0 & 0 & 0 & 0 & 0 \\ 55371 & 41605 & 12863 & 898.1 & 0 & 0 & 0 & 0 & 0 \\ 725 & 0 & 0 & 0 & 0 & 362.5 & 362.5 & 0 & 0 \\ 8630 & 0 & 0 & 0 & 0 & 638 & 4023 & 3949.42 & 0 \\ 49547 & 0 & 0 & 0 & 0 & 0 & 0 & 4952.7 & 44574.3 \\ 2750 & 0 & 0 & 0 & 0 & 476.7 & 325.6 & 1947.83 & 0 \\ 21649 & 0 & 0 & 0 & 0 & 10824.5 & 10824.5 & 0 & 0 \\ 11000 & 0 & 0 & 0 & 0 & 5500 & 5500 & 0 & 0 \\ 880 & 0 & 0 & 0 & 0 & 826.2 & 0 & 53.8 & 0 \end{bmatrix}$$

MACHINES

	TOTAL	#1	#2	#3	#4	#5	#6	#7	#8
	0	0	0	0	0	0	0	0	0
	0	0	0	0	0	0	0	0	0
	0	0	0	0	0	0	0	0	0
	0	0	0	0	0	0	0	0	0
IP9 =	0	0	0	0	0	0	0	0	0
	16.5	0	0	0	0	2.86	1.954	11.687	0
	12.989	0	0	0	0	6.495	6.495	0	0
	55	0	0	0	0	27.5	27.5	0	0
	0	0	0	0	0	0	0	0	0

ANALYSIS: NW-3A CONTAINED 16.5 TOTAL LBS OF FORMALDEHYDE, OF WHICH 2.86 LBS WERE USED IN MACHINE 5, 1.954 LBS WERE USED IN MACHINE 6 AND 11.687 LBS WERE USED IN MACHINE 7. RW-41 CONTAINED A TOTAL OF 12.989 LBS FORMALDEHYDE, 6.4955 LBS WERE USED IN MACHINE 5 AND 6.495 LBS WERE USED IN MACHINE 6. TGC CONTAINED 55 LBS FORMALDEHYDE, 27.5 LBS WERE USED IN MACHINE 5 AND 27.5 LBS WERE USED IN MACHINE 6.

TO CALCULATE FORMALDEHYDE USAGE FOR BOTH 1995 AND 1996, WE PROCEED AS FOLLOWS:

$$\text{IPFORM9596} := \text{diag}(\text{FORM}) \cdot \text{INV}^T$$

					1995						1996								
	tot95	#1	#2	#3	#4	#5	#6	#7	#8	Tot96	#1	#2	#3	#4	#5	#6	#7	#8
	0	0	0	0	0	0	0	0	0	0	0	0	0	0	0	0	0	0
	0	0	0	0	0	0	0	0	0	0	0	0	0	0	0	0	0	0
	0	0	0	0	0	0	0	0	0	0	0	0	0	0	0	0	0	0
	0	0	0	0	0	0	0	0	0	0	0	0	0	0	0	0	0	0
IPFORM9596=	0	0	0	0	0	0	0	0	0	0	0	0	0	0	0	0	0	0
	16.5	0	0	0	0	2.86	1.954	11.687	0	16.68	0	0	0	0	2.891	1.975	11.815	0
	12.989	0	0	0	0	6.495	6.495	0	0	13.65	0	0	0	0	6.825	6.825	0	0
	55	0	0	0	0	27.5	27.5	0	0	60	0	0	0	0	30	30	0	0
	0	0	0	0	0	0	0	0	0	0	0	0	0	0	0	0	0	0

TO CALCULATE THE ITEMPAGE MATRIX FOR ISOPROPANOL:

$$\text{ISOP} := \text{DB}^{<3>}$$

$$\text{IP3} := \text{diag}(\text{ISOP}) \cdot \text{INV}^T$$

TOTAL95

$IP3 =$

	#1	#2	#3	#4	#5	#6	#7	#8	TOT96	#1	#2	#3	#4	#5	#6	#7	#8
0	0	0	0	0	0	0	0	0	0	0	0	0	0	0	0	0	0
2214.84	1664.2	514.52	35.924	0	0	0	0	0	2259.24	1697.56	523.732	36.644	0	0	0	0	0
14.5	0	0	0	0	7.25	7.25	0	0	16	0	0	0	0	8	8	0	0
0	0	0	0	0	0	0	0	0	0	0	0	0	0	0	0	0	0
0	0	0	0	0	0	0	0	0	0	0	0	0	0	0	0	0	0
0	0	0	0	0	0	0	0	0	0	0	0	0	0	0	0	0	0
0	0	0	0	0	0	0	0	0	0	0	0	0	0	0	0	0	0
0	0	0	0	0	0	0	0	0	0	0	0	0	0	0	0	0	0

ANALYSIS: FOR 1996, CSS CONTAINED A TOTAL OF 2259.24 LBS ISOPROPANOL, 1697.56 LBS USED IN MACHINE 1, 523.732 LBS USED IN MACHINE 2 AND 36.644 LBS USED IN MACHINE 3. DGP CONTAINED 16 LBS ISOPROPANOL, 8 LBS USED IN MACHINE 5 AND USED 8 LBS IN MACHINE6.

LABELING THE MATRICES

Lets use MathCad to manipulate the labels inside the CUI Matrix PG. 90

INVENTORY := (0 AP CSS DGP FC - 37 KFR - 18 NW - 3 A RW - 41 TGC ZON)

The spreadsheet Matrix looks like this, but we have a problem, if we multiply in this form the labels will attach to the values, so we must leave the top row except position 1,1 as zeros and position 1,1 = 1

$INV1995 :=$

0	AP	CSS	DGP	FC — 37	KFR — 18	NW — 3A	RW — 41	TGC	ZON
0	5590	55371	725	8630	49547	2750	21649	11000	880
0	5590	41605	0	0	0	0	0	0	0
0	0	12863	0	0	0	0	0	0	0
0	0	898.1	0	0	0	0	0	0	0
0	0	0	0	0	0	0	0	0	0
0	0	0	362.5	638	0	476.7	10824.5	5500	826.2
0	0	0	362.5	4023	0	325.6	10824.5	5500	0
0	0	0	0	3949.42	4952.7	1947.83	0	0	53.8
0	0	0	0	0	44574.3	0	0	0	0

The inventory matrix (spreadsheet matrix) now looks like this:

$$
\text{SPRDSHT1995} := \begin{bmatrix}
1 & 0 & 0 & 0 & 0 & 0 & 0 & 0 & 0 & 0 \\
0 & 5590 & 55371 & 725 & 8630 & 49547 & 2750 & 21649 & 11000 & 880 \\
0 & 5590 & 41605 & 0 & 0 & 0 & 0 & 0 & 0 & 0 \\
0 & 0 & 12683 & 0 & 0 & 0 & 0 & 0 & 0 & 0 \\
0 & 0 & 898.1 & 0 & 0 & 0 & 0 & 0 & 0 & 0 \\
0 & 0 & 0 & 0 & 0 & 0 & 0 & 0 & 0 & 0 \\
0 & 0 & 0 & 362.5 & 638 & 0 & 476.7 & 10824.5 & 5500 & 826.2 \\
0 & 0 & 0 & 362.5 & 4023 & 0 & 325.6 & 108244.5 & 5500 & 0 \\
0 & 0 & 0 & 0 & 3949.42 & 49952.7 & 1947.83 & 0 & 0 & 53.8 \\
0 & 0 & 0 & 0 & 0 & 44572.3 & 0 & 0 & 0 & 0
\end{bmatrix}
$$

Now we must create the database matrix, but because it is too big we must make two separate matrices and augment them into a bigger matrix.

$$
I := \begin{bmatrix}
0 & .02 & .06 & 0 & 0 & 0 & 0 & 0 & 0 & 0 \\
0 & 0 & 0 & .12 & .04 & .10 & .03 & 0 & 0 & 0 \\
0 & 0 & 0 & 0 & .02 & 0 & 0 & 0 & 0 & 0 \\
0 & 0 & 0 & 0 & 0 & 0 & 0 & .0001 & .001 & 0 \\
0 & 0 & 0 & 0 & 0 & 0 & 0 & 0 & 0 & .01 \\
0 & 0 & 0 & 0 & 0 & 0 & 0 & 0 & 0 & 0 \\
0 & 0 & 0 & 0 & 0 & 0 & 0 & 0 & 0 & 0 \\
0 & 0 & 0 & 0 & 0 & 0 & 0 & 0 & 0 & 0 \\
0 & 0 & 0 & 0 & 0 & 0 & 0 & 0 & .04 & 0
\end{bmatrix}
\qquad
J := \begin{bmatrix}
0 & 0 & 0 & 0 & 0 & 0 \\
0 & 0 & 0 & 0 & 0 & 0 \\
0 & 0 & 0 & 0 & 0 & 0 \\
0 & 0 & 0 & 0 & 0 & 0 \\
0 & 0 & 0 & 0 & 0 & 0 \\
.006 & 0 & 0 & 0 & 0 & 0 \\
.001 & .0002 & .0002 & .0001 & 0 & 0 \\
.005 & 0 & 0 & 0 & .02 & 0 \\
0 & 0 & 0 & 0 & 0 & .075
\end{bmatrix}
$$

DB1 := augment (I, J)

$$
\text{DB1} = \begin{bmatrix}
0 & 0.02 & 0.06 & 0 & 0 & 0 & 0 & 0 & 0 & 0 & 0 & 0 & 0 & 0 & 0 & 0 \\
0 & 0 & 0 & 0.12 & 0.04 & 0.1 & 0.03 & 0 & 0 & 0 & 0 & 0 & 0 & 0 & 0 & 0 \\
0 & 0 & 0 & 0 & 0.02 & 0 & 0 & 0 & 0 & 0 & 0 & 0 & 0 & 0 & 0 & 0 \\
0 & 0 & 0 & 0 & 0 & 0 & 0 & 0.0001 & 0.001 & 0 & 0 & 0 & 0 & 0 & 0 & 0 \\
0 & 0 & 0 & 0 & 0 & 0 & 0 & 0 & 0 & 0.01 & 0 & 0 & 0 & 0 & 0 & 0 \\
0 & 0 & 0 & 0 & 0 & 0 & 0 & 0 & 0 & 0 & 0.006 & 0 & 0 & 0 & 0 & 0 \\
0 & 0 & 0 & 0 & 0 & 0 & 0 & 0 & 0 & 0 & 0.001 & 0.0002 & 0.0002 & 0.0001 & 0 & 0 \\
0 & 0 & 0 & 0 & 0 & 0 & 0 & 0 & 0 & 0 & 0.005 & 0 & 0 & 0 & 0.02 & 0 \\
0 & 0 & 0 & 0 & 0 & 0 & 0 & 0 & 0.04 & 0 & 0 & 0 & 0 & 0 & 0 & 0.075
\end{bmatrix}
$$

Now we will label the individual chemicals that are in the various mixtures by using the stack function over the database matrix.

CHEMICALS := (0 APE DPE NAP ISOP 2 EH oDCB 2 EE EG AMM FORM 2 BA DIOX ACRYL DEGG ACET)

DB := stack (CHEMICALS, DB1)

DB →

	APE	DPE	NAP	ISOP	2·EH	oDCB	2·EE	EG	AMM	FORM	2·BA	DIOX	ACRYL	DEGG	ACET
0	$2{\cdot}10^{-2}$	$6{\cdot}10^{-2}$	0	0	0	0	0	0	0	0	0	0	0	0	0
0	0	0	.12	$4{\cdot}10^{-2}$	.10	$3{\cdot}10^{-2}$	0	0	0	0	0	0	0	0	0
0	0	0	0	$2{\cdot}10^{-2}$	0	0	0	0	0	0	0	0	0	0	0
0	0	0	0	0	0	0	$1{\cdot}10^{-4}$	$1{\cdot}10^{-3}$	0	0	0	0	0	0	0
0	0	0	0	0	0	0	0	0	$1{\cdot}10^{-2}$	0	0	0	0	0	0
0	0	0	0	0	0	0	0	0	0	$6{\cdot}10^{-3}$	0	0	0	0	0
0	0	0	0	0	0	0	0	0	0	$1{\cdot}10^{-3}$	$2{\cdot}10^{-4}$	$2{\cdot}10^{-4}$	$1{\cdot}10^{-4}$	0	0
0	0	0	0	0	0	0	0	0	0	$5{\cdot}10^{-3}$	0	0	0	$2{\cdot}10^{-2}$	0
0	0	0	0	0	0	0	$4{\cdot}10^{-2}$	0	0	0	0	0	0	0	$7.5{\cdot}10^{-2}$

Shrinking this matrix down so we can see it we have:

	APE	DPE	NAP	**ISOP**	**2EH**	oDCB	**2EE**	EG	AMM	FORM	**2BA**	DIOX	ACRYL	DEGG	ACET
0	0.02	0.06	0	0	0	0	0	0	0	0	0	0	0	0	0
0	0	0	0.12	0.04	0.1	0.03	0	0	0	0	0	0	0	0	0
0	0	0	0	0.02	0	0	0	0	0	0	0	0	0	0	0
0	0	0	0	0	0	0	0	0.001	0	0	0	0	0	0	0
0	0	0	0	0	0	0	0	0	0.01	0	0	0	0	0	0
0	0	0	0	0	0	0	0	0	0	0.006	0	0	0	0	0
0	0	0	0	0	0	0	0	0	0	0.001	.0002	.0002	.0001	0	0
0	0	0	0	0	0	0	0	0	0	0.005	0	0	0	0.02	0
0	0	0	0	0	0	0	0.04	0	0	0	0	0	0	0	0.075

CUI := SPRDSHT1995·DB

SPRDSHT1995·DB →

	APE	DPE	NAP	ISOP	2·EH	oDCB	2·EE	EG	AMM	FORM	2·BA	DIOX	ACRYL	DEGG	ACET
0	111.80	335.40	6644.52	2229.34	5537.10	1661.13	.8630	43.830	495.47	93.149	4.3298	4.3298	2.1649	220.00	66.000
0	111.80	335.40	4992.60	1664.20	4160.50	1248.15	0	0	0	0	0	0	0	0	0
0	0	0	1521.96	507.32	1268.30	380.49	0	0	0	0	0	0	0	0	0
0	0	0	107.772	35.924	89.810	26.943	0	0	0	0	0	0	0	0	0
0	0	0	0	0	0	0	0	0	0	0	0	0	0	0	0
0	0	0	0	7.250	0	0	$6.38{\cdot}10^{-2}$	33.686	0	41.1847	2.16490	2.16490	1.08245	110.00	61.9650
0	0	0	0	7.250	0	0	.4023	4.023	0	137.6981	21.64890	21.64890	10.82445	110.00	0
0	0	0	0	0	0	0	.394942	6.10142	499.527	11.68698	0	0	0	0	4.0350
0	0	0	0	0	0	0	0	0	445.723	0	0	0	0	0	0

CUI →

	APE	DPE	NAP	ISOP	2·EH	oDCB	2·EE	EG	AMM	FORM	2·BA	DIOX	ACRYL	DEGG	ACET
0	111.80	335.40	6644.52	2229.34	5537.10	1661.13	.8630	43.830	495.47	93.149	4.3298	4.3298	2.1649	220.00	66.000
0	111.80	335.40	4992.60	1664.20	4160.50	1248.15	0	0	0	0	0	0	0	0	0
0	0	0	1521.96	507.32	1268.30	380.49	0	0	0	0	0	0	0	0	0
0	0	0	107.772	35.924	89.810	26.943	0	0	0	0	0	0	0	0	0
0	0	0	0	0	0	0	0	0	0	0	0	0	0	0	0
0	0	0	0	7.250	0	0	$6.38{\cdot}10^{-2}$	33.686	0	41.1847	2.16490	2.16490	1.08245	110.00	61.9650
0	0	0	0	7.250	0	0	.4023	4.023	0	137.6981	21.64890	21.64890	10.82445	110.00	0
0	0	0	0	0	0	0	.394942	6.10142	499.527	11.68698	0	0	0	0	4.0350
0	0	0	0	0	0	0	0	0	445.723	0	0	0	0	0	0

Now we want to label the values, so we must make the label matrix by augmenting:

$$
A := \begin{bmatrix}
0 & 0 & 0 & 0 & 0 & 0 & 0 & 0 & 0 & 0 \\
0 & 0 & 0 & 0 & 0 & 0 & 0 & 0 & 0 & 0 \\
0 & 0 & 0 & 0 & 0 & 0 & 0 & 0 & 0 & 0 \\
0 & 0 & 0 & 0 & 0 & 0 & 0 & 0 & 0 & 0 \\
0 & 0 & 0 & 0 & 0 & 0 & 0 & 0 & 0 & 0 \\
0 & 0 & 0 & 0 & 0 & 0 & 0 & 0 & 0 & 0 \\
0 & 0 & 0 & 0 & 0 & 0 & 0 & 0 & 0 & 0 \\
0 & 0 & 0 & 0 & 0 & 0 & 0 & 0 & 0 & 0 \\
0 & 0 & 0 & 0 & 0 & 0 & 0 & 0 & 0 & 0 \\
0 & 0 & 0 & 0 & 0 & 0 & 0 & 0 & 0 & 0
\end{bmatrix}
\qquad
B := \begin{bmatrix}
0 & 0 & 0 & 0 & 0 \\
0 & 0 & 0 & 0 & 0 \\
0 & 0 & 0 & 0 & 0 \\
0 & 0 & 0 & 0 & 0 \\
0 & 0 & 0 & 0 & 0 \\
0 & 0 & 0 & 0 & 0 \\
0 & 0 & 0 & 0 & 0 \\
0 & 0 & 0 & 0 & 0 \\
0 & 0 & 0 & 0 & 0 \\
0 & 0 & 0 & 0 & 0
\end{bmatrix}
$$

LABEL1 := augment (A, B)

$$
\text{LABEL1} = \begin{bmatrix}
0 & 0 & 0 & 0 & 0 & 0 & 0 & 0 & 0 & 0 & 0 & 0 & 0 & 0 & 0 \\
0 & 0 & 0 & 0 & 0 & 0 & 0 & 0 & 0 & 0 & 0 & 0 & 0 & 0 & 0 \\
0 & 0 & 0 & 0 & 0 & 0 & 0 & 0 & 0 & 0 & 0 & 0 & 0 & 0 & 0 \\
0 & 0 & 0 & 0 & 0 & 0 & 0 & 0 & 0 & 0 & 0 & 0 & 0 & 0 & 0 \\
0 & 0 & 0 & 0 & 0 & 0 & 0 & 0 & 0 & 0 & 0 & 0 & 0 & 0 & 0 \\
0 & 0 & 0 & 0 & 0 & 0 & 0 & 0 & 0 & 0 & 0 & 0 & 0 & 0 & 0 \\
0 & 0 & 0 & 0 & 0 & 0 & 0 & 0 & 0 & 0 & 0 & 0 & 0 & 0 & 0 \\
0 & 0 & 0 & 0 & 0 & 0 & 0 & 0 & 0 & 0 & 0 & 0 & 0 & 0 & 0 \\
0 & 0 & 0 & 0 & 0 & 0 & 0 & 0 & 0 & 0 & 0 & 0 & 0 & 0 & 0 \\
0 & 0 & 0 & 0 & 0 & 0 & 0 & 0 & 0 & 0 & 0 & 0 & 0 & 0 & 0
\end{bmatrix}
\qquad
C := \begin{bmatrix}
0 \\
AP \\
CSS \\
DGP \\
FC37 \\
KFR18 \\
NW3A \\
RW41 \\
TGC \\
ZON
\end{bmatrix}
$$

And now we join the labels to the matrix:

LABEL := augment (C, LABEL1)

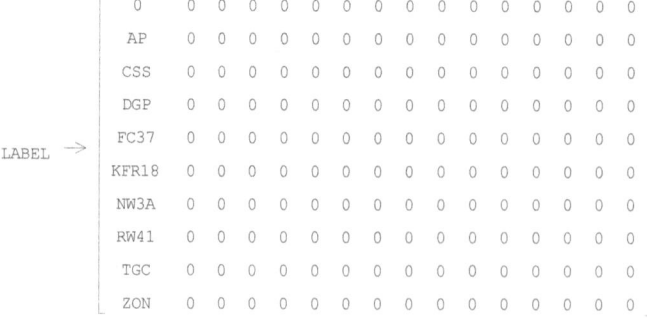

Finally we add the Chemical Usage Inventory Matrix to the Label
Matrix to get the final form we want.:

CUI + LABEL →

0	APE	DPE	NAP	ISOP	2·EH	oDCB	2·EE	EG	AMM	FORM	2·BA	DIOX	ACRYL	DEGG	ACET
AP	111.80	335.40	6644.52	2229.34	5537.10	1661.13	.8630	43.830	495.47	93.149	4.3298	4.3298	2.1649	220.00	66.000
CSS	111.80	335.40	4992.60	1664.20	4160.50	1248.15	0	0	0	0	0	0	0	0	0
DGP	0	0	1521.96	507.32	1268.30	380.49	0	0	0	0	0	0	0	0	0
FC37	0	0	107.772	35.924	89.810	26.943	0	0	0	0	0	0	0	0	0
KFR18	0	0	0	0	0	0	0	0	0	0	0	0	0	0	0
NW3A	0	0	0	7.250	0	0	$6.38 \cdot 10^{-2}$	33.686	0	41.1847	2.16490	2.16490	1.08245	110.00	61.9650
RW41	0	0	0	7.250	0	0	.4023	4.023	0	137.6981	21.64890	21.64890	10.82445	110.00	0
TGC	0	0	0	0	0	0	.394942	6.10142	499.527	11.68698	0	0	0	0	4.0350
ZON	0	0	0	0	0	0	0	0	445.723	0	0	0	0	0	0

Suppose we are a company who wishes to keep our inventory secret
in case hackers break into our computer. Also suppose we are
paranoid of any commercial encryption software out on the market
(anyone who can understand the program can possibly decode our
information). Lets encrypt the information ourselves. We must
remember to save the key on a disk and delete all info as to
the identity of the key from mathcad. (Of course, we can just
keep the CUI on disk and not in the computer if we wish too)
As you will see, if we wish to encrypt our data, we really
should do it with our unlabled computations and add the labels
in after de-encryption.

A := CUI + LABEL

A →

0	APE	DPE	NAP	ISOP	2·EH	oDCB	2·EE	EG	AMM	FORM	2·BA	DIOX	ACRYL	DEGG	ACET
AP	111.80	335.40	6644.52	2229.34	5537.10	1661.13	.8630	43.830	495.47	93.149	4.3298	4.3298	2.1649	220.00	66.000
CSS	111.80	335.40	4992.60	1664.20	4160.50	1248.15	0	0	0	0	0	0	0	0	0
DGP	0	0	1521.96	507.32	1268.30	380.49	0	0	0	0	0	0	0	0	0
FC37	0	0	107.772	35.924	89.810	26.943	0	0	0	0	0	0	0	0	0
KFR18	0	0	0	0	0	0	0	0	0	0	0	0	0	0	0
NW3A	0	0	0	7.250	0	0	$6.38 \cdot 10^{-2}$	33.686	0	41.1847	2.16490	2.16490	1.08245	110.00	61.9650
RW41	0	0	0	7.250	0	0	.4023	4.023	0	137.6981	21.64890	21.64890	10.82445	110.00	0
TGC	0	0	0	0	0	0	.394942	6.10142	499.527	11.68698	0	0	0	0	4.0350
ZON	0	0	0	0	0	0	0	0	445.723	0	0	0	0	0	0

First we will scramble just the rows, then we will scramble the
columns. This simple CUI has 10 rows and 16 columns, we must
make a 10x10 matrix to interchange the rows. Note that there
is only a single 1 in any row or column, we can scramble the
ones as we wish.

$$
CODE := \begin{bmatrix}
0 & 0 & 0 & 1 & 0 & 0 & 0 & 0 & 0 & 0 \\
0 & 1 & 0 & 0 & 0 & 0 & 0 & 0 & 0 & 0 \\
0 & 0 & 0 & 0 & 0 & 0 & 0 & 1 & 0 & 0 \\
0 & 0 & 0 & 0 & 0 & 1 & 0 & 0 & 0 & 0 \\
0 & 0 & 1 & 0 & 0 & 0 & 0 & 0 & 0 & 0 \\
1 & 0 & 0 & 0 & 0 & 0 & 0 & 0 & 0 & 0 \\
0 & 0 & 0 & 0 & 1 & 0 & 0 & 0 & 0 & 0 \\
0 & 0 & 0 & 0 & 0 & 0 & 0 & 0 & 0 & 1 \\
0 & 0 & 0 & 0 & 0 & 0 & 0 & 0 & 1 & 0 \\
0 & 0 & 0 & 0 & 0 & 0 & 1 & 0 & 0 & 0
\end{bmatrix}
$$

Now premultiplying on the data matrix we get:

CODE·A →

	0	APE	DPE	NAP	ISOP	2·EH	oDCB	2·EE	EG	AMM	FORM	2·BA	DIOX	ACRYL	DEGG	ACET
DGP	0	0	1521.96	507.32	1268.30	380.49	0	0	0	0	0	0	0	0	0	0
AP	111.80	335.40	6644.52	2229.34	5537.10	1661.13	.8630	43.830	495.47	93.149	4.3298	4.3298	2.1649	220.00	66.000	
RW41	0	0	0	7.250	0	0	.4023	4.023	0	137.6981	21.64890	21.64890	10.82445	110.00	0	
KFR18	0	0	0	0	0	0	0	0	0	0	0	0	0	0	0	0
CSS	111.80	335.40	4992.60	1664.20	4160.50	1248.15	0	0	0	0	0	0	0	0	0	0
0	APE	DPE	NAP	ISOP	2·EH	oDCB	2·EE	EG	AMM	FORM	2·BA	DIOX	ACRYL	DEGG	ACET	
FC37	0	0	107.772	35.924	89.810	26.943	0	0	0	0	0	0	0	0	0	0
ZON	0	0	0	0	0	0	0	0	445.723	0	0	0	0	0	0	0
TGC	0	0	0	0	0	0	.394942	6.10142	499.527	11.68698	0	0	0	0		4.0350
NW3A	0	0	0	7.250	0	0	$6.30 \cdot 10^{-2}$	33.686	0	41.1847	2.16490	2.16490	1.08245	110.00	61.9650	

Now we will scramble the columns, since this is a 10x16 matrix, we need a 16x16 matrix, but Mathcad can only form single matrices of 100 elements. so we must create the matrices of less than 100 elements and stack them together.

$$
CODEB1 := \begin{bmatrix}
0 & 0 & 1 & 0 & 0 & 0 & 0 & 0 & 0 & 0 & 0 & 0 & 0 & 0 & 0 & 0 \\
0 & 0 & 0 & 0 & 1 & 0 & 0 & 0 & 0 & 0 & 0 & 0 & 0 & 0 & 0 & 0 \\
1 & 0 & 0 & 0 & 0 & 0 & 0 & 0 & 0 & 0 & 0 & 0 & 0 & 0 & 0 & 0 \\
0 & 0 & 0 & 1 & 0 & 0 & 0 & 0 & 0 & 0 & 0 & 0 & 0 & 0 & 0 & 0 \\
0 & 0 & 0 & 0 & 0 & 1 & 0 & 0 & 0 & 0 & 0 & 0 & 0 & 0 & 0 & 0 \\
0 & 1 & 0 & 0 & 0 & 0 & 0 & 0 & 0 & 0 & 0 & 0 & 0 & 0 & 0 & 0
\end{bmatrix}
$$

$$
CODEB2 := \begin{bmatrix}
0 & 0 & 0 & 0 & 0 & 0 & 1 & 0 & 0 & 0 & 0 & 0 & 0 & 0 & 0 & 0 \\
0 & 0 & 0 & 0 & 0 & 0 & 0 & 0 & 0 & 0 & 1 & 0 & 0 & 0 & 0 & 0 \\
0 & 0 & 0 & 0 & 0 & 0 & 0 & 1 & 0 & 0 & 0 & 0 & 0 & 0 & 0 & 0 \\
0 & 0 & 0 & 0 & 0 & 0 & 0 & 0 & 0 & 0 & 0 & 1 & 0 & 0 & 0 & 0 \\
0 & 0 & 0 & 0 & 0 & 0 & 0 & 0 & 1 & 0 & 0 & 0 & 0 & 0 & 0 & 0 \\
0 & 0 & 0 & 0 & 0 & 0 & 0 & 0 & 0 & 0 & 0 & 0 & 1 & 0 & 0 & 0
\end{bmatrix}
$$

$$
CODEB3 := \begin{bmatrix}
0 & 0 & 0 & 0 & 0 & 0 & 0 & 0 & 0 & 0 & 0 & 0 & 0 & 0 & 0 & 1 \\
0 & 0 & 0 & 0 & 0 & 0 & 0 & 0 & 0 & 0 & 0 & 0 & 0 & 0 & 1 & 0 \\
0 & 0 & 0 & 0 & 0 & 0 & 0 & 0 & 0 & 0 & 0 & 0 & 0 & 0 & 0 & 1 \\
0 & 0 & 0 & 0 & 0 & 0 & 0 & 0 & 0 & 0 & 0 & 0 & 0 & 1 & 0 & 0
\end{bmatrix}
$$

CODEB4 := stack(CODEB1,CODEB2)

CODEB := stack(CODEB4,CODEB3)

This is the scrambler matrix for the column scramble

$$
CODEB = \begin{bmatrix}
0 & 0 & 1 & 0 & 0 & 0 & 0 & 0 & 0 & 0 & 0 & 0 & 0 & 0 & 0 & 0 \\
0 & 0 & 0 & 0 & 1 & 0 & 0 & 0 & 0 & 0 & 0 & 0 & 0 & 0 & 0 & 0 \\
1 & 0 & 0 & 0 & 0 & 0 & 0 & 0 & 0 & 0 & 0 & 0 & 0 & 0 & 0 & 0 \\
0 & 0 & 0 & 1 & 0 & 0 & 0 & 0 & 0 & 0 & 0 & 0 & 0 & 0 & 0 & 0 \\
0 & 0 & 0 & 0 & 0 & 1 & 0 & 0 & 0 & 0 & 0 & 0 & 0 & 0 & 0 & 0 \\
0 & 1 & 0 & 0 & 0 & 0 & 0 & 0 & 0 & 0 & 0 & 0 & 0 & 0 & 0 & 0 \\
0 & 0 & 0 & 0 & 0 & 0 & 1 & 0 & 0 & 0 & 0 & 0 & 0 & 0 & 0 & 0 \\
0 & 0 & 0 & 0 & 0 & 0 & 0 & 0 & 0 & 1 & 0 & 0 & 0 & 0 & 0 & 0 \\
0 & 0 & 0 & 0 & 0 & 0 & 0 & 1 & 0 & 0 & 0 & 0 & 0 & 0 & 0 & 0 \\
0 & 0 & 0 & 0 & 0 & 0 & 0 & 0 & 0 & 0 & 1 & 0 & 0 & 0 & 0 & 0 \\
0 & 0 & 0 & 0 & 0 & 0 & 1 & 0 & 0 & 0 & 0 & 0 & 0 & 0 & 0 & 0 \\
0 & 0 & 0 & 0 & 0 & 0 & 0 & 0 & 0 & 0 & 0 & 1 & 0 & 0 & 0 & 0 \\
0 & 0 & 0 & 0 & 0 & 0 & 0 & 0 & 0 & 0 & 0 & 0 & 0 & 0 & 0 & 1 \\
0 & 0 & 0 & 0 & 0 & 0 & 0 & 0 & 0 & 0 & 0 & 0 & 0 & 1 & 0 & 0 \\
0 & 0 & 0 & 0 & 0 & 0 & 0 & 0 & 0 & 0 & 0 & 0 & 0 & 0 & 1 & 0 \\
0 & 0 & 0 & 0 & 0 & 0 & 0 & 0 & 0 & 0 & 0 & 0 & 1 & 0 & 0 & 0
\end{bmatrix}
\blacksquare
$$

C := CODE·A·CODEB

This matrix has it's rows and columns interchanged.

To decode this and get our information back again we just multiply the coded matrix by the transpose of the encoding matrices i.e.:

Which is the unscrambled inventory.

CALCULATION OF A WATER BILL

I worked for a water meter reader in Butner, North Carolina for awhile and wrote this just for the heck of it. I will not elaborate any on the math, just an example to see another way to use inventories and accounting equations.

WATER BILL

METER READING THIS MONTH	—	METER READING LAST MONTH =	GALLONS WATER USED

$$\begin{bmatrix} 7100 \\ 107000 \\ 126700 \\ 11200 \end{bmatrix} \qquad \begin{bmatrix} 6100 \\ 106500 \\ 126000 \\ 11000 \end{bmatrix} \qquad \begin{bmatrix} 1000 \\ 500 \\ 700 \\ 200 \end{bmatrix}$$

$$MRTM := \begin{bmatrix} 7100 \\ 107000 \\ 126700 \\ 11200 \end{bmatrix} \qquad MRLM := \begin{bmatrix} 6100 \\ 106500 \\ 126000 \\ 11000 \end{bmatrix}$$

$$GWU := MRTM - MRLM$$

$$GWU = \begin{bmatrix} 1000 \\ 500 \\ 700 \\ 200 \end{bmatrix}$$

TO COMPUTE THE CHARGE, SUPPOSE THE COST OF WATER IS 3¢ PER GALLON, THEN $[GWU]_{4,1} \circ c[1]_{4,1}$, WHERE C = CHARGE PER GALLON.

$$FOUR1 := \begin{bmatrix} 1 \\ 1 \\ 1 \\ 1 \end{bmatrix} \qquad cFOUR1 := .03 \cdot FOUR1$$

CHARGEPERCUSTOMER $:= \overrightarrow{(\text{GWU} \cdot \text{cFOUR1})}$

$$
\begin{bmatrix} 1000 \\ 500 \\ 700 \\ 200 \end{bmatrix} \circ \begin{bmatrix} 0.03 \\ 0.03 \\ 0.03 \\ 0.03 \end{bmatrix} = \begin{bmatrix} 30 \\ 15 \\ 21 \\ 6 \end{bmatrix}
$$

THE HOLLOW DOT MEANS TO MULTIPLY STRAIGHT ACROSS.

$$
\text{CHARGEPERCUSTOMER} = \begin{bmatrix} 30 \\ 15 \\ 21 \\ 6 \end{bmatrix}
$$

NOTE: THESE VALUES ARE IN DOLLARS.

OR WE CAN COMPUTE THIS BY:

$$
\text{ANTI} - \text{LOG}(\text{LOG} \begin{bmatrix} 1000 \\ 500 \\ 700 \\ 200 \end{bmatrix} \blacksquare + \text{LOG} \begin{bmatrix} 0.03 \\ 0.03 \\ 0.03 \\ 0.03 \end{bmatrix}) =
$$

PROGRAM:

CPC $:= \overrightarrow{(\log(\text{GWU}) + \log(\text{cFOUR1}))}$ Here we take the log of each individual element and add them together

$$
\text{CPC} = \begin{bmatrix} 1.477 \\ 1.176 \\ 1.322 \\ 0.778 \end{bmatrix}
$$
(THIS REPRESENTS THE SUMMED LOGS OF THE GALLONS + PRICE MATRICES)

CHARGEPERCUSTOMERANTILOG $:= \overrightarrow{10^{\text{CPC}}}$ (THIS TAKES THE ANTI-LOG OF THE ABOVE COMPUTATION.)

$$\text{CHARGEPERCUSTOMERANTILOG} = \begin{bmatrix} 30 \\ 15 \\ 21 \\ 6 \end{bmatrix}$$

THE TOTAL AMOUNT OWED BY ALL CUSTOMERS TO THE UTILITY IS:

$$\text{GWU}^T \cdot \text{cFOUR1} = 72$$

$$\text{GWU}^T = (\begin{matrix} 1000 & 500 & 700 & 200 \end{matrix})$$

$$(\begin{matrix} 1000 & 500 & 700 & 200 \end{matrix}) \cdot \begin{bmatrix} 0.03 \\ 0.03 \\ 0.03 \\ 0.03 \end{bmatrix} = 72$$

THE TOTAL OF MONTHLY BILLS PER CUSTOMER TOTALED FOR THE ENTIRE YEAR TO DATE = LAST MONTHS BILL(JAN. + THIS MONTHS BILL(FEB) = TWO MONTHS TOTAL PER CUSTOMER

$$\begin{bmatrix} 30 \\ 20 \\ 16 \\ 10 \end{bmatrix} + \begin{bmatrix} 30 \\ 15 \\ 21 \\ 6 \end{bmatrix} = \begin{bmatrix} 60 \\ 35 \\ 37 \\ 16 \end{bmatrix}$$

BUT SUPPOSE THE CHARGES FOR THE FIRST 500 GALLONS = 3¢/GAL AND THE CHARGE FOR THE NEXT 1000 GALLONS RISES TO 5¢/GALLON, WE HAVE (WE IGNORE ZERO SIGNS):

$$\begin{bmatrix} 500 & -500 \\ 500 & -500 \\ 500 & -500 \\ 200 & -500 \end{bmatrix} + \begin{bmatrix} 0 & 1000 \\ 0 & 500 \\ 0 & 700 \\ 0 & 200 \end{bmatrix} = \begin{bmatrix} 500 & 500 \\ 500 & 0 \\ 500 & 200 \\ 200 & -300 \end{bmatrix}$$ (CHANGE THE NEGATIVE NUMBERS TO ZERO)

$$\text{CHRGDIFFERENTIAL} = \begin{bmatrix} 500 & 500 \\ 500 & 0 \\ 500 & 200 \\ 200 & 0 \end{bmatrix}$$

$$\text{PRICE} := \begin{bmatrix} .03 & .05 \\ .03 & .05 \\ .03 & .05 \\ .03 & .05 \end{bmatrix}$$

$$\overrightarrow{\text{TOTALDIFFERENTIALCHARGES} := (\text{CHRGDIFFERENTIAL} \cdot \text{PRICE})}$$

$$\text{TOTALDIFFERENTIALCHARGES} = \begin{bmatrix} 15 & 25 \\ 15 & 0 \\ 15 & 10 \\ 6 & 0 \end{bmatrix}$$

NOW WE NEED TO ADD THE CHARGES TOGETHER.

$$\text{TWO1} := \begin{pmatrix} 1 \\ 1 \end{pmatrix}$$

$$\text{TOTALCHARGE} := \text{TOTALDIFFERENTIALCHARGES} \cdot \text{TWO1}$$

$$\begin{bmatrix} 15 & 25 \\ 15 & 0 \\ 15 & 10 \\ 6 & 0 \end{bmatrix} \cdot \begin{pmatrix} 1 \\ 1 \end{pmatrix} = \begin{bmatrix} 40 \\ 15 \\ 25 \\ 6 \end{bmatrix}$$

THE FIRST CUSTOMER USED $40 WORTH OF WATER, THE SECOND CUSTOMER USED $15 WORTH OF WATER, THE THIRD CUSTOMER USED $25 WORTH OF WATER AND THE FOURTH CUSTOMER ONLY USED $6 WORTH OF WATER.

OR WE CAN USE LOGARITHMS:

$$\text{DIFFCHRGS} := \begin{bmatrix} 500 & 500 \\ 500 & 10^{-100} \\ 500 & 200 \\ 200 & 10^{-100} \end{bmatrix}$$

(NOTE: MUST CHANGE ZERO'S TO DEFAULT $0 \approx 10^{-100}$)

$$\overrightarrow{\text{LOGDIFFWATERBILL} := (\log(\text{DIFFCHRGS}) + \log(\text{PRICE}))}$$

$$\text{TOTALDIFFERENTIALCHARGES} := 10^{\overrightarrow{\text{LOGDIFFWATERBILL}}} \cdot \text{TWO1}$$

$$\text{TOTALDIFFERENTIALCHARGES} = \begin{bmatrix} 40 \\ 15 \\ 25 \\ 6 \end{bmatrix}$$

With MathCad, we don't have to take the logs of the elements of the matrices, we are just multiplying the first element in matrix one by the first element in matrix two; the second element in matrix one by its corresponding element in matrix two, etc. But computers cannot do the matrix multiplication's this way (we can with MathCad through the vectorize operation) so we have to use logarithms to get our solutions.

To find out who paid their bill, and who is behind in their payments, we proceed as follows:

We need an accounts paid matrix where each element represents the amount paid to the utility by the water user.

Owed   paid

$$\begin{vmatrix} \$40 \\ \$15 \\ \$25 \\ \$6 \end{vmatrix} - \begin{vmatrix} \$40 \\ 0 \\ \$35 \\ \$6 \end{vmatrix} = \begin{vmatrix} 0 \\ \$15 \\ -\$10 \\ 0 \end{vmatrix} \quad \begin{array}{l} \text{paid up through February} \\ \text{owes \$15 from February} \\ \text{credit \$10 for February} \\ \text{paid up through February} \end{array}$$

Suppose the utility charges 1% per month penalty for the unpaid portion of the bill. We will ignore the \$10 credit:

$$\begin{vmatrix} 0 \\ 15 \\ 0 \\ 0 \end{vmatrix} \circ \begin{vmatrix} .01 \\ .01 \\ .01 \\ .01 \end{vmatrix} = \begin{vmatrix} 0 \\ .15 \\ 0 \\ 0 \end{vmatrix} \quad \text{Then} \quad \begin{vmatrix} 0 \\ .15 \\ 0 \\ 0 \end{vmatrix} + \begin{vmatrix} 0 \\ \$15 \\ -\$10 \\ 0 \end{vmatrix} = \begin{vmatrix} 0 \\ 15.15 \\ -10 \\ 0 \end{vmatrix} + \begin{vmatrix} \text{MARCH'S} \\ \text{NEW} \\ \text{BILL} \end{vmatrix}$$

Or we could compute the taxes, say at 6%, for the charges in the same manner:

$$\text{TAX} = \begin{vmatrix} 40 \\ 15 \\ 25 \\ 6 \end{vmatrix} \circ \begin{vmatrix} .06 \\ .06 \\ .06 \\ .06 \end{vmatrix} + \begin{vmatrix} 40 \\ 15 \\ 25 \\ 6 \end{vmatrix} = \begin{vmatrix} 2.40 \\ .90 \\ 1.50 \\ .36 \end{vmatrix} + \begin{vmatrix} 40 \\ 15 \\ 25 \\ 6 \end{vmatrix} = \begin{vmatrix} 42.40 \\ 15.90 \\ 26.50 \\ 6.36 \end{vmatrix}$$

If a business gets a tax break, we just put in the value charged in place of the .06 and proceed as above.

One more example to show how to extend the calculations even further. Suppose after the first 1500 gallons, the charge increases to 10¢/gallon. Let's add home 5 who used 1599 gallons of water last month. The first row will always be 500 gallons unless less water is consumed. Our matrix now becomes:

$$
\begin{vmatrix}
1000 \\
500 \\
700 \\
200 \\
1599
\end{vmatrix}
$$

And the problem becomes:

$$
\begin{bmatrix}
500 & -500 & -1500 \\
500 & -500 & -1500 \\
500 & -500 & -1500 \\
200 & -500 & -1500 \\
500 & -500 & -1500
\end{bmatrix}
+
\begin{bmatrix}
0 & 1000 & 1000 \\
0 & 500 & 500 \\
0 & 700 & 700 \\
0 & 200 & 200 \\
0 & 1599 & 1599
\end{bmatrix}
=
\begin{bmatrix}
500 & 500 & -500 \\
500 & 0 & -1000 \\
500 & 200 & -800 \\
200 & -300 & -1300 \\
500 & 1099 & 99
\end{bmatrix}
$$

SINCE THE MAXIMUM FOR THE SECOND COLUMN IS 1000 GALLONS, ANYTHING OVER BELONGS IN THE THIRD COLUMN, SO LET'S SEE IF THERE ARE ANY VALUES ABOVE 1000 IN THE SECOND COLUMN. (HERE WE CAN SEE IT, BUT IF THIS WERE THE WATER BILL FOR NYC OR LA THERE WOULD BE MILLIONS OF LISTINGS, TOO MANY TO LOOK FOR BY HAND). IF ANY THREE VALUES IN ANY ROW ARE ALL POSITIVE, OVER 1500 GALLONS OF WATER WERE USED. A CORRECTION MATRIX CAN BE DERIVED AS FOLLOWS: REPLACE EVERY NEGATIVE VALUE WITH ZERO.

$$
\begin{bmatrix}
500 & 500 & 0 \\
500 & 0 & 0 \\
500 & 200 & 0 \\
200 & 0 & 0 \\
500 & 1099 & 99
\end{bmatrix}
\cdot
\begin{pmatrix}
0 \\
1 \\
-1
\end{pmatrix}
=
\begin{bmatrix}
500 \\
0 \\
200 \\
0 \\
1000
\end{bmatrix}
$$

$$\begin{bmatrix} 500 & 500 & 0 \\ 500 & 0 & 0 \\ 500 & 200 & 0 \\ 200 & 0 & 0 \\ 500 & 1000 & 99 \end{bmatrix}$$

MATHCAD WILL LET US DELETE THE MIDDLE COLUMN, BUT IT WON'T LET US INSERT THE CORRECTED MATRIX IN IT'S PLACE. WE MUST SELECT THE 500 AT THE TOP OF THE MIDDLE COLUMN. CLICK THE MATRIX PALLET AND FOR ROWS INSERT A ZERO, AND IN COLUMNS INSERT A ONE. THEN CLICK THE DELETE BUTTON. THEN CLICK NEXT TO THE 500 IN THE FIRST COLUMN, BUT DON'T PUT IT IN A SELECTION BOX, JUST PUT THE CURSOR TO THE RIGHT OF IT AND PRESS INSERT. WE MUST RE-TYPE IN THE NEW VALUES. WE CAM EXTRACT A SINGLE COLUMN FROM A MATRIX, BUT WE CAN'T PUT ONE IN.

THE PROBLEM NOW BECOMES, AFTER SETTING UP THE COST MATRIX:

$$\overrightarrow{\begin{bmatrix} 500 & 500 & 0 \\ 500 & 0 & 0 \\ 500 & 200 & 0 \\ 200 & 0 & 0 \\ 500 & 1000 & 99 \end{bmatrix} \cdot \begin{bmatrix} .03 & .05 & .10 \\ .03 & .05 & .10 \\ .03 & .05 & .10 \\ .03 & .05 & .10 \\ .03 & .05 & .10 \end{bmatrix}} = \begin{bmatrix} 15 & 25 & 0 \\ 15 & 0 & 0 \\ 15 & 10 & 0 \\ 6 & 0 & 0 \\ 15 & 50 & 9.9 \end{bmatrix} \qquad \text{THREE1} := \begin{pmatrix} 1 \\ 1 \\ 1 \end{pmatrix}$$

AND THE TOTAL BILL IS:

$$\text{INV} := \begin{bmatrix} 500 & 500 & 0 \\ 500 & 0 & 0 \\ 500 & 200 & 0 \\ 200 & 0 & 0 \\ 500 & 1000 & 99 \end{bmatrix} \qquad \text{COST} := \begin{bmatrix} .03 & .05 & .10 \\ .03 & .05 & .10 \\ .03 & .05 & .10 \\ .03 & .05 & .10 \\ .03 & .05 & .10 \end{bmatrix}$$

$$\overrightarrow{(\text{INV} \cdot \text{COST})} \cdot \text{THREE1} = \begin{bmatrix} 40 \\ 15 \\ 25 \\ 6 \\ 74.9 \end{bmatrix}$$

(THE VECTORIZED INV x COST MULTIPLIES THE MATRICES ONE ON ONE.)

HONEOWNER ONE'S BILL IS $40, TWO'S BILL IS $15, THREE'S BILL
IS $25, FOUR'S BILL IS $6 AND FIVE'S BILL IS $74.90.

PROGRAM: ([THIS MO. READING]$_{5,1}$-[LAST MO. READING]$_{5,1}$)∘[.03]$_{5,1}$=
[BILL FOR SEPARATE HOMES]$_{5,1}$

COMPLEX BILL: ANTI-LOG(LOG[SEP. INV.]$_{5,3}$)+LOG([COSTS]$_{5,3}$))[1]$_{3,1}$=
[COMPLEX BILL FOR SEPARATE HOMES]$_{5,1}$

TO KEEP TRACK OF WHOLE AREA, STATE, USA OR WORLD ALL AT THE SAME TIME

I am not going to solve this in detail like I did for the problem above, I'm just going to set this and solve for a simple system.

Suppose Butner charges 2¢/first 500 gal, 5¢/next 1000gal 11¢/ all use above
Suppose Cozart charges 3¢/first 500 gal, 5¢/next 1000gal 10¢/ all use above
Suppose Creedmore charges 4¢/first 500 gal, 6¢/next 1000gal 12¢/all use above

After reading the meters and calculating gallons used:

$$
\text{INV} = \begin{bmatrix}
500 & 300 & 0 \\
500 & 1000 & 100 \\
300 & 0 & 0 \\
500 & 1000 & 10000 \\
500 & 500 & 0 \\
500 & 0 & 0 \\
500 & 200 & 0 \\
200 & 0 & 0 \\
500 & 1000 & 99 \\
500 & 1000 & 2000 \\
400 & 0 & 0 \\
500 & 300 & 0 \\
300 & 0 & 0
\end{bmatrix}
\quad
\text{COST} = \begin{bmatrix}
.02 & .05 & .11 \\
.02 & .05 & .11 \\
.02 & .05 & .11 \\
.02 & .05 & .11 \\
.03 & .05 & .10 \\
.03 & .05 & .10 \\
.03 & .05 & .10 \\
.03 & .05 & .10 \\
.03 & .05 & .10 \\
.04 & .06 & .12 \\
.04 & .06 & .12 \\
.04 & .06 & .12 \\
.04 & .06 & .12
\end{bmatrix}
\quad
\text{BILL} = \begin{bmatrix}
25 \\
71 \\
6 \\
1160 \\
40 \\
15 \\
25 \\
6 \\
74.9 \\
320 \\
16 \\
38 \\
12
\end{bmatrix}
$$

TO COMPUTE THE TOTAL BILL:

$$\text{BILL} = (\overrightarrow{\text{INV} \cdot \text{COST}}) \cdot \text{THREE1}$$

BUTNER: PRINT 1 - 4
COZART: PRINT 5 - 9
CREEDMORE: PRINT 10 - 13

$$\text{OR: } \quad \text{INVA} := \begin{bmatrix} 500 & 300 & 10^{-100} \\ 500 & 1000 & 100 \\ 300 & 10^{-100} & 10^{-100} \\ 500 & 1000 & 10000 \\ 500 & 500 & 10^{-100} \\ 500 & 10^{-100} & 10^{-100} \\ 500 & 200 & 10^{-100} \\ 200 & 10^{-100} & 10^{-100} \\ 500 & 1000 & 99 \\ 500 & 1000 & 2000 \\ 400 & 10^{-100} & 10^{-100} \\ 500 & 300 & 10^{-100} \\ 300 & 10^{-100} & 10^{-100} \end{bmatrix}$$

$$\text{LOGTOTALINVA} := \overrightarrow{\log(\text{INVA})} + \overrightarrow{\log(\text{COST})}$$

$$\text{INVAREA} := \overrightarrow{10^{\text{LOGTOTALINVA}} \cdot \text{THREE1}} \qquad \text{INVAREA} = \begin{bmatrix} 25 \\ 71 \\ 6 \\ 1160 \\ 40 \\ 15 \\ 25 \\ 6 \\ 74.9 \\ 320 \\ 16 \\ 38 \\ 12 \end{bmatrix}$$

THIS IS HOW TO ACHIEVE THE SOLUTION USING LOGS.

Now for one final example on how this method works. Suppose the government wishes to keep track of all the public utilities: water, gas and electricity. In the following example, this is the order that I will put in each value. Suppose the meter readings have been taken and subtracted. The cost for water is as above, the cost for gas is .001¢/cubic foot and the cost for electricity is 7¢/kilowatthour. The inventory and cost matrices are now:

$$
A := \begin{bmatrix}
500 & 300 \\
5500 & 0 \\
125 & 0 \\
500 & 1000 \\
6000 & 0 \\
150 & 0 \\
300 & 0 \\
7500 & 0 \\
250 & 0 \\
500 & 1000 \\
8000 & 0 \\
2000 & 0 \\
500 & 500 \\
10000 & 0 \\
450 & 0 \\
500 & 0 \\
12500 & 0 \\
550 & 0 \\
500 & 200 \\
4650 & 0 \\
125 & 0 \\
200 & 0 \\
9900 & 0 \\
325 & 0 \\
500 & 1000 \\
8750 & 0 \\
225 & 0 \\
500 & 1000 \\
6750 & 0 \\
330 & 0 \\
400 & 0 \\
7500 & 0 \\
150 & 0 \\
500 & 300 \\
8550 & 0 \\
260 & 0 \\
300 & 0 \\
11000 & 0 \\
550 & 0
\end{bmatrix}
\qquad
B := \begin{bmatrix}
0 \\
0 \\
0 \\
100 \\
0 \\
0 \\
0 \\
0 \\
0 \\
10000 \\
0 \\
0 \\
0 \\
0 \\
0 \\
0 \\
0 \\
0 \\
0 \\
0 \\
0 \\
0 \\
0 \\
0 \\
99 \\
0 \\
0 \\
2000 \\
0 \\
0 \\
0 \\
0 \\
0 \\
0 \\
0 \\
0 \\
0 \\
0 \\
0
\end{bmatrix}
$$

INVUSA := augment(A , B)

$$
INVUSA = \begin{bmatrix}
500 & 300 & 0 \\
5500 & 0 & 0 \\
125 & 0 & 0 \\
500 & 1000 & 100 \\
6000 & 0 & 0 \\
150 & 0 & 0 \\
300 & 0 & 0 \\
7500 & 0 & 0 \\
250 & 0 & 0 \\
500 & 1000 & 10000 \\
8000 & 0 & 0 \\
2000 & 0 & 0 \\
500 & 500 & 0 \\
10000 & 0 & 0 \\
450 & 0 & 0 \\
500 & 0 & 0 \\
12500 & 0 & 0 \\
550 & 0 & 0 \\
500 & 200 & 0 \\
4650 & 0 & 0 \\
125 & 0 & 0 \\
200 & 0 & 0 \\
9900 & 0 & 0 \\
325 & 0 & 0 \\
500 & 1000 & 99 \\
8750 & 0 & 0 \\
225 & 0 & 0 \\
500 & 1000 & 2000 \\
6750 & 0 & 0 \\
330 & 0 & 0 \\
400 & 0 & 0 \\
7500 & 0 & 0 \\
150 & 0 & 0 \\
500 & 300 & 0 \\
8550 & 0 & 0 \\
260 & 0 & 0 \\
300 & 0 & 0 \\
11000 & 0 & 0 \\
550 & 0 & 0
\end{bmatrix}
\qquad
COSTUSA = \begin{bmatrix}
.02 & .05 & .11 \\
.001 & 0 & 0 \\
.07 & 0 & 0 \\
.12 & .05 & .11 \\
.001 & 0 & 0 \\
.07 & 0 & 0 \\
.02 & .05 & .11 \\
.001 & 0 & 0 \\
.07 & 0 & 0 \\
.02 & .05 & .11 \\
.001 & 0 & 0 \\
.07 & 0 & 0 \\
.03 & .05 & .10 \\
.001 & 0 & 0 \\
.07 & 0 & 0 \\
.03 & .05 & .10 \\
.001 & 0 & 0 \\
.07 & 0 & 0 \\
.03 & .05 & .10 \\
.001 & 0 & 0 \\
.07 & 0 & 0 \\
.03 & .05 & .10 \\
.001 & 0 & 0 \\
.07 & 0 & 0 \\
.03 & .05 & .10 \\
.001 & 0 & 0 \\
.07 & 0 & 0 \\
.04 & .06 & .12 \\
.001 & 0 & 0 \\
.07 & 0 & 0 \\
.04 & .06 & .12 \\
.001 & 0 & 0 \\
.07 & 0 & 0 \\
.04 & .06 & .12 \\
.001 & 0 & 0 \\
.07 & 0 & 0 \\
.04 & .06 & .12 \\
.001 & 0 & 0 \\
.07 & 0 & 0
\end{bmatrix}
$$

THEN:

$$(\overrightarrow{INVUSA \cdot COSTUSA}) \cdot THREE1$$

$$TOTALWGEUSA := (\overrightarrow{INVUSA \cdot COSTUSA}) \cdot THREE1$$

$$TOTALWGEUSA = \begin{bmatrix} 25 \\ 5.5 \\ 8.75 \\ 121 \\ 6 \\ 10.5 \\ 6 \\ 7.5 \\ 17.5 \\ 1160 \\ 8 \\ 140 \\ 40 \\ 10 \\ 31.5 \\ 15 \\ 12.5 \\ 38.5 \\ 25 \\ 4.65 \\ 8.75 \\ 6 \\ 9.9 \\ 22.75 \\ 74.9 \\ 8.75 \\ 15.75 \\ 320 \\ 6.75 \\ 23.1 \\ 16 \\ 7.5 \\ 10.5 \\ 38 \\ 8.55 \\ 18.2 \\ 12 \\ 11 \\ 38.5 \end{bmatrix}$$

$$
\text{TAXDB} = \begin{bmatrix}
.06 \\
.06 \\
.06 \\
.06 \\
.06 \\
.06 \\
.06 \\
.06 \\
.06 \\
.06 \\
.06 \\
.06 \\
.06 \\
.06 \\
.06 \\
.06 \\
.06 \\
.06 \\
.06 \\
.06 \\
.06 \\
.06 \\
.06 \\
.06 \\
.06 \\
.06 \\
.06 \\
.06 \\
.06 \\
.06 \\
.06 \\
.06 \\
.06 \\
.06 \\
.06 \\
.06 \\
.06 \\
.06 \\
.06 \\
.06
\end{bmatrix}
$$

ONE39 :=(1 1 1 1 1 1 1 1 1 1 1 1 1 1 1 1 1 1 1 1 1 1 1 1 1 1 1 1 1 1 1 1 1 1 1 1 1 1 1)

THEN THE TOTAL BILL =:

$$\text{TOTALBILLWGEUSA} := (\overrightarrow{\text{TOTALWGEUSA} \cdot \text{TAXDB}}) + \text{TOTALWGEUSA}$$

$$\text{TOTALTAXES} := \text{ONE39} \cdot (\overrightarrow{\text{TOTALWGEUSA} \cdot \text{TAXDB}})$$

$$\text{TOTALTAXES} = 140.988$$

TOTAL ACCOUNTPAGE FOR TAXES   TOTAL BILL SENT TO CUSTOMERS

$$(\overrightarrow{\text{TOTALWGEUSA} \cdot \text{TAXDB}}) = \begin{bmatrix} 1.5 \\ 0.33 \\ 0.525 \\ 7.26 \\ 0.36 \\ 0.63 \\ 0.36 \\ 0.45 \\ 1.05 \\ 69.6 \\ 0.48 \\ 8.4 \\ 2.4 \\ 0.6 \\ 1.89 \\ 0.9 \\ 0.75 \\ 2.31 \\ 1.5 \\ 0.279 \\ 0.525 \\ 0.36 \\ 0.594 \\ 1.365 \\ 4.494 \\ 0.525 \\ 0.945 \\ 19.2 \\ 0.405 \\ 1.386 \\ 0.96 \\ 0.45 \\ 0.63 \\ 2.28 \\ 0.513 \\ 1.092 \\ 0.72 \\ 0.66 \\ 2.31 \end{bmatrix} \qquad \text{TOTALBILLWGEUSA} = \begin{bmatrix} 26.5 \\ 5.83 \\ 9.275 \\ 128.26 \\ 6.36 \\ 11.13 \\ 6.36 \\ 7.95 \\ 18.55 \\ 1229.6 \\ 8.48 \\ 148.4 \\ 42.4 \\ 10.6 \\ 33.39 \\ 15.9 \\ 13.25 \\ 40.81 \\ 26.5 \\ 4.929 \\ 9.275 \\ 6.36 \\ 10.494 \\ 24.115 \\ 79.394 \\ 9.275 \\ 16.695 \\ 339.2 \\ 7.155 \\ 24.486 \\ 16.96 \\ 7.95 \\ 11.13 \\ 40.28 \\ 9.063 \\ 19.292 \\ 12.72 \\ 11.66 \\ 40.81 \end{bmatrix}$$

Household #1 received a bill of $26.50 for water, $5.83 for gas and $9.27 for electric with taxes of $1.50 for water, $0.33 tax on gas and $0.53 tax on electricity. Household #4 is a business, and paid $1229.6 for water, $8.48 for gas and $148.40 for electricity, etc.

TWO FACTOR MIXED DESIGN: REPEATED MEASURES ON ONE FACTOR FOR STATISTICS AND ACCOUNTING/INVENTORY SYSTEMS

COMPUTATIONAL HANDBOOK OF STATISTICS, 2ᴺᴰ EDITION; SECTION 2.7, PG. 55-61.

EXAMPLE: A man has three hot-dog stands which he operates downtown. He sells only three products, two hot-dogs for $1, a burrito for $1.25 and a polish sausage for $1.50.

These prices include the tax. He operates the stands 5½ hours and records the total sales of each item every 1/2 hour. Each employee makes $5/hr or $27.50 each day. We will find the total sales, the sales per cart, the sales by item, the taxes on the sales, the employees wages and federal, state and Social Security taxes. We shall then do a two factor mixed design analysis of variance on the sales to see if there is any statistical correlation between location of the carts, sales as the day goes by and if there is any statistical significance as to whether any item sold more than any other.

Note: This is problem 2.7 Analysis of Variance in Statistics section of book)

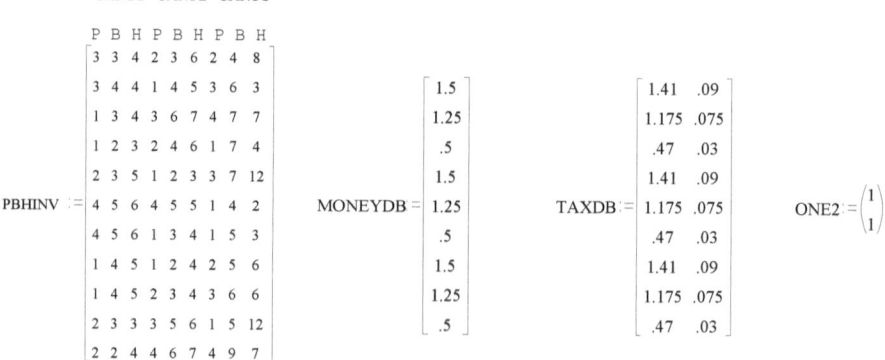

$$TAXDB \cdot ONE2 = \begin{bmatrix} 1.5 \\ 1.25 \\ 0.5 \\ 1.5 \\ 1.25 \\ 0.5 \\ 1.5 \\ 1.25 \\ 0.5 \end{bmatrix}$$ NOTE: TAXDB with the rows summed equals MONEYDB

PROGRAM: $[A]_{11,9}$ $[MONEY]_{9,1}$ = MONEY MADE EVERY HALF HOUR
MONEYEACHHALFHOUR := PBHINV·MONEYDB

$$MONEYEACHHALFHOUR = \begin{bmatrix} 32 \\ 34 \\ 41 \\ 28.75 \\ 34 \\ 37.5 \\ 31.75 \\ 27.25 \\ 32.75 \\ 35.75 \\ 45.25 \end{bmatrix}$$

In the first half-hour he brought in $32, in the second half-hour he brought in $34, etc.

TOTAL SALES SEPARATED BY CART

PROGRAM: MONEYEACHHALFHOUR(MONEYDBoDB1)

$$DB1 := \begin{bmatrix} 1 & 0 & 0 \\ 1 & 0 & 0 \\ 1 & 0 & 0 \\ 0 & 1 & 0 \\ 0 & 1 & 0 \\ 0 & 1 & 0 \\ 0 & 0 & 1 \\ 0 & 0 & 1 \\ 0 & 0 & 1 \end{bmatrix}$$

THE DB1 MATRIX SEPARATES THE COSTS OF THE PRODUCE INTO 3 SEPARATE COLUMNS INSTEAD OF JUST ONE.

MONEYoDB1 $:=$ diag(MONEYDB)·DB1

$$
\text{MONEYoDB1} =
\begin{bmatrix}
1.5 & 0 & 0 \\
1.25 & 0 & 0 \\
0.5 & 0 & 0 \\
0 & 1.5 & 0 \\
0 & 1.25 & 0 \\
0 & 0.5 & 0 \\
0 & 0 & 1.5 \\
0 & 0 & 1.25 \\
0 & 0 & 0.5
\end{bmatrix}
$$

SALESPERCART $:=$ PBHINV·MONEYoDB1

$$
\begin{bmatrix}
3 & 3 & 4 & 2 & 3 & 6 & 2 & 4 & 8 \\
3 & 4 & 4 & 1 & 4 & 5 & 3 & 6 & 3 \\
1 & 3 & 4 & 3 & 6 & 7 & 4 & 7 & 7 \\
1 & 2 & 3 & 2 & 4 & 6 & 1 & 7 & 4 \\
2 & 3 & 5 & 1 & 2 & 3 & 3 & 7 & 12 \\
4 & 5 & 6 & 4 & 5 & 5 & 1 & 4 & 2 \\
4 & 5 & 6 & 1 & 3 & 4 & 1 & 5 & 3 \\
1 & 4 & 5 & 1 & 2 & 4 & 2 & 5 & 6 \\
1 & 4 & 5 & 2 & 3 & 4 & 3 & 6 & 6 \\
2 & 3 & 3 & 3 & 5 & 6 & 1 & 5 & 12 \\
2 & 2 & 4 & 4 & 6 & 7 & 4 & 9 & 7
\end{bmatrix}
\cdot
\begin{bmatrix}
1.5 & 0 & 0 \\
1.25 & 0 & 0 \\
0.5 & 0 & 0 \\
0 & 1.5 & 0 \\
0 & 1.25 & 0 \\
0 & 0.5 & 0 \\
0 & 0 & 1.5 \\
0 & 0 & 1.25 \\
0 & 0 & 0.5
\end{bmatrix}
=
\begin{bmatrix}
10.25 & 9.75 & 12 \\
11.5 & 9 & 13.5 \\
7.25 & 15.5 & 18.25 \\
5.5 & 11 & 12.25 \\
9.25 & 5.5 & 19.25 \\
15.25 & 14.75 & 7.5 \\
15.25 & 7.25 & 9.25 \\
9 & 6 & 12.25 \\
9 & 8.75 & 15 \\
8.25 & 13.75 & 13.75 \\
7.5 & 17 & 20.75
\end{bmatrix}
$$

$$
\text{SALESPERCART} =
\begin{bmatrix}
10.25 & 9.75 & 12 \\
11.5 & 9 & 13.5 \\
7.25 & 15.5 & 18.25 \\
5.5 & 11 & 12.25 \\
9.25 & 5.5 & 19.25 \\
15.25 & 14.75 & 7.5 \\
15.25 & 7.25 & 9.25 \\
9 & 6 & 12.25 \\
9 & 8.75 & 15 \\
8.25 & 13.75 & 13.75 \\
7.5 & 17 & 20.75
\end{bmatrix}
$$

In the first half-hour, cart 1 had a total of $10.25 in sales, cart 2 sold $9.75 and cart 3 sold a total of $12 worth of polish sausage, burritos and hotdogs, etc.

To find the total sales for each cart, we proceed as follows:

ONE11 $:=$ (1 1 1 1 1 1 1 1 1 1 1)

TOTALSALESFORDAY $:=$ ONE11·SALESPERCART

$$
(1\ 1\ 1\ 1\ 1\ 1\ 1\ 1\ 1\ 1\ 1) \cdot
\begin{array}{ccc}
C1 & C2 & C3 \\
\begin{bmatrix}
10.25 & 9.75 & 12 \\
11.5 & 9 & 13.5 \\
7.25 & 15.5 & 18.25 \\
5.5 & 11 & 12.25 \\
9.25 & 5.5 & 19.25 \\
15.25 & 14.75 & 7.5 \\
15.25 & 7.25 & 9.25 \\
9 & 6 & 12.25 \\
9 & 8.75 & 15 \\
8.25 & 13.75 & 13.75 \\
7.5 & 17 & 20.75
\end{bmatrix}
\end{array}
= (\ 108\quad 118.25\quad 153.75\ )
$$

Cart 1 had a total of \$108 in sales, cart 2 had a total of \$118.25 and cart 3 had a total of \$153.75 for the day.

The totals for polish sausage, burritos and hotdogs separated are:

PBHINV(MONEYDBoDB3)

Or it can be written as:

$[\text{PBINV}]_{11,9}\,([\text{MONEYDB}]_{9,1}\ o\,[\text{DB3}]_{9,3})$

Since we can't compute matrices in parenthesis, we solve the equation as follows (using diagonal matrices):

$$
\text{DB3} :=
\begin{bmatrix}
1 & 0 & 0 \\
0 & 1 & 0 \\
0 & 0 & 1 \\
1 & 0 & 0 \\
0 & 1 & 0 \\
0 & 0 & 1 \\
1 & 0 & 0 \\
0 & 1 & 0 \\
0 & 0 & 1
\end{bmatrix}
$$

$\text{MONEYoDB3} := \text{diag}(\text{MONEYDB}) \cdot \text{DB3}$

$$MONEYoDB3 = \begin{bmatrix} 1.5 & 0 & 0 \\ 0 & 1.25 & 0 \\ 0 & 0 & 0.5 \\ 1.5 & 0 & 0 \\ 0 & 1.25 & 0 \\ 0 & 0 & 0.5 \\ 1.5 & 0 & 0 \\ 0 & 1.25 & 0 \\ 0 & 0 & 0.5 \end{bmatrix}$$

SALESPERITEM $:=$ PBHINV·MONEYoDB3

Here we are just adding the cost of each polish sausage, each burrito and each hot-dog. The math looks like:

$$\begin{bmatrix} 3 & 3 & 4 & 2 & 3 & 6 & 2 & 4 & 8 \\ 3 & 4 & 4 & 1 & 4 & 5 & 3 & 6 & 3 \\ 1 & 3 & 4 & 3 & 6 & 7 & 4 & 7 & 7 \\ 1 & 2 & 3 & 2 & 4 & 6 & 1 & 7 & 4 \\ 2 & 3 & 5 & 1 & 2 & 3 & 3 & 7 & 12 \\ 4 & 5 & 6 & 4 & 5 & 5 & 1 & 4 & 2 \\ 4 & 5 & 6 & 1 & 3 & 4 & 1 & 5 & 3 \\ 1 & 4 & 5 & 1 & 2 & 4 & 2 & 5 & 6 \\ 1 & 4 & 5 & 2 & 3 & 4 & 3 & 6 & 6 \\ 2 & 3 & 3 & 3 & 5 & 6 & 1 & 5 & 12 \\ 2 & 2 & 4 & 4 & 6 & 7 & 4 & 9 & 7 \end{bmatrix} \cdot \begin{bmatrix} 1.5 & 0 & 0 \\ 0 & 1.25 & 0 \\ 0 & 0 & 0.5 \\ 1.5 & 0 & 0 \\ 0 & 1.25 & 0 \\ 0 & 0 & 0.5 \\ 1.5 & 0 & 0 \\ 0 & 1.25 & 0 \\ 0 & 0 & 0.5 \end{bmatrix} = \begin{bmatrix} 10.5 & 12.5 & 9 \\ 10.5 & 17.5 & 6 \\ 12 & 20 & 9 \\ 6 & 16.25 & 6.5 \\ 9 & 15 & 10 \\ 13.5 & 17.5 & 6.5 \\ 9 & 16.25 & 6.5 \\ 6 & 13.75 & 7.5 \\ 9 & 16.25 & 7.5 \\ 9 & 16.25 & 10.5 \\ 15 & 21.25 & 9 \end{bmatrix}$$

$$SALESPERITEM = \begin{bmatrix} 10.5 & 12.5 & 9 \\ 10.5 & 17.5 & 6 \\ 12 & 20 & 9 \\ 6 & 16.25 & 6.5 \\ 9 & 15 & 10 \\ 13.5 & 17.5 & 6.5 \\ 9 & 16.25 & 6.5 \\ 6 & 13.75 & 7.5 \\ 9 & 16.25 & 7.5 \\ 9 & 16.25 & 10.5 \\ 15 & 21.25 & 9 \end{bmatrix}$$

In the first half-hour, he sold $10.50 in polish sausages, $12.50 in burritos and $9 in hotdogs, in the 6th half-hour, he sold $13.50 in Polish sausages, $17.50 in burritos and $6.50 worth of hot-dogs, etc.

    TOTAL SALES OF POLISH SAUSAGE, BURRITOS AND HOTDOGS

TOTALSALESPERITEM $:=$ ONE11·SALESPERITEM

$$(1\ 1\ 1\ 1\ 1\ 1\ 1\ 1\ 1\ 1\ 1) \cdot \begin{bmatrix} 10.5 & 12.5 & 9 \\ 10.5 & 17.5 & 6 \\ 12 & 20 & 9 \\ 6 & 16.25 & 6.5 \\ 9 & 15 & 10 \\ 13.5 & 17.5 & 6.5 \\ 9 & 16.25 & 6.5 \\ 6 & 13.75 & 7.5 \\ 9 & 16.25 & 7.5 \\ 9 & 16.25 & 10.5 \\ 15 & 21.25 & 9 \end{bmatrix} = (\ 109.5 \quad 182.5 \quad 88\ )$$

For the day, he sold $109.50 in polish sausage, $182.50 in burritos and $88 in hotdogs.

THE TAX DATABASE

Suppose taxes are 6%. There are two ways of computing the taxes.

METHOD #1: [PBHINV]([MONEYDB]o[TAX]) (This computes the taxes directly from the money taken in that day)

$$TAXA = \begin{bmatrix} .06 \\ .06 \\ .06 \\ .06 \\ .06 \\ .06 \\ .06 \\ .06 \\ .06 \end{bmatrix} \qquad TAXB = \begin{bmatrix} .06 \\ .06 \\ .06 \\ .06 \\ .06 \\ .06 \\ .06 \\ .06 \\ .06 \\ .06 \\ .06 \end{bmatrix}$$

The math from the equation looks like:

$$
\begin{bmatrix} 32 \\ 34 \\ 41 \\ 28.75 \\ 34 \\ 37.5 \\ 31.75 \\ 27.25 \\ 32.75 \\ 35.75 \\ 45.25 \end{bmatrix} \cdot \begin{bmatrix} .06 \\ .06 \\ .06 \\ .06 \\ .06 \\ .06 \\ .06 \\ .06 \\ .06 \\ .06 \\ .06 \end{bmatrix} = \begin{bmatrix} 1.92 \\ 2.04 \\ 2.46 \\ 1.725 \\ 2.04 \\ 2.25 \\ 1.905 \\ 1.635 \\ 1.965 \\ 2.145 \\ 2.715 \end{bmatrix}
$$

$$
\mathrm{MONEYEACHHALFHOUR} = \begin{bmatrix} 32 \\ 34 \\ 41 \\ 28.75 \\ 34 \\ 37.5 \\ 31.75 \\ 27.25 \\ 32.75 \\ 35.75 \\ 45.25 \end{bmatrix} \qquad \mathrm{TAX} = \begin{bmatrix} 1.92 \\ 2.04 \\ 2.46 \\ 1.725 \\ 2.04 \\ 2.25 \\ 1.905 \\ 1.635 \\ 1.965 \\ 2.145 \\ 2.715 \end{bmatrix}
$$

THIS IS THE TAX ON EACH HALF-HOUR OF SALES. THE TOTAL TAX WILL BE $[1]_{1,11}$ x $[\mathrm{TAX}]_{11,1}$

$$
(1\ 1\ 1\ 1\ 1\ 1\ 1\ 1\ 1\ 1\ 1) \cdot \begin{bmatrix} 1.92 \\ 2.04 \\ 2.46 \\ 1.725 \\ 2.04 \\ 2.25 \\ 1.905 \\ 1.635 \\ 1.965 \\ 2.145 \\ 2.715 \end{bmatrix} = 22.8
\qquad
\begin{bmatrix} 32 \\ 34 \\ 41 \\ 28.75 \\ 34 \\ 37.5 \\ 31.75 \\ 27.25 \\ 32.75 \\ 35.75 \\ 45.25 \end{bmatrix} - \begin{bmatrix} 1.92 \\ 2.04 \\ 2.46 \\ 1.725 \\ 2.04 \\ 2.25 \\ 1.905 \\ 1.635 \\ 1.965 \\ 2.145 \\ 2.715 \end{bmatrix} = \begin{bmatrix} 30.08 \\ 31.96 \\ 38.54 \\ 27.025 \\ 31.96 \\ 35.25 \\ 29.845 \\ 25.615 \\ 30.785 \\ 33.605 \\ 42.535 \end{bmatrix}
$$

THIS IS THE AMOUNT MADE AFTER TAXES

Summing the profits up to a single total, we find he made
$357.20 after taxes.

$$(1\ 1\ 1\ 1\ 1\ 1\ 1\ 1\ 1\ 1\ 1) \cdot \begin{bmatrix} 30.08 \\ 31.96 \\ 38.54 \\ 27.025 \\ 31.96 \\ 35.25 \\ 29.845 \\ 25.615 \\ 30.785 \\ 33.605 \\ 42.535 \end{bmatrix} = 357.2$$

TOTAL AMOUNT HE MADE AFTER TAXES

The government makes $22.80 for the total day, he made $357.20.
Or we can compute it with the TAXDB matrix: (This is computed
directly from the inventory with no accounting necessary)

$$\text{TAXDB} := \begin{bmatrix} 1.41 & .09 \\ 1.175 & .075 \\ .47 & .03 \\ 1.41 & .09 \\ 1.175 & .075 \\ .47 & .03 \\ 1.41 & .09 \\ 1.175 & .075 \\ .47 & .03 \end{bmatrix}$$

$$\text{TAXES2} := \text{PBHINV} \cdot \text{TAXDB}$$

$$
\begin{bmatrix}
3 & 3 & 4 & 2 & 3 & 6 & 2 & 4 & 8 \\
3 & 4 & 4 & 1 & 4 & 5 & 3 & 6 & 3 \\
1 & 3 & 4 & 3 & 6 & 7 & 4 & 7 & 7 \\
1 & 2 & 3 & 2 & 4 & 6 & 1 & 7 & 4 \\
2 & 3 & 5 & 1 & 2 & 3 & 3 & 7 & 12 \\
4 & 5 & 6 & 4 & 5 & 5 & 1 & 4 & 2 \\
4 & 5 & 6 & 1 & 3 & 4 & 1 & 5 & 3 \\
1 & 4 & 5 & 1 & 2 & 4 & 2 & 5 & 6 \\
1 & 4 & 5 & 2 & 3 & 4 & 3 & 6 & 6 \\
2 & 3 & 3 & 3 & 5 & 6 & 1 & 5 & 12 \\
2 & 2 & 4 & 4 & 6 & 7 & 4 & 9 & 7
\end{bmatrix}
\cdot
\begin{bmatrix}
1.41 & .09 \\
1.175 & .075 \\
.47 & .03 \\
1.41 & .09 \\
1.175 & .075 \\
.47 & .03 \\
1.41 & .09 \\
1.175 & .075 \\
.47 & .03
\end{bmatrix}
=
\begin{bmatrix}
30.08 & 1.92 \\
31.96 & 2.04 \\
38.54 & 2.46 \\
27.025 & 1.725 \\
31.96 & 2.04 \\
35.25 & 2.25 \\
29.845 & 1.905 \\
25.615 & 1.635 \\
30.785 & 1.965 \\
33.605 & 2.145 \\
42.535 & 2.715
\end{bmatrix}
\qquad \text{TAXES2} =
\begin{bmatrix}
30.08 & 1.92 \\
31.96 & 2.04 \\
38.54 & 2.46 \\
27.025 & 1.725 \\
31.96 & 2.04 \\
35.25 & 2.25 \\
29.845 & 1.905 \\
25.615 & 1.635 \\
30.785 & 1.965 \\
33.605 & 2.145 \\
42.535 & 2.715
\end{bmatrix}
$$

Out of the $32 made in the first half-hour, we keep $30.08 and Uncle Sam gets 1.92. Net income = NETINCOME := ONE11·TAXES2

$$
(1\ 1\ 1\ 1\ 1\ 1\ 1\ 1\ 1\ 1\ 1) \cdot
\begin{bmatrix}
30.08 & 1.92 \\
31.96 & 2.04 \\
38.54 & 2.46 \\
27.025 & 1.725 \\
31.96 & 2.04 \\
35.25 & 2.25 \\
29.845 & 1.905 \\
25.615 & 1.635 \\
30.785 & 1.965 \\
33.605 & 2.145 \\
42.535 & 2.715
\end{bmatrix}
$$

NETINCOME = (357.2   22.8)

For the day, we made $357.20, and paid the government $22.80. This is out of a total of $380 total brought in.

To check TAXES2 for error, multiply TAXES2 by TWO1. This sums the rows and they should equal to MONEYEACHHALFHOUR.

$$\text{TAXES2}\cdot\text{ONE2} = \begin{bmatrix} 32 \\ 34 \\ 41 \\ 28.75 \\ 34 \\ 37.5 \\ 31.75 \\ 27.25 \\ 32.75 \\ 35.75 \\ 45.25 \end{bmatrix} \qquad \text{MONEYEACHHALFHOUR} = \begin{bmatrix} 32 \\ 34 \\ 41 \\ 28.75 \\ 34 \\ 37.5 \\ 31.75 \\ 27.25 \\ 32.75 \\ 35.75 \\ 45.25 \end{bmatrix}$$

Now we subtract the two and get zero.

$$\text{TAXES2}\cdot\text{ONE2} - \text{MONEYEACHHALFHOUR} = \begin{bmatrix} 0 \\ 0 \\ 0 \\ 3.553\cdot10^{-15} \\ 0 \\ 0 \\ 3.553\cdot10^{-15} \\ 3.553\cdot10^{-15} \\ 0 \\ 0 \\ 0 \end{bmatrix} = 0$$

Which checks.

EMPLOYEE WAGES AND TAXES

$$C := \begin{bmatrix} 3 & 3 & 4 & 2 & 3 & 6 & 2 & 4 & 8 \\ 3 & 4 & 4 & 1 & 4 & 5 & 3 & 6 & 3 \\ 1 & 3 & 4 & 3 & 6 & 7 & 4 & 7 & 7 \\ 1 & 2 & 3 & 2 & 4 & 6 & 1 & 7 & 4 \\ 2 & 3 & 5 & 1 & 2 & 3 & 3 & 7 & 12 \\ 4 & 5 & 6 & 4 & 5 & 5 & 1 & 4 & 2 \\ 4 & 5 & 6 & 1 & 3 & 4 & 1 & 5 & 3 \\ 1 & 4 & 5 & 1 & 2 & 4 & 2 & 5 & 6 \\ 1 & 4 & 5 & 2 & 3 & 4 & 3 & 6 & 6 \\ 2 & 3 & 3 & 3 & 5 & 6 & 1 & 5 & 12 \\ 2 & 2 & 4 & 4 & 6 & 7 & 4 & 9 & 7 \end{bmatrix} \qquad D := \begin{matrix} \text{E1} & \text{E2} & \text{E3} \\ \begin{bmatrix} -1 & -1 & -1 \\ -1 & -1 & -1 \\ -1 & -1 & -1 \\ -1 & -1 & -1 \\ -1 & -1 & -1 \\ -1 & -1 & -1 \\ -1 & -1 & -1 \\ -1 & -1 & -1 \\ -1 & -1 & -1 \\ -1 & -1 & -1 \\ -1 & -1 & -1 \end{bmatrix} \end{matrix}$$

TOTALCOSTINVENTORY :=augment(C,D) E1 = employee # 1, etc. The -1 represents that that employee worked that half-hour period and that it is computed as a cost rather than income. To simplify matters, we will just add the employees wage matrix to the sales matrix.

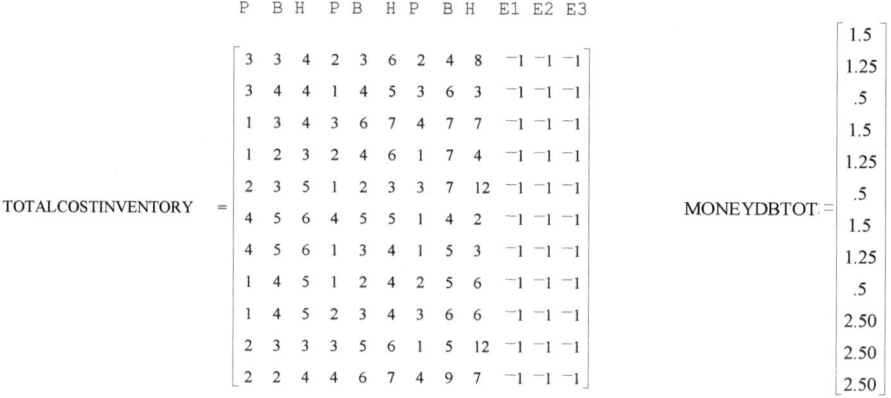

TOTALINCOME :=TOTALCOSTINVENTORY ·MONEYDBTOT

$$
\begin{bmatrix}
3 & 3 & 4 & 2 & 3 & 6 & 2 & 4 & 8 & -1 & -1 & -1 \\
3 & 4 & 4 & 1 & 4 & 5 & 3 & 6 & 3 & -1 & -1 & -1 \\
1 & 3 & 4 & 3 & 6 & 7 & 4 & 7 & 7 & -1 & -1 & -1 \\
1 & 2 & 3 & 2 & 4 & 6 & 1 & 7 & 4 & -1 & -1 & -1 \\
2 & 3 & 5 & 1 & 2 & 3 & 3 & 7 & 12 & -1 & -1 & -1 \\
4 & 5 & 6 & 4 & 5 & 5 & 1 & 4 & 2 & -1 & -1 & -1 \\
4 & 5 & 6 & 1 & 3 & 4 & 1 & 5 & 3 & -1 & -1 & -1 \\
1 & 4 & 5 & 1 & 2 & 4 & 2 & 5 & 6 & -1 & -1 & -1 \\
1 & 4 & 5 & 2 & 3 & 4 & 3 & 6 & 6 & -1 & -1 & -1 \\
2 & 3 & 3 & 3 & 5 & 6 & 1 & 5 & 12 & -1 & -1 & -1 \\
2 & 2 & 4 & 4 & 6 & 7 & 4 & 9 & 7 & -1 & -1 & -1
\end{bmatrix}
\cdot
\begin{bmatrix}
1.5 \\ 1.25 \\ .5 \\ 1.5 \\ 1.25 \\ .5 \\ 1.5 \\ 1.25 \\ .5 \\ 2.50 \\ 2.50 \\ 2.50
\end{bmatrix}
=
\begin{bmatrix}
24.5 \\ 26.5 \\ 33.5 \\ 21.25 \\ 26.5 \\ 30 \\ 24.25 \\ 19.75 \\ 25.25 \\ 28.25 \\ 37.75
\end{bmatrix}
\blacksquare
$$

CHECK:

24.5 + 7.50 = 32
26.5 + 7.50 = 34

$$\text{TOTALINCOME} = \begin{bmatrix} 24.5 \\ 26.5 \\ 33.5 \\ 21.25 \\ 26.5 \\ 30 \\ 24.25 \\ 19.75 \\ 25.25 \\ 28.25 \\ 37.75 \end{bmatrix}$$

After paying his employees, he made \$24.50 the first half-hour, \$30 the 6th half-hour and \$37.75 the last half-hour.

Suppose the first employee was 1/2 hour late, the second employee had to take an hour off for lunch and the third employee stayed an extra 1/2 hour for cleanup. The inventory matrix now looks like:

$$\text{ADJTOTALCOSTINVENTORY} := \begin{bmatrix} 3 & 3 & 4 & 2 & 3 & 6 & 2 & 4 & 8 & 0 & -1 & -1 \\ 3 & 4 & 4 & 1 & 4 & 5 & 3 & 6 & 3 & -1 & -1 & -1 \\ 1 & 3 & 4 & 3 & 6 & 7 & 4 & 7 & 7 & -1 & -1 & -1 \\ 1 & 2 & 3 & 2 & 4 & 6 & 1 & 7 & 4 & -1 & -1 & -1 \\ 2 & 3 & 5 & 1 & 2 & 3 & 3 & 7 & 12 & -1 & -1 & -1 \\ 4 & 5 & 6 & 4 & 5 & 5 & 1 & 4 & 2 & -1 & 0 & -1 \\ 4 & 5 & 6 & 1 & 3 & 4 & 1 & 5 & 3 & -1 & 0 & -1 \\ 1 & 4 & 5 & 1 & 2 & 4 & 2 & 5 & 6 & -1 & -1 & -1 \\ 1 & 4 & 5 & 2 & 3 & 4 & 3 & 6 & 6 & -1 & -1 & -1 \\ 2 & 3 & 3 & 3 & 5 & 6 & 1 & 5 & 12 & -1 & -1 & -1 \\ 2 & 2 & 4 & 4 & 6 & 7 & 4 & 9 & 7 & -1 & -1 & -2 \end{bmatrix}$$

ADJUSTEDNETINCOME := ADJTOTALCOSTINVENTORY ·MONEYDBTOT

$$
\begin{bmatrix}
3 & 3 & 4 & 2 & 3 & 6 & 2 & 4 & 8 & 0 & -1 & -1 \\
3 & 4 & 4 & 1 & 4 & 5 & 3 & 6 & 3 & -1 & -1 & -1 \\
1 & 3 & 4 & 3 & 6 & 7 & 4 & 7 & 7 & -1 & -1 & -1 \\
1 & 2 & 3 & 2 & 4 & 6 & 1 & 7 & 4 & -1 & -1 & -1 \\
2 & 3 & 5 & 1 & 2 & 3 & 3 & 7 & 12 & -1 & -1 & -1 \\
4 & 5 & 6 & 4 & 5 & 5 & 1 & 4 & 2 & -1 & 0 & -1 \\
4 & 5 & 6 & 1 & 3 & 4 & 1 & 5 & 3 & -1 & 0 & -1 \\
1 & 4 & 5 & 1 & 2 & 4 & 2 & 5 & 6 & -1 & -1 & -1 \\
1 & 4 & 5 & 2 & 3 & 4 & 3 & 6 & 6 & -1 & -1 & -1 \\
2 & 3 & 3 & 3 & 5 & 6 & 1 & 5 & 12 & -1 & -1 & -1 \\
2 & 2 & 4 & 4 & 6 & 7 & 4 & 9 & 7 & -1 & -1 & -2
\end{bmatrix}
\cdot
\begin{bmatrix}
1.5 \\ 1.25 \\ .5 \\ 1.5 \\ 1.25 \\ .5 \\ 1.5 \\ 1.25 \\ .5 \\ 2.50 \\ 2.50 \\ 2.50
\end{bmatrix}
=
\begin{bmatrix}
27 \\ 26.5 \\ 33.5 \\ 21.25 \\ 26.5 \\ 32.5 \\ 26.75 \\ 19.75 \\ 25.25 \\ 28.25 \\ 35.25
\end{bmatrix} .
$$

$$
\text{ADJUSTEDNETINCOME} =
\begin{bmatrix}
27 \\ 26.5 \\ 33.5 \\ 21.25 \\ 26.5 \\ 32.5 \\ 26.75 \\ 19.75 \\ 25.25 \\ 28.25 \\ 35.25
\end{bmatrix}
$$

With some employees missing some time and another working an extra half-hour without overtime, in the first half-hour he made $27, in the last half-hour he made $32.25.

CHECK:

27+5=32
26.5+7.50=34, etc.

TOTAL WAGES PER EMPLOYEE

$$
\text{THREE1} := \begin{pmatrix} 1 \\ 1 \\ 1 \end{pmatrix}
$$

$$
\begin{bmatrix}
3 & 3 & 4 & 2 & 3 & 6 & 2 & 4 & 8 & 0 & -1 & -1 \\
3 & 4 & 4 & 1 & 4 & 5 & 3 & 6 & 3 & -1 & -1 & -1 \\
1 & 3 & 4 & 3 & 6 & 7 & 4 & 7 & 7 & -1 & -1 & -1 \\
1 & 2 & 3 & 2 & 4 & 6 & 1 & 7 & 4 & -1 & -1 & -1 \\
2 & 3 & 5 & 1 & 2 & 3 & 3 & 7 & 12 & -1 & -1 & -1 \\
4 & 5 & 6 & 4 & 5 & 5 & 1 & 4 & 2 & -1 & 0 & -1 \\
4 & 5 & 6 & 1 & 3 & 4 & 1 & 5 & 3 & -1 & 0 & -1 \\
1 & 4 & 5 & 1 & 2 & 4 & 2 & 5 & 6 & -1 & -1 & -1 \\
1 & 4 & 5 & 2 & 3 & 4 & 3 & 6 & 6 & -1 & -1 & -1 \\
2 & 3 & 3 & 3 & 5 & 6 & 1 & 5 & 12 & -1 & -1 & -1 \\
2 & 2 & 4 & 4 & 6 & 7 & 4 & 9 & 7 & -1 & -1 & -2
\end{bmatrix}
\cdot
\begin{bmatrix}
0 & 0 & 0 \\
0 & 0 & 0 \\
0 & 0 & 0 \\
0 & 0 & 0 \\
0 & 0 & 0 \\
0 & 0 & 0 \\
0 & 0 & 0 \\
0 & 0 & 0 \\
0 & 0 & 0 \\
2.5 & 0 & 0 \\
0 & 2.5 & 0 \\
0 & 0 & 2.5
\end{bmatrix}
=
\begin{bmatrix}
0 & -2.5 & -2.5 \\
-2.5 & -2.5 & -2.5 \\
-2.5 & -2.5 & -2.5 \\
-2.5 & -2.5 & -2.5 \\
-2.5 & -2.5 & -2.5 \\
-2.5 & 0 & -2.5 \\
-2.5 & 0 & -2.5 \\
-2.5 & -2.5 & -2.5 \\
-2.5 & -2.5 & -2.5 \\
-2.5 & -2.5 & -2.5 \\
-2.5 & -2.5 & -5
\end{bmatrix}
$$

$$
\text{EMPLOYEEPAYPERHALFHOUR} :=
\begin{bmatrix}
0 & 0 & 0 \\
0 & 0 & 0 \\
0 & 0 & 0 \\
0 & 0 & 0 \\
0 & 0 & 0 \\
0 & 0 & 0 \\
0 & 0 & 0 \\
0 & 0 & 0 \\
0 & 0 & 0 \\
2.5 & 0 & 0 \\
0 & 2.5 & 0 \\
0 & 0 & 2.5
\end{bmatrix}
$$

Here we are just computing the wages, not costs.

$$
\begin{bmatrix}
0 & -2.5 & -2.5 \\
-2.5 & -2.5 & -2.5 \\
-2.5 & -2.5 & -2.5 \\
-2.5 & -2.5 & -2.5 \\
-2.5 & -2.5 & -2.5 \\
-2.5 & 0 & -2.5 \\
-2.5 & 0 & -2.5 \\
-2.5 & -2.5 & -2.5 \\
-2.5 & -2.5 & -2.5 \\
-2.5 & -2.5 & -2.5 \\
-2.5 & -2.5 & -5
\end{bmatrix}
\cdot
\begin{pmatrix}
1 \\
1 \\
1
\end{pmatrix}
=
\begin{bmatrix}
-5 \\
-7.5 \\
-7.5 \\
-7.5 \\
-7.5 \\
-5 \\
-5 \\
-7.5 \\
-7.5 \\
-7.5 \\
-10
\end{bmatrix}
$$

THE NEGATIVE SIGNS SHOW THAT THIS IS MONEY PAID OUT RATHER THAN
INCOME RECEIVED.

EMPLOYEEPAY $:=$ ADJTOTALCOSTINVENTORY $\cdot$ EMPLOYEEPAYPERHALFHOUR $\cdot$ THREE1

$$
\text{EMPLOYEEPAY} =
\begin{bmatrix}
-5 \\
-7.5 \\
-7.5 \\
-7.5 \\
-7.5 \\
-5 \\
-5 \\
-7.5 \\
-7.5 \\
-7.5 \\
-10
\end{bmatrix}
$$

$$
(1\ 1\ 1\ 1\ 1\ 1\ 1\ 1\ 1\ 1\ 1) \cdot
\begin{bmatrix}
0 & -2.5 & -2.5 \\
-2.5 & -2.5 & -2.5 \\
-2.5 & -2.5 & -2.5 \\
-2.5 & -2.5 & -2.5 \\
-2.5 & -2.5 & -2.5 \\
-2.5 & 0 & -2.5 \\
-2.5 & 0 & -2.5 \\
-2.5 & -2.5 & -2.5 \\
-2.5 & -2.5 & -2.5 \\
-2.5 & -2.5 & -2.5 \\
-2.5 & -2.5 & -5
\end{bmatrix}
= (-25\ \ -22.5\ \ -30)
$$

The first employee made $25, the second employee made $22.50
and the third employee made $30.

And finally, suppose the Social Security administration
takes out 5%, Uncle Sam takes out 10% and the state takes out
6% of the employees wages:

ONE11 $:=(1\ 1\ 1\ 1\ 1\ 1\ 1\ 1\ 1\ 1\ 1)$

$$
\text{ADJTAXDB} :=
\begin{pmatrix}
.05 & .1 & .06 \\
.05 & .1 & .06 \\
.05 & .1 & .06
\end{pmatrix}
$$

EMPINDIVIDUALWAGES $:=(-25\ -22.5\ -30\ )$

DIAGEMPINDIVWAGES $:=$ diag $\left(\text{EMPINDIVIDUALWAGES}^{T}\right)$

TOTALADJUSTEDTAXES $:=$ DIAGEMPINDIVWAGES ADJTAXDB

$$\text{TOTALADJUSTEDTAXES} = \begin{pmatrix} -1.25 & ^-2.5 & ^-1.5 \\ ^-1.125 & ^-2.25 & ^-1.35 \\ ^-1.5 & ^-3 & ^-1.8 \end{pmatrix}$$

SUMTOTALADJUSTEDTAXES $:=$ TOTALADJUSTEDTAXES $\cdot$ THREE1

$$\text{SUMTOTALADJUSTEDTAXES} = \begin{pmatrix} ^-5.25 \\ ^-4.725 \\ ^-6.3 \end{pmatrix}$$

NETPAY $:=$ EMPINDIVIDUALWAGES$-$ SUMTOTALADJUSTEDTAXES ^{T}

NETPAY $=(^-19.75\ \ ^-17.775\ \ ^-23.7\ )$

Employee 1 takes home a net $19.75, Employee 2 takes home a net of $17.78 and Employee 3 takes home a net of $23.70 after all taxes and deductions.

To check the solution:

$(-25\ \ -22.5\ \ -30\ )$

$-25\cdot.05 = ^-1.25$ $-25\cdot.1 = ^-2.5$ $-25\cdot.06 = ^-1.5$ $-1.25 +^-2.5 +^-1.5 +25 = 19.75$

$22.5\cdot.05 = 1.125$ $22.5\cdot.1 = 2.25$ $22.5\cdot.06 = 1.35$ $22.5 - 1.125 - 2.25 - 1.35 = 17.775$

$-30\cdot.05 = ^-1.5$ $-30\cdot.1 = ^-3$ $-30\cdot.06 = ^-1.8$ $30 - 1.5 - 3 - 1.8 = 23.7$

Suppose now he wishes to check out how much money he made in the 6th half-hour and the first half-hour:

$$PBHINV = \begin{bmatrix} 3 & 3 & 4 & 2 & 3 & 6 & 2 & 4 & 8 \\ 3 & 4 & 4 & 1 & 4 & 5 & 3 & 6 & 3 \\ 1 & 3 & 4 & 3 & 6 & 7 & 4 & 7 & 7 \\ 1 & 2 & 3 & 2 & 4 & 6 & 1 & 7 & 4 \\ 2 & 3 & 5 & 1 & 2 & 3 & 3 & 7 & 12 \\ 4 & 5 & 6 & 4 & 5 & 5 & 1 & 4 & 2 \\ 4 & 5 & 6 & 1 & 3 & 4 & 1 & 5 & 3 \\ 1 & 4 & 5 & 1 & 2 & 4 & 2 & 5 & 6 \\ 1 & 4 & 5 & 2 & 3 & 4 & 3 & 6 & 6 \\ 2 & 3 & 3 & 3 & 5 & 6 & 1 & 5 & 12 \\ 2 & 2 & 4 & 4 & 6 & 7 & 4 & 9 & 7 \end{bmatrix} \qquad MONEYDB = \begin{bmatrix} 1.5 \\ 1.25 \\ 0.5 \\ 1.5 \\ 1.25 \\ 0.5 \\ 1.5 \\ 1.25 \\ 0.5 \end{bmatrix}$$

$PBHINVTRN := PBHINV^T$

$a := PBHINVTRN^{<6>}$ Removes the 6th column from the transposed inventory matrix

$b := PBHINVTRN^{<1>}$ Removes the 1st column from the transposed inventory matrix

$$a = \begin{bmatrix} 4 \\ 5 \\ 6 \\ 4 \\ 5 \\ 5 \\ 1 \\ 4 \\ 2 \end{bmatrix} \qquad b = \begin{bmatrix} 3 \\ 3 \\ 4 \\ 2 \\ 3 \\ 6 \\ 2 \\ 4 \\ 8 \end{bmatrix}$$

$DIAGa := diag(a)$ $AP6 := DIAGa \, MONEYDB$

$DIAGb := diag(b)$ $AP1 := DIAGb \, MONEYDB$

$$AP6 = \begin{bmatrix} 6 \\ 6.25 \\ 3 \\ 6 \\ 6.25 \\ 2.5 \\ 1.5 \\ 5 \\ 1 \end{bmatrix} \qquad AP1 = \begin{bmatrix} 4.5 \\ 3.75 \\ 2 \\ 3 \\ 3.75 \\ 3 \\ 3 \\ 5 \\ 4 \end{bmatrix}$$

In the sixth half-hour period, he sold $6 worth of polish in the first cart, 6.25 worth of burritos in the second cart and $1 worth of hot-dogs in the third cart, etc. [AP1] says $4.50 worth of polish was sold in the first half-hour in the first cart, $3.75 worth of burritos in the second cart and $4 worth of hot-dogs in the third cart, etc. Let's find the taxes for the first and sixth half-hours:

$$TAXDB := \begin{bmatrix} 1.41 & .09 \\ 1.175 & .075 \\ .47 & .03 \\ 1.41 & .09 \\ 1.175 & .075 \\ .47 & .03 \\ 1.41 & .09 \\ 1.175 & .075 \\ .47 & .03 \end{bmatrix}$$

$$DIAGa\,TAXDB = \begin{bmatrix} 5.64 & 0.36 \\ 5.875 & 0.375 \\ 2.82 & 0.18 \\ 5.64 & 0.36 \\ 5.875 & 0.375 \\ 2.35 & 0.15 \\ 1.41 & 0.09 \\ 4.7 & 0.3 \\ 0.94 & 0.06 \end{bmatrix}$$

$$ONE9 := (1\ 1\ 1\ 1\ 1\ 1\ 1\ 1\ 1)$$

$$\text{DIAG}b\text{ TAXDB} = \begin{bmatrix} 4.23 & 0.27 \\ 3.525 & 0.225 \\ 1.88 & 0.12 \\ 2.82 & 0.18 \\ 3.525 & 0.225 \\ 2.82 & 0.18 \\ 2.82 & 0.18 \\ 4.7 & 0.3 \\ 3.76 & 0.24 \end{bmatrix}$$

FOR THE AP MATRICES, FOR THE [SOL] MATRIX:

AP6Tx $\doteq$ DIAGb TAXDB

TOTALHALFHOUR6 $\doteq$ ONE9 AP6Tx

TOTALHALFHOUR6 = (30.08   1.92)

USING THE PROPERTIES OF THE HALF-MULTIPLIER OPERATOR

We will just summarize the results here and check them against what we have done above.

$$\text{PBHINVTRN} = \begin{bmatrix} 3 & 3 & 1 & 1 & 2 & 4 & 4 & 1 & 1 & 2 & 2 \\ 3 & 4 & 3 & 2 & 3 & 5 & 5 & 4 & 4 & 3 & 2 \\ 4 & 4 & 4 & 3 & 5 & 6 & 6 & 5 & 5 & 3 & 4 \\ 2 & 1 & 3 & 2 & 1 & 4 & 1 & 1 & 2 & 3 & 4 \\ 3 & 4 & 6 & 4 & 2 & 5 & 3 & 2 & 3 & 5 & 6 \\ 6 & 5 & 7 & 6 & 3 & 5 & 4 & 4 & 4 & 6 & 7 \\ 2 & 3 & 4 & 1 & 3 & 1 & 1 & 2 & 3 & 1 & 4 \\ 4 & 6 & 7 & 7 & 7 & 4 & 5 & 5 & 6 & 5 & 9 \\ 8 & 3 & 7 & 4 & 12 & 2 & 3 & 6 & 6 & 12 & 7 \end{bmatrix}$$

$([\text{INV}]_{9,11}[1]_{11,1})$ o $[\text{MONEYDB}]_{9,1}$

$$
\begin{bmatrix}
3 & 3 & 1 & 1 & 2 & 4 & 4 & 1 & 1 & 2 & 2 \\
3 & 4 & 3 & 2 & 3 & 5 & 5 & 4 & 4 & 3 & 2 \\
4 & 4 & 4 & 3 & 5 & 6 & 6 & 5 & 5 & 3 & 4 \\
2 & 1 & 3 & 2 & 1 & 4 & 1 & 1 & 2 & 3 & 4 \\
3 & 4 & 6 & 4 & 2 & 5 & 3 & 2 & 3 & 5 & 6 \\
6 & 5 & 7 & 6 & 3 & 5 & 4 & 4 & 4 & 6 & 7 \\
2 & 3 & 4 & 1 & 3 & 1 & 1 & 2 & 3 & 1 & 4 \\
4 & 6 & 7 & 7 & 7 & 4 & 5 & 5 & 6 & 5 & 9 \\
8 & 3 & 7 & 4 & 12 & 2 & 3 & 6 & 6 & 12 & 7
\end{bmatrix}
\cdot
\begin{bmatrix}
1 \\ 1 \\ 1 \\ 1 \\ 1 \\ 1 \\ 1 \\ 1 \\ 1 \\ 1 \\ 1
\end{bmatrix}
=
\begin{bmatrix}
24 \\ 38 \\ 49 \\ 24 \\ 43 \\ 57 \\ 25 \\ 65 \\ 70
\end{bmatrix}
$$

Here we sum the rows of the transposed inventory matrix.

$$
\text{PBHINVTRN} \cdot \text{ELEVEN1} =
\begin{bmatrix}
24 \\ 38 \\ 49 \\ 24 \\ 43 \\ 57 \\ 25 \\ 65 \\ 70
\end{bmatrix}
$$

$$
\begin{bmatrix}
24 \\ 38 \\ 49 \\ 24 \\ 43 \\ 57 \\ 25 \\ 65 \\ 70
\end{bmatrix}
\cdot
\begin{bmatrix}
1.5 \\ 1.25 \\ .5 \\ 1.5 \\ 1.25 \\ .5 \\ 1.5 \\ 1.25 \\ .5
\end{bmatrix}
=
\begin{bmatrix}
36 \\ 47.5 \\ 24.5 \\ 36 \\ 53.75 \\ 28.5 \\ 37.5 \\ 81.25 \\ 35
\end{bmatrix}
$$

Here we hollow dot multiply the summed rows by MONEYDB.

He sold 24 polish sausages in cart one in 5½ hours for a
total of $36, he sold 43 burritos in cart 2 for a total of $53.75
and in cart 3, he sold 70 hot-dogs for a total of $35.

$$ONE9 \cdot \begin{bmatrix} 36 \\ 47.5 \\ 24.5 \\ 36 \\ 53.75 \\ 28.5 \\ 37.5 \\ 81.25 \\ 35 \end{bmatrix} = 380$$

CHECK: The sum total still sums to $380.

Now lets reverse multiply the transposed inventory matrix by the MONEYDB matrix.

[MONEYDB] ○ [PBHINVTRN]

DIAGMONEYDB = diag(MONEYDB)
ITEMPAGEMONEY = DIAGMONEYDB PBHINVTRN

$$DIAGMONEYDB = \begin{bmatrix} 1.5 & 0 & 0 & 0 & 0 & 0 & 0 & 0 & 0 \\ 0 & 1.25 & 0 & 0 & 0 & 0 & 0 & 0 & 0 \\ 0 & 0 & 0.5 & 0 & 0 & 0 & 0 & 0 & 0 \\ 0 & 0 & 0 & 1.5 & 0 & 0 & 0 & 0 & 0 \\ 0 & 0 & 0 & 0 & 1.25 & 0 & 0 & 0 & 0 \\ 0 & 0 & 0 & 0 & 0 & 0.5 & 0 & 0 & 0 \\ 0 & 0 & 0 & 0 & 0 & 0 & 1.5 & 0 & 0 \\ 0 & 0 & 0 & 0 & 0 & 0 & 0 & 1.25 & 0 \\ 0 & 0 & 0 & 0 & 0 & 0 & 0 & 0 & 0.5 \end{bmatrix}$$

$$\begin{bmatrix} 1.5 & 0 & 0 & 0 & 0 & 0 & 0 & 0 & 0 \\ 0 & 1.25 & 0 & 0 & 0 & 0 & 0 & 0 & 0 \\ 0 & 0 & 0.5 & 0 & 0 & 0 & 0 & 0 & 0 \\ 0 & 0 & 0 & 1.5 & 0 & 0 & 0 & 0 & 0 \\ 0 & 0 & 0 & 0 & 1.25 & 0 & 0 & 0 & 0 \\ 0 & 0 & 0 & 0 & 0 & 0.5 & 0 & 0 & 0 \\ 0 & 0 & 0 & 0 & 0 & 0 & 1.5 & 0 & 0 \\ 0 & 0 & 0 & 0 & 0 & 0 & 0 & 1.25 & 0 \\ 0 & 0 & 0 & 0 & 0 & 0 & 0 & 0 & 0.5 \end{bmatrix} \cdot \begin{bmatrix} 3 & 3 & 1 & 1 & 2 & 4 & 4 & 1 & 1 & 2 & 2 \\ 3 & 4 & 3 & 2 & 3 & 5 & 5 & 4 & 4 & 3 & 2 \\ 4 & 4 & 4 & 3 & 5 & 6 & 6 & 5 & 5 & 3 & 4 \\ 2 & 1 & 3 & 2 & 1 & 4 & 1 & 1 & 2 & 3 & 4 \\ 3 & 4 & 6 & 4 & 2 & 5 & 3 & 2 & 3 & 5 & 6 \\ 6 & 5 & 7 & 6 & 3 & 5 & 4 & 4 & 4 & 6 & 7 \\ 2 & 3 & 4 & 1 & 3 & 1 & 1 & 2 & 3 & 1 & 4 \\ 4 & 6 & 7 & 7 & 7 & 4 & 5 & 5 & 6 & 5 & 9 \\ 8 & 3 & 7 & 4 & 12 & 2 & 3 & 6 & 6 & 12 & 7 \end{bmatrix} =$$

$$ITEMPAGEMONEY = \begin{bmatrix} 4.5 & 4.5 & 1.5 & 1.5 & 3 & 6 & 6 & 1.5 & 1.5 & 3 & 3 \\ 3.75 & 5 & 3.75 & 2.5 & 3.75 & 6.25 & 6.25 & 5 & 5 & 3.75 & 2.5 \\ 2 & 2 & 2 & 1.5 & 2.5 & 3 & 3 & 2.5 & 2.5 & 1.5 & 2 \\ 3 & 1.5 & 4.5 & 3 & 1.5 & 6 & 1.5 & 1.5 & 3 & 4.5 & 6 \\ 3.75 & 5 & 7.5 & 5 & 2.5 & 6.25 & 3.75 & 2.5 & 3.75 & 6.25 & 7.5 \\ 3 & 2.5 & 3.5 & 3 & 1.5 & 2.5 & 2 & 2 & 2 & 3 & 3.5 \\ 3 & 4.5 & 6 & 1.5 & 4.5 & 1.5 & 1.5 & 3 & 4.5 & 1.5 & 6 \\ 5 & 7.5 & 8.75 & 8.75 & 8.75 & 5 & 6.25 & 6.25 & 7.5 & 6.25 & 11.25 \\ 4 & 1.5 & 3.5 & 2 & 6 & 1 & 1.5 & 3 & 3 & 6 & 3.5 \end{bmatrix}$$

We can see at a glance that in the second half-hour, he sold $4.50 worth of polish sausage in cart one, $5 worth of burritos and $2 worth of hotdogs. In the eighth half-hour in cart 3, he sold $3 worth of polish, $6.25 worth of burritos and $3 worth of hot-dogs, etc.

Lets see how much tax he paid for each sum of items each half-hour. We can do this two ways, there is a tax of 9¢ on every polish sausage, .075¢ on each burrito and 3¢ on every hot-dog. We can multiply this by the inventory, or we can take the accountpage which has the sales in terms of money and multiply each value by .06 for 6%. i.e.

$$ITEMTAX = TAXDB^{<2>}$$

$$ITEMTAX = \begin{bmatrix} 0.09 \\ 0.075 \\ 0.03 \\ 0.09 \\ 0.075 \\ 0.03 \\ 0.09 \\ 0.075 \\ 0.03 \end{bmatrix} \qquad TAXA = \begin{bmatrix} 0.06 \\ 0.06 \\ 0.06 \\ 0.06 \\ 0.06 \\ 0.06 \\ 0.06 \\ 0.06 \\ 0.06 \end{bmatrix}$$

[ITEMTAX]○[PBHINVTRN] and [TAXA]○[ITEMPAGEMONEY]

DIAGITEMTAX = diag(ITEMTAX)   ITEMTAXES = DIAGITEMTAX·PBHINVTRN
DIAGTAXA = diag(TAXA)   TAXESACCT = DIAGTAXA·ITEMPAGEMONEY

$$
ITEMTAXES = \begin{bmatrix}
0.27 & 0.27 & 0.09 & 0.09 & 0.18 & 0.36 & 0.36 & 0.09 & 0.09 & 0.18 & 0.18 \\
0.225 & 0.3 & 0.225 & 0.15 & 0.225 & 0.375 & 0.375 & 0.3 & 0.3 & 0.225 & 0.15 \\
0.12 & 0.12 & 0.12 & 0.09 & 0.15 & 0.18 & 0.18 & 0.15 & 0.15 & 0.09 & 0.12 \\
0.18 & 0.09 & 0.27 & 0.18 & 0.09 & 0.36 & 0.09 & 0.09 & 0.18 & 0.27 & 0.36 \\
0.225 & 0.3 & 0.45 & 0.3 & 0.15 & 0.375 & 0.225 & 0.15 & 0.225 & 0.375 & 0.45 \\
0.18 & 0.15 & 0.21 & 0.18 & 0.09 & 0.15 & 0.12 & 0.12 & 0.12 & 0.18 & 0.21 \\
0.18 & 0.27 & 0.36 & 0.09 & 0.27 & 0.09 & 0.09 & 0.18 & 0.27 & 0.09 & 0.36 \\
0.3 & 0.45 & 0.525 & 0.525 & 0.525 & 0.3 & 0.375 & 0.375 & 0.45 & 0.375 & 0.675 \\
0.24 & 0.09 & 0.21 & 0.12 & 0.36 & 0.06 & 0.09 & 0.18 & 0.18 & 0.36 & 0.21
\end{bmatrix}
$$

$$
TAXESACCT = \begin{bmatrix}
0.27 & 0.27 & 0.09 & 0.09 & 0.18 & 0.36 & 0.36 & 0.09 & 0.09 & 0.18 & 0.18 \\
0.225 & 0.3 & 0.225 & 0.15 & 0.225 & 0.375 & 0.375 & 0.3 & 0.3 & 0.225 & 0.15 \\
0.12 & 0.12 & 0.12 & 0.09 & 0.15 & 0.18 & 0.18 & 0.15 & 0.15 & 0.09 & 0.12 \\
0.18 & 0.09 & 0.27 & 0.18 & 0.09 & 0.36 & 0.09 & 0.09 & 0.18 & 0.27 & 0.36 \\
0.225 & 0.3 & 0.45 & 0.3 & 0.15 & 0.375 & 0.225 & 0.15 & 0.225 & 0.375 & 0.45 \\
0.18 & 0.15 & 0.21 & 0.18 & 0.09 & 0.15 & 0.12 & 0.12 & 0.12 & 0.18 & 0.21 \\
0.18 & 0.27 & 0.36 & 0.09 & 0.27 & 0.09 & 0.09 & 0.18 & 0.27 & 0.09 & 0.36 \\
0.3 & 0.45 & 0.525 & 0.525 & 0.525 & 0.3 & 0.375 & 0.375 & 0.45 & 0.375 & 0.675 \\
0.24 & 0.09 & 0.21 & 0.12 & 0.36 & 0.06 & 0.09 & 0.18 & 0.18 & 0.36 & 0.21
\end{bmatrix}
$$

27¢ of the $4.50 in polish sausage sales in the first half hour went to taxes. The highest taxes are in the 3rd, 4th, 5th and 11th half-hours in cart 3 for burritos, which tie at 52.5¢ and a high of 67.5¢ in the 11th half-hour.

The accounting is the same, it all depends upon whether it is easier to tax the inventory or tax the accounting.

CHECK: 4 polish sausage were sold in cart one in the first half-hour, there is a tax of 9¢ on each sausage. So we have 4x.09= 27¢. In the 11th half-hour in cart 3, 9 burritos were sold for a total tax of 9x.075 = 67.5¢.

$4.50 worth polish sausage were sold in cart one in the first half-hour, there is a tax of 6¢ on each sausage. So we have 4.50x.06= 27¢. In the 11th half-hour in cart 3, $11.25 worth of burritos were sold for a total tax of 11.25x.06= 67.5¢.

STATISTICAL ANALYSIS OF SALES

There is a lot more we can do to improve this inventory, so I am going to go into the Analysis of Variance for the sales between the three carts. Note that the Database Matrices I use in the statistical analysis are the same which are used to separate the sales by cart (or by store) and to separate the sales to each individual item (like only the sale of hot dogs, only the total of polish sausages sold and only the sales total of burritos (these could also represent every item sold in a store or in stores controlled by chains). This is the exact same statistical analysis you'll find in Section 2.7 in the chapters on statistics in the Handbook of Statistical Computations, 2nd edition by Bruning and Kertz. If you do not understand what I've done here, please read their book dealing with Statistics for a basic background.

STEP 1: Table the data:

$$[A]_{11,9} = \begin{array}{|ccc\ ccc\ ccc|}
P & B & H & P & B & H & P & B & H \\
3 & 3 & 4 & 2 & 3 & 6 & 2 & 4 & 8 \\
3 & 4 & 4 & 1 & 4 & 5 & 3 & 6 & 3 \\
1 & 3 & 4 & 3 & 6 & 7 & 4 & 7 & 7 \\
1 & 2 & 3 & 2 & 4 & 6 & 1 & 7 & 4 \\
2 & 3 & 5 & 1 & 2 & 3 & 3 & 7 & 12 \\
4 & 5 & 6 & 4 & 5 & 5 & 1 & 4 & 2 \\
4 & 5 & 6 & 1 & 3 & 4 & 1 & 5 & 3 \\
1 & 4 & 5 & 1 & 2 & 4 & 2 & 5 & 6 \\
1 & 4 & 5 & 2 & 3 & 4 & 3 & 6 & 6 \\
2 & 3 & 3 & 3 & 5 & 6 & 1 & 5 & 12 \\
2 & 2 & 4 & 4 & 6 & 7 & 4 & 9 & 7
\end{array}$$

VARIABLES:

$ONE11 := (1 \ 1 \ 1 \ 1 \ 1 \ 1 \ 1 \ 1 \ 1 \ 1 \ 1)$ $\qquad NINE1 := ONE9^{T}$

$THREE1 := \begin{pmatrix} 1 \\ 1 \\ 1 \end{pmatrix}$ $\quad ONE9 := (1 \ 1 \ 1 \ 1 \ 1 \ 1 \ 1 \ 1 \ 1)$

$$A_{11,9} := \begin{bmatrix} 3 & 3 & 4 & 2 & 3 & 6 & 2 & 4 & 8 \\ 3 & 4 & 4 & 1 & 4 & 5 & 3 & 6 & 3 \\ 1 & 3 & 4 & 3 & 6 & 7 & 4 & 7 & 7 \\ 1 & 2 & 3 & 2 & 4 & 6 & 1 & 7 & 4 \\ 2 & 3 & 5 & 1 & 2 & 3 & 3 & 7 & 12 \\ 4 & 5 & 6 & 4 & 5 & 5 & 1 & 4 & 2 \\ 4 & 5 & 6 & 1 & 3 & 4 & 1 & 5 & 3 \\ 1 & 4 & 5 & 1 & 2 & 4 & 2 & 5 & 6 \\ 1 & 4 & 5 & 2 & 3 & 4 & 3 & 6 & 6 \\ 2 & 3 & 3 & 3 & 5 & 6 & 1 & 5 & 12 \\ 2 & 2 & 4 & 4 & 6 & 7 & 4 & 9 & 7 \end{bmatrix} \quad DB1 := \begin{bmatrix} 1 & 0 & 0 \\ 1 & 0 & 0 \\ 1 & 0 & 0 \\ 0 & 1 & 0 \\ 0 & 1 & 0 \\ 0 & 1 & 0 \\ 0 & 0 & 1 \\ 0 & 0 & 1 \\ 0 & 0 & 1 \end{bmatrix} \quad DB3 := \begin{bmatrix} 1 & 0 & 0 \\ 0 & 1 & 0 \\ 0 & 0 & 1 \\ 1 & 0 & 0 \\ 0 & 1 & 0 \\ 0 & 0 & 1 \\ 1 & 0 & 0 \\ 0 & 1 & 0 \\ 0 & 0 & 1 \end{bmatrix}$$

$$SS := \begin{bmatrix} 467 \\ 181 \\ 39 \\ 142 \\ 286 \\ 170 \\ 19 \\ 97 \end{bmatrix} \quad ERROR := \begin{bmatrix} 142 \\ 30 \\ 142 \\ 30 \\ 142 \\ 30 \\ 142 \\ 30 \\ 97 \\ 60 \\ 97 \\ 60 \\ 97 \\ 60 \\ 97 \\ 60 \end{bmatrix} \quad df := \begin{bmatrix} 98 \\ 32 \\ 2 \\ 30 \\ 66 \\ 2 \\ 4 \\ 60 \end{bmatrix}$$

STEP 2: Add the scores in each column for each cart:

SUM1CARTS $= [1]_{1,11} [A]_{11,9}$

SUM1CARTS $:=$ ONE11·$A_{11,9}$

$$\text{SUM1CARTS} := (1 \quad 1 \quad 1 \quad 1 \quad 1 \quad 1 \quad 1 \quad 1 \quad 1 \quad 1 \quad 1) \cdot \begin{bmatrix} 3 & 3 & 4 & 2 & 3 & 6 & 2 & 4 & 8 \\ 3 & 4 & 4 & 1 & 4 & 5 & 3 & 6 & 3 \\ 1 & 3 & 4 & 3 & 6 & 7 & 4 & 7 & 7 \\ 1 & 2 & 3 & 2 & 4 & 6 & 1 & 7 & 4 \\ 2 & 3 & 5 & 1 & 2 & 3 & 3 & 7 & 12 \\ 4 & 5 & 6 & 4 & 5 & 5 & 1 & 4 & 2 \\ 4 & 5 & 6 & 1 & 3 & 4 & 1 & 5 & 3 \\ 1 & 4 & 5 & 1 & 2 & 4 & 2 & 5 & 6 \\ 1 & 4 & 5 & 2 & 3 & 4 & 3 & 6 & 6 \\ 2 & 3 & 3 & 3 & 5 & 6 & 1 & 5 & 12 \\ 2 & 2 & 4 & 4 & 6 & 7 & 4 & 9 & 7 \end{bmatrix}$$

SUM1CARTS = (24   38   49   24   43   57   25   65   70)

HP-48G PROGRAM:

```
RCL ONE11
RCL A
x
STO SUM1CARTS
```

STEP 3: Obtain the sum of each cart by adding the sums of the
individual sales.

PROGRAM: GROUPSUMSCARTS = $[1]_{1,11}[A]_{11,9}[DB1]_{9,3}$

GROUPSUMSCARTS := ONE11·$A_{11,9}$·DB1

$$\text{GROUPSUMSCARTS} := (1 \; 1 \; 1 \; 1 \; 1 \; 1 \; 1 \; 1 \; 1 \; 1 \; 1) \cdot \begin{bmatrix} 3 & 3 & 4 & 2 & 3 & 6 & 2 & 4 & 8 \\ 3 & 4 & 4 & 1 & 4 & 5 & 3 & 6 & 3 \\ 1 & 3 & 4 & 3 & 6 & 7 & 4 & 7 & 7 \\ 1 & 2 & 3 & 2 & 4 & 6 & 1 & 7 & 4 \\ 2 & 3 & 5 & 1 & 2 & 3 & 3 & 7 & 12 \\ 4 & 5 & 6 & 4 & 5 & 5 & 1 & 4 & 2 \\ 4 & 5 & 6 & 1 & 3 & 4 & 1 & 5 & 3 \\ 1 & 4 & 5 & 1 & 2 & 4 & 2 & 5 & 6 \\ 1 & 4 & 5 & 2 & 3 & 4 & 3 & 6 & 6 \\ 2 & 3 & 3 & 3 & 5 & 6 & 1 & 5 & 12 \\ 2 & 2 & 4 & 4 & 6 & 7 & 4 & 9 & 7 \end{bmatrix} \cdot \begin{bmatrix} 1 & 0 & 0 \\ 1 & 0 & 0 \\ 1 & 0 & 0 \\ 0 & 1 & 0 \\ 0 & 1 & 0 \\ 0 & 1 & 0 \\ 0 & 0 & 1 \\ 0 & 0 & 1 \\ 0 & 0 & 1 \end{bmatrix}$$

GROUPSUMSCARTS = (111   124   160)

HP-48G PROGRAM:

RCL SUM1

RCL DB1

x

STO GROUPSUMSCARTS   [111 124 160]

STEP 4:   Add the scores for each item in each Cart:

PROGRAM: ITEMSUMS = $[A]_{11,9}[DB1]_{9,3}$

$$\text{ITEMSUMS} := A_{11,9} \cdot DB1$$

$$\text{ITEMSUMS} :=
\begin{bmatrix}
3 & 3 & 4 & 2 & 3 & 6 & 2 & 4 & 8 \\
3 & 4 & 4 & 1 & 4 & 5 & 3 & 6 & 3 \\
1 & 3 & 4 & 3 & 6 & 7 & 4 & 7 & 7 \\
1 & 2 & 3 & 2 & 4 & 6 & 1 & 7 & 4 \\
2 & 3 & 5 & 1 & 2 & 3 & 3 & 7 & 12 \\
4 & 5 & 6 & 4 & 5 & 5 & 1 & 4 & 2 \\
4 & 5 & 6 & 1 & 3 & 4 & 1 & 5 & 3 \\
1 & 4 & 5 & 1 & 2 & 4 & 2 & 5 & 6 \\
1 & 4 & 5 & 2 & 3 & 4 & 3 & 6 & 6 \\
2 & 3 & 3 & 3 & 5 & 6 & 1 & 5 & 12 \\
2 & 2 & 4 & 4 & 6 & 7 & 4 & 9 & 7
\end{bmatrix}
\cdot
\begin{bmatrix}
1 & 0 & 0 \\
1 & 0 & 0 \\
1 & 0 & 0 \\
1 & 0 & 0 \\
0 & 1 & 0 \\
0 & 1 & 0 \\
0 & 1 & 0 \\
0 & 0 & 1 \\
0 & 0 & 1 \\
0 & 0 & 1
\end{bmatrix}
\qquad
\text{ITEMSUMS} =
\begin{bmatrix}
10 & 11 & 14 \\
11 & 10 & 12 \\
8 & 16 & 18 \\
6 & 12 & 12 \\
10 & 6 & 22 \\
15 & 14 & 7 \\
15 & 8 & 9 \\
10 & 7 & 13 \\
10 & 9 & 15 \\
8 & 14 & 18 \\
8 & 17 & 20
\end{bmatrix}$$

HP-48G PROGRAM:

RCL A

RCL DB1

x

STO ITEMSUMS

STEP 5:   Square each term in the table and sum:

PROGRAM: GRANDSUMSQ = $[1]_{1,9}$ (diagonal$[A]^{T}[A])[1]_{9,1}$ or $[1]_{1,11}([A]_{11,9} \otimes [A]_{11,9})[1]_{9,1}$ =

$$\text{GRANDSUMSQ} := \text{ONE11} \cdot \overrightarrow{\left(A_{11,9} \cdot A_{11,9} \right)} \cdot \text{NINE1}$$

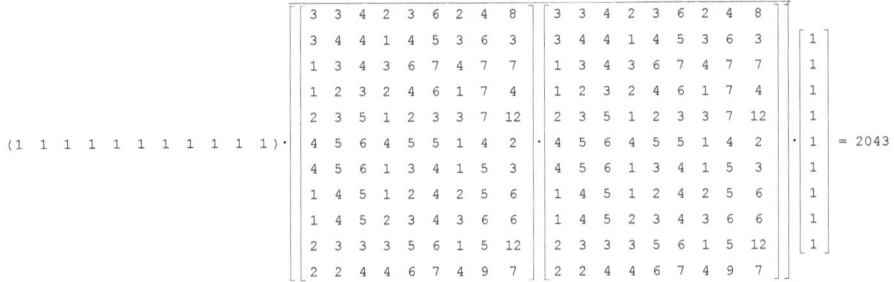

GRANDSUMSQ = 2043

$$[1\ 1\ 1\ 1\ 1\ 1\ 1\ 1] \begin{vmatrix} 66 & & & & & & & \\ & 142 & & & & & & \\ & & 229 & & & & & \\ & & & 66 & & 0 & & \\ & & & & 189 & & & \\ & & & & & 313 & & \\ & 0 & & & & & 71 & \\ & & & & & & & 407 \\ & & & & & & & & 560 \end{vmatrix} \begin{vmatrix} 1 \\ 1 \\ 1 \\ 1 \\ 1 \\ 1 \\ 1 \\ 1 \end{vmatrix} = 2043$$

HP-48G PROGRAM:

```
RCL ONE9
↑
TRN
SWAP
RCL A
↑
TRN
SWAP
x
MTH,MATR,NXT →DIAG
9
↑
DIAG→
x
SWAP
x
STO GNDSUMSQ
```

STEP 6: Sum each score in the table for the Grand Sum.

PROGRAM: GRANDSUM = $[1]_{1,11}[A]_{11,9}[1]_{9,1}$

GRANDSUM := ONE11·$A_{11,9}$·NINE1

$$
\text{GRANDSUM} := (1\ \ 1\ \ 1\ \ 1\ \ 1\ \ 1\ \ 1\ \ 1\ \ 1\ \ 1\ \ 1) \cdot
\begin{bmatrix}
3 & 3 & 4 & 2 & 3 & 6 & 2 & 4 & 8 \\
3 & 4 & 4 & 1 & 4 & 5 & 3 & 6 & 3 \\
1 & 3 & 4 & 3 & 6 & 7 & 4 & 7 & 7 \\
1 & 2 & 3 & 2 & 4 & 6 & 1 & 7 & 4 \\
2 & 3 & 5 & 1 & 2 & 3 & 3 & 7 & 12 \\
4 & 5 & 6 & 4 & 5 & 5 & 1 & 4 & 2 \\
4 & 5 & 6 & 1 & 3 & 4 & 1 & 5 & 3 \\
1 & 4 & 5 & 1 & 2 & 4 & 2 & 5 & 6 \\
1 & 4 & 5 & 2 & 3 & 4 & 3 & 6 & 6 \\
2 & 3 & 3 & 3 & 5 & 6 & 1 & 5 & 12 \\
2 & 2 & 4 & 4 & 6 & 7 & 4 & 9 & 7
\end{bmatrix}
\cdot
\begin{bmatrix}
1 \\ 1 \\ 1 \\ 1 \\ 1 \\ 1 \\ 1 \\ 1 \\ 1
\end{bmatrix}
$$

GRANDSUM = 395

HP-48G PROGRAM:

```
RCL ONE9TRN
RCL ONE11
RCL A
x
SWAP
x
STO GRNDSUM   395
```

STEP 7:   Computation of the Correction Factor CORRFACT: Square the Grand Sum and divide by the total # of measures in the table. (N_T = 99)

PROGRAM: CORRFACT = $1/N_T\ ([1]_{1,11}[A]_{11,9}[1]_{9,1})^2$

$$
\text{CORRFACT} := \left(\frac{1}{N_T}\right) \cdot \left(\text{ONE11} \cdot A_{11,9} \cdot \text{NINE1}\right)^2
$$

$$
\text{CORRFACT} := \left(\frac{1}{99}\right) \cdot \text{GRANDSUM}^2
$$

$$\text{CORRFACT} := \left(\frac{1}{99}\right) \cdot 395^2$$

CORRFACT = 1576.01

$(1/99)(395)(395) = \mathbf{1576.0101} = \mathbf{CORRFACT}$

HP-48G PROGRAM: or:

RCL GRNDSUM	395	
SQ		
RCL N_T = 99		
/		
STO CORRFACT	1576.0101	

RCL ONE11
RCL A
x
RCL ONE9TRN
x
SQ
RCL N_T = 99
/
STO CORRFACT

STEP 8: Computation of the Total Sum of Squares SS_T. Subtract the Correction Factor from the Grand Sum of Squares (GRNDSUMSQ)

SST = GRNDSUMSQ - CORRFACT = 2043 - 1576 = **467**

$$\text{SST} := \text{ONE11} \cdot \overrightarrow{\left(A_{11,9} \cdot A_{11,9}\right)} \cdot \text{NINE1} - \left(\frac{1}{N_T}\right) \cdot \left(\text{ONE11} \cdot A_{11,9} \cdot \text{NINE1}\right)^2 \square$$

SST := GRANDSUMSQ − CORRFACT
SST = 466.99

BETWEEN SUBJECTS EFFECTS

STEP 9: Computation of Between Subjects Sum of Squares SS_b. N_B = # Items in each cart: N_b = 3.

PROGRAM: SSb = $(1/N_b)$ $[1]_{1,3}$**diagonal**$([A]_{11,9}[DB1]_{9,3})^2[1]_{3,1}$ - **CORRFACT** = SS_b.

$$\text{SSb} := \left(\frac{1}{N_b}\right) \cdot \left[\text{ONE11} \cdot \overrightarrow{\left(\left[\left(A_{11,9} \cdot DB1\right) \cdot \left(A_{11,9} \cdot DB1\right)\right]\right)}\right] \cdot \text{NINE1} - \text{CORRFACT} \square$$

$$SSb := \left(\frac{1}{3}\right) \cdot \left[ONE11 \cdot \left(\overrightarrow{\left[\left(A_{11,9} \cdot DB1\right) \cdot \left(A_{11,9} \cdot DB1\right) \right]} \right) \right] \cdot THREE1 - CORRFACT\square$$

Again we must separate the problem into parts.

$$SUM1A := \left(A_{11,9} \cdot DB1\right)$$

$$SUM1A = \begin{bmatrix} 10 & 11 & 14 \\ 11 & 10 & 12 \\ 8 & 16 & 18 \\ 6 & 12 & 12 \\ 10 & 6 & 22 \\ 15 & 14 & 7 \\ 15 & 8 & 9 \\ 10 & 7 & 13 \\ 10 & 9 & 15 \\ 8 & 14 & 18 \\ 8 & 17 & 20 \end{bmatrix}$$

$$SSb := \left(\frac{1}{3}\right) \cdot \left[ONE11 \cdot \left(\overrightarrow{\left((SUM1A) \cdot (SUM1A) \right)} \right) \right] \cdot THREE1 - CORRFACT\square$$

$$SSb := \left(\frac{1}{3}\right) \cdot \left[(1\ 1\ 1\ 1\ 1\ 1\ 1\ 1\ 1\ 1\ 1) \cdot \overrightarrow{\left(\begin{bmatrix} 10 & 11 & 14 \\ 11 & 10 & 12 \\ 8 & 16 & 18 \\ 6 & 12 & 12 \\ 10 & 6 & 22 \\ 15 & 14 & 7 \\ 15 & 8 & 9 \\ 10 & 7 & 13 \\ 10 & 9 & 15 \\ 8 & 14 & 18 \\ 8 & 17 & 20 \end{bmatrix} \cdot \begin{bmatrix} 10 & 11 & 14 \\ 11 & 10 & 12 \\ 8 & 16 & 18 \\ 6 & 12 & 12 \\ 10 & 6 & 22 \\ 15 & 14 & 7 \\ 15 & 8 & 9 \\ 10 & 7 & 13 \\ 10 & 9 & 15 \\ 8 & 14 & 18 \\ 8 & 17 & 20 \end{bmatrix} \right)} \cdot \begin{pmatrix} 1 \\ 1 \\ 1 \end{pmatrix} \right] - CORRFACT$$

$$SSb = 180.99$$

$\Sigma Tr1\ \Sigma Tr2\ \Sigma tR3$

```
| 3 3 4  2 3 6  2 4  8 |   | 1 0 0 |       | 10  11  14 |
| 3 4 4  1 4 5  3 6  3 |   | 1 0 0 |       | 11  10  12 |
| 1 3 4  3 6 7  4 7  7 |   | 1 0 0 |       |  8  16  18 |
| 1 2 3  2 4 6  1 7  4 |   | 0 1 0 |       |  6  12  22 |
| 2 3 5  1 2 3  3 7 12 |   | 0 1 0 |  =    | 10   6  22 |
| 4 5 6  4 5 5  1 4  2 |   | 0 1 0 |       | 15  14   7 |
| 4 5 6  1 3 4  1 5  3 |   | 0 0 1 |       | 15   8   9 |
| 1 4 5  1 2 4  2 5  6 |   | 0 0 1 |       | 10   7  13 |
| 1 4 5  2 3 4  3 6  6 |   | 0 0 1 |       | 10   9  15 |
| 2 3 3  3 5 6  1 5 12 |                   |  8  14  18 |
| 2 2 4  4 6 7  4 9  7 |                   |  8  17  20 |
```

$C^2 = C^T C$ (Multiplying to get smallest possible square matrix, in this case a 3 x 3, we proceed as follows.)

```
       | 10 11  8  6 10 15 15 10 10  8  8 |   | 10 11 14 |
(1/Nᴮ) | 11 10 16 12  6 14  8  7  9 14 17 |   | 11 10 12 |
       | 14 12 18 22 22  7  9 13 15 18 20 |   |  8 16 18 |
                                              |  6 12 22 |
                                              | 10  6 22 |       | 1199 1218 1532 |
                                              | 15 14  7 |   =   | 1218 1532 1826 |
                                              | 15  8  9 |       | 1532 1826 2540 |
                                              | 10  7 13 |
                                              | 10  9 15 |
                                              |  8 14 18 |
                                              |  8 17 20 |
```

And summing the diagonal we get

$$SS_b = (1/3)\ [1\ 1\ 1]\ \begin{vmatrix} 1199 & & \\ & 1532 & \\ & & 2540 \end{vmatrix}\ \begin{vmatrix} 1 \\ 1 \\ 1 \end{vmatrix} - CORRFACT = 5271/3 - 1576 = \mathbf{181}$$

HP-48G PROGRAM:

```
RCL ONE3      [1 1 1]
↑
TRN
SWAP
RCL A
RCL DB1
x
↑
TRN
SWAP
x
MTH,MATR,NEXT →DIAG
```

3

↑

DIAG→

x

SWAP

x

RCL N_b = 3

/

RCL CORRFACT

−

STO SSb   181

STEP 10: Computation of the total sales between the three carts. (Performance of each cart in sales) (The meaningfulness effects).

$$SS_{CARTS} \cdot N_{CARTS} = N_T/\text{\# columns in DB1} = 3.$$
$$N_{CARTS} = 99/3 = 33.$$

PROGRAM: SSCARTS = $1/N_c([1]_{1,11} [A]_{11,9} [DB1]_{9,3})^2$ − CORRFACT =
$1/33(SUM1CARTS \times DB1)^2$−CORRFACT

$$SS_{CARTS} := \left(\frac{1}{N_{CARTS}}\right) \cdot \left(ONE11 \cdot A_{11,9} \cdot DB1\right)^2 - CORRFACT\square$$

$$SS_{CARTS} := \left(\frac{1}{33}\right) \cdot \left(ONE11 \cdot A_{11,9} \cdot DB1\right)^2 - CORRFACT\square$$

$$SUM1B := \left(ONE11 \cdot A_{11,9} \cdot DB1\right)$$

$$SUM1B := (1\ 1\ 1\ 1\ 1\ 1\ 1\ 1\ 1\ 1\ 1) \cdot \begin{bmatrix} 3 & 3 & 4 & 2 & 3 & 6 & 2 & 4 & 8 \\ 3 & 4 & 4 & 1 & 4 & 5 & 3 & 6 & 3 \\ 1 & 3 & 4 & 3 & 6 & 7 & 4 & 7 & 7 \\ 1 & 2 & 3 & 2 & 4 & 6 & 1 & 7 & 4 \\ 2 & 3 & 5 & 1 & 2 & 3 & 3 & 7 & 12 \\ 4 & 5 & 6 & 4 & 5 & 5 & 1 & 4 & 2 \\ 4 & 5 & 6 & 1 & 3 & 4 & 1 & 5 & 3 \\ 1 & 4 & 5 & 1 & 2 & 4 & 2 & 5 & 6 \\ 1 & 4 & 5 & 2 & 3 & 4 & 3 & 6 & 6 \\ 2 & 3 & 3 & 3 & 5 & 6 & 1 & 5 & 12 \\ 2 & 2 & 4 & 4 & 6 & 7 & 4 & 9 & 7 \end{bmatrix} \cdot \begin{bmatrix} 1 & 0 & 0 \\ 1 & 0 & 0 \\ 1 & 0 & 0 \\ 0 & 1 & 0 \\ 0 & 1 & 0 \\ 0 & 1 & 0 \\ 0 & 0 & 1 \\ 0 & 0 & 1 \\ 0 & 0 & 1 \end{bmatrix}$$

$$\text{SUM1B} = (\ 111 \quad 124 \quad 160\ )$$

$$\text{SS}_{\text{CARTS}} := \left(\frac{1}{33}\right) \cdot \left(\text{SUM1B} \cdot \text{SUM1B}^T\right) - \text{CORRFACT}$$

$$\text{SS}_{\text{CARTS}} := \left(\frac{1}{33}\right) \cdot \left[(111 \quad 124 \quad 160) \cdot \begin{pmatrix} 111 \\ 124 \\ 160 \end{pmatrix} \right] - \text{CORRFACT}$$

$$\text{SS}_{\text{CARTS}} = 39.051$$

HP–48 PROGRAM:

```
RCL ONE11
RCLA
x
RCL DB1
x
↑
TRN
x
RCL N_CARTS
/
RCL CORRFACT
−
STO SSCARTS
```

STEP 11: Computation of Between Subjects Error Term, SS_{Errb}.

$$\text{SS}_{\text{Errb}} = \text{SSb} - \text{SSC} = 181 - 39 = \mathbf{142}$$

$$\text{SSERRb} := \text{SSb} - \text{SS}_{\text{CARTS}}$$

The math for the Error Term is,

$$\text{SSERRb} := \left[\left(\frac{1}{N_b}\right) \cdot \left[\text{ONE11} \cdot \left(\overrightarrow{\left[\left(A_{11,9} \cdot \text{DB1}\right) \cdot \left(A_{11,9} \cdot \text{DB1}\right) \right]} \right) \right] \cdot \text{NINE1} - \text{CORRFACT} \right] - \left[\left(\frac{1}{N_C}\right) \cdot \left(\text{ONE11} \cdot A_{11,9} \cdot \text{DB1}\right)^2 - \text{CORRFACT} \right] \square$$

$$\text{SSERRb} = 141.939$$

WITHIN SUBJECTS EFFECTS

STEP 12: Computation of Within Subjects Sum of Squares, SS_w.

$$\mathbf{SSW} = SST - SSb = 467 - 181 = \mathbf{286}$$

STEP 13: Computation for Sum of Squares for Polish, Burritos and Hot-dogs, SS_{PBH} N_{PBH} = N_T/# columns in DB3 = 3.

$$N_{PBH} = 99/3 = 33.$$

PROGRAM: SSPBH = $(1/N_{Tr})$ $([1]_{1,11}$ $[A]_{11,9}$ $[DB3]_{9,3+})^2$ − **CORRFACT**
$$= (1/33) \ (\mathbf{SUM1 \ x \ DB3})^2 - \mathbf{CORRFACT}$$

$$SS_{PBH} := \left(\frac{1}{N_{PBH}}\right) \cdot (SUM1 \cdot DB3)^2 - CORRFACT\square$$

$$SUM1C := SUM1CARTS \cdot DB3$$

$$SUM1C := (24 \quad 38 \quad 49 \quad 24 \quad 43 \quad 57 \quad 25 \quad 65 \quad 70) \cdot \begin{bmatrix} 1 & 0 & 0 \\ 0 & 1 & 0 \\ 0 & 0 & 1 \\ 1 & 0 & 0 \\ 0 & 1 & 0 \\ 0 & 0 & 1 \\ 1 & 0 & 0 \\ 0 & 1 & 0 \\ 0 & 0 & 1 \end{bmatrix}$$

$$SUM1C = (73 \quad 146 \quad 176)$$

$$SS_{PBH} := \left(\frac{1}{33}\right) \cdot \left(SUM1C \cdot SUM1C^T\right) - CORRFACT\square$$

$$SS_{PBH} := \left(\frac{1}{33}\right) \cdot \left[(73 \quad 146 \quad 176) \cdot \begin{pmatrix} 73 \\ 146 \\ 176 \end{pmatrix} \right] - CORRFACT$$

$$SS_{PBH} = 170.081$$

HP-48G PROGRAM:

RCL SUM1CARTS
RCL DB3
x
↑
TRN
x
RCL N_{PBH}
/
RCL CORRFACT
-
STO SSPBH

STEP 14: Computation of the Sum of Squares for the squaring of SUM1:
SSTrxC. $N_{TrxC} = N_T/\#$ elements in SUM1 = 9. $N_{TrxC} = 99/9 = 11$.

**PROGRAM: SSPBHxCARTS = $(1/N_{PBHxCARTS})$ $([1]_{1,11}$ $[A]_{11,9})^2$ - CORRFACT -
SSCARTS - SSPBH =(1/11) $([SUM1CARTS]_{1,9}$ $[SUM1CARTS]_{1,9}^T)$ -
CORRFACT - SSCARTS—SSPBH**

$$SS_{PBHxCARTS} := \left(\frac{1}{N_{PBHxCARTS}}\right) \cdot \left(SUM1 \cdot SUM1^T\right) - CORRFACT - SS_{CARTS} - SS_{PBH}$$

$$SS_{PBHxCARTS} := \left(\frac{1}{11}\right) \cdot \left((24\ 38\ 49\ 24\ 43\ 57\ 25\ 65\ 70) \cdot \begin{bmatrix} 24 \\ 38 \\ 49 \\ 24 \\ 43 \\ 57 \\ 25 \\ 65 \\ 70 \end{bmatrix} \right) - CORRFACT - SS_{CARTS} - SS_{PBH}$$

$$SS_{PBHxCARTS} = 18.949$$

HP-48G PROGRAM:

RCL SUM1CARTS
↑
TRN
x
RCL NPBHxCARTS

/

RCL CORRFACT

–

RCL SSC   ARTS

–

RCL SSPBH

–

STO SSPBHxCARTS

STEP 15: Computation of the Within Subjects error term, SS_{Errw}.

PROGRAM AND SOLUTION: SSErrw = SSW – SSTr – SSTrxC = 286 – 170 – 19 = **97**.

STEP 16: Tabling of data and calculation of degrees of freedom df.

SST = 467	df = # measures – 1 = 99 – 1 =	98
SSb = 181	df = # subjects – 1 = 33 – 1 =	32
SSCARTS = 39	df = # groups – 1 = 3 – 1 =	2
SSErrb = 142	df = dfSSb – dfSSC = 32 – 2 =	30
SSW = 286	df = dfSST – dfSSb = 98 – 32 =	66
SSPBH = 170	df = # trials – 1 = 3 – 1 =	2
SSPBHxCARTS = 19	df = dfSSTr x dfSSC = 2 x 2 =	4
SSErrw = 97	df = dfSSW – dfSSTr – dfSSTrxC = 66 – 2 – 4 =	60

STEP 17: Computation of **F.**

F = SS o df^{-1} o ERROR^{-1}

$$F := \overrightarrow{\left(SS \cdot df^{-1} \cdot ERROR^{-1} \right)}$$

$$
SS = \begin{bmatrix} 467 \\ 181 \\ 39 \\ 142 \\ 286 \\ 170 \\ 19 \\ 97 \end{bmatrix}
\quad
df = \begin{bmatrix} 98 \\ 32 \\ 2 \\ 30 \\ 66 \\ 2 \\ 4 \\ 60 \end{bmatrix}
\quad
ERROR = \begin{bmatrix} 4.733 \\ 4.733 \\ 4.733 \\ 4.733 \\ 1.617 \\ 1.617 \\ 1.617 \\ 1.617 \end{bmatrix}
$$

$$F := \begin{bmatrix} 467 \\ 181 \\ 39 \\ 142 \\ 286 \\ 170 \\ 19 \\ 97 \end{bmatrix} \cdot \begin{bmatrix} 98 \\ 32 \\ 2 \\ 30 \\ 66 \\ 2 \\ 4 \\ 60 \end{bmatrix}^{-1} \cdot \begin{bmatrix} 4.733 \\ 4.733 \\ 4.733 \\ 4.733 \\ 1.617 \\ 1.617 \\ 1.617 \\ 1.617 \end{bmatrix}^{-1} \qquad F = \begin{bmatrix} 1.007 \\ 1.195 \\ 4.12 \\ 1 \\ 2.68 \\ 52.566 \\ 2.938 \\ 1 \end{bmatrix}$$

In conventional mathematics, the operation above is computed as,

$$\begin{vmatrix} 467 \\ \quad 181 \\ \qquad 39 \\ \qquad\quad 142 \\ \qquad\qquad 286 \\ \qquad\qquad\quad 170 \\ \qquad\qquad\qquad 19 \\ \qquad\qquad\qquad\quad 97 \end{vmatrix} \begin{Vmatrix} 1/98 \\ \quad 1/32 \\ \qquad 1/2 \\ \qquad\quad 1/30 \\ \qquad\qquad 1/66 \\ \qquad\qquad\quad 1/2 \\ \qquad\qquad\qquad 1/4 \\ \qquad\qquad\qquad\quad 1/60 \end{Vmatrix} \begin{vmatrix} .2113 \\ \quad .2113 \\ \qquad .2113 \\ \qquad\quad .2113 \\ \qquad\qquad .6186 \\ \qquad\qquad\quad .6186 \\ \qquad\qquad\qquad .6186 \\ \qquad\qquad\qquad\quad .6186 \end{vmatrix} [1]_{8,1} =$$

F		P
1.0068		–
1.1950		–
4.1197		<.05
1.0000	;	–
2.6806		–
52.5773		<.001
2.9381		<.05
1.0000		–

Hence it is concluded:
1. The location of the carts significantly affected the overall amount sold
2. Sales improved as the day went by

3. The items sold at different rates.

HP-48G PROGRAM:

```
RCL SS
8
↑
MATH,MATR,NXT,DIAG→
RCL df
8
↑
DIAG→
INV
x
RCL ERROR
<<INV>>
ααTEACH
EXAM,PRG,APLY
x
STO F
```

POSTSCRIPT

Suppose the cost of each hot-dog is 15¢, the cost for each burrito is 35¢ and the cost for each polish sausage is 50¢. Then the total profits can be computed as:

$$
\begin{bmatrix}
3 & 3 & 4 & 2 & 3 & 6 & 2 & 4 & 8 & 0 & -1 & -1 \\
3 & 4 & 4 & 1 & 4 & 5 & 3 & 6 & 3 & -1 & -1 & -1 \\
1 & 3 & 4 & 3 & 6 & 7 & 4 & 7 & 7 & -1 & -1 & -1 \\
1 & 2 & 3 & 2 & 4 & 6 & 1 & 7 & 4 & -1 & -1 & -1 \\
2 & 3 & 5 & 1 & 2 & 3 & 3 & 7 & 12 & -1 & -1 & -1 \\
4 & 5 & 6 & 4 & 5 & 5 & 1 & 4 & 2 & -1 & 0 & -1 \\
4 & 5 & 6 & 1 & 3 & 4 & 1 & 5 & 3 & -1 & 0 & -1 \\
1 & 4 & 5 & 1 & 2 & 4 & 2 & 5 & 6 & -1 & -1 & -1 \\
1 & 4 & 5 & 2 & 3 & 4 & 3 & 6 & 6 & -1 & -1 & -1 \\
2 & 3 & 3 & 3 & 5 & 6 & 1 & 5 & 12 & -1 & -1 & -1 \\
2 & 2 & 4 & 4 & 6 & 7 & 4 & 9 & 7 & -1 & -1 & -2
\end{bmatrix}
\cdot
\begin{bmatrix}
1.5 & .50 \\
1.25 & .35 \\
.5 & .15 \\
1.5 & .5 \\
1.25 & .35 \\
.5 & .15 \\
1.5 & .50 \\
1.25 & .35 \\
.5 & .15 \\
2.50 & 0 \\
2.50 & 0 \\
2.50 & 0
\end{bmatrix}
=
\begin{bmatrix}
27 & 9.7 \\
26.5 & 10.2 \\
33.5 & 12.3 \\
21.25 & 8.5 \\
26.5 & 10.2 \\
32.5 & 11.35 \\
26.75 & 9.5 \\
19.75 & 8.1 \\
25.25 & 9.8 \\
28.25 & 10.7 \\
35.25 & 13.65
\end{bmatrix}
$$

$$
\begin{bmatrix}
27 & 9.7 \\
26.5 & 10.2 \\
33.5 & 12.3 \\
21.25 & 8.5 \\
26.5 & 10.2 \\
32.5 & 11.35 \\
26.75 & 9.5 \\
19.75 & 8.1 \\
25.25 & 9.8 \\
28.25 & 10.7 \\
35.25 & 13.65
\end{bmatrix}
\cdot
\begin{pmatrix}
1 \\
-1
\end{pmatrix}
=
\begin{bmatrix}
17.3 \\
16.3 \\
21.2 \\
12.75 \\
16.3 \\
21.15 \\
17.25 \\
11.65 \\
15.45 \\
17.55 \\
21.6
\end{bmatrix}
$$

But since the values in the second column are costs, it might be more illustrative to write the problem as:

$$\begin{bmatrix} 3 & 3 & 4 & 2 & 3 & 6 & 2 & 4 & 8 & 0 & -1 & -1 \\ 3 & 4 & 4 & 1 & 4 & 5 & 3 & 6 & 3 & -1 & -1 & -1 \\ 1 & 3 & 4 & 3 & 6 & 7 & 4 & 7 & 7 & -1 & -1 & -1 \\ 1 & 2 & 3 & 2 & 4 & 6 & 1 & 7 & 4 & -1 & -1 & -1 \\ 2 & 3 & 5 & 1 & 2 & 3 & 3 & 7 & 12 & -1 & -1 & -1 \\ 4 & 5 & 6 & 4 & 5 & 5 & 1 & 4 & 2 & -1 & 0 & -1 \\ 4 & 5 & 6 & 1 & 3 & 4 & 1 & 5 & 3 & -1 & 0 & -1 \\ 1 & 4 & 5 & 1 & 2 & 4 & 2 & 5 & 6 & -1 & -1 & -1 \\ 1 & 4 & 5 & 2 & 3 & 4 & 3 & 6 & 6 & -1 & -1 & -1 \\ 2 & 3 & 3 & 3 & 5 & 6 & 1 & 5 & 12 & -1 & -1 & -1 \\ 2 & 2 & 4 & 4 & 6 & 7 & 4 & 9 & 7 & -1 & -1 & -2 \end{bmatrix} \cdot \begin{bmatrix} 1.5 & -.50 \\ 1.25 & -.35 \\ .5 & -.15 \\ 1.5 & -.5 \\ 1.25 & -.35 \\ .5 & -.15 \\ 1.5 & -.50 \\ 1.25 & -.35 \\ .5 & -.15 \\ 2.50 & 0 \\ 2.50 & 0 \\ 2.50 & 0 \end{bmatrix} = \begin{bmatrix} 27 & -9.7 \\ 26.5 & -10.2 \\ 33.5 & -12.3 \\ 21.25 & -8.5 \\ 26.5 & -10.2 \\ 32.5 & -11.35 \\ 26.75 & -9.5 \\ 19.75 & -8.1 \\ 25.25 & -9.8 \\ 28.25 & -10.7 \\ 35.25 & -13.65 \end{bmatrix}$$

$$\begin{bmatrix} 27 & -9.7 \\ 26.5 & -10.2 \\ 33.5 & -12.3 \\ 21.25 & -8.5 \\ 26.5 & -10.2 \\ 32.5 & -11.35 \\ 26.75 & -9.5 \\ 19.75 & -8.1 \\ 25.25 & -9.8 \\ 28.25 & -10.7 \\ 35.25 & -13.65 \end{bmatrix} \cdot \begin{pmatrix} 1 \\ 1 \end{pmatrix} = \begin{bmatrix} 17.3 \\ 16.3 \\ 21.2 \\ 12.75 \\ 16.3 \\ 21.15 \\ 17.25 \\ 11.65 \\ 15.45 \\ 17.55 \\ 21.6 \end{bmatrix}$$

CHECK: In the first half-hour, he brought in $32, there were
$5 in wages paid out, and the total cost is:

$3 \cdot .5 + 3 \cdot .35 + 4 \cdot .15 + 2 \cdot .5 + 3 \cdot .35 + 6 \cdot .15 + 2 \cdot .5 + 4 \cdot .35 + 8 \cdot .15 = 9.7$

So his total profit is 32-5-9.7= $17.30.

One more check, in the fifth half-hour, he brought in

$34 - 7.50 - (2 \cdot .5 + 3 \cdot .35 + 5 \cdot .15 + 1 \cdot .5 + 2 \cdot .35 + 3 \cdot .15 + 3 \cdot .5 + 7 \cdot .35 + 12 \cdot .15) = 16.3$

Which also checks out.

THE PHILOSOPHICAL MEANING WHEN
$\Sigma A_{IJ}{}^T \circ B_{JK} = \infty$

3-26/27-97

Suppose we take every atom in existence in all the galaxies of the Universe and we give each of them a place in a Grand Matrix. The zeros represent empty space, and everything in creation has its unique matrix element or position in the Grand Matrix, even every Soul has to be somewhere. This is the Matrix A_{ij}, the Datastream Matrix. it is the positioning and accounting Matrix for everything in creation. When the columns of the Datastream Matrix $A_{ij}{}^T$ are half-multiplied, we get the sum of All of creation at the moment of multiplication. We can identify position and account for everything in all the galaxies of all the stars of all the worlds and planets, their proper motions and directions through space, even to the vagaries that live on and between worlds (if we had a way to keep track of them, that is). So you see, with this equation we can keep track of Everything, but only at the moment of half-multiplication. We could continuously re-half-multiply by $[1]_{1i}$ and get a 'picture' or more likely a series of pictures (like motion picture film) of the changes in motion for any given particle. Even waves in space are positioned somewhere! Even electrons in their orbits are accounted for, for they have their place in the Grand Matrix. Suppose the speed of the half-multiplier operator approaches the speed at which reality is perceived, (say 60 multiplications per second, the speed in which the eye detects the motion in films for movies) we can keep up to date track of EVERYTHING in Existence! As it happens! (at least up to the last half-multiplication). And we don't need calculus, nor algebra, or differential equations, all we need to know is how to add, subtract, multiply & divide! (and maybe a logarithm or two, just to satisfy the mathematics), and Accounting/Inventories is directly related to Statistics! Let the grand Sum of all creation be:

Grand Sum of everything in creation. $[1]_{1,i}[A]_{ij}[1]_{j,1}$

$[1]_{1,i}[A]_{ij}$ = The sum of the columns of A_{ij}, here we keep track of each individual element.

$[\mathbf{A}]_{ij}$ $[1]_{j,1}$ = Sum of the rows of the Matrix These are the coordinates for each individual item in the Grand Matrix.

Lets limit the size of the Matrix and for every row in the Matrix we will put the name of every person in the world. In each column we put everything that is up for sale in the world. One's. twos, threes, whatever quantity anyone buys or sells, the amounts exchanged, from whom to whom (buyer/seller) and for how much can easily be calculated for each person. The only practical restraint is memory capacity of the computer. For each new item added to the matrix we must add 3 billion rows, one for each person in the world. That's 3 Gigabytes of memory for each item for sale to everyone in the world.

This Grand Matrix of all things is $[\mathbf{A}]^{T}_{IJ}$ o $[\ \mathbf{B}]_{JK} = \Sigma\ [\mathbf{C}]_{JK}$

Then: $[1]_{J1}$ o $[\mathbf{C}_{JK}]$ Accounts for everything bought and sold, keeping account of each individual row.

And: $[1]_{1I}$ o $[A]_{IJ}$ gives total sum of everything bought & sold and whatever data you may wish to include. Sums every column.

$[1]_{J1}$ o $[\mathbf{A}]^{T}_{ij}$ Sums a total inventory of parts to a piece by piece amount for every single item bought/sold in the world.

$[1]_{1i}$ x $[A]_{iJ}$ Total summing of the columns, total sum of everything bought/sold in the world.

$[1]_{iJ}$ x $[A]_{ij}$ x $[DB1]_{jk}$ (Elementary statistics) Sums which groups you wish to add together.

$1/N$ $([1]_{iJ}$ x $[A]_{ij}[DB]_{jk})^{2}$ Analysis of variance of the chosen groups under study. We may be limited to the statistics we can do here, i.e., we can compare sales of brown shoes in comparison to sales of black shoes (even to a particular brand verses a similar brand item) and compare the sales. Or we can do this comparing sales of all brands and comparable items and their volume (inventory) and sales of each compared to the other. But comparing sales of apples and oranges or shoes Vs refrigerators isn't practical unless we are looking for statistical relationships between the sales (data mining).

EXAMPLE: FARMLAND INDUSTRIES RECIPIE/ EMPLOYEE HOUR/WAGE INVENTORY

Suppose we need to run a daily inventory for a company that makes sausages and bologna. The recipes are given below. All we wish to know is how much meat and spices are used daily. Note: This example is full of matrices that run off the page. When this is the case, I will try to place the complete matrix underneath the incomplete display. These are real recipies, try them, they are really good.

THE RECIPIES

The basic equation for equating all ingredients in the above recipes to lbs is:

Nx1 lb = (1/N)oz 1 lb/16 oz. I assume 3 t =1/2 oz or 2 T = 1 oz

BEEF LOG
1 LBS HAMB =
2.5t PEP = 2.5/5X6X16=.0052   lbs
2.5 t TENDERIZER = .0052lb = oz/(5x6x16) = 480
3.75 t salt = .0078
1.25 t garlic = .0052/2 = .0026
2.5 t mustard seed = .0052
2 t celery seed = 2/5x6x16 = .00417
1 t dill = 1/480 = .00208
2.5 t liqsmk = .0052

BEEF BOLOGNA
lb = oz/(16x25) = oz/400
2.5 LBS BEEF = 1 lb
tenderizer = 1/25 = .04 lbs
1 oz pep. = .0625/5 = .0025
1/2 cup sugar = 4/16 = .25/25 = .0100
3 T liq smk = 1.5oz/25 = = .00375
2 t saltpeter = .333/400 = .0008333

SUMMER SAUSAGE
1 LB HMB
    1.25 t garlic = .0026 lbs
    3.75 t salt = .0078
    2.5 t must seed = .0052
    2.5 t pepper = .0052
    2.5 tsp liqsmk = .0052
    2.5 t tenderizer = .0052

HERTERS SAUSAGE
1lb = 1/(34x16) = 1/544
water 4 oz = .00735 lbs
pepper 1.5 oz = 1.5/544 = .00276
ginger 1 oz = 1/544 = .00184
nutmeg 1.5 oz = .00276 lbs
allspice 1/2 oz = .000919
paprika 1/2 oz = .000919
garlic 1/3 oz = .000613
onion 1/3 oz = .000613 lbs
salt 12 oz = .02205
dry milk 8 oz = .0147
liqsmk 1.5 oz = .00276 lbs

THE DATABASE MATRIX

All ingredients are listed in pounds.

HMB	PRK	LQSMK	PEP	GIN	NTMG	ALSP	PAP	GAR	ONION	SALT	DMLK	TNDR	MSTD	SUG	CLSD	WT	DIL	SP
.5	.5	.00276	.00276	.00184	.00276	.000919	.000919	.000613	.000613	.0221	.0147	0	0	0	0	.0074	0	0
1	0	.0052	.0052	0	0	0	0	.0026	0	.0078	0	.0052	.0052	0	.00417	0	.00208	0
1	0	.0052	.0052	0	0	0	0	.0026	0	.0078	0	.0052	.0052	0	0	0	0	0
1	0	.00375	.0025	0	0	0	0	0	0	0	0	.04	0	.01	0	0	0	.00083

$R :=$ (matrix above)

The first row represents Herters sausage, the second row represents the beef log, the third row represents the summer sausage and the fourth row represents the beef bologna.

We now need to sum each row of the matrix to find the correction factors (we are going to normalize the Matrix so that one pound of ingredients equals one pound of product). We need a column matrix of 19 rows and one column:

$$\text{NINETEEN1} := \begin{bmatrix} 1 \\ 1 \\ 1 \\ 1 \\ 1 \\ 1 \\ 1 \\ 1 \\ 1 \\ 1 \\ 1 \\ 1 \\ 1 \\ 1 \\ 1 \\ 1 \\ 1 \\ 1 \\ 1 \end{bmatrix}.$$

$$R \cdot \text{NINETEEN1} = \begin{bmatrix} 1.057 \\ 1.037 \\ 1.031 \\ 1.057 \end{bmatrix}$$

To compute the corrected matrix for final product, diagonalize this matrix, take it's inverse and multiply to R:

CLINTON L. HOLT

$$T := \begin{bmatrix} \dfrac{1}{1.05739} & 0 & 0 & 0 \\[2ex] 0 & \dfrac{1}{1.03745} & 0 & 0 \\[2ex] 0 & 0 & \dfrac{1}{1.0312} & 0 \\[2ex] 0 & 0 & 0 & \dfrac{1}{1.0571} \end{bmatrix}.$$

Multiplying we get,

T·R =

	HMB	PRK	LQSMK	PEP	GIN	NTMG	ALSP	PAP	GAR	ONION	SALT	DMLK	TNDR	MSTD	SUG	CLSD	WT	DIL	SP
	0.473	0.473	0.003	0.003	0.002	0.003	0.001	0.001	0.001	0.001	0.021	0.014	0	0	0	0	0.007	0	0
	0.964	0	0.005	0.005	0	0	0	0	0.003	0	0.008	0	0.005	0.005	0	0.004	0	0.002	0
	0.97	0	0.005	0.005	0	0	0	0	0.003	0	0.008	0	0.005	0.005	0	0	0	0	0
	0.946	0	0.004	0.002	0	0	0	0	0	0	0	0	0.038	0	0.009	0	0	0	0.001

CORRECTEDTOTAL =

0.473	0.473	0.003	0.003	0.002	0.003	0.001	0.001	0.001	0.001	0.021	0.014	0	0	0	0	0.007	0	0
0.964	0	0.005	0.005	0	0	0	0	0.003	0	0.008	0	0.005	0.005	0	0.004	0	0.002	0
0.97	0	0.005	0.005	0	0	0	0	0.003	0	0.008	0	0.005	0.005	0	0	0	0	0
0.946	0	0.004	0.002	0	0	0	0	0	0	0	0	0.038	0	0.009	0	0	0	0.001

THIS PROCESS CHECKS TO MAKE SURE THE VALUES ARE CORRECTED SO THAT ONE LB OF FINISHED SAUSAGE OR BOLOGNA EQUALS ONE POUND OF INGREDIENTS. The values of the sums of the rows should equal one.

$$T \cdot R \cdot NINETEEN1 = \begin{bmatrix} 1 \\ 1 \\ 1 \\ 1 \end{bmatrix}$$

TO FIND OUT HOW MUCH INGREDIENTS WERE USED IN ONE DAY: Suppose the final account for the day gives

50,000 LBS HERTERS SAUSAGE
45,000 LBS BEEF LOG
62,000 LBS SUMMER SAUSAGE
24,000 LBS BEEF BOLOGNA
DAILYTOTAL := (50000 45000 52000 24000)

FINALINVENTORY := DAILYTOTAL·CORRECTEDTOTAL

FINALINVENTORY = (140174 23650 731 683 100 150 50 50 341 50 1826 700 1397 485 216 180 350 90 24)

THE SOLUTION SAYS THAT FOR A TOTAL OF 171,000 POUNDS OF FINISHED
PRODUCT, WE USED 140,174 LBS OF HAMBURGER, 23,650 LBS OF PORK, 731
LBS OF LIQUID SMOKE, 683 LBS OF PEPPER, 100 LBS OF GINGER, ETC.

LET'S DO A ROUGH CHECK ON THE AMOUNTS OF HAMBURGER AND PORK TO
CHECK TO SEE IF THE SOLUTION IS CORRECT:

50,000 x .5 = 25,000 LBS HAMBURGER
45,000 x   1 = 45,000 LBS HAMBURGER
52,000 X   1 = 52,000 LBS HAMBURGER
24,000 x   1 = 24,000 LBS HAMBURGER
25,000 + 45,000 + 52,000 + 24,000 = 146,000 LBS HAMBURGER

NOW LET'S CHECK THE ROUGH AMOUNT OF PORK:

.5x50,000 = 25,000 LBS PORK.

$$\frac{25000}{23650} = 1.057082$$

THE CORRECT ANSWER FOR PORK IS 23,643 LBS WHICH IS WITHIN
ROUNDING ERROR OF THE SPREADSHEET MATRIX I'M USING.

Of course we can include the cost of each of the ingredients,
federal and state taxes, and employee wages if we wish (computed
later).

PROGRAM CONDENSED

DAILY TOTAL = [50,000 45,000 52,000 24,000]   changes daily
R = [DATABASE]   stays the same unless recipes change
NINETEEN1 = $[1]_{19,1}$ stays the same unless recipes change

SUMR = R x NINETEEN1   THESE 5 STEPS ARE THE PROGRAM
T = diag(SUMR)$^{-1}$
CORRECTEDTOTAL = T x R
FINALINVENTORY = DAILYTOTAL x CORRECTEDTOTAL
PRINT

Suppose that instead of just one department, the sausages
and bologna are made in four different departments. We wish to

keep track of not only the total inventory, but the amounts of ingredients used by each Department. The total usage for the day is:

$$\text{DAILYTOTAL} = \begin{pmatrix} \text{H.S} & \text{B.L} & \text{S.S.} & \text{BOL} \\ 50000 & 45000 & 52000 & 24000 \end{pmatrix}$$

Suppose Dept. 1 made 20,000 lbs Herter's sausage, 10,000 lbs beef log, 12,000 lbs summer sausage and no bologna.
Dept. 2 made 5,000 lbs Herter's sausage, 0 lbs beef logs, 25,000 lbs summer sausage and 6,000 lbs bologna.
Dept. 3 made 25,000 lbs Herter's sausage, 15,000 lbs beef logs, 0 lbs summer sausage and 12,000 lbs bologna.
Dept. 4 made 0 lbs Herter's sausage, 20,000 lbs beef logs, 15,000 lbs summer sausage and 6,000 lbs bologna.

Then the inventory matrix becomes:

$$\begin{bmatrix} \text{H.S.} & \text{B.L.} & \text{S.S.} & \text{BOL} \\ 50,000 & 45,000 & 52,000 & 24,000 \\ 20,000 & 10,000 & 12,000 & 0 \\ 5,000 & 0 & 25,000 & 6,000 \\ 25,000 & 15,000 & 0 & 12,000 \\ 0 & 20,000 & 15,000 & 6,000 \end{bmatrix}$$

$$\text{DEPARTMENTTOTALS} = \begin{bmatrix} \text{H.S.} & \text{B.L.} & \text{S.S.} & \text{BOL} \\ 50000 & 45000 & 52000 & 24000 \\ 20000 & 10000 & 12000 & 0 \\ 5000 & 0 & 25000 & 6000 \\ 25000 & 15000 & 0 & 12000 \\ 0 & 20000 & 15000 & 6000 \end{bmatrix}$$

COMPLETEDEPARTMENTINV = DEPARTMENTTOTALS · CORRECTEDTOTAL

HAMBURGER PORK LQ.SMK PEP GING NTMG ALSP PAP GAR ON SALT DRYMLK TNDR MSTD SUG CELERY WTR DILL SP.

	HAMBURGER	PORK	LQ.SMK	PEP	GING	NTMG	ALSP	PAP	GAR	ON	SALT	DRYMLK	TNDR	MSTD	SUG	CELERY	WTR	DILL	SP.
	140174	23650	731	683	100	150	50	50	341	50	1826	700	1397	485	216	180	350	90	24
	30740	9460	170	170	40	60	20	20	86	20	596	280	110	110	0	40	140	20	0
COMPLETEDEPARTMENTINV =	32291	2365	164	152	10	15	5	5	80	5	305	70	353	125	54	0	35	0	6
	37637	11825	198	174	50	75	25	25	70	25	645	350	531	75	108	60	175	30	12
	39506	0	199	187	0	0	0	0	105	0	280	0	403	175	54	80	0	40	6

DEPT. 1 used 30,740 lbs hamburger, DEPT. 2 used 32,291 lbs hamburger, DEPT. 3 used 11,825 lbs pork and 645 lbs salt. DEPT. 4

used 39,506 lbs hamburger, 199 lbs liq. smoke, 175 lbs mustard and 6 lbs salt peter.

Just for ease in the computations, suppose that out of the 50,000 lbs of Herter's sausage produced, 1,000 lbs was waste, of the 45,000 lbs beef logs, 800 lbs was waste, of the 52,000 lbs summer sausage, 1,500 lbs was waste and of the 24,000 lbs bologna, 500 lbs was discarded. What is the new total inventory? To see what I am doing, I will use just the DAILY TOTAL matrix, not the DEPARTMENT TOTALS matrix.

$$
\begin{array}{cccc}
\text{H.S.} & \text{B.L.} & \text{S.S.} & \text{BOL} \\
\end{array}
$$
$$
\begin{pmatrix}
50000 & 45000 & 52000 & 24000 \\
-1000 & -800 & -1500 & -500
\end{pmatrix}
$$

$$
\text{TOTALWITHDISCARDS} =
\begin{array}{cccc}
\text{H.S.} & \text{B.L.} & \text{S.S.} & \text{BOL} \\
\end{array}
\begin{pmatrix}
50000 & 45000 & 52000 & 24000 \\
-1000 & -800 & -1500 & -500
\end{pmatrix}
$$

$\text{ONE2} = (1 \quad 1)$

$\text{TOTALAFTERDISCARDS} := \text{TOTALWITHDISCARDS} \cdot \text{CORRECTEDTOTAL}$

$\text{TOTALAFTERDISCARDS} =$

	HMB	PRK	LQSMK	PEP	GIN	NTMG	ALSP	PAP	GAR	ONION	SALT	DMLK	TNDR	MSTD	SUG	CLSD	WT	DIL	SP
	140174	23650	731	683	100	150	50	50	341	50	1826	700	1397	485	216	180	350	90	24
	-3172.2	-473	-16.5	-15.5	-2	-3	-1	-1	-7.9	-1	-39.4	-14	-30.5	-11.5	-4.5	-3.2	-7	-1.6	-0.5

To find the total used that was good we add the two rows together:

$\text{GRANDTOTAL} := \text{ONE2} \cdot \text{TOTALAFTERDISCARDS}$

	HMB	PRK	LQSMK	PEP	GIN	NTMG	ALSP	PAP	GAR	ONION	SALT	DMLK	TNDR	MSTD	SUG	CLSD	WT	DIL	SP
GRANDTOTAL = (	137001.8	23177	714.5	667.5	98	147	49	49	333.1	49	1786.6	686	1366.5	473.5	211.5	176.8	343	88.4	23.5)

For the day, we passed 137,001.8 lbs hamburger, 23,177 lbs of pork, 1366.5 lbs of salt and 23.5 lbs salt peter, etc.

Let's look at this problem in another light. Suppose that instead of computing the inventory at the end of the day, we wish to manufacture 50,000 lbs of Herter's sausage, 45,000 lbs of beef logs, 52,000 lbs of summer sausage and 24,000 lbs of bologna. We want to calculate the recipes. We will compute the recipes for all four departments as well as the total amount of ingredients. We will use the corrected database and

DEPARTMENTTOTALS matrix, but with a twist, we will use the account page matrix to obtain the desired proportions. i.e.

$$DAILYTOTAL^T = \begin{bmatrix} 50000 \\ 45000 \\ 52000 \\ 24000 \end{bmatrix}$$

$$DAILYTOTALDIAG = diag\left(DAILYTOTAL^T\right)$$

$$DAILYTOTALDIAG = \begin{bmatrix} 50000 & 0 & 0 & 0 \\ 0 & 45000 & 0 & 0 \\ 0 & 0 & 52000 & 0 \\ 0 & 0 & 0 & 24000 \end{bmatrix}$$

$$AMOUNTINGREDIENTS = DAILYTOTALDIAG \cdot CORRECTEDTOTAL$$

	HMB	PRK	LQSMK	PEP	GIN	NTMG	ALSP	PAP	GAR	ONION	SALT	DMLK	TNDR	MSTD	SUG	CLSD	WT	DIL	SP
AMOUNTINGREDIENTS =	23650	23650	150	150	100	150	50	50	50	50	1050	700	0	0	0	0	350	0	0
	43380	0	225	225	0	0	0	0	135	0	360	0	225	225	0	180	0	90	0
	50440	0	260	260	0	0	0	0	156	0	416	0	260	260	0	0	0	0	0
	22704	0	96	48	0	0	0	0	0	0	0	0	912	0	216	0	0	0	24

THE INGREDIENTS FOR BOLOGNA (TO MAKE 24,000 LBS FINAL PRODUCT) ARE 22,704 LBS HAMBURGER, 96 LBS LIQUID SMOKE, 48 LBS PEPPER, 912 LBS TENDERIZER, 216 LBS SUGAR AND 24 LBS SALT PETER. IN THIS RECIPIE, THE TENDERIZER CONTAINS THE SALT, SO I DO NOT INCLUDE SALT IN THE RECIPE. THIS IS THE TOTAL FOR ALL DEPARTMENTS. LET'S SEE WHAT INGREDIENTS ARE NEEDED FOR DEPARTMENT ONE.

MATHCAD CANNOT REMOVE A ROW OF A MATRIX, BUT IT CAN REMOVE A COLUMN. SINCE DEPARTMENT ONE IS REPRESENTED IN ROW 2, WE MUST TRANSPOSE THE DEPARTMENTTOTALS MATRIX AND REMOVE THE SECOND COLUMN, DIAGONALIZE IT AND MULTIPLY TO THE CORRECTED TOTAL MATRIX. WE MUST CHANGE THE ORIGIN OF THE MATRIX FROM 0 TO 1. i.e.

$$DEPARTMENTTOTALS^T = \begin{bmatrix} 50000 & 20000 & 5000 & 25000 & 0 \\ 45000 & 10000 & 0 & 15000 & 20000 \\ 52000 & 12000 & 25000 & 0 & 15000 \\ 24000 & 0 & 6000 & 12000 & 6000 \end{bmatrix}$$

$$\left(\text{DEPARTMENTTOTALS}^{\text{T}}\right)^{<2>} = \begin{bmatrix} 20000 \\ 10000 \\ 12000 \\ 0 \end{bmatrix}$$

$$\text{DEPTONETOTAL} := \text{diag}\left[\left(\text{DEPARTMENTTOTALS}^{\text{T}}\right)^{<2>}\right]$$

$$\text{DEPTONETOTAL} = \begin{bmatrix} \text{H.S.} & \text{S.S} & \text{B.L.} & \text{BOL} \\ 20000 & 0 & 0 & 0 \\ 0 & 10000 & 0 & 0 \\ 0 & 0 & 12000 & 0 \\ 0 & 0 & 0 & 0 \end{bmatrix}$$

$$\text{AMOUNTINGREDIENTSDEPT1} := \text{DEPTONETOTAL} \cdot \text{CORRECTEDTOTAL}$$

AMOUNTINGREDIENTSDEPT1 =

	HMB	PRK	LQSMK	PEP	GIN	NTMG	ALSP	PAP	GAR	ONON	SLT	DMLK	TNDR	MSTD	SUG	CLSD	WT	DIL	SP
	9460	9460	60	60	40	60	20	20	20	20	420	280	0	0	0	0	140	0	0
	9640	0	50	50	0	0	0	0	30	0	80	0	50	50	0	40	0	20	0
	11640	0	60	60	0	0	0	0	36	0	96	0	60	60	0	0	0	0	0
	0	0	0	0	0	0	0	0	0	0	0	0	0	0	0	0	0	0	0

FOR THE INGREDIENTS NEEDED BY DEPT. 1 TO MAKE 12,000 LBS OF SUMMER SAUSAGE, WE NEED 11,640 LBS HAMBURGER, 60 LBS LIQ. SMOKE, 60 LBS PEPPER, 36 LBS GARLIC, 96 LBS SALT, 60 LBS TENDERIZER AND 60 LBS MUSTARD.

LET'S NOW LOOK AT THE COST TO MAKE THE FINISHED PRODUCT. THE PRICES OF THE INGREDIENTS MUST BE IN DOLLARS PER POUND. SUPPOSE

HAMBURGER   = .50/LB
PORK   = .55/LB
LIQUID SMOKE = 1.00/LB
PEPPER   = 2.50/LB
GINGER   = .75/LB
NUTMEG   = 2.75/LB
ALLSPICE   = 2.35/LB
PAPRIKA   = .35/LB
GARLIC   = 1.00/LB
ONION   = 1.00/LB
SALT   = .35/LB
DRIED MILK   = 2.50/LB
TENDERIZER   = 1.25/LB
MUSTARD   = .75/LB

```
SUGAR          =   .40/LB
CELERY SEED    = 1.00/LB
WATER          =   .01/LB
DILL           = 1.00/LB
SALT PETER     = 1.50/LB
```

$$
\text{COST} :=
\begin{bmatrix}
.50 \\
.55 \\
1.00 \\
2.50 \\
.75 \\
2.75 \\
2.35 \\
.35 \\
1.00 \\
1.00 \\
.35 \\
2.50 \\
1.25 \\
.75 \\
.40 \\
1.00 \\
.01 \\
1.00 \\
1.50
\end{bmatrix}
$$

COSTTOMAKE := FINALINVENTORY·COST

COSTTOMAKE = 91441.5

IF WE WISH AN ACCOUNTING OF EVERY INDIVIDUAL COST, WE CAN COMPUTE
THE ACCOUNT PAGE MATRIX.

ACCOUNTPAGECOSTS := $\mathrm{diag}\left(\text{FINALINVENTORY}^{T}\right)$·COST

$$
\text{ACCOUNTPAGECOSTS} = \begin{bmatrix}
70087 \\
13007.5 \\
731 \\
1707.5 \\
75 \\
412.5 \\
117.5 \\
17.5 \\
341 \\
50 \\
639.1 \\
1750 \\
1746.25 \\
363.75 \\
86.4 \\
180 \\
3.5 \\
90 \\
36
\end{bmatrix}
$$

LET'S CHECK THE SOLUTIONS OF THE PRODUCTS THAT COST $1.00/LB. THESE ARE LIQUID SMOKE, GARLIC, ONION, CELERY SEED AND DILL. THE COST SHOULD EQUAL THE NUMBER OF POUNDS USED.

```
LIQUID SMOKE = 731 LBS = $731
GARLIC       = 341 LBS = $341
ONION        = 50  LBS = $50
CELERY SEED  = 180 LBS = $180
DILL         = 90  LBS = $90
THE ANSWERS ALL CHECK
```

TO CHECK THE TOTAL, WE SUM THE ACCOUNTPAGECOSTS COLUMN:

$\text{CHECK} := \text{NINETEEN1}^{\text{T}} \cdot \text{ACCOUNTPAGECOSTS}$
$\text{CHECK} = 91441.5$

BUT SOME OF THE PRODUCT WAS QA'd OUT, HOW MUCH MONEY WAS LOST?

$\text{ZERO1} := (0 \quad 1)$
$\text{POUNDSLOST} := \text{ZERO1} \cdot \text{TOTALAFTERDISCARDS}$
$\text{POUNDSLOST} = (^{-}3172.2 \;\; ^{-}473 \;\; ^{-}16.5 \;\; ^{-}15.5 \;\; ^{-}2 \;\; ^{-}3 \;\; ^{-}1 \;\; ^{-}1 \;\; ^{-}7.9 \;\; ^{-}1 \;\; ^{-}39.4 \;\; ^{-}14 \;\; ^{-}30.5 \;\; ^{-}11.5 \;\; ^{-}4.5 \;\; ^{-}3.2 \;\; ^{-}7 \;\; ^{-}1.6 \;\; ^{-}0.5)$

MONEYLOST := POUNDSLOST·COST
MONEYLOST = ⁻2025.81

PROGRAMS CONDENSED

TO FIND POUNDS USED
SUMR := R·NINETEEN1

$T := diag(SUMR)^{(-1)}$.
CORRECTEDTOTAL := T·R

FINALINVENTORY := DAILYTOTAL·CORRECTEDTOTAL

TO FIND POUNDAGE PER DEPARTMENT

COMPLETEDEPARTMENTINV := DEPARTMENTTOTALS·CORRECTEDTOTAL

TO FIND FINAL POUNDAGE PLUS WASTE POUNDAGE

TOTALAFTERDISCARDS := TOTALWITHDISCARDS·CORRECTEDTOTAL

TO FIND THE TOTAL USED THAT WAS GOOD MINUS THE WASTE

GRANDTOTAL := ONE2·TOTALAFTERDISCARDS

TO FIND INGREDIENTS NEEDED

$$DAILYTOTAL^T = \begin{bmatrix} 50000 \\ 45000 \\ 52000 \\ 24000 \end{bmatrix}$$

$DAILYTOTALDIAG := diag(DAILYTOTAL^T)$

AMOUNTINGREDIENTS := DAILYTOTALDIAG·CORRECTEDTOTAL

COST TO MAKE FINISHED PRODUCT

COSTTOMAKE := FINALINVENTORY·COST

ACCOUNTPAGE FOR COSTS

$$\text{ACCOUNTPAGECOSTS} := \text{diag}\left(\text{FINALINVENTORY}^{T}\right) \cdot \text{COST}$$

TO FIND MONEY LOST

ZERO1 := (0 1)

POUNDSLOST := ZERO1·TOTALAFTERDISCARDS

MONEYLOST := POUNDSLOST·COST

TO FIND INGREDIENTS NEEDED

$$\text{DAILYTOTAL}^{T} = \begin{bmatrix} 50000 \\ 45000 \\ 52000 \\ 24000 \end{bmatrix}$$

$$\text{DAILYTOTALDIAG} := \text{diag}\left(\text{DAILYTOTAL}^{T}\right)$$

AMOUNTINGREDIENTS := DAILYTOTALDIAG·CORRECTEDTOTAL

COST TO MAKE FINISHED PRODUCT

COSTTOMAKE := FINALINVENTORY·COST

ACCOUNTPAGE FOR COSTS

$$\text{ACCOUNTPAGECOSTS} := \text{diag}\left(\text{FINALINVENTORY}^{T}\right) \cdot \text{COST}$$

TO FIND MONEY LOST

ZERO1 := (0 1)

POUNDSLOST := ZERO1·TOTALAFTERDISCARDS

MONEYLOST := POUNDSLOST·COST

LET'S LOOK AT OUR EMPLOYEES AND THEIR WAGES. SUPPOSE DEPT. 1 THROUGH DEPT. 4 HAVE TWO EMPLOYEES EACH (THIS IS JUST TO KEEP THE EXAMPLE TO A MANAGABLE SIZE). THE EMPLOYEES AND THEIR WAGES ARE:

```
DEPARTMENT 1:   CHRIS   $10/HR;      $15/HR OT
                FRAN    $ 9/HR;      $13.50 OT
DEPARTMENT 2:   TINA    $11/HR;      $16.50 OT
                BRUCE   $8.50/HR;    $12.75 OT
DEPARTMENT 3:   JEANIE  $10.50/HR;   $15.75 OT
                RAY     $9.50/HR;    $14.25 OT
DEPARTMENT 4:   PENNY   $12.00/HR;   $18.00 OT
                HELEN   $11.50/HR;   $17.25 OT
```

THE COMPANY WORKS 8 HOUR SHIFTS 6 DAYS A WEEK. WE WILL COMPUTE THE PAY MATRIX FIRST AS A SEPARATE ENTITY, THEN TRY TO CONSOLIDATE IT INTO THE COSTS MATRIX.

OR

$$\text{STRAIGHTHOURS2} := (8 \quad 8 \quad 8 \quad 8 \quad 8 \quad 8 \quad 8 \quad 8)$$

$$\text{STRAIGHTTIME} := \begin{bmatrix} 10 \\ 9 \\ 11 \\ 8.5 \\ 10.5 \\ 9.5 \\ 12 \\ 11.5 \end{bmatrix} \qquad \text{TIMEWITHOT} := \begin{bmatrix} 10 & 15 \\ 9 & 13.5 \\ 11 & 16.5 \\ 8.5 & 12.75 \\ 10.5 & 15.75 \\ 9.5 & 14.25 \\ 12 & 18 \\ 11.5 & 17.25 \end{bmatrix}$$

I'LL JUST GO AHEAD AND COMPUTE WAGES PLUS OVERTIME; TOTAL PAID OUT AND TOTAL TO EACH EMPLOYEE.

TOTAL PAID OUT BY COMPANY:

$$\text{ONE2} := \begin{pmatrix} 1 \\ 1 \end{pmatrix}$$ Below we are adding the straight hours to the ingredients inventory.

$$\text{FINALINVENTORYPLUSHOURS} := (140174 \quad 23650 \quad 731 \quad 683 \quad 100 \quad 150 \quad 50 \quad 50 \quad 341 \quad 50 \quad 1826 \quad 700 \quad 1397 \quad 485 \quad 216 \quad 180 \quad 350 \quad 90 \quad 24 \quad 8 \quad 8 \quad 8 \quad 8 \quad 8 \quad 8 \quad 8 \quad 8)$$

$$\text{TWO1} := \begin{pmatrix} 1 \\ 1 \end{pmatrix}$$

$$\text{TOTALCOSTTOMAKEPRODUCEWITHOT} := \text{FINALINVENTORYPLUSHOURS·COSTS}$$

TOTALCOSTTOMAKEPRODUCEWITHOT $= (92097.5 \quad 984)$

I'LL JUST GO AHEAD AND COMPUTE WAGES PLUS OVERTIME; TOTAL PAID OUT AND TOTAL TO EACH EMPLOYEE.

TOTAL PAID OUT BY COMPANY:

ONE2 $:= (1 \quad 1)$

ZERO1 $:= \begin{pmatrix} 0 \\ 1 \end{pmatrix}$

ONEZERO $:= \begin{pmatrix} 1 \\ 0 \end{pmatrix}$

TOTALWAGESPAIDITEMIZED $:=$ STRAIGHTHOURS2 $\cdot$ TIMEWITHOT
TOTALWAGESPAIDITEMIZED $= (656 \quad 984)$

IF THE EMPLOYEES WORKED A REGULAR WORKDAY, THEIR WAGESWOULD BE $656, IF THEY WORKED OVERTIME, THEIR WAGES FOR THE DAY WOULD BE $984.

TOTALWAGESPAID $:=$ TOTALWAGESPAIDITEMIZED $\cdot$ ONEZERO
TOTALWAGESPAID $= 656$
TOTALOTPAID $:=$ TOTALWAGESPAIDITEMIZED $\cdot$ ZERO1
TOTALOTPAID $= 984$
EMPLOYEESINDIVIDUALWAGES $:= \mathrm{diag}\left(\text{STRAIGHTHOURS2}^{\mathrm{T}} \right) \cdot$ TIMEWITHOT

$$
\text{EMPLOYEESINDIVIDUALWAGES} = \begin{bmatrix} 80 & 120 \\ 72 & 108 \\ 88 & 132 \\ 68 & 102 \\ 84 & 126 \\ 76 & 114 \\ 96 & 144 \\ 92 & 138 \end{bmatrix}
$$

WHEN I ADD THIS TO THE COSTS MATRIX, I WILL NOT DEDUCT FOR THE EMPLOYEES TAXES. THE EMPLOYEES WAGES WILL GO TOWARD THE COST OF MANUFACTURING THE SAUSAGES AND BOLOGNA. BUT HERE WE CAN COMPUTE THE TAXES AND WITHHOLDINGS. I DO NOT KNOW THE FEDERAL INCOME TAX WITHHOLDING FORMULAS, SO I WILL MAKE THE PERCENT DEDUCTION

THE SAME PERCENTAGE AS THEIR OVERTIME WAGES. SOCIAL SECURITY
WILL BE 10% AND STATE TAXES WILL BE 6%.

$$\text{THREE1} := \begin{pmatrix} 1 \\ 1 \\ 1 \end{pmatrix}$$

$$\text{TAX} := \begin{bmatrix} -.15 & -.1 & -.06 \\ -.135 & -.1 & -.06 \\ -.165 & -.1 & -.06 \\ -.1275 & -.1 & -.06 \\ -.1575 & -.1 & -.06 \\ -.1425 & -.1 & -.06 \\ -.18 & -.1 & -.06 \\ -.1725 & -.1 & -.06 \end{bmatrix}$$

EMPLOYEESINDIVIDUALWAGESNOOT := EMPLOYEESINDIVIDUALWAGES ONEZERO

TAXES := diag(EMPLOYEESINDIVIDUALWAGESNOOT)·TAX·THREE1

$$\text{TAXES} = \begin{bmatrix} -24.8 \\ -21.24 \\ -28.6 \\ -19.55 \\ -26.67 \\ -22.99 \\ -32.64 \\ -30.59 \end{bmatrix}$$

TO ITEMIZE FOR FEDERAL, STATE AND SOCIAL SECURITY, LEAVE OUT
THE MULTIPLICATION BY THREE1.

ITEMIZEDTAXES := diag(EMPLOYEESINDIVIDUALWAGESNOOT)·TAX

$$
\text{ITEMIZEDTAXES} =
\begin{bmatrix}
-12 & -8 & -4.8 \\
-9.72 & -7.2 & -4.32 \\
-14.52 & -8.8 & -5.28 \\
-8.67 & -6.8 & -4.08 \\
-13.23 & -8.4 & -5.04 \\
-10.83 & -7.6 & -4.56 \\
-17.28 & -9.6 & -5.76 \\
-15.87 & -9.2 & -5.52
\end{bmatrix}
$$

TAKEHOMEPAY := EMPLOYEESINDIVIDUALWAGES ONEZERO + TAXES

$$
\text{TAKEHOMEPAY} =
\begin{bmatrix}
55.2 \\
50.76 \\
59.4 \\
48.45 \\
57.33 \\
53.01 \\
63.36 \\
61.41
\end{bmatrix}
$$

OVERTIMETAKEHOMEPAY := EMPLOYEESINDIVIDUALWAGES ZERO1 + TAXES

$$
\text{OVERTIMETAKEHOMEPAY} =
\begin{bmatrix}
95.2 \\
86.76 \\
103.4 \\
82.45 \\
99.33 \\
91.01 \\
111.36 \\
107.41
\end{bmatrix}
$$

FOR THE PURPOSES OF COMPUTING THE TOTAL COSTS OF PRODUCING THE SAUSAGES AND BOLOGNA, WE WILL ASSUME THE EMPLOYEES ARE PART OF THE INGREDIENTS IN MAKING THE PRODUCE.

$$
\text{COSTS} := \begin{bmatrix}
.50 & 0 \\
.55 & 0 \\
1.00 & 0 \\
2.50 & 0 \\
.75 & 0 \\
2.75 & 0 \\
2.35 & 0 \\
.35 & 0 \\
1.00 & 0 \\
1.00 & 0 \\
.35 & 0 \\
2.50 & 0 \\
1.25 & 0 \\
.75 & 0 \\
.40 & 0 \\
1.00 & 0 \\
.01 & 0 \\
1.00 & 0 \\
1.50 & 0 \\
10 & 15 \\
9 & 13.5 \\
11 & 16.5 \\
8.5 & 12.75 \\
10.5 & 15.75 \\
9.5 & 14.25 \\
12 & 18 \\
11.5 & 17.25
\end{bmatrix}
$$

FINALINVENTORYPLUSHOURS := (140174 23650 731 683 100 150 50 50 341 50 1826 700 1397 485 216 180 350 90 24 8 8 8 8 8 8 8 8)

$$
\text{FINALINVENTORYPLUSOTHOURS} := \begin{pmatrix}
140174 & 23650 & 731 & 683 & 100 & 150 & 50 & 50 & 341 & 50 & 1826 & 700 & 1397 & 485 & 216 & 180 & 350 & 90 & 24 & 0 & 0 & 0 & 0 & 0 & 0 & 0 & 0 \\
0 & 0 & 0 & 0 & 0 & 0 & 0 & 0 & 0 & 0 & 0 & 0 & 0 & 0 & 0 & 0 & 0 & 0 & 0 & 8 & 8 & 8 & 8 & 8 & 8 & 8 & 8
\end{pmatrix}
$$

$$
\text{TWO1} := \begin{pmatrix} 1 \\ 1 \end{pmatrix}
$$

TOTALCOSTTOMAKEPRODUCEWITHOUTOT := FINALINVENTORYPLUSHOURS·COSTS·ONEZERO

TOTALCOSTTOMAKEPRODUCEWITHOUTOT = 92097.5

TOTALCOSTTOMAKEPRODUCEWITHOT1 := FINALINVENTORYPLUSOTHOURS·COSTS

$$I2 := \begin{pmatrix} 1 & 0 \\ 0 & 1 \end{pmatrix}$$

TOTALCOSTTOMAKEPRODUCEWITHOT2 $:=\overrightarrow{(\text{TOTALCOSTTOMAKEPRODUCEWITHOT1} \cdot I2)}$

$$\text{TOTALCOSTTOMAKEPRODUCEWITHOT2} = \begin{pmatrix} 91441.5 & 0 \\ 0 & 984 \end{pmatrix}$$

TOTALCOSTTOMAKEPRODUCEWITHOT $:=$ONE2·TOTALCOSTTOMAKEPRODUCEWITHOT2 ·TWO1

TOTALCOSTTOMAKEPRODUCEWITHOT $= 92425.5$

NOW LET'S THROW AWAY EVERYTHING WE HAVE DONE ABOVE THIS POINT AND COMPUTE EVERYTHING ALL AT THE SAME TIME, INCLUDING LOSSES. BECAUSE WE DO NOT WISH TO INCLUDE EMPLOYEE WAGES IN THE FIRST MULTIPLICATION, WE WILL INTRODUCE WHAT I WILL CALL TEMPLATE MATRICES.

CORRTEMP1 $:= \begin{pmatrix} .473 & .473 & .003 & .003 & .002 & .003 & .001 & .001 & .001 & .001 & .021 & .014 & 0 & 0 & 0 & 0 & .007 & 0 & 0 & 0 & 0 & 0 & 0 & 0 & 0 & 0 \\ .964 & 0 & .005 & .005 & 0 & 0 & 0 & 0 & .003 & 0 & .008 & 0 & .005 & .005 & 0 & .004 & 0 & .002 & 0 & 0 & 0 & 0 & 0 & 0 & 0 & 0 \\ .97 & 0 & .005 & .005 & 0 & 0 & 0 & 0 & .003 & 0 & .008 & 0 & .005 & .005 & 0 & 0 & 0 & 0 & 0 & 0 & 0 & 0 & 0 & 0 & 0 & 0 \end{pmatrix}$

CORRTEMP2 $:=(.946 \ 0 \ .004 \ .002 \ 0 \ 0 \ 0 \ 0 \ 0 \ 0 \ 0 \ 0 \ .038 \ 0 \ .009 \ 0 \ 0 \ 0 \ .001 \ 0 \ 0 \ 0 \ 0 \ 0 \ 0 \ 0)$

CORRTEMP $:=$stack(CORRTEMP1 , CORRTEMP2)

CORRTEMP $= \begin{bmatrix} 0.473 & 0.473 & 0.003 & 0.003 & 0.002 & 0.003 & 0.001 & 0.001 & 0.001 & 0.001 & 0.021 & 0.014 & 0 & 0 & 0 & 0 & 0.007 & 0 & 0 & 0 & 0 & 0 & 0 & 0 & 0 & 0 \\ 0.964 & 0 & 0.005 & 0.005 & 0 & 0 & 0 & 0 & 0.003 & 0 & 0.008 & 0 & 0.005 & 0.005 & 0 & 0.004 & 0 & 0.002 & 0 & 0 & 0 & 0 & 0 & 0 & 0 & 0 \\ 0.97 & 0 & 0.005 & 0.005 & 0 & 0 & 0 & 0 & 0.003 & 0 & 0.008 & 0 & 0.005 & 0.005 & 0 & 0 & 0 & 0 & 0 & 0 & 0 & 0 & 0 & 0 & 0 & 0 \\ 0.946 & 0 & 0.004 & 0.002 & 0 & 0 & 0 & 0 & 0 & 0 & 0 & 0 & 0.038 & 0 & 0.009 & 0 & 0 & 0 & 0.001 & 0 & 0 & 0 & 0 & 0 & 0 & 0 \end{bmatrix}$

This Matrix does not fit the page, this is the full Matrix below.

$\begin{bmatrix} 0.473 & 0.473 & 0.003 & 0.003 & 0.002 & 0.003 & 0.001 & 0.001 & 0.001 & 0.001 & 0.021 & 0.014 & 0 & 0 & 0 & 0 & 0.007 & 0 & 0 & 0 & 0 & 0 & 0 & 0 & 0 & 0 \\ 0.964 & 0 & 0.005 & 0.005 & 0 & 0 & 0 & 0 & 0.003 & 0 & 0.008 & 0 & 0.005 & 0.005 & 0 & 0.004 & 0 & 0.002 & 0 & 0 & 0 & 0 & 0 & 0 & 0 & 0 \\ 0.97 & 0 & 0.005 & 0.005 & 0 & 0 & 0 & 0 & 0.003 & 0 & 0.008 & 0 & 0.005 & 0.005 & 0 & 0 & 0 & 0 & 0 & 0 & 0 & 0 & 0 & 0 & 0 & 0 \\ 0.946 & 0 & 0.004 & 0.002 & 0 & 0 & 0 & 0 & 0 & 0 & 0 & 0 & 0.038 & 0 & 0.009 & 0 & 0 & 0 & 0.001 & 0 & 0 & 0 & 0 & 0 & 0 & 0 \end{bmatrix}$

DEPTTOTALSPLUSWASTE $:= \begin{bmatrix} 50000 & 45000 & 52000 & 24000 \\ -1000 & -800 & -1500 & -500 \\ 20000 & 10000 & 12000 & 0 \\ 5000 & 0 & 25000 & 6000 \\ 25000 & 15000 & 0 & 12000 \\ 0 & 20000 & 15000 & 6000 \end{bmatrix}$

TOTALPOUNDSPLUSWASTE $:=$DEPTTOTALSPLUSWASTE ·CORRTEMP

TOTALPOUNDSPLUSWASTE =

$$
\begin{bmatrix}
140174 & 23650 & 731 & 683 & 100 & 150 & 50 & 50 & 341 & 50 & 1826 & 700 & 1397 & 485 & 216 & 180 & 350 & 90 & 24 & 0 & 0 & 0 & 0 & 0 & 0 & 0 & 0 \\
-3172.2 & -473 & -16.5 & -15.5 & -2 & -3 & -1 & -1 & -7.9 & -1 & -39.4 & -14 & -30.5 & -11.5 & -4.5 & -3.2 & -7 & -1.6 & -0.5 & 0 & 0 & 0 & 0 & 0 & 0 & 0 & 0 \\
30740 & 9460 & 170 & 170 & 40 & 60 & 20 & 20 & 86 & 20 & 596 & 280 & 110 & 110 & 0 & 40 & 140 & 20 & 0 & 0 & 0 & 0 & 0 & 0 & 0 & 0 & 0 \\
32291 & 2365 & 164 & 152 & 10 & 15 & 5 & 5 & 80 & 5 & 305 & 70 & 353 & 125 & 54 & 0 & 35 & 0 & 6 & 0 & 0 & 0 & 0 & 0 & 0 & 0 & 0 \\
37637 & 11825 & 198 & 174 & 50 & 75 & 25 & 25 & 70 & 25 & 645 & 350 & 531 & 75 & 108 & 60 & 175 & 30 & 12 & 0 & 0 & 0 & 0 & 0 & 0 & 0 & 0 \\
39506 & 0 & 199 & 187 & 0 & 0 & 0 & 0 & 105 & 0 & 280 & 0 & 403 & 175 & 54 & 80 & 0 & 40 & 6 & 0 & 0 & 0 & 0 & 0 & 0 & 0 & 0
\end{bmatrix}
$$

The complete Matrix is:

$$
\begin{bmatrix}
140174 & 23650 & 731 & 683 & 100 & 150 & 50 & 50 & 341 & 50 & 1826 & 700 & 1397 & 485 & 216 & 180 & 350 & 90 & 24 & 0 & 0 & 0 & 0 & 0 & 0 & 0 & 0 \\
-3172 & -473 & -17 & -16 & -2 & -3 & -1 & -1 & -8 & -1 & -39 & -14 & -31 & -12 & -5 & -3 & -7 & -2 & -1 & 0 & 0 & 0 & 0 & 0 & 0 & 0 & 0 \\
30740 & 9460 & 170 & 170 & 40 & 60 & 20 & 20 & 86 & 20 & 596 & 280 & 110 & 110 & 0 & 40 & 140 & 20 & 0 & 0 & 0 & 0 & 0 & 0 & 0 & 0 & 0 \\
32291 & 2365 & 164 & 152 & 10 & 15 & 5 & 5 & 80 & 5 & 305 & 70 & 353 & 125 & 54 & 0 & 35 & 0 & 6 & 0 & 0 & 0 & 0 & 0 & 0 & 0 & 0 \\
37637 & 11825 & 198 & 174 & 50 & 75 & 25 & 25 & 70 & 25 & 645 & 350 & 531 & 75 & 108 & 60 & 175 & 30 & 12 & 0 & 0 & 0 & 0 & 0 & 0 & 0 & 0 \\
39506 & 0 & 199 & 187 & 0 & 0 & 0 & 0 & 105 & 0 & 280 & 0 & 403 & 175 & 54 & 80 & 0 & 40 & 6 & 0 & 0 & 0 & 0 & 0 & 0 & 0 & 0
\end{bmatrix}
$$

HOURTEMP :=

$$
\begin{bmatrix}
0 & 0 & 0 & 0 & 0 & 0 & 0 & 0 & 0 & 0 & 0 & 0 & 0 & 0 & 0 & 0 & 0 & 0 & 0 & 8 & 8 & 8 & 8 & 8 & 8 & 8 & 8 \\
0 & 0 & 0 & 0 & 0 & 0 & 0 & 0 & 0 & 0 & 0 & 0 & 0 & 0 & 0 & 0 & 0 & 0 & 0 & 0 & 0 & 0 & 0 & 0 & 0 & 0 & 0 \\
0 & 0 & 0 & 0 & 0 & 0 & 0 & 0 & 0 & 0 & 0 & 0 & 0 & 0 & 0 & 0 & 0 & 0 & 0 & 8 & 8 & 0 & 0 & 0 & 0 & 0 & 0 \\
0 & 0 & 0 & 0 & 0 & 0 & 0 & 0 & 0 & 0 & 0 & 0 & 0 & 0 & 0 & 0 & 0 & 0 & 0 & 0 & 0 & 8 & 8 & 0 & 0 & 0 & 0 \\
0 & 0 & 0 & 0 & 0 & 0 & 0 & 0 & 0 & 0 & 0 & 0 & 0 & 0 & 0 & 0 & 0 & 0 & 0 & 0 & 0 & 0 & 0 & 8 & 8 & 0 & 0 \\
0 & 0 & 0 & 0 & 0 & 0 & 0 & 0 & 0 & 0 & 0 & 0 & 0 & 0 & 0 & 0 & 0 & 0 & 0 & 0 & 0 & 0 & 0 & 0 & 0 & 8 & 8
\end{bmatrix}
$$

TOTALACCOUNTMATRIX := HOURTEMP + TOTALPOUNDSPLUSWASTE

Remember there are 2 employees in each Department, so there are 2 employee columns for each Dept.

HMB PRK LQSMK PEP GIN NTMG ALSP PAP GAR ONION SALT DMLK TNDR MSTD SUG CLSD WT DIL SP HSemp BLemp SSemp boloemp

TOTALACCOUNTMATRIX =

HMB	PRK	LQSMK	PEP	GIN	NTMG	ALSP	PAP	GAR	ONION	SALT	DMLK	TNDR	MSTD	SUG	CLSD	WT	DIL	SP	HSemp	BLemp		SSemp			boloemp	
140174	23650	731	683	100	150	50	50	341	50	1826	700	1397	485	216	180	350	90	24	8	8	8	8	8	8	8	8
-3172.2	-473	-16.5	-15.5	-2	-3	-1	-1	-7.9	-1	-39.4	-14	-30.5	-11.5	-4.5	-3.2	-7	-1.6	-0.5	0	0	0	0	0	0	0	0
30740	9460	170	170	40	60	20	20	86	20	596	280	110	110	0	40	140	20	0	8	8	0	0	0	0	0	0
32291	2365	164	152	10	15	5	5	80	5	305	70	353	125	54	0	35	0	6	0	0	8	8	0	0	0	0
37637	11825	198	174	50	75	25	25	70	25	645	350	531	75	108	60	175	30	12	0	0	0	0	8	8	0	0
39506	0	199	187	0	0	0	0	105	0	280	0	403	175	54	80	0	40	6	0	0	0	0	0	0	8	8

NOTE: THE HOURTEMP MATRIX WILL ALWAYS BE THE SAME UNLESS WE HIRE MORE EMPLOYEES; THE CORRTEMP MATRIX WILL ALWAYS BE THE SAME UNLESS WE CHANGE THE RECIPES OR ADD NEW ONES OR DELETE OLD ONES. THE **DEPTTOTALSPLUSWASTE** MATRIX WILL CHANGE DAILY. THE COSTS MATRIX WILL ALWAYS BE THE SAME UNLESS THERE IS A CHANGE IN PRICE OF THE INGREDIENTS OR AN EMPLOYEE GETS A RAISE. WE JUST CHANGE THOSE ITEMS AND PROCEED AS FOLLOWS. IN OTHER WORDS,

ONCE THE DATABASES ARE ENTERED, WE NEED ONLY TO KEEP TRACK OF
WHAT IS PRODUCED AND WASTED EACH DAY.

NOW THE ABOVE MATRIX IS FINE, BUT I AM NOT SATISFIED WITH IT.
WE CANNOT DO SIMPLE STATISTICS AS IT STANDS, SO I AM GOING TO
PUT THE WASTE PER DEPARTMENT IN THE **DEPTTOTALSPLUSWASTE** MATRIX.
WE'LL LET THE COMPUTER DO ALL THE EXTRA MATH FOR US.

$$
\text{COMPLETEDEPTTOTALSPLUSWASTE} :=
\begin{array}{cccc}
\text{H.S.} & \text{B.L.} & \text{S.S.} & \text{BOL} \\
\end{array}
$$

$$
\begin{bmatrix}
50000 & 45000 & 52000 & 24000 \\
-1000 & -800 & -1500 & -500 \\
20000 & 10000 & 12000 & 0 \\
-400 & -200 & -600 & 0 \\
5000 & 0 & 25000 & 6000 \\
-300 & 0 & -400 & -100 \\
25000 & 15000 & 0 & 12000 \\
-300 & -300 & 0 & -200 \\
0 & 20000 & 15000 & 6000 \\
0 & -300 & -500 & -200
\end{bmatrix}
$$

COMPINV := COMPLETEDEPTTOTALSPLUSWASTE · CORRTEMP

$$
\text{COMPINV} =
\begin{bmatrix}
140174 & 23650 & 731 & 683 & 100 & 150 & 50 & 50 & 341 & 50 & 1826 & 700 & 1397 & 485 & 216 & 180 & 350 & 90 & 24 & 0 & 0 & 0 & 0 & 0 & 0 & 0 \\
-3172.2 & -473 & -16.5 & -15.5 & -2 & -3 & -1 & -1 & -7.9 & -1 & -39.4 & -14 & -30.5 & -11.5 & -4.5 & -3.2 & -7 & -1.6 & -0.5 & 0 & 0 & 0 & 0 & 0 & 0 & 0 \\
30740 & 9460 & 170 & 170 & 40 & 60 & 20 & 20 & 86 & 20 & 596 & 280 & 110 & 110 & 0 & 40 & 140 & 20 & 0 & 0 & 0 & 0 & 0 & 0 & 0 & 0 \\
-964 & -189.2 & -5.2 & -5.2 & -0.8 & -1.2 & -0.4 & -0.4 & -2.8 & -0.4 & -14.8 & -5.6 & -4 & -4 & 0 & -0.8 & -2.8 & -0.4 & 0 & 0 & 0 & 0 & 0 & 0 & 0 & 0 \\
32291 & 2365 & 164 & 152 & 10 & 15 & 5 & 5 & 80 & 5 & 305 & 70 & 353 & 125 & 54 & 0 & 35 & 0 & 6 & 0 & 0 & 0 & 0 & 0 & 0 & 0 \\
-624.5 & -141.9 & -3.3 & -3.1 & -0.6 & -0.9 & -0.3 & -0.3 & -1.5 & -0.3 & -9.5 & -4.2 & -5.8 & -2 & -0.9 & 0 & -2.1 & 0 & -0.1 & 0 & 0 & 0 & 0 & 0 & 0 & 0 \\
37637 & 11825 & 198 & 174 & 50 & 75 & 25 & 25 & 70 & 25 & 645 & 350 & 531 & 75 & 108 & 60 & 175 & 30 & 12 & 0 & 0 & 0 & 0 & 0 & 0 & 0 \\
-620.3 & -141.9 & -3.2 & -2.8 & -0.6 & -0.9 & -0.3 & -0.3 & -1.2 & -0.3 & -8.7 & -4.2 & -9.1 & -1.5 & -1.8 & -1.2 & -2.1 & -0.6 & -0.2 & 0 & 0 & 0 & 0 & 0 & 0 & 0 \\
39506 & 0 & 199 & 187 & 0 & 0 & 0 & 0 & 105 & 0 & 280 & 0 & 403 & 175 & 54 & 80 & 0 & 40 & 6 & 0 & 0 & 0 & 0 & 0 & 0 & 0 \\
-963.4 & 0 & -4.8 & -4.4 & 0 & 0 & 0 & 0 & -2.4 & 0 & -6.4 & 0 & -11.6 & -4 & -1.8 & -1.2 & 0 & -0.6 & -0.2 & 0 & 0 & 0 & 0 & 0 & 0 & 0
\end{bmatrix}
$$

BEFORE WE ADD IN THE EMPLOYEES HOURS, I'LL COMPUTE BOTH THE
WASTE AND THE TOTAL POUNDAGE THAT IS GOOD. FOR SIMPLICITY,
ALTHOUGH WE DO NOT HAVE TO DO THIS, I'LL GET RID OF THE TOTAL
POUNDS AND LEAVE ONLY THE WASTE POUNDAGE.

ONE10 := (0 1 0 1 0 1 0 1 0 1)
WASTEONLY := diag$\left(\text{ONE10}^T\right)$ · COMPINV

$$
\text{WASTEONLY} =
\begin{bmatrix}
0 & 0 & 0 & 0 & 0 & 0 & 0 & 0 & 0 & 0 & 0 & 0 & 0 & 0 & 0 & 0 & 0 & 0 & 0 & 0 & 0 & 0 & 0 & 0 & 0 & 0 \\
-3172.2 & -473 & -16.5 & -15.5 & -2 & -3 & -1 & -1 & -7.9 & -1 & -39.4 & -14 & -30.5 & -11.5 & -4.5 & -3.2 & -7 & -1.6 & -0.5 & 0 & 0 & 0 & 0 & 0 & 0 & 0 \\
0 & 0 & 0 & 0 & 0 & 0 & 0 & 0 & 0 & 0 & 0 & 0 & 0 & 0 & 0 & 0 & 0 & 0 & 0 & 0 & 0 & 0 & 0 & 0 & 0 & 0 \\
-964 & -189.2 & -5.2 & -5.2 & -0.8 & -1.2 & -0.4 & -0.4 & -2.8 & -0.4 & -14.8 & -5.6 & -4 & -4 & 0 & -0.8 & -2.8 & -0.4 & 0 & 0 & 0 & 0 & 0 & 0 & 0 & 0 \\
0 & 0 & 0 & 0 & 0 & 0 & 0 & 0 & 0 & 0 & 0 & 0 & 0 & 0 & 0 & 0 & 0 & 0 & 0 & 0 & 0 & 0 & 0 & 0 & 0 & 0 \\
-624.5 & -141.9 & -3.3 & -3.1 & -0.6 & -0.9 & -0.3 & -0.3 & -1.5 & -0.3 & -9.5 & -4.2 & -5.8 & -2 & -0.9 & 0 & -2.1 & 0 & -0.1 & 0 & 0 & 0 & 0 & 0 & 0 & 0 \\
0 & 0 & 0 & 0 & 0 & 0 & 0 & 0 & 0 & 0 & 0 & 0 & 0 & 0 & 0 & 0 & 0 & 0 & 0 & 0 & 0 & 0 & 0 & 0 & 0 & 0 \\
-620.3 & -141.9 & -3.2 & -2.8 & -0.6 & -0.9 & -0.3 & -0.3 & -1.2 & -0.3 & -8.7 & -4.2 & -9.1 & -1.5 & -1.8 & -1.2 & -2.1 & -0.6 & -0.2 & 0 & 0 & 0 & 0 & 0 & 0 & 0 \\
0 & 0 & 0 & 0 & 0 & 0 & 0 & 0 & 0 & 0 & 0 & 0 & 0 & 0 & 0 & 0 & 0 & 0 & 0 & 0 & 0 & 0 & 0 & 0 & 0 & 0 \\
-963.4 & 0 & -4.8 & -4.4 & 0 & 0 & 0 & 0 & -2.4 & 0 & -6.4 & 0 & -11.6 & -4 & 0 & -1.8 & -1.2 & 0 & -0.6 & -0.2 & 0 & 0 & 0 & 0 & 0 & 0 \\
\end{bmatrix}
$$

THE COST IN DOLLARS FOR THE WASTEIS GIVEN BY:

COSTWASTE := WASTEONLY·COSTS

$$
\text{COSTWASTE} =
\begin{bmatrix}
0 & 0 \\
-2025.81 & 0 \\
0 & 0 \\
-640.85 & 0 \\
0 & 0 \\
-429.99 & 0 \\
0 & 0 \\
-432.52 & 0 \\
0 & 0 \\
-522.46 & 0 \\
\end{bmatrix}
$$

THE COSTS OF THE DEPARTMENT WASTES SHOULD EQUAL THE TOTAL WASTE ($2025.81):

$$
\text{CHECKWASTE} := (0\ \ 0\ \ 0\ \ 1\ \ 0\ \ 1\ \ 0\ \ 1\ \ 0\ \ 1) \cdot
\begin{bmatrix}
0 & 0 \\
-2025.81 & 0 \\
0 & 0 \\
-640.85 & 0 \\
0 & 0 \\
-429.99 & 0 \\
0 & 0 \\
-432.52 & 0 \\
0 & 0 \\
-522.46 & 0 \\
\end{bmatrix}
$$

CHECKWASTE $= (-2025.82\ \ 0\ )$

LET'S NOW LOOK AT THE TOTAL POUNDAGE THAT PASSED. DB1 IS A STATISTICAL DATABASE MATRIX THAT IS USED FOR ACCOUNTING PURPOSES:

$$DB1 := \begin{bmatrix} 1 & 1 & 0 & 0 & 0 & 0 & 0 & 0 & 0 & 0 \\ 0 & 0 & 1 & 1 & 0 & 0 & 0 & 0 & 0 & 0 \\ 0 & 0 & 0 & 0 & 1 & 1 & 0 & 0 & 0 & 0 \\ 0 & 0 & 0 & 0 & 0 & 0 & 1 & 1 & 0 & 0 \\ 0 & 0 & 0 & 0 & 0 & 0 & 0 & 0 & 1 & 1 \end{bmatrix}$$

POUNDSPASSED := DB1·COMPINV

Sorry, this Matrix runs off the page, so we'll copy it below
without the variable name.

POUNDSPASSED =

137001.8	23177	714.5	667.5	98	147	49	49	333.1	49	1786.6	686	1366.5	473.5	211.5	176.8	343	88.4	23.5	0	0	0	0	0	0	0
29776	9270.8	164.8	164.8	39.2	58.8	19.6	19.6	83.2	19.6	581.2	274.4	106	106	0	39.2	137.2	19.6	0	0	0	0	0	0	0	0
31666.5	2223.1	160.7	148.9	9.4	14.1	4.7	4.7	78.5	4.7	295.5	65.8	347.2	123	53.1	0	32.9	0	5.9	0	0	0	0	0	0	0
37016.7	11683.1	194.8	171.2	49.4	74.1	24.7	24.7	68.8	24.7	636.3	345.8	521.9	73.5	106.2	58.8	172.9	29.4	11.8	0	0	0	0	0	0	0
38542.6	0	194.2	182.6	0	0	0	0	102.6	0	273.6	0	391.4	171	52.2	78.8	0	39.4	5.8	0	0	0	0	0	0	0

HMB	PRK	LQSMK	PEP	GIN	NTMG	ALS	PAP	GAR	ON	SALT	DMLK	TNDR	MSTD	SUG	CLSD	WT	DIL	SP 1	2	3	4	5	6	7	8
137002	23177	715	668	98	147	49	49	333	49	1787	686	1367	474	211	177	343	88	24 0	0	0	0	0	0	0	0
29776	9271	165	165	39	59	20	20	83	20	581	274	106	106	0	39	137	20	0 0	0	0	0	0	0	0	0
31667	2223	161	149	9	14	5	5	79	5	296	66	347	123	53	0	33	0	6 0	0	0	0	0	0	0	0
37017	11683	195	171	49	74	25	25	69	25	636	346	522	74	106	59	173	29	12 0	0	0	0	0	0	0	0
38543	0	194	183	0	0	0	0	103	0	274	0	391	171	52	79	0	39	6 0	0	0	0	0	0	0	0

LET'S LOOK AT THE % LOSSES. % ERROR = 100x (AMOUNT PASSED – TOTAL
AMOUNT)/(TOTAL AMOUNT)

TEN1 := (1 0 1 0 1 0 1 0 1 0)

TOTALPOUNDS1 := diag$\left(\text{TEN1}^T\right)$·COMPINV

TOTALPOUNDS1 =

HMB	PRK	LQSMK	PEP	GIN	NTMG	ALSP	PAP	GAR	ON	SALT	DMLK	TNDR	MSTD	SUG	CLSD	WT	DIL	SP							
140174	23650	731	683	100	150	50	50	341	50	1826	700	1397	485	216	180	350	90	24	0	0	0	0	0	0	0
0	0	0	0	0	0	0	0	0	0	0	0	0	0	0	0	0	0	0	0	0	0	0	0	0	0
30740	9460	170	170	40	60	20	20	86	20	596	280	110	110	0	40	140	20	0	0	0	0	0	0	0	0
0	0	0	0	0	0	0	0	0	0	0	0	0	0	0	0	0	0	0	0	0	0	0	0	0	0
32291	2365	164	152	10	15	5	5	80	5	305	70	353	125	54	0	35	0	6	0	0	0	0	0	0	0
0	0	0	0	0	0	0	0	0	0	0	0	0	0	0	0	0	0	0	0	0	0	0	0	0	0
37637	11825	198	174	50	75	25	25	70	25	645	350	531	75	108	60	175	30	12	0	0	0	0	0	0	0
0	0	0	0	0	0	0	0	0	0	0	0	0	0	0	0	0	0	0	0	0	0	0	0	0	0
39506	0	199	187	0	0	0	0	105	0	280	0	403	175	54	80	0	40	6	0	0	0	0	0	0	0
0	0	0	0	0	0	0	0	0	0	0	0	0	0	0	0	0	0	0	0	0	0	0	0	0	0

TOTALPOUNDS := DB1·TOTALPOUNDS1

TOTALPOUNDS =

140174	23650	731	683	100	150	50	50	341	50	1826	700	1397	485	216	180	350	90	24	0	0	0	0	0	0	0
30740	9460	170	170	40	60	20	20	86	20	596	280	110	110	0	40	140	20	0	0	0	0	0	0	0	0
32291	2365	164	152	10	15	5	5	80	5	305	70	353	125	54	0	35	0	6	0	0	0	0	0	0	0
37637	11825	198	174	50	75	25	25	70	25	645	350	531	75	108	60	175	30	12	0	0	0	0	0	0	0
39506	0	199	187	0	0	0	0	105	0	280	0	403	175	54	80	0	40	6	0	0	0	0	0	0	0

$$
\text{TOTALLBSINV} := \begin{bmatrix}
\frac{1}{140174} & \frac{1}{23650} & \frac{1}{731} & \frac{1}{683} & \frac{1}{100} & \frac{1}{150} & \frac{1}{50} & \frac{1}{50} & \frac{1}{341} & \frac{1}{50} & \frac{1}{1826} & \frac{1}{700} & \frac{1}{1397} & \frac{1}{485} & \frac{1}{216} & \frac{1}{180} & \frac{1}{350} & \frac{1}{90} & \frac{1}{24} & 0 & 0 & 0 & 0 & 0 & 0 & 0 \\
\frac{1}{30740} & \frac{1}{9460} & \frac{1}{170} & \frac{1}{170} & \frac{1}{40} & \frac{1}{60} & \frac{1}{20} & \frac{1}{20} & \frac{1}{86} & \frac{1}{20} & \frac{1}{596} & \frac{1}{280} & \frac{1}{110} & \frac{1}{110} & 0 & \frac{1}{40} & \frac{1}{140} & \frac{1}{20} & 0 & 0 & 0 & 0 & 0 & 0 & 0 & 0 \\
\frac{1}{32291} & \frac{1}{2365} & \frac{1}{164} & \frac{1}{152} & \frac{1}{10} & \frac{1}{15} & \frac{1}{5} & \frac{1}{5} & \frac{1}{80} & \frac{1}{5} & \frac{1}{305} & \frac{1}{70} & \frac{1}{353} & \frac{1}{125} & \frac{1}{54} & 0 & \frac{1}{35} & 0 & \frac{1}{6} & 0 & 0 & 0 & 0 & 0 & 0 & 0 \\
\frac{1}{37637} & \frac{1}{11825} & \frac{1}{198} & \frac{1}{174} & \frac{1}{50} & \frac{1}{75} & \frac{1}{25} & \frac{1}{25} & \frac{1}{70} & \frac{1}{25} & \frac{1}{645} & \frac{1}{350} & \frac{1}{531} & \frac{1}{75} & \frac{1}{108} & \frac{1}{60} & \frac{1}{175} & \frac{1}{30} & \frac{1}{12} & 0 & 0 & 0 & 0 & 0 & 0 & 0 \\
\frac{1}{39506} & 0 & \frac{1}{199} & \frac{1}{187} & 0 & 0 & 0 & 0 & \frac{1}{105} & 0 & \frac{1}{280} & 0 & \frac{1}{403} & \frac{1}{175} & \frac{1}{54} & \frac{1}{80} & 0 & \frac{1}{40} & \frac{1}{6} & 0 & 0 & 0 & 0 & 0 & 0 & 0
\end{bmatrix}
$$

NOTE: MATHCAD CANNOT TAKE THE INDIVIDUAL INVERSES OF THE ABOVE MATRIX BECAUSE IT CANNOT DIVIDE BY ZERO, SO I HAD TO COMPUTE THE INVERSES BY HAND. THE EASIEST WAY WOULD NOT BE TO ITEMIZE THE % ERROR BUT TO COMPUTE THE TOTAL ERROR.

$$\text{PCTERROR1} := 100 \cdot (\text{TOTALPOUNDS} - \text{POUNDSPASSED})$$

$$\text{PCTERROR} := \overrightarrow{(\text{PCTERROR1} \cdot \text{TOTALLBSINV})}$$

$$
\text{PCTERROR} = \begin{bmatrix}
2.26 & 2 & 2.26 & 2.27 & 2 & 2 & 2 & 2 & 2.32 & 2 & 2.16 & 2 & 2.18 & 2.37 & 2.08 & 1.78 & 2 & 1.78 & 2.08 & 0 & 0 & 0 & 0 & 0 & 0 & 0 \\
3.14 & 2 & 3.06 & 3.06 & 2 & 2 & 2 & 2 & 3.26 & 2 & 2.48 & 2 & 3.64 & 3.64 & 0 & 2 & 2 & 2 & 0 & 0 & 0 & 0 & 0 & 0 & 0 & 0 \\
1.93 & 6 & 2.01 & 2.04 & 6 & 6 & 6 & 6 & 1.88 & 6 & 3.11 & 6 & 1.64 & 1.6 & 1.67 & 0 & 6 & 0 & 1.67 & 0 & 0 & 0 & 0 & 0 & 0 & 0 \\
1.65 & 1.2 & 1.62 & 1.61 & 1.2 & 1.2 & 1.2 & 1.2 & 1.71 & 1.2 & 1.35 & 1.2 & 1.71 & 2 & 1.67 & 2 & 1.2 & 2 & 1.67 & 0 & 0 & 0 & 0 & 0 & 0 & 0 \\
2.44 & 0 & 2.41 & 2.35 & 0 & 0 & 0 & 0 & 2.29 & 0 & 2.29 & 0 & 2.88 & 2.29 & 3.33 & 1.5 & 0 & 1.5 & 3.33 & 0 & 0 & 0 & 0 & 0 & 0 & 0
\end{bmatrix}
$$

DEPT. 2 HAD THE HIGHEST PERCENT WASTE, IT WASTED 6% OF THE PORK. DEPT. 3 HAD THE BEST SHOWING.

$$
\text{ONE27} := \begin{bmatrix} 1 \\ 1 \\ 1 \\ 1 \\ 1 \\ 1 \\ 1 \\ 1 \\ 1 \\ 1 \\ 1 \\ 1 \\ 1 \\ 1 \\ 1 \\ 1 \\ 1 \\ 1 \\ 1 \\ 1 \\ 1 \\ 1 \\ 1 \\ 1 \\ 1 \\ 1 \\ 1 \end{bmatrix}
$$

LBSPASSEDTOT := POUNDSPASSED·ONE27

TOTALLBS := TOTALPOUNDS·ONE27

$$
\text{LBSPASSEDTOT} = \begin{bmatrix} 167441.7 \\ 40880 \\ 35238.7 \\ 51288.8 \\ 40034.2 \end{bmatrix}
$$

CLINTON L. HOLT

$$TOTALLBS = \begin{bmatrix} 171247 \\ 42082 \\ 36040 \\ 52090 \\ 41035 \end{bmatrix}$$

$$TOTALLBSINV := \overrightarrow{TOTALLBS^{-1}}$$

$$TOTALLBSINV = \begin{bmatrix} 5.839518 \cdot 10^{-6} \\ 0.000024 \\ 0.000028 \\ 0.000019 \\ 0.000024 \end{bmatrix}$$

$$TOTALPCTERROR := \overrightarrow{(100 \cdot (LBSPASSEDTOT - TOTALLBS) \cdot TOTALLBSINV)}$$

$$TOTALPCTERROR = \begin{bmatrix} ^{-}2.22 \\ ^{-}2.86 \\ ^{-}2.22 \\ ^{-}1.54 \\ ^{-}2.44 \end{bmatrix}$$

THE TOTAL % WASTE = 2.22%; THE % WASTE FOR DEPT.1 = 2.86%; THE % WASTE FOR DEPT. 2 = 2.22%, ETC.

NOW LET'S GO AHEAD AND ADD IN THE EMPLOYEE HOURS. NOTE THAT THIS IS THE EASIEST WAY TO COMPUTE WAGES, BUT IT IS NOT THE BEST WAY. I MEAN, SUPPOSE ONE EMPLOYEE WORKED 8 HOURS A DAY IN THE REGULAR WORKWEEK, ANOTHER WORKED 7 HOURS THAT DAY, BUT BOTH ONLY WORKED FOUR HOURS SATURDAY? THIS EXAMPLE IS FOR DEMONSTRATION. THE OVERTIME PROBLEM IS SIMPLE BUT I WON'T GO INTO THAT HERE YET. I'LL JUST DEMONSTRATE THE PROBLEM AS DESCRIBED ABOVE.

$$HOURSTEMP := \begin{bmatrix} 0 & 0 & 0 & 0 & 0 & 0 & 0 & 0 & 0 & 0 & 0 & 0 & 0 & 0 & 0 & 0 & 0 & 0 & 0 & 0 & 8 & 8 & 8 & 8 & 8 & 8 & 8 \\ 0 & 0 & 0 & 0 & 0 & 0 & 0 & 0 & 0 & 0 & 0 & 0 & 0 & 0 & 0 & 0 & 0 & 0 & 0 & 0 & 8 & 8 & 0 & 0 & 0 & 0 & 0 \\ 0 & 0 & 0 & 0 & 0 & 0 & 0 & 0 & 0 & 0 & 0 & 0 & 0 & 0 & 0 & 0 & 0 & 0 & 0 & 0 & 8 & 8 & 0 & 0 & 0 & 0 & 0 \\ 0 & 0 & 0 & 0 & 0 & 0 & 0 & 0 & 0 & 0 & 0 & 0 & 0 & 0 & 0 & 0 & 0 & 0 & 0 & 0 & 0 & 8 & 8 & 0 & 0 \\ 0 & 0 & 0 & 0 & 0 & 0 & 0 & 0 & 0 & 0 & 0 & 0 & 0 & 0 & 0 & 0 & 0 & 0 & 0 & 0 & 0 & 0 & 0 & 8 & 8 \end{bmatrix}$$

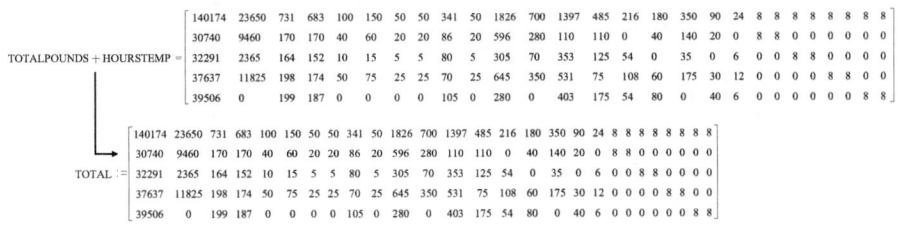

$$\text{TOTALPOUNDS} + \text{HOURSTEMP} = \begin{bmatrix} 140174 & 23650 & 731 & 683 & 100 & 150 & 50 & 50 & 341 & 50 & 1826 & 700 & 1397 & 485 & 216 & 180 & 350 & 90 & 24 & 8 & 8 & 8 & 8 & 8 & 8 & 8 & 8 \\ 30740 & 9460 & 170 & 170 & 40 & 60 & 20 & 20 & 86 & 20 & 596 & 280 & 110 & 110 & 0 & 40 & 140 & 20 & 0 & 8 & 8 & 0 & 0 & 0 & 0 & 0 & 0 \\ 32291 & 2365 & 164 & 152 & 10 & 15 & 5 & 5 & 80 & 5 & 305 & 70 & 353 & 125 & 54 & 0 & 35 & 0 & 6 & 0 & 0 & 8 & 8 & 0 & 0 & 0 & 0 \\ 37637 & 11825 & 198 & 174 & 50 & 75 & 25 & 25 & 70 & 25 & 645 & 350 & 531 & 75 & 108 & 60 & 175 & 30 & 12 & 0 & 0 & 0 & 0 & 8 & 8 & 0 & 0 \\ 39506 & 0 & 199 & 187 & 0 & 0 & 0 & 0 & 105 & 0 & 280 & 0 & 403 & 175 & 54 & 80 & 0 & 40 & 6 & 0 & 0 & 0 & 0 & 0 & 0 & 8 & 8 \end{bmatrix}$$

$$\text{TOTAL} := \begin{bmatrix} 140174 & 23650 & 731 & 683 & 100 & 150 & 50 & 50 & 341 & 50 & 1826 & 700 & 1397 & 485 & 216 & 180 & 350 & 90 & 24 & 8 & 8 & 8 & 8 & 8 & 8 & 8 & 8 \\ 30740 & 9460 & 170 & 170 & 40 & 60 & 20 & 20 & 86 & 20 & 596 & 280 & 110 & 110 & 0 & 40 & 140 & 20 & 0 & 8 & 8 & 0 & 0 & 0 & 0 & 0 & 0 \\ 32291 & 2365 & 164 & 152 & 10 & 15 & 5 & 5 & 80 & 5 & 305 & 70 & 353 & 125 & 54 & 0 & 35 & 0 & 6 & 0 & 0 & 8 & 8 & 0 & 0 & 0 & 0 \\ 37637 & 11825 & 198 & 174 & 50 & 75 & 25 & 25 & 70 & 25 & 645 & 350 & 531 & 75 & 108 & 60 & 175 & 30 & 12 & 0 & 0 & 0 & 0 & 8 & 8 & 0 & 0 \\ 39506 & 0 & 199 & 187 & 0 & 0 & 0 & 0 & 105 & 0 & 280 & 0 & 403 & 175 & 54 & 80 & 0 & 40 & 6 & 0 & 0 & 0 & 0 & 0 & 0 & 8 & 8 \end{bmatrix}$$

EVERYTHING := TOTAL ·COSTS

$$\text{EVERYTHING} = \begin{bmatrix} 92097.5 & 984 \\ 22865 & 228 \\ 19141.2 & 234 \\ 28495.2 & 240 \\ 21596.1 & 282 \end{bmatrix}$$

THIS IS GOOD, BUT I DON'T LIKE IT, WHAT IF WE WERE WORKING SATURDAY? LET'S RE-DEFINE THE COSTS MATRIX AS:

$$
\text{COSTSPLUSOT} := \begin{bmatrix}
.50 & .50 \\
.55 & .55 \\
1.00 & 1.00 \\
2.50 & 2.50 \\
.75 & .75 \\
2.75 & 2.75 \\
2.35 & 2.35 \\
.35 & .35 \\
1.00 & 1.00 \\
1.00 & 1.00 \\
.35 & .35 \\
2.50 & 2.50 \\
1.25 & 1.25 \\
.75 & .75 \\
.40 & .40 \\
1.00 & 1.00 \\
.01 & .01 \\
1.00 & 1.00 \\
1.50 & 1.50 \\
10 & 15 \\
9 & 13.5 \\
11 & 16.5 \\
8.5 & 12.75 \\
10.5 & 15.75 \\
9.5 & 14.25 \\
12 & 18 \\
11.5 & 17.25
\end{bmatrix}
$$

EVERYTHING $:=$ TOTAL$\cdot$COSTSPLUSOT

$$
\text{EVERYTHING} = \begin{bmatrix}
92097.5 & 92425.5 \\
22865 & 22941 \\
19141.2 & 19219.2 \\
28495.2 & 28575.2 \\
21596.1 & 21690.1
\end{bmatrix}
$$

IF WE RAN THIS ON A WEEKDAY, THE PRODUCTION COSTS WOULD BE:

EVERYTHINGNORMALDAY $:=$ TOTAL$\cdot$COSTSPLUSOT$\cdot$ONEZERO

$$\text{EVERYTHINGNORMALDAY} = \begin{bmatrix} 92097.5 \\ 22865 \\ 19141.2 \\ 28495.2 \\ 21596.1 \end{bmatrix}$$

THE TOTAL COSTS FOR THE WORKDAY WOULD BE $92,097.50 FOR THE DAILY
TOTAL, $22,865 FOR DEPT. 1; 19,141.20 FOR DEPT. 2; $28,495.20
FOR DEPT. 3 AND $21,596.10 FOR DEPT. 4.

IF THIS PRODUCTION WERE RUN ON SATURDAY, THE COSTS WOULD BE:

EVERYTHINGSATURDAY $=$ TOTAL·COSTSPLUSOT·ZERO1

$$\text{EVERYTHINGSATURDAY} = \begin{bmatrix} 92425.5 \\ 22941 \\ 19219.2 \\ 28575.2 \\ 21690.1 \end{bmatrix}$$

THE TOTAL COSTS FOR THE SATURDAY WITH OVERTIME WOULD BE
$92,425.50 FOR THE DAILY PRODUCTION TOTAL; $22,941 FOR DEPT. 1;
$19,219.20 FOR DEPT. 2; $28,575.20 FOR DEPT. 3 AND $21,690.10
FOR DEPT. 4.

NOW THIS WOULDN'T BE A PROPER INVENTORY SYSTEM IF WE COULD
NOT KEEP TRACK OF OUR TOTAL INVENTORY ALSO. WITH DECIMALS, THE
MATRICES ARE TOO LARGE TO FIT ON THIS PAGE, SO I AM GOING TO
ROUND OFF TO THE NEAREST POUND.

HMB PRK LQSMK PEP GIN NTMG ALSP PAP GAR ON SLT DMLK TNDR MSTD
SUG CLSD WTR DIL SP
SUPPOSE IN-HOUSE WE HAVE:

```
1,000,000 LBS HAMBURGER
  500,000 LBS PORK
  100,000 LBS LIQUID SMOKE
   50,000 LBS PEPPER
   25,000 LBS GINGER
   30,000 LBS NUTMEG
   20,000 LBS ALLSPICE
   75,000 LBS PAPRIKA
```

```
 80,000 LBS GARLIC POWDER
100,000 LBS ONION POWDER
150,000 LBS SALT
 55,000 LBS DRY MILK
125,000 LBS TENDERIZER
100,000 LBS MUSTARD
 20,000 LBS SUGAR
 40,000 LBS CELERY SEED
400,000 LBS WATER
 15,000 LBS DILL SEED
 99,000 LBS SALT PETER
```

INHOUSEINV := (1000000 500000 100000 50000 25000 30000 20000 75000 80000 100000 150000 55000 125000 100000 20000 40000 400000 15000 99000)

FINALINVTEMPLATE := (1 0 0 0 0)

COMPLETEDEPARTMENTINV :=

HMB	PRK	LQSMK	PEP	GIN	NTMG	ALSP	PAP	GAR	ON	SLT	DMLK	TNDR	MSTD	SUG	CLSD	WTR	DIL	SP
140150	23643	703	675	87	131	43	43	273	29	1777	695	1396	488	227	181	350	90	19
30733	9457	163	163	35	52	17	17	67	12	584	278	111	111	0	40	140	20	0
32284	2364	160	153	9	13	4	4	66	3	294	70	353	126	57	0	35	0	5
37632	11822	183	169	44	65	22	22	52	14	635	348	529	75	114	60	175	30	9
39500	0	197	190	0	0	0	0	88	0	264	0	403	176	57	80	0	40	5

FINALINV := FINALINVTEMPLATE·COMPLETEDEPARTMENTINV

FINALINV = (140150   23643   703   675   87   131   43   43   273   29   1777   695   1396   488   227   181   350   90   19)

INHOUSEINV− FINALINV

INHOUSEINV102697 := INHOUSEINV− FINALINV

INHOUSEINV102697 = (859850   476357   99297   49325   24913   29869   19957   74957   79727   99971   148223   54305   123604   99512   19773   39819   399650   14910   98981)

```
Oops, off the page again, the complete Matrix is:
```

(859850   476357   99297   49325   24913   29869   19957   74957   79727   99971   148223   54305   123604   99512   19773   39819   399650   14910   98981)

EXAMPLE: ARGONIA STORE INVENTORY

SUPPOSE WE HAVE A STORE THAT SELLS 9 ITEMS. THE STORE IS IN A TOWN THAT HAS 10 RESIDENTS. THE 9 ITEMS, THEIR PRICES, THE SHOPKEEPERS COST TO BUY THE ITEMS, AND THE SALES TAX ARE LISTED BELOW. THE SHOPKEEPER WISHES TO KEEP A LIST OF WHAT IS BOUGHT AND WHO BUYS IT, HOW MUCH MONEY SHE BROUGHT IN, HOW MUCH TAX WAS COLLECTED, EMPLOYEE WAGES AND TOTAL PROFITS FOR THE DAY. SHE ALSO WISHES TO KNOW HOW MUCH OF YESTERDAYS INVENTORY IS LEFT AFTER TODAY'S SALES.

	LET	TOM	CEL	BEEF	CHICK	PORK	MILK	EGGS	BEER
LARRY	2	3	1	0	0	0	2	1	0
HELEN	0	1	0	2	4	0	1	1	0
TANYA	1	1	0	1	2	0	1	0	0
BECKY	0	2	0	2	3	1	0	1	1
DON	0	0	0	4	2	2	1	0	0
SHANNA	2	1	1	1.5	1	0	2	1	0
JEANIE	1	0	1	2	2	0	1	1	1
LORI	1	0	0	3	3	2	1	0	1
BOBBYJO	0	1	0	2	0	2	2	1	0
PAULA	1	0	1	3	1	0	1	1	0

		PRICE	COST	TAX
L = LETTUCE	(HEADS)	$.99	$.75	.06
T = TOMATO'S	(POUNDS)	$1.09	$.50	.06
C = CELERY	(STALKS)	$.69	$.35	.06
Bf = BEEF	(POUNDS)	$2.29	$1.45	.06
C = CHICKEN	(POUNDS)	$1.59	$.85	.06
P = PORK	(POUNDS)	$2.25	$1.65	.06
M = MILK	(HALF-GALLONS)	$1.59	$1.00	.06
E = EGGS	(DOZEN)	$1.09	$.55	.06
B = BEER	(SIX-PACK)	$2.50	$1.25	.06

WE PUT THE SALES INVENTORY INTO MATRIX FORM AND NAME THE MATRIX SALES. WE THEN COMPUTE THE 6% TAX ON THE SALE PRICE AND ENTER IT INTO THE THIRD COLUMN. THE COST TO BUY THE ITEMS ARE ENTERED AS NEGATIVE VALUES IN THE SECOND COLUMN. WE NAME THIS MATRIX DB FOR DATABASE.

$$
\text{SALES} :=
\begin{bmatrix}
L & T & C & Bf & C & P & M & E & B \\
2 & 3 & 1 & 0 & 0 & 0 & 2 & 1 & 0 \\
0 & 1 & 0 & 2 & 4 & 0 & 1 & 1 & 0 \\
1 & 1 & 0 & 1 & 2 & 0 & 1 & 0 & 0 \\
0 & 2 & 0 & 2 & 3 & 1 & 0 & 1 & 1 \\
0 & 0 & 0 & 4 & 2 & 2 & 1 & 0 & 0 \\
2 & 1 & 1 & 1.5 & 1 & 0 & 2 & 1 & 0 \\
1 & 0 & 1 & 2 & 2 & 0 & 1 & 1 & 1 \\
1 & 0 & 0 & 3 & 3 & 2 & 1 & 0 & 1 \\
0 & 1 & 0 & 2 & 0 & 2 & 2 & 1 & 0 \\
1 & 0 & 1 & 3 & 1 & 0 & 1 & 1 & 0
\end{bmatrix}
\qquad
\text{DB} :=
\begin{bmatrix}
Pr & C & Tx \\
.99 & -.75 & .0594 \\
1.09 & -.50 & .0654 \\
.69 & -.35 & .0414 \\
2.29 & -1.45 & .1374 \\
1.59 & -.85 & .0954 \\
2.25 & -1.65 & .135 \\
1.59 & -1.00 & .0954 \\
1.09 & -.55 & .0654 \\
2.50 & -1.25 & .15
\end{bmatrix}
$$

SALESPERPERSON :=SALES·DB

$$
\begin{bmatrix}
2 & 3 & 1 & 0 & 0 & 0 & 2 & 1 & 0 \\
0 & 1 & 0 & 2 & 4 & 0 & 1 & 1 & 0 \\
1 & 1 & 0 & 1 & 2 & 0 & 1 & 0 & 0 \\
0 & 2 & 0 & 2 & 3 & 1 & 0 & 1 & 1 \\
0 & 0 & 0 & 4 & 2 & 2 & 1 & 0 & 0 \\
2 & 1 & 1 & 1.5 & 1 & 0 & 2 & 1 & 0 \\
1 & 0 & 1 & 2 & 2 & 0 & 1 & 1 & 1 \\
1 & 0 & 0 & 3 & 3 & 2 & 1 & 0 & 1 \\
0 & 1 & 0 & 2 & 0 & 2 & 2 & 1 & 0 \\
1 & 0 & 1 & 3 & 1 & 0 & 1 & 1 & 0
\end{bmatrix}
\cdot
\begin{bmatrix}
.99 & -.75 & .0594 \\
1.09 & -.50 & .0654 \\
.69 & -.35 & .0414 \\
2.29 & -1.45 & .1374 \\
1.59 & -.85 & .0954 \\
2.25 & -1.65 & .135 \\
1.59 & -1.00 & .0954 \\
1.09 & -.55 & .0654 \\
2.50 & -1.25 & .15
\end{bmatrix}
= \text{SALESPERPERSON} =
\begin{bmatrix}
Pr & C & Tx \\
10.21 & -5.9 & 0.613 \\
14.71 & -8.35 & 0.883 \\
9.14 & -5.4 & 0.548 \\
17.37 & -9.9 & 1.042 \\
18.43 & -11.8 & 1.106 \\
13.055 & -7.925 & 0.783 \\
14.62 & -8.5 & 0.877 \\
21.22 & -13.2 & 1.273 \\
14.44 & -9.25 & 0.866 \\
12.82 & -7.85 & 0.769
\end{bmatrix}
$$

THE TOTAL SALE TO LARRY IS $10.21, SALES TAX IS 61CENTS, YOUR
COST IS $5.90 FOR A PROFIT OF 10.21 - 5.90 = $4.31. THE TOTAL
SALE TO BOB IS $21.22, HE PAID 1.27 IN TAXES AND YOU MADE A
PROFIT OF 21.22 - 13.20 = $8.02. LET'S FIND YOUR TOTAL PROFITS
PER CUSTOMER:

$$
\text{PROFITS} := \begin{pmatrix} 1 \\ 1 \\ 0 \end{pmatrix}
\qquad
\text{SALESPLUSTAX} := \begin{pmatrix} 1 \\ 0 \\ 1 \end{pmatrix}
$$

PROFITSPERCUSTOMER ¦=SALESPERPERSON·PROFITS

$$
\begin{bmatrix}
10.21 & -5.9 & 0.613 \\
14.71 & -8.35 & 0.883 \\
9.14 & -5.4 & 0.548 \\
17.37 & -9.9 & 1.042 \\
18.43 & -11.8 & 1.106 \\
13.055 & -7.925 & 0.783 \\
14.62 & -8.5 & 0.877 \\
21.22 & -13.2 & 1.273 \\
14.44 & -9.25 & 0.866 \\
12.82 & -7.85 & 0.769
\end{bmatrix}
\cdot \begin{pmatrix} 1 \\ 1 \\ 0 \end{pmatrix}
= \text{PROFITSPERCUSTOMER} =
\begin{bmatrix}
4.31 \\
6.36 \\
3.74 \\
7.47 \\
6.63 \\
5.13 \\
6.12 \\
8.02 \\
5.19 \\
4.97
\end{bmatrix}
$$

TOTALCASHREGISTERRECEIPT ¦=SALESPERPERSON·SALESPLUSTAX

$$
\begin{bmatrix}
10.21 & -5.9 & 0.613 \\
14.71 & -8.35 & 0.883 \\
9.14 & -5.4 & 0.548 \\
17.37 & -9.9 & 1.042 \\
18.43 & -11.8 & 1.106 \\
13.055 & -7.925 & 0.783 \\
14.62 & -8.5 & 0.877 \\
21.22 & -13.2 & 1.273 \\
14.44 & -9.25 & 0.866 \\
12.82 & -7.85 & 0.769
\end{bmatrix}
\cdot \begin{pmatrix} 1 \\ 0 \\ 1 \end{pmatrix}
= \text{TOTALCASHREGISTERRECEIPT} =
\begin{bmatrix}
10.823 \\
15.593 \\
9.688 \\
18.412 \\
19.536 \\
13.838 \\
15.497 \\
22.493 \\
15.306 \\
13.589
\end{bmatrix}
$$

THESE VALUES SHOULD MATCH THE CASHREGISTER RECEIPTS.

BESIDES YOU, YOU HAVE TWO EMPLOYEES, ONE WORKS PART-TIME, THE
OTHER WORKS FULL-TIME. THE PART-TIME WORKER MAKES $6.00/HR AND
WORKED 20 HOURS THIS WEEK, THE FULL-TIME WORKER MAKES $7.50/HR
AND WORKED 40 HOURS. BUT SUPPOSE THE ABOVE SALES INVENTORY WAS
A ONE DAY TOTAL, AND THE FULL-TIME WORKER WORKED 8 HOURS WHILE
YOUR PART-TIME HELP WORKED 3 HOURS.

$$\text{DBWITHHOURS} := \begin{bmatrix} .99 & -.75 & .0594 & 0 \\ 1.09 & -.50 & .0654 & 0 \\ .69 & -.35 & .0414 & 0 \\ 2.29 & -1.45 & .1374 & 0 \\ 1.59 & -.85 & .0954 & 0 \\ 2.25 & -1.65 & .135 & 0 \\ 1.59 & -1.00 & .0954 & 0 \\ 1.09 & -.55 & .0654 & 0 \\ 2.50 & -1.25 & .15 & 0 \\ 0 & 0 & 0 & -7.50 \\ 0 & 0 & 0 & -6.00 \end{bmatrix}$$

$$\text{INVWITHHOURSA} := \begin{bmatrix} 2 & 3 & 1 & 0 & 0 & 0 & 2 & 1 & 0 & 0 \\ 0 & 1 & 0 & 2 & 4 & 0 & 1 & 1 & 0 & 0 \\ 1 & 1 & 0 & 1 & 2 & 0 & 1 & 0 & 0 & 0 \\ 0 & 2 & 0 & 2 & 3 & 1 & 0 & 1 & 1 & 0 \\ 0 & 0 & 0 & 4 & 2 & 2 & 1 & 0 & 0 & 0 \\ 2 & 1 & 1 & 1.5 & 1 & 0 & 2 & 1 & 0 & 0 \\ 1 & 0 & 1 & 2 & 2 & 0 & 1 & 1 & 1 & 0 \\ 1 & 0 & 0 & 3 & 3 & 2 & 1 & 0 & 1 & 0 \\ 0 & 1 & 0 & 2 & 0 & 2 & 2 & 1 & 0 & 0 \\ 1 & 0 & 1 & 3 & 1 & 0 & 1 & 1 & 0 & 0 \end{bmatrix}$$

$$B := \begin{pmatrix} 0 & 0 & 0 & 0 & 0 & 0 & 0 & 0 & 0 & 8 \\ 0 & 0 & 0 & 0 & 0 & 0 & 0 & 0 & 0 & 0 \end{pmatrix}$$

THE MATRIX IS TOO BIG FOR MATHCAD TO ALLOW US TO ADD NEW ROWS AND COLUMNS, SO WE MUST USE THE AUGMENT AND STACK FUNCTIONS TO INCREASE THE SIZE OF OUR INVENTORY MATRIX:

INVWITHHOURSB := stack(INVWITHHOURSA, B)

$$\text{INVWITHHOURSB} = \begin{bmatrix} 2 & 3 & 1 & 0 & 0 & 0 & 2 & 1 & 0 & 0 \\ 0 & 1 & 0 & 2 & 4 & 0 & 1 & 1 & 0 & 0 \\ 1 & 1 & 0 & 1 & 2 & 0 & 1 & 0 & 0 & 0 \\ 0 & 2 & 0 & 2 & 3 & 1 & 0 & 1 & 1 & 0 \\ 0 & 0 & 0 & 4 & 2 & 2 & 1 & 0 & 0 & 0 \\ 2 & 1 & 1 & 1.5 & 1 & 0 & 2 & 1 & 0 & 0 \\ 1 & 0 & 1 & 2 & 2 & 0 & 1 & 1 & 1 & 0 \\ 1 & 0 & 0 & 3 & 3 & 2 & 1 & 0 & 1 & 0 \\ 0 & 1 & 0 & 2 & 0 & 2 & 2 & 1 & 0 & 0 \\ 1 & 0 & 1 & 3 & 1 & 0 & 1 & 1 & 0 & 0 \\ 0 & 0 & 0 & 0 & 0 & 0 & 0 & 0 & 0 & 8 \\ 0 & 0 & 0 & 0 & 0 & 0 & 0 & 0 & 0 & 0 \end{bmatrix}$$

$$A := \begin{bmatrix} 0 \\ 0 \\ 0 \\ 0 \\ 0 \\ 0 \\ 0 \\ 0 \\ 0 \\ 0 \\ 0 \\ 3 \end{bmatrix}$$

INVWITHHOURS $:=$ augment(INVWITHHOURSB , A)

$$INVWITHHOURS = \begin{bmatrix} 2 & 3 & 1 & 0 & 0 & 0 & 2 & 1 & 0 & 0 & 0 \\ 0 & 1 & 0 & 2 & 4 & 0 & 1 & 1 & 0 & 0 & 0 \\ 1 & 1 & 0 & 1 & 2 & 0 & 1 & 0 & 0 & 0 & 0 \\ 0 & 2 & 0 & 2 & 3 & 1 & 0 & 1 & 1 & 0 & 0 \\ 0 & 0 & 0 & 4 & 2 & 2 & 1 & 0 & 0 & 0 & 0 \\ 2 & 1 & 1 & 1.5 & 1 & 0 & 2 & 1 & 0 & 0 & 0 \\ 1 & 0 & 1 & 2 & 2 & 0 & 1 & 1 & 1 & 0 & 0 \\ 1 & 0 & 0 & 3 & 3 & 2 & 1 & 0 & 1 & 0 & 0 \\ 0 & 1 & 0 & 2 & 0 & 2 & 2 & 1 & 0 & 0 & 0 \\ 1 & 0 & 1 & 3 & 1 & 0 & 1 & 1 & 0 & 0 & 0 \\ 0 & 0 & 0 & 0 & 0 & 0 & 0 & 0 & 0 & 8 & 0 \\ 0 & 0 & 0 & 0 & 0 & 0 & 0 & 0 & 0 & 0 & 3 \end{bmatrix}$$

WE PLACE THE EMPLOYEES HOURS IN THIS MANNER BECAUSE WE DO NOT WISH
TO INTERFERE WITH CUSTOMER SALES INVENTORY OR SALES ACCOUNTING.

TOTALINCOMEEXPENSEMATRIX $:=$ INVWITHHOURS·DBWITHHOURS

$$\begin{bmatrix} 2 & 3 & 1 & 0 & 0 & 0 & 2 & 1 & 0 & 0 & 0 \\ 0 & 1 & 0 & 2 & 4 & 0 & 1 & 1 & 0 & 0 & 0 \\ 1 & 1 & 0 & 1 & 2 & 0 & 1 & 0 & 0 & 0 & 0 \\ 0 & 2 & 0 & 2 & 3 & 1 & 0 & 1 & 1 & 0 & 0 \\ 0 & 0 & 0 & 4 & 2 & 2 & 1 & 0 & 0 & 0 & 0 \\ 2 & 1 & 1 & 1.5 & 1 & 0 & 2 & 1 & 0 & 0 & 0 \\ 1 & 0 & 1 & 2 & 2 & 0 & 1 & 1 & 1 & 0 & 0 \\ 1 & 0 & 0 & 3 & 3 & 2 & 1 & 0 & 1 & 0 & 0 \\ 0 & 1 & 0 & 2 & 0 & 2 & 2 & 1 & 0 & 0 & 0 \\ 1 & 0 & 1 & 3 & 1 & 0 & 1 & 1 & 0 & 0 & 0 \\ 0 & 0 & 0 & 0 & 0 & 0 & 0 & 0 & 0 & 8 & 0 \\ 0 & 0 & 0 & 0 & 0 & 0 & 0 & 0 & 0 & 0 & 3 \end{bmatrix} \cdot \begin{bmatrix} .99 & -.75 & .0594 & 0 \\ 1.09 & -.50 & .0654 & 0 \\ .69 & -.35 & .0414 & 0 \\ 2.29 & -1.45 & .1374 & 0 \\ 1.59 & -.85 & .0954 & 0 \\ 2.25 & -1.65 & .135 & 0 \\ 1.59 & -1.00 & .0954 & 0 \\ 1.09 & -.55 & .0654 & 0 \\ 2.50 & -1.25 & .15 & 0 \\ 0 & 0 & 0 & -7.50 \\ 0 & 0 & 0 & -6.00 \end{bmatrix} = $$

INVWITHHOURS·DBWITHHOURS $=$

$$\begin{bmatrix} 10.21 & -5.9 & 0.613 & 0 \\ 14.71 & -8.35 & 0.883 & 0 \\ 9.14 & -5.4 & 0.548 & 0 \\ 17.37 & -9.9 & 1.042 & 0 \\ 18.43 & -11.8 & 1.106 & 0 \\ 13.055 & -7.925 & 0.783 & 0 \\ 14.62 & -8.5 & 0.877 & 0 \\ 21.22 & -13.2 & 1.273 & 0 \\ 14.44 & -9.25 & 0.866 & 0 \\ 12.82 & -7.85 & 0.769 & 0 \\ 0 & 0 & 0 & -60 \\ 0 & 0 & 0 & -18 \end{bmatrix}$$

THE FULL-TIME EMPLOYEE EARNED $60, AND THE PART-TIME EMPLOYEE
EARNED $18. SO NOW, WHAT IS YOUR TOTAL PROFIT FOR THE DAY?

$$TOTALPROFITMATRIX := \begin{bmatrix} 1 \\ 1 \\ 0 \\ 1 \end{bmatrix}$$

ONE12 $:=$ (1 1 1 1 1 1 1 1 1 1 1 1)

MONEYMADETODAY1 $:=$ TOTALINCOMEEXPENSEMATRIX·TOTALPROFITMATRIX

$$
\begin{bmatrix}
10.21 & -5.9 & 0.613 & 0 \\
14.71 & -8.35 & 0.883 & 0 \\
9.14 & -5.4 & 0.548 & 0 \\
17.37 & -9.9 & 1.042 & 0 \\
18.43 & -11.8 & 1.106 & 0 \\
13.055 & -7.925 & 0.783 & 0 \\
14.62 & -8.5 & 0.877 & 0 \\
21.22 & -13.2 & 1.273 & 0 \\
14.44 & -9.25 & 0.866 & 0 \\
12.82 & -7.85 & 0.769 & 0 \\
0 & 0 & 0 & -60 \\
0 & 0 & 0 & -18
\end{bmatrix}
\cdot
\begin{bmatrix}
1 \\ 1 \\ 0 \\ 1
\end{bmatrix}
=
\begin{bmatrix}
4.31 \\
6.36 \\
3.74 \\
7.47 \\
6.63 \\
5.13 \\
6.12 \\
8.02 \\
5.19 \\
4.97 \\
-60 \\
-18
\end{bmatrix}
$$

MONEYMADETODAY := ONE12·MONEYMADETODAY1

$$
(1\ 1\ 1\ 1\ 1\ 1\ 1\ 1\ 1\ 1\ 1\ 1) \cdot
\begin{bmatrix}
4.31 \\
6.36 \\
3.74 \\
7.47 \\
6.63 \\
5.13 \\
6.12 \\
8.02 \\
5.19 \\
4.97 \\
-60 \\
-18
\end{bmatrix}
= -20.06
$$

MONEYMADETODAY = -20.06

TODAY WE MADE NO PROFIT, WE DID NOT LOSE ANY MONEY BECAUSE WE BROUGHT IN (MINUS WAGES):

$$
\text{TOTALINCOMEMATRIX} :=
\begin{bmatrix}
1 \\ 0 \\ 0 \\ 1
\end{bmatrix}
$$

TOTAL := TOTALINCOMEEXPENSEMATRIX·TOTALINCOMEMATRIX

$$
\begin{bmatrix}
10.21 & -5.9 & 0.613 & 0 \\
14.71 & -8.35 & 0.883 & 0 \\
9.14 & -5.4 & 0.548 & 0 \\
17.37 & -9.9 & 1.042 & 0 \\
18.43 & -11.8 & 1.106 & 0 \\
13.055 & -7.925 & 0.783 & 0 \\
14.62 & -8.5 & 0.877 & 0 \\
21.22 & -13.2 & 1.273 & 0 \\
14.44 & -9.25 & 0.866 & 0 \\
12.82 & -7.85 & 0.769 & 0 \\
0 & 0 & 0 & -60 \\
0 & 0 & 0 & -18
\end{bmatrix}
\cdot
\begin{bmatrix}
1 \\ 0 \\ 0 \\ 1
\end{bmatrix}
=
\begin{bmatrix}
10.21 \\
14.71 \\
9.14 \\
17.37 \\
18.43 \\
13.055 \\
14.62 \\
21.22 \\
14.44 \\
12.82 \\
{}^-60 \\
{}^-18
\end{bmatrix}
$$

$$
\text{TOTALINCOMEEXPENSEMATRIX} \cdot \text{TOTALINCOMEMATRIX} =
\begin{bmatrix}
10.21 \\
14.71 \\
9.14 \\
17.37 \\
18.43 \\
13.055 \\
14.62 \\
21.22 \\
14.44 \\
12.82 \\
{}^-60 \\
{}^-18
\end{bmatrix}
$$

$\text{AMOUNTAHEAD} := \text{ONE12} \cdot \text{TOTAL}$

$$
(1 \ \ 1 \ \ 1 \ \ 1 \ \ 1 \ \ 1 \ \ 1 \ \ 1 \ \ 1 \ \ 1 \ \ 1 \ \ 1) \cdot
\begin{bmatrix}
10.21 \\
14.71 \\
9.14 \\
17.37 \\
18.43 \\
13.055 \\
14.62 \\
21.22 \\
14.44 \\
12.82 \\
-60 \\
-18
\end{bmatrix}
= 68.015
$$

AMOUNTAHEAD = 68.015

WE HAVE $68 LEFT IN THE TILL, BUT MADE A PROFIT OF -$20.06 FOR THE DAY.

SUPPOSE WE ARE HAVING A SALE ON BEER AND TOMATOES, AND TO SAVE ON COSTS, WE WISH TO MAIL FLYERSOUT ONLY TO THOSE CUSTOMERS WHO BUY BEER AND TOMATOES. (Rememer, we have kept track of who buys what items so we can do this.) WE PROCEED TO HALF-MULTIPLY AS FOLLOWS:

$$
\text{SALES} :=
\begin{bmatrix}
2 & 3 & 1 & 0 & 0 & 0 & 2 & 1 & 0 \\
0 & 1 & 0 & 2 & 4 & 0 & 1 & 1 & 0 \\
1 & 1 & 0 & 1 & 2 & 0 & 1 & 0 & 0 \\
0 & 2 & 0 & 2 & 3 & 1 & 0 & 1 & 1 \\
0 & 0 & 0 & 4 & 2 & 2 & 1 & 0 & 0 \\
2 & 1 & 1 & 1.5 & 1 & 0 & 2 & 1 & 0 \\
1 & 0 & 1 & 2 & 2 & 0 & 1 & 1 & 1 \\
1 & 0 & 0 & 3 & 3 & 2 & 1 & 0 & 1 \\
0 & 1 & 0 & 2 & 0 & 2 & 2 & 1 & 0 \\
1 & 0 & 1 & 3 & 1 & 0 & 1 & 1 & 0
\end{bmatrix}
\qquad
\text{SALES}^{<2>} =
\begin{bmatrix}
3 \\ 1 \\ 1 \\ 2 \\ 0 \\ 1 \\ 0 \\ 0 \\ 1 \\ 0
\end{bmatrix}
\qquad
\text{SALES}^{<9>} =
\begin{bmatrix}
0 \\ 0 \\ 0 \\ 1 \\ 0 \\ 0 \\ 1 \\ 1 \\ 0 \\ 0
\end{bmatrix}
$$

HERE WE REMOVE THE SECOND ROW (TOMATOES) AND THE NINETH COLUMN (BEER) FROM THE SALES MATRIX:

$$
\text{TOMATOES} :=
\begin{bmatrix}
3 \\ 1 \\ 1 \\ 2 \\ 0 \\ 1 \\ 0 \\ 0 \\ 1 \\ 0
\end{bmatrix}
\qquad
\text{BEER} :=
\begin{bmatrix}
0 \\ 0 \\ 0 \\ 1 \\ 0 \\ 0 \\ 1 \\ 1 \\ 0 \\ 0
\end{bmatrix}
$$

ADVERTISE := TOMATOES + BEER

THE PROOFS SAY WE CAN ADD BOTH COLUMNS TOGETHER AND OBTAIN THE SUM OF BOTH INSTEAD OF FINDING THE SOLUTIONS FOR EACH COLUMN SEPARATELY AND ADDING THE TWO MATRICES. WE GET THE SAME ANSWER.

$$
\text{ADVERTISE} =
\begin{bmatrix}
3 \\ 1 \\ 1 \\ 3 \\ 0 \\ 1 \\ 1 \\ 1 \\ 1 \\ 0
\end{bmatrix}
$$

NOW WE MUST HALF-MULTIPLY, BUT SINCE NO ONE HAS EVER DISCOVERED THIS TYPE OF MULTIPLICATION BEFORE, IT CAN NOT BE DONE ON ANY COMPUTER ON EARTH TODAY, SO WE MUST DIAGONALIZE THE COLUMN MATRIX AND MULTIPLY TO THE SALES MATRIX TO HALF-MULTIPLY.

DIAGONALADVERTISE := diag(ADVERTISE)

$$
\text{DIAGONALADVERTISE} =
\begin{bmatrix}
3 & 0 & 0 & 0 & 0 & 0 & 0 & 0 & 0 & 0 \\
0 & 1 & 0 & 0 & 0 & 0 & 0 & 0 & 0 & 0 \\
0 & 0 & 1 & 0 & 0 & 0 & 0 & 0 & 0 & 0 \\
0 & 0 & 0 & 3 & 0 & 0 & 0 & 0 & 0 & 0 \\
0 & 0 & 0 & 0 & 0 & 0 & 0 & 0 & 0 & 0 \\
0 & 0 & 0 & 0 & 0 & 1 & 0 & 0 & 0 & 0 \\
0 & 0 & 0 & 0 & 0 & 0 & 1 & 0 & 0 & 0 \\
0 & 0 & 0 & 0 & 0 & 0 & 0 & 1 & 0 & 0 \\
0 & 0 & 0 & 0 & 0 & 0 & 0 & 0 & 1 & 0 \\
0 & 0 & 0 & 0 & 0 & 0 & 0 & 0 & 0 & 0
\end{bmatrix}
$$

THOSETOWHOMTOSENDTHEADS := DIAGONALADVERTISE SALES

$$
\begin{bmatrix}
3 & 0 & 0 & 0 & 0 & 0 & 0 & 0 & 0 & 0 \\
0 & 1 & 0 & 0 & 0 & 0 & 0 & 0 & 0 & 0 \\
0 & 0 & 1 & 0 & 0 & 0 & 0 & 0 & 0 & 0 \\
0 & 0 & 0 & 3 & 0 & 0 & 0 & 0 & 0 & 0 \\
0 & 0 & 0 & 0 & 0 & 0 & 0 & 0 & 0 & 0 \\
0 & 0 & 0 & 0 & 0 & 1 & 0 & 0 & 0 & 0 \\
0 & 0 & 0 & 0 & 0 & 0 & 1 & 0 & 0 & 0 \\
0 & 0 & 0 & 0 & 0 & 0 & 0 & 1 & 0 & 0 \\
0 & 0 & 0 & 0 & 0 & 0 & 0 & 0 & 1 & 0 \\
0 & 0 & 0 & 0 & 0 & 0 & 0 & 0 & 0 & 0
\end{bmatrix}
\cdot
\begin{bmatrix}
2 & 3 & 1 & 0 & 0 & 0 & 2 & 1 & 0 \\
0 & 1 & 0 & 2 & 4 & 0 & 1 & 1 & 0 \\
1 & 1 & 0 & 1 & 2 & 0 & 1 & 0 & 0 \\
0 & 2 & 0 & 2 & 3 & 1 & 0 & 1 & 1 \\
0 & 0 & 0 & 4 & 2 & 2 & 1 & 0 & 0 \\
2 & 1 & 1 & 1.5 & 1 & 0 & 2 & 1 & 0 \\
1 & 0 & 1 & 2 & 2 & 0 & 1 & 1 & 1 \\
1 & 0 & 0 & 3 & 3 & 2 & 1 & 0 & 1 \\
0 & 1 & 0 & 2 & 0 & 2 & 2 & 1 & 0 \\
1 & 0 & 1 & 3 & 1 & 0 & 1 & 1 & 0
\end{bmatrix}
=
\begin{bmatrix}
6 & 9 & 3 & 0 & 0 & 0 & 6 & 3 & 0 \\
0 & 1 & 0 & 2 & 4 & 0 & 1 & 1 & 0 \\
1 & 1 & 0 & 1 & 2 & 0 & 1 & 0 & 0 \\
0 & 6 & 0 & 6 & 9 & 3 & 0 & 3 & 3 \\
0 & 0 & 0 & 0 & 0 & 0 & 0 & 0 & 0 \\
2 & 1 & 1 & 1.5 & 1 & 0 & 2 & 1 & 0 \\
1 & 0 & 1 & 2 & 2 & 0 & 1 & 1 & 1 \\
1 & 0 & 0 & 3 & 3 & 2 & 1 & 0 & 1 \\
0 & 1 & 0 & 2 & 0 & 2 & 2 & 1 & 0 \\
0 & 0 & 0 & 0 & 0 & 0 & 0 & 0 & 0
\end{bmatrix}
$$

$$\text{THOSETOWHOMTOSENDTHEADS} = \begin{bmatrix} 6 & 9 & 3 & 0 & 0 & 0 & 6 & 3 & 0 \\ 0 & 1 & 0 & 2 & 4 & 0 & 1 & 1 & 0 \\ 1 & 1 & 0 & 1 & 2 & 0 & 1 & 0 & 0 \\ 0 & 6 & 0 & 6 & 9 & 3 & 0 & 3 & 3 \\ 0 & 0 & 0 & 0 & 0 & 0 & 0 & 0 & 0 \\ 2 & 1 & 1 & 1.5 & 1 & 0 & 2 & 1 & 0 \\ 1 & 0 & 1 & 2 & 2 & 0 & 1 & 1 & 1 \\ 1 & 0 & 0 & 3 & 3 & 2 & 1 & 0 & 1 \\ 0 & 1 & 0 & 2 & 0 & 2 & 2 & 1 & 0 \\ 0 & 0 & 0 & 0 & 0 & 0 & 0 & 0 & 0 \end{bmatrix}$$

WE MAIL ADS FOR TOMATOES AND BEER TO THOSE PEOPLE WHO DO NOT HAVE A ROW OF ZEROS. SUPPOSE WE JUST HAVE A SPECIAL ON BEER?

DIAGBEER $=$ diag(BEER)

$$\text{DIAGBEER} = \begin{bmatrix} 0 & 0 & 0 & 0 & 0 & 0 & 0 & 0 & 0 & 0 \\ 0 & 0 & 0 & 0 & 0 & 0 & 0 & 0 & 0 & 0 \\ 0 & 0 & 0 & 0 & 0 & 0 & 0 & 0 & 0 & 0 \\ 0 & 0 & 0 & 1 & 0 & 0 & 0 & 0 & 0 & 0 \\ 0 & 0 & 0 & 0 & 0 & 0 & 0 & 0 & 0 & 0 \\ 0 & 0 & 0 & 0 & 0 & 0 & 0 & 0 & 0 & 0 \\ 0 & 0 & 0 & 0 & 0 & 0 & 1 & 0 & 0 & 0 \\ 0 & 0 & 0 & 0 & 0 & 0 & 0 & 1 & 0 & 0 \\ 0 & 0 & 0 & 0 & 0 & 0 & 0 & 0 & 0 & 0 \\ 0 & 0 & 0 & 0 & 0 & 0 & 0 & 0 & 0 & 0 \end{bmatrix}$$

THOSETOWHOMTOSENDTHEBEERADS $=$ DIAGBEER·SALES

$$\begin{bmatrix} 0 & 0 & 0 & 0 & 0 & 0 & 0 & 0 & 0 & 0 \\ 0 & 0 & 0 & 0 & 0 & 0 & 0 & 0 & 0 & 0 \\ 0 & 0 & 0 & 0 & 0 & 0 & 0 & 0 & 0 & 0 \\ 0 & 0 & 0 & 1 & 0 & 0 & 0 & 0 & 0 & 0 \\ 0 & 0 & 0 & 0 & 0 & 0 & 0 & 0 & 0 & 0 \\ 0 & 0 & 0 & 0 & 0 & 0 & 0 & 0 & 0 & 0 \\ 0 & 0 & 0 & 0 & 0 & 0 & 1 & 0 & 0 & 0 \\ 0 & 0 & 0 & 0 & 0 & 0 & 0 & 1 & 0 & 0 \\ 0 & 0 & 0 & 0 & 0 & 0 & 0 & 0 & 0 & 0 \\ 0 & 0 & 0 & 0 & 0 & 0 & 0 & 0 & 0 & 0 \end{bmatrix} \cdot \begin{bmatrix} 2 & 3 & 1 & 0 & 0 & 0 & 2 & 1 & 0 \\ 0 & 1 & 0 & 2 & 4 & 0 & 1 & 1 & 0 \\ 1 & 1 & 0 & 1 & 2 & 0 & 1 & 0 & 0 \\ 0 & 2 & 0 & 2 & 3 & 1 & 0 & 1 & 1 \\ 0 & 0 & 0 & 4 & 2 & 2 & 1 & 0 & 0 \\ 2 & 1 & 1 & 1.5 & 1 & 0 & 2 & 1 & 0 \\ 1 & 0 & 1 & 2 & 2 & 0 & 1 & 1 & 1 \\ 1 & 0 & 0 & 3 & 3 & 2 & 1 & 0 & 1 \\ 0 & 1 & 0 & 2 & 0 & 2 & 2 & 1 & 0 \\ 1 & 0 & 1 & 3 & 1 & 0 & 1 & 1 & 0 \end{bmatrix} = \begin{bmatrix} 0 & 0 & 0 & 0 & 0 & 0 & 0 & 0 & 0 \\ 0 & 0 & 0 & 0 & 0 & 0 & 0 & 0 & 0 \\ 0 & 0 & 0 & 0 & 0 & 0 & 0 & 0 & 0 \\ 0 & 2 & 0 & 2 & 3 & 1 & 0 & 1 & 1 \\ 0 & 0 & 0 & 0 & 0 & 0 & 0 & 0 & 0 \\ 0 & 0 & 0 & 0 & 0 & 0 & 0 & 0 & 0 \\ 1 & 0 & 1 & 2 & 2 & 0 & 1 & 1 & 1 \\ 1 & 0 & 0 & 3 & 3 & 2 & 1 & 0 & 1 \\ 0 & 0 & 0 & 0 & 0 & 0 & 0 & 0 & 0 \\ 0 & 0 & 0 & 0 & 0 & 0 & 0 & 0 & 0 \end{bmatrix}$$

$$\text{THOSETOWHOMTOSENDTHEBEERADS} = \begin{bmatrix} 0 & 0 & 0 & 0 & 0 & 0 & 0 & 0 & 0 \\ 0 & 0 & 0 & 0 & 0 & 0 & 0 & 0 & 0 \\ 0 & 0 & 0 & 0 & 0 & 0 & 0 & 0 & 0 \\ 0 & 2 & 0 & 2 & 3 & 1 & 0 & 1 & 1 \\ 0 & 0 & 0 & 0 & 0 & 0 & 0 & 0 & 0 \\ 0 & 0 & 0 & 0 & 0 & 0 & 0 & 0 & 0 \\ 1 & 0 & 1 & 2 & 2 & 0 & 1 & 1 & 1 \\ 1 & 0 & 0 & 3 & 3 & 2 & 1 & 0 & 1 \\ 0 & 0 & 0 & 0 & 0 & 0 & 0 & 0 & 0 \\ 0 & 0 & 0 & 0 & 0 & 0 & 0 & 0 & 0 \end{bmatrix}$$

WE SEND ADS ONLY TO BECKY, JEANIE AND BOB.

FOR THE INVENTORY LEFT IN THE STORE, WE PROCEED AS FOLLOWS:

INVENTORY102597 := (25 100 32 75 35 42 20 25 12)   INVENTORY AS OF YESTERDAY
ONE10 := (1 1 1 1 1 1 1 1 1)
TODAYSINVENTORY := ONE10·SALES   SUMS THE COLUMNS IN THE SALES MATRIX
INVENTORY102697 := INVENTORY102597 − TODAYSINVENTORY   HERE WE FIND THE NEW
                                                        IN-STORE INVENTORY
INVENTORY102697 = (17 91 28 54.5 17 35 8 18 9)   NEW INVENTORY

    WE HAVE ONLY 8 HALF-GALLONS OF MILK LEFT, IT IS TIME TO
ORDER MORE. WE MIGHT AS WELL ORDER SOME MORE BEER SINCE WE ARE
RUNNING LOW ON IT TOO. WE HAVE ENOUGH OF THE REST TO LAST FOR
AWHILE. i.e. 17 HEADS OF LETTUCE LEFT, 91 TOMATOES LEFT, 28
STALKS OF CELERY, 54.5 LBS OF BEEF, 17 POUNDS OF CHICKEN, 35
POUNDS OF PORK AND 18 DOZEN EGGS.

LOVE BOX COMPANY BASIC INVENTORY PROGRAM

The following is an inventory system that keeps track of the different parts that go into making-up a single finished product. It is sort of simple, sort of complex, but if we do not have to keep track of everything entered here, just the inventory, the system would become much simpler to compute. The matrix SHIFT123 gives the total products completed by each shift. The first row is first shifts production, the second row is 2nd shifts daily production and the third row the 3rd shifts production. The matrix SHIFT123T gives the maximum total of items that should have been produces with no losses. The DATABASE matrix gives each product (the rows) and the columns give how many parts of each are needed to complete each item. The last column gives the total area in square feet for the finished item. We want to know the total square feet produced, square feet lost, the number of items produced per hour and the cost as applied to worker hours to produce each piece.

4	3	4	3
0	2	0	2
0	1	0	0
5	5	5	5
0	0	4	4
0	0	0	0
0	0	1	1

Note: These columns are not Matrices, but the production codes for the items eing manufactured. They are just too long to put in rows in the Matrix.

$$SHIFT123 := \begin{pmatrix} 0 & 485 & 495 & 500 & 492 \\ 0 & 490 & 500 & 498 & 495 \\ 0 & 500 & 495 & 490 & 493 \end{pmatrix}$$

$$SHIFT123T := \begin{pmatrix} 0 & 500 & 500 & 500 & 500 \\ 0 & 500 & 500 & 500 & 500 \\ 0 & 500 & 500 & 500 & 500 \end{pmatrix}$$

4	4	3	3	4	4	3	3	S
0	0	2	2	0	0	2	2	Q
0	0	1	1	0	0	0	0	U
/	/	/	/	/	/	/	/	A
5	5	5	5	5	5	5	5	R
0	0	0	0	4	4	4	4	E
0	0	0	0	0	0	0	0	F
0	0	0	0	1	1	1	1	E
A	B	A		A2	A1	A		E
								T

$$DATABASE := \begin{bmatrix} 3000 & 1500 & 7500 & 1500 & 3000 & 1500 & 3000 & 1500 & 52687.65 \\ 2 & 1 & 0 & 0 & 0 & 0 & 0 & 0 & 6.5078 \\ 0 & 0 & 5 & 1 & 0 & 0 & 0 & 0 & 13.0156 \\ 0 & 0 & 0 & 0 & 2 & 1 & 0 & 0 & 10.5023 \\ 0 & 0 & 0 & 0 & 0 & 0 & 2 & 1 & 5.0994 \end{bmatrix}$$

Total Produced	3000	1500	7500	1500	3000	1500	3000	1500	52687.65
400-5000	2	1	0	0	0	0	0	0	6.5078
321-5000	0	0	5	1	0	0	0	0	13.0156
400-5401	0	0	0	0	2	1	0	0	10.5023
320-5401	0	0	0	0	0	0	2	1	5.0994

The columns represent the number of pieces needed to make each item. i.e. item 321-5000 needs 5 pieces of item 321-5000A and

1 piece of item 321-5000. Of the items 321-5000A there were
7500 made and 1500 pieces of 321-5000.

$$
\text{SHIFT123} \cdot \text{DATABASE} =
\begin{pmatrix}
970 & 485 & 2475 & 495 & 1000 & 500 & 984 & 492 & 17359.0598 \\
980 & 490 & 2500 & 500 & 996 & 498 & 990 & 495 & 17450.9704 \\
1000 & 500 & 2475 & 495 & 980 & 490 & 986 & 493 & 17356.7532
\end{pmatrix}
$$

$$
\text{NINE1} :=
\begin{bmatrix}
1 \\ 1 \\ 1 \\ 1 \\ 1 \\ 1 \\ 1 \\ 1 \\ 0
\end{bmatrix}
\qquad
\text{INVENTORY1} :=
\begin{pmatrix}
970 & 485 & 2475 & 495 & 1000 & 500 & 984 & 492 & 17359.06 \\
980 & 490 & 2500 & 500 & 996 & 498 & 990 & 495 & 17450.97 \\
1000 & 500 & 2475 & 495 & 980 & 490 & 986 & 493 & 17356.753
\end{pmatrix}
$$

$$
\text{INVENTORY1} \cdot \text{NINE1} =
\begin{pmatrix}
7401 \\ 7449 \\ 7419
\end{pmatrix}
$$

This gives the total number of pieces completed in one shift
for all three shifts.

The following list gives the total number of pieces made by
the corrugator for the three shifts.

CORRUGATOR $:=$ (3000   1500   7500   1500   3000   1500   3000   1500   52687.65)
ONE3 $:=$ (1   1   1)
ONE3 $\cdot$ INVENTORY1 = (2950   1475   7450   1490   2976   1488   2960   1480   52166.783)

**This computation defines the total number of pieces for all
three shifts.**

DAILYUSE $:=$ (2950   1475   7450   1490   2976   1488   2960   1480   52166.783)
CORRUGATOR $-$ DAILYUSE = (50   25   50   10   24   12   40   20   520.867)

**This is the loss in pieces and square footage for all three
shifts combined.**

This is the theoretical inventory assuming no pieces are lost.

$$\text{SHIFT123T} \cdot \text{DATABASE} = \begin{pmatrix} 1000 & 500 & 2500 & 500 & 1000 & 500 & 1000 & 500 & 17562.55 \\ 1000 & 500 & 2500 & 500 & 1000 & 500 & 1000 & 500 & 17562.55 \\ 1000 & 500 & 2500 & 500 & 1000 & 500 & 1000 & 500 & 17562.55 \end{pmatrix}$$

$$\text{DAILYUSET} := \begin{pmatrix} 1000 & 500 & 2500 & 500 & 1000 & 500 & 1000 & 500 & 17562.55 \\ 1000 & 500 & 2500 & 500 & 1000 & 500 & 1000 & 500 & 17562.55 \\ 1000 & 500 & 2500 & 500 & 1000 & 500 & 1000 & 500 & 17562.55 \end{pmatrix}$$

$$\text{ONE3} \cdot \text{DAILYUSET} = (\; 3000 \quad 1500 \quad 7500 \quad 1500 \quad 3000 \quad 1500 \quad 3000 \quad 1500 \quad 52687.65 \;)$$

To compute the losses for each shift we proceed as follows:

$$\text{INVENTORY1} := \begin{pmatrix} 970 & 485 & 2475 & 495 & 1000 & 500 & 984 & 492 & 17359.06 \\ 980 & 490 & 2500 & 500 & 996 & 498 & 990 & 495 & 17450.97 \\ 1000 & 500 & 2475 & 495 & 980 & 490 & 986 & 493 & 17356.753 \end{pmatrix}$$

We must divide the totals of the corrugator by three to get the total for each shift.

$$\text{LOSSSHIFT123} := \begin{bmatrix} \dfrac{1}{3} & -485 & -495 & -500 & -492 \\[2mm] \dfrac{1}{3} & -490 & -500 & -498 & -495 \\[2mm] \dfrac{1}{3} & -500 & -495 & -490 & -493 \end{bmatrix}$$

$$\text{LOSSSHIFT123} \cdot \text{DATABASE} = \begin{pmatrix} 30 & 15 & 25 & 5 & 0 & 0 & 16 & 8 & 203.4902 \\ 20 & 10 & 0 & 0 & 4 & 2 & 10 & 5 & 111.5796 \\ 0 & 0 & 25 & 5 & 20 & 10 & 14 & 7 & 205.7968 \end{pmatrix}$$

$$(50 \quad 25 \quad 50 \quad 10 \quad 24 \quad 12 \quad 40 \quad 20 \quad 520.867\,)$$

50 pieces lost of 400-5000A, 24 pieces lost of 400-5401A2 were lost and first shift lost a total of 203.49 square feet. First shift lost 30 pieces of 400-5000A and second shift lost 20 pieces while third shift lost none.

To check if this correct, we sum each column and compare to the total losses for all three shifts.

$$
\text{LOSSPERSHIFT} := \begin{pmatrix} 30 & 15 & 25 & 5 & 0 & 0 & 16 & 8 & 203.49 \\ 20 & 10 & 0 & 0 & 4 & 2 & 10 & 5 & 111.58 \\ 0 & 0 & 25 & 5 & 20 & 10 & 14 & 7 & 205.797 \end{pmatrix}
$$

CHECK := ONE3·LOSSPERSHIFT

CHECK = (50   25   50   10   24   12   40   20   520.867)
QED

Suppose we wish to keep track of the employees time per task:

$$
\text{SHIFT123} := \begin{pmatrix} 0 & 485 & 495 & 500 & 492 \\ 0 & 490 & 500 & 498 & 495 \\ 0 & 500 & 495 & 490 & 493 \end{pmatrix}
$$

$$
\text{INVENTORYANDTIME} := \begin{pmatrix} 0 & 485 & 495 & 500 & 492 & 1.75 & 2.5 & 1 & 1.75 \\ 0 & 490 & 500 & 498 & 495 & 1.75 & 2.5 & 1 & 1.75 \\ 0 & 500 & 495 & 490 & 493 & 1.75 & 2.5 & 1 & 1.75 \end{pmatrix}
$$

First shift took 1.75 hours to complete the first project,
second shift took 2.5 hours to complete the second project and
third shift took 1 hour to complete the third project, etc.

$$
\text{DATABASEANDTIME} := \begin{bmatrix} 3000 & 1500 & 7500 & 1500 & 3000 & 1500 & 3000 & 1500 & 52687.65 \\ 2 & 1 & 0 & 0 & 0 & 0 & 0 & 0 & 6.5078 \\ 0 & 0 & 5 & 1 & 0 & 0 & 0 & 0 & 13.0156 \\ 0 & 0 & 0 & 0 & 2 & 1 & 0 & 0 & 10.5023 \\ 0 & 0 & 0 & 0 & 0 & 0 & 2 & 1 & 5.0994 \\ 1 & 1 & 0 & 0 & 0 & 0 & 0 & 0 & 0 \\ 0 & 0 & 1 & 1 & 0 & 0 & 0 & 0 & 0 \\ 0 & 0 & 0 & 0 & 1 & 1 & 0 & 0 & 0 \\ 0 & 0 & 0 & 0 & 0 & 0 & 1 & 1 & 0 \end{bmatrix}
$$

$$
B := \begin{bmatrix} 0 & 0 \\ 0 & 0 \\ 0 & 0 \\ 0 & 0 \\ 0 & 0 \\ 0 & 0 \\ 0 & 0 \\ 1 & 0 \\ 0 & 1 \end{bmatrix}
$$

$$A := \begin{bmatrix} 3000 & 1500 & 7500 & 1500 & 3000 & 1500 & 3000 & 1500 & 52687.65 & 0 & 0 \\ 2 & 1 & 0 & 0 & 0 & 0 & 0 & 0 & 6.5078 & 0 & 0 \\ 0 & 0 & 5 & 1 & 0 & 0 & 0 & 0 & 13.0156 & 0 & 0 \\ 0 & 0 & 0 & 0 & 2 & 1 & 0 & 0 & 10.5023 & 0 & 0 \\ 0 & 0 & 0 & 0 & 0 & 0 & 2 & 1 & 5.0994 & 0 & 0 \\ 0 & 0 & 0 & 0 & 0 & 0 & 0 & 0 & 0 & 1 & 0 \\ 0 & 0 & 0 & 0 & 0 & 0 & 0 & 0 & 0 & 0 & 1 \\ 0 & 0 & 0 & 0 & 0 & 0 & 0 & 0 & 0 & 0 & 0 \\ 0 & 0 & 0 & 0 & 0 & 0 & 0 & 0 & 0 & 0 & 0 \end{bmatrix}$$

DATABASEANDTIME := augment (A,B)

$$DATABASEANDTIME = \begin{bmatrix} 3000 & 1500 & 7500 & 1500 & 3000 & 1500 & 3000 & 1500 & 52687.65 & 0 & 0 & 0 & 0 \\ 2 & 1 & 0 & 0 & 0 & 0 & 0 & 0 & 6.5078 & 0 & 0 & 0 & 0 \\ 0 & 0 & 5 & 1 & 0 & 0 & 0 & 0 & 13.0156 & 0 & 0 & 0 & 0 \\ 0 & 0 & 0 & 0 & 2 & 1 & 0 & 0 & 10.5023 & 0 & 0 & 0 & 0 \\ 0 & 0 & 0 & 0 & 0 & 0 & 2 & 1 & 5.0994 & 0 & 0 & 0 & 0 \\ 0 & 0 & 0 & 0 & 0 & 0 & 0 & 0 & 0 & 1 & 0 & 0 & 0 \\ 0 & 0 & 0 & 0 & 0 & 0 & 0 & 0 & 0 & 0 & 1 & 0 & 0 \\ 0 & 0 & 0 & 0 & 0 & 0 & 0 & 0 & 0 & 0 & 0 & 1 & 0 \\ 0 & 0 & 0 & 0 & 0 & 0 & 0 & 0 & 0 & 0 & 0 & 0 & 1 \end{bmatrix}$$

INVENTORYPLUSTIME := INVENTORYANDTIME·DATABASEANDTIME

$$INVENTORYPLUSTIME = \begin{pmatrix} 970 & 485 & 2475 & 495 & 1000 & 500 & 984 & 492 & 17359.0598 & 1.75 & 2.5 & 1 & 1.75 \\ 980 & 490 & 2500 & 500 & 996 & 498 & 990 & 495 & 17450.9704 & 1.75 & 2.5 & 1 & 1.75 \\ 1000 & 500 & 2475 & 495 & 980 & 490 & 986 & 493 & 17356.7532 & 1.75 & 2.5 & 1 & 1.75 \end{pmatrix}$$

Now lets compute the pieces per hour:
First let's add the total pieces for each product:

$$ADD1 := \begin{bmatrix} 1 & 0 & 0 & 0 \\ 1 & 0 & 0 & 0 \\ 0 & 1 & 0 & 0 \\ 0 & 1 & 0 & 0 \\ 0 & 0 & 1 & 0 \\ 0 & 0 & 1 & 0 \\ 0 & 0 & 0 & 1 \\ 0 & 0 & 0 & 1 \\ 0 & 0 & 0 & 0 \\ 0 & 0 & 0 & 0 \\ 0 & 0 & 0 & 0 \\ 0 & 0 & 0 & 0 \\ 0 & 0 & 0 & 0 \end{bmatrix} \qquad ADD2 := \begin{bmatrix} 0 & 0 & 0 & 0 & 0 \\ 0 & 0 & 0 & 0 & 0 \\ 0 & 0 & 0 & 0 & 0 \\ 0 & 0 & 0 & 0 & 0 \\ 0 & 0 & 0 & 0 & 0 \\ 0 & 0 & 0 & 0 & 0 \\ 0 & 0 & 0 & 0 & 0 \\ 0 & 0 & 0 & 0 & 0 \\ 0 & 0 & 0 & 0 & 0 \\ 0 & 1 & 0 & 0 & 0 \\ 0 & 0 & 1 & 0 & 0 \\ 0 & 0 & 0 & 1 & 0 \\ 0 & 0 & 0 & 0 & 1 \end{bmatrix}$$

ADD := augment(ADD1,ADD2)

PIECESADDED := INVENTORYPLUSTIME·ADD

Here we sum the total number of separate pieces for each order.

$$\text{PIECESADDED} = \begin{pmatrix} 1455 & 2970 & 1500 & 1476 & 0 & 1.75 & 2.5 & 1 & 1.75 \\ 1470 & 3000 & 1494 & 1485 & 0 & 1.75 & 2.5 & 1 & 1.75 \\ 1500 & 2970 & 1470 & 1479 & 0 & 1.75 & 2.5 & 1 & 1.75 \end{pmatrix}$$

$$\text{PIECESADDED1} := \begin{pmatrix} 1455 & 2970 & 1500 & 1476 & 10^{-100} & 1.75 & 2.5 & 1 & 1.75 \\ 1470 & 3000 & 1494 & 1485 & 10^{-100} & 1.75 & 2.5 & 1 & 1.75 \\ 1500 & 2970 & 1470 & 1479 & 10^{-100} & 1.75 & 2.5 & 1 & 1.75 \end{pmatrix}$$

$$\text{LOGPIECESADDED1} := \overrightarrow{\log(\text{PIECESADDED1})}$$

$$\text{LOGPIECESADDED1} = \begin{pmatrix} 3.163 & 3.473 & 3.176 & 3.169 & ^-100 & 0.243 & 0.398 & 0 & 0.243 \\ 3.167 & 3.477 & 3.174 & 3.172 & ^-100 & 0.243 & 0.398 & 0 & 0.243 \\ 3.176 & 3.473 & 3.167 & 3.17 & ^-100 & 0.243 & 0.398 & 0 & 0.243 \end{pmatrix}$$

$$\text{DB1} := \begin{bmatrix} 1 & 0 & 0 & 0 \\ 0 & 1 & 0 & 0 \\ 0 & 0 & 1 & 0 \\ 0 & 0 & 0 & 1 \\ 0 & 0 & 0 & 0 \\ -1 & 0 & 0 & 0 \\ 0 & -1 & 0 & 0 \\ 0 & 0 & -1 & 0 \\ 0 & 0 & 0 & -1 \end{bmatrix}$$

LOGPIECESPERHOUR := LOGPIECESADDED1·DB1

$$\text{LOGPIECESPERHOUR} = \begin{pmatrix} 2.919825 & 3.074816 & 3.176091 & 2.926048 \\ 2.924279 & 3.079181 & 3.174351 & 2.928688 \\ 2.933053 & 3.074816 & 3.167317 & 2.92693 \end{pmatrix}$$

$$\text{PIECESPERHOUR} := \overrightarrow{10^{\text{LOGPIECESPERHOUR}}}$$

$$\text{PIECESPERHOUR} = \begin{pmatrix} 831.428571 & 1188 & 1500 & 843.428571 \\ 840 & 1200 & 1494 & 848.571429 \\ 857.142857 & 1188 & 1470 & 845.142857 \end{pmatrix}$$

Therefore for Kawasaki 400-5000, first shift did 831.43 pieces per hour, for Kawasaki 321-5000, second shift did 1200 pieces per hour, for Kawasaki 400-5401, third shift did 1470 pieces per hour and for Kawasaki 320-5401, second shift did 848.57 pieces per hour.

Let's look at employee wages and the cost to produce each piece.

Now let's compute the cost per piece.

$$\text{INVENTORYANDTIME} := \begin{pmatrix} 0 & 485 & 495 & 500 & 492 & 1.75 & 2.5 & 1 & 1.75 \\ 0 & 490 & 500 & 498 & 495 & 1.75 & 2.5 & 1 & 1.75 \\ 0 & 500 & 495 & 490 & 493 & 1.75 & 2.5 & 1 & 1.75 \end{pmatrix}$$

Combined pay second shift = \$18.06
combined pay first shift = \$20.75
Combined pay third shift = \$17.06

$$\text{PAY} := \begin{pmatrix} 0 & 1 & 1 & 1 & 1 & 20.75 & 20.75 & 20.75 & 20.75 \\ 0 & 1 & 1 & 1 & 1 & 18.06 & 18.06 & 18.06 & 18.06 \\ 0 & 1 & 1 & 1 & 1 & 17.06 & 17.06 & 17.06 & 17.06 \end{pmatrix}$$

$$\text{INVENTORYANDTIME} := \begin{pmatrix} 0 & 485 & 495 & 500 & 492 & 1.75 & 2.5 & 1 & 1.75 \\ 0 & 490 & 500 & 498 & 495 & 1.75 & 2.5 & 1 & 1.75 \\ 0 & 500 & 495 & 490 & 493 & 1.75 & 2.5 & 1 & 1.75 \end{pmatrix}$$

$$\text{COSTTOMAKE} := \overrightarrow{(\text{INVENTORYANDTIME} \cdot \text{PAY})}$$

$$\text{COSTTOMAKE} = \begin{pmatrix} 0 & 485 & 495 & 500 & 492 & 36.3125 & 51.875 & 20.75 & 36.3125 \\ 0 & 490 & 500 & 498 & 495 & 31.605 & 45.15 & 18.06 & 31.605 \\ 0 & 500 & 495 & 490 & 493 & 29.855 & 42.65 & 17.06 & 29.855 \end{pmatrix}$$

$$\text{COSTPERPIECE1} := \begin{pmatrix} 10^{-100} & 485 & 495 & 500 & 492 & 36.3125 & 51.875 & 20.75 & 36.3125 \\ 10^{-100} & 490 & 500 & 498 & 495 & 31.605 & 45.15 & 18.06 & 31.605 \\ 10^{-100} & 500 & 495 & 490 & 493 & 29.855 & 42.65 & 17.06 & 29.855 \end{pmatrix}$$

$$\text{LOGCOSTPERPIECE} := \overrightarrow{\log(\text{COSTPERPIECE1})}$$

$$DB2 := \begin{bmatrix} 1 & 0 & 0 & 0 & 0 \\ 0 & -1 & 0 & 0 & 0 \\ 0 & 0 & -1 & 0 & 0 \\ 0 & 0 & 0 & -1 & 0 \\ 0 & 0 & 0 & 0 & -1 \\ 0 & 1 & 0 & 0 & 0 \\ 0 & 0 & 1 & 0 & 0 \\ 0 & 0 & 0 & 1 & 0 \\ 0 & 0 & 0 & 0 & 1 \end{bmatrix}$$

COSTPERPIECE2 := LOGCOSTPERPIECE·DB2

$$COSTPERPIECE := 10^{\overrightarrow{COSTPERPIECE2}}$$

$$COSTPERPIECE = \begin{pmatrix} 0 & 0.074871 & 0.104798 & 0.0415 & 0.073806 \\ 0 & 0.0645 & 0.0903 & 0.036265 & 0.063848 \\ 0 & 0.05971 & 0.086162 & 0.034816 & 0.060558 \end{pmatrix}$$

Let's check to see if the above problem is correct. First shift produced 485 pieces in 1.75 hours with a combined pay of $20.75. The cost is 20.75 x 1.75 = $36.21. So cost per piece would equal 36.3125/485 = .074871.

MATRIX SOLUTION FOR GAUSSIAN REDUCTION

In this section, I will show how I discovered this operator. Mathematicians already know about this method, but I do not think anyone has ever derived it from a mathematical basis or that anyone has ever proved Gaussian Reduction. Suppose we wish to reduce the following matrix:

$$A_{ij} = \begin{vmatrix} a_{11} & a_{12} & a_{13} & . & . & . & a_{1n} \\ a_{21} & a_{22} & a_{23} & . & . & . & a_{2n} \\ a_{31} & a_{32} & a_{33} & . & . & . & a_{3n} \\ . & . & . & . & . & . & . \\ . & . & . & . & . & . & . \\ a_{n1} & a_{n2} & a_{n3} & . & . & . & a_{nn} \end{vmatrix}$$

Now normally to perform the Gaussian Reduction, we write out the top row, then multiply the first row of the matrix by the element in another row that we are trying to reduce to zero. We multiply this second row by the first element and subtract so that the number in the row below row one equals zero under the diagonal. Instead of multiplying and subtracting by hand, one day I put zero's in the places I wasn't interested in and formed a column matrix from them. After playing around with the idea for awhile, this is what I came up with:

$$\begin{vmatrix} 1 \\ 0 \\ 0 \\ . \\ . \\ 0 \end{vmatrix} \begin{vmatrix} a_{11} & a_{12} & a_{13} & . & . & . & a_{1n} \\ a_{21} & a_{22} & a_{23} & . & . & . & a_{2n} \\ a_{31} & a_{32} & a_{33} & . & . & . & a_{3n} \\ . & . & . & . & . & . & . \\ . & . & . & . & . & . & . \\ a_{n1} & a_{n2} & a_{n3} & . & . & . & a_{nn} \end{vmatrix} + \begin{vmatrix} a_{21} \\ -a_{11} \\ 0 \\ . \\ . \\ 0 \end{vmatrix} \begin{vmatrix} a_{11} & a_{12} & a_{13} & . & . & . & a_{1n} \\ a_{21} & a_{22} & a_{23} & . & . & . & a_{2n} \\ a_{31} & a_{32} & a_{33} & . & . & . & a_{3n} \\ . & . & . & . & . & . & . \\ a_{n1} & a_{n2} & a_{n3} & . & . & . & a_{nn} \end{vmatrix} + \, . \, . \, . \, +$$

$$\begin{vmatrix} a_{11} \\ 0 \\ 0 \\ -a_{11} \\ . \\ 0 \end{vmatrix} \circ \begin{vmatrix} a_{11} & a_{12} & a_{13} & . & . & . & a_{1n} \\ a_{21} & a_{22} & a_{23} & . & . & . & a_{2n} \\ . & . & . & . & . & . & . \\ a_{11} & a_{12} & a_{13} & . & . & . & a_{1n} \\ . & . & . & . & . & . & . \\ a_{n1} & a_{n2} & a_{n3} & . & . & . & a_{nn} \end{vmatrix} + \, . \, . \, . \, + \begin{vmatrix} a_{n1} \\ 0 \\ 0 \\ . \\ -a_{11} \end{vmatrix} \circ \begin{vmatrix} a_{11} & a_{12} & a_{13} & . & . & . & a_{1n} \\ a_{21} & a_{22} & a_{23} & . & . & . & a_{2n} \\ a_{31} & a_{32} & a_{33} & . & . & . & a_{3n} \\ . & . & . & . & . & . & . \\ a_{n1} & a_{n2} & a_{n3} & . & . & . & a_{nn} \end{vmatrix}$$

Cross multiplying the matrices and adding we get:

$$
\begin{vmatrix}
a_{11} & a_{12} & a_{13} & & & a_{1n} \\
a_{21}a_{11}-a_{11}a_{21} & a_{21}a_{12}-a_{11}a_{22} & a_{21}a_{13}-a_{11}a_{23} & \cdot & \cdot & \cdot & a_{21}a_{1n}-a_{11}a_{2n} \\
a_{31}a_{11}-a_{11}a_{31} & a_{31}a_{12}-a_{11}a_{32} & a_{31}a_{13}-a_{11}a_{33} & \cdot & \cdot & \cdot & a_{31}a_{1n}-a_{11}a_{3n} \\
\cdot & \cdot & \cdot & \cdot & \cdot & \cdot & \cdot \\
a_{i1}a_{11}-a_{11}a_{i1} & a_{i1}a_{12}-a_{11}a_{i2} & a_{i1}a_{13}-a_{11}a_{i3} & \cdot & \cdot & \cdot & a_{i1}a_{1n}-a_{11}a_{in} \\
\cdot & \cdot & \cdot & \cdot & \cdot & \cdot & \cdot \\
a_{n1}a_{11}-a_{11}a_{n1} & a_{n1}a_{12}-a_{11}a_{n2} & a_{n1}a_{13}-a_{11}a_{n3} & \cdot & \cdot & \cdot & a_{n1}a_{1n}-a_{11}a_{nn}
\end{vmatrix}
$$

All the terms under a_{11} are equal to zero. To help keep things simple, I'm going to define $b_{11} = a_{21}a_{12}-a_{11}a_{22}$; $b_{12} = a_{21}a_{13}-a_{11}a_{23}$; $b_{in} = a_{i1}a_{1n}-a_{11}a_{in}$; $b_{nn} = a_{n1}a_{1n}-a_{11}a_{nn}$; etc. But first, I'm going to collect all the pre-multipliers of $[A]_{nn}$ and form them into a matrix in the order I multiplied them.:

$$
\begin{bmatrix}
1 & a_{21} & a_{31} & \cdot & \cdot & a_{11} & \cdot & \cdot & a_{n-1,1} & a_{n1} \\
0 & -a11 & 0 & \cdot & \cdot & \cdot & 0 & \cdot & \cdot & 0 & 0 \\
0 & 0 & -a_{11} & \cdot & \cdot & \cdot & 0 & \cdot & \cdot & 0 & 0 \\
0 & 0 & 0 & \cdot & \cdot & \cdot & \cdot & \cdot & \cdot & 0 & 0 \\
\cdot & \cdot & \cdot & \cdot & \cdot & \cdot & -a_{11} & \cdot & \cdot & \cdot & \cdot \\
0 & 0 & 0 & \cdot & \cdot & \cdot & 0 & \cdot & \cdot & \cdot & 0 \\
\cdot & \cdot & \cdot & \cdot & \cdot & \cdot & \cdot & \cdot & -a_{11} & 0 \\
0 & 0 & 0 & \cdot & \cdot & \cdot & 0 & \cdot & \cdot & 0 & -a_{11}
\end{bmatrix}
$$

And re-writing the new Matrix we get:

$$
G_1 :=
\begin{bmatrix}
a11 & a_{12} & a_{13} & \cdot & \cdot & \cdot & a_{1n} \\
0 & b_{11} & b_{12} & \cdot & \cdot & \cdot & b_{1,n-1} \\
0 & b_{21} & b_{22} & \cdot & \cdot & \cdot & b_{2,n-1} \\
\cdot & \cdot & \cdot & \cdot & \cdot & \cdot & \cdot \\
0 & b_{i1} & b_{i1} & \cdot & \cdot & \cdot & b_{i,n-1} \\
\cdot & \cdot & \cdot & \cdot & \cdot & \cdot & \cdot \\
0 & b_{n-1,1} & b_{n-1,2} & \cdot & \cdot & \cdot & b_{n-1,n-1}
\end{bmatrix}
$$

Now we will get all the terms under b_{11} to equal zero: Note: we are multiplying the same as in Gaussian Reduction except we multiply by zeros in those terms in which we are not currently interested.

$$
\begin{vmatrix} 1 \\ 0 \\ 0 \\ \cdot \\ 0 \\ \cdot \\ 0 \end{vmatrix} \circ
\begin{vmatrix} a_{11} & a_{12} & a_{13} & \cdots & a_{1n} \\ 0 & b_{11} & b_{12} & \cdots & b_{1n-1} \\ 0 & b_{21} & b_{22} & \cdots & b_{2n-1} \\ \cdot & & & & \cdot \\ 0 & b_{i1} & b_{i2} & \cdots & b_{in-1} \\ \cdot & & & & \cdot \\ 0 & b_{n-1,1} & b_{n-1,2} & \cdots & b_{nn-1} \end{vmatrix} +
\begin{vmatrix} 0 \\ 1 \\ 0 \\ 0 \\ \cdot \\ 0 \end{vmatrix} \circ
\begin{vmatrix} a_{11} & a_{12} & a_{13} & \cdots & a_{1n} \\ 0 & b_{11} & b_{12} & \cdots & b_{1n-1} \\ 0 & b_{21} & b_{22} & \cdots & b_{2n-1} \\ \cdot & & & & \cdot \\ 0 & b_{i1} & b_{i2} & \cdots & b_{in-1} \\ \cdot & & & & \cdot \\ 0 & b_{n-1,1} & b_{n-1,2} & \cdots & b_{nn-1} \end{vmatrix} +
$$

$$
\begin{vmatrix} 0 \\ b_{21} \\ -b_{11} \\ 0 \\ 0 \\ \cdot \\ 0 \end{vmatrix} \circ
\begin{vmatrix} a_{11} & a_{12} & a_{13} & \cdots & a_{1n} \\ 0 & b_{11} & b_{12} & \cdots & b_{1n-1} \\ 0 & b_{21} & b_{22} & \cdots & b_{2n-1} \\ \cdot & & & & \cdot \\ 0 & b_{i1} & b_{i2} & \cdots & b_{in-1} \\ \cdot & & & & \cdot \\ 0 & b_{n-1,1} & b_{n-1,2} & \cdots & b_{nn-1} \end{vmatrix} +
\begin{vmatrix} 0 \\ b_{i1} \\ 0 \\ \cdot \\ -b_{11} \\ \cdot \\ 0 \end{vmatrix} \circ
\begin{vmatrix} a_{11} & a_{12} & a_{13} & \cdots & a_{1n} \\ 0 & b_{11} & b_{12} & \cdots & b_{1n-1} \\ 0 & b_{21} & b_{22} & \cdots & b_{2n-1} \\ 0 & & & & \\ 0 & b_{i1} & b_{i2} & \cdots & b_{in-1} \\ \cdot & & & & \\ 0 & b_{n-1,1} & b_{n-1,2} & \cdots & b_{nn-1} \end{vmatrix} + \cdots +
$$

$$
\begin{vmatrix} 0 \\ b_{n-1} \\ 0 \\ 0 \\ \cdot \\ \cdot \\ -b_{11} \end{vmatrix} \circ
\begin{vmatrix} a_{11} & a_{12} & a_{13} & \cdots & a_{1n} \\ 0 & b_{11} & b_{12} & \cdots & b_{1n-1} \\ 0 & b_{21} & b_{22} & \cdots & b_{2n-1} \\ \cdot & & & & \cdot \\ 0 & b_{i1} & b_{i2} & \cdots & b_{in-1} \\ \cdot & & & & \cdot \\ 0 & b_{n-1,1} & b_{n-1,2} & \cdots & b_{nn-1} \end{vmatrix} =
$$

$$
\begin{vmatrix}
a_{11} & a_{12} & a_{13} & & a_{1n} \\
0 & b_{11} & b_{12} & & b_{1,n-1} \\
0 & b_{21}b_{11}-b_{11}b_{21} & b_{21}b_{12}-b_{11}b_{22} & \cdots & b_{21}b_{1n-1}-b_{11}b_{2,n-1} \\
\cdot & \cdot & \cdot & & \cdot \\
\cdot & \cdot & \cdot & & \cdot \\
0 & b_{i1}b_{11}-b_{11}b_{i1} & b_{i1}b_{12}-b_{11}b_{i2} & \cdots & b_{i1}b_{11}-b_{11}b_{i,n-1} \\
\cdot & \cdot & \cdot & & \cdot \\
0 & b_{n-1,1}b_{11}-b_{11}b_{n-1,1} & b_{n-1,1}b_{12}-b_{11}b_{n-1,2} & & b_{n-1,1}b_{11}-b_{11}b_{n-1,n-1}
\end{vmatrix} .
$$

And, again putting all the pre-multiplier column matrices into a single square matrix, and noting all the elements below b_{11} are equal to zero, we let $c_{11} = b_{21}b_{12}-b_{11}b_{22}$, etc., we get:

pre-multiplier for b

$$
\begin{bmatrix}
1 & a_{21} & a_{31} & \cdots & a_{11} & \cdots & a_{n-1,1} & a_{n1} \\
0 & 1 & 0 & \cdots & 0 & \cdots & 0 & 0 \\
0 & 0 & b_{21} & \cdots & 0 & \cdots & 0 & 0 \\
0 & 0 & -b_{11} & \cdots & & \cdots & 0 & 0 \\
\cdot & \cdot & \cdot & \cdots & -b_{11} & \cdots & \cdot & \cdot \\
0 & 0 & 0 & \cdots & 0 & \cdots & \cdot & 0 \\
\cdot & \cdot & \cdot & & \cdot & \cdots & -b_{11} & 0 \\
0 & 0 & 0 & \cdots & 0 & \cdots & 0 & -b_{11}
\end{bmatrix}
$$

$$
G_2 := \begin{bmatrix}
a11 & a_{12} & a_{13} & \cdot & \cdot & \cdot & a_{1n} \\
0 & b_{11} & b_{12} & \cdot & \cdot & \cdot & b_{1,n-1} \\
0 & 0 & c_{11} & \cdot & \cdot & \cdot & c_{1,n-1} \\
\cdot & \cdot & \cdot & \cdot & \cdot & \cdot & \cdot \\
0 & 0 & c_{i1} & \cdot & \cdot & \cdot & c_{i,n-1} \\
\cdot & \cdot & \cdot & \cdot & \cdot & \cdot & \cdot \\
0 & 0 & c_{n-2,2} & \cdot & \cdot & \cdot & c_{n-2,n-\blacksquare}
\end{bmatrix} \quad \square
$$

Note the inner Matrix is getting smaller by one row and column
with each successive operation. The new c sub-matrix is 2 rows
and two columns smaller than the size of the original a matrix.

Following the pattern, the pre-multiplier matrix for c is:

Pre-multiplier to reduce c:

$$
\begin{vmatrix}
1 & 0 & 0 & 0 & 0 & \cdot & 0 & \cdot & 0 \\
0 & 1 & 0 & 0 & 0 & \cdot & 0 & \cdot & 0 \\
0 & 0 & 1 & 0 & 0 & \cdot & \cdot & \cdot & 0 \\
0 & 0 & 0 & c_{21} & c_{31} & \cdot & \cdot & \cdot & c_{n-3,1} \\
0 & 0 & 0 & -c_{11} & 0 & \cdot & \cdot & \cdot & 0 \\
0 & 0 & 0 & 0 & -c_{11} & \cdot & \cdot & \cdot & \cdot \\
\cdot & \cdot & \cdot & & \cdot & & \cdot & \cdot & \cdot \\
0 & 0 & 0 & 0 & 0 & & -c_{11} & & 0 \\
0 & 0 & 0 & 0 & 0 & & 0 & & -c_{11}
\end{vmatrix}
$$

This goes on in this pattern and we get to the I'th row:

$$
G^I = \begin{vmatrix}
a_{11} & a_{12} & a_{13} & \cdot & \cdot & \cdot & a_{1n} \\
0 & b_{11} & b_{12} & \cdot & \cdot & \cdot & b_{1n-1} \\
& 0 & c_{11} & \cdot & \cdot & \cdot & b_{1,n-2} \\
0 & & \cdot & \cdot & & & \cdot \\
0 & 0 & I_{11} & \cdot & \cdot & \cdot & I_{i,n-i} \\
& & \cdot & & & & \cdot \\
0 & 0 & I_{n-i,1} & \cdot & \cdot & I_{n-i,n-i}
\end{vmatrix}
$$

$$
\begin{vmatrix}
1 \\ 0 \\ 0 \\ \cdot \\ 0 \\ \cdot \\ 0
\end{vmatrix} \circ
\begin{vmatrix}
a_{11} & a_{12} & a_{13} & \cdot & \cdot & \cdot & a_{1n} \\
0 & b_{11} & b_{12} & \cdot & \cdot & \cdot & b_{1n-1} \\
0 & 0 & c_{11} & \cdot & \cdot & \cdot & b_{1,n-2} \\
\cdot & & \cdot & \cdot & & & \cdot \\
0 & 0 & I_{11} & \cdot & \cdot & \cdot & I_{i,n-i} \\
\cdot & & \cdot & & & & \cdot \\
0 & 0 & I_{n-i,1} & \cdot & \cdot & I_{n-i,n-I}
\end{vmatrix} +
\begin{vmatrix}
0 \\ 1 \\ 0 \\ \cdot \\ 0 \\ \cdot \\ 0
\end{vmatrix} \circ
\begin{vmatrix}
a_{11} & a_{12} & a_{13} & \cdot & \cdot & \cdot & a_{1n} \\
0 & b_{11} & b_{12} & \cdot & \cdot & \cdot & b_{1n-1} \\
0 & 0 & c_{11} & \cdot & \cdot & \cdot & b_{1,n-2} \\
\cdot & & \cdot & \cdot & & & \cdot \\
0 & 0 & I_{11} & \cdot & \cdot & \cdot & I_{i,n-i} \\
\cdot & & \cdot & & & & \cdot \\
0 & 0 & I_{n-i,1} & \cdot & \cdot & I_{n-i,n-I}
\end{vmatrix} + \cdot \cdot \cdot +
$$

$$
\begin{vmatrix} 0 \\ 0 \\ 0 \\ \cdot \\ 1 \\ \cdot \\ 0 \end{vmatrix} \circ
\begin{vmatrix} a_{11} & a_{12} & a_{13} & \cdot \cdot \cdot & a_{1n} \\ 0 & b_{11} & b_{12} & \cdot \cdot \cdot & b_{1n-1} \\ 0 & 0 & c_{11} & \cdot \cdot \cdot & b_{1,n-2} \\ \cdot & \cdot & \cdot & \cdot \cdot \cdot & \cdot \\ 0 & 0 & I_{11} & \cdot \cdot \cdot & I_{i,n-I} \\ \cdot & \cdot & \cdot & \cdot \cdot \cdot & \cdot \\ 0 & 0 & I_{n-i,1} & \cdot \cdot & I_{n-i,n-I} \end{vmatrix} +
\begin{vmatrix} 0 \\ 0 \\ 0 \\ \cdot \\ I_{21} \\ -I_{11} \\ 0 \end{vmatrix} \circ
\begin{vmatrix} a_{11} & a_{12} & a_{13} & \cdot \cdot \cdot & a_{1n} \\ 0 & b_{11} & b_{12} & \cdot \cdot \cdot & b_{1n-1} \\ 0 & 0 & c_{11} & \cdot \cdot \cdot & b_{1,n-2} \\ \cdot & \cdot & \cdot & \cdot \cdot \cdot & \cdot \\ 0 & 0 & I_{11} & \cdot \cdot \cdot & I_{i,n-i} \\ \cdot & \cdot & \cdot & \cdot \cdot \cdot & \cdot \\ 0 & 0 & I_{n-i,1} & \cdot \cdot & I_{n-i,n-I} \end{vmatrix} + \cdot \cdot \cdot +
$$

$$
\begin{vmatrix} 0 \\ 0 \\ 0 \\ \cdot \\ I_{n-1} \\ 0 \\ -I_{11} \end{vmatrix} \circ
\begin{vmatrix} a_{11} & a_{12} & a_{13} & \cdot \cdot \cdot & a_{1n} \\ 0 & b_{11} & b_{12} & \cdot \cdot \cdot & b_{1n-1} \\ 0 & 0 & c_{11} & \cdot \cdot \cdot & b_{1,n-2} \\ \cdot & \cdot & \cdot & \cdot \cdot \cdot & \cdot \\ 0 & 0 & I_{11} & \cdot \cdot \cdot & I_{i,n-I} \\ \cdot & \cdot & \cdot & \cdot \cdot \cdot & \cdot \\ 0 & 0 & I_{n-i,1} & \cdot \cdot & I_{n-i,n-I} \end{vmatrix} =
$$

The pre-multiplier matrix is:

$$
\begin{vmatrix}
1 & 0 & 0 & \cdot & \cdot & 0 & & 0 \\
0 & 1 & 0 & \cdot & \cdot & 0 & & 0 \\
0 & 0 & 1 & \cdot & \cdot & 0 & & 0 \\
\cdot & \cdot & \cdot & & & \cdot & & \cdot \\
0 & 0 & 0 & \cdot & \cdot & 1 & \cdot & \cdot \\
0 & 0 & 0 & \cdot & \cdot & I_{21} & I_{31} & I_{n-i,1} \\
0 & 0 & 0 & \cdot & \cdot & -I_{11} & 0 & \cdot \cdot & 0 \\
0 & 0 & 0 & \cdot & \cdot & & -I_{11} & \cdot \cdot & 0 \\
0 & 0 & 0 & \cdot & \cdot & \cdot & & \cdot \cdot \cdot \\
0 & 0 & 0 & \cdot & \cdot & & 0 & & -I_{11}
\end{vmatrix}
$$

And the last row (which is now reduced to a 2x2 submatrix) is calculated by:

$$
G_{n-1} := \begin{bmatrix}
A_{11} & a_{12} & a_{13} & \cdot & \cdot & & \cdot & a_{1n} \\
0 & b_{11} & b_{12} & \cdot & \cdot & & \cdot & b_{1,n-1} \\
0 & 0 & c_{11} & \cdot & \cdot & & \cdot & c_{1,n-2} \\
\cdot & \cdot & \cdot & \cdot & \cdot & & \cdot & \cdot \\
0 & 0 & 0 & \cdot & I_{11} & & \cdot & I_{1,n-i} \\
\cdot & \cdot & \cdot & \cdot & \cdot & & \cdot & \cdot \\
\cdot & 0 & 0 & \cdot & 0 & N_{11} & & N_{12} \\
0 & 0 & 0 & \cdot & 0 & N_{21} & & N_{22}
\end{bmatrix} \quad \square
$$

The pre-multiplier matrix ix:

$$
\begin{bmatrix}
1 & 0 & 0 & 0 & . & . & . & 0 & 0 \\
0 & 1 & 0 & 0 & . & . & . & 0 & 0 \\
0 & 0 & 1 & 0 & . & . & . & 0 & 0 \\
. & . & . & . & . & . & . & . & 0 \\
0 & 0 & 0 & 0 & . & . & . & 1 & 0 \\
0 & 0 & 0 & 0 & . & . & . & N_{21} & -N_{11}
\end{bmatrix}
$$

And $G_n =$

$$
G_n :=
\begin{bmatrix}
A_{11} & a_{12} & a_{13} & . & . & . & a_{1n} \\
0 & b_{11} & b_{12} & . & . & . & b_{1,n-1} \\
0 & 0 & c_{11} & . & . & . & c_{1,n-2} \\
. & . & . & . & . & . & . \\
0 & 0 & 0 & . & I_{11} & . & I_{1,n-i} \\
. & . & . & . & 0 & . & . \\
. & 0 & 0 & . & 0 & N_{11} & N_{1,n-m} \\
0 & 0 & 0 & . & 0 & 0 & M_{11}
\end{bmatrix}
\quad \square
$$

Before I go on, let me recap what We've done. We did the
Gaussian Reduction on an nxn square matrix (or an nxm augmented
matrix, with the solutions = to zero). We multiplied two rows
at a time and added such that the value below the diagonal
under the top element would equal zero. We kept track of each
hand calculated multiplication and addition process by putting
them into a column matrix. When we were finished, we combined
each column into a square nxn matrix. For an nxn matrix, there
are n-1 cofactor matrices to set up. Rather than keep on going
on an infinite nxn matrix, we will let n = 3 and we will solve
for a general 3x4 augmented matrix. According to the proofs of
the Half-Multiplier Operator, we just take the pre-multiplier
matrices developed by hand, transpose them and multiply. We
will get the same answers.

$$\begin{pmatrix} a11 & a12 & a13 & A \\ a21 & a22 & a23 & B \\ a31 & a23 & a33 & C \end{pmatrix} \quad \text{The 3 x 4 augmented matrix.}$$

$$\begin{pmatrix} 1 & 0 & 0 \\ 0 & 1 & 0 \\ 0 & b21 & -b11 \end{pmatrix} \cdot \begin{pmatrix} 1 & 0 & 0 \\ a21 & -a11 & 0 \\ a31 & 0 & -a11 \end{pmatrix} \cdot \begin{pmatrix} a11 & a12 & a13 & A \\ a21 & a22 & a23 & B \\ a31 & a23 & a33 & C \end{pmatrix}$$

We must first find the values to put in for the b matrix.

$$\begin{pmatrix} 1 & 0 & 0 \\ a21 & -a11 & 0 \\ a31 & 0 & -a11 \end{pmatrix} \cdot \begin{pmatrix} a11 & a12 & a13 & A \\ a21 & a22 & a23 & B \\ a31 & a23 & a33 & C \end{pmatrix} \rightarrow \begin{pmatrix} a11 & a12 & a13 & A \\ 0 & a21{\cdot}a12 - a11{\cdot}a22 & a21{\cdot}a13 - a11{\cdot}a23 & a21{\cdot}A - a11{\cdot}B \\ 0 & a31{\cdot}a12 - a11{\cdot}a23 & a31{\cdot}a13 - a11{\cdot}a33 & a31{\cdot}A - a11{\cdot}C \end{pmatrix}$$

We use only the values under a12 of the first multiplication and ignore the rest.

b11 = a21 x a12 - a11 x a22
b21 = a31 x a12 - a11 x a23

$$\begin{pmatrix} 1 & 0 & 0 \\ 0 & 1 & 0 \\ 0 & b21 & -b11 \end{pmatrix} \begin{pmatrix} a11 & a12 & a13 & A \\ 0 & a21{\cdot}a12 - a11{\cdot}a22 & a21{\cdot}a13 - a11{\cdot}a23 & a21{\cdot}A - a11{\cdot}B \\ 0 & a31{\cdot}a12 - a11{\cdot}a23 & a31{\cdot}a13 - a11{\cdot}a33 & a31{\cdot}A - a11{\cdot}C \end{pmatrix}$$

Note all the values under the principle diagonal are zero.

$$\begin{pmatrix} a11 & a12 & a13 & A \\ 0 & a21{\cdot}a12 - a11{\cdot}a22 & a21{\cdot}a13 - a11{\cdot}a23 & a21{\cdot}A - a11{\cdot}B \\ 0 & 0 & -a31{\cdot}a12{\cdot}a11{\cdot}a23 - a11{\cdot}a23{\cdot}a21{\cdot}a13 + a11^2{\cdot}a23^2 + a21{\cdot}a12{\cdot}a11{\cdot}a33 + a11{\cdot}a22{\cdot}a31{\cdot}a13 - a11^2{\cdot}a22{\cdot}a33 & -a31{\cdot}a12{\cdot}a11{\cdot}B - a11{\cdot}a23{\cdot}a21{\cdot}A + a11^2{\cdot}a23{\cdot}B + a21{\cdot}a12{\cdot}a11{\cdot}C + a11{\cdot}a22{\cdot}a31{\cdot}A - a11^2{\cdot}a22{\cdot}C \end{pmatrix}$$

For a 4x5 augmented matrix, evaluated symbolically by MathCad, we get:

$$\begin{bmatrix} 1 & 0 & 0 & 0 \\ A21 & -A11 & 0 & 0 \\ A31 & 0 & -A11 & 0 \\ A41 & 0 & 0 & -A11 \end{bmatrix} \begin{bmatrix} A11 & A12 & A13 & A14 & A \\ A21 & A22 & A23 & A24 & B \\ A31 & A32 & A33 & A34 & C \\ A41 & A42 & A43 & A44 & D \end{bmatrix} \rightarrow \begin{bmatrix} A11 & A12 & A13 & A14 & A \\ 0 & A21{\cdot}A12 - A11{\cdot}A22 & A21{\cdot}A13 - A11{\cdot}A23 & A21{\cdot}A14 - A11{\cdot}A24 & A21{\cdot}A - A11{\cdot}B \\ 0 & A31{\cdot}A12 - A11{\cdot}A32 & A31{\cdot}A13 - A11{\cdot}A33 & A31{\cdot}A14 - A11{\cdot}A34 & A31{\cdot}A - A11{\cdot}C \\ 0 & A41{\cdot}A12 - A11{\cdot}A42 & A41{\cdot}A13 - A11{\cdot}A43 & A41{\cdot}A14 - A11{\cdot}A44 & A41{\cdot}A - A11{\cdot}D \end{bmatrix}$$

$$\begin{bmatrix} A11 & A12 & A13 & A14 & A \\ 0 & A21{\cdot}A12 - A11{\cdot}A22 & A21{\cdot}A13 - A11{\cdot}A23 & A21{\cdot}A14 - A11{\cdot}A24 & A21{\cdot}A - A11{\cdot}B \\ 0 & A31{\cdot}A12 - A11{\cdot}A32 & A31{\cdot}A13 - A11{\cdot}A33 & A31{\cdot}A14 - A11{\cdot}A34 & A31{\cdot}A - A11{\cdot}C \\ 0 & A41{\cdot}A12 - A11{\cdot}A42 & A41{\cdot}A13 - A11{\cdot}A43 & A41{\cdot}A14 - A11{\cdot}A44 & A41{\cdot}A - A11{\cdot}D \end{bmatrix}$$

$$\begin{bmatrix} 1 & 0 & 0 & 0 \\ 0 & 1 & 0 & 0 \\ 0 & B21 & -B11 & 0 \\ 0 & B31 & 0 & -B11 \end{bmatrix} \cdot \begin{bmatrix} a11 & a12 & a13 & a14 & A \\ 0 & B11 & B12 & B13 & B1 \\ 0 & B21 & B22 & B23 & C1 \\ 0 & B31 & B32 & B33 & D1 \end{bmatrix} \rightarrow \begin{bmatrix} a11 & a12 & a13 & a14 & A \\ 0 & B11 & B12 & B13 & B1 \\ 0 & 0 & B21\,B12-\,B11\,B22 & B21\,B13-\,B11\,B23 & B21\,B1-\,B11\,C1 \\ 0 & 0 & B31\,B12-\,B11\,B32 & B31\,B13-\,B11\,B33 & B31\,B1-\,B11\,D1 \end{bmatrix}$$

$$\begin{bmatrix} 1 & 0 & 0 & 0 \\ 0 & 1 & 0 & 0 \\ 0 & 0 & 1 & 0 \\ 0 & 0 & C21 & -C11 \end{bmatrix} \cdot \begin{bmatrix} a11 & a12 & a13 & a14 & A \\ 0 & B11 & B12 & B13 & B1 \\ 0 & 0 & C11 & C12 & C2 \\ 0 & 0 & C21 & C22 & D2 \end{bmatrix} \rightarrow \begin{bmatrix} a11 & a12 & a13 & a14 & A \\ 0 & B11 & B12 & B13 & B1 \\ 0 & 0 & C11 & C12 & C2 \\ 0 & 0 & C21\cdot C11-\,C11\cdot C21 & C21\cdot C12-\,C11\,C22 & C21\cdot C2-\,C11\cdot D2 \end{bmatrix}$$

$C21 \cdot C11 - C11 \cdot C21 = 0$, MathCad doesn't seem to want to do the subtraction.

SO OUR MATRIX HAS NOW BEEN REDUCED TO:

$$\begin{bmatrix} a11 & a12 & a13 & a14 & A \\ 0 & B11 & B12 & B13 & B1 \\ 0 & 0 & C11 & C12 & C2 \\ 0 & 0 & 0 & D11 & D3 \end{bmatrix}$$

WHICH IS WHERE WE WANT IT TO BE. THE VALUES FOR THE B'S, C'S AND D'S ARE: (THE EXPRESSIONS ARE WAY TOO LONG TO PUT INTO TERMS OF a). The values for the B's, C's and D's are:

```
B11 = A21 x A12 - A11 x A22    C11 = B21 x B12 - B11 x B22    D11 = C21 x C12 - C11 x C22
B21 = A31 x A12 - A11 x A32    C21 = B31 x B12 - B11 x B32    D11 = C21 x C2 - C11 x D2
B31 = A41 x A12 - A11 x A42    C12 = B21 x B13 - B11 x B23
B12 = A21 x A13 - A11 x A23    C22 = B31 x B13 - B11 x B33
B22 = A31 x A13 - A11 x A33    C2 = B21 x B1 - B11 x C1
B32 = A41 x A13 - A11 x A43    D2 = B31 x B1 - B11 x D1
B13 = A21 x A14 - A11 x A24
B23 = A31 x A14 - A11 x A34
B33 = A41 x A14 - A11 x A44
B1 = A21 x A - A11 x B
C1 = A31 x A - A11 x C
D1 = A41 x A - A11 x D
```

MathCad cannot define the above variables, so I've written them in Word 7.

The pre-multiplier matrices used to get zero's below the diagonal will be denoted as ½H↓. Now let's try to complete the solution to Gaussian reduction by eliminating all off-diagonal elements going up. This set of matrices will be denoted as ½H↑. I am not going to go through the problem like I did for the ½H↓ operator. I'm just going to derive the first Gaussian Matrix and extrapolate from there following the pattern.

$$\begin{vmatrix} -D_{11} \\ 0 \\ 0 \\ a_{14} \end{vmatrix} \circ \begin{vmatrix} a_{11} & a_{12} & a_{13} & a_{14} & A \\ 0 & B_{11} & B_{12} & B_{13} & B1 \\ 0 & 0 & C_{11} & C_{12} & C2 \\ 0 & 0 & 0 & d_{11} & D3 \end{vmatrix} + \begin{vmatrix} 0 \\ -D_{11} \\ 0 \\ B_{13} \end{vmatrix} \circ \begin{vmatrix} a_{11} & a_{12} & a_{13} & a_{14} & A \\ 0 & B_{11} & B_{12} & B_{13} & B1 \\ 0 & 0 & C_{11} & C_{12} & C2 \\ 0 & 0 & 0 & d_{11} & D3 \end{vmatrix} + \begin{vmatrix} 0 \\ 0 \\ -D_{11} \\ C_{13} \end{vmatrix} \circ \begin{vmatrix} a_{11} & a_{12} & a_{13} & a_{14} & A \\ 0 & B_{11} & B_{12} & B_{13} & B1 \\ 0 & 0 & C_{11} & C_{12} & C2 \\ 0 & 0 & 0 & d_{11} & D3 \end{vmatrix} +$$

$$\begin{vmatrix} 0 \\ 0 \\ 0 \\ 1 \end{vmatrix} \circ \begin{vmatrix} a_{11} & a_{12} & a_{13} & a_{14} & A \\ 0 & B_{11} & B_{12} & B_{13} & B1 \\ 0 & 0 & C_{11} & C_{12} & C2 \\ 0 & 0 & 0 & d_{11} & D3 \end{vmatrix} =$$

Putting the pre-multipliers together and transposing we get:

$$\begin{bmatrix} -D11 & 0 & 0 & 0 \\ 0 & -D11 & 0 & 0 \\ 0 & 0 & -D11 & 0 \\ a14 & B13 & C12 & 1 \end{bmatrix}^T \rightarrow \begin{bmatrix} -D11 & 0 & 0 & a14 \\ 0 & -D11 & 0 & B13 \\ 0 & 0 & -D11 & C12 \\ 0 & 0 & 0 & 1 \end{bmatrix}$$

Likewise, we can just write out the other two Gaussian pre-multipliers, so that the total ½H↑ operator becomes:

$$\begin{bmatrix} -B11 & B12 & 0 & 0 \\ 0 & 1 & 0 & 0 \\ 0 & 0 & 1 & 0 \\ 0 & 0 & 0 & 1 \end{bmatrix} \cdot \begin{bmatrix} -C11 & 0 & a13 & 0 \\ 0 & -C11 & B12 & 0 \\ 0 & 0 & 1 & 0 \\ 0 & 0 & 0 & 1 \end{bmatrix} \cdot \begin{bmatrix} -D11 & 0 & 0 & a14 \\ 0 & -D11 & 0 & B13 \\ 0 & 0 & -D11 & C12 \\ 0 & 0 & 0 & 1 \end{bmatrix} \cdot \begin{bmatrix} a11 & a12 & a13 & a14 & A \\ 0 & B11 & B12 & B13 & B2 \\ 0 & 0 & C11 & C12 & C3 \\ 0 & 0 & 0 & D11 & D4 \end{bmatrix}$$

To find the values for the second pre-multiplier, we must first multiply the first pre-multiplier by the reduced Gaussian Matrix:

$$\begin{bmatrix} -D11 & 0 & 0 & a14 \\ 0 & -D11 & 0 & B13 \\ 0 & 0 & -D11 & C12 \\ 0 & 0 & 0 & 1 \end{bmatrix} \begin{bmatrix} a11 & a12 & a13 & a14 & A \\ 0 & B11 & B12 & B13 & B2 \\ 0 & 0 & C11 & C12 & C3 \\ 0 & 0 & 0 & D11 & D4 \end{bmatrix} \rightarrow \begin{bmatrix} -D11 \cdot a11 & -D11 \cdot a12 & -D11 \cdot a13 & 0 & -D11 \cdot A + a14 \cdot D4 \\ 0 & -D11 \cdot B11 & -D11 \cdot B12 & 0 & -D11 \cdot B2 + B13 \cdot D4 \\ 0 & 0 & -D11 \cdot C11 & 0 & -D11 \cdot C3 + C12 \cdot D4 \\ 0 & 0 & 0 & D11 & D4 \end{bmatrix}$$

$$\begin{bmatrix} -D11 \cdot a11 & -D11 \cdot a12 & -D11 \cdot a13 & 0 & -D11 \cdot A + a14 \cdot D4 \\ 0 & -D11 \cdot B11 & -D11 \cdot B12 & 0 & -D11 \cdot B2 + B13 \cdot D4 \\ 0 & 0 & -D11 \cdot C11 & 0 & -D11 \cdot C3 + C12 \cdot D4 \\ 0 & 0 & 0 & D11 & D4 \end{bmatrix}$$

The transformations are at the bottom of the problem, if I put them here, MathCad will put the values in the matrix and the algebra would be too long to put the solution on the same page much less the same line. To find the values for the third pre-multiplier matrix, we must now multiply by the second pre-multiplier to the fourth Gaussian Matrix:

$$
\begin{bmatrix}
-F11 & 0 & H13 & 0 \\
0 & -F11 & G12 & 0 \\
0 & 0 & 1 & 0 \\
0 & 0 & 0 & 1
\end{bmatrix}
\begin{bmatrix}
H11 & H12 & H13 & 0 & A1 \\
0 & G11 & G12 & 0 & B3 \\
0 & 0 & F11 & 0 & C4 \\
0 & 0 & 0 & D11 & D4
\end{bmatrix}
\rightarrow
\begin{bmatrix}
-F11{\cdot}H11 & -F11{\cdot}H12 & 0 & 0 & -F11{\cdot}A1+H13{\cdot}C4 \\
0 & -F11{\cdot}G11 & 0 & 0 & -F11{\cdot}B3+G12{\cdot}C4 \\
0 & 0 & F11 & 0 & C4 \\
0 & 0 & 0 & D11 & D4
\end{bmatrix}
$$

We now fill in for the final pre-multiplier matrix times the fifth Gaussian Matrix:

$$
\begin{bmatrix}
-I11 & J12 & 0 & 0 \\
0 & 1 & 0 & 0 \\
0 & 0 & 1 & 0 \\
0 & 0 & 0 & 1
\end{bmatrix}
\begin{bmatrix}
J11 & J12 & 0 & 0 & -F11{\cdot}A1+H13{\cdot}C4 \\
0 & I11 & 0 & 0 & -F11{\cdot}B3+G12{\cdot}C4 \\
0 & 0 & F11 & 0 & C4 \\
0 & 0 & 0 & D11 & D4
\end{bmatrix}
\rightarrow
\begin{bmatrix}
-I11{\cdot}J11 & 0 & 0 & 0 & I11{\cdot}F11{\cdot}A1-I11{\cdot}H13{\cdot}C4-J12{\cdot}F11{\cdot}B3+J12{\cdot}G12{\cdot}C4 \\
0 & I11 & 0 & 0 & -F11{\cdot}B3+G12{\cdot}C4 \\
0 & 0 & F11 & 0 & C4 \\
0 & 0 & 0 & D11 & D4
\end{bmatrix}
$$

And we now have the total solution to the system of equations.

The variable substitutions for the matrices are:

```
F11 = D11 x C11        I11= F11 x G11           D11 = D4
G11 = D11 x B11        J11= F11 x H11           F11 = C4
G12 = D11 x B12        J12= F11 x H12           I11 = -F11 x B3 + G12 x C4
H11 = D11 x a11        A2= F11 x A1 + H13 x C4
H12 = D11 x a12        B4= F11 x B3 + G12 x C4
H13 = D11 x a13
A1 = D11 x A + a14 x D4
B3 = D11 x B2 + B13 x D4
C4 = D11 x C3 + C12 x D4
-I11xJ11 = I11xF11xA1 - I11xH13xC4 - J12xF11xB3 + J12xG12xC 4
```

Note that the solutions for the ½H Matrices, whether you are multiplying up or down, are pre-calculated for us by the solution of the previous Gaussian Matrix. i.e. The first solution for the ½H↓ gives a column of zero's under the position a_{11}. All the numbers needed for the solution of the next Gaussian Reducer Matrix now lie under the position a_{22}. We must get all numbers below this to equal zero. The topmost number (it is on the next diagonal) is given a negative value and completes the rest of the diagonal.

The numbers just below it are placed in position under the new 1 (identity) matrix occupying the place (row) of the row just reduced. Their signs are unchanged. All other values in the Gaussian Reduction Matrix are made equal to zero. We continue in this manner until all elements under the principle diagonal are equal to zero. Then we repeat the process, but instead of listing all elements below the a_{11} column, we list them above the a_{nn} position. A one is placed in the a_{nn} position for the row one which is not to be changed. The negative of the value occupying the a_{nn} position is put along the rest of the Gaussian Reduction Matrix diagonal. Then we pre-multiply to the previous solution. In all instances the previous calculation calculates the next needed values for us. **The matrix itself, in reduction, determines it's own solution.** This is why this method should be so powerful for computers, we don't have to calculate the later values, just read the column under the diagonal of the Gaussian Reduction Matrix just computed, plug them into the next pre-multiplier matrix in the sequence and multiply on, repeating the process until the solution is complete.

Let's try solving a real problem. This will be a 3x3 matrix both for brevity and illustration.

We will define the augmented matrix $[A]_{34}$ as:

$$A]_{34} = \begin{vmatrix} 3 & -1 & -2 & -1 \\ -1 & 6 & -3 & 0 \\ -2 & -3 & 6 & -6 \end{vmatrix}$$

which represents the set of equations:

$$\begin{array}{l} 3x -y-2z = 1 \\ -x+6y-3z = 0 \\ -2x-3y+6z = 6 \end{array} \quad \text{Which is equivalent to:} \quad \begin{array}{l} 3x -y-2z-1 = 0 \\ -x+6y-3z = 0 \\ -2x-3y+6z-6 = 0 \end{array}$$

Then ½H↓ is equal to:

$$\begin{pmatrix} 1 & 0 & 0 \\ 0 & 1 & 0 \\ 0 & b21 & -b11 \end{pmatrix} \cdot \begin{pmatrix} 1 & 0 & 0 \\ -1 & -3 & 0 \\ -2 & 0 & -3 \end{pmatrix}$$

Or if we wish to start with ½H↑, we use the following Pre-multipliers:

$$\begin{pmatrix} -b11 & b12 & 0 \\ 0 & 1 & 0 \\ 0 & 0 & 1 \end{pmatrix} \cdot \begin{pmatrix} -6 & 0 & 2 \\ 0 & -6 & -3 \\ 0 & 0 & 1 \end{pmatrix}$$

The values for the b's are computed during the first multiplication. So ½H↓ is equal to:

$$
\begin{pmatrix} 1 & 0 & 0 \\ 0 & 1 & 0 \\ 0 & b21 & -b11 \end{pmatrix} \cdot \begin{pmatrix} 1 & 0 & 0 \\ -1 & -3 & 0 \\ -2 & 0 & -3 \end{pmatrix} \cdot \begin{pmatrix} 3 & -1 & -2 & -1 \\ -1 & 6 & -3 & 0 \\ -2 & -3 & 6 & -6 \end{pmatrix}
$$

The first multiplication is:

$$
\begin{pmatrix} 1 & 0 & 0 \\ -1 & -3 & 0 \\ -2 & 0 & -3 \end{pmatrix} \cdot \begin{pmatrix} 3 & -1 & -2 & -1 \\ -1 & 6 & -3 & 0 \\ -2 & -3 & 6 & -6 \end{pmatrix} = \begin{pmatrix} 3 & -1 & -2 & -1 \\ 0 & -17 & 11 & 1 \\ 0 & 11 & -14 & 20 \end{pmatrix}
$$

The number -17 lies on the diagonal, and is represented by b11, and the number just below it is 11, which represents b21. We plug these values into the b pre-multiplier and multiply to the first Gaussian reduction matrix. Remember, we take the negative of -17.

$$
\begin{pmatrix} 1 & 0 & 0 \\ 0 & 1 & 0 \\ 0 & 11 & 17 \end{pmatrix} \cdot \begin{pmatrix} 3 & -1 & -2 & -1 \\ 0 & -17 & 11 & 1 \\ 0 & 11 & -14 & 20 \end{pmatrix} = \begin{pmatrix} 3 & -1 & -2 & -1 \\ 0 & -17 & 11 & 1 \\ 0 & 0 & -117 & 351 \end{pmatrix}
$$

We now have all zero's under the principle diagonal. It is now time to reduce the matrix going up: The number on the principal diagonal is -117. We take it's negative and place it on the principle diagonal. The two numbers above -117 are 11 and -2, they are placed above the 1 occupying the a33 position on the pre-multiplier matrix. i.e.

So ½H↑ is equal to:

$$
\begin{pmatrix} -b11 & b12 & 0 \\ 0 & 1 & 0 \\ 0 & 0 & 1 \end{pmatrix} \cdot \begin{pmatrix} 117 & 0 & -2 \\ 0 & 117 & 11 \\ 0 & 0 & 1 \end{pmatrix} \cdot \begin{pmatrix} 3 & -1 & -2 & -1 \\ 0 & -17 & 11 & 1 \\ 0 & 0 & -117 & 351 \end{pmatrix}
$$

The first multiplication is:

$$\begin{pmatrix} 117 & 0 & -2 \\ 0 & 117 & 11 \\ 0 & 0 & 1 \end{pmatrix} \cdot \begin{pmatrix} 3 & -1 & -2 & -1 \\ 0 & -17 & 11 & 1 \\ 0 & 0 & -117 & 351 \end{pmatrix} = \begin{pmatrix} 351 & -117 & 0 & -819 \\ 0 & -1989 & 0 & 3978 \\ 0 & 0 & -117 & 351 \end{pmatrix}$$

The number -1989 is the next number up on the principle diagonal with -117 above it. -b11 = -1989 and -117 = b12. Substituting and multiplying we get:

$$\begin{pmatrix} 1989 & -117 & 0 \\ 0 & 1 & 0 \\ 0 & 0 & 1 \end{pmatrix} \cdot \begin{pmatrix} 351 & -117 & 0 & -819 \\ 0 & -1989 & 0 & 3978 \\ 0 & 0 & -117 & 351 \end{pmatrix} = \begin{pmatrix} 698139 & 0 & 0 & -2094417 \\ 0 & -1989 & 0 & 3978 \\ 0 & 0 & -117 & 351 \end{pmatrix}$$

Even though it is illegal (although mathematicians do it all the time) let's simplify by dividing each row by the value along it's diagonal. i.e.

$$-\frac{2094417}{698139} = -3$$

$$\frac{3978}{-1989} = -2$$

$$\frac{351}{-117} = -3$$

We can do this totally by matrix by using the following method (which is legal, by the way):

$$G := \begin{pmatrix} 698139 & 0 & 0 & -2094417 \\ 0 & -1989 & 0 & 3978 \\ 0 & 0 & -117 & 351 \end{pmatrix}$$

First we isolate the diagonal and then take the inverse of each of it's elements

$$\begin{pmatrix} 698139 & 0 & 0 & -2094417 \\ 0 & -1989 & 0 & 3978 \\ 0 & 0 & -117 & 351 \end{pmatrix} \cdot \begin{bmatrix} 1 & 0 & 0 \\ 0 & 1 & 0 \\ 0 & 0 & 1 \\ 0 & 0 & 0 \end{bmatrix} = \begin{pmatrix} 698139 & 0 & 0 \\ 0 & -1989 & 0 \\ 0 & 0 & -117 \end{pmatrix} ,$$

I3 := identity (3)

$$A1 := \begin{pmatrix} 698139 & 0 & 0 \\ 0 & -1989 & 0 \\ 0 & 0 & -117 \end{pmatrix}$$ We've isolated the diagonal.

$$A2 := \overline{\left((I3 \cdot A1) \right)}^{-1}$$ We're taking the inverse of each element along the diagonal.

$$A2 = \begin{bmatrix} 1.432 \cdot 10^{-6} & 0 & 0 \\ 0 & -0.001 & 0 \\ 0 & 0 & -0.009 \end{bmatrix}$$ We've taken the inverse of each element along the diagonal

Areducedsolution := A2·G We're multiplying each element in A2 by the inverse of the diagonal.

$$\text{Areducesolution} = \begin{pmatrix} 1 & 0 & 0 & -3 \\ 0 & 1 & 0 & -2 \\ 0 & 0 & 1 & -3 \end{pmatrix}$$ The solution of the augmented matrix

The equations are:

X-3=0 ; x = 3
y-2=0 ; y = 2
z-3=0 ; z = 3

Now that the four pre-multipliers are computed for ½H↓ and ½H↑, we can multiply them together and get a single matrix for ½H↓ and a single matrix for ½H↑. We can also multiply (½H↓) (½H↑) or (½H↑) (½H↓) to get a single matrix that will also solve the augmented matrix. Let's see how this works.

The ½H↑, =
Gaussian 2 up x Gaussian 1 up = ½H↑,

$$\begin{pmatrix} 1989 & -117 & 0 \\ 0 & 1 & 0 \\ 0 & 0 & 1 \end{pmatrix} \cdot \begin{pmatrix} 117 & 0 & -2 \\ 0 & 117 & 11 \\ 0 & 0 & 1 \end{pmatrix} = \begin{pmatrix} 232713 & -13689 & -5265 \\ 0 & 117 & 11 \\ 0 & 0 & 1 \end{pmatrix}$$

The ½H↓ matrix =

$$\begin{pmatrix} 1 & 0 & 0 \\ 0 & 1 & 0 \\ 0 & 11 & 17 \end{pmatrix} \cdot \begin{pmatrix} 1 & 0 & 0 \\ -1 & -3 & 0 \\ -2 & 0 & -3 \end{pmatrix} = \begin{pmatrix} 1 & 0 & 0 \\ -1 & -3 & 0 \\ -45 & -33 & -51 \end{pmatrix}$$

Then (½H↑) (½H↓) =

$$\begin{pmatrix} 232713 & -13689 & -5265 \\ 0 & 117 & 11 \\ 0 & 0 & 1 \end{pmatrix} \cdot \begin{pmatrix} 1 & 0 & 0 \\ -1 & -3 & 0 \\ -45 & -33 & -51 \end{pmatrix} = \begin{pmatrix} 483327 & 214812 & 268515 \\ -612 & -714 & -561 \\ -45 & -33 & -51 \end{pmatrix}$$

And

(½H↑) (½H↓)　　　　　　x　　　　　A　　　　=

$$\begin{pmatrix} 483327 & 214812 & 268515 \\ -612 & -714 & -561 \\ -45 & -33 & -51 \end{pmatrix} \cdot \begin{pmatrix} 3 & -1 & -2 & -1 \\ -1 & 6 & -3 & 0 \\ -2 & -3 & 6 & -6 \end{pmatrix} = \begin{pmatrix} 698139 & 0 & 0 & -2094417 \\ 0 & -1989 & 0 & 3978 \\ 0 & 0 & -117 & 351 \end{pmatrix}$$

Note: These operators do not commute. The operator's are computed starting with the down operation. Commuting them with these values of ½H↑ first gives invalid results. i.e.

$$\begin{pmatrix} 1 & 0 & 0 \\ -1 & -3 & 0 \\ -45 & -33 & -51 \end{pmatrix} \cdot \begin{pmatrix} 232713 & -13689 & -5265 \\ 0 & 117 & 11 \\ 0 & 0 & 1 \end{pmatrix} = \begin{bmatrix} 232713 & -13689 & -5265 \\ -232713 & 13338 & 5232 \\ -1.047 \cdot 10^7 & 612144 & 236511 \end{bmatrix}$$

$$\begin{bmatrix} 232713 & -13689 & -5265 \\ -232713 & 13338 & 5232 \\ -1.047 \cdot 10^7 & 612144 & 236511 \end{bmatrix} \cdot \begin{pmatrix} 3 & -1 & -2 & -1 \\ -1 & 6 & -3 & 0 \\ -2 & -3 & 6 & -6 \end{pmatrix} = \begin{bmatrix} 722358 & -299052 & -455949 & -201123 \\ -721941 & 297045 & 456804 & 201321 \\ -3.25 \cdot 10^7 & 1.343 \cdot 10^7 & 2.052 \cdot 10^7 & 9050934 \end{bmatrix}$$

But if we start reducing the augmented matrix by getting all zero's above the diagonal first, and then below the diagonal to finish the problem, we get the two Gaussian reduction matrices for the up first operation as:

Gaussian 2 up x Gaussian 1 up

$$
\begin{vmatrix} 27 & 12 & 0 \\ 0 & 1 & 0 \\ 0 & 0 & 1 \end{vmatrix} \cdot \begin{vmatrix} -6 & 0 & -2 \\ 0 & -6 & -3 \\ 0 & 0 & 1 \end{vmatrix} = \begin{vmatrix} -162 & -72 & -90 \\ 0 & -6 & -3 \\ 0 & 0 & 1 \end{vmatrix}
$$

And Gaussian 2 down x Gaussian 1 down =

$$
\begin{vmatrix} 1 & 0 & 0 \\ 0 & 1 & 0 \\ 0 & -702 & 6318 \end{vmatrix} \cdot \begin{vmatrix} 1 & 0 & 0 \\ 12 & 234 & 0 \\ -2 & 0 & 234 \end{vmatrix} = \begin{vmatrix} 1 & 0 & 0 \\ 12 & 234 & 0 \\ -21060 & -164268 & 1478412 \end{vmatrix}
$$

And multiplying the two together we obtain:

$$
\begin{vmatrix} 1 & 0 & 0 \\ 12 & 234 & 0 \\ -21060 & -164268 & 1478412 \end{vmatrix} \cdot \begin{vmatrix} -162 & -72 & -90 \\ 0 & -6 & -3 \\ 0 & 0 & 1 \end{vmatrix} \cdot \begin{vmatrix} 3 & -1 & -2 & -1 \\ -1 & 6 & -3 & 0 \\ -2 & -3 & 6 & -6 \end{vmatrix} = \begin{vmatrix} -234 & 0 & 0 & 702 \\ 0 & -6318 & 0 & 12636 \\ 0 & 0 & 8870472 & -26611416 \end{vmatrix}
$$

Or:

$$
\begin{vmatrix} 1 & 0 & 0 \\ 12 & 234 & 0 \\ -21060 & -164268 & 1478412 \end{vmatrix} \cdot \begin{vmatrix} -162 & -72 & -90 \\ 0 & -6 & -3 \\ 0 & 0 & 1 \end{vmatrix} = \begin{vmatrix} -162 & -72 & -90 \\ -1944 & -2268 & -1782 \\ 3411720 & 2501928 & 3866616 \end{vmatrix}
$$

$$
\begin{vmatrix} -162 & -72 & -90 \\ -1944 & -2268 & -1782 \\ 3411720 & 2501928 & 3866616 \end{vmatrix} \cdot \begin{vmatrix} 3 & -1 & -2 & -1 \\ -1 & 6 & -3 & 0 \\ -2 & -3 & 6 & -6 \end{vmatrix} = \begin{vmatrix} -234 & 0 & 0 & 702 \\ 0 & -6318 & 0 & 12636 \\ 0 & 0 & 8870472 & -26611416 \end{vmatrix}
$$

Simplifying:

$$\begin{pmatrix} -234 & 0 & 0 & 702 \\ 0 & -6318 & 0 & 12636 \\ 0 & 0 & 8870472 & -26611416 \end{pmatrix} \cdot \begin{bmatrix} 1 & 0 & 0 \\ 0 & 1 & 0 \\ 0 & 0 & 1 \\ 0 & 0 & 0 \end{bmatrix} = \begin{pmatrix} -234 & 0 & 0 \\ 0 & -6318 & 0 \\ 0 & 0 & 8870472 \end{pmatrix}$$

$$\text{Gup} := \begin{pmatrix} -234 & 0 & 0 \\ 0 & -6318 & 0 \\ 0 & 0 & 8870472 \end{pmatrix}$$

$$\text{INVGup} := \text{Gup}^{-1}$$

$$\text{INVGup} = \begin{pmatrix} -0.004 & 0 & 0 \\ 0 & 0 & 0 \\ 0 & 0 & 0 \end{pmatrix}$$ These are not actually zero's along the diagonal, the numbers are too small for MathCad to show.

$$\text{Aup} := \begin{pmatrix} -234 & 0 & 0 & 702 \\ 0 & -6318 & 0 & 12636 \\ 0 & 0 & 8870472 & -26611416 \end{pmatrix}$$

$$\text{UPSOLUTION} := \text{INVGup} \, \text{Aup}$$

$$\text{UPSOLUTION} = \begin{pmatrix} 1 & 0 & 0 & -3 \\ 0 & 1 & 0 & -2 \\ 0 & 0 & 1 & -3 \end{pmatrix}$$

Note: I did this differently than from above because I just figured it out. Especially the part of reducing the 3 x 4 augmented matrix to its diagonal. (7-8-97).

This operator notation is confusing, even to me, so I am going to change the look of the operator a little to make it a little less confusing. If we start with a down operation, the down operators will be

$$(\tfrac{1}{2}H\!\uparrow)_d \, (\tfrac{1}{2}H\!\downarrow)_d$$

and the up operator that begins after the down operation is complete will be

$$(½H\downarrow)_{du} (½H\uparrow)_{du}$$

If we begin with the up operators, the notation will be

$$(½H\downarrow)_{u} (½H\uparrow)_{u}$$

and the down operator that begins after the up operation is complete will be

$$(½H\uparrow)_{ud} (½H\downarrow)_{ud}$$

And

$$\mathbf{H\uparrow} = (½H\uparrow)_{ud} (½H\downarrow)_{ud} (½H\downarrow)_{u} (½H\uparrow)_{u}$$

We can now define

$$\mathbf{H\downarrow} = (½H\downarrow)_{du} (½H\uparrow)_{du} (½H\uparrow)_{d} (½H\downarrow)_{d}$$

as the solution which begins with a down multiplication

And

$$\mathbf{H\uparrow} = (½H\uparrow)_{ud} (½H\downarrow)_{ud} (½H\downarrow)_{u} (½H\uparrow)_{u}$$

as the solution which begins with an up multiplication

Then

$$\mathbf{H\uparrow \neq H\downarrow}$$

but

$$(H\uparrow)A = (H\downarrow)A$$

SIMPLIFYING

Note that even with a 3x3 matrix, the elements in the matrix get very large for **H↑** & **H↓**. By a rough calculation, a 9x9 matrix could give a value for **H** larger than the memory capacity of my calculator (10^{499}). So like mathematicians have done inductively with Gaussian reduction for the past few hundred years or so, we will divide out the larger numbers as we come across them.

Lets look at the down operation again

$$(\tfrac{1}{2}H\uparrow)_d (\tfrac{1}{2}H\downarrow)_d = \begin{pmatrix} 1 & 0 & 0 \\ 0 & 1 & 0 \\ 0 & 11 & 17 \end{pmatrix} \begin{pmatrix} 1 & 0 & 0 \\ -1 & -3 & 0 \\ -2 & 0 & -3 \end{pmatrix}$$

$$\begin{vmatrix} 1 & 0 & 0 \\ -1 & -3 & 0 \\ -2 & 0 & -3 \end{vmatrix} \cdot \begin{vmatrix} 3 & -1 & -2 & -1 \\ -1 & 6 & -3 & 0 \\ -2 & -3 & 6 & -6 \end{vmatrix} = \begin{vmatrix} 3 & -1 & -2 & -1 \\ 0 & -17 & 11 & 1 \\ 0 & 11 & -14 & 20 \end{vmatrix} \blacksquare$$

No simplification needed here, let's go to the next step:

$$\begin{vmatrix} 1 & 0 & 0 \\ 0 & 1 & 0 \\ 0 & 11 & 17 \end{vmatrix} \cdot \begin{vmatrix} 3 & -1 & -2 & -1 \\ 0 & -17 & 11 & 1 \\ 0 & 11 & -14 & 20 \end{vmatrix} = \begin{vmatrix} 3 & -1 & -2 & -1 \\ 0 & -17 & 11 & 1 \\ 0 & 0 & -117 & 351 \end{vmatrix} \blacksquare$$

Let's get rid of the large numbers in the last row by dividing by the number on the diagonal:

$$\begin{vmatrix} 1 & & \\ & 1 & \\ & & 1/117 \end{vmatrix} \circ \begin{vmatrix} 3 & -1 & -2 & -1 \\ 0 & -17 & 11 & 1 \\ 0 & 0 & -117 & 351 \end{vmatrix} \overset{1/117_{R3}}{=} \begin{vmatrix} 3 & -1 & -2 & -1 \\ 0 & -17 & 11 & 1 \\ 0 & 0 & -1 & 3 \end{vmatrix}$$

So the new and different $(\tfrac{1}{2}H\uparrow)_{du}$ operator matrix is denoted as $1/117_{R3}$

$$\begin{vmatrix} 1 & 0 & -2 \\ 0 & 1 & 11 \\ 0 & 0 & 1 \end{vmatrix}$$ and the first up multiplication becomes $1/117_{R3}$

$$\begin{pmatrix} 1 & 0 & -2 \\ 0 & 1 & 11 \\ 0 & 0 & 1 \end{pmatrix} \cdot \begin{pmatrix} 3 & -1 & -2 & -1 \\ 0 & -17 & 11 & 1 \\ 0 & 0 & -1 & 3 \end{pmatrix} = \begin{pmatrix} 3 & -1 & 0 & -7 \\ 0 & -17 & 0 & 34 \\ 0 & 0 & -1 & 3 \end{pmatrix} \blacksquare$$

Here in the second row we can divide by 17

$$\begin{vmatrix} 1 \\ 1/17 \\ 1 \end{vmatrix} \circ \begin{vmatrix} 3 & -1 & 0 & -7 \\ 0 & -17 & 0 & 34 \\ 0 & 0 & -1 & 3 \end{vmatrix} \overset{1/17_{R2}}{=} \begin{vmatrix} 3 & -1 & 0 & -7 \\ 0 & -1 & 0 & 2 \\ 0 & 0 & -1 & 3 \end{vmatrix}$$

So the final multiplication becomes:

$$\begin{pmatrix} 1 & -1 & 0 \\ 0 & 1 & 0 \\ 0 & 0 & 1 \end{pmatrix} \cdot \begin{pmatrix} 3 & -1 & 0 & -7 \\ 0 & -1 & 0 & 2 \\ 0 & 0 & -1 & 3 \end{pmatrix} = \begin{pmatrix} 3 & 0 & 0 & -9 \\ 0 & -1 & 0 & 2 \\ 0 & 0 & -1 & 3 \end{pmatrix} \blacksquare$$

And our solution is complete.

Note: We write the divisor in the upper left hand corner of
the matrix mainly for two reasons, the first so that it won't
be mistaken for an exponent, the second is to remind us of
what steps We took to simplify the results and when and where,
just in case We make a mistake and need to know where to look
to correct it.

INVERSE OF A MATRIX

Suppose we extend A by adding on to it the identity matrix such that

$$A = \begin{vmatrix} a_{11} & a_{12} & a_{13} & 1 & 0 & 0 \\ a_{21} & a_{22} & a_{23} & 0 & 1 & 0 \\ a_{31} & a_{32} & a_{33} & 0 & 0 & 1 \end{vmatrix} \; ; \quad \text{Then } (\,{}^{1\!/\!2}H{\uparrow})_d (\,{}^{1\!/\!2}H{\downarrow})_d A = \begin{vmatrix} 1 & 0 & 0 & \text{INVERSE} \\ 0 & 1 & 0 & \text{OF} \\ 0 & 0 & 1 & \text{A} \end{vmatrix}$$

We've already calculated **H↓**, so let's see how this works.

$$\begin{pmatrix} 483327 & 214812 & 268515 \\ -612 & -714 & -561 \\ -45 & -33 & -51 \end{pmatrix} \cdot \begin{pmatrix} 3 & -1 & -2 & 1 & 0 & 0 \\ -1 & 6 & -3 & 0 & 1 & 0 \\ -2 & -3 & 6 & 0 & 0 & 1 \end{pmatrix} = \begin{pmatrix} 698139 & 0 & 0 & 483327 & 214812 & 268515 \\ 0 & -1989 & 0 & -612 & -714 & -561 \\ 0 & 0 & -117 & -45 & -33 & -51 \end{pmatrix}$$

$$\begin{pmatrix} 698139 & 0 & 0 & 483327 & 214812 & 268515 \\ 0 & -1989 & 0 & -612 & -714 & -561 \\ 0 & 0 & -117 & -45 & -33 & -51 \end{pmatrix} \cdot \begin{bmatrix} 1 & 0 & 0 \\ 0 & 1 & 0 \\ 0 & 0 & 1 \\ 0 & 0 & 0 \\ 0 & 0 & 0 \\ 0 & 0 & 0 \end{bmatrix} = \begin{pmatrix} 698139 & 0 & 0 \\ 0 & -1989 & 0 \\ 0 & 0 & -117 \end{pmatrix}$$

$$R2 := \begin{bmatrix} 0 & 0 & 0 \\ 0 & 0 & 0 \\ 0 & 0 & 0 \\ 1 & 0 & 0 \\ 0 & 1 & 0 \\ 0 & 0 & 1 \end{bmatrix}$$

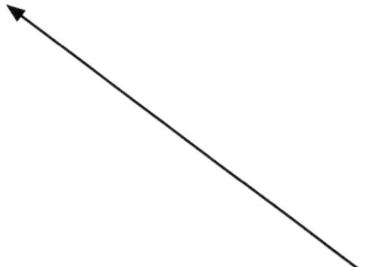

R2 stands for reducer matrix 2. Reducer matrix 1 is here. This operator takes a larger matrix to legally makes a smaller matrix out of it. Multiplying by R2 will return the inverse

$$I := \begin{pmatrix} 698139 & 0 & 0 \\ 0 & -1989 & 0 \\ 0 & 0 & -117 \end{pmatrix}$$

$$INV := I^{-1}$$

$$
INVERSE1 = INV \cdot \begin{pmatrix} 698139 & 0 & 0 & 483327 & 214812 & 268515 \\ 0 & -1989 & 0 & -612 & -714 & -561 \\ 0 & 0 & -117 & -45 & -33 & -51 \end{pmatrix}
$$

In mathematics, we can't legally multiply a single row by a single number. When we multiply a matrix by a number, we multiply every element in that matrix by that number. By separating the left diagonal, taking it's inverse and multiplying, we are doing this operation, but are doing it legally. The operation everyone has really been doing all these centuries and thinking it illegal is

$$
\begin{vmatrix} 1/698139 \\ -1/1989 \\ -1/117 \end{vmatrix} \circ \begin{vmatrix} 483{,}327 & 214{,}812 & 268{,}515 \\ -612 & -714 & -561 \\ -45 & -33 & -51 \end{vmatrix} = \begin{vmatrix} .692 & .308 & .385 \\ .308 & .359 & .282 \\ .385 & .282 & .436 \end{vmatrix} =
$$

Or it is legally equal to:

$$
\begin{pmatrix} 0.00000143 & 0 & 0 \\ 0 & -0.00050277 & 0 \\ 0 & 0 & -0.00854701 \end{pmatrix} \cdot \begin{pmatrix} 698139 & 0 & 0 & 483327 & 214812 & 268515 \\ 0 & -1989 & 0 & -612 & -714 & -561 \\ 0 & 0 & -117 & -45 & -33 & -51 \end{pmatrix} = \begin{pmatrix} 0.998 & 0 & 0 & 0.691 & 0.307 & 0.384 \\ 0 & 1 & 0 & 0.308 & 0.359 & 0.282 \\ 0 & 0 & 1 & 0.385 & 0.282 & 0.436 \end{pmatrix}
$$

$$
INVERSE1 = \begin{pmatrix} 1 & 0 & 0 & 0.692 & 0.308 & 0.385 \\ 0 & 1 & 0 & 0.308 & 0.359 & 0.282 \\ 0 & 0 & 1 & 0.385 & 0.282 & 0.436 \end{pmatrix}
$$

INVERSE = INVERSE1R2

$$
INVERSE = \begin{pmatrix} 0.692 & 0.308 & 0.385 \\ 0.308 & 0.359 & 0.282 \\ 0.385 & 0.282 & 0.436 \end{pmatrix}
$$

To check this answer, let's multiply by the un-augmented matrix A

$$
A := \begin{pmatrix} 3 & -1 & -2 \\ -1 & 6 & -3 \\ -2 & -3 & 6 \end{pmatrix}
$$

CHECK := A·INVERSE   Here we're multiplying A by it's inverse

$$\text{CHECK} = \begin{pmatrix} 1 & 0 & 0 \\ 0 & 1 & 0 \\ 0 & 0 & 1 \end{pmatrix}$$

CHECKSUM = INVERSEA   Here we're multiplying in reverse order.

$$\text{CHECKSUM} = \begin{pmatrix} 1 & 0 & 0 \\ 0 & 1 & 0 \\ 0 & 0 & 1 \end{pmatrix}$$

The matrix A and it's inverse are commutative.

SYMMETRY

This section is for that Mathematician or Physicist who finds it unnecessary to learn two solution sets when one will do. So let's go ahead and solve the same augmented matrix using only the down operator:

$$\begin{pmatrix} 1 & 0 & 0 \\ 0 & 1 & 0 \\ 0 & 11 & 17 \end{pmatrix} \cdot \begin{pmatrix} 1 & 0 & 0 \\ -1 & -3 & 0 \\ -2 & 0 & -3 \end{pmatrix} \cdot \begin{pmatrix} 3 & -1 & -2 & -1 \\ -1 & 6 & -3 & 0 \\ -2 & -3 & 6 & -6 \end{pmatrix} = \begin{pmatrix} 3 & -1 & -2 & -1 \\ 0 & -17 & 11 & 1 \\ 0 & 0 & -117 & 351 \end{pmatrix} \cdot \blacksquare$$

Now, since we are only using the ½H↓ operator, we must rotate the matrix along it's central element a_{22} (in this case -17) 180°. In other words, we must rotate the first Gaussian solution matrix such that a_{11} and a_{33} change places, likewise a_{31} and a_{13} change places. We leave the solutions with their original rows, that part of the math does not change.

First I am going to divide row three by 117. The rotated matrix now becomes:

$$\begin{vmatrix} -1 & 0 & 0 & 3 \\ 11 & -17 & 0 & 1 \\ -2 & -1 & 3 & -1 \end{vmatrix} \quad \text{or} \quad \begin{array}{r} -z +0y+0x = -3 \\ 11z-17y+0x = -1 \\ -2z - y+3x = 1 \end{array}$$

So let's calculate down again:

$$\begin{pmatrix} 1 & 0 & 0 \\ 11 & 1 & 0 \\ -2 & 0 & 1 \end{pmatrix} \cdot \begin{pmatrix} -1 & 0 & 0 & 3 \\ 11 & -17 & 0 & 1 \\ -2 & -1 & 3 & -1 \end{pmatrix} = \begin{pmatrix} -1 & 0 & 0 & 3 \\ 0 & -17 & 0 & 34 \\ 0 & -1 & 3 & -7 \end{pmatrix}$$

Now we need to get the -1 under the -17 to be zero, we accomplish this as follows:

$$\begin{pmatrix} 1 & 0 & 0 \\ 0 & 1 & 0 \\ 0 & -1 & 17 \end{pmatrix} \cdot \begin{pmatrix} -1 & 0 & 0 & 3 \\ 0 & -17 & 0 & 34 \\ 0 & -1 & 3 & -7 \end{pmatrix} = \begin{pmatrix} -1 & 0 & 0 & 3 \\ 0 & -17 & 0 & 34 \\ 0 & 0 & 51 & -153 \end{pmatrix}$$

Now we divide by the numbers along the principle diagonal to simplify:

$$\begin{pmatrix} -1 & 0 & 0 & 3 \\ 0 & -17 & 0 & 34 \\ 0 & 0 & 51 & -153 \end{pmatrix} \cdot \begin{bmatrix} 1 & 0 & 0 \\ 0 & 1 & 0 \\ 0 & 0 & 1 \\ 0 & 0 & 0 \end{bmatrix} = \begin{pmatrix} -1 & 0 & 0 \\ 0 & -17 & 0 \\ 0 & 0 & 51 \end{pmatrix}$$

$$\begin{pmatrix} -1 & 0 & 0 \\ 0 & -17 & 0 \\ 0 & 0 & 51 \end{pmatrix}^{-1} \cdot \begin{pmatrix} -1 & 0 & 0 & 3 \\ 0 & -17 & 0 & 34 \\ 0 & 0 & 51 & -153 \end{pmatrix} = \begin{pmatrix} 1 & 0 & 0 & -3 \\ 0 & 1 & 0 & -2 \\ 0 & 0 & 1 & -3 \end{pmatrix}$$

Which is the solution we are looking for.

NESTED ARRAYS AND GAUSSIAN REDUCTION

Sometimes, especially in physics and engineering we have a matrix's characteristic equation or determinant for which we have solved the roots or eigenvalues, but we need to plug each root back into the matrix and solve the system of equations for each root. Usually this is done one root at a time. I am going to use nested arrays to show how to solve all these matrix solutions at the same time. We will solve for the wave functions of butadiene all in one set of operations.

$$\begin{bmatrix} X & 1 & 0 & 0 \\ 1 & X & 1 & 0 \\ 0 & 1 & X & 1 \\ 0 & 0 & 1 & X \end{bmatrix}$$ The butadiene secular matrix

$x^4 - 3 \cdot x^2 + 1$ The symbolic determinant of the butadiene matrix

$$U := \begin{bmatrix} 1 \\ 0 \\ -3 \\ 0 \\ 1 \end{bmatrix}$$

The coefficients of the symbolic determinant in order from constant to X^4

$$\text{polyroots}(U) = \begin{bmatrix} -1.61803 \\ -0.61803 \\ 0.61803 \\ 1.61803 \end{bmatrix}$$ The roots of the symbolic determinant.

Normally, we would put -1.618 in the matrix for X and reduce the matrix to it's simplest form, then put in -.618 in for X and reduce. We do this for each of the roots, and will end up with four solutions. We will do this all at the same time, although we could use row, column transformations, we will transform the transposed Nested Array into a diagonal. It takes up more memory, but shows what we are doing better.

$$
\text{TOTALSOLMATRIX} =
\begin{bmatrix}
-1.618 & 1 & 0 & 0 & 0 & 0 & 0 & 0 & 0 & 0 & 0 & 0 & 0 & 0 & 0 & 0 \\
1 & -1.618 & 1 & 0 & 0 & 0 & 0 & 0 & 0 & 0 & 0 & 0 & 0 & 0 & 0 & 0 \\
0 & 1 & -1.618 & 1 & 0 & 0 & 0 & 0 & 0 & 0 & 0 & 0 & 0 & 0 & 0 & 0 \\
0 & 0 & 1 & -1.618 & 0 & 0 & 0 & 0 & 0 & 0 & 0 & 0 & 0 & 0 & 0 & 0 \\
0 & 0 & 0 & 0 & -0.618 & 1 & 0 & 0 & 0 & 0 & 0 & 0 & 0 & 0 & 0 & 0 \\
0 & 0 & 0 & 0 & 1 & -0.618 & 1 & 0 & 0 & 0 & 0 & 0 & 0 & 0 & 0 & 0 \\
0 & 0 & 0 & 0 & 0 & 1 & -0.618 & 1 & 0 & 0 & 0 & 0 & 0 & 0 & 0 & 0 \\
0 & 0 & 0 & 0 & 0 & 0 & 1 & -0.618 & 0 & 0 & 0 & 0 & 0 & 0 & 0 & 0 \\
0 & 0 & 0 & 0 & 0 & 0 & 0 & 0 & 0.618 & 1 & 0 & 0 & 0 & 0 & 0 & 0 \\
0 & 0 & 0 & 0 & 0 & 0 & 0 & 0 & 1 & 0.618 & 1 & 0 & 0 & 0 & 0 & 0 \\
0 & 0 & 0 & 0 & 0 & 0 & 0 & 0 & 0 & 1 & 0.618 & 1 & 0 & 0 & 0 & 0 \\
0 & 0 & 0 & 0 & 0 & 0 & 0 & 0 & 0 & 0 & 1 & 0.618 & 0 & 0 & 0 & 0 \\
0 & 0 & 0 & 0 & 0 & 0 & 0 & 0 & 0 & 0 & 0 & 0 & 1.618 & 1 & 0 & 0 \\
0 & 0 & 0 & 0 & 0 & 0 & 0 & 0 & 0 & 0 & 0 & 0 & 1 & 1.618 & 1 & 0 \\
0 & 0 & 0 & 0 & 0 & 0 & 0 & 0 & 0 & 0 & 0 & 0 & 0 & 1 & 1.618 & 1 \\
0 & 0 & 0 & 0 & 0 & 0 & 0 & 0 & 0 & 0 & 0 & 0 & 0 & 0 & 1 & 1.618
\end{bmatrix}
$$

The first Gaussian Reduction Matrix is:

$$
\text{GRM1} =
\begin{bmatrix}
1 & 0 & 0 & 0 & 0 & 0 & 0 & 0 & 0 & 0 & 0 & 0 & 0 & 0 & 0 & 0 \\
1 & 1.618 & 0 & 0 & 0 & 0 & 0 & 0 & 0 & 0 & 0 & 0 & 0 & 0 & 0 & 0 \\
0 & 0 & -1.618 & 0 & 0 & 0 & 0 & 0 & 0 & 0 & 0 & 0 & 0 & 0 & 0 & 0 \\
0 & 0 & 0 & 1.618 & 0 & 0 & 0 & 0 & 0 & 0 & 0 & 0 & 0 & 0 & 0 & 0 \\
0 & 0 & 0 & 0 & 1 & 0 & 0 & 0 & 0 & 0 & 0 & 0 & 0 & 0 & 0 & 0 \\
0 & 0 & 0 & 0 & 1 & 0.618 & 0 & 0 & 0 & 0 & 0 & 0 & 0 & 0 & 0 & 0 \\
0 & 0 & 0 & 0 & 0 & 0 & .618 & 0 & 0 & 0 & 0 & 0 & 0 & 0 & 0 & 0 \\
0 & 0 & 0 & 0 & 0 & 0 & 0 & .618 & 0 & 0 & 0 & 0 & 0 & 0 & 0 & 0 \\
0 & 0 & 0 & 0 & 0 & 0 & 0 & 0 & 1 & 0 & 0 & 0 & 0 & 0 & 0 & 0 \\
0 & 0 & 0 & 0 & 0 & 0 & 0 & 0 & 1 & -.618 & 0 & 0 & 0 & 0 & 0 & 0 \\
0 & 0 & 0 & 0 & 0 & 0 & 0 & 0 & 0 & 0 & -.618 & 0 & 0 & 0 & 0 & 0 \\
0 & 0 & 0 & 0 & 0 & 0 & 0 & 0 & 0 & 0 & 0 & -.618 & 0 & 0 & 0 & 0 \\
0 & 0 & 0 & 0 & 0 & 0 & 0 & 0 & 0 & 0 & 0 & 0 & 1 & 0 & 0 & 0 \\
0 & 0 & 0 & 0 & 0 & 0 & 0 & 0 & 0 & 0 & 0 & 0 & 1 & -1.618 & 0 & 0 \\
0 & 0 & 0 & 0 & 0 & 0 & 0 & 0 & 0 & 0 & 0 & 0 & 0 & 0 & -1.618 & 0 \\
0 & 0 & 0 & 0 & 0 & 0 & 0 & 0 & 0 & 0 & 0 & 0 & 0 & 0 & 0 & -1.618
\end{bmatrix}
$$

In the next computation the matrix is too big to fit on the page, so I am redefining TOTALSOLMATRIX as TSM (for this one problem). Multiplying the Nested Array by the first Gaussian Reduction Matrix we get:

TSM := TOTALSOLMATRIX

GRM1·TSM =

-1.618	1	0	0	0	0	0	0	0	0	0	0	0	0	0	0
0	-1.618	1.618	0	0	0	0	0	0	0	0	0	0	0	0	0
0	-1.618	2.618	-1.618	0	0	0	0	0	0	0	0	0	0	0	0
0	0	1.618	-2.618	0	0	0	0	0	0	0	0	0	0	0	0
0	0	0	0	-0.618	1	0	0	0	0	0	0	0	0	0	0
0	0	0	0	0	0.618	0.618	0	0	0	0	0	0	0	0	0
0	0	0	0	0	0.618	-0.382	0.618	0	0	0	0	0	0	0	0
0	0	0	0	0	0	0.618	-0.382	0	0	0	0	0	0	0	0
0	0	0	0	0	0	0	0	0.618	1	0	0	0	0	0	0
0	0	0	0	0	0	0	0	0	0.618	-0.618	0	0	0	0	0
0	0	0	0	0	0	0	0	0	-0.618	-0.382	-0.618	0	0	0	0
0	0	0	0	0	0	0	0	0	0	-0.618	-0.382	0	0	0	0
0	0	0	0	0	0	0	0	0	0	0	0	1.618	1	0	0
0	0	0	0	0	0	0	0	0	0	0	0	0	-1.618	-1.618	0
0	0	0	0	0	0	0	0	0	0	0	0	0	-1.618	-2.618	-1.618
0	0	0	0	0	0	0	0	0	0	0	0	0	0	-1.618	-2.618

RM1 := GRM1·TOTALSOLMATRIX

Note that there are all zero's under the diagonal in the 1st, 5th, 9th and 13th columns. Now to get zero's under the diagonal in the 2nd, 6th, 10th and 14th columns:

GRM2 :=

1	0	0	0	0	0	0	0	0	0	0	0	0	0	0	0
0	1	0	0	0	0	0	0	0	0	0	0	0	0	0	0
0	-1.618	1.618	0	0	0	0	0	0	0	0	0	0	0	0	0
0	0	0	1.618	0	0	0	0	0	0	0	0	0	0	0	0
0	0	0	0	1	0	0	0	0	0	0	0	0	0	0	0
0	0	0	0	0	1	0	0	0	0	0	0	0	0	0	0
0	0	0	0	0	.618	-.618	0	0	0	0	0	0	0	0	0
0	0	0	0	0	0	0	-.618	0	0	0	0	0	0	0	0
0	0	0	0	0	0	0	0	1	0	0	0	0	0	0	0
0	0	0	0	0	0	0	0	0	1	0	0	0	0	0	0
0	0	0	0	0	0	0	0	0	-.618	-.618	0	0	0	0	0
0	0	0	0	0	0	0	0	0	0	0	-.618	0	0	0	0
0	0	0	0	0	0	0	0	0	0	0	0	1	0	0	0
0	0	0	0	0	0	0	0	0	0	0	0	0	1	0	0
0	0	0	0	0	0	0	0	0	0	0	0	0	-1.618	1.618	0
0	0	0	0	0	0	0	0	0	0	0	0	0	0	0	1.618

Now we multiply the first product RM1 by the second Gaussian Reduction Matrix:

$$
\text{GRM2·RM1} =
\begin{bmatrix}
-1.618 & 1 & 0 & 0 & 0 & 0 & 0 & 0 & 0 & 0 & 0 & 0 & 0 & 0 & 0 & 0 \\
0 & -1.618 & 1.618 & 0 & 0 & 0 & 0 & 0 & 0 & 0 & 0 & 0 & 0 & 0 & 0 & 0 \\
0 & 0 & 1.618 & -2.618 & 0 & 0 & 0 & 0 & 0 & 0 & 0 & 0 & 0 & 0 & 0 & 0 \\
0 & 0 & 2.618 & -4.236 & 0 & 0 & 0 & 0 & 0 & 0 & 0 & 0 & 0 & 0 & 0 & 0 \\
0 & 0 & 0 & 0 & -0.618 & 1 & 0 & 0 & 0 & 0 & 0 & 0 & 0 & 0 & 0 & 0 \\
0 & 0 & 0 & 0 & 0 & 0.618 & 0.618 & 0 & 0 & 0 & 0 & 0 & 0 & 0 & 0 & 0 \\
0 & 0 & 0 & 0 & 0 & 0 & 0.618 & -0.382 & 0 & 0 & 0 & 0 & 0 & 0 & 0 & 0 \\
0 & 0 & 0 & 0 & 0 & 0 & -0.382 & 0.236 & 0 & 0 & 0 & 0 & 0 & 0 & 0 & 0 \\
0 & 0 & 0 & 0 & 0 & 0 & 0 & 0 & 0.618 & 1 & 0 & 0 & 0 & 0 & 0 & 0 \\
0 & 0 & 0 & 0 & 0 & 0 & 0 & 0 & 0 & 0.618 & -0.618 & 0 & 0 & 0 & 0 & 0 \\
0 & 0 & 0 & 0 & 0 & 0 & 0 & 0 & 0 & 0 & 0.618 & 0.382 & 0 & 0 & 0 & 0 \\
0 & 0 & 0 & 0 & 0 & 0 & 0 & 0 & 0 & 0 & 0.382 & 0.236 & 0 & 0 & 0 & 0 \\
0 & 0 & 0 & 0 & 0 & 0 & 0 & 0 & 0 & 0 & 0 & 0 & 1.618 & 1 & 0 & 0 \\
0 & 0 & 0 & 0 & 0 & 0 & 0 & 0 & 0 & 0 & 0 & 0 & 0 & -1.618 & -1.618 & 0 \\
0 & 0 & 0 & 0 & 0 & 0 & 0 & 0 & 0 & 0 & 0 & 0 & 0 & 0 & -1.618 & -2.618 \\
0 & 0 & 0 & 0 & 0 & 0 & 0 & 0 & 0 & 0 & 0 & 0 & 0 & 0 & -2.618 & -4.236 \\
\end{bmatrix}
$$

RM2 := GRM2·RM1   (This is the computer program, believe it or not)

Now we want to get zero's under the diagonal in the 3rd, 7th, 11th and 15th columns:

$$
\text{GRM3} :=
\begin{bmatrix}
1 & 0 & 0 & 0 & 0 & 0 & 0 & 0 & 0 & 0 & 0 & 0 & 0 & 0 & 0 & 0 \\
0 & 1 & 0 & 0 & 0 & 0 & 0 & 0 & 0 & 0 & 0 & 0 & 0 & 0 & 0 & 0 \\
0 & 0 & 1 & 0 & 0 & 0 & 0 & 0 & 0 & 0 & 0 & 0 & 0 & 0 & 0 & 0 \\
0 & 0 & 2.618 & -1.618 & 0 & 0 & 0 & 0 & 0 & 0 & 0 & 0 & 0 & 0 & 0 & 0 \\
0 & 0 & 0 & 0 & 1 & 0 & 0 & 0 & 0 & 0 & 0 & 0 & 0 & 0 & 0 & 0 \\
0 & 0 & 0 & 0 & 0 & 1 & 0 & 0 & 0 & 0 & 0 & 0 & 0 & 0 & 0 & 0 \\
0 & 0 & 0 & 0 & 0 & 0 & 1 & 0 & 0 & 0 & 0 & 0 & 0 & 0 & 0 & 0 \\
0 & 0 & 0 & 0 & 0 & 0 & -.382 & -.618 & 0 & 0 & 0 & 0 & 0 & 0 & 0 & 0 \\
0 & 0 & 0 & 0 & 0 & 0 & 0 & 0 & 1 & 0 & 0 & 0 & 0 & 0 & 0 & 0 \\
0 & 0 & 0 & 0 & 0 & 0 & 0 & 0 & 0 & 1 & 0 & 0 & 0 & 0 & 0 & 0 \\
0 & 0 & 0 & 0 & 0 & 0 & 0 & 0 & 0 & 0 & 1 & 0 & 0 & 0 & 0 & 0 \\
0 & 0 & 0 & 0 & 0 & 0 & 0 & 0 & 0 & 0 & .382 & -.618 & 0 & 0 & 0 & 0 \\
0 & 0 & 0 & 0 & 0 & 0 & 0 & 0 & 0 & 0 & 0 & 0 & 1 & 0 & 0 & 0 \\
0 & 0 & 0 & 0 & 0 & 0 & 0 & 0 & 0 & 0 & 0 & 0 & 0 & 1 & 0 & 0 \\
0 & 0 & 0 & 0 & 0 & 0 & 0 & 0 & 0 & 0 & 0 & 0 & 0 & 0 & 1 & 0 \\
0 & 0 & 0 & 0 & 0 & 0 & 0 & 0 & 0 & 0 & 0 & 0 & 0 & 0 & -2.618 & 1.618 \\
\end{bmatrix}
$$

Now we multiply the second reduced matrix by the third Gaussian Reduction Matrix:

$$\text{GRM3}\cdot\text{RM2} = \begin{bmatrix}
-1.618 & 1 & 0 & 0 & 0 & 0 & 0 & 0 & 0 & 0 & 0 & 0 & 0 & 0 & 0 & 0 \\
0 & -1.618 & 1.618 & 0 & 0 & 0 & 0 & 0 & 0 & 0 & 0 & 0 & 0 & 0 & 0 & 0 \\
0 & 0 & 1.618 & -2.618 & 0 & 0 & 0 & 0 & 0 & 0 & 0 & 0 & 0 & 0 & 0 & 0 \\
0 & 0 & 0 & 0 & 0 & 0 & 0 & 0 & 0 & 0 & 0 & 0 & 0 & 0 & 0 & 0 \\
0 & 0 & 0 & 0 & -0.618 & 1 & 0 & 0 & 0 & 0 & 0 & 0 & 0 & 0 & 0 & 0 \\
0 & 0 & 0 & 0 & 0 & 0.618 & 0.618 & 0 & 0 & 0 & 0 & 0 & 0 & 0 & 0 & 0 \\
0 & 0 & 0 & 0 & 0 & 0 & 0.618 & -0.382 & 0 & 0 & 0 & 0 & 0 & 0 & 0 & 0 \\
0 & 0 & 0 & 0 & 0 & 0 & 0 & 0 & 0 & 0 & 0 & 0 & 0 & 0 & 0 & 0 \\
0 & 0 & 0 & 0 & 0 & 0 & 0 & 0 & 0.618 & 1 & 0 & 0 & 0 & 0 & 0 & 0 \\
0 & 0 & 0 & 0 & 0 & 0 & 0 & 0 & 0 & 0.618 & -0.618 & 0 & 0 & 0 & 0 & 0 \\
0 & 0 & 0 & 0 & 0 & 0 & 0 & 0 & 0 & 0 & 0.618 & 0.382 & 0 & 0 & 0 & 0 \\
0 & 0 & 0 & 0 & 0 & 0 & 0 & 0 & 0 & 0 & 0 & 0 & 0 & 0 & 0 & 0 \\
0 & 0 & 0 & 0 & 0 & 0 & 0 & 0 & 0 & 0 & 0 & 0 & 1.618 & 1 & 0 & 0 \\
0 & 0 & 0 & 0 & 0 & 0 & 0 & 0 & 0 & 0 & 0 & 0 & 0 & -1.618 & -1.618 & 0 \\
0 & 0 & 0 & 0 & 0 & 0 & 0 & 0 & 0 & 0 & 0 & 0 & 0 & 0 & -1.618 & -2.618 \\
0 & 0 & 0 & 0 & 0 & 0 & 0 & 0 & 0 & 0 & 0 & 0 & 0 & 0 & 0 & 0
\end{bmatrix}$$

Woah! They're all zero's! This means we can't reduce the matrix to simpler terms. We can play with it if we want, but no further reductions are really needed. Let's put all values in terms of the diagonal:

I16 = identity(16)

$$\text{DIAG} := \overrightarrow{\big((\text{I16}\cdot\text{RM3})\big)}$$

$$\text{INVDIAG} := \overrightarrow{\text{DIAG}^{-1}}$$

$$\text{SOL} := \text{INVDIAG}\,\text{RM3}$$

MATHCAD CAN'T TAKE THE INVERSE BECAUSE OF THE ZERO'S ALONG THE DIAGONAL, SO I'LL HAVE TO DO IT BY HAND:

$$INVDIAG = \begin{bmatrix}
-.618 & 0 & 0 & 0 & 0 & 0 & 0 & 0 & 0 & 0 & 0 & 0 & 0 & 0 & 0 & 0 \\
0 & -.618 & 0 & 0 & 0 & 0 & 0 & 0 & 0 & 0 & 0 & 0 & 0 & 0 & 0 & 0 \\
0 & 0 & .618 & 0 & 0 & 0 & 0 & 0 & 0 & 0 & 0 & 0 & 0 & 0 & 0 & 0 \\
0 & 0 & 0 & 0 & 0 & 0 & 0 & 0 & 0 & 0 & 0 & 0 & 0 & 0 & 0 & 0 \\
0 & 0 & 0 & 0 & -1.618 & 0 & 0 & 0 & 0 & 0 & 0 & 0 & 0 & 0 & 0 & 0 \\
0 & 0 & 0 & 0 & 0 & 1.618 & 0 & 0 & 0 & 0 & 0 & 0 & 0 & 0 & 0 & 0 \\
0 & 0 & 0 & 0 & 0 & 0 & 1.618 & 0 & 0 & 0 & 0 & 0 & 0 & 0 & 0 & 0 \\
0 & 0 & 0 & 0 & 0 & 0 & 0 & 0 & 0 & 0 & 0 & 0 & 0 & 0 & 0 & 0 \\
0 & 0 & 0 & 0 & 0 & 0 & 0 & 0 & 1.618 & 0 & 0 & 0 & 0 & 0 & 0 & 0 \\
0 & 0 & 0 & 0 & 0 & 0 & 0 & 0 & 0 & 1.618 & 0 & 0 & 0 & 0 & 0 & 0 \\
0 & 0 & 0 & 0 & 0 & 0 & 0 & 0 & 0 & 0 & 1.618 & 0 & 0 & 0 & 0 & 0 \\
0 & 0 & 0 & 0 & 0 & 0 & 0 & 0 & 0 & 0 & 0 & 0 & 0 & 0 & 0 & 0 \\
0 & 0 & 0 & 0 & 0 & 0 & 0 & 0 & 0 & 0 & 0 & 0 & .618 & 0 & 0 & 0 \\
0 & 0 & 0 & 0 & 0 & 0 & 0 & 0 & 0 & 0 & 0 & 0 & 0 & -.618 & 0 & 0 \\
0 & 0 & 0 & 0 & 0 & 0 & 0 & 0 & 0 & 0 & 0 & 0 & 0 & 0 & -.618 & 0 \\
0 & 0 & 0 & 0 & 0 & 0 & 0 & 0 & 0 & 0 & 0 & 0 & 0 & 0 & 0 & 0
\end{bmatrix}$$

I also clicked out the zero's on the diagonal and replaced them with real zero's. Due to rounding errors, the zero's in RM3 are very small numbers. HOW TO READ: For the first wave function a11=.618a2; a2=a3; a3=1.618a4. For the second wave function, a1=1.618a2; a2=-a3; a3=.618a4. For the third wave function, a1=-1.618a2; a2=a3; a3=-.618a4. For the fourth and last wave function, a1=-.618a2; a2=-a3; a3=-1.618a4

$$INVDIAGRM3 = \begin{bmatrix}
1 & -0.618 & 0 & 0 & 0 & 0 & 0 & 0 & 0 & 0 & 0 & 0 & 0 & 0 & 0 & 0 \\
0 & 1 & -1 & 0 & 0 & 0 & 0 & 0 & 0 & 0 & 0 & 0 & 0 & 0 & 0 & 0 \\
0 & 0 & 1 & -1.618 & 0 & 0 & 0 & 0 & 0 & 0 & 0 & 0 & 0 & 0 & 0 & 0 \\
0 & 0 & 0 & 0 & 0 & 0 & 0 & 0 & 0 & 0 & 0 & 0 & 0 & 0 & 0 & 0 \\
0 & 0 & 0 & 0 & 1 & -1.618 & 0 & 0 & 0 & 0 & 0 & 0 & 0 & 0 & 0 & 0 \\
0 & 0 & 0 & 0 & 0 & 1 & 1 & 0 & 0 & 0 & 0 & 0 & 0 & 0 & 0 & 0 \\
0 & 0 & 0 & 0 & 0 & 0 & 1 & -0.618 & 0 & 0 & 0 & 0 & 0 & 0 & 0 & 0 \\
0 & 0 & 0 & 0 & 0 & 0 & 0 & 0 & 0 & 0 & 0 & 0 & 0 & 0 & 0 & 0 \\
0 & 0 & 0 & 0 & 0 & 0 & 0 & 0 & 1 & 1.618 & 0 & 0 & 0 & 0 & 0 & 0 \\
0 & 0 & 0 & 0 & 0 & 0 & 0 & 0 & 0 & 1 & -1 & 0 & 0 & 0 & 0 & 0 \\
0 & 0 & 0 & 0 & 0 & 0 & 0 & 0 & 0 & 0 & 1 & 0.618 & 0 & 0 & 0 & 0 \\
0 & 0 & 0 & 0 & 0 & 0 & 0 & 0 & 0 & 0 & 0 & 0 & 0 & 0 & 0 & 0 \\
0 & 0 & 0 & 0 & 0 & 0 & 0 & 0 & 0 & 0 & 0 & 0 & 1 & 0.618 & 0 & 0 \\
0 & 0 & 0 & 0 & 0 & 0 & 0 & 0 & 0 & 0 & 0 & 0 & 0 & 1 & 1 & 0 \\
0 & 0 & 0 & 0 & 0 & 0 & 0 & 0 & 0 & 0 & 0 & 0 & 0 & 0 & 1 & 1.618 \\
0 & 0 & 0 & 0 & 0 & 0 & 0 & 0 & 0 & 0 & 0 & 0 & 0 & 0 & 0 & 0
\end{bmatrix}$$

GENERAL MATRIX SOLUTION SET FOR ½H↓

$$
\begin{vmatrix}
1 & 0 & 0 & . & . & . & 0 & 0 \\
0 & 1 & 0 & . & . & . & 0 & 0 \\
0 & 0 & 1 & . & . & . & 0 & 0 \\
. & . & . & . & . & . & . & . \\
. & . & . & . & . & . & . & . \\
0 & 0 & 0 & . & . & . & 1 & 0 \\
0 & 0 & 0 & . & . & . & M_{21} & -M_{11}
\end{vmatrix}
\begin{vmatrix}
1 & 0 & 0 & . & . & . & 0 & 0 & 0 \\
0 & 1 & 0 & . & . & . & 0 & 0 & 0 \\
0 & 0 & 1 & . & . & . & 0 & 0 & 0 \\
. & . & . & . & . & . & . & . & . \\
. & . & . & . & . & . & . & . & . \\
0 & 0 & 0 & . & . & . & 1 & 0 & 0 \\
0 & 0 & 0 & . & . & . & L_{21} & -L_{11} & 0 \\
0 & 0 & 0 & . & . & . & L_{31} & 0 & -L_{11}
\end{vmatrix}
\begin{vmatrix}
1 & 0 & 0 & . & . & . & 0 & 0 & 0 & 0 \\
0 & 1 & 0 & . & . & . & 0 & 0 & 0 & 0 \\
0 & 0 & 1 & . & . & . & 0 & 0 & 0 & 0 \\
. & . & . & . & . & . & . & . & . & . \\
. & . & . & . & . & . & . & . & . & . \\
0 & 0 & 0 & . & . & . & 1 & 0 & 0 & 0 \\
0 & 0 & 0 & . & . & . & K_{21} & -K_{11} & 0 & 0 \\
0 & 0 & 0 & . & . & . & K_{31} & 0 & -K_{11} & 0 \\
0 & 0 & 0 & . & . & . & K_{41} & 0 & 0 & -K_{11}
\end{vmatrix}
\; . \; . \; .
$$

$$
\begin{vmatrix}
1 & 0 & 0 & . & . & . & 0 & 0 & 0 & . & . & . & 0 & 0 \\
0 & 1 & 0 & . & . & . & 0 & 0 & 0 & . & . & . & 0 & 0 \\
0 & 0 & 1 & . & . & . & 0 & 0 & 0 & . & . & . & 0 & 0 \\
. & . & . & . & . & . & . & . & . & . & . & . & . & . \\
. & . & . & . & . & . & . & . & . & . & . & . & . & . \\
0 & 0 & 0 & . & . & . & 1 & 0 & 0 & . & . & . & 0 & 0 \\
0 & 0 & 0 & . & . & . & I_{21} & -I_{11} & 0 & 0 & . & . & . & 0 & 0 \\
0 & 0 & 0 & . & . & . & I_{31} & 0 & -I_{11} & 0 & 0 & . & . & . & 0 & 0 \\
0 & 0 & 0 & . & . & . & I_{41} & 0 & -I_{11} & 0 & 0 & . & . & . & 0 & 0 \\
. & . & . & . & . & . & . & . & . & . & . & . & . & . \\
. & . & . & . & . & . & . & . & . & . & . & . & . & . \\
0 & 0 & 0 & . & . & . & I_{11} & 0 & 0 & -I_{11} & . & 0 & 0 \\
. & . & . & . & . & . & . & . & . & . & . & . & . & . \\
. & . & . & . & . & . & . & . & . & . & . & . & . & . \\
0 & 0 & 0 & . & . & . & I_{(n-I),1} & 0 & 0 & . & . & . & -I_{11}
\end{vmatrix}
\; . \; . \; .
$$

$$
\begin{vmatrix}
1 & . & . & . & 0 & 0 & 0 & . & . & . & 0 & 0 \\
0 & . & . & . & 1 & 0 & 0 & . & . & . & 0 & 0 \\
0 & . & . & . & B_{21} & -B_{11} & 0 & 0 & . & . & . & 0 & 0 \\
0 & . & . & . & B_{31} & 0 & -B_{11} & 0 & 0 & . & . & . & 0 & 0 \\
0 & . & . & . & B_{41} & 0 & -B_{11} & 0 & 0 & . & . & . & 0 & 0 \\
. & . & . & . & . & . & . & . & . & . & . & . \\
. & . & . & . & . & . & . & . & . & . & . & . \\
0 & . & . & . & B_{i1} & 0 & 0 & -B_{11} & 0 & 0 \\
0 & . & . & . & B_{i1} & 0 & 0 & . & -B_{11} & 0 \\
0 & . & . & . & B_{(n-I),1} & 0 & 0 & . & . & . & -B_{11}
\end{vmatrix}
$$

$$
\begin{vmatrix}
1 & 0 & 0 & . & . & . & 0 & 0 \\
A_{21} & -A_{11} & 0 & 0 & . & . & . & 0 & 0 \\
A_{31} & 0 & -A_{11} & 0 & 0 & . & . & . & 0 & 0 \\
A_{41} & 0 & -A_{11} & 0 & . & . & . & 0 & 0 \\
. & . & . & . & . & . & . \\
. & . & . & . & . & . & . \\
A_{i1} & 0 & 0 & -A_{11} & . & 0 & 0 \\
A_{i1} & 0 & 0 & . & -A_{11} & 0 \\
A_{(n-I),1} & 0 & 0 & . & . & . & -A_{11}
\end{vmatrix}
$$

GENERAL MATRIX SOLUTION SET FOR ½H↑

$$
\begin{vmatrix}
A_{21} & -A_{11} & 0 & . & . & .0 \\
0 & 1 & 0 & . & . & .0 \\
0 & 0 & 0 & . & . & .0 \\
. & . & . & .. & . & .. \\
. & . & . & .. & . & .. \\
0 & 0 & 0 & 1 & 0 & \\
0 & 0 & 0 & 0 & 1 &
\end{vmatrix}
\begin{vmatrix}
-B_{11} & 0 & B_{13} & 0 & . & . & . & 0 \\
0_{21} & -B_{11} & B_{12} & 0 & . & . & . & 0 \\
0 & 0 & 1 & 0 & . & . & . & 0 \\
. & . & . & . & .. & .. & & \\
. & . & . & . & .. & .. & & \\
0 & 0 & 0 & 0. & .1. & 0 & & \\
0 & 0 & 0 & 0. & .0. & 1 & &
\end{vmatrix}
\begin{vmatrix}
-C_{11} & 0 & 0 & C_{14} & . & 0 & . & . & . & 0 \\
0 & -C_{11} & 0 & C_{13} & 0 & . & . & . & & 0 \\
0 & 0 & -C_{11} & C_{12} & 0 & . & . & . & & 0 \\
0 & 0 & 0 & 1 & 0 & . & . & . & & 0 \\
0 & 0 & 0 & 0 & 1 & . & . & . & & 0 \\
. & . & . & . & . & . & . & . & & \\
. & . & . & . & . & . & . & . & & \\
0 & 0 & 0 & 0 & 0 & 1 & 0 & & & \\
0 & 0 & 0 & 0 & 0 & 0 & 1 & & &
\end{vmatrix}
$$

$$
\begin{vmatrix}
-I_{11} & 0 & 0 & . & . & . & . & I_{(n-1),1} & 0 & . & . & . & . & . & 0 \\
0 & -I_{11} & 0 & . & . & . & . & & 0 & . & . & . & . & . & . \\
. & . & . & . & . & . & . & & . & . & . & . & . & . & . \\
. & . & . & . & . & . & . & & . & . & . & . & . & . & . \\
0 & . & 0 & -I_{11} & . & . & . & I_{14} & 0 & . & . & . & . & . & 0 \\
0 & . & 0 & & .-I_{11}. & & . & I_{13} & 0 & . & . & . & . & . & 0 \\
0 & . & 0 & & . & . & .-I_{11} & I_{12} & 0. & & & & . & 0. \\
0 & . & 0 & & . & . & . & 0 & 1 & 0. & & . & . & . & 0 \\
. & . & . & & . & & . & & . & . & . & . & . & & . \\
. & . & . & & . & & . & & . & . & . & . & . & & . \\
0 & . & 0 & . & . & . & 0 & 0 & 1. & & . & . & . & . & 0 \\
0 & . & 0 & . & . & . & 0 & 0 & 0. & 1 & 0 & 0 & 0 & 0 & \\
0 & . & 0 & . & . & . & 0 & 0 & 0. & 0 & 1 & 0 & 0 & 0 & \\
0 & . & 0 & . & . & . & 0 & 0 & 0. & 0 & 0 & 1 & 0 & 0 & \\
0 & . & 0 & . & . & . & 0 & 0 & 0. & 0 & 0 & 0 & 1 & 0 & \\
0 & . & 0 & . & . & . & 0 & 0 & 0. & 0 & 0 & 0 & 0 & 1 &
\end{vmatrix}
\; . \; .
\begin{vmatrix}
-B_{11} & 0 & 0 & . & . & . & . & B_{(n-2),1} & 0 \\
0 & -B_{11} & 0 & . & . & . & . & & 0 \\
0 & . & -B_{11} & & . & & . & B_{14} & 0 \\
0 & . & 0 & -B_{11} & & . & & B_{13} & 0 \\
0 & . & 0 & & -B_{11} & & & B_{12} & 0 \\
0 & . & 0 & & & -B_{11} & & B_{12} & 0 \\
0 & . & 0 & & & & & 1 & 0 \\
0 & . & 0 & & & & & 0 & 1
\end{vmatrix}
$$

$$
\begin{vmatrix}
-A_{11} & 0 & 0 & . & . & . & . & A_{(n-1),1} \\
0 & -A_{11} & 0 & . & . & . & . & \\
. & . & . & . & . & . & . & \\
. & . & . & . & . & . & . & \\
0 & . & -A_{11} & & . & & & A_{14} \\
0 & . & 0 & -A_{11} & & . & & A_{13} \\
0 & . & 0 & & -A_{11} & & & A_{12} \\
0 & . & 0 & & & & & 1
\end{vmatrix}
$$

ELEMENTARY MOLECULAR ORBITAL METHODS

(PART ONE)
[1]HÜCKEL MO THEORY

In elementary Hückel Molecular calculations, there are three assumptions made:

The wave function can be written $\Psi = A(\Sigma)(\Pi)$ where Σ is an anti-symmetric function of σ electrons, and Π is an anti-symmetric function of π electrons.

Σ and Π are separately normalized to unity.

Σ and Π can each be expanded in terms of a set of orthonormal Slater determinants. In Hückel theory, Σ is ignored and Π is crudely solved. The basis set is limited to 2p orbitals.[1]

It is also assumed that an eigenfunction can be used to represent the Π values of an electron in a molecule just as an electron in an atom can be described by a similar function. But in a molecule, one has to also worry about the electrostatic fields of the surrounding nuclei.[2]

Beyond the hydrogen atom, it is impossible to obtain an exact solution of the Schrodinger equation (hydrogen, with six vectors to describe the x, y and z positions of both the nucleus and electrons, forms a system of six equations in six unknowns, which we can solve. Helium, on the other hand, has the same six equations but nine unknowns, which we cannot solve exactly). Approximate methods of calculating these energies must be used. One method is the Theory of Variations.

THE THEORY OF VARIATIONS
From the Schrodinger Equation

$$H\Psi = E\Psi$$

---

[1]   Turner, Methods in MO Theory, pp. 107

[2]   Liberales, Introduction to MO Theory, pp. 82-85

Multiplying both sides by Ψ* (the complex conjugate of Ψ) we get

$$\Psi*H\Psi = \Psi*E\Psi$$

But E (energy) is a constant, so the equation transforms to:

$$\Psi*H\Psi = E\Psi^2$$

or

$$E = \frac{\int \Psi*H\Psi \, d\tau}{\int \Psi^2 \, d\tau}$$

Suppose that $\Psi = (C_1P_1 + C_2P_2)$ is a solution to the above equation. We just plug it in and simplify the result.

$$E = \frac{\int (C_1P_1 + C_2P_2)*H(C_1P_1 + C_2P_2) \, d\tau}{\int (C_1P_1 + C_2P_2)^2 \, d\tau}$$

$$= \frac{\int (C_1*P_1*HC_1P_1 + C_1*P_1*HC_2P_2 + C_1*P_2*HC_1P_1 + C_2*P_2*HC_2P_2) \, d\tau}{\int (C_1^2 P_1^2 + 2C_1*C_2P_1*P_2 + C_2^2 P_2^2) \, d\tau}$$

Where $\int (P_1*HP_2) \, d\tau = \int (P_2HP_1*) \, d\tau$ (The properties if the integrals are symmetric) Simplifying:

$H_{11} = \int (P_1*HP_1) \, d\tau$; $H_{12} = H_{21} = \int (P_1*HP_1) \, d\tau = \int (P_1*HP_2) \, d\tau = \int (P_2HP_1*) \, d\tau$

$H_{22} = \int (P_2*HP_2) \, d\tau$; $S_{11} = \int P_1^2 \, d\tau$; $S_{12} = S_{21} = \int (P_1*P_2) \, d\tau = \int (P_2*P_1) \, d\tau$; $S_{22} = \int P_2^2 \, d\tau$

Therefore

$$E = \frac{C_1^2 H_{11} + 2C_1*C_2H_{12} + C_2^2 H_{22}}{C_1^2 S_{11} + 2C_1*C_2S_{12} + C_2^2 S_{22}} \qquad \text{EQ. 1}$$

Suppose $\Psi = \Sigma a_i \phi_i$ where a_i are real algebraic numbers, and ϕ is a set of arbitrary real well-behaved (conforms to physical reality) functions from which we seek to choose parameters for a_i such that Ψ approximates most closely to Ψ_0.

Now we will substitute $\Psi = \Sigma a_i \phi_I$ into the Schrodinger Equation:

$$E = \frac{\int (\Sigma a_i \phi_I) H (\Sigma a_j \phi_j) \ d\tau}{\int (\Sigma a_i \phi_I)(\Sigma a_j \phi_j) \ d\tau}$$

Now simplify $H_{ij} = \int (\phi_I H \phi_j) \ d\tau$; $S_{ij} = \int (\phi_I \phi_j) \ d\tau$ and the equation becomes:

$$E = \frac{a_i a_j H_{ij}}{A_{ij} S_{ij}} = f_1/f_2 \qquad \text{EQUATION 2.}$$

where f_1/f_2 is a constant and the integrals H_{ij} and S_{ij} do not involve the parameters a_i.

Now we must use the Calculus of Variations to solve this equation:

$$\partial E/\partial a_i = 0$$
$$\partial E/\partial a_i = (1/f_2) \cdot \partial f_1/\partial a_i - (f_1/f_2) \cdot \partial f_2/\partial a_i = 0$$

and using Equation 2:

$$\partial E/\partial a_i = \partial f_1/\partial a_i - E \cdot \partial f_2/\partial a_i \quad (E = (f_1/f_2)$$
$$E = (f_1/f_2) = 2a_j H_{ij} - E 2a_j S_{ij}$$
$$\text{or}$$
$$\Sigma 2a_j (H_{ij} - E S_{ij}) = 0 \ {}^{3}$$

To demonstrate, suppose $\Psi = a_1 P_1 + a_2 P_2$: Lets solve for a_2 first. $a_i = a_2$, then $\partial E/\partial a_2$ and Eq. 1 become:

$$\partial E/\partial a_2 = M \partial N/M^2 - N \partial M/M^2 = 0$$

Where $M = a_1^2 S_{11} + 2a_1 a_2 S_{12} + a_2^2 S_{22}$
and
$N = a_1^2 H_{11} + 2a_1 a_2 H_{12} + a_2^2 H_{22}$

---

3   Dewar, MO Theory of Organic Chemistry

but $\partial E/\partial a_2 = 0$, therefore

$$M\partial N/M^2 = N\partial M/M^2$$

$$\partial N = (N/M)\,\partial M,$$

but $N/M = E$ (EQ.1)

therefore

$$\partial N = E\partial M$$

or

$$2a_1H_{12} + 2a_2H_{22} = E(2a_1H_{12} + 2a_2H_{22})\ \text{the 2's cancel leaving}$$

$$a_1H_{12} + a_2H_{22} = E(a_1H_{12} + a_2H_{22}),$$

$$a_1(H_{12} - ES_{12}) + a_2(H_{22} - ES_{22}) = \partial E/\partial a_2$$

by similar reasoning, we find

$$a_1(H_{11} - ES_{12}) + a_2(H_{22} - ES_{22}) = \partial E/\partial a_2 \text{ [4]}$$

These are the Slater determinants that we will use in our calculations. Even if they are not understandable how they are derived, it is important to at least see where they come from.

BUTADIENE

The Hückel method chooses eigenfunctions which obey the rules of AO's (Atomic Orbitals) and the Aufbau principle. With use of the above Slater determinant, energies of an electron in various possible MO's (Molecular Orbitals) can be calculated where each AO accommodates electrons of opposite spins. Orbitals are filled starting from the lowest energies until the electrons are used up. When dealing with more than two electrons, two are placed in the most stable orbit, two in the next, then two more and so on until all the electrons are accounted for. In cases of degenerate orbitals, electrons with parallel spins go into each of the orbitals (with equal probability of occupation) with only one electron per orbital until each orbital is half-filled. Then, if there are enough electrons left over, we begin filling the half-filled orbitals until all electrons are used. i.e. in the case of butadiene, orbitals are filled as $E_1 = A-2B$, $E_2 = A = E_3$ and $E_4 = A+2B$. And the orbitals are filled as indicated:

---

[4]   Roberts, MO Calculations, pp. 27-28

$E_1 = A-2B$ $E_2 = A =$ E_3 $E_4 = A+2B$

where ↑ represents electrons with parallel + spins, & ↓ represents electrons with − spins.

One serious drawback to the Hückel MO method is that overlap is generally neglected (a simple method is included on pg. 110-111) and the energy associated with surrounding nuclei and orbitals is assumed to remain unchanged even in the presence of other electrons. To help visualize this a little better, suppose we have a molecule of ethylene before its π bonds are formed:

$$CH_3=CH_3 \equiv \quad C = C$$

where the electron is in one of the p orbitals. Suppose we find it is in p_1. Then $Hp_1 = Ep_1$. Multiplying both sides by p_1 and integrating, we have:

$$E = \int p_1 Hp_1 d\tau = \int p_1 Ep_1 d\tau$$

where E is a constant, and

$$\int p_1 p_1 d\tau = 1$$

Therefore

$$\int p_1 Hp_1 d\tau = E5$$

This is the coulomb integral representing the energy of an electron in a 2p orbital of carbon. It's value must be negative since the electron is attracted to the nucleus. This particular integral is denoted as H_{11} or A. Similarly $\int p_2 Hp_2 d\tau = H_{22} = A$.

B represents the energy of an electron when it is between carbon atoms 1 and 2. It is also a negative quantity and is called the resonance integral. It is defined by:

$$\int p_1 Hp_2 d\tau = H_{12} = B = \int p_2 Hp_1 d\tau = H_{21}$$

To find the most probable location of an electron in a molecule, we describe a function Ψ such that $\Psi = a_1 p_1 + a_2 p_2$ (which represents the π bonding of ethylene having the energy A + B). The more negative the energy, the more stable the bond.

ETHYLENE

We are now in a position to solve the energy states of ethylene. First we set up the secular determinant for ethylene with the conditions:

$$S_{ij} = \begin{array}{l} i=j = 1 \\ i \neq j = 0 \end{array} \quad \begin{array}{l} \text{coulomb computation} \\ \text{resonance with overlap neglected} \end{array} \quad \begin{array}{l} \text{and } \Psi = a_1p_1 + a_2p_2. \\ \text{and } E_{11} = E_{22} \end{array}$$

$$\begin{vmatrix} H_{11} - S_{11}E_{11} & H_{12} - S_{12}E_{12} \\ H_{21} - S_{21}E_{21} & H_{22} - S_{22}E_{22} \end{vmatrix} = \begin{vmatrix} H_{11} - E_{11} & H_{12} \\ H_{21} & H_{22} - E_{22} \end{vmatrix} = 0 \equiv \begin{vmatrix} A - E & B \\ B & A - E \end{vmatrix}$$

We divide both rows by B and let $(A-E)/B = X$ (EQ. 3), which gives us the simplified Slater determinant

$\begin{vmatrix} X & 1 \\ 1 & X \end{vmatrix} = 0$. The value of the determinant is $(X^2 - 1) = 0$ The roots of this 1 X equation are X=1, X=-1. Putting each value of X into EQ. 3 and solving for E_1 (defined as the most negative or highest energy or bonding energy, and E_2 as the least negative, lowest energy or non-bonding energy, we get:

$$(A-E)/B = 1; \quad -E = B-A; \quad E_2 = A - B$$
$$(A-E)/B = -1; \quad -E = -B-A; \quad E_1 = A + B$$

Or we can write the equation out as: $(X+1)(X-1) = (A-E_1+B)(A-E_2-B)$ and solve for E.

(This proves our assumption above that the energy for ethylene is A + B.)

Since A & B are negative quantities, E_1 is the most stable orbital and E_2 is the least. E_1 is called a bonding orbital (it is occupied) and E_2 is called an anti-bonding orbital (it's orbitals are unoccupied). There are two orbitals to fill (in the non-bonding orbital, p_1 & p_2), so the total π energy (E_π) is 2A + 2B which is twice the energy of a single occupied orbital.

We have just computed E_π = 2A + 2B as the total de-localized energy of ethylene. To determine the energy of localized ethylene (for resonance energy calculations), we [5]assume that the carbon

_____

[5] Liberles, Intro. To MO Theory, pp. 119-121

atoms are separated in space (the electrons occupy orbitals on the carbon atoms and not in between). In this case, $H_{12} = H_{21} = 0$; and $H_{11} = H_{22} = A$. The secular determinant now becomes

$$\begin{vmatrix} A - E & 0 \\ 0 & A - E \end{vmatrix} = 0 = (A - E)^2. \qquad \begin{matrix} E_1 = A \\ E_2 = A \end{matrix}$$

Which is degenerate (the roots are equal and of the same sign). This is interpreted to mean each of the carbon atoms contain one electron each in their p orbitals and $E_\pi = 2A$. The resonance energy $DE_\pi = E_\pi$ (de-localized) $- E_\pi$ (localized) $= 2A + 2B - 2A = 2B$. This difference is considered to be the energy difference between the most stable classical structure and the compound itself. It is the energy of the π bond formation. [5]

A more exact energy diagram of ethylene may be constructed if overlap is included. [6] In this case, for S_{ij}, $i \neq j \neq 0$; $H_{11} = H_{22} = A$; $S_{12} = S_{21} = S$ and $H_{12} = H_{21} = \gamma$. The Slater determinant becomes:

$$\begin{vmatrix} H_{11} - S_{11}E_{11} & H_{12} - S_{12}E_{12} \\ H_{21} - S_{21}E_{21} & H_{22} - S_{22}E_{22} \end{vmatrix} = \begin{vmatrix} H_{11} - E_{11} & \gamma - S_{12}E_{12} \\ \gamma - S_{21}E_{21} & H_{22} - E_{22} \end{vmatrix} = 0 \equiv \begin{vmatrix} A - E & \gamma - SE \\ \gamma - SE & A - E \end{vmatrix}$$

This looks like a more complex determinant to solve, but if we let $B = \gamma - SE$, the determinant changes to

$$\begin{vmatrix} A-E & B \\ B & A-E \end{vmatrix}$$

Which is the same form as above, dividing by B, and letting $X = (A-E)/B$ or $X = (A-E)/(\gamma-SE)$, we get (EQ.4)

$$\begin{vmatrix} X & 1 \\ 1 & X \end{vmatrix} = 0.$$

Solving the same way as outlined above, the solution to the determinant is $(X^2 - 1)$ and $X = \pm 1$. Now we must solve for E_1 and E_2:

---

[6]   Ibid, pp. 114-116

ACCOUNTING FOR OVERLAP INTERGAL

Suppose $X = (A-E)/(\gamma-SE) = -1$; $A - E_1 = -\gamma + SE_1$
$$A + \gamma = E_1 + SE_1$$
$$A + \gamma = E_1(1 + S)$$
$$\mathbf{E_1 = (A + \gamma)/(1 + S)}$$

Suppose $X = (A-E_2)/(\gamma-SE_2) = 1$; $A - E_2 = \gamma - SE_2$

then

$$\mathbf{E_2 = (A - \gamma)/(1 - S)}$$

Where S is not a constant, but is a function of the distance separating the carbon atoms. For ethylene, experiment gives this distance as 1.33Å or about a 25% overlap. Therefore, $S = .25$. Plugging this value for S into both equations, we get:

$$\mathbf{E_1} = (A + \gamma)/(1 + .25) = (A + \gamma)/(1.25) = \mathbf{.8(A + \gamma)}$$
$$\mathbf{E_2} = (A - \gamma)/(1 - .25) = (A - \gamma)/(.75) = \mathbf{1.33(A - \gamma)}$$

$\mathbf{E_\pi}$ (de-localized) $= 2E_1 = 2x.8(A + \gamma) = \mathbf{1.6(A + \gamma)}$
$\mathbf{E_\pi}$ (localized) $= (A - E)^2$, $E_1 = E_2 = A = 2A$ (A for one orbital, twice this for both orbitals).
$\mathbf{DE_\pi}$ (resonance energy) $= 1.6A + 1.6\gamma -2A = \mathbf{-.4A + 1.6\gamma}$.

To find the coefficients for $\Psi = a_1p_1 + a_2p_2$, we will reduce the secular determinant by use of Gaussian Reduction. Solving first for X=1, we plug this value for X into the determinant

$$\begin{vmatrix} X & 1 \\ 1 & X \end{vmatrix} = 0 = \begin{vmatrix} 1 & 1 \\ 1 & 1 \end{vmatrix}$$

This is simple, just subtract row 2 from row one, and we get (remembering the first column represents the values of a_1 and the second column represents the values of a_2)

$\begin{matrix} a_1 & a_2 \\ \begin{vmatrix} 1 & 1 \\ 0 & 0 \end{vmatrix} \end{matrix}$ $a_1 + a_2 = 0$ or $\mathbf{a_1 = -a_2}$.

We will now write a_2 in terms of a_1 and normalize,: remembering $a_1 = 1$ and $a_2 = 1$:

a_1 (normalized) $= 1/(a_1^2 + (-a_1^2)) = 1/(1^2 + 1^2)^{1/2} = 1/(2)^{1/2}$ therefore
$a_1 = 1/\sqrt{2}$, and $\Psi_2 = 1/\sqrt{2}\ P_1 - 1/\sqrt{2}\ P_2 = \mathbf{1/\sqrt{2}\ (P_1 - P_2)}$

Similarly, for $X = -1$, the determinant becomes

$\begin{vmatrix} -1 & 1 \\ 1 & -1 \end{vmatrix} = 0 =$ (adding row 2 to row 1) $\begin{vmatrix} -1 & 1 \\ 0 & 0 \end{vmatrix}$ or $a_1 = a_2$.   Setting in terms of a_1
and normalizing:

a_1 (normalized) $= 1/(1^2 + 1^2)^{1/2} = 1/\sqrt{2}$ and $\mathbf{\Psi_1 = 1/\sqrt{2}\ (P_1 + P_2)}$.

The total wave function describing ethylene is:

$\Psi_1 = \mathbf{1/\sqrt{2}\ (P_1 + P_2)}$ **bonding**

$\Psi_2 = \mathbf{1/\sqrt{2}\ (P_1 - P_2)}$ **anti-bonding**

Since only Ψ_1 is occupied, each electron contributes $(1/\sqrt{2})^2 = 1/2$ of the total electron density at each carbon, or $1/2 + 1/2 = 1 = \Psi_1^1$ for the total molecule. But at some point on Ψ_2, the coefficient of a_1 will equal a_2, and $a_1 - a_2 = 0$, meaning that at some point the molecule will experience a state of zero electron density (a node). But Ψ_1 will always equal a negative quantity such that there is a non-vanishing electron density. Since in nature it is assumed the orbitals of an atom or molecule never vanish, Ψ_1 is chosen as the function to represent the bonding state of ethylene, Ψ_1 is therefore a well behaved function.[7]

---

[7]   Goodrich, A Primer in Quantum Chemistry, pp. 35-36

BOND ORDER

The bond order of a substance is an indication of the location of the π bonds in a molecule. It is calculated from the occupied orbitals and is defined as

$$P_{ij} = \sum^{\Psi occ} N a_i a_j$$

Since we are calculating between two carbons (not the interaction between three or four), $N = 2$ (N will always be two as long as we are calculating for adjacent carbons). So for ethylene,

$$P_{12} = \sum^{\Psi occ} 2\,(1/\sqrt{2})\,(1/\sqrt{2}) = 2\,(1/2) = 1$$

It is defined that the bond order of ethylene is the maximum strength of a π bond, with all other π bonds in other, larger molecules either equal or less than this value.

Note here the above Ψ function can be represented as a matrix:

Ψ_1
Ψ_2
$\quad \begin{vmatrix} \sqrt{2} & \sqrt{2} \\ \sqrt{2} & -\sqrt{2} \end{vmatrix}$ and the bond order can be calculated by squaring the first row of the matrix. We can do this as follows: $\Psi_{1occ}\Psi_1{}^T{}_{occ}$

$$P_{12} = 2\left(\begin{vmatrix} 1 & 0 \\ 0 & 0 \end{vmatrix}\begin{vmatrix} \sqrt{2} & \sqrt{2} \\ \sqrt{2} & -\sqrt{2} \end{vmatrix}\right)\left(\begin{vmatrix} 1 & 0 \\ 0 & 0 \end{vmatrix}\begin{vmatrix} \sqrt{2} & \sqrt{2} \\ \sqrt{2} & -\sqrt{2} \end{vmatrix}\right)^T = 2\,[\sqrt{2}\ \ \sqrt{2}]\begin{vmatrix} \sqrt{2} \\ \sqrt{2} \end{vmatrix} = 2\,(1/2) = 1$$

Lets take a quantum leap in our mathematics and run a comparison of butadiene and cyclobutadiene and compare their relative stability's using the Hückel method.

BUTADIENE

Butadiene has the structure $CH_2=CH-CH=CH_2$. To set up the Slater determinant, we number the carbons in order from left to right
1   2   3   4, and set up the math such that $H_{11}=H_{22}=H_{33}=H_{44}=$ A (each
C- C- C- C
carbon atom has an orbital to fill). Also suppose the determinant of A is symmetric, then $H_{12}=H_{21}=H_{23}=H_{32}=H_{34}=H_{43}$ = B (the electron can exist equally in the regions between any two adjacent carbon atoms), and $H_{13}=H_{31}=H_{14}=H_{41}=H_{24}=H_{42}$ = 0 (there is assumed to be no inter-electronic interaction between non-adjacent carbon atoms). These assumptions are not really correct, for it is assumed here that $B_{12}=B_{21}=B_{23}=B_{32}=B_{34}=B_{43}$ = B and $S_{12}=S_{21}=S_{23}=S_{32}=S_{34}=S_{43}$ = 0 which is not true as the bond lengths in butadiene are not the same. But since S=.25, overlap is usually ignored.[8]

So the Slater determinant for butadiene becomes

$$\begin{vmatrix} A-E & B & 0 & 0 \\ B & A-E & B & 0 \\ 0 & B & A-E & B \\ 0 & 0 & B & A-E \end{vmatrix} = 0 \quad \text{Letting X=(A-E)/B, the determinant becomes} \quad \begin{vmatrix} X & 1 & 0 & 0 \\ 1 & X & 1 & 0 \\ 0 & 1 & X & 1 \\ 0 & 0 & 1 & X \end{vmatrix} = 0$$

To solve the determinant and find it's characteristic equation, we compute the following steps using Cramer's rule:

$$(X \begin{vmatrix} X & 1 & 0 \\ 1 & X & 1 \\ 0 & 1 & X \end{vmatrix} - 1 \begin{vmatrix} 1 & 1 & 0 \\ 0 & X & 1 \\ 0 & 1 & X \end{vmatrix}) = X(X \begin{vmatrix} X & 1 \\ 1 & X \end{vmatrix} - 1 \begin{vmatrix} 1 & 1 \\ 0 & X \end{vmatrix}) - 1(1 \begin{vmatrix} X & 1 \\ 1 & X \end{vmatrix} - 1 \begin{vmatrix} 0 & 1 \\ 0 & X \end{vmatrix}) =$$

$X(X(X^2-1)-X)-((X^2-1)-X) = X(X^3-X-X)-X^2+1 = X^4-2X^2-X^2+1= \mathbf{X^4-3X^2+1=0}$.

Note: This is one of the reasons I discovered the half-multiplier operator. I was trying to obtain the characteristic equation using Gaussian Reduction (which I am still unable to do) when it occurred to me to prove it, as I had never been shown the proof in any math or physics class I had ever taken. The rest is mathematical history, as from it's proof is derived the Unified Field Equation. It also puts the final touches on the definition of number and the operations of addition, subtraction, multiplication and division. Mathematicians and physicists have

---

[8]   Flurry, MO Theory of Bonding in Organic Molecules, pp. 54-55

used Gaussian Reduction for over 200 years, and not one of them ever set about to prove it, it's usefulness was just intuitive, never proven, it was used because it worked, and thus lost the chance to discover the missing link in mathematics.

The roots of this equation are $\pm(3/2 \pm 1/2\sqrt{2})$ or ±1.618 and . ±618. The energy diagram is as follows:

$E_1 = A + 1.618B$ ⥮   occupied bonding orbital

$E_2 = A + .618B$ ⥮   occupied bonding orbital

$E_3 = A - .618B$ ———   un-occupied anti-bonding orbital

$E_4 = A - 1.618B$ ———   un-occupied anti-bonding orbital

We will now calculate the a_i coefficients for the Ψ functions. First we write the secular determinant of butadiene:

$a_1\ a_2\ a_3\ a_4$ Setting X=-1.618

a1 a2 a3 a4

$$
\begin{vmatrix} X & 1 & 0 & 0 \\ 1 & X & 1 & 0 \\ 0 & 1 & X & 1 \\ 0 & 0 & 1 & X \end{vmatrix} = 0
\qquad
\begin{vmatrix} -1.628 & 1 & 0 & 0 \\ 1 & -1.618 & 1 & 0 \\ 0 & 1 & -1.618 & 1 \\ 0 & 0 & 1 & -1.618 \end{vmatrix} = 0
$$

Normally, in all the books, this determinant is solved using Cramer's rule, but the math takes a quarter page per a_i, or about an entire page to solve the wave function Ψ_i. Note also that in this method of solution, the above equation should be an augmented matrix of 4 rows and 5 columns, the fifth column being the negative of the matrix to the right of the equal sign. But since here this value is 0, I'm just going to ignore it, but remember that it's there. i.e.

$a_1\ a_2\ a_3\ a_4$ Setting X=-1.618

$$
\begin{array}{l} \Psi_1 \\ \Psi_2 \\ \Psi_3 \\ \Psi_4 \end{array}
\begin{vmatrix} X & 1 & 0 & 0 & -0 \\ 1 & X & 1 & 0 & -0 \\ 0 & 1 & X & 1 & -0 \\ 0 & 0 & 1 & X & -0 \end{vmatrix} = 0
\qquad
\begin{vmatrix} -1.618 & 1 & 0 & 0 & -0 \\ 1 & -1.618 & 1 & 0 & -0 \\ 0 & 1 & -1.618 & 1 & -0 \\ 0 & 0 & 1 & -1.618 & -0 \end{vmatrix} = 0
$$

To solve, we do the following steps:

1. Divide row 1 by 1.618, and add to row two.
   Put solution in row 2, leaving row 1 unchanged.

$$1/1.618R1+R2$$

$$\begin{vmatrix} -1 & .618 & 0 & 0 \\ 0 & -1 & 1 & 0 \\ 0 & 1 & -1.618 & 1 \\ 0 & 0 & 1 & -1.618 \end{vmatrix}$$

2. Add row 2 to row 3
   Put solution in row 3, leaving rows 1&2 unchanged.

$$R2+R3$$

$$\begin{vmatrix} -1 & .618 & 0 & 0 \\ 0 & -1 & 1 & 0 \\ 0 & 0 & -.618 & 1 \\ 0 & 0 & 1 & -1.618 \end{vmatrix}$$

3. Divide row 4 by 1.618 and add to row 3. Put solution in row 4, leaving rows 1,2&3 unchanged.

$$1/1.618R4+R3$$

$$\begin{vmatrix} -1 & .618 & 0 & 0 \\ 0 & -1 & 1 & 0 \\ 0 & 0 & -.618 & 1 \\ 0 & 0 & 0 & 0 \end{vmatrix}$$

This is as simple as we can get this reduction. Remembering that this is actually an augmented matrix, we will subtract the terms that are negative to get them on the other side of the equal sign. This gives us

$a_1 = .618a_2$
$a_2 = a_3$
$.618a_3 = a_4$

To make the calculations simpler, we will put all these values in terms of a_1.

$a_1 = a_1$
$a_2 = 1.618a_1$
$a_3 = 1.618a_1$
$a_4 = a_1$

Therefore, $\Psi_1 = P_1 + 1.618P_2 + 1.618P_3 + P_4$ Now we must normalize Ψ_1:

$\Psi_1(N) = (a_1^2 + a_2^2 + a_3^2 + a_4^2)^{1/2} = (1^2 + 1.618^2 + 1.618^2 + 1^2)^{1/2} =$
$(1+2.618+2.168+1)^{1/2} = (7.75)^{1/2} = \mathbf{2.69}$

This is also equal to $(\Psi_1\Psi_1^T)^{1/2} = ([1\ 1.618\ 1.618\ 1] \begin{vmatrix} 1 \\ 1.618 \\ 1.618 \\ 1 \end{vmatrix})^{1/2} = 2.69$

So the equation in general is: $\Psi_i/(\Psi_i\Psi_i^T)^{1/2}$

and for $\Psi_1 = 1/2.69[1\ 1.618\ 1.618\ 1] = \mathbf{[.3717\ .6015\ .6015\ .3717]}$ or
$\Psi_1 = \mathbf{.3717P_1 + .6015P_2 + .6015P_3 + .3717P_4}$

Lets look at another way to solve the secular determinant when X = -.618, also using Gaussian reduction, but in its pure matrix form (see section in paper about Gaussian Reduction for details, theory and proofs). The matrix for $\Psi_2 =$

```
      a₁ a₂ a₃ a₄   Setting X=-.618
Ψ₁  | X  1  0  0 |                      | -.618    1      0      0   |
Ψ₂  | 1  X  1  0 | = 0                  |   1   -.618    1      0   | = 0
Ψ₃  | 0  1  X  1 |                      |   0      1   -.618    1   |
Ψ₄  | 0  0  1  X |                      |   0      0      1   -.618 |
```

To get zero's under the -.618 in the first column:

```
       G1                        PSI                              PSI1
| 1   0    0    0  |     | -.618    1      0      0  |      | -.618    1      0      0   |
| 1 .618   0    0  |     |   1   -.618    1      0  |  =   |   0    .618   .618    0   |
| 0   0  .618   0  |     |   0      1   -.618    1  |      |   0    .618  -.382   .618 |
| 0   0    0  .618 |     |   0      0      1   -.618 |      |   0      0    .618  -.382 |
```

Using the newly calculated matrix, to get zero's under the +.618, multiply by:

```
       G1                        PSI1                             PSI2
| 1    0     0     0  |   | -.618    1      0      0  |   | -.618    1      0      0   |
| 0    1     0     0  |   |   0    .618   .618    0  | = |   0    .618   .618    0   |
| 0  .618 -.618    0  |   |   0    .618  -.382   .618 |   |   0      0    .618  -.382 |
| 0    0     0  -.618 |   |   0      0    .618  -.382 |   |   0      0   -.382   .236 |
```

Now we want a zero under the diagonal in the third column, so we multiply by:

$$
\begin{array}{c}
G2 \\
\begin{vmatrix}
1 & 0 & 0 & 0 \\
0 & 1 & 0 & 0 \\
0 & 0 & 1 & 0 \\
0 & 0 & -.382 & .618
\end{vmatrix}
\end{array}
\quad
\begin{array}{c}
PSI2 \\
\begin{vmatrix}
-.618 & 1 & 0 & 0 \\
0 & .618 & .618 & 0 \\
0 & 0 & .618 & -.382 \\
0 & 0 & -.382 & .236
\end{vmatrix}
\end{array}
=
\begin{array}{c}
PSI3 \\
\begin{vmatrix}
-.618 & 1 & 0 & 0 \\
0 & .618 & .618 & 0 \\
0 & 0 & .618 & -.382 \\
0 & 0 & 0 & 0
\end{vmatrix}
\end{array}
$$

We are done with the Gaussian reduction, the coefficients a_i are:

$a_1 = a_1$
$.618a_1 = a_2$
$a_2 = -a_3$
$.618a_3 = .382a_4$

Putting these all in terms of a_1 we have:

$a_1 = a_1$
$a_2 = .618a_1$
$a_3 = -.618a_1$
$a_4 = (.618)(-.618a_1)/.382 = -1a_1$

Before we finish, we are going to write out the MathCad +6 program for this Gaussian Reduction. G1 is the first Gaussian Multiplier, G2 is the second and G3 is the third. PSI is the Slater determinant, PSI1 is the determinant after being multiplied by G1. PSI2 is PSI1 after being multiplied by G2 and PSI3 is the solution, after multiplying PSI2 by G3.

MATHCAD +6 PROGRAM:

```
PSI1 = G1xPSI
PSI2 = G2xPSI1
PSI3 = G3xPSI2        The math is done for Ψ₁
a₁ = a₁
.618a₁ = a₂
a₂ = -a₃
.618a₃ = .382a₄
```

$PSI3 = G3xPSI2$ The math is done for Ψ_1
$a_1 = a_1$
$.618a_1 = a_2$
$a_2 = -a_3$
$.618a_3 = .382a_4$

Putting these all in terms of a_1 we have:

$a_1 = a_1$
$a_2 = .618a_1$
$a_3 = -.618a_1$
$a_4 = (.618)(-.618a_1)/.382 = -1a_1$

A2 = [1 1.618 -1.618 -1]   Starting Ψ_2
A2SQ = A2 x A2^T

ATWO = $\overrightarrow{(A2SQ)^{1/2}}$
ATWO = 1.662
PSITWO = ATWO^{-1} x A2
PSITWO = [.6015 .3717 -.3717 -.6015]

Note, in MathCad, the zero's are mostly in decimals of 10^{-5} or 10^{-6}. This is due to rounding error, just look at them as zero's.

Using any method (even Cramer's rule, if you prefer) let the remaining X's = .618 and 1.618 respectively. The four wave functions for butadiene can be expressed as:

Ψ_1 = .3717P$_1$ + .6015P$_2$ + .6015P$_3$ + .3717P$_4$ bonding
Ψ_2 = .6015P$_1$ + .3717P$_2$ - .3717P$_3$ - .6015P$_4$ bonding
Ψ_3 = .6015P$_1$ - .3717P$_2$ - .3717P$_3$ + .6015P$_4$ anti-bonding
Ψ_4 = .3717P$_1$ +-.6015P$_2$ + .6015P$_3$ - .3717P$_4$ anti-bonding

or

$$\Psi = \begin{vmatrix} .3717 & .6015 & .6015 & .3717 \\ .6015 & .3717 & -.3717 & -.6015 \\ .6015 & -.3717 & -.3717 & .6015 \\ .3717 & -.6015 & .6015 & -.3717 \end{vmatrix}$$

Diagramatically, these functions look like:

Ψ_1 is the most stable state where all the orbitals are reinforcing each other additively. In Ψ_2 the bonding between P_2 and P_3 cancels out at a node, reinforcing the bonding between P_1-P_2 and P_3-P_4. The relative strengths of these bonds can be found via the bond order. The bond order P_{ij} is defined as:

$$P_{ij} = \Sigma Na_i a_j$$

(Which is the same as for ethylene in the previous example).

$$P_{12} = (Na_1 a_2)\Psi_1 + (Na_1 a_2)\Psi_2$$
$$= 2(.6015)(.3717) + 2(.6015)(.3717) = .8942$$
$$P_{23} = (Na_2 a_3)\Psi_1 + (Na_2 a_3)\Psi_2$$
$$= = 2(.6015)(.6015) + 2(.3717)(-.3717) = .4473$$
$$P_{34} = P_{12} = .8942 \quad (9)$$

Remember we are using only the occupied orbitals here.

To calculate these values using matrices, we proceed as follows: $\Psi^T_{occ} \Psi_{occ}$.

$$2 \left(\begin{vmatrix} .3717 & .6015 \\ .6015 & .3717 \\ .6015 & -.3717 \\ .3717 & -.6015 \end{vmatrix} \begin{vmatrix} .3717 & .6015 & .6015 & .3717 \\ .6015 & .3717 & -.3717 & -.6015 \end{vmatrix} \right) =$$

$$\begin{vmatrix} 1 & 0.894 & 0 & -0.447 \\ 0.894 & 1 & 0.447 & 0 \\ 0 & 0.447 & 1 & 0.894 \\ -0.447 & 0 & 0.894 & 1 \end{vmatrix}$$

We ignore the values along the diagonal, and of course P_{14} and P_{41}. Note, the in-between values are not necessarily all zero's, it just happened to come out this way in this case.

These are defined as the mobile bond orders. To get the total bond order, the bond order of the σ bonds must be added. For simplicity, the hydrogen bonds will be defined as: $\sigma = 1$. Therefore, $P_{12} = P_{21} = P_{34} = P_{43} = 1.8942$; and $P_{23} = P_{32} = 1.4473$

The molecular diagram at this point is:

```
   1.8942     1.4473   1.8942
CH₂  -   CH  -  CH  -  CH₂
```

This shows us that there is more double bond character in the 1-2, 3-4 positions than in the 2-3 position. So in the butadiene molecule, the π bonds should favor $CH_2=CH-CH=CH_2$.

The Free-Valence Index [10] of a compound is the measure of the degree that the atoms in a molecule are bonded to adjacent atoms relative to their maximum bonding power. If particular atoms are not bonded much compared to this maximum, they are said to occupy especially reactive positions.

$$\text{Free-Valence Index } (\mathbf{F}) = \text{max. bonding energy} - \Sigma P_{ij}$$

EXERCISE: Using trimethylene methane, verify $\Sigma P_{ijMax} = 4.732$. (or 1.732 ignoring σ bonds)

$$CH_2 - C \begin{matrix} CH_2 \\ \\ CH_2 \end{matrix}$$

$A = A^T$ (Matrix is symmetric); $H_{11}=H_{22}=H_{33}=H_{44} = A$; $H_{12}=H_{21}=H_{23}=H_{32}=H_{34}=H_{43} = B$ and $H_{13}=H_{31}=H_{14}=H_{41}=H_{24}=H_{42} = 0$;

The Secular determinant is:
$$\begin{matrix} 1 & 2 & 3 \\ C & - C & - C \\ & | & \\ & C\ 4 & \end{matrix}$$

$$\begin{vmatrix} A-E & B & 0 & 0 \\ B & A-E & B & B \\ 0 & B & A-E & 0 \\ 0 & B & 0 & A-E \end{vmatrix} = 0$$ Letting $X=(A-E)/B$, the determinant becomes $$\begin{vmatrix} X & 1 & 0 & 0 \\ 1 & X & 1 & 1 \\ 0 & 1 & X & 0 \\ 0 & 1 & 0 & X \end{vmatrix} = 0$$

The determinant is equal to:

$$X \begin{vmatrix} X & 1 & 1 \\ 1 & X & 0 \\ 1 & 0 & X \end{vmatrix} - 1 \begin{vmatrix} 1 & 1 & 1 \\ 0 & X & 0 \\ 0 & 0 & X \end{vmatrix} = X(1 \begin{vmatrix} 1 & 1 \\ X & 0 \end{vmatrix} + X \begin{vmatrix} X & 1 \\ 1 & X \end{vmatrix}) - 1(1 \begin{vmatrix} X & 0 \\ 0 & X \end{vmatrix}) =$$

$$X[(-X)+X(X^2-1)]-X^2 = X(-X+X^3-X)-X^2 = -X^2+X^4-X^2-X^2 = X^4-3X^2 = \mathbf{X^2(X^2-3)=0}.$$

We don't need any special math to determine the roots of this equation. From inspection, we can easily see $X = 0$, 0, $\sqrt{2}$, $-\sqrt{3}$. Therefore:

$E_1 = A+\sqrt{3}B$

$E_2 = A$

$E_3 = A$

$E_4 = A-\sqrt{3}B$

To solve for the coefficients a_i, let $X = -\sqrt{3} = -1.732$, putting this value for X into the secular determinant, we get:

$$\begin{vmatrix} -1.732 & 1 & 0 & 0 \\ 1 & -1.732 & 1 & 1 \\ 0 & 1 & -1.732 & 0 \\ 0 & 1 & 0 & -1.732 \end{vmatrix} = 0$$

$1/1.732 R_1 + R_2$

$$\begin{vmatrix} -1 & .577 & 0 & 0 \\ 0 & -1.155 & 1 & 1 \\ 0 & 1 & -1.732 & 0 \\ 0 & 1 & 0 & -1.732 \end{vmatrix} = 0$$

$1/1.732 (R_3 + R_2)$

$$\begin{vmatrix} -1 & .577 & 0 & 0 \\ 0 & -1.155 & 1 & 1 \\ 0 & .577 & -1 & 0 \\ 0 & .577 & 0 & -1 \end{vmatrix} = 0$$

$-R_3 + R_1$

$$\begin{vmatrix} -1 & 0 & 1 & 0 \\ 0 & -1.155 & 1 & 1 \\ 0 & .577 & -1 & 0 \\ 0 & .577 & 0 & -1 \end{vmatrix} = 0$$

$R_3 + R_2$

$$\begin{vmatrix} -1 & 0 & 1 & 0 \\ 0 & -.578 & 0 & 1 \\ 0 & .577 & -1 & 0 \\ 0 & .577 & 0 & -1 \end{vmatrix} = 0$$

$R_2 + R_4$

$$\begin{vmatrix} -1 & 0 & 1 & 0 \\ 0 & -.578 & 0 & 1 \\ 0 & .577 & -1 & 0 \\ 0 & 0 & 0 & 0 \end{vmatrix} = 0$$

$R_3 + R_2$

$$\begin{vmatrix} -1 & 0 & 1 & 0 \\ 0 & 0 & -1 & 1 \\ 0 & .577 & -1 & 0 \\ 0 & 0 & 0 & 0 \end{vmatrix} = 0$$

Note: Even though the problem is solved in the step above this one, it is necessary to reduce the determinant to its simplest terms to obtain the correct Ψ function.

$a_1 = a_3$

$a_3 = a_4$

$.577 a_2 = a_3$

To be different, I am going to put these values in terms of a_3 instead of a_1, just to show how to do it and that we don't have to do it for a_1 all the time.

$a_3 = a_1$
$a_3 = a_4$
$.577a_2 = a_3$; $1.732a_3 = a_2$
$a_3 = a_3$

Therefore, $(\Psi^T\Psi)^{1/2} = (1^1 + (3^{1/2})^2 \ 1^2 + 1^2)^{1/2} = (6)^{1/2}$ and

$\Psi = 1/(6)^{1/2}P_1 + (3)^{1/2}/(6)^{1/2}P_3 \ 1/(6)^{1/2}P_1 + 1/(6)^{1/2}P_4$
$\Psi = .4082P_1 + .4082P_2 + .7071P_1 + .4082P_4$

$$P_{ij} = \Sigma Na_i a_j = \Psi^T_{Occ}\Psi_{Occ} = \begin{vmatrix} .4082 \\ .7071 \\ .4082 \\ .4082 \end{vmatrix} [.4082 \ .7071 \ .4082 \ .4082] =$$

Note: With this method $(\Psi^T_{Occ}\Psi_{Occ})$, the value of N is automatically taken care of in the math, so we do not need to include it in our mathematics. To find the maximum energy, we must also include the values of the σ bonds for $\sigma_{12} = \sigma_{23} = \sigma_{24} = 1$, or for the total σ energy, we have $\sigma_{12} + \sigma_{23} + \sigma_{24} = 3$, and the total energy $\Sigma P_{ijMax} = \Psi^T\Psi + \sigma_{ij}$.

The MathCad program and mathematics are as follows:

$$\Psi\text{TRN} = \begin{vmatrix} .4082 \\ .7071 \\ .4082 \\ .4082 \end{vmatrix} \quad \Psi = [.4082 \ .7071 \ .4082 \ .4082] \quad \text{ONE4} = [1\ 1\ 1\ 1]$$

$$\text{FOUR1} = \begin{vmatrix} 1 \\ 1 \\ 1 \\ 1 \end{vmatrix}$$

$$\text{LOGPij} = \begin{vmatrix} -100 & 0 & -100 & -100 \\ 0 & -100 & 0 & 0 \\ -100 & 0 & -100 & -100 \\ -100 & 0 & -100 & -100 \end{vmatrix} \quad P_{ij} = \begin{vmatrix} 0 & 1 & 0 & 0 \\ 1 & 0 & 1 & 1 \\ 0 & 1 & 0 & 0 \\ 0 & 1 & 0 & 0 \end{vmatrix}$$

Where the one's in P_{ij} occupy the positions for $P_{12} = P_{21} = P_{23} = P_{32} = P_{24} = P_{42} = 1$. All other values in the matrix are identical to zero. This way, in a one to one multiplication, we retrieve only the values in these positions and ignore the rest. Compare with the Slater determinant, these values all occupy the positions where the one's are located, with X=0. This is the easiest way to set up the P_{ij} and LOGPij matrix templates.

Note: The log of 0 is undefined, so to be able to use it in a problem, I give 0 a default value of 10^{-100}. MathCad can only go as low as 10^{-336}, so the default value has to be at least half that value, because as we add 0 + 0, we will get 10^{-200} which the computer can handle. I am not including the HP-48G program here, mainly because I cannot multiply the matrices one on one like I can in MathCad.

$\Psi SQ := \Psi TRN \cdot \Psi$

$$\Psi SQ = \begin{bmatrix} 0.167 & 0.289 & 0.167 & 0.167 \\ 0.289 & 0.5 & 0.289 & 0.289 \\ 0.167 & 0.289 & 0.167 & 0.167 \\ 0.167 & 0.289 & 0.167 & 0.167 \end{bmatrix}$$

$LOC\Psi SQ := \overrightarrow{\log(\Psi SQ)}$

Here we take the log of each individual element in the matrix ΨSQ.

$Pij := \overrightarrow{10^{(LOGPij + LOG\Psi SQ)}}$

Here we add the two log matrices (in effect multiplying each element one on one) and take the anti-log of each element to return the numbers to our real number system.

$$Pij = \begin{bmatrix} 0 & 0.289 & 0 & 0 \\ 0.289 & 0 & 0.289 & 0.289 \\ 0 & 0.289 & 0 & 0 \\ 0 & 0.289 & 0 & 0 \end{bmatrix}$$

$SUMPij := ONE4 \cdot Pij \cdot FOUR1$ Here we sum all the values in the matrix.

$SUMPij = 1.732$

To complete the calculation: $\Sigma P_{ijMax} = \Psi^T\Psi + \sigma_{ij} = 1.732 + 3 = $ **4.732**.

Because of having to now subtract the σ bonds during the calculations, we do not furthermore use the value of 4.732, we just use the 1.732 value (SUMPij) because it is much easier to do. The value 4.732 is for advanced quantum chemistry users.

So for butadiene, $F_i = 1.732 - \Sigma P_{ij}$

$F_1 = F_4 = 1.732 - .8942 = $ **.8378**
$F_2 = F_3 = 1.732 - (.8942 + .4473) = $ **.391**

Or to do this totally by matrices, we proceed as follows:

$F_i = [1.73]_{4,1} - ([P]_{4,4}[1]_{1,4})$

(Remembering to use only those values that correspond to the positions of the one's in the original Slater determinant).

$$\begin{vmatrix} 1.732 \\ 1.732 \\ 1.732 \\ 1.732 \end{vmatrix} - \begin{vmatrix} 0 & 0.894 & 0 & 0 \\ 0.894 & 0 & 0.447 & 0 \\ 0 & 0.447 & 0 & 0.894 \\ 0 & 0 & 0.894 & 0 \end{vmatrix} \begin{vmatrix} 1 \\ 1 \\ 1 \\ 1 \end{vmatrix} = \begin{vmatrix} 1.732 \\ 1.732 \\ 1.732 \\ 1.732 \end{vmatrix} - \begin{vmatrix} .8942 \\ 1.341 \\ 1.341 \\ .894 \end{vmatrix} = \begin{vmatrix} .8378 \\ .3910 \\ .3910 \\ .8378 \end{vmatrix}$$

Our molecular diagram now looks like:

The charge distribution q_i (the deviation from the 'normal' electron density at a given π bonded atom) is obtained by summing the electron's probabilities corresponding to the contribution of that particular AO (Atomic Orbital) to the other various **occupied** orbitals. Corrections have to be made concerning formal charges resulting from the σ bonds to obtain the overall charge. Therefore, if a carbon forms 3 σ bonds and is π bonded, it will be neutral if there is an average of one electron in it's 2p π bonded orbitals.

If q_i is defined as the deviation from neutrality of such a carbon, then

$$q_i = 1.000 - \Sigma Na_i{}^2\Psi_I \quad {}^9$$

For butadiene, $q_1 = q_4$

$\begin{aligned} q &= 1.000 - \Sigma Na_1{}^2\Psi_I - \Sigma Na_1{}^2\Psi_2 = \\ &1.000 - 2(.3717)^2 - 2(.6015)^2 = \\ &1.000 - 2(.1382) - 2(.3618) = 1.000 - .2763 - .7236 = 0 \end{aligned}$

for $q_2 = q_3$ $q = 1.000 - 2(.6015)^2 - 2(.3717)^2 = 0$

Now this seems quite simple to do by hand, much more simple than the following MathCad programs. But suppose we were dealing with a protein that has a matrix of dimensions 100,000x100,000, then we would need a computer because it would be impossible to add these values by hand. I am going to show two ways, the first by calculating each value of q separately, then by using the method of Nested Arrays. Remember, I'm just going to use the Ψ_{Occ} matrix here, not the whole Ψ matrix.

MATHCAD +6 PROGRAM # 1

$$B := \begin{pmatrix} .3717 & .6015 \\ .6015 & .3717 \\ .6015 & -.6015 \\ .3717 & -.3717 \end{pmatrix} \qquad A1 := \begin{pmatrix} 1 & 0 \\ 0 & 0 \end{pmatrix} \qquad A2 := \begin{pmatrix} 0 & 0 \\ 0 & 1 \end{pmatrix} \qquad TWO1 := \begin{pmatrix} 1 \\ 1 \end{pmatrix}$$

$ONE4 := (1 \quad 1 \quad 1 \quad 1) \qquad FOUR1 := ONE4^T$

Solving for q_1: (Am using only the values for the occupied orbitals.)

$B1 := B \cdot A1 \cdot TWO1$

$$B1 := \begin{pmatrix} .3717 & .6015 \\ .6015 & .3717 \\ .6015 & -.6015 \\ .3717 & -.3717 \end{pmatrix} \cdot \begin{pmatrix} 1 & 0 \\ 0 & 0 \end{pmatrix} \cdot \begin{pmatrix} 1 \\ 1 \end{pmatrix} \qquad B1 = \begin{pmatrix} 0.372 \\ 0.602 \\ 0.602 \\ 0.372 \end{pmatrix}$$

$B1SQ := \overrightarrow{B1^2}$

---

[9]   Liberles, pp.59

$$B1SQ = \begin{pmatrix} 0.138 \\ 0.362 \\ 0.362 \\ 0.138 \end{pmatrix}$$

$q_1 := 1.000 - ONE4 \cdot B1SQ$

$q_1 = 0$

Now to solve for q_2:

$B2 := B \cdot A2 \cdot TWO1$

$$B2 := \begin{pmatrix} .3717 & .6015 \\ .6015 & .3717 \\ .6015 & -.6015 \\ .3717 & -.3717 \end{pmatrix} \cdot \begin{pmatrix} 0 & 0 \\ 0 & 1 \end{pmatrix} \cdot \begin{pmatrix} 1 \\ 1 \end{pmatrix} \qquad B2 = \begin{pmatrix} 0.602 \\ 0.372 \\ -0.602 \\ -0.372 \end{pmatrix}$$

$B2SQ := \overrightarrow{B2^2}$

$$B2SQ = \begin{pmatrix} 0.362 \\ 0.138 \\ 0.362 \\ 0.138 \end{pmatrix}$$

$q_2 := 1.000 - ONE4 \cdot B2SQ$

$q_2 = 0$

(See part 2, Bond order for Acrolein for simpler method)

NESTED ARRAYS

Here we will compute q_1 and q_2 at the same time using nested arrays:

Use the variable definitions above plus: (See below how we obtain NESTARRAYB)

$$EIGHT1 := \begin{bmatrix} 1 \\ 1 \\ 1 \\ 1 \\ 1 \\ 1 \\ 1 \\ 1 \end{bmatrix} \qquad NESTARRAYB := \begin{bmatrix} .3717 \\ .6015 \\ .6015 \\ .3717 \\ .6015 \\ .3717 \\ -.6015 \\ -.3717 \end{bmatrix} \qquad DB1 := \begin{pmatrix} 1 & 1 & 1 & 1 & 0 & 0 & 0 & 0 \\ 0 & 0 & 0 & 0 & 1 & 1 & 1 & 1 \end{pmatrix}$$

$B1 := B \cdot A1 \cdot TWO1$

$B2 := B \cdot A2 \cdot TWO1$

$$q1q2 := 1.000 - DB1 \cdot diag\left(\overrightarrow{NESTARRAYB^2}\right) \cdot EIGHT1$$

$$q1q2 := 1 - \begin{pmatrix} 1 & 1 & 1 & 1 & 0 & 0 & 0 & 0 \\ 0 & 0 & 0 & 0 & 1 & 1 & 1 & 1 \end{pmatrix} \cdot \overrightarrow{\begin{bmatrix} 0.372 & 0 & 0 & 0 & 0 & 0 & 0 & 0 \\ 0 & 0.602 & 0 & 0 & 0 & 0 & 0 & 0 \\ 0 & 0 & 0.602 & 0 & 0 & 0 & 0 & 0 \\ 0 & 0 & 0 & 0.372 & 0 & 0 & 0 & 0 \\ 0 & 0 & 0 & 0 & 0.602 & 0 & 0 & 0 \\ 0 & 0 & 0 & 0 & 0 & 0.372 & 0 & 0 \\ 0 & 0 & 0 & 0 & 0 & 0 & -0.602 & 0 \\ 0 & 0 & 0 & 0 & 0 & 0 & 0 & -0.372 \end{bmatrix}^2} \cdot \begin{bmatrix} 1 \\ 1 \\ 1 \\ 1 \\ 1 \\ 1 \\ 1 \\ 1 \end{bmatrix}$$

$$q1q2 := 1 - \begin{pmatrix} 1 & 1 & 1 & 1 & 0 & 0 & 0 & 0 \\ 0 & 0 & 0 & 0 & 1 & 1 & 1 & 1 \end{pmatrix} \cdot \begin{bmatrix} 0.138 & 0 & 0 & 0 & 0 & 0 & 0 & 0 \\ 0 & 0.362 & 0 & 0 & 0 & 0 & 0 & 0 \\ 0 & 0 & 0.362 & 0 & 0 & 0 & 0 & 0 \\ 0 & 0 & 0 & 0.138 & 0 & 0 & 0 & 0 \\ 0 & 0 & 0 & 0 & 0.362 & 0 & 0 & 0 \\ 0 & 0 & 0 & 0 & 0 & 0.138 & 0 & 0 \\ 0 & 0 & 0 & 0 & 0 & 0 & 0.362 & 0 \\ 0 & 0 & 0 & 0 & 0 & 0 & 0 & 0.138 \end{bmatrix} \cdot \begin{bmatrix} 1 \\ 1 \\ 1 \\ 1 \\ 1 \\ 1 \\ 1 \\ 1 \end{bmatrix}$$

$$q1q2 = \begin{pmatrix} 0 \\ 0 \end{pmatrix}$$

First we must compute the half-multiplication the hard way, mainly because there is no computer program that can accomplish this (programmers don't even know this mathematical operator exists yet). The neat thing about this operator is that there is a regular matrix method that can accomplish the same results, but with a lot more mathematical steps, so instead of the elegant [b] o [1], we must use the following method:

B1 = [B][A1][TWO1]
B2 = [B][A2][TWO1]

Then

$$
[B] \; o \; [1] = \begin{vmatrix} .3717 & .6015 \\ .6015 & .3717 \\ .6015 & -.6015 \\ .3717 & -.3717 \end{vmatrix} o \begin{vmatrix} 1 \\ 1 \\ 1 \\ 1 \end{vmatrix} = \begin{Vmatrix} \begin{vmatrix} .3717 \\ .6015 \\ .6015 \\ .3717 \end{vmatrix} & \begin{vmatrix} .6015 \\ .3717 \\ -.6015 \\ -.3717 \end{vmatrix} \end{Vmatrix}
$$

Now to transform this so we can use it in our regular mathematics, we transpose the nested array, remembering that when we transpose, we transpose the matrices (not the elements) and drop the inner brackets (we partition the Nested Array).

The equation now to solve is: **[DB1]([NESTARRAYB]o[I8])²** ([DB1] is used to keep the solutions q1 and q2 separate)

$$
\text{NESTARRAYB} = \begin{vmatrix} .3717 \\ .6015 \\ .6015 \\ .3717 \\ .6015 \\ .3717 \\ -.6015 \\ -.3717 \end{vmatrix}
$$

PROGRAM:

DIAGONALNESTARRAYB   = diag(NESTARRAYB)
DIAGONALNESTARRAYBSQ = [DIAGONALNESTARRAYB]²
q = 1.000 - [DB1][DIAGONALNESTARRAYBSQ][EIGHT1] = 0

or due to rounding errors is equal to:

$$
q = \begin{vmatrix} 7.372EEX-5 \\ 7.372EEX-5 \end{vmatrix} = \begin{vmatrix} 0 \\ 0 \end{vmatrix}
$$

We have completed our basic calculations. The complete molecular diagram for butadiene is:

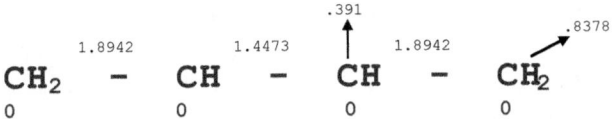

The bond order on the 1,4 carbons deviates the most from maximum, they are predicted to be the most reactive positions. Since $q_i = 0$, butadiene is designated as a molecule with a self-consistent field.

CYCLOBUTADIENE

We shall now do the molecule cyclobutadiene. We are going to consolidate the math to save space, for explanation of the steps and computations, see the solution for butadiene above. We start by numbering the carbons, which we do in consecutive order, starting from left and counting to the right.

C_1 —— C_2 $H_{11} = H_{22} = H_{33} = H_{44} = A$

| | $H_{12} = H_{21} = H_{23} = H_{32} = H_{34} = H_{43} = H_{14} = H_{41} = B$

C_4 —— C_3 $H_{13} = H_{31} = H_{24} = H_{42} = 0$

The secular determinant is:

$$\begin{vmatrix} X & 1 & 0 & 1 \\ 1 & X & 1 & 0 \\ 0 & 1 & X & 1 \\ 1 & 0 & 1 & X \end{vmatrix} = 0$$

And its roots (determined by Cramer's rule) are:

$X^4 - 4X^2 = X^2(X^2 - 4)$; and the roots are $(0, 0, +2, -2)$

$X = -2;\ E_1 = A + 2B$

$X = 0;\ E_2 = A$

$X = 0;\ E_3 = A$

$X = 2;\ E_4 = A - 2B$

At X=2:

$$\begin{vmatrix} -2 & 1 & 0 & 1 \\ 1 & -2 & 1 & 0 \\ 0 & 1 & -2 & 1 \\ 1 & 0 & 1 & -2 \end{vmatrix} \begin{vmatrix} 1 & 0 & 1 & -2 \\ 0 & 1 & -2 & 1 \\ 0 & -2 & 0 & 2 \\ 1 & 0 & 2 & -2 \end{vmatrix} \begin{vmatrix} 1 & 0 & 1 & -2 \\ 0 & 1 & -2 & 1 \\ 0 & -2 & 0 & 2 \\ 0 & 0 & 2 & -3 \end{vmatrix} \begin{vmatrix} 1 & 0 & 0 & 0 \\ 0 & 1 & 0 & 0 \\ 0 & 0 & 1 & 0 \\ 0 & 0 & 0 & 1 \end{vmatrix}$$

$a_1 = a_1$
$a_1 = a_2$
$a_1 = a_3$
$a_1 = a_4$

Normalize: $(1^2+1^2+1^2+1^2)^{1/2} = (4)^{1/2} = 2;\ a_1 = 1/2$ and $\Psi_1 = 1/2$ $(P_1+P_2+P_3+P_4)$

At X=0:

$$\begin{vmatrix} 0 & 1 & 0 & 1 \\ 1 & 0 & 1 & 0 \\ 0 & 1 & 0 & 1 \\ 1 & 0 & 1 & 0 \end{vmatrix} \begin{vmatrix} 1 & 0 & 1 & 0 \\ 0 & 1 & 0 & 1 \\ 0 & 0 & 0 & 0 \\ 0 & 0 & 0 & 0 \end{vmatrix}$$

$a1 = -a3$ Putting in terms of a1: $a1 = a1$

$a2 = -a4$ $a2 = 0$

                                       $a1 = -a3$

                                       $a4 = 0$

Normalize: $(1^2+(-1^2))^{1/2} = (2)^{1/2} = 2;\ a_2 = a_3 = 1/\sqrt{2}$ and Ψ_2 and $\Psi_3 = 1/\sqrt{2}(P_1-P_3)$

310

At X=2:

$$\begin{vmatrix} 2 & 1 & 0 & 0 \\ 1 & 2 & 1 & 0 \\ 0 & 1 & 2 & 1 \\ 1 & 0 & 1 & 2 \end{vmatrix} \quad \begin{vmatrix} 1 & 0 & 1 & 2 \\ 0 & 1 & 2 & 1 \\ 2 & 1 & 0 & 0 \\ 1 & 2 & 1 & 0 \end{vmatrix} \quad \begin{vmatrix} 1 & 0 & 0 & 0 \\ 0 & -1 & 0 & 0 \\ 0 & 0 & 1 & 0 \\ 0 & 0 & 0 & -1 \end{vmatrix} \quad \begin{matrix} a_1 = a_1 \\ a_1 = -a_2 \\ a_1 = a_3 \\ a_1 = -a_4 \end{matrix}$$

Normalize: $(1^2 + (-1)^2 + 1^2 + (-1)^2)^{1/2} = (4)^{1/2} = 2$; $a_1 = 1/2$ and
$\Psi_4 = 1/2(P_1 + P_2 + P_3 + P_4)$

The wave functions are:

$\Psi_1 = 1/2(P_1 + P_2 + P_3 + P_4)$
$\Psi_2 = 1/\sqrt{2}(P_1 - P_3)$
$\Psi_3 = 1/\sqrt{2}(P_2 - P_4)$
$\Psi_4 = 1/2(P_1 + P_2 + P_3 + P_4)$

And in matrix form:

$$\Psi_I = \begin{vmatrix} .5 & .5 & .5 & .5 \\ 1.4142 & 0 & -1.4142 & 0 \\ 0 & 1.4142 & 0 & -1.4142 \\ .5 & .5 & .5 & .5 \end{vmatrix}$$

Bond Order: $P_{12} = P_{23} = P_{34} = P_{14} = 2(1/2)(1/2)\Psi_1 + 2(1/\sqrt{2})(0)\Psi_2 + 2(1/\sqrt{2})(0)\Psi_3 = 1/2$

F_i: $a_{12} = a_{14}$; $\quad P_{12} + P_{14} = (1/2) + (1/2) = 1$, $\sigma = 3$; $\Sigma P_{ij} = 1 + 3 = 4$. F = 4.7318 - 4 = .7318
   or $\Sigma P_{ij} = 1$, F = 1.7318 - 1 = .7318 ignoring the σ bonds.

There is no need to compute q_i since cyclobutadiene is a molecule which contains a self-consistent field, thus $q_i = 0$.

The molecular diagram is:

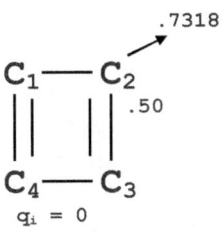

.7318

C_1 — C_2

.50

C_4 — C_3

$q_i = 0$

This molecule has never been isolated, presumably due to it's instability in it's resonance stabilization.[10]

E_1 (BUTADIENE)   = A + 1.618B
E_1 (CYCLOBUTADIENE) = A + 2B

Because E_1 (BUTADIENE) has a lower energy than E_1 (CYCLOBUTADIENE), it would be predicted to be more stable than cyclobutadiene.

---

[10]   Lederle, pp. 98-102 Note: I wrote this paper in a shorter form 20 years ago & didn't write down the book or publisher.

OZONE

Lets look at a non-carbon molecule ozone and see if we can mathematically determine if ozone is a linear molecule, or if all it's bonds are connected forming a triangular structure. We'll look at the linear model first.

```
1   2   3
O — O — O
```

$H_{11}=H_{22}=H_{33} = A+2B$

$H_{12}=H_{21}=H_{23}=H_{32} = \sqrt{2}B$

$H_{13}=H_{31} = 0$

$S_{ij} = \quad i=j = 1$

$\qquad\quad i\neq j = 0$

And

$$\begin{vmatrix} A+2B-E & \sqrt{2}B & 0 \\ \sqrt{2}B & A+2B-E & \sqrt{2}B \\ 0 & \sqrt{2}B & A+2B-E \end{vmatrix} = 0$$

Let $X = (A+2B-E)/\sqrt{2}B$, then the Slater determinant becomes

$$\begin{vmatrix} X & 1 & 0 \\ 1 & X & 1 \\ 0 & 1 & X \end{vmatrix} = 0$$

The determinant is equal to: $X(\begin{vmatrix} X & 1 \\ 1 & X \end{vmatrix}) - 1(\begin{vmatrix} 1 & 1 \\ 0 & X \end{vmatrix}) =$

$$X(X^2-1)-X = X^3-X-X = X^3-2X = X(X^2-2)$$

and the roots are (by inspection): $(0, \sqrt{2}, -\sqrt{2})$

$E_1 = (A+2B-E)/\sqrt{2}B = -\sqrt{2} = (A+2B-E)= -2B; \quad E_1 = A+4B$

$E_2 = (A+2B-E)/\sqrt{2}B = 0; \quad E_2 = A+2B$

$E_3 = (A+2B-E)/\sqrt{2}B = \sqrt{2}; =(A+2B-E)= 2B; \quad E_3 = A+2B-2B = A$

$E_1 = A+4B \qquad \underline{\quad\uparrow\downarrow\quad}$

$E_2 = A+2B \qquad \underline{\quad\uparrow\quad}$

$E_3 = A \qquad\qquad \underline{\qquad\qquad}$

Now we'll look at the triangular model for ozone:

```
        1
        O
       / \
  3 O ——— O 2
```

$H_{11}=H_{22}=H_{33} = A+2B$

$H_{12}=H_{21}=H_{23}=H_{32}=H_{13}=H_{31} = \sqrt{2}B$

$H_{13}=H_{31} = 0$

$S_{ij} = \quad i=j = 1; \quad i\neq j = 0$

and

$$\begin{vmatrix} A+2B-E & \sqrt{2}B & \sqrt{2}B \\ \sqrt{2}B & A+2B-E & \sqrt{2}B \\ \sqrt{2}B & \sqrt{2}B & A+2B-E \end{vmatrix}$$

Let X = (A+2B-E)/ $\sqrt{2}$B, then the Slater determinant becomes:

$$\begin{vmatrix} X & 1 & 1 \\ 1 & X & 1 \\ 1 & 1 & X \end{vmatrix} = 0;$$ The equation for this determinant is X^3-3X+2; The roots are (1, 1, -2).

E_1 = (A+2B-E)/ $\sqrt{2}$B = -2 = (A+2B-E)= -2$\sqrt{2}$B; E_1 = A+4.83B
E_2 = (A+2B-E)/ $\sqrt{2}$B = 1; E_2 = A+(2-$\sqrt{2}$)B = A+.586B

E_1 = A+4.83B —↑↓—

E_2 = A+.568B —↑—

E_3 = A+.586B —↑—

    Because E_1 is smaller in the linear form of ozone, it is predicted to be the most stable. Just for the heck of it, let's compute the Ψ function for ozone. The Slater determinant is:

Normalizing: $(1^2+0+(-1)^2)^{1/2}$ = $\sqrt{2}$ and Ψ_3 = 1/$\sqrt{2}$ (P$_1$-P$_3$) = [.7071 0 -.7071]

$$\begin{vmatrix} X & 1 & 0 \\ 1 & X & 1 \\ 0 & 1 & X \end{vmatrix} =0 \; ; \text{ At X=}\sqrt{2} \text{ we get} \begin{vmatrix} 1.41 & 1 & 0 \\ 1 & 1.41 & 1 \\ 0 & 1 & 1.41 \end{vmatrix} = \overset{R_1-\sqrt{2}R_2}{\begin{vmatrix} \sqrt{2} & 1 & 0 \\ 0 & -1 & -\sqrt{2} \\ 0 & 1 & \sqrt{2} \end{vmatrix}} \overset{R_2+R_3}{\begin{vmatrix} \sqrt{2} & 1 & 0 \\ 0 & -1 & -\sqrt{2} \\ 0 & 0 & 0 \end{vmatrix}}$$

$\sqrt{2}a_1$ = $-a_2$; a_1 = -.707a_2
a_2 = -$\sqrt{2}a_3$ (-$\sqrt{2}a_1$) = -$\sqrt{2}a_3$; a_1=a_3
a_1 = a_1 a_1 = a_1

Normalizing: $(1^2+(-\sqrt{2})^2+1^2)^{1/2}$ = $\sqrt{4}$ = 2; a = 1/2 ; Ψ_2 = 1/2 (P$_1$-$\sqrt{2}$ P$_2$+P$_3$) = [.5 -.7071 .5]

For X = $-\sqrt{2}$

$$\begin{vmatrix} X & 1 & 0 \\ 1 & X & 1 \\ 0 & 1 & X \end{vmatrix} =0 \; ; \text{ At X=}-\sqrt{2} \text{ we get} \begin{vmatrix} 1.41 & 1 & 0 \\ 1 & 1.41 & 1 \\ 0 & 1 & 1.41 \end{vmatrix} = \overset{1/\sqrt{2}R_1+R_2}{\begin{vmatrix} -1 & 1/\sqrt{2} & 0 \\ 1 & -\sqrt{2} & 1 \\ 0 & 1 & -\sqrt{2} \end{vmatrix}} \overset{\sqrt{2}R_2+R_1}{\begin{vmatrix} -1 & 0 & 1 \\ 0 & -1 & -\sqrt{2} \\ 0 & 0 & 0 \end{vmatrix}}$$

a_1=a_3
a_2=$\sqrt{2}a_3$

Putting in terms of a_1:

$a_1 = a_1$
$a_1 = 1/\sqrt{2}a_2$
$a_2 = \sqrt{2}a_3$

Normalizing: $(1^2 + (\sqrt{2})^2 + 1^2)^{1/2} = \sqrt{4} = 2$; $a = 1/2$; $\Psi_1 = 1/2(P_1 + \sqrt{2}$
$P_2 + P_3) = [.5 \ .7071 \ .5]$

$\Psi_1 = 1/2(P_1 + \sqrt{2}P_2 + P_3)$
$\Psi_2 = 1/2(P_1 - \sqrt{2}P_2 + P_3)$
$\Psi_3 = 1/\sqrt{2}(P_1 - P_3)$

or

$$\Psi = \begin{vmatrix} .5 & .7071 & .5 \\ .5 & -.7071 & .5 \\ .7071 & 0 & -.7071 \end{vmatrix}$$

Let's use the matrix solution of Gaussian Reduction to solve all three states of ozone at the same time. We will solve them in the order of $X = -\sqrt{2}$, 0 and $\sqrt{2}$. But since putting 0 in the diagonal for $X = 0$ gives us a diagonal of zero, I am going to switch rows two and one. This makes the calculations easier. Let B equal the matrix of matrices where the roots replace X, and let A be the first Gaussian Reduction matrix needed to get zero's under the values of the elements in the topmost corner of the diagonal.

$$A := \begin{bmatrix} 1 & 0 & 0 & 0 & 0 & 0 & 0 & 0 & 0 \\ 1 & \sqrt{2} & 0 & 0 & 0 & 0 & 0 & 0 & 0 \\ 0 & 0 & \sqrt{2} & 0 & 0 & 0 & 0 & 0 & 0 \\ 0 & 0 & 0 & 1 & 0 & 0 & 0 & 0 & 0 \\ 0 & 0 & 0 & 0 & 1 & 0 & 0 & 0 & 0 \\ 0 & 0 & 0 & 0 & 0 & 1 & 0 & 0 & 0 \\ 0 & 0 & 0 & 0 & 0 & 0 & 1 & 0 & 0 \\ 0 & 0 & 0 & 0 & 0 & 0 & 1 & -\sqrt{2} & 0 \\ 0 & 0 & 0 & 0 & 0 & 0 & 0 & 0 & -\sqrt{2} \end{bmatrix}$$

$$B := \begin{bmatrix} -\sqrt{2} & 1 & 0 & 0 & 0 & 0 & 0 & 0 & 0 \\ 1 & -\sqrt{2} & 1 & 0 & 0 & 0 & 0 & 0 & 0 \\ 0 & 1 & -\sqrt{2} & 0 & 0 & 0 & 0 & 0 & 0 \\ 0 & 0 & 0 & 1 & 0 & 1 & 0 & 0 & 0 \\ 0 & 0 & 0 & 0 & 1 & 0 & 0 & 0 & 0 \\ 0 & 0 & 0 & 0 & 1 & 0 & 0 & 0 & 0 \\ 0 & 0 & 0 & 0 & 0 & 0 & \sqrt{2} & 1 & 0 \\ 0 & 0 & 0 & 0 & 0 & 0 & 1 & \sqrt{2} & 1 \\ 0 & 0 & 0 & 0 & 0 & 0 & 0 & 1 & \sqrt{2} \end{bmatrix}$$

Then multiplying AxB will give zeros below each of the a_{11} elements. Note the matrix for X=0 already has all zero's under it, so I just multiply by identity and leave it unchanged for this computation.

A x B = C

$$
C = \begin{bmatrix}
-1.414 & 1 & 0 & 0 & 0 & 0 & 0 & 0 & 0 \\
0 & -1 & 1.414 & 0 & 0 & 0 & 0 & 0 & 0 \\
0 & 1.414 & -2 & 0 & 0 & 0 & 0 & 0 & 0 \\
0 & 0 & 0 & 1 & 0 & 1 & 0 & 0 & 0 \\
0 & 0 & 0 & 0 & 1 & 0 & 0 & 0 & 0 \\
0 & 0 & 0 & 0 & 1 & 0 & 0 & 0 & 0 \\
0 & 0 & 0 & 0 & 0 & 0 & 1.414 & 1 & 0 \\
0 & 0 & 0 & 0 & 0 & 0 & 0 & -1 & -1.414 \\
0 & 0 & 0 & 0 & 0 & 0 & 0 & -1.414 & -2
\end{bmatrix}
$$

Now we want to make the elements under the a_{22} positions of the matrices zero, so we multiply C by the matrix A2:

$$
A2 := \begin{bmatrix}
1 & 0 & 0 & 0 & 0 & 0 & 0 & 0 & 0 \\
0 & 1 & 0 & 0 & 0 & 0 & 0 & 0 & 0 \\
0 & \sqrt{2} & 1 & 0 & 0 & 0 & 0 & 0 & 0 \\
0 & 0 & 0 & 1 & 0 & 0 & 0 & 0 & 0 \\
0 & 0 & 0 & 0 & 1 & 0 & 0 & 0 & 0 \\
0 & 0 & 0 & 0 & 1 & -1 & 0 & 0 & 0 \\
0 & 0 & 0 & 0 & 0 & 0 & 1 & 0 & 0 \\
0 & 0 & 0 & 0 & 0 & 0 & 0 & 1 & 0 \\
0 & 0 & 0 & 0 & 0 & 0 & 0 & -\sqrt{2} & 1
\end{bmatrix}
$$

Then A2 x C = D

$$D = \begin{bmatrix} -1.414 & 1 & 0 & 0 & 0 & 0 & 0 & 0 & 0 \\ 0 & -1 & 1.414 & 0 & 0 & 0 & 0 & 0 & 0 \\ 0 & 0 & 0 & 0 & 0 & 0 & 0 & 0 & 0 \\ 0 & 0 & 0 & 1 & 0 & 1 & 0 & 0 & 0 \\ 0 & 0 & 0 & 0 & 1 & 0 & 0 & 0 & 0 \\ 0 & 0 & 0 & 0 & 0 & 0 & 0 & 0 & 0 \\ 0 & 0 & 0 & 0 & 0 & 0 & 1.414 & 1 & 0 \\ 0 & 0 & 0 & 0 & 0 & 0 & 0 & -1 & -1.414 \\ 0 & 0 & 0 & 0 & 0 & 0 & 0 & 0 & 0 \end{bmatrix}$$

We are almost done, all we need to do is get rid of the 1 in the a_{12} position in the matrix where $X = -\sqrt{2}$. We do not wish to change anything else, so the rest of the sub-matrices equal identity.

$$A3 := \begin{bmatrix} 1 & 1 & 0 & 0 & 0 & 0 & 0 & 0 & 0 \\ 0 & 1 & 0 & 0 & 0 & 0 & 0 & 0 & 0 \\ 0 & 0 & 1 & 0 & 0 & 0 & 0 & 0 & 0 \\ 0 & 0 & 0 & 1 & 0 & 0 & 0 & 0 & 0 \\ 0 & 0 & 0 & 0 & 1 & 0 & 0 & 0 & 0 \\ 0 & 0 & 0 & 0 & 0 & 1 & 0 & 0 & 0 \\ 0 & 0 & 0 & 0 & 0 & 0 & 1 & 0 & 0 \\ 0 & 0 & 0 & 0 & 0 & 0 & 0 & 1 & 0 \\ 0 & 0 & 0 & 0 & 0 & 0 & 0 & 0 & 1 \end{bmatrix}$$

And A3 x D = E

$$E = \begin{bmatrix} -1.414 & 0 & 1.414 & 0 & 0 & 0 & 0 & & 0 & 0 \\ 0 & -1 & 1.414 & 0 & 0 & 0 & 0 & & 0 & 0 \\ 0 & 0 & 0 & & 0 & 0 & 0 & 0 & 0 & 0 \\ 0 & 0 & 0 & & 1 & 0 & 1 & 0 & & 0 & 0 \\ 0 & 0 & 0 & & 0 & 1 & 0 & 0 & & 0 & 0 \\ 0 & 0 & 0 & & 0 & 0 & 0 & 0 & & 0 & 0 \\ 0 & 0 & 0 & & 0 & 0 & 0 & 1.414 & 1 & 0 \\ 0 & 0 & 0 & & 0 & 0 & 0 & 0 & & -1 & -1.414 \\ 0 & 0 & 0 & & 0 & 0 & 0 & 0 & & 0 & 0 \end{bmatrix}$$

So lets see what we've got:

For the sub-matrix representing the roots of X, and putting in terms of a_1:

$X = - \sqrt{2}$:
$a_1 = a_3$
$a_2 = \sqrt{2}a_3$; $a_2 = \sqrt{2}a_1$; $a_1 = 1/\sqrt{2}a_2$
$a_1 = a_1$

$X = 0$:
$a_1 = -a_3$
$a_2 = 0$

$X = \sqrt{2}$:
$a_1 = a_1$
$\sqrt{2}a_1 = -a_2$; $a_1 = -1/\sqrt{2}a_2$
$a_2 = -\sqrt{2}a_3$; $\sqrt{2}a_1 = -\sqrt{2}a_3$; $a_1 = -a_3$

Which are the same values computed above.

UTILIZING SYMMETRY IN BUTADIENE[11]

We can take advantage of the symmetry of butadiene to write:

$C_1=C_2-C_3=C_4$ $C_1=C_4$; $C_2=C_3$. The Ψ function can be written as:

$$\Psi = a_1/\sqrt{2}\,(P_1 \pm P_4) + a_2/\sqrt{2}\,(P_2 \pm P_3)$$

Let's choose the positive values to find those roots:

$$\Psi = a_1/\sqrt{2}\,(P_1 + P_4) + a_2/\sqrt{2}\,(P_2 + P_3) =$$

$$\begin{vmatrix} \frac{1}{2}\int (P_1 + P_4)H(P_1 + P_4)d\tau - E & \frac{1}{2}\int (P_1 + P_4)H(P_2 + P_3)d\tau \\ \frac{1}{2}\int (P_2 + P_3)H(P_1 + P_4)d\tau & \frac{1}{2}\int (P_2 + P_3)H(P_2 + P_3)d\tau - E \end{vmatrix} = 0 = \text{(Multiply \& add terms)}$$

$$= \begin{vmatrix} \frac{1}{2}\int (P_1HP_1 + P_1HP_4 + P_4HP_1 + P_4HP_4)d\tau - E & \frac{1}{2}\int (P_1HP_2 + P_1HP_3 + P_4HP_2 + P_4HP_3)d\tau \\ \frac{1}{2}\int (P_2HP_1 + P_2HP_4 + P_3HP_1 + P_3HP_4)d\tau & \frac{1}{2}\int (P_2HP_2 + P_2HP_3 + P_3HP_2 + P_3HP_3)d\tau - E \end{vmatrix} = 0$$

$$= \begin{vmatrix} 1/2\,(H_{11} + 2H_{14} + H_{44}) - E & 1/2\,(H_{12} + H_{13} + H_{42} + H_{43}) \\ 1/2\,(H_{21} + H_{31} + H_{24} + H_{34}) & 1/2\,(H_{22} + 2H_{23} + H_{33}) - E \end{vmatrix} = 0$$

$$= \begin{vmatrix} 1/2\,(A+0+A) - E & 1/2\,(B+0+0+B) \\ 1/2\,(B+0+0+B) & 1/2\,(A+2B+A) - E \end{vmatrix} = \begin{vmatrix} A-E & B \\ B & A+B-E \end{vmatrix} = \begin{vmatrix} X & 1 \\ 1 & X+1 \end{vmatrix} = X_2 + X - 1 = 0$$

$$X = (-1 \pm \sqrt{5})/2 = -1.618;\ .618$$

Before we compute for the negative values of these roots, let's see how we might solve this problem using Nested Arrays. For convenience, I'm going to ignore the integral sign

---

[11] Turner, Methods in MO Theory, pp. 108

$$\begin{vmatrix} P_2 \\ P_3 \end{vmatrix} H \ [P_1 \ P_4] = \begin{vmatrix} P_2H \\ P_3H \end{vmatrix} [P_1 \ P_4] = \begin{vmatrix} P2HP1 & P2HP4 \\ P3HP1 & P3HP4 \end{vmatrix}$$

$$\begin{vmatrix} P_1 \\ P_4 \end{vmatrix} H \ [P_1 \ P_4] = \begin{vmatrix} P_1H \\ P_4H \end{vmatrix} [P_1 \ P_4] = \begin{vmatrix} P1HP1 & P1HP4 \\ P4HP1 & P4HP4 \end{vmatrix}$$

$$\begin{vmatrix} P_2 \\ P_3 \end{vmatrix} H \ [P_2 \ P_3] = \begin{vmatrix} P_2H \\ P_3H \end{vmatrix} [P_2 \ P_3] = \begin{vmatrix} P2HP2 & P2HP3 \\ P3HP2 & P3HP3 \end{vmatrix}$$

$$\begin{vmatrix} P_1 \\ P_4 \end{vmatrix} H \ [P_2 \ P_3] = \begin{vmatrix} P_1H \\ P_4H \end{vmatrix} [P_2 \ P_3] = \begin{vmatrix} P1HP2 & P1HP3 \\ P4HP2 & P4HP3 \end{vmatrix}$$

So the problem becomes:

$$\begin{Vmatrix} \begin{vmatrix} P_1H \\ P_4H \end{vmatrix} [P_1 \ P_4]-E & \begin{vmatrix} P_1H \\ P_4H \end{vmatrix} [P_2 \ P_3] \\ \begin{vmatrix} P_2H \\ P_3H \end{vmatrix} [P_1 \ P_4] & \begin{vmatrix} P_2H \\ P_3H \end{vmatrix} [P_2 \ P_3]-E \end{Vmatrix} = \begin{Vmatrix} \begin{vmatrix} P1HP1 & P1HP4 \\ P4HP1 & P4HP4 \end{vmatrix}-E & \begin{vmatrix} P1HP2 & P1HP3 \\ P4HP2 & P4HP3 \end{vmatrix} \\ \begin{vmatrix} P2HP1 & P2HP4 \\ P3HP1 & P3HP4 \end{vmatrix} & \begin{vmatrix} P2HP2 & P2HP3 \\ P3HP2 & P3HP3 \end{vmatrix}-E \end{Vmatrix}$$

Now E is single valued and we do not want to drag it along when we transpose, so we will separate it from the nested array and re-combine it later. i.e.

$$\begin{Vmatrix} \begin{vmatrix} P1HP1 & P1HP4 \\ P4HP1 & P4HP4 \end{vmatrix} & \begin{vmatrix} P1HP2 & P1HP3 \\ P4HP2 & P4HP3 \end{vmatrix} \\ \begin{vmatrix} P2HP1 & P2HP4 \\ P3HP1 & P3HP4 \end{vmatrix} & \begin{vmatrix} P2HP2 & P2HP3 \\ P3HP2 & P3HP3 \end{vmatrix} \end{Vmatrix} + \begin{vmatrix} -E & 0 \\ 0 & -E \end{vmatrix}$$

Now we will transpose the nested array so that we can use it mathematically. In this case, we wish to compute each sub-matrix separately. We do not want to operate on them all at the same time, so the transposed matrix will be a diagonal rather than a column matrix. We wish to sum the four elements in each sub-matrix, so we will multiply by $[DB1]_{4,16}$.

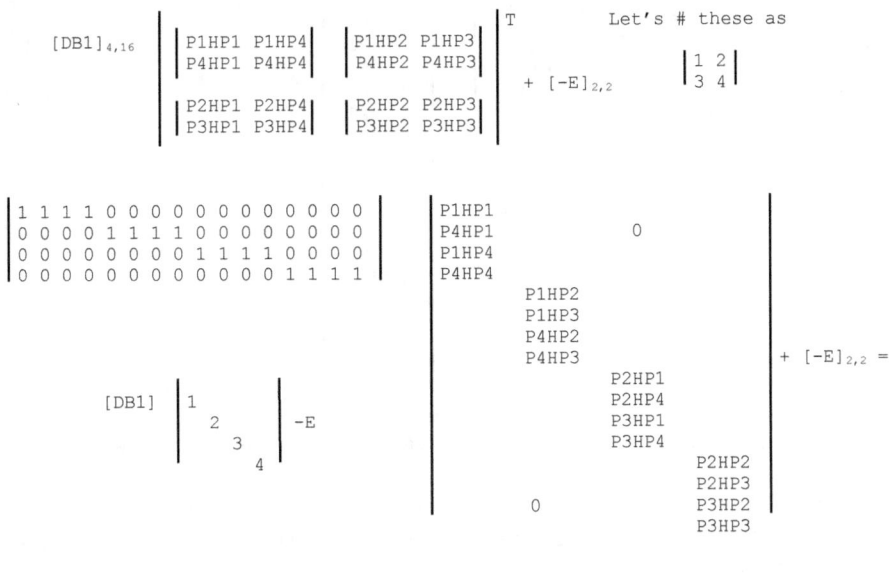

$$
\begin{vmatrix}
\text{P1HP1+P4HP1+P1HP4+P4HP4} & & & 0 \\
& \text{P1HP2+P1HP3+P4HP2+P4HP3} & & \\
& & \text{P2HP1+P2HP4+P3HP1+P3HP4} & \\
0 & & & \text{P2HP2+P2HP3+P3HP2+P3HP3}
\end{vmatrix}
$$

Now we re-transpose, returning the matrix to it's original form, and add [-E] to complete the computation.

$$
\begin{vmatrix}
\text{P1HP1+P4HP1+P1HP4+P4HP4 -E} & \text{P1HP2+P1HP3+P4HP2+P4HP3} \\
\text{P2HP1+P2HP4+P3HP1+P3HP4} & \text{P2HP2+P2HP3+P3HP2+P3HP3 -E}
\end{vmatrix}
\quad ; \quad
\begin{vmatrix}
\text{1-E} & 2 \\
3 & \text{4-E}
\end{vmatrix}
$$

Of course the numbers inside the brackets are matrices, but since they sum to a single number, it is superfluous whether we add them in or not.

Lets also solve this when we put the transposed nested array into a single column:

CLINTON L. HOLT

$$
\begin{vmatrix}
1 & 1 & 1 & 1 & 0 & 0 & 0 & 0 & 0 & 0 & 0 & 0 & 0 & 0 & 0 & 0 \\
0 & 0 & 0 & 0 & 1 & 1 & 1 & 1 & 0 & 0 & 0 & 0 & 0 & 0 & 0 & 0 \\
0 & 0 & 0 & 0 & 0 & 0 & 0 & 0 & 1 & 1 & 1 & 1 & 0 & 0 & 0 & 0 \\
0 & 0 & 0 & 0 & 0 & 0 & 0 & 0 & 0 & 0 & 0 & 0 & 1 & 1 & 1 & 1
\end{vmatrix}
\begin{vmatrix}
P1HP1 \\
P4HP1 \\
P1HP4 \\
P4HP4 \\
P1HP2 \\
P1HP3 \\
P4HP2 \\
P4HP3 \\
P2HP1 \\
P2HP4 \\
P3HP1 \\
P3HP4 \\
P2HP2 \\
P2HP3 \\
P3HP2 \\
P3HP3
\end{vmatrix}
+ [-E]_{2,2} =
\qquad [DB1]
\begin{vmatrix}
1 \\ 2 \\ 3 \\ 4
\end{vmatrix}
$$

$$
\begin{array}{cccc}
1 & 2 & 3 & 4
\end{array}
$$
[P1HP1+P4HP1+P1HP4+P4HP4 \quad P1HP2+P1HP3+P4HP2+P4HP3 \quad P2HP1+P2HP4+P3HP1+P3HP4 \quad P2HP2+P2HP3+P3HP2+P3HP3]

Which gives us the same answer, but not in the matrix form we need. Of course, this is the first time we've encountered a Nested Array whose dimensions were greater than Nx1 or 1xN. But if we follow the rules, the resulting array re-formed from this row matrix would be the first two elements form the first row, and the second two would form the second row. i.e.

$$
\begin{vmatrix}
1 & 2 \\
3 & 4
\end{vmatrix}
$$

Hey, it works! So I guess we can use nested arrays both as diagonals and columns and most probably as a row also, but we will not get into that here, maybe the next example. When I started this example, I did not know this column method would work, but I'm going to keep my doubts and accomplishments as they were written. The exploration of mathematics is a beautiful wilderness, we learn as we explore it's beauty. Sometimes, even when I think something can't work, the math proves me wrong. The above example is such a case.

Letting $\Psi = a_1/\sqrt{2}(P_1 - P_4) + a_2/\sqrt{2}(P_2 - P_3) =$

$$
\begin{vmatrix}
1/2(H_{11} - 2H_{14} + H_{44}) - E & 1/2(H_{12} - H_{13} - H_{24} - H_{43}) \\
1/2(H_{21} - H_{31} - H_{42} - H_{34}) & 1/2(H_{22} - 2H_{23} + H_{33}) - E
\end{vmatrix} = 0
$$

$$
\begin{vmatrix}
1/2(A-0+A)-E & 1/2(B-0-0+B) \\
1/2(B-0-0+B) & 1/2(A-2B+A)-E
\end{vmatrix}
=
\begin{vmatrix}
A-E & B \\
B & A-B-E
\end{vmatrix}
=
\begin{vmatrix}
X & 1 \\
1 & X-1
\end{vmatrix}
= X^2-X-1 = 0
$$

$X = (1 \pm \sqrt{5})/2 = 1.618; \; -.618$

COEFFICIENTS BY PROPERTIES OF NON-BONDING MOLECULAR ORBITALS (NBMO's)[12]

Let's look at methyl hexane:

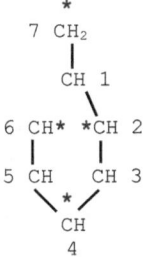

```
   7 CH₂
      |
     CH 1
    /    \
 6 CH    CH 2
    |     |
 5 CH    CH 3
    \    /
      CH
      4
```

First we number the carbons, starting with the longest continuous chain. Then the positions on the molecule are alternately starred to get the greatest number of starred positions.

```
      *
   7 CH₂
      |
     CH 1
        \
 6 CH*  *CH 2
    |     |
 5 CH    CH 3
    \  * /
      CH
      4
```

Then the sum of the coefficients of the AO's of the starred atoms directly linked to a given un-starred atom is zero. Thus the sum of $a_2 + a_4$ attached to the un-starred atom 3 must equal zero, etc. . . . Thus

$$a_2 + a_4 = 0$$
$$a_6 + a_4 = 0$$
$$a_2 + a_6 + a_7 = 0$$

---

[12]   Roberts, MO Calculations, pp. 106-107

If we set $a_4 = 1$, then $a_2 = a_6 = -1$ and $a_7 = 2$. These coefficients are not normalized. To normalize, we must first determine the energies of each carbon.

$$
\begin{vmatrix} X&1&0&0&0&1&1\\ 1&X&1&0&0&0&0\\ 0&1&X&1&0&0&0\\ 0&0&1&X&1&0&0\\ 0&0&0&1&X&1&0\\ 1&0&0&0&1&X&0\\ 1&0&0&0&0&0&X \end{vmatrix} = (1(\begin{vmatrix} 1&0&0&0&1&1\\ X&1&0&0&0&0\\ 1&X&1&0&0&0\\ 0&1&X&1&0&0\\ 0&0&1&X&1&0\\ 0&0&0&1&X&0 \end{vmatrix}) + (X(\begin{vmatrix} X&1&0&0&0&1\\ 1&X&1&0&0&0\\ 0&1&X&1&0&0\\ 0&0&1&X&1&0\\ 0&0&0&1&X&0\\ 1&0&0&0&1&X \end{vmatrix}) =
$$

$$
1[-1(\begin{vmatrix} 1&0&0&1&1\\ X&1&0&0&0\\ 1&X&1&0&0\\ 0&1&X&0&0\\ 0&0&1&1&0 \end{vmatrix} + X(\begin{vmatrix} 1&0&0&0&1\\ X&1&0&0&0\\ 1&X&1&0&0\\ 0&1&X&1&0\\ 0&0&1&X&0 \end{vmatrix}] + X[X(\begin{vmatrix} X&1&0&0&0\\ 1&X&1&0&0\\ 0&1&X&1&0\\ 0&0&1&X&0\\ 0&0&0&1&X \end{vmatrix})-1(\begin{vmatrix} 1&X&1&0&0\\ 0&1&X&1&0\\ 0&0&1&X&1\\ 0&0&0&1&X\\ 1&0&0&0&1 \end{vmatrix})] =
$$

$$
1[1(1[\begin{vmatrix} 1&0&0&1\\ X&1&0&0\\ 1&X&1&0\\ 0&1&X&1 \end{vmatrix} + -1[1(\begin{vmatrix} 1&0&0&1\\ X&1&0&0\\ 1&X&1&0\\ 0&1&X&0 \end{vmatrix} + X(1(\begin{vmatrix} 1&0&0&1\\ X&1&0&0\\ 1&X&0&0\\ 0&1&1&0 \end{vmatrix})] +X[-X([\begin{vmatrix} 1&0&0&1\\ X&1&0&0\\ 1&X&1&0\\ 0&1&X&1 \end{vmatrix}
$$

We are going to stop here, find the determinant is quite complex, as you can see. The MathCad +6 program can find it easily for us, and also the roots. Proceed as follows:

click SYMBOLIC
choose MATRIX OPERATIONS
click DETERMINANT OF MATRIX. =
$$= -7 \cdot X + 13 \cdot X^3 - 7 \cdot X^5 + X^7$$

To find the roots, click matrix and set for 8 rows, 1 column. Put the coefficients, starting from the constant and going on up to the coefficient of X^7. Type in polyroots(P) and press the = key.

$$P = \begin{vmatrix} 0 \\ -7 \\ 0 \\ 13 \\ 0 \\ -7 \\ 0 \\ 1 \end{vmatrix} \quad ; \quad \text{polyroots}(P) = \begin{vmatrix} -2.101 \\ -1.259 \\ -1.000 \\ 0 \\ 1.000 \\ 1.259 \\ 2.101 \end{vmatrix}$$

$E_1 = (A-E)/B = -2.101; \quad A-E=-2.101B; \quad E_1 = A+2.101B$
$E_2 = A+1.259B$
$E_3 = A+B$
$E_4 = A$
$E_5 = A-B$
$E_6 = A-1.259B$
$E_7 = A-2.101B$

$E_\pi = 7A + 8.72B$

To normalize: $(a_2^2 + a_4^2 + a_6^2 + a_7^2)^{1/2} = (1^2+1^2+1^2+2^2)^{1/2} = \sqrt{7}$.

$\Psi = 1/\sqrt{7}(-P_2+P_4-P_6+2P_7$

This alternate carbon structure has a self-consistent field, $q_i = 0$. Therefore, if an electron is removed from the NBMO to obtain a benzyl cation, the positive charge will be distributed solely over those atoms whose orbital coefficients are not zero for the NBMO. The same is true if we add an electron to get a benzyl anion.

CYCLOBUTADIENE by NBMO's:

First we must find the coefficients a_i for cyclobutadiene.

First we star the greatest number of non-adjacent points on the molecule (in this case we choose C_1 and C_3). Then for the point about C_2, the starred atoms must equal 0. $C_1 + C_3 = 0$. Since there are tow NBMO's about C_3 (C_2 and C_4), then $C_2 + C_4 = 0$ also. We arbitrarily choose $C_1 = 1$ and $C_2 = -1$ ($C_1 + C_3 = 0$; $1 + (-1) = 0$). Then Ψ_2 (the point between C_1 and C_3) $= 1/\sqrt{2}(P_1 - P_3)$ and rotating in the starred direction, $\Psi_3 = 1/\sqrt{2}(P_2 - P_4)$. And, of course, Ψ_1 is in general always positive, so $\Psi_1 = 1/\sqrt{4}(P_1 + P_2 + P_3 + P_4)$. Ψ_4 is determined by, starting at $C_1 = 1$, (the first adjacent atom) then $C_2 = -1$, $C_3 = 1$ and $C_4 = -1$ as $\Psi_4 = 1/2(P_1 - P_2 + P_3 - P_4)$. These correspond to the earlier equations. This works on molecules where $q_i = 0$.

HETEROCYCLIC COMPOUNDS[13]

Wieland and Pauli adjusted the values of A_N (the A value for nitrogen) and A_0 for oxygen by expressing them in terms of A plus some multiple of B.

$A_N = A + \lambda_N B$ $\quad$ $A_N(1) = A + 1/2B$ $\qquad$ $A_N(2) = A + 3/2B$
and $\qquad\qquad$ where $\qquad\qquad\qquad\qquad$ and
$A_0 = A + \lambda_0 B$ $\quad$ $A_0(1) = A + 3/2B$ $\qquad$ $A_0(2) = A + 5/2B$

where the one in parenthesis denotes one electron has been donated, and the two in parenthesis denotes that 2 electrons have been donated. Contribution of two electrons by the heteroatom decreases shielding at the position. When π bonding involves the P_2 orbital, the electrons in the system are under the influence of a more attractive potential, and the energy of the electron in the $2P_z$ orbital decreases. Starting with a molecule of pyrrole:

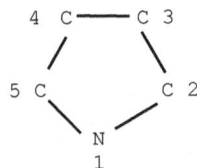

$(A_N(2) - E)$

The Slater determinant is: $\begin{vmatrix} A+3/2B-E & B & 0 & 0 & B \\ B & A-E & B & 0 & 0 \\ 0 & B & A-E & B & 0 \\ 0 & 0 & B & A-E & B \\ B & 0 & 0 & B & A-E \end{vmatrix}$ $\begin{vmatrix} X+3/2 & 1 & 0 & 0 & 0 \\ 1 & X & 1 & 0 & 0 \\ 0 & 1 & X & 1 & 0 \\ 0 & 0 & 1 & X & 1 \\ 1 & 0 & 0 & 1 & X \end{vmatrix} = 0$

The determinant is equal to: $X^5 + 3/2X^4 + -5X^3 - 9/2X^2 + 5X + 7/2 = 0$

And has roots: $X = (-2.55, -1.15, -.618, 1.20, 1.62)$

Substituting each root for X in the Slater determinant, we determine the wave function to be:

$\Psi_1 = .749P_1 + .393P_2 + .254P_3 + .254P_4 + .393P_5$
$\Psi_2 = .503P_1 - .089P_2 - .605P_3 - .605P_4 - .089P_5$

---

[13] Liberles, Introduction to MO Theory, pp. 170-174

$$\Psi_3 = .602P_2 - .372P_3 - .372P_4 - .602P_5$$
$$\Psi_4 = .430P_1 - .580P_2 + .267P_3 + .267P_4 - .580P_5$$
$$\Psi_5 = .372P_2 - .602P_3 + .602P_4 - .372P_5$$

For this exercise, we will check the correctness of the coefficients. We proceed as follows:

Checking for Ψ_1.

$$H\Psi_1 = E\Psi_1$$

Where $E = A + 2.55B$

Multiply both sides by Ψ_1 and integrate

$$\int \Psi_1 H\Psi_1 d\tau = \int \Psi_1 E\Psi_1 d\tau$$

$$E = \frac{\int \Psi_1 H\Psi_1 d\tau}{\int \Psi_1 \Psi_1 d\tau}$$

but $\int \Psi_1 \Psi_1 d\tau = 1$

Therefore

$$\mathbf{E = \int \Psi_1 H\Psi_1 d\tau}$$

$\Psi_1 = .749P_1 + .393P_2 + .254P_3 + .254P_4 + .393P_5$ and $H_{AB} = H_{BA}$.

$$H_{11} = A + 3/2B; \quad H_{22} = H_{33} = H_{44} = H_{55} = A$$
$$H_{12} = H_{21} = H_{15} = H_{51} = H_{23} = H_{32} = H_{34} = H_{43} = H_{45} = H_{54} = B$$
$$H_{13} = H_{31} = H_{14} = H_{41} = H_{24} = H_{42} = H_{25} = H_{52} = H_{35} = H_{53} = 0$$

I am going to do this problem using MathCad again, mainly to reduce the amount of typing needed to illustrate this problem. I can't double up on lines imported from MathCad, so I'll have to put in these items one line at a time which really takes up space, but first we'll solve this by hand.

$E = (.749)^2 H_{11} + 2(.749)(.393)H_{12} + (H_{13,31}$ & $H_{14,41} = 0) + 2(.749)$
$(.393)H_{15} + (.393)^2 H_{22} + 2(.393)(.254)H_{23} + (H_{24,42}$ & $H_{25,52} = 0) +$
$(.254)^2 H_{33} + 2(.254)(.254)H_{34} + (.254)^2 H_{44} + (H_{35,53} = 0) + 2(.254)$
$(.393)H_{45} + (.393)^2 H_{55} =$

$H_{11} = A + 3/2B$; $H_{22}=H_{33}=H_{44}=H_{55}=A$

$H_{12}=H_{21}=H_{15}=H_{51}=H_{23}=H_{32}=H_{34}=H_{43}=H_{45}=H_{54}=B$

$\mathbf{E} = (.749)^2(A+3/2B) + (.393)^2A + (.254)^2A + (.254)^2A + (.393)^2A +$
$4(.294)B + 4(.1)B + 2(.065)B = .999A + .842B + 1.18B +$
$.400B + .130B = \mathbf{.999A + 2.552B}$

MATHCAD +6 PROGRAM

$$\Psi^T = \begin{vmatrix} .749 \\ .393 \\ .254 \\ .254 \\ .393 \end{vmatrix} \qquad \Psi = [.749\ .393\ .254\ .254\ .393] \qquad ONE5 = [1\ 1\ 1\ 1\ 1]$$

$$FIVE1 = \begin{vmatrix} 1 \\ 1 \\ 1 \\ 1 \\ 1 \end{vmatrix} \qquad I5 = identity(5) \quad (5x5\ identity\ matrix)$$

$$\Psi SQ = \Psi^T\Psi \quad = \quad \begin{vmatrix} .561 & .294 & .190 & .190 & .294 \\ .294 & .154 & .100 & .100 & .154 \\ .190 & .100 & .065 & .065 & .100 \\ .190 & .100 & .065 & .065 & .100 \\ .294 & .154 & .100 & .100 & .154 \end{vmatrix}$$

We want only the real values (the values where the carbons
are connected) so we must make a template matrix. Go to the
Slater determinant and for every element that isn't zero, put
in a one in it's place. Put zero's in for zero's. But we want
the log of this matrix, and computers can't take the log of
zero, in fact it is undefined, so we must make it ourselves. I
will define $0 \approx 10^{-100}$ so the log $0 \approx -100$ which is close enough to
zero for the purposes of this problem. So to write the log of
the template matrix, for every non-zero element in the Slater
determinant we will put a zero, and for every zero, we will
replace it with -100.

$$TEMPLATE1 = \begin{vmatrix} 0 & 0 & -100 & -100 & 0 \\ 0 & 0 & 0 & -100 & -100 \\ -100 & 0 & 0 & 0 & -100 \\ -100 & -100 & 0 & 0 & 0 \\ 0 & -100 & -100 & 0 & 0 \end{vmatrix}$$

We must take the log of each individual element in ΨSQ, in MathCad, this accomplished by:

$$\text{LOG}\overrightarrow{\Psi\text{SQ}} = \log(\Psi\text{SQ})$$

$$\text{LOGENERGY} = \text{LOG}\Psi\text{SQ} + \text{TEMPLATE1}$$

$$\text{ENERGY} = 10^{\overrightarrow{\text{LOGENERGY}}}$$

This takes the anti-log of every element in LOGENERGY.

$$\text{ENERGY} = \begin{vmatrix} .561 & .294 & 0 & 0 & .294 \\ .294 & .154 & .100 & 0 & 0 \\ 0 & .100 & .065 & .065 & 0 \\ 0 & 0 & .065 & .065 & .100 \\ .294 & 0 & 0 & .100 & .154 \end{vmatrix}$$

Now A is equal to the sum of the diagonal, and B is equal to the sum of the elements of the matrix minus the diagonal. Since this is a heteroatom, we must correct the B term in H_{11}. MathCad can make a diagonal from a column matrix, but it cannot remove one, a big problem that they must correct, so we must construct the template to remove the diagonal. The log if the 5x5 identity matrix is:

$$\text{LOGI5} = \log(\overrightarrow{\text{I5}})$$

$$\text{LOGI5} = \begin{vmatrix} 0 & -100 & -100 & -100 & -100 \\ -100 & 0 & -100 & -100 & -100 \\ -100 & -100 & 0 & -100 & -100 \\ -100 & -100 & -100 & 0 & -100 \\ -100 & -100 & -100 & -100 & 0 \end{vmatrix}$$

$$\text{ENERGYA1} = \text{LOGI5} + \text{ENERGY}$$

$$\text{ENERGYA2} = 10^{\overrightarrow{\text{ENERGYA1}}}$$

$$\text{ENERGYA2} = \begin{vmatrix} .561 & & & & \\ & .154 & & & \\ & & .065 & & \\ & & & .065 & \\ & & & & .154 \end{vmatrix}$$

ENERGYA = ONE5*ENERGYA2*FIVE1 = **.999**

To compute B, we must first correct for the Nitrogen term in a_{11}:

$$
\text{ENERGYCORR} =
\begin{vmatrix}
3/2 & 0 & 0 & 0 & 0 \\
0 & 0 & 0 & 0 & 0 \\
0 & 0 & 0 & 0 & 0 \\
0 & 0 & 0 & 0 & 0 \\
0 & 0 & 0 & 0 & 0
\end{vmatrix}
$$

ENERGYB1 = ENERGYA2*ENERGYCORR

$$
\text{ENERGYB1} =
\begin{vmatrix}
.842 & 0 & 0 & 0 & 0 \\
0 & 0 & 0 & 0 & 0 \\
0 & 0 & 0 & 0 & 0 \\
0 & 0 & 0 & 0 & 0 \\
0 & 0 & 0 & 0 & 0
\end{vmatrix}
$$

ENERGYBCORRECTED = ENERGY - ENERGYA2 + ENERGYB1

$$
\begin{vmatrix}
.561 & .294 & 0 & 0 & .294 \\
.294 & .154 & .100 & 0 & 0 \\
0 & .100 & .065 & .065 & 0 \\
0 & 0 & .065 & .065 & .100 \\
.294 & 0 & 0 & .100 & .154
\end{vmatrix}
-
\begin{vmatrix}
.561 & & & & \\
& .154 & & & \\
& & .065 & & \\
& & & .065 & \\
& & & & .154
\end{vmatrix}
+
\begin{vmatrix}
.842 & 0 & 0 & 0 & 0 \\
0 & 0 & 0 & 0 & 0 \\
0 & 0 & 0 & 0 & 0 \\
0 & 0 & 0 & 0 & 0 \\
0 & 0 & 0 & 0 & 0
\end{vmatrix}
=
$$

ENERGYB = ONE5*ENERGYBCORRECTED*FIVE1

ENERGYB = 2.547

E = **.999A + 2.547B** which matches.

PERTURBATION THEORY[14]

This will be the largest section in this part of my paper, mainly because of the importance of perturbation theory to the chemist. The majority of chemical problems are rarely concerned with the absolute energies of molecules, but more with the relative energies of pairs of molecules that are nearly identical. Since chemical reactions rarely involve the formation/breaking of more than one bond, the products and reactants in reactions are necessarily closely related in structure. Even in the examples involving localized and de-localized forms of the same molecule, the calculated resonance energy depends only upon a slight change in the classical structure of the molecule. Thus if one molecule of naphthalene (for instance) is resolved by the Hückel method, the relative energies of mono and di substituted naphthalene compounds can readily be calculated by setting up the un-perturbed matrix and the perturbation matrix for the perturbed state of naphthalene. Since naphthalene is known (we presumably have already calculated the un-perturbed wave function), the perturbed state can easily be found by multiplying the perturbation matrix by the wave function found for the un-perturbed molecule.

Approximate solutions to many problems may be calculated if it is assumed that it's solution differs only slightly from one that is already known. Suppose $H\Psi = E\Psi$ and that H is the sum of two symmetric matrices $H = H^0 + \varepsilon H^1$, where ε is a scalar, εH^1 is small, and $H^0\Psi^0 = E^0\Psi^0$ is known. So if εH^1 is small, the eigenvalues E of the perturbed system should not differ much from Ψ^0 and E^0.

Suppose some E^0_K is non-degenerate with eigenvector Ψ^0_K and a perturbation of εH^1. Then $E = E^0_K + \varepsilon E_K^1 + O(\varepsilon^2) + \ldots$ where $O(\varepsilon^2)$ means all subsequent terms are multiplied by powers of ε greater than or equal to 2. We shall determine E_K^1 and the coefficients of the higher powers of ε. Similarly, $\Psi = \Psi^0_K + \varepsilon\Psi^1_K + O(\varepsilon^2) + \ldots$ where Ψ^1_K and E are to be determined. Substituting these values of Ψ and E into $H\Psi=E\Psi$ we get:

$$(H^0 + \varepsilon H^1)(\Psi^0_K + \varepsilon\Psi^1_K + O(\varepsilon^2) + \ldots) = (E^0_K + \varepsilon E_K^1 + O(\varepsilon^2) + \ldots)$$
$$(\Psi^0_K + \varepsilon\Psi^1_K + O(\varepsilon^2) + \ldots)$$

---

[14] Goodrich, A Primer of Quantum Chemistry, pp. 98-103

or

$$H^\circ\Psi_K^\circ + E[H^\circ\Psi_K^1 + H^1\Psi_K^\circ + \ldots] = E^\circ_K\Psi_K^\circ + \varepsilon[E^\circ_K\Psi_K^1 + E_K^1\Psi_K^\circ + \ldots]$$

and equating coefficients of like powers of ε, we have for the zeroth order term

$$H^\circ\Psi_K^\circ = E^\circ_K\Psi_k^\circ$$

which is where we started. The first order correction (from the coefficients of ε) is:

$$H^\circ\Psi_K^1 + H^1\Psi_k^\circ = E^\circ_K\Psi_K^1 + E_K^1\Psi_K^\circ,$$

with E_K^1 and Ψ_K^1 to be determined. H° forms a basis in a vector space of n dimensions, so we can expand Ψ_K^1 in terms of Ψ_j°.

$$\Psi_K^1 = \sum_j c_{jk}\Psi_j^\circ, \text{ and } H^1\Psi_k^\circ = \sum_j b_{jk}\Psi_j^\circ$$

c_{jk} is unknown, b_{jk} is known: $b_{jk} = (\Psi_j^\circ)^T H^1\Psi_K^\circ$. Since the matrix is symmetric, $b_{jk} = b_{kj}$. These values can now be substituted into the above equation:

$$H^\circ\Psi_K^1 = H^\circ\sum c_{jk}\Psi_j^\circ = \sum c_{jk}E^\circ_j\Psi_j^\circ$$

and the first order equation becomes:

$$\sum c_{jk}E^\circ_j\Psi_j^\circ + \sum b_{jk}\Psi_j^\circ = E^\circ_K\sum c_{jk}\Psi_j^\circ + E_K^1\Psi_K^2.$$

Again, Ψ_j° is a set of n linearly independent vectors in n dimensions, so the above equation can be solved only if the coefficients of identical vectors on either side are separately equal. From the coefficients of Ψ_K° we have $b_{KK} = E_K^1$. which determines the first order correction to the eigenvalue. From the coefficient of Ψ_j° i≠k, we get

$$c_{jk} = b_{jk}/(E^\circ_K - E^\circ_j),$$

the values of which determine the first order correction to the eigenvector. So we now have the equation:

$$\Psi = \Psi_k^\circ + (\varepsilon\sum b_{jk}\Psi_j^\circ)/(E^\circ_K - E^\circ_J) + O(\varepsilon)^2$$

and it's associated eigenvalue

$$E = E^{\circ}_{K} + \varepsilon (\Psi^{\circ}_{k})^{T} H^{1} \Psi^{\circ}_{K} + O(\varepsilon)^{2}$$

EXAMPLE: FORMALDEHYDE

In this example, $H_2C=O$ is considered to be a perturbation of ethylene where a CH_2 is replaced by an oxygen atom. Since oxygen is more electronegative than carbon, the effect of such a change will make the diagonal element corresponding to oxygen more negative. Assume H_{22} for oxygen $= A + B$, then the determinant corresponding to $H_2C=O$ is

$$\begin{vmatrix} A & B \\ B & A+B \end{vmatrix} = 0 \; ; \qquad H = H^{\circ} + \varepsilon H^{1} \text{ and } \varepsilon = 1. \quad \text{(If } H_{22} = A + .5B, \text{ then } \varepsilon = .5\text{)}$$

So in this case, $H^{\circ} = \begin{vmatrix} A & B \\ B & A \end{vmatrix}$ and $H^{1} = \begin{vmatrix} 0 & 0 \\ 0 & B \end{vmatrix}$.

The solution for H° (the un-perturbed function for ethylene) is:

$$E_1 = A + B \text{ and } \Psi_1 = 1/\sqrt{2}(P_1 + P_2)$$
$$E_2 = A - B \text{ and } \Psi_2 = 1/\sqrt{2}(P_1 - P_2)$$

The coefficients b_{jk} are calculated from the zero order eigenvectors

$$b_{jk} = (\Psi^{\circ}_{j})^{T} H^{1} \Psi^{\circ}_{K}$$

$$\Psi^{\circ}_1 \text{ (in matrix form)} = [1/\sqrt{2}, \ 1/\sqrt{2}]$$
$$\Psi^{\circ}_2 \text{ (in matrix form)} = [1/\sqrt{2}, \ -1/\sqrt{2}]$$

$$b_{11} = [1/\sqrt{2}, \ 1/\sqrt{2}] \begin{vmatrix} 0 & 0 \\ 0 & B \end{vmatrix} \begin{vmatrix} 1/\sqrt{2} \\ 1/\sqrt{2} \end{vmatrix} = [0 \ \ 1/\sqrt{2}B] \begin{vmatrix} 1/\sqrt{2} \\ 1/\sqrt{2} \end{vmatrix} = 1/2B$$

$$b_{22} = [1/\sqrt{2}, \ -1/\sqrt{2}] \begin{vmatrix} 0 & 0 \\ 0 & B \end{vmatrix} \begin{vmatrix} 1/\sqrt{2} \\ -1/\sqrt{2} \end{vmatrix} = [0 \ \ -1/\sqrt{2}B] \begin{vmatrix} 1/\sqrt{2} \\ -1/\sqrt{2} \end{vmatrix} = 1/2B$$

$$b_{12} = [1/\sqrt{2}, \ -1/\sqrt{2}] \begin{vmatrix} 0 & 0 \\ 0 & B \end{vmatrix} \begin{vmatrix} 1/\sqrt{2} \\ 1/\sqrt{2} \end{vmatrix} = [0 \ \ -1/\sqrt{2}B] \begin{vmatrix} 1/\sqrt{2} \\ 1/\sqrt{2} \end{vmatrix} = -1/2B$$

$$\text{PSI1} := \begin{pmatrix} 1 & 1 \\ 1 & -1 \end{pmatrix}$$

$$H1 = \begin{vmatrix} 0 & 0 \\ 0 & B \end{vmatrix}$$

BE SURE THAT IN THE MATH MENU AUTOMATIC MODE AND LIVE SYMBOLICS
ARE CHECKED. TYPE IN THE FUNCTION AND CLICK TO PUT EXPRESSION
IN A SELECTION BOX, THEN PRESS CONTROL PERIOD (CTRL .), THIS
WILL GIVE THE SYMBOLIC EQUAL SIGN. THEN CLICK ANYWHERE OUTSIDE
THE EXPRESSION TO GET THE SOLUTION.

$$\left(\frac{1}{\sqrt{2}}\right) \cdot \left(\frac{1}{\sqrt{2}}\right) \cdot PSI1^{T} \cdot H1 \cdot PSI1 \rightarrow \begin{bmatrix} \frac{1}{2} \cdot B & \frac{-1}{2} \cdot B \\ \frac{-1}{2} \cdot B & \frac{1}{2} \cdot B \end{bmatrix}$$

OR WE CAN TAKE THE ROW FROM THE PSI1 MATRIX THAT CORRESPONDS
TO THE ROW THAT THE B IS IN AND SQUARE IT AS PSI1SQ =
PSI[2]^TxPSI[2]. (IN THIS CASE, IT IS IN ROW 2) i.e.

$$\left(\frac{1}{\sqrt{2}}\right)^{2} \cdot B \cdot \begin{pmatrix} 1 \\ -1 \end{pmatrix} \cdot (1 \quad -1) \rightarrow \begin{bmatrix} \frac{1}{2} \cdot B & \frac{-1}{2} \cdot B \\ \frac{-1}{2} \cdot B & \frac{1}{2} \cdot B \end{bmatrix}$$

WHICH GIVES US THE SAME ANSWER (SEE PART TWO OF THIS PAPER).

The un-normalized perturbed ground state function is given by:

$\Psi_1 = \Psi^\circ_1 + b_{21}/(E^\circ_1 - E^\circ_2) = \Psi^\circ_1 - 1/2B/(A+B)-(A-B) = \Psi^\circ_1 - 1/2B/2B$
$\Psi^\circ_2 = \Psi^\circ_1 - 1/4\Psi^\circ_2$

Normalizing Ψ_1

$\Psi_{1NORM} = 1/[1^1 + (-1/4)^2]^{1/2} = 1/1.033$

Therefore $\Psi_{1NORM} = 1/1.033(\Psi^\circ_1 - 1/4\Psi^\circ_2) = \Psi_1 = .968\Psi^\circ_1 - .243\Psi^\circ_2 =$

(To see how to normalize using matrices, see example in part
two of this paper.)

$.968[\ 1/\sqrt{2}\ \ 1/\sqrt{2}]\ -\ .243\ [1/\sqrt{2}\ -1/\sqrt{2}]\ \equiv\ [1\ 1]\ \begin{vmatrix} .968 & 0 \\ 0 & -.243 \end{vmatrix}\ \begin{vmatrix} 1/\sqrt{2} & 1/\sqrt{2} \\ 1/\sqrt{2} & -1/\sqrt{2} \end{vmatrix}\ =$

$$\Psi\text{norm} = \begin{pmatrix} 0.68448 & 0.68448 \\ -0.17183 & 0.17183 \end{pmatrix}$$

$\Psi_{1\text{NORM}}\ =\ \ \ [1\ 1]\ \begin{vmatrix} .68448 & .68448 \\ -.17183 & .17183 \end{vmatrix} =$

The program for the rest of this problem is:

$$\Psi\text{norm} = \begin{pmatrix} 0.68448 & 0.68448 \\ -0.17183 & 0.17183 \end{pmatrix}$$

$\Psi 1\text{norm} := \text{ONE2}\ \Psi\text{norm}$

$\Psi 1\text{norm} = (\ 0.51265\ \ 0.85631\ )$

And now we will solve for Ψ_2.

$E_2\ =\ E^\circ_2\ +\varepsilon b_{22}\ =\ A-B\ +1/2B\ =\ \mathbf{A-1/2B}$

$\Psi_2\ =\ \Psi^\circ_2\ +\ b_{12}/(E^\circ_2\ -\ E^\circ_1)\ =\ \Psi^\circ_2\ -\ 1/2B/(A-B)-(A+B)\ =\ \Psi^\circ_1\ -\ 1/2B/-2B$

$\Psi^\circ_1\ =\ \mathbf{\Psi^\circ_2\ +1/4\Psi^\circ_1}$

Normalizing Ψ_2

$\Psi_{2\text{NORM}}\ =\ 1/[1^1\ +\ (1/4)^2]^{1/2}\ =\ 1/1.033$

$\Psi_{2\text{NORM}}\ =\ -.968\Psi^\circ_1\ -\ .243\Psi^\circ_2\ =$

    $-.968[\ 1/\sqrt{2}\ \ 1/\sqrt{2}]\ -\ .243\ [1/\sqrt{2}\ -1/\sqrt{2}]\ \equiv\ [1\ 1]\ \begin{vmatrix} -.968 & 0 \\ 0 & -.243 \end{vmatrix}\ \begin{vmatrix} 1/\sqrt{2} & 1/\sqrt{2} \\ 1/\sqrt{2} & -1/\sqrt{2} \end{vmatrix}\ =$

The rest I will solve using MathCad +6:

$$\Psi 2\text{norm} = \begin{pmatrix} -0.68448 & -0.68448 \\ -0.17183 & 0.17183 \end{pmatrix}$$

$\Psi 2 := \text{ONE2}\ \Psi 2\text{norm}$

$\Psi 2 = (-0.85631 \ -0.51265)$

or

$$\Psi = \begin{vmatrix} .514 & .858 \\ -.858 & -.514 \end{vmatrix}$$

The two electrons occupy the ground state Ψ, so E_1 = A + 1.5B and E_π = 2(A + 1.5B) = 2A + 3B. If the bond is broken, the electrons occupy the separate atoms, A for carbon, A + B for oxygen, or E = A + (A + B) = 2A + B. So DE_π = 2A = 3B - 2A - B = 2B, which is the same DE_π of un-perturbed ethylene.

The charge distribution q_i for carbon is 2(.514)^2 = .528. The charge distribution q_2 for oxygen is 2(.858)^2 = 1.476. Subtracting the residual charge of 1 from each value we get the molecular charge distribution. C = .472; O = -.472.

The dipole moment μ is defined as the product of the charge concentration about the positive pole times the inter-atomic distance.

For this example, μ = (.472)e$^-$(1.21x10^{-5}cm) = (.472) (4.803x10^{-10}esu)(1.21x10^{-5}cm) = 2.74 Debye units. Experimental values for μ for formaldehyde are found to be from 2.29 - 2.34 Debye units. But note, the perturbation for ethylene covers 50% of the molecule, which is a very large percentage (50%). If the molecule had 100 carbons, the perturbation of adding one oxygen would be 1%, so the larger the molecule, the closer to the actual approximated value the answers will be.

PART 2. PERTURBATION THEORY

Around October 1995, I discovered what seems to be a new matrix operator. This operator solves problems in statistics, quantum chemistry, the solution of equations (Gaussian reduction) and can create a universal accounting/inventory system that can keep track of **everything** that is made and the parts it is made of, everything that is bought and sold, and who bought it and from whom and for how much, and also can keep track of taxes, all in four program steps. Not millions, like in Excel or Lotus databases, but 4 steps, six steps if we want specialized information from the system. I say this because I proved we are able to multiply matrices differently from the ordinary dot product and still come up with the correct answer. In the Half-Multiplier mode, though, we can compute the intermediate values of matrix multiplication before we sum, and many of these values are of interest and some are very important to understanding the system we are working on.

The Half-Multiplier Operator works the same as the normal matrix multiplier, but it 'freezes' the operation of multiplication before summation is done, creating Nested Arrays as intermediates. Information about a system not available through accepted methods of multiplication are easily accessible through utilization of this operator. The sub-matrices calculated each represent one summed row, or one summed column in the final matrix.

Although not necessarily a linear combination, the sub-matrices seem more like nested arrays that await our use. We may add/subtract them as in a linear combination of matrices, cross multiply them, or re-define the dimensions of the matrix, transpose and by transposing, drop the inner brackets to produce one large grand matrix (partitioning) that follows the rules of our regular mathematics, and use this to solve many different problems all at the same time, keeping them all separate, or adding them all together in one lump. One of the really neat properties of this operator is if a computer is not set up to do the math, there is a one to one mathematical correspondence with regular matrices, but the normal way uses many more steps than by using the Half-Multiplier. For instance, $[A]^T_{31} \circ [B]_{33}$ needs nine multiplication's to get the

answer, but the equivalent matrix operation is to change [A]
(a column matrix) into a square diagonal matrix of 9 elements.
The first row needs 9 multiplication's, as does the second and
third row, which makes 27 multiplication's, even though 6 of
every 9 is a multiplication by zero. Plus we must add the 27
numbers three at a time to obtain the final solution. In other
words, with the Half-Multiplier, we get the answer in 3^2 or
9 steps, but with regular matrix multiplication, the answer
needs $2(3)^3$ or 54 mathematical steps to get the same result. Or
better yet, suppose we are working with the anti-AIDS medicine
AZT which has about 100,000 carbons and various other atoms
composing it. I will show that for a single perturbation,
we can calculate b_{ij} in $100,000^2$ steps, or 10^{10} mathematical
operations. The accepted way is $b_{ij} = \Psi^T H \Psi$. $\Psi^T H$ needs $2(10^5)^3$
or $2x10^{15}$ mathematical steps. Then $(\Psi^T H)\Psi$ also needs $2(10^5)^3$
steps for a total of $4(10^{15})$ mathematical operations, which is
a factor of 40 thousand more than with the Half-Multiplier. Of
course, with $b_{ij} = \Psi^T H \Psi$ we can calculate all the perturbations
at once, with the Half Multiplier though, we must compute
one perturbation at a time and add the solutions. So for
two perturbations, rather than one, we need $2x10^{10}$ steps to
multiply and add another 100,000 to add them, for a total of
$3x10^{10}$ steps, still much fewer steps than by computing b_{ij} the
accepted way.

Although the original derivation of the Half-Multiplier
Operator had nothing to do with science, one of the field
equations I developed $(c\Sigma(A_{ij}^T \circ T_{jk}) = G_{jk}$ (see derivation of
Einstein's First Field Equation for Gravity and the derivation
for the wave equation for gravity at the beginning of this
paper) reminded me so much of Einstein's First Field Equation
$(8\pi\rho T_{jk} = G_{jk})$ that I thought the operator might be applicable to
physics as well as accounting/inventory systems and the solving
of linear equations. If in the equation

$$(c\Sigma(A_{ij}^T \circ T_{jk}) = G_{jk}$$

$A_{ij} = I$ and $c = 8\pi\rho$, the equation reduces to Einstein's equation.
In Einstein's equation, T_{jk} is a square matrix, if in general, B_{jk}
is not a square matrix, $[I]_{ij}^T \circ [B]_{jk} = [B]_{jk}$. Therefore $[I]_{ij}^T$ can
also be defined as the identity matrix in the half-multiplier
mode. i.e.

$$
\begin{vmatrix}
1 & o & a_{11} & a_{12} \\
1 & o & a_{21} & a_{22} \\
. & o & . & . \\
1 & o & a_{i1} & a_{i2} \\
. & o & . & . \\
. & o & . & . \\
1 & o & a_{n1} & a_{n2}
\end{vmatrix}
=
\begin{vmatrix}
a_{11} & a_{12} \\
a_{21} & a_{22} \\
. & . \\
a_{i1} & a_{i2} \\
. & . \\
a_{n1} & a_{n2}
\end{vmatrix}
\equiv
\begin{vmatrix}
1 & & & & \\
& 1 & & & \\
& & 1 & & \\
& & & . & \\
& & & & 1
\end{vmatrix}
\begin{vmatrix}
a_{11} & a_{12} \\
a_{21} & a_{22} \\
. & . \\
a_{i1} & a_{i2} \\
. & . \\
a_{n1} & a_{n2}
\end{vmatrix}
$$

There is no new physics presented here, all the derivations were obtained from books on Quantum Chemistry and my notes from Quantum Mechanics I & II I took while in college at Fort Hays State in Hays, KS. What this paper covers is how to simplify the calculations so I could program the problem on my HP-48G calculator and MathCad +6 computer program.

So long as an exact or as close to exact solution can be obtained by hand or computer program first, no matter how much calculus or differential equations are used, the perturbation calculation can be used to further approximate solutions for tiny changes in the system using only regular addition, subtraction, multiplication and division, with some miscellaneous logarithms thrown in for good measure. I remember virtually nothing about matrix mechanics since, when I graduated 29 years ago, I have forgotten all the math I was ever taught. When I wanted to see if the operator worked in physics, I had to find a simple problem that even I could understand and solve. Non-degenerate perturbation theory stood out from all other matrix mechanics problems for it's simplicity and ease of comprehension. Everything else was beyond my mathematical abilities. This part of the paper is long, but that is because I cover every step with few or no shortcuts. To conserve on the length, therefore, there is more math than words of explanation. The math is so simple it should be understood with little trouble (as long as you understand how matrices are multiplied). Perturbation theory is not only applicable to chemistry, but also to electrical networks, mechanical vibrations, statistics and everything else where matrices are used to calculate linear systems.

EXAMPLE: COMPUTING THE WAVE FUNCTION
OF ACROLEIN FROM BUTADIENE

In Quantum Mechanics, the perturbation calculation is given as

$$\Psi^1_{i \text{ PERTURBED}} = (C_{ij})_{\text{NORM}} \Psi^0_i$$

with

$$C_{ij} = b_{ij}/(E^0_I - E^0_j) \; ; \; b_{ij} = (\Psi^0_i)^T H^1 \Psi^0_i$$

Because there is no onto mapping of matrices for matric multiplication defined, only into mappings for multiplication, (onto mappings are defined only for addition and subtraction), I get around this difficulty by taking the log of the b_{ij} matrix and the log of the ΔE_{ij} matrix defining any zero's as 10^{-100}. Since we are dividing the b_{ij} by ΔE_{ij}, we just subtract the logs. It is now a simple subtraction problem to divide each element individually without manually placing the divisions by hand. The half-multiplier operator takes care of this automatically. I have also extended the theory of multiple perturbations later in this paper. In this example, the energy for oxygen = A + εB where ε = 1. ε is not necessarily = 1, it depends upon the value determined by experiment. For butadiene,

$$
\Psi^0_1 =
\begin{array}{cccc}
\Psi_1 & \Psi_2 & \Psi_3 & \Psi_4 \\
\end{array}
\begin{vmatrix}
.3717 & .6015 & .6015 & .3717 \\
.6015 & .3717 & -.3717 & -.6015 \\
.6015 & -.3717 & -.3717 & .6015 \\
.3717 & -.6015 & .6015 & -.3717
\end{vmatrix}
$$

Suppose we wish now to solve for the Ψ function for acrolein (C=C-C=O). Instead of solving exactly, perturbation theory may be used. NOTE: For purposes of this paper, Ψ_I is defined as the I'th row of the Ψ_{ij} matrix; Ψ^0_i is defined as the whole Ψ matrix; Ψ_j is the j'th column of the Ψ_{ij} matrix. Ψ_I may also be denoted as Ψ_{1j} for the first row, Ψ_{2j} for the second row, etc. . . Ψ_{ij} = one single element of the matrix Ψ^0_i. i.e. Ψ_{23} is the element in Ψ^0_I corresponding to the third row, second column, in this case, it equals -.3717.

$\Psi^1_K = \Sigma C_{jk}\Psi^o_j$ and $b_{jk} = (\Psi^o_j)^T H^1 \Psi^o_K$; with E for oxygen = A + B. The off-diagonal elements are given by $C_{jk} = b_{jk}/(E^o_K - E^o_j)$, and $E_i = b_{ii}$. The Slater determinant is given by:

$$
\begin{vmatrix}
A-E & B & 0 & 0 \\
B & A-E & B & 0 \\
0 & B & A-E & B \\
0 & 0 & B & (A+B)-E
\end{vmatrix} = 0 \;;\; \text{letting } X = (A-E)/B;
\text{ we get }
\begin{vmatrix}
X & 1 & 0 & 0 \\
1 & X & 1 & 0 \\
0 & 1 & X & 1 \\
0 & 0 & 1 & X+1
\end{vmatrix}
\overset{H_0}{=}
\begin{vmatrix}
X & 1 & 0 & 0 \\
1 & X & 1 & 0 \\
0 & 1 & X & 1 \\
0 & 0 & 1 & X
\end{vmatrix}
+
\overset{H_1}{
\begin{vmatrix}
0 & 0 & 0 & 0 \\
0 & 0 & 0 & 0 \\
0 & 0 & 0 & 0 \\
0 & 0 & 0 & 1
\end{vmatrix}}
$$

First we will calculate b_{jk}:

$$ b_{jk} = (\Psi^o_j)^T H^1 \Psi^o_K $$

$$
b_{jk} =
\begin{vmatrix}
.3717 & .6015 & .6015 & .3717 \\
.6015 & .3717 & -.3717 & -.6015 \\
.6015 & -.3717 & -.3717 & .6015 \\
.3717 & -.6015 & .6015 & -.3717
\end{vmatrix}
\begin{vmatrix}
0 & 0 & 0 & 0 \\
0 & 0 & 0 & 0 \\
0 & 0 & 0 & 0 \\
0 & 0 & 0 & 1
\end{vmatrix}
\begin{vmatrix}
.3717 & .6015 & .6015 & .3717 \\
.6015 & .3717 & -.3717 & -.6015 \\
.6015 & -.3717 & -.3717 & .6015 \\
.3717 & -.6015 & .6015 & -.3717
\end{vmatrix} =
$$

$$
\begin{vmatrix}
.1382 & -.2236 & .2236 & -.1382 \\
-.2236 & .3618 & -.3618 & .2236 \\
.2236 & -.3618 & .3618 & -.2236 \\
-.1382 & .2236 & -.2236 & .1382
\end{vmatrix}
$$

The energy correction lies along the principle diagonal b_{ii}.

HP-48G PROGRAM:

```
RCL PSI01
TRN
RCL H1
x
RCL PSI01
x
STO bij
```

MATHCAD PROGRAM:

bij: $\Psi01^T * H1 * \Psi01$

To calculate the energy for acrolein, we proceed as follows:

BY HAND:

$E_1 = E^o_1 + E^1_1 = (A + 1.618B) + .1382B = \mathbf{A + 1.7562B}$

$E_2 = E^o_2 + E^1_2 = (A + .618B) + .3618B = \mathbf{A + .9798B}$

$$\mathbf{E_3} = E^0_3 + E^1_3 = (A - .618B) + .3618B = \mathbf{A - .2562B}$$
$$\mathbf{E_4} = E^0_4 + E^1_4 = (A - 1.618B) + .1382B = \mathbf{A - 1.4798B}$$

Or in matrix form:

$$\begin{vmatrix} A+1.618B & & & \\ & A+.618B & & \\ & & A-.618B & \\ & & & A-1.618B \end{vmatrix} + \begin{vmatrix} .1382B & & & \\ & .3618B & & \\ & & .3618B & \\ & & & .1382B \end{vmatrix} =$$

$$\begin{bmatrix} A+1.7562B & & & \\ & A+.9798B & & \\ & & A-.2562B & \\ & & & A-1.4798B \end{bmatrix}$$

Or

$$\begin{vmatrix} (A + 1.618B) \\ (A + .618B) \\ (A - .618B) \\ (A - 1.618B) \end{vmatrix} + \begin{vmatrix} .1382B \\ .3618B \\ .3618B \\ .1382B \end{vmatrix} = \begin{vmatrix} A + 1.7562B \\ A + .9798B \\ A - .2562B \\ A - 1.4798B \end{vmatrix}$$

HP-48G PROGRAM: (2 methods)

```
RCL bij            RCL bij
→DIAG              →DIAG
RCL E              4
+                  ↑
STO EACROLIN       DIAG→
                   RCL E
                   4
                   ↑
                   DIAG→
                   +
                   STO EACROLIN
```

To get MathCad to add symbolically, we must first divide by B:

$$\begin{bmatrix} \frac{A}{B}+1.618 & 0 & 0 & 0 \\ 0 & \frac{A}{B}+.618 & 0 & 0 \\ 0 & 0 & \frac{A}{B}-.618 & 0 \\ 0 & 0 & 0 & \frac{A}{B}-1.618 \end{bmatrix} + \begin{bmatrix} .1382 & 0 & 0 & 0 \\ 0 & .3618 & 0 & 0 \\ 0 & 0 & .3618 & 0 \\ 0 & 0 & 0 & .1382 \end{bmatrix} \rightarrow \begin{bmatrix} \frac{A}{B}+1.7562 & 0 & 0 & 0 \\ 0 & \frac{A}{B}+.9798 & 0 & 0 \\ 0 & 0 & \frac{A}{B}-.2562 & 0 \\ 0 & 0 & 0 & \frac{A}{B}-1.4798 \end{bmatrix}$$

We cannot do the following computations symbolically, so we must just use the values of B to calculate the energy.

$$bjk := \begin{bmatrix} .1382 & -.2236 & .2236 & -.1382 \\ -.2236 & .3618 & -.3618 & .2236 \\ .2236 & -.3618 & .3618 & -.2236 \\ -.1382 & .2236 & -.2236 & .1382 \end{bmatrix} \qquad E := \begin{bmatrix} 1.618 \\ .618 \\ -.618 \\ -1.618 \end{bmatrix}$$

$I4 := identity(4)$ creates a 4x4 identity matrix

$DIAGbjk := \overrightarrow{(I4 \cdot bjk)}$ multiplies onto b_{jk} to give diagonal of b_{jk}

$DIAGE := diag(E)$ diagonalizes the E matrix

$EACROLEIN = DIAGbjk + DIAGE$ adds the two diagonalized matrices giving us the B values.

$$EACROLEIN = \begin{bmatrix} 1.756 & 0 & 0 & 0 \\ 0 & 0.98 & 0 & 0 \\ 0 & 0 & -0.256 & 0 \\ 0 & 0 & 0 & -1.48 \end{bmatrix}$$

Another, more elegant method to compute the Energy of acrolein is to square by Half- Multiplier the row in $\Psi^0{}_1$ that corresponds to the row in which the perturbation occurs:

$$E^1{}_I = E_i + \varepsilon B (\Psi_4{}^T)^{02}$$

$$E^1{}_I = \begin{array}{c} E_i \\ \begin{vmatrix} 1.618 \\ .618 \\ -.618 \\ -1.618 \end{vmatrix} \end{array} + \begin{array}{c} \Psi_4{}^T \\ \begin{vmatrix} .3717 \\ -.6015 \\ .6015 \\ -.3717 \end{vmatrix} \end{array} \circ \begin{array}{c} \Psi_4{}^T \\ \begin{vmatrix} .3717 \\ -.6015 \\ .6015 \\ -.3717 \end{vmatrix} \end{array} = \begin{vmatrix} 1.618 \\ .618 \\ -.618 \\ -1.618 \end{vmatrix} + \begin{vmatrix} .1382 \\ .3618 \\ .3618 \\ .1382 \end{vmatrix} = \begin{vmatrix} 1.7562 \\ .9798 \\ -.2562 \\ -1.4798 \end{vmatrix}$$

To calculate all the non-diagonal elements, for all $i \neq j$; $C_{jk} = b_{jk} / (E^0{}_i - E^0{}_j)$. For a solution to the operator **using the Half-Multiplier Operator**, we will calculate the b_{jk} matrix as follows:

$$b_{jk} = (\Psi^0{}_j)^T H^1 \Psi^0{}_K = \Sigma (((\Psi^0{}_j)^T \circ H^1)^T \circ \Psi^0{}_K$$

$(\Psi^{o}_{j})^{T}$ **o** $\mathbf{H^{1}})^{T}$ **=** For Ψ_{1}, take the first row of (Ψ^{o}_{j}) [.3717 .6015 .6015 .3717], transpose, and o multiply into H^{1} We will have four solutions to this first part. For Ψ_{2}, take the second row of (Ψ^{o}_{j}) [.6015 .3717 -.3717 -.6015], transpose, and o multiply into (Ψ^{o}_{j}). Repeat for Ψ_{3} and Ψ_{4}.

$$b_{jk} = \begin{vmatrix} .3717 & .6015 & .6015 & .3717 \\ .6015 & .3717 & -.3717 & -.6015 \\ .6015 & -.3717 & -.3717 & .6015 \\ .3717 & -.6015 & .6015 & -.3717 \end{vmatrix} \mathbf{o} \begin{vmatrix} 0 & 0 & 0 & 0 \\ 0 & 0 & 0 & 0 \\ 0 & 0 & 0 & 0 \\ 0 & 0 & 0 & 1 \end{vmatrix}$$

$$= \left[\begin{vmatrix} .3717 \\ .6015 \\ .6015 \\ .3717 \end{vmatrix} \mathbf{o} \begin{vmatrix} 0 & 0 & 0 & 0 \\ 0 & 0 & 0 & 0 \\ 0 & 0 & 0 & 0 \\ 0 & 0 & 0 & 1 \end{vmatrix} \quad \begin{vmatrix} .6015 \\ .3717 \\ -.3717 \\ -.6015 \end{vmatrix} \mathbf{o} \begin{vmatrix} 0 & 0 & 0 & 0 \\ 0 & 0 & 0 & 0 \\ 0 & 0 & 0 & 0 \\ 0 & 0 & 0 & 1 \end{vmatrix} \quad \begin{vmatrix} .6015 \\ -.3717 \\ -.3717 \\ .6015 \end{vmatrix} \mathbf{o} \begin{vmatrix} 0 & 0 & 0 & 0 \\ 0 & 0 & 0 & 0 \\ 0 & 0 & 0 & 0 \\ 0 & 0 & 0 & 1 \end{vmatrix} \quad \begin{vmatrix} .3717 \\ -.6015 \\ .6015 \\ -.3717 \end{vmatrix} \mathbf{o} \begin{vmatrix} 0 & 0 & 0 & 0 \\ 0 & 0 & 0 & 0 \\ 0 & 0 & 0 & 0 \\ 0 & 0 & 0 & 1 \end{vmatrix} \right]$$

$$\begin{vmatrix} 0 & 0 & 0 & 0 \\ 0 & 0 & 0 & 0 \\ 0 & 0 & 0 & 0 \\ 0 & 0 & 0 & .3717 \end{vmatrix} \begin{vmatrix} 0 & 0 & 0 & 0 \\ 0 & 0 & 0 & 0 \\ 0 & 0 & 0 & 0 \\ 0 & 0 & 0 & -.6015 \end{vmatrix} \begin{vmatrix} 0 & 0 & 0 & 0 \\ 0 & 0 & 0 & 0 \\ 0 & 0 & 0 & 0 \\ 0 & 0 & 0 & .6015 \end{vmatrix} \begin{vmatrix} 0 & 0 & 0 & 0 \\ 0 & 0 & 0 & 0 \\ 0 & 0 & 0 & 0 \\ 0 & 0 & 0 & -.3717 \end{vmatrix}$$

Then using the Cross Product property of the Half-Multiplier, we sum the rows to form a single matrix from the Nested Array.

$$\begin{vmatrix} 0 & 0 & 0 & 0 \\ 0 & 0 & 0 & 0 \\ 0 & 0 & 0 & 0 \\ .3717 & -.6015 & .6015 & -.3717 \end{vmatrix}$$

Now we o multiply this value into Ψ^{o}_{1}: Note: the pre-multiplier matrix is already transposed, so we do not have to transpose it again.

$$\begin{vmatrix} 0 & 0 & 0 & 0 \\ 0 & 0 & 0 & 0 \\ 0 & 0 & 0 & 0 \\ .3717 & -.6015 & .6015 & -.3717 \end{vmatrix} \mathbf{o} \begin{vmatrix} .3717 & .6015 & .6015 & .3717 \\ .6015 & .3717 & -.3717 & -.6015 \\ .6015 & -.3717 & -.3717 & .6015 \\ .3717 & -.6015 & .6015 & -.3717 \end{vmatrix} =$$

$$\begin{vmatrix} 0 \\ 0 \\ 0 \\ .3717 \end{vmatrix} \mathbf{o} \begin{vmatrix} .3717 & .6015 & .6015 & .3717 \\ .6015 & .3717 & -.3717 & -.6015 \\ .6015 & -.3717 & -.3717 & .6015 \\ .3717 & -.6015 & .6015 & -.3717 \end{vmatrix} \quad \begin{vmatrix} 0 \\ 0 \\ 0 \\ -.6015 \end{vmatrix} \mathbf{o} \begin{vmatrix} .3717 & .6015 & .6015 & .3717 \\ .6015 & .3717 & -.3717 & -.6015 \\ .6015 & -.3717 & -.3717 & .6015 \\ .3717 & -.6015 & .6015 & -.3717 \end{vmatrix}$$

$$\begin{vmatrix} 0 \\ 0 \\ 0 \\ .6015 \end{vmatrix} \mathbf{o} \begin{vmatrix} .3717 & .6015 & .6015 & .3717 \\ .6015 & .3717 & -.3717 & -.6015 \\ .6015 & -.3717 & -.3717 & .6015 \\ .3717 & -.6015 & .6015 & -.3717 \end{vmatrix} \quad \begin{vmatrix} 0 \\ 0 \\ 0 \\ -.3717 \end{vmatrix} \mathbf{o} \begin{vmatrix} .3717 & .6015 & .6015 & .3717 \\ .6015 & .3717 & -.3717 & -.6015 \\ .6015 & -.3717 & -.3717 & .6015 \\ .3717 & -.6015 & .6015 & -.3717 \end{vmatrix} =$$

Note: these matrices should be in a single row, I just do not have the room on this page to place them so. Now we o multiply all the sub-matrices and keep them in a single row:

$$
\left\|
\begin{array}{cccc}
0 & 0 & 0 & 0 \\
0 & 0 & 0 & 0 \\
0 & 0 & 0 & 0 \\
.1382 & -.2236 & .2236 & -.1382
\end{array}
\right\|
\left\|
\begin{array}{cccc}
0 & 0 & 0 & 0 \\
0 & 0 & 0 & 0 \\
0 & 0 & 0 & 0 \\
-.2236 & .3618 & -.3618 & .2236
\end{array}
\right\|
\left\|
\begin{array}{cccc}
0 & 0 & 0 & 0 \\
0 & 0 & 0 & 0 \\
0 & 0 & 0 & 0 \\
.2236 & -.3618 & .3618 & -.2236
\end{array}
\right\|
\left\|
\begin{array}{cccc}
0 & 0 & 0 & 0 \\
0 & 0 & 0 & 0 \\
0 & 0 & 0 & 0 \\
.1382 & .2236 & -.2236 & .1382
\end{array}
\right\|
$$

Now we wish to sum the columns and create a new matrix of these summed columns, so lets transpose the Nested Array, keeping each sub-matrix as a 4x4 square of elements.

$$
\left\|
\begin{array}{cccc}
0 & 0 & 0 & 0 \\
0 & 0 & 0 & 0 \\
0 & 0 & 0 & 0 \\
.1382 & -.2236 & .2236 & -.1382
\end{array}
\right\|
\left\|
\begin{array}{cccc}
0 & 0 & 0 & 0 \\
0 & 0 & 0 & 0 \\
0 & 0 & 0 & 0 \\
-.2236 & .3618 & -.3618 & .2236
\end{array}
\right\|
\left\|
\begin{array}{cccc}
0 & 0 & 0 & 0 \\
0 & 0 & 0 & 0 \\
0 & 0 & 0 & 0 \\
.2236 & -.3618 & .3618 & -.2236
\end{array}
\right\|
\left\|
\begin{array}{cccc}
0 & 0 & 0 & 0 \\
0 & 0 & 0 & 0 \\
0 & 0 & 0 & 0 \\
-.1382 & .2236 & -.2236 & .1382
\end{array}
\right\|^{\text{T}} =
$$

Now we remove the inner brackets, partitioning the Nested Array into a single Matrix and multiply.

$$
\left\|
\begin{array}{cccccccccccccccc}
1 & 1 & 1 & 1 & 0 & 0 & 0 & 0 & 0 & 0 & 0 & 0 & 0 & 0 & 0 & 0 \\
0 & 0 & 0 & 0 & 1 & 1 & 1 & 1 & 0 & 0 & 0 & 0 & 0 & 0 & 0 & 0 \\
0 & 0 & 0 & 0 & 0 & 0 & 0 & 0 & 1 & 1 & 1 & 1 & 0 & 0 & 0 & 0 \\
0 & 0 & 0 & 0 & 0 & 0 & 0 & 0 & 0 & 0 & 0 & 0 & 1 & 1 & 1 & 1
\end{array}
\right\|
\left\|
\begin{array}{cccc}
0 & 0 & 0 & 0 \\
0 & 0 & 0 & 0 \\
0 & 0 & 0 & 0 \\
.1382 & -.2236 & .2236 & -.1382 \\
0 & 0 & 0 & 0
\end{array}
\right\|
$$

Now we remove the inner brackets, partitioning the Nested Array into a single Matrix and multiply.

$$
\left\|
\begin{array}{cccccccccccccccc}
1 & 1 & 1 & 1 & 0 & 0 & 0 & 0 & 0 & 0 & 0 & 0 & 0 & 0 & 0 & 0 \\
0 & 0 & 0 & 0 & 1 & 1 & 1 & 1 & 0 & 0 & 0 & 0 & 0 & 0 & 0 & 0 \\
0 & 0 & 0 & 0 & 0 & 0 & 0 & 0 & 1 & 1 & 1 & 1 & 0 & 0 & 0 & 0 \\
0 & 0 & 0 & 0 & 0 & 0 & 0 & 0 & 0 & 0 & 0 & 0 & 1 & 1 & 1 & 1
\end{array}
\right\|
\left\|
\begin{array}{cccc}
0 & 0 & 0 & 0 \\
0 & 0 & 0 & 0 \\
0 & 0 & 0 & 0 \\
.1382 & -.2236 & .2236 & -.1382 \\
0 & 0 & 0 & 0 \\
0 & 0 & 0 & 0 \\
-.2236 & .3618 & -.3618 & .2236 \\
0 & 0 & 0 & 0 \\
0 & 0 & 0 & 0 \\
0 & 0 & 0 & 0 \\
.2236 & -.3618 & .3618 & -.2236 \\
0 & 0 & 0 & 0 \\
0 & 0 & 0 & 0 \\
0 & 0 & 0 & 0 \\
-.1382 & .2236 & -.2236 & .1382
\end{array}
\right\|
=
\left\|
\begin{array}{cccc}
.1382 & -.2236 & .2236 & -.1382 \\
-.2236 & .3618 & -.3618 & .2236 \\
.2236 & -.3618 & .3618 & -.2236 \\
-.1382 & .2236 & -.2236 & .1382
\end{array}
\right\|
$$

Which is the same solution we got for $b_{jk} = (\Psi^{\circ}{}_{j})^{\text{T}} H^{1} \Psi^{\circ}{}_{\kappa}$. But what a lot of math to go through to get the same answer. There is a pattern, let's see if it can take us anywhere. All values in the problem after the first Half-Multiplication are between the elements in row four and row four itself. i.e.

$$
b_{jk} =
\left\|
\begin{array}{c}
\Psi_{41} \circ \Psi_4 \\
\Psi_{42} \circ \Psi_4 \\
\Psi_{43} \circ \Psi_4 \\
\Psi_{44} \circ \Psi_4
\end{array}
\right\|
=
\left\|
\left\|
\begin{array}{c}
.3717 \\
-.6015 \\
.6015 \\
-.3717
\end{array}
\right\|
\circ
\left\|
\begin{array}{cccc}
.3717 & -.6015 & .6015 & -.3717 \\
.3717 & -.6015 & .6015 & -.3717 \\
.3717 & -.6015 & .6015 & -.3717 \\
.3717 & -.6015 & .6015 & -.3717
\end{array}
\right\|
\right\|
=
$$

$$
\left\|
\begin{array}{cccc}
.1382 & -.2236 & .2236 & -.1382 \\
-.2236 & .3618 & -.3618 & .2236 \\
.2236 & -.3618 & .3618 & -.2236 \\
-.1382 & .2236 & -.2236 & .1382
\end{array}
\right\|
= b_{jk} = \varepsilon B \sum_{1}^{4} \Psi_{4k} \circ \Psi_4
$$

Hmmm, 16 mathematical steps, this is a step in the right direction. Since the matrix is symmetric, $b_{jk} = \varepsilon B\sum \Psi_{j4} \circ \Psi_4$ also works. The energy E is equal to the diagonal, Or $E = \varepsilon B\Psi_{jj}$.

GENERAL PROOF

To generalize, suppose we have

$$H^1 = \begin{vmatrix} 0_{11} & & & & & \\ & 0_{22} & & & & \\ & & 0_{33} & & & \\ & & & \cdot & & \\ & & & & \varepsilon B_{jj} & \\ & & & & & \cdot \\ & & & & & & 0_{nn} \end{vmatrix} \quad \text{and } \Psi_{\tau} = \Psi_j^{\ T} \quad \text{(matrices are symmetric)}$$

then

$$b_{jk} = \begin{vmatrix} \Psi_{j1} \circ \Psi_j \\ \Psi_{j2} \circ \Psi_j \\ \Psi_{j3} \circ \Psi_j \\ \cdot \\ \Psi_{jk} \circ \Psi_j \\ \cdot \\ \Psi_{jn} \circ \Psi_j \end{vmatrix} \varepsilon B = \varepsilon B\sum \Psi_{jk} \circ \Psi_{\tau} = \varepsilon B\sum (\Psi_{kj})^T \circ \Psi_{\tau}$$

$$E = \varepsilon B\varepsilon B\sum \Psi_{jk} \circ [\Psi_{\tau}]_{jj}$$

We put $[\Psi_{\tau}]_{jj}$ in brackets to show it is a square diagonal matrix, not the jj'th element of the matrix.

But there is a simpler way to express this for a perturbation of the fourth row, we take the original multiplication and get rid of all the zero's, leaving only the rows with numbers. i.e. For a perturbation at the fourth carbon, we can write [.3717 -.6015 .6015 -.3717] $\circ$[.3717 -.6015 .6015 -.3717] = $\Psi_4 \circ \Psi_4 = \Psi_4^{\circ 2}$ which is different from $(\Psi_4^T)^{\circ 2}$ which gives the energy correction term. We are not going to solve $\Psi_4 \circ \Psi_4$ because We have already solved it above. Remember, the Half-Multiplier has a one-to-one relationship with regular matrix multiplication, all we need to do is transpose the first matrix and multiply, so lets try it.

$$\Psi_4^{\circ 2} = \Psi_4 \circ \Psi_4 = \Psi_4^T \Psi_4 = \begin{vmatrix} .3717 \\ -.6015 \\ .6015 \\ -.3717 \end{vmatrix} [.3717 \ -.6015 \ .6015 \ -.3717] =$$

$$\begin{vmatrix} .1382 & -.2236 & .2236 & -.1382 \\ -.2236 & .3618 & -.3618 & .2236 \\ .2236 & -.3618 & .3618 & -.2236 \\ -.1382 & .2236 & -.2236 & .1382 \end{vmatrix} = b_{jk} = \varepsilon B\sum_1^4 \Psi_{4x} \circ \Psi_4 = (\Psi^{\circ}_j)^T H^1 \Psi^{\circ}_K = \Psi_4^{\circ 2} = \Psi_4 \circ \Psi_4 = \Psi_4^T \Psi_4$$

For convenience, I'm going to call the row in which the perturbation occurs as Ψ_B, in this case, $\Psi_B = \Psi_4$. I am now going to attempt to prove $(\Psi^{\circ}_j)^T H^1 \Psi^{\circ}_K = \Psi_4^{\circ 2}$.

$$
\begin{vmatrix}
a_{11} & a_{21} & \cdots & a_{j1} & \cdots & a_{n1} & a_{n+1,1} \\
a_{12} & a_{22} & \cdots & a_{j2} & \cdots & a_{n2} & a_{n+1,2} \\
\cdot & \cdot & & \cdot & & \cdot & \cdot \\
a_{1k} & a_{2k} & \cdots & a_{jk} & \cdots & a_{nk} & a_{n+1,k} \\
\cdot & \cdot & & \cdot & & \cdot & \cdot \\
a_{1n} & a_{2n} & \cdots & a_{jn} & \cdots & a_{nn} & a_{n+1,n} \\
a_{1,n+1} & a_{2,n+1} & \cdots & a_{j,n+1} & \cdots & a_{n,n+1} & a_{n+1,n+1}
\end{vmatrix}
$$

$$
\begin{vmatrix}
a_{11} & a_{21} & \cdots & a_{j1} & \cdots & a_{n1} & a_{n+1,1} \\
a_{12} & a_{22} & \cdots & a_{j2} & \cdots & a_{n2} & a_{n+1,2} \\
\cdot & \cdot & & \cdot & & \cdot & \cdot \\
a_{1k} & a_{2k} & \cdots & a_{jk} & \cdots & a_{nk} & a_{n+1,k} \\
\cdot & \cdot & & \cdot & & \cdot & \cdot \\
a_{1n} & a_{2n} & \cdots & a_{jn} & \cdots & a_{nn} & a_{n+1,n} \\
a_{1,n+1} & a_{2,n+1} & \cdots & a_{j,n+1} & \cdots & a_{n,n+1} & a_{n+1,n+1}
\end{vmatrix}
\begin{vmatrix}
0 & & & & & \\
& 0 & & & & \\
& & 0 & & & \\
& & & \cdot & & \\
& & & & \varepsilon B_{jj} & \\
& & & & \cdot & \\
& & & & & 0_{n+1,n+1}
\end{vmatrix}
$$

$$
\begin{vmatrix}
a_{11} & a_{21} & \cdots & a_{j1} & \cdots & a_{n1} & a_{n+1,1} \\
a_{12} & a_{22} & \cdots & a_{j2} & \cdots & a_{n2} & a_{n+1,2} \\
\cdot & \cdot & & \cdot & & \cdot & \cdot \\
a_{1k} & a_{2k} & \cdots & a_{jk} & \cdots & a_{nk} & a_{n+1,k} \\
\cdot & \cdot & & \cdot & & \cdot & \cdot \\
a_{1n} & a_{2n} & \cdots & a_{jn} & \cdots & a_{nn} & a_{n+1,n} \\
a_{1,n+1} & a_{2,n+1} & \cdots & a_{j,n+1} & \cdots & a_{n,n+1} & a_{n+1,n+1}
\end{vmatrix}
\begin{vmatrix}
0 & & & & & \\
& 0 & & & & \\
& & 0 & & & \\
& & & \cdot & & \\
& & & & \varepsilon B_{jj} & \\
& & & & \cdot & \\
& & & & & 0_{n+1,n+1}
\end{vmatrix}
\begin{vmatrix}
a_{11} & a_{12} & \cdots & a_{1k} & \cdots & a_{1n} & a_{1,n+1} \\
a_{21} & a_{22} & \cdots & a_{2k} & \cdots & a_{2n} & a_{2,n+1} \\
\cdot & \cdot & & \cdot & & \cdot & \cdot \\
a_{j1} & a_{j2} & \cdots & a_{jk} & \cdots & a_{jn} & a_{j,n+1} \\
\cdot & \cdot & & \cdot & & \cdot & \cdot \\
a_{n1} & a_{n2} & \cdots & a_{nk} & \cdots & a_{nn} & a_{n+1,n} \\
a_{n+1,1} & a_{n+1,2} & \cdots & a_{n+1,k} & \cdots & a_{n+1,n} & a_{n+1,n+1}
\end{vmatrix}
$$

STEP 1: Multiply $(\Psi^o_j)^T H^1$

$$
\begin{vmatrix}
a_{11} & a_{21} & \cdots & a_{j1} & \cdots & a_{n1} & a_{n+1,1} \\
a_{12} & a_{22} & \cdots & a_{j2} & \cdots & a_{n2} & a_{n+1,2} \\
\cdot & \cdot & & \cdot & & \cdot & \cdot \\
a_{1k} & a_{2k} & \cdots & a_{jk} & \cdots & a_{nk} & a_{n+1,k} \\
\cdot & \cdot & & \cdot & & \cdot & \cdot \\
a_{1n} & a_{2n} & \cdots & a_{jn} & \cdots & a_{nn} & a_{n+1,n} \\
a_{1,n+1} & a_{2,n+1} & \cdots & a_{j,n+1} & \cdots & a_{n,n+1} & a_{n+1,n+1}
\end{vmatrix}
\begin{vmatrix}
0 & & & & & \\
& 0 & & & & \\
& & 0 & & & \\
& & & \cdot & & \\
& & & & 1 & \\
& & & & \cdot & \\
& & & & & 0_{n+1,n+1}
\end{vmatrix}
\varepsilon B =\varepsilon B
\begin{vmatrix}
0 & 0 & 0 & \cdot & \cdot & a_{j1} & \cdot & 0 \\
0 & 0 & 0 & \cdot & \cdot & a_{j2} & \cdot\cdot & 0 \\
\cdot & \cdot & \cdot & & & \cdot & & 0 \\
\cdot & \cdot & \cdot & \cdot & & a_{jk} & \cdot\cdot & 0 \\
\cdot & \cdot & \cdot & \cdot & & \cdot & & 0 \\
0 & 0 & 0 & \cdot & \cdot & a_{jn} & \cdot\cdot & 0 \\
0 & 0 & 0 & \cdot & \cdot & a_{j,n+1} & \cdot & 0
\end{vmatrix}
$$

STEP 2: Multiply Ψ^o_j by the product of $(\Psi^o_j)^T H^1$:

$$
\begin{vmatrix}
0 & 0 & 0 & \cdot & \cdot & a_{j1} & \cdot & 0 \\
0 & 0 & 0 & \cdot & \cdot & a_{j2} & \cdot & 0 \\
\cdot & \cdot & \cdot & & & \cdot & & 0 \\
\cdot & \cdot & \cdot & & \cdot & \cdot & & 0 \\
\cdot & \cdot & \cdot & \cdot & & a_{jk} & \cdot\cdot & 0 \\
\cdot & \cdot & \cdot & & \cdot & \cdot & & 0 \\
0 & 0 & 0 & \cdot & \cdot & a_{jn} & \cdot\cdot & 0 \\
0 & 0 & 0 & \cdot & \cdot & a_{j,n+1} & \cdot & 0
\end{vmatrix}
\varepsilon B
\begin{vmatrix}
a_{11} & a_{12} & \cdots & a_{1k} & \cdots & a_{1n} & a_{1,n+1} \\
a_{21} & a_{22} & \cdots & a_{2k} & \cdots & a_{2n} & a_{2,n+1} \\
\cdot & \cdot & & \cdot & & \cdot & \cdot \\
a_{j1} & a_{j2} & \cdots & a_{jk} & \cdots & a_{jn} & a_{j,n+1} \\
\cdot & \cdot & & \cdot & & \cdot & \cdot \\
a_{n1} & a_{n2} & \cdots & a_{nk} & \cdots & a_{nn} & a_{n+1,n} \\
a_{n+1,1} & a_{n+1,2} & \cdots & a_{n+1,k} & \cdots & a_{n+1,n} & a_{n+1,n+1}
\end{vmatrix}
=
$$

$$
\begin{vmatrix}
a_{j1}a_{j1} & a_{j1}a_{j2} & \cdots & a_{j1}a_{jk} & \cdots & a_{j1}a_{jn} & a_{j1}a_{j,n+1} \\
a_{j1}a_{j2} & a_{j2}a_{j2} & \cdots & a_{j2}a_{jk} & \cdots & a_{j2}a_{jn} & a_{j2}a_{j,n+1} \\
\cdot & \cdot & & \cdot & & \cdot & \cdot \\
a_{j1}a_{jk} & a_{j2}a_{jk} & \cdots & a_{jk}a_{jk} & \cdots & a_{jk}a_{jn} & a_{jk}a_{j,n+1} \\
\cdot & \cdot & & \cdot & & \cdot & \cdot \\
a_{j1}a_{jn} & a_{j2}a_{jn} & \cdots & a_{jk}a_{jn} & \cdots & a_{jn}a_{jn} & a_{jn}a_{j,n+1} \\
a_{j1}a_{j,n+1} & a_{j2}a_{j,n+1} & \cdots & a_{jk}a_{j,n+1} & \cdots & a_{jn}a_{j,n+1} & a_{j,n+1}a_{j,n+1}
\end{vmatrix}
\qquad \varepsilon B = (\Psi^o_j)^T H^1 \Psi^o_K
$$

Now we'll multiply $\Psi_B^T * \Psi_B$ but I'll do this by MathCad symbolic:

$$
\begin{bmatrix} aj1 \\ aj2 \\ ajk \\ ajn \\ ajn+1 \end{bmatrix} \cdot \begin{pmatrix} aj1 & aj2 & ajk & ajn & ajn+1 \end{pmatrix} \rightarrow
\begin{bmatrix}
aj1^2 & aj1\cdot aj2 & aj1\cdot ajk & aj1\cdot ajn & aj1\cdot(ajn+1) \\
aj1\cdot aj2 & aj2^2 & aj2\cdot ajk & aj2\cdot ajn & aj2\cdot(ajn+1) \\
aj1\cdot ajk & aj2\cdot ajk & ajk^2 & ajk\cdot ajn & ajk\cdot(ajn+1) \\
aj1\cdot ajn & aj2\cdot ajn & ajk\cdot ajn & ajn^2 & ajn\cdot(ajn+1) \\
aj1\cdot(ajn+1) & aj2\cdot(ajn+1) & ajk\cdot(ajn+1) & ajn\cdot(ajn+1) & (ajn+1)^2
\end{bmatrix}
$$

NOTE: In MathCad symbolic, the program won't take commas in the subscripts, nor will it take the . . . notation, so to rewrite it in proper form we get:

$$
\left|
\begin{matrix}
a_{j1}a_{j1} & a_{j1}a_{j2}. & . & . & a_{j1}a_{jk} & . & . & . & a_{j1}a_{jn} & a_{j1}a_{j,n+1} \\
a_{j1}a_{j2} & a_{j2}a_{j2}. & . & . & a_{j2}a_{jk} & . & . & . & a_{j2}a_{jn} & a_{j2}a_{j,n+1} \\
. & . & . & . & . & . & . & . & . & . \\
a_{j1}a_{jk} & a_{j2}a_{jk}. & . & . & a_{jk}a_{jk} & . & . & . & a_{jk}a_{jn} & a_{jk}a_{j,n+1} \\
. & . & . & . & . & . & . & . & . & . \\
a_{j1}a_{jn} & a_{j2}a_{jn}. & . & . & a_{jk}a_{jn} & . & . & . & a_{jn}a_{jn} & a_{jn}a_{j,n+1} \\
a_{j1}a_{j,n+1} & a_{j2}a_{j,n+1} & . & . & a_{jk}a_{j,n+1} & . & . & a_{jn}a_{j,n+1} & a_{j,n+1}a_{j,n+1}
\end{matrix}
\right|
\quad \varepsilon B = (\Psi^o_j)^T {}^1_H \Psi^o_K = \Psi_B{}^{o2} = \Psi_B^T * \Psi_B = b_{jk}
$$

QED

Now we will compute $C_{jk} = b_{jk}/(E_j - E_k)$

First I am going to **re-define the [1] operator.** Up until now I have been using it as a row or a column matrix, but now I want to define it as an n x m matrix, or in the case of Quantum Chemistry, as an n x n matrix.

Define $\delta_{jj}[1]_{jj}$ j=k=1; and ${}^{-1}\delta_{jj}[1]_{jj}$ j=k=0
$\phantom{Define \delta_{jj}[1]_{jj}}$ j≠k=0 $\phantom{; and   {}^{-1}\delta_{jj}[1]_{jj}}$ j≠k=1

such that $\delta_{jj}[1]_{jj} + {}^{-1}\delta_{jj}[1]_{jj} = [1]_{jj}$. To illustrate for j=k=4:

$\delta_{jj}[1]_{jj} \; + \; {}^{-1}\delta_{jj}[1]_{jj} = \; [1]_{jj}$

$$
\begin{vmatrix} 1 & 0 & 0 & 0 \\ 0 & 1 & 0 & 0 \\ 0 & 0 & 1 & 0 \\ 0 & 0 & 0 & 1 \end{vmatrix} +
\begin{vmatrix} 0 & 1 & 1 & 1 \\ 1 & 0 & 1 & 1 \\ 1 & 1 & 0 & 1 \\ 1 & 1 & 1 & 0 \end{vmatrix} =
\begin{vmatrix} 1 & 1 & 1 & 1 \\ 1 & 1 & 1 & 1 \\ 1 & 1 & 1 & 1 \\ 1 & 1 & 1 & 1 \end{vmatrix}
$$

Define $B_{jk} = (b_{jk} - E_{jj} + \delta_{jj}[1]_{jj}) = (\mathbf{b_{jk}} - \mathbf{b_{jj}} + \boldsymbol{\delta_{jj}}[\mathbf{1}]_{\mathbf{jj}})$

Define $\Delta E_{jk} - \delta_{jj}[1]_{jj} = (E_1)^T \circ ({}^{-1}\delta_{jj}[1]_{jj})$

Define Subtraction as $\Delta E^-_{jk} - \delta_{jj}[1]_{jj} = (E_1)^T \circ ({}^{-1}\delta_{jj}[1]_{jj}) - ((E_1)^T \circ$
$({}^{-1}\delta_{jj}[1]_{jj}))^T$

Define addition as $\Delta E^+_{jk} - \delta_{jj}[1]_{jj} = (E_1)^T \circ ({}^{-1}\delta_{jj}[1]_{jj}) + ((E_1)^T \circ$
$({}^{-1}\delta_{jj}[1]_{jj}))^T$

Note: This mess is not actually necessary if we only wish to perform subtraction, the subtraction process itself puts zero's in the principle diagonal. But if for some reason we wish to add the energies, the above equation is the only way it will work.

Again

$$C_{jk} = b_{jk}/(E_j - E_k)$$

Therefore, $\text{Log } B_{jk} - \text{Log } \Delta E^-_{jk} = \text{Log } C_{jk}$

and thus

$$\text{Anti-log } (C_{jk})_{\text{NORM}} \Psi^0_1 = \Psi^1_{i \text{ PERTURBED}}$$

Then for the perturbation at Ψ_4:

$$b_4 = \Psi_4^{\circ 2} = \Psi_4^T \Psi_4$$

$$b_4 = \Psi_4^{\circ 2} = \Psi_4 \circ \Psi_4 = \Psi_4^T \Psi_4 =$$

$$[.3717 \; -.6015 \; .6015 \; -.3717] = \begin{vmatrix} .3717 \\ -.6015 \\ .6015 \\ -.3717 \end{vmatrix} = \begin{vmatrix} .1382 & -.2236 & .2236 & -.1382 \\ -.2236 & .3618 & -.3618 & .2236 \\ .2236 & -.3618 & .3618 & -.2236 \\ -.1382 & .2236 & -.2236 & .1382 \end{vmatrix}$$

$$B_{jk} = (b_{jk} - E_{jj} + \delta_{jj}[1]_{jj}) = (b_{jk} - b_{jj} + \delta_{jj}[1]_{jj})$$

$$B_4 = \begin{vmatrix} .1382 & -.2236 & .2236 & -.1382 \\ -.2236 & .3618 & -.3618 & .2236 \\ .2236 & -.3618 & .3618 & -.2236 \\ -.1382 & .2236 & -.2236 & .1382 \end{vmatrix} - \begin{vmatrix} .1382 & & & \\ & .3618 & & \\ & & .3618 & \\ & & & .1382 \end{vmatrix} + \begin{vmatrix} 1 & 0 & 0 & 0 \\ 0 & 1 & 0 & 0 \\ 0 & 0 & 1 & 0 \\ 0 & 0 & 0 & 1 \end{vmatrix} =$$

$$B_4 = \begin{vmatrix} 1 & -.2236 & .2236 & -.1382 \\ -.2236 & 1 & -.3618 & .2236 \\ .2236 & -.3618 & 1 & -.2236 \\ -.1382 & .2236 & -.2236 & 1 \end{vmatrix}$$

COMPUTING ΔE

Now we've computed b_{ij} and put ones in the diagonal. Now we must divide each element by it's $E_j - E_k$ value. We don't have to do it this way, because when we subtract the E matrix and it's transpose, the diagonal will be all zero's anyway. But I'm going to show the other way, just in case we ever have to add the energies rather than subtract them.

The subtraction matrix is (in general)

$$\begin{pmatrix} E_1 & E_2 & E_3 & E_4 \\ -E_1 & -E_2 & -E_3 & -E_4 \end{pmatrix}$$

Then $\Delta E^-_{jk} - \delta_{jj}[1]_{jj} = (E_1)^T \circ ({}^{-1}\delta_{jj}[1]_{jj}) - ((E_1)^T \circ ({}^{-1}\delta_{jj}[1]_{jj}))^T$

We have derived an wxpression for the energy in a single equation.

Let's translate so we can understand what this equation means:

$$\begin{vmatrix} E_1 & -E_1 \\ E_2 & -E_2 \\ E_3 & -E_3 \\ E_4 & -E_4 \end{vmatrix} \circ \begin{vmatrix} 0 & 1 & 1 & 1 \\ 1 & 0 & 1 & 1 \\ 1 & 1 & 0 & 1 \\ 1 & 1 & 1 & 0 \end{vmatrix} = \begin{vmatrix} E_1 \\ E_2 \\ E_3 \\ E_4 \end{vmatrix} \circ \begin{vmatrix} 0 & 1 & 1 & 1 \\ 1 & 0 & 1 & 1 \\ 1 & 1 & 0 & 1 \\ 1 & 1 & 1 & 0 \end{vmatrix} - \left(\begin{vmatrix} E_1 \\ E_2 \\ E_3 \\ E_4 \end{vmatrix} \circ \begin{vmatrix} 0 & 1 & 1 & 1 \\ 1 & 0 & 1 & 1 \\ 1 & 1 & 0 & 1 \\ 1 & 1 & 1 & 0 \end{vmatrix} \right)^T + \delta_{jj}[1]_{jj}$$

$$\Delta E^-_{jk} = \begin{vmatrix} 0 & E_1 & E_1 & E_1 \\ E_2 & 0 & E_2 & E_2 \\ E_3 & E_3 & 0 & E_3 \\ E_4 & E_4 & E_4 & 0 \end{vmatrix} - \begin{vmatrix} 0 & E_2 & E_3 & E_4 \\ E_1 & 0 & E_3 & E_4 \\ E_1 & E_2 & 0 & E_4 \\ E_1 & E_2 & E_3 & 0 \end{vmatrix} + \begin{vmatrix} 1 & 0 & 0 & 0 \\ 0 & 1 & 0 & 0 \\ 0 & 0 & 1 & 0 \\ 0 & 0 & 0 & 1 \end{vmatrix} = \begin{vmatrix} 1 & E_1-E_2 & E_1-E_3 & E_1-E_4 \\ E_2-E_1 & 1 & E_2-E_3 & E_2-E_4 \\ E_3-E_1 & E_3-E_2 & 1 & E_3-E_4 \\ E_4-E_1 & E_4-E_2 & E_4-E_3 & 1 \end{vmatrix}$$

Now $\Delta E^-_{jk} = \begin{vmatrix} 1.618 & .618 & -.618 & -1.618 \\ -1.618 & -.618 & .618 & 1.618\backslash \end{vmatrix}$,

Transposing ΔE^-_{jk} and multiplying we get:

$$\begin{vmatrix} 1.618 & -1.618 \\ .618 & -1.618 \\ -.618 & .618 \\ -1.618 & 1.618 \end{vmatrix} \circ \begin{vmatrix} 0 & 1 & 1 & 1 \\ 1 & 0 & 1 & 1 \\ 1 & 1 & 0 & 1 \\ 1 & 1 & 1 & 0 \end{vmatrix} = \begin{vmatrix} 1.618 \\ .618 \\ -.618 \\ -1.618 \end{vmatrix} \circ \begin{vmatrix} 0 & 1 & 1 & 1 \\ 1 & 0 & 1 & 1 \\ 1 & 1 & 0 & 1 \\ 1 & 1 & 1 & 0 \end{vmatrix} + \left(\begin{vmatrix} -1.618 \\ -.618 \\ .618 \\ 1.618 \end{vmatrix} \circ \begin{vmatrix} 0 & 1 & 1 & 1 \\ 1 & 0 & 1 & 1 \\ 1 & 1 & 0 & 1 \\ 1 & 1 & 1 & 0 \end{vmatrix} \right)^T + \delta_{jj}[1]_{jj} =$$

$$
\Delta E := \begin{bmatrix} 0 & 1.618 & 1.618 & 1.618 \\ .618 & 0 & .618 & .618 \\ -.618 & -.618 & 0 & -.618 \\ -1.618 & -1.618 & -1.618 & 0 \end{bmatrix} - \begin{bmatrix} 0 & .618 & -.618 & -1.618 \\ 1.618 & 0 & -.618 & -1.618 \\ 1.618 & .618 & 0 & -1.618 \\ 1.618 & .618 & -.618 & 0 \end{bmatrix} + \begin{bmatrix} 1 & 0 & 0 & 0 \\ 0 & 1 & 0 & 0 \\ 0 & 0 & 1 & 0 \\ 0 & 0 & 0 & 1 \end{bmatrix} *
$$

$$
\Delta E = \begin{bmatrix} 1 & 1 & 2.236 & 3.236 \\ -1 & 1 & 1.236 & 2.236 \\ -2.236 & -1.236 & 1 & 1 \\ -3.236 & -2.236 & -1 & 1 \end{bmatrix}
$$

$\text{LOG}\Delta E := \overrightarrow{\log(\Delta E)}$ We use the vectorize operator to take the log of ΔE.

$$
\text{LOG}\Delta E = \begin{bmatrix} 0 & 0 & 0.349 & 0.51 \\ 1.364i & 0 & 0.092 & 0.349 \\ 0.349 + 1.364i & 0.092 + 1.364i & 0 & 0 \\ 0.51 + 1.364i & 0.349 + 1.364i & 1.364i & 0 \end{bmatrix}
$$

$$
\text{B4} := \begin{bmatrix} 1 & -.2236 & .2236 & -.1382 \\ -.2236 & 1 & -.3618 & .2236 \\ .2236 & -.3618 & 1 & -.2236 \\ -.1382 & .2236 & -.2236 & 1 \end{bmatrix} *
$$

$\text{LOGB4} := \overrightarrow{\log(\text{B4})}$ We use the vectorize operator to take the log of B_4.

$$
\text{LOGB4} = \begin{bmatrix} 0 & -0.651 + 1.364i & -0.651 & -0.859 + 1.364i \\ -0.651 + 1.364i & 0 & -0.442 + 1.364i & -0.651 \\ -0.651 & -0.442 + 1.364i & 0 & -0.651 + 1.364i \\ -0.859 + 1.364i & -0.651 & -0.651 + 1.364i & 0 \end{bmatrix}
$$

$\text{LOGCjk} := \text{LOGB4} - \text{LOG}\Delta E_*$ We subtract the logs, effectively dividing each element in B_4 by it's corresponding element in ΔE.

$$
\text{LOGCjk} = \begin{bmatrix} 0 & -0.651 + 1.364i & -1 & -1.37 + 1.364i \\ -0.651 & 0 & -0.534 + 1.364i & -1 \\ -1 - 1.364i & -0.534 & 0 & -0.651 + 1.364i \\ -1.37 & -1 - 1.364i & -0.651 & 0 \end{bmatrix}
$$

$Cjk := 10^{\overrightarrow{LOGCjk}}$ We subtract the logs, effectively dividing each element in B_4 by it's corresponding element in ΔE.

$$Cjk = \begin{bmatrix} 1 & -0.224 & 0.1 & -0.043 \\ 0.224 & 1 & -0.293 & 0.1 \\ -0.1 & 0.293 & 1 & -0.224 \\ 0.043 & -0.1 & 0.224 & 1 \end{bmatrix}$$

This is the un-normalized form of C_{jk}. We will use this when we compute the second and higher order energy corrections. But for the purpose of finding the wave function of acrolein, we must normalize this matrix. Let's calculate them all at the same time. All we need to do is square $C_{jkUN-NORM}$, the diagonal is the sum of the separate functions squared. We must then invert the values and take the square root of the values. Then we can o multiply the normalization matrix into $C_{jkUN-NORM}$, and multiply by the original un-perturbed wave function to obtain the corrected wave function of acrolein.

To find the normalization factors: (I hope my notation here is correct)

γ_{jj} = diagonal $(C_{jk}{}^{T}C_{jk})$; then

$$(\gamma_{jj}{}^{-1})^{1/2} C_{jk} \Psi^0{}_1 = \Psi^1{}_1$$

$$\gamma jj = \begin{bmatrix} 1.062 & 0 & 0 & 0 \\ 0 & 1.146 & 0 & 0 \\ 0 & 0 & 1.146 & 0 \\ 0 & 0 & 0 & 1.062 \end{bmatrix}$$

MATHCAD CANNOT TAKE THE SQUARE ROOT OF THIS MATRIX, SO WE'LL HAVE TO MANIPULATE IT A LITTLE SO WE CAN TAKE THE SQUARE ROOT. FIRST WE MUST GET RID OF THE ZERO'S BY MAKING γ_{JJ} INTO A COLUMN MATRIX.

$$\text{FOUR1} := \begin{bmatrix} 1 \\ 1 \\ 1 \\ 1 \end{bmatrix} *$$

$$\gamma_{jj}1 := \gamma_{jj} \cdot \text{FOUR1} *$$

$$\gamma_{jj}1 = \begin{bmatrix} 1.062 \\ 1.146 \\ 1.146 \\ 1.062 \end{bmatrix}$$

$$\gamma_{jj}2 := \overrightarrow{\left(\gamma_{jj}1^{-1} \right)^{\frac{1}{2}}}$$

$$\gamma_{jj}2 = \begin{bmatrix} 0.97 \\ 0.934 \\ 0.934 \\ 0.97 \end{bmatrix}$$

$$\gamma_{jj}\text{NORM} := \text{diag}(\gamma_{jj}2)$$

$$\gamma_{jj}\text{NORM} = \begin{bmatrix} 0.97 & 0 & 0 & 0 \\ 0 & 0.934 & 0 & 0 \\ 0 & 0 & 0.934 & 0 \\ 0 & 0 & 0 & 0.97 \end{bmatrix}$$

We have computed the square root. Now we are ready to finish the calculation $(\gamma_{jj}^{-1})^{1/2} C_{jk} \Psi^0{}_1 = \Psi^1{}_1$

$$\Psi^1{}_1 = \begin{vmatrix} .9705 & & & \\ & .9343 & & \\ & & .9343 & \\ & & & .9705 \end{vmatrix} \begin{vmatrix} 1 & -.2236 & .1000 & -.0427 \\ .2236 & 1 & -.2928 & .1000 \\ -.1000 & .2928 & 1 & -.2236 \\ .0427 & -.1000 & -.2236 & 1 \end{vmatrix} \begin{vmatrix} .3717 & .6015 & .6015 & .3717 \\ .6015 & .3717 & -.3717 & -.6015 \\ .6015 & -.3717 & -.3717 & .6015 \\ .3717 & -.6015 & .6015 & -.3717 \end{vmatrix} =$$

$$\begin{vmatrix} .2732 & .4919 & .6034 & .5650 \\ .5098 & .5184 & -.0637 & -.6835 \\ .6142 & -.1761 & -.6307 & .4403 \\ .1873 & -.5142 & .7254 & -.4175 \end{vmatrix} = \Psi^1{}_1$$

We are finished! The rest of the calculations we have already
done, except for the second order and higher energy corrections.
Let's proceed and look at the energy diagrams:

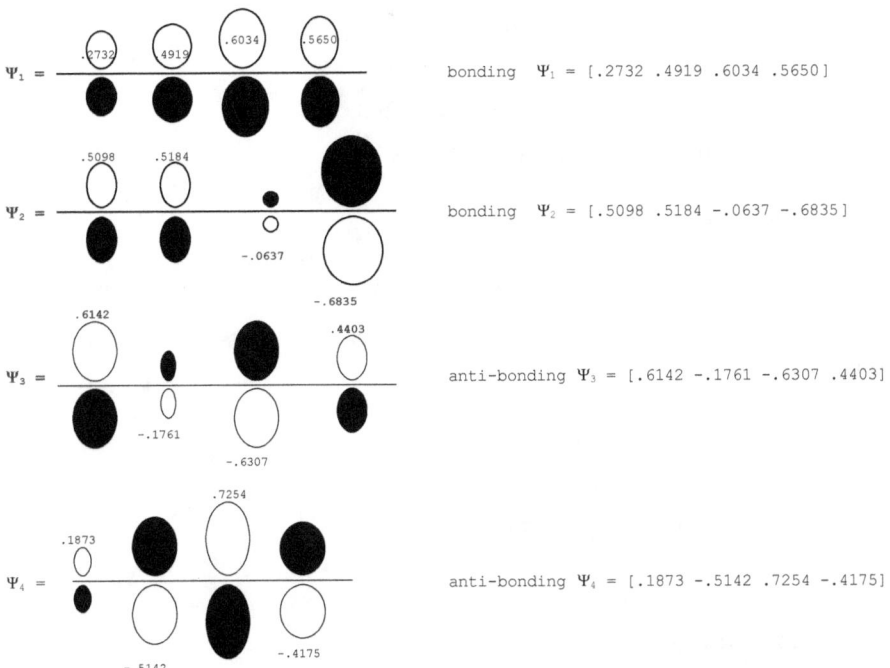

bonding $\Psi_1 = [.2732\ .4919\ .6034\ .5650]$

bonding $\Psi_2 = [.5098\ .5184\ -.0637\ -.6835]$

anti-bonding $\Psi_3 = [.6142\ -.1761\ -.6307\ .4403]$

anti-bonding $\Psi_4 = [.1873\ -.5142\ .7254\ -.4175]$

Let's look at this function as the Half-Multiplier acts as
a "**Picture Function**".

We will freeze the computation of Ψ^1_1 just before summing.
First we multiply C_{jk} by the normalization matrix:

PICTFUNCT $:= \gamma jjNORM \cdot Cjk$

$$
\text{PICTFUNCT} =
\begin{bmatrix}
0.97045 & -0.21699 & 0.09705 & -0.04145 \\
0.2089 & 0.93426 & -0.27348 & 0.09343 \\
-0.09343 & 0.27348 & 0.93426 & -0.2089 \\
0.04145 & -0.09705 & 0.21699 & 0.97045
\end{bmatrix}
$$

Now we take the first row, diagonalize it, and multiply
into Ψ^0_1:

FOUR11 $:=($ 1   0   0   0 $)$

PICTFUNCT $:= \gamma jj NORM \cdot Cjk$

$$\text{PICTFUNCT} = \begin{bmatrix} 0.97045 & -0.21699 & 0.09705 & -0.04145 \\ 0.2089 & 0.93426 & -0.27348 & 0.09343 \\ -0.09343 & 0.27348 & 0.93426 & -0.2089 \\ 0.04145 & -0.09705 & 0.21699 & 0.97045 \end{bmatrix}$$

PFΨ1A $:=$ FOUR11$\cdot$PICTFUNCT

PFΨ1A $= ($ 0.97045 -0.21699 0.09705 $-0.04145)$

PFΨ11 $:= \mathrm{diag}\left(\text{PF}\Psi\text{1A}^T \right)$

$$\text{PF}\Psi11 = \begin{bmatrix} 0.97045 & 0 & 0 & 0 \\ 0 & -0.21699 & 0 & 0 \\ 0 & 0 & 0.09705 & 0 \\ 0 & 0 & 0 & -0.04145 \end{bmatrix}$$

$$\Psi := \begin{bmatrix} .3717 & .6015 & .6015 & .3717 \\ .6015 & .3717 & -.3717 & -.6015 \\ .6015 & -.3717 & -.3717 & .6015 \\ .3717 & -.6015 & .6015 & -.3717 \end{bmatrix}$$

PFΨ1 $:=$ PFΨ11$\cdot\Psi$

$$\text{PF}\Psi1 = \begin{bmatrix} 0.36072 & 0.58373 & 0.58373 & 0.36072 \\ -0.13052 & -0.08066 & 0.08066 & 0.13052 \\ 0.05837 & -0.03607 & -0.03607 & 0.05837 \\ -0.01541 & 0.02493 & -0.02493 & 0.01541 \end{bmatrix}$$

Now we'll do the same for the second row:

FOUR12 $:=($ 0   1   0   0 $)$

PFΨ1B $:=$ FOUR12$\cdot$PICTFUNCT

$PF\Psi 1B = (\; 0.2089 \quad 0.93426 \; {}^-0.27348 \; 0.09343 \;)$

$PF\Psi 12 := diag\left(PF\Psi 1B^T\right)$

$$PF\Psi 12 = \begin{bmatrix} 0.2089 & 0 & 0 & 0 \\ 0 & 0.93426 & 0 & 0 \\ 0 & 0 & {}^-0.27348 & 0 \\ 0 & 0 & 0 & 0.09343 \end{bmatrix}$$

$PF\Psi 2 := PF\Psi 12 \cdot \Psi$

$$PF\Psi 2 = \begin{bmatrix} 0.07765 & 0.12565 & 0.12565 & 0.07765 \\ 0.56196 & 0.34726 & {}^-0.34726 & {}^-0.56196 \\ {}^-0.1645 & 0.10165 & 0.10165 & {}^-0.1645 \\ 0.03473 & {}^-0.0562 & 0.0562 & {}^-0.03473 \end{bmatrix}$$

Let's graph the picture function for Ψ_1 diagramming all the functions:

$$PF\Psi 1 = \begin{bmatrix} 0.36072 & 0.58373 & 0.58373 & 0.36072 \\ {}^-0.13052 & {}^-0.08066 & 0.08066 & 0.13052 \\ 0.05837 & {}^-0.03607 & {}^-0.03607 & 0.05837 \\ {}^-0.01541 & 0.02493 & {}^-0.02493 & 0.01541 \end{bmatrix}$$

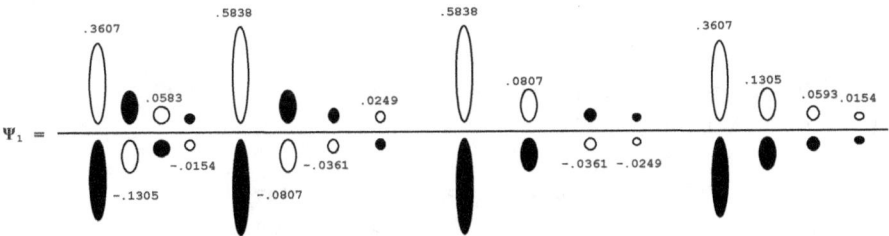

$\Psi_1 =$

Each group of four represents the first column, the second column, etc. . . When these are summed, we get the functions calculated for on page 41. These represent the interactions of the individual Atomic Orbitals.

The picture function for Ψ_2 is represented by:

$$PF\,\Psi2 = \begin{bmatrix} 0.07765 & 0.12565 & 0.12565 & 0.07765 \\ 0.56196 & 0.34726 & -0.34726 & -0.56196 \\ -0.1645 & 0.10165 & 0.10165 & -0.1645 \\ 0.03473 & -0.0562 & 0.0562 & -0.03473 \end{bmatrix}$$

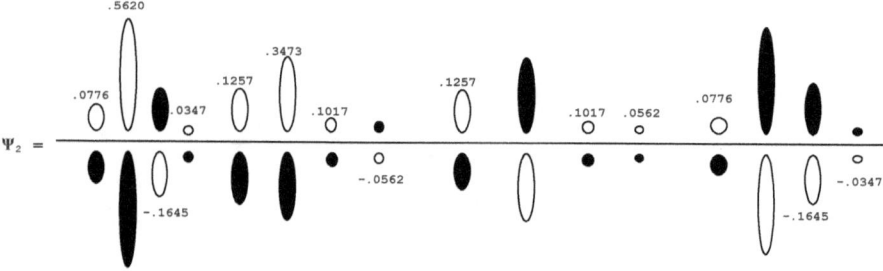

$\Psi_2 =$

Now to finish the problem by computing the **bond order**, **free-valence index** and the **charge density**. As you remember, the bond order is given by:

$$P_{ij} \;=\; \Sigma Na_i a_j \;=\Psi^T{}_{Occ}\Psi_{Occ} \;=$$

$$\begin{vmatrix} .2732 & .4919 \\ .4919 & .5184 \\ .6034 & -.0637 \\ .5650 & -.6835 \end{vmatrix}\begin{vmatrix} .2732 & .4919 & .6034 & .5650 \\ .4919 & .5184 & -.0637 & -.6835 \end{vmatrix} \;=\; \begin{vmatrix} .6691 & .7973 & .2647 & -.3882 \\ .7973 & 1.0214 & .5276 & -.1528 \\ .2647 & .5276 & .7363 & .7689 \\ -.3882 & -.1528 & .7689 & 1.5728 \end{vmatrix}$$

and $\quad q_i \;=\; 1.000 \,-\, \Sigma Na_i{}^2\Psi_i$

but $\Sigma Na_i{}^2\Psi_I$ = the diagonal of the product of = $2\,(\Psi^T{}_{Occ}\Psi_{Occ})$. We must subtract this value from 1.

$q_1 \;=\; 1 \,-\; .6691 \;=\; .3309 \qquad\qquad P_{12} = P_{21} = .7973$

$q_2 \;=\; 1 \,-\; 1.0214 \;=\; -.0214 \qquad\quad P_{23} = P_{32} = .5276$

$q_3 \;=\; 1 \,-\; .7363 \;=\; .2637 \qquad\qquad P_{34} = P_{43} = .7689$

$q_4 \;=\; 1 \,-\; 1.5728 \;=\; -.5728$

Hey! That's even easier to compute q_i than in the section on butadiene! Let's see if we can write a program for this.

$$\Psi := \begin{pmatrix} .2732 & .4919 & .6034 & .5650 \\ .5098 & .5184 & -.0637 & -.6835 \end{pmatrix} \quad \text{This is the } \Psi_{occ} \text{ matrix.}$$

I4 := identity(4) This is a 4x4 identity matrix.

ΨSQ := $2 \cdot \Psi^T \cdot \Psi$ The value is 2 times the square of Ψ.

$$\Psi SQ = \begin{bmatrix} 0.6691 & 0.7973 & 0.2647 & -0.3882 \\ 0.7973 & 1.0214 & 0.5276 & -0.1528 \\ 0.2647 & 0.5276 & 0.7363 & 0.7689 \\ -0.3882 & -0.1528 & 0.7689 & 1.5728 \end{bmatrix}$$

Ψbondorder := $I4 - \overrightarrow{(I4 \cdot \Psi SQ)}$ The second term removes the diagonal
from ΨSQ, then it is subtracted from 1.000.

$$\Psi bondorder = \begin{bmatrix} 0.3309 & 0 & 0 & 0 \\ 0 & -0.0214 & 0 & 0 \\ 0 & 0 & 0.2637 & 0 \\ 0 & 0 & 0 & -0.5728 \end{bmatrix}$$

So the equation is $q_i = I_{44} - \mathbf{diag}(2\,(\Psi_{24})^T\Psi_{24}) = q_{44}$

Now to calculate the **Free-valence Index**. We make the template
matrix by taking the secular determinant and replacing all the
X's with zero's. We then take the log of this matrix, letting
all zero's equal 10^{-100} (this is so we can take the log, there
is no definition for the log of zero.)

$$FVI1 := \begin{bmatrix} 0 & 1 & 0 & 0 \\ 1 & 0 & 1 & 0 \\ 0 & 1 & 0 & 1 \\ 0 & 0 & 1 & 0 \end{bmatrix} \quad LOGFVI := \begin{bmatrix} -100 & 0 & -100 & -100 \\ 0 & -100 & 0 & -100 \\ -100 & 0 & -100 & 0 \\ -100 & -100 & 0 & -100 \end{bmatrix} \quad EMAX := \begin{bmatrix} 1.73 \\ 1.73 \\ 1.73 \\ 1.73 \end{bmatrix} \quad FOUR1 := \begin{bmatrix} 1 \\ 1 \\ 1 \\ 1 \end{bmatrix}$$

LOGFVI := LOGFVI $\overrightarrow{+ \log(\Psi SQ)}$

FVI := EMAX $-\ \overrightarrow{10^{LOGFVI}} \cdot$ FOUR1 $FVI = \begin{bmatrix} 0.9327 \\ 0.4051 \\ 0.4335 \\ 0.9611 \end{bmatrix}$ The molecular diagram
thus becomes:

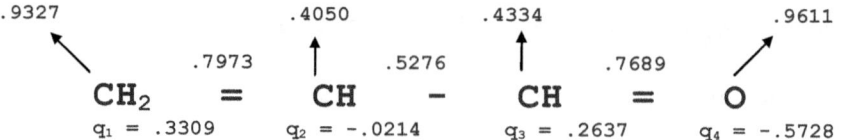

The most reactive positions are at the CH-CH bond (free-radical attack), the C-O bond undergoes acid attack while the CH_2 -CH undergoes alkali attacks most readily than the rest of the molecule. (Remember, it has been 20 years of scientific exile since I took any chemistry classes, so I do not really remember very well if the interpretations above for the reactivity of acrolein are correct).

Before I get to the section on multiple perturbations and the higher order energy corrections, let me show you how we come up with the bond order and P_{ij} and charge density q_i.

$$
\begin{vmatrix}
.2732 & .4919 & .6034 & .5650 \\
.5098 & .5184 & -.0637 & -.6835 \\
.6142 & -.1761 & -.6307 & .4403 \\
.1873 & -.5142 & .7254 & -.4175
\end{vmatrix} = \Psi^1_1
$$

By hand

$P_{12} = P_{21} = 2(.2731)(.4919) + 2(.5098)(.5185) = .7973$

$P_{23} = P_{32} = 2(.4919)(.6035) + 2(.5185)(-.0637) = .5277$

$P_{34} = P_{43} = 2(.6035)(.5649) + 2(-.0637)(-.6836) = .7689$

Putting the pre-multipliers in a separate matrix and half-multiplying we get:

$$\begin{vmatrix} .2732 \\ .5098 \end{vmatrix} \circ \begin{vmatrix} .2732 \\ .5098 \end{vmatrix} \quad = \quad q_1 = 2(.0746 + .2599) = .6691$$

$$\begin{vmatrix} .2732 \\ .5098 \end{vmatrix} \circ \begin{vmatrix} .2732 & .4919 \\ .5098 & .5185 \end{vmatrix} \quad = \quad \begin{matrix} 2(& .0747 & .1344 &) \\ + & .2599 & .2634 \\ q_1 = & .6691 & P_{12} = .7974 \end{matrix}$$

$$\begin{vmatrix} .4919 \\ .5185 \end{vmatrix} \circ \begin{vmatrix} .4919 & .6035 \\ .5185 & -.0637 \end{vmatrix} \quad = \quad \begin{matrix} 2(& .2420 & .2969 &) \\ + & .2688 & -.0330 \\ q_2 = & 1.0216 & P_{23} = .5278 \end{matrix}$$

$$\begin{vmatrix} .6035 \\ -.0637 \end{vmatrix} \circ \begin{vmatrix} .6035 & .5650 \\ -.0637 & -.6835 \end{vmatrix} \quad = \quad \begin{matrix} 2(& .3641 & .3409 &) \\ + & .0041 & .0435 \\ q_3 = & .7364 & P_{34} = .7688 \end{matrix}$$

$$\begin{vmatrix} .5650 \\ -.6835 \end{vmatrix} \circ \begin{vmatrix} .5650 \\ -.6835 \end{vmatrix} \quad = \quad 2(.3192 + .4672) = q_4 = 1.5728$$

All the other element values are don't cares, but we cannot replace their values with zero. Now according to the properties of the Half-Multiplier Operator, we take the pre-multipliers and form a matrix out of them. When this matrix is transposed, we multiply it to the post-multiplier to get the same solution. i.e.

$$\begin{vmatrix} .2732 & .4919 & .6034 & .5650 \\ .5098 & .5185 & -.0637 & -.6835 \end{vmatrix} \circ \begin{vmatrix} .2732 & .4919 & .6035 & .5650 \\ .5098 & .5185 & -.0637 & -.6835 \end{vmatrix} \quad =$$

$$\begin{vmatrix} .2732 & .4919 & .6034 & .5650 \\ .5098 & .5185 & -.0637 & -.6835 \end{vmatrix}^T \begin{vmatrix} .2732 & .4919 & .6035 & .5650 \\ .5098 & .5185 & -.0637 & -.6835 \end{vmatrix} \quad =$$

$$\begin{vmatrix} .2732 & .4919 \\ .4919 & .5184 \\ .6034 & -.0637 \\ .5650 & -.6835 \end{vmatrix} \begin{vmatrix} .2732 & .4919 & .6034 & .5650 \\ .4919 & .5184 & -.0637 & -.6835 \end{vmatrix} = \begin{vmatrix} .6691 & .7973 & .2647 & -.3882 \\ .7973 & 1.0214 & .5276 & -.1528 \\ .2647 & .5276 & .7363 & .7689 \\ -.3882 & -.1528 & .7689 & 1.5728 \end{vmatrix}$$

Before I get into multiple perturbations, let's check the energy of the perturbed acrolein against the direct calculation from the Slater determinant.

TYPE IN THE MATRIX, SELECT IT (ENCLOSE IT IN A BOX), CLICK SYMBOLIC ON THE TOOL BAR, CLICK MATRIX OPERATIONS, CLICK SYMBOLIC DETERMINANT).

$$\begin{bmatrix} x & 1 & 0 & 0 \\ 1 & x & 1 & 0 \\ 0 & 1 & x & 1 \\ 0 & 0 & 1 & x+1 \end{bmatrix}$$

$$x^4 + x^3 - 3 \cdot x^2 - 2 \cdot x + 1$$

PUT THE COEFFICIENTS IN A COLUMN MATRIX, STARTING WITH THE CONSTANT, THEN THE COEFFICIENT OF X AND ON TO THE HIGHEST POWER OF X. THE MATRIX MUST BE DEFINED WITH A VARIABLE NAME. THEN TYPE IN polyroots(ACROLEIN). NOTE: POLYROOTS MUST BE LOWER CASE. THEN PRESS THE = SIGN TO GET ROOTS.

$$\text{ACROLEIN} := \begin{bmatrix} 1 \\ -2 \\ -3 \\ 1 \\ 1 \end{bmatrix}$$

$$\text{polyroots(ACROLEIN)} = \begin{bmatrix} -1.87939 \\ -1 \\ 0.3473 \\ 1.53209 \end{bmatrix}$$

TO COMPARE WITH THE PERTURBATION CALCULATIONS, WE SUBTRACT THE ENERGY WE OBTAINED FROM THE PERTURBATION CALCULATIONS FROM THE ENERGY CALCULATED FROM THE SLATER DETERMINANT, THEN DIVIDE THE DIFFERENCE BY THE ENERGY FROM THE SLATER DETERMINANT x 100 TO PUT THE ANSWER IN PERCENTS. SINCE IT IS DEFINED bjj = E, WE TAKE THE DIAGONAL FROM THE bjk MATRIX AND ADD IT TO THE UNPERTURBED E OF BUTADIENE.

$$\begin{bmatrix} \dfrac{A}{B} + 1.87939 \\ \dfrac{A}{B} + 1 \\ \dfrac{A}{B} - .3473 \\ \dfrac{A}{B} - 1.53209 \end{bmatrix} - \begin{bmatrix} \dfrac{A}{B} + 1.7562 \\ \dfrac{A}{B} + .9798 \\ \dfrac{A}{B} - .2562 \\ \dfrac{A}{B} - 1.4798 \end{bmatrix} \rightarrow \begin{bmatrix} .12319 \\ 2.02 \cdot 10^{-2} \\ -9.11 \cdot 10^{-2} \\ -5.229 \cdot 10^{-2} \end{bmatrix} = \begin{bmatrix} 0.12319 \\ 0.0202 \\ -0.0911 \\ -0.05229 \end{bmatrix}$$

$$
\begin{pmatrix} \dfrac{A}{B} + 1.618 \\[1em] \dfrac{A}{B} + .618 \\[1em] \dfrac{A}{B} - .618 \\[1em] \dfrac{A}{B} - 1.618 \end{pmatrix} \blacksquare + \begin{pmatrix} .1382 \\ .3618 \\ .3618 \\ .1382 \end{pmatrix} \rightarrow \begin{pmatrix} \dfrac{A}{B} + 1.7562 \\[1em] \dfrac{A}{B} + .9798 \\[1em] \dfrac{A}{B} - .2562 \\[1em] \dfrac{A}{B} - 1.4798 \end{pmatrix} \blacksquare
$$

This is the energy for the perturbed Ethylene.

$$
\text{ENERGYDIFFERENCE} := \begin{bmatrix} 0.12319 \\ 0.0202 \\ -0.0911 \\ -0.05229 \end{bmatrix}
$$

$$
\text{ENERGYACROLEIN} := \begin{bmatrix} -1.87939 \\ -1 \\ 0.3473 \\ 1.53209 \end{bmatrix}
$$

INVERSE := ENERGYACROLEIN^{-1} Here I take the inverse of the ENERGYACROLEIN matrix.

PERCENTERROR := (INVERSE·ENERGYDIFFERENCE)·100   Here I Half-Multiply the inverse to the difference in energy, effectively dividing each element in the ENERGYDIFFERENCE matrix by each element in the ENERGYACROLEIN matrix.

% error = 100x(correct answer-your answer)/correct answer

$$
\text{PERCENTERROR} = \begin{bmatrix} -6.55479 \\ -2.02 \\ -26.23092 \\ -3.41298 \end{bmatrix}
$$

Not too bad an approximation for a perturbation that perturbs 25% of the molecule, but the third value is quite high, perhaps we can correct it a little bit see if it makes a better fit.

SECOND ORDER ENERGY CORRECTION

In Quantum Mechanics, the term for the second order energy correction (for diagonal elements only) is given by:

$$E^2_{N',N} = \sum_{N' \neq N} \frac{\left| H_{N',N} \right|^2}{E^0_{n'} - E^0_n}$$

Note, the 2 in $E^2_{N',N}$ is not E^2, but a notation stating that this is a second order perturbation. It is an index number only. This equation looks like gibberish, let's see if we can simplify it a little bit. The energy computed for acrolein is the sum of the four energies of butadiene, the molecule we started out with, plus the values of the diagonal of the un-normalized b_{jk} matrix, placed in order of the most negative energy first to the most positive energy last for E^0_1. Then the energy correction term for Ψ_1 is

$$E^2_{1n} = \frac{H_{12}{}^2 \, \Psi^0_2}{E^0_{n'} - E^0_n} + \frac{H_{13}{}^2 \, \Psi^0_3}{E^0_{n'} - E^0_n} + \frac{H_{14}{}^2 \, \Psi^0_4}{E^0_{n'} - E^0_n} = \frac{(b_{12})^2 \, \Psi^0_2}{E^0_{n'} - E^0_n} + \frac{(b_{13})^2 \, \Psi^0_3}{E^0_{n'} - E^{0n}} + \frac{(b_{14})^2 \, \Psi^0_4}{E^0_{n'} - E^0_n}$$

We've got to remember that b_{21} is not a matrix, but is the number that occupies the position of row 2, column 1 in the b_{jk} matrix. Calculating likewise for the energy terms E^2_{2n}, E^2_{3n} and E^2_{4n} and putting in matrix form, we get:

	Ψ^0_1	Ψ^0_2	Ψ^0_3	Ψ^0_4
$= E^2_{1n}$	0	$\dfrac{(b_{12})^2}{E^0_1 - E^0_2}$	$\dfrac{(b_{13})^2}{E^0_1 - E^0_3}$	$\dfrac{(b_{14})^2}{E^0_1 - E^0_4}$
E^2_{2n}	$\dfrac{(b_{21})^2}{E^0_2 - E^0_1}$	0	$\dfrac{(b_{23})^2}{E^0_2 - E^0_3}$	$\dfrac{(b_{24})^2}{E^0_2 - E^0_4}$
E^2_{3n}	$\dfrac{(b_{31})^2}{E^0_3 - E^0_1}$	$\dfrac{(b_{32})^2}{E^0_3 - E^0_2}$	0	$\dfrac{(b_{34})^2}{E^0_3 - E^0_4}$
E^2_{4n}	$\dfrac{(b_{41})^2}{E^0_4 - E^0_1}$	$\dfrac{(b_{42})^2}{E^0_4 - E^0_2}$	$\dfrac{(b_{43})^2}{E^0_4 - E^0_3}$	0

The second order energy correction matrix is exactly the same in form as the un-normalized C_{jk} matrix except that the non-diagonal b_{jk} elements are squared. So all we need to do to compute the second order correction is multiply the non-diagonal

b_{jk} matrix by the un-normalized matrix. i.e. for a perturbation at location Ψ_4 we have:

$$(B_4 - B_{jj})\,(C_{jk})_{\text{UN-NORM}} =$$

SECOND ORDER CORRECTION

$$Cjk := \begin{bmatrix} 1 & -.2236 & .1 & -.0427 \\ .2236 & 1 & -.2927 & .1 \\ -.1 & .2927 & 1 & -.2234 \\ .0427 & -.1 & -.2236 & 1 \end{bmatrix} \qquad B4MINUSBjj := \begin{bmatrix} 0 & -.2236 & .2236 & -.1382 \\ -.2236 & 0 & -.3618 & .2236 \\ .2236 & -.3618 & 0 & -.2236 \\ -.1382 & .2236 & -.2236 & 0 \end{bmatrix}$$

$$Cjk \cdot B4MINUSBjj = \begin{bmatrix} 0.07826 & -0.26933 & 0.31405 & -0.21056 \\ -0.30287 & 0.07826 & -0.33416 & 0.25815 \\ 0.18903 & -0.38939 & -0.07831 & -0.14433 \\ -0.16584 & 0.29495 & -0.17787 & 0.02174 \end{bmatrix}$$

So, taking the diagonal of this matrix and adding it to the first-order energy computation of acrolein, we have:

$$
\begin{aligned}
E^2_1 &= 1.618 + .1382 + .0783 = 1.8345 = 2.39\% \text{ ERROR} \\
E^2_2 &= .618 + .3618 + .0783 = 1.0581 = 5.8\% \text{ ERROR} \\
E^2_3 &= -.618 + .3618 - .0783 = -.3662 = 3.2\% \text{ ERROR} \\
E^2_4 &= -1.618 + .1382 + .0217 = -1.5015 = 1.99\% \text{ ERROR}
\end{aligned}
$$

HIGHER ORDER ENERGY CORRECTIONS

The higher order energy corrections are given by:

$$H = H^0 + bH^1 + b^2H^2 + b^3H^3 + \ldots$$

Where b^2 and so on represent the square of the element of b, and $H^{2,3,4,}$ . . . are still indexes, not squares. So lets take b_i, replace the diagonal with ones, take the log of the elements and store in location LOGb. Take C_{jk} un-normalized, replace the diagonal with zero's, (10^{-100} in this case) take the log of it's elements and store in location LOGC.

HP-48G PROGRAM:

```
RCL LOGb
RCL LOGC
+
<<10ˣ>>
↑
ααTEACH,EXAM,PRGS,APLY, ααTEACH
RCL FOUR1
x
STO E2
```

$$Cjk := \begin{bmatrix} 1 & -.2236 & .1 & -.0427 \\ .2236 & 1 & -.2927 & .1 \\ -.1 & .2927 & 1 & -.2236 \\ .0427 & -.1 & -.2236 & 1 \end{bmatrix}$$

$$bi := \begin{bmatrix} 10^{-100} & -.2236 & .2236 & -.1382 \\ -.2236 & 10^{-100} & -.3618 & .2236 \\ .2236 & -.3618 & 10^{-100} & -.2236 \\ -.1382 & .2236 & -.2236 & 10^{-100} \end{bmatrix}$$

$$FOUR1 := \begin{bmatrix} 1 \\ 1 \\ 1 \\ 1 \end{bmatrix} \qquad I4 := identity(4)$$

$$E := \begin{bmatrix} 1.618 \\ .618 \\ -.618 \\ -1.618 \end{bmatrix} \qquad E1 := \begin{bmatrix} .1382 \\ .3618 \\ .3618 \\ .1382 \end{bmatrix}$$

$$E+E1 = \begin{bmatrix} 1.7562 \\ 0.9798 \\ -0.2562 \\ -1.4798 \end{bmatrix} \qquad \text{(FIRST ORDER CORRECTION)}$$

SECOND ORDER CORRECTION

METHOD 1   **METHOD 2**

$b2 := Cjk \cdot bi$

$LOGE2M2 := \overrightarrow{(\log(bi) + \log(Cjk))}$ $E2M2 := 10^{\overrightarrow{LOGE2M2}} \cdot FOUR1$

$$b2 = \begin{bmatrix} 0.07826 & -0.26933 & 0.31405 & -0.21056 \\ -0.30287 & 0.07826 & -0.33416 & 0.25815 \\ 0.18905 & -0.38944 & -0.07826 & -0.14433 \\ -0.16584 & 0.29495 & -0.17787 & 0.02174 \end{bmatrix}$$

$$E2M2 = \begin{bmatrix} 0.07826 \\ 0.07826 \\ -0.07826 \\ 0.02174 \end{bmatrix}$$

$E2 := \overrightarrow{(I4 \cdot b2) \cdot FOUR1}$

$$E2 = \begin{bmatrix} 0.07826 \\ 0.07826 \\ -0.07826 \\ 0.02174 \end{bmatrix}$$

$E2CORR := E + E1 + E2$

PERCENT ERROR FOR SECOND ORDER ENERGY CORRECTION

$$E2CORR = \begin{bmatrix} 1.83446 \\ 1.05806 \\ -0.33446 \\ -1.45806 \end{bmatrix} \qquad ENERGYACROLEIN = \begin{bmatrix} 1.87939 \\ 1 \\ -0.3473 \\ -1.53209 \end{bmatrix}$$

$ENERGYDIFFE2CORR := ENERGYACROLEIN - E2CORR$

$PCTERRORE2 := \overrightarrow{(INVERSE \cdot ENERGYDIFFE2CORR)} \cdot 100$

$$PCTERRORE2 = \begin{bmatrix} -2.39077 \\ 5.80619 \\ -3.69654 \\ -4.83169 \end{bmatrix}$$

THIRD ORDER CORRECTION

METHOD1: MATRIC   METHOD 2: LOGARITHMIC

$$LOGE3M2 := \overrightarrow{(2 \cdot (\log(bi)) + \log(Cjk))}$$

$b3 := Cjk \cdot \overrightarrow{bi}^2$ HERE WE ARE SQUARING EACH INDIVIDUAL ELEMENT IN b AND PRE-MULTIPLYING IT BY C_{jk}. THE EQUATION IS GIVEN BY:

$$E3M2 := \overrightarrow{\left(10^{\overrightarrow{LOGE3M2}} \right)} \cdot FOUR1$$

$$b3 = \begin{bmatrix} -0.007 & 0.06095 & 0.01859 & 0.01292 \\ 0.03727 & -0.02214 & 0.14708 & 0.03963 \\ 0.06036 & 0.11472 & 0.02214 & 0.06272 \\ 0.00292 & 0.02286 & 0.03904 & -0.01536 \end{bmatrix} \qquad E3M2 = \begin{bmatrix} -0.007 \\ -0.02214 \\ 0.02214 \\ -0.01536 \end{bmatrix}$$

$E3 := \overrightarrow{(I4 \cdot b3)} \cdot FOUR1$ HERE WE ARE REMOVING THE DIAGONAL OF b3 AND MAKING A COLUMN MATRIX OUT OF IT

$$E3 = \begin{bmatrix} -0.007 \\ -0.0221 \\ 0.0221 \\ -0.0154 \end{bmatrix}$$

PERCENT ERROR FOR THIRD ORDER CORRECTION

$$ENERGYACROLEIN := \begin{bmatrix} 1.87939 \\ 1 \\ -0.3473 \\ -1.53209 \end{bmatrix}$$

$E3CORR := E2CORR + E3$

$$E3CORR = \begin{bmatrix} 1.82746 \\ 1.03593 \\ -0.31233 \\ -1.47343 \end{bmatrix}$$

ENERGYDIFFE3CORR $:=$ ENERGYACROLEIN$-$ E3CORR HERE WE ARE COMPUTING THE % ERROR FOR THE THIRD ORDER ENERGY CORRECTION

PERCENTERRORE3CORR $:=\overrightarrow{(\text{INVERSE} \cdot \text{ENERGYDIFFE3CORR})} \cdot 100$

$$\text{PERCENTERRORE3CORR} = \begin{bmatrix} -2.76297 \\ 3.59267 \\ -10.07005 \\ -3.82891 \end{bmatrix}$$

IT LOOKS LIKE THE SECOND ORDER CORRECTION WILL BE THE BEST, BUT LETS TRY A FOURTH ORDER CORRECTION JUST FOR THE HECK OF IT, AND ALSO THIS MAY HELP TO SEE THE PATTERN BETTER OF WHAT WE ARE TRYING TO DO.

FOURTH ORDER ENERGY CORRECTION

METHOD 1: MATRIC   METHOD 2: LOGARITHMIC

$b4 := \text{Cjk} \cdot \overrightarrow{\text{bi}}^{3}$ $\text{LOGE4M2} := \overrightarrow{(3 \cdot (\log(\text{bi})) + \log(\text{Cjk}))}$

HERE WE ARE CUBING EACH INDIVIDUAL ELEMENT IN b AND PRE—MULTIPLYING BY Cjk.

$E4 := \overrightarrow{(I4 \cdot b4)} \cdot \text{FOUR1}$ $\text{E4M2} := \overrightarrow{\left(10^{\text{LOGE4M2}}\right)} \cdot \text{FOUR1}$

E4CORR $:=$ E3CORR $+$ E4

$$\text{E4M2} = \begin{bmatrix} 0.00373 \\ 0.01248 \\ -0.01248 \\ 0.00127 \end{bmatrix}$$

E4CORR $:=$ E3CORR $+$ E4M2

$$E4CORR = \begin{bmatrix} 1.83119 \\ 1.04841 \\ -0.32481 \\ -1.47216 \end{bmatrix}$$

PCTERRORE4A $:=$**ENERGYACROLEIN**$-$**E4CORR**   HERE   WE   ARE   COMPUTING   THE PERCENT ERROR FOR THE FOURTH ORDER CORRECTION.

PCTERRORE4 $:=(\overrightarrow{\text{INVERSE} \cdot \text{PCTERRORE4A}}) \cdot 100$

$$PCTERRORE4 = \begin{bmatrix} -2.56449 \\ 4.8407 \\ -6.47653 \\ -3.91174 \end{bmatrix}$$

SO IT LOOKS LIKE THE SECOND ORDER CORRECTION WAS THE BEST FOR THIS PROBLEM.

MULTIPLE PERTURBATIONS

Finally, suppose we have perturbed two or more carbons on a molecule rather than just the one we've calculated for:

Suppose:

$$H^1 = \begin{vmatrix} 0 & 0 & 0 & 0 \\ 0 & 0 & 0 & 0 \\ 0 & 0 & 1 & 0 \\ 0 & 0 & 0 & 1 \end{vmatrix}$$

$$b_{jk} = \begin{vmatrix} .3717 & .6015 & .6015 & .3717 \\ .6015 & .3717 & -.3717 & -.6015 \\ .6015 & -.3717 & -.3717 & .6015 \\ .3717 & -.6015 & .6015 & -.3717 \end{vmatrix} \begin{vmatrix} 0 & 0 & 0 & 0 \\ 0 & 0 & 0 & 0 \\ 0 & 0 & 1 & 0 \\ 0 & 0 & 0 & 1 \end{vmatrix} \begin{vmatrix} .3717 & .6015 & .6015 & .3717 \\ .6015 & .3717 & -.3717 & -.6015 \\ .6015 & -.3717 & -.3717 & .6015 \\ .3717 & -.6015 & .6015 & -.3717 \end{vmatrix} =$$

$$\begin{vmatrix} .5000 & -.4472 & 0 & .2236 \\ -.4472 & .5000 & -.2236 & 0 \\ 0 & -.2236 & .5000 & -.4472 \\ .2236 & 0 & -.4472 & .5000 \end{vmatrix}$$

but this is equal to $\Psi_3^T\Psi_3 + \Psi_4^T\Psi_4 =$

$$\begin{vmatrix} .6015 \\ -.3717 \\ -.3717 \\ .6015 \end{vmatrix} [.6015 \ -.3717 \ -.3717 \ .6015] + \begin{vmatrix} .3717 \\ -.6015 \\ .6015 \\ -.3717 \end{vmatrix} [.3717 \ -.6015 \ .6015 \ -.3717] =$$

$$\Psi := \begin{bmatrix} .3717 & .6015 & .6015 & .3717 \\ .6015 & .3717 & -.3717 & -.6015 \\ .6015 & -.3717 & -.3717 & .6015 \\ .3717 & -.6015 & .6015 & -.3717 \end{bmatrix} \qquad H1 := \begin{bmatrix} 0 & 0 & 0 & 0 \\ 0 & 0 & 0 & 0 \\ 0 & 0 & 1 & 0 \\ 0 & 0 & 0 & 1 \end{bmatrix}$$

$$\Psi3 := (.6015 \ -.3717 \ -.3717 \ .6015) \qquad \Psi4 := (.3717 \ -.6015 \ .6015 \ -.3717)$$

$$b3 := \Psi3^T \cdot \Psi3$$

$$b3 = \begin{bmatrix} 0.3618 & -0.2236 & -0.2236 & 0.3618 \\ -0.2236 & 0.1382 & 0.1382 & -0.2236 \\ -0.2236 & 0.1382 & 0.1382 & -0.2236 \\ 0.3618 & -0.2236 & -0.2236 & 0.3618 \end{bmatrix}$$

b4 $= \Psi 4^{T} \cdot \Psi 4$

$$b4 = \begin{bmatrix} 0.1382 & -0.2236 & 0.2236 & -0.1382 \\ -0.2236 & 0.3618 & -0.3618 & 0.2236 \\ 0.2236 & -0.3618 & 0.3618 & -0.2236 \\ -0.1382 & 0.2236 & -0.2236 & 0.1382 \end{bmatrix}$$

$$b3 + b4 = \begin{bmatrix} 0.5 & -0.4472 & 0 & 0.2236 \\ -0.4472 & 0.5 & -0.2236 & 0 \\ 0 & -0.2236 & 0.5 & -0.4472 \\ 0.2236 & 0 & -0.4472 & 0.5 \end{bmatrix}$$

But b3 + b4 = b34

b34 $= \Psi^{T} \cdot H1 \cdot \Psi$

$$b34 = \begin{bmatrix} 0.5 & -0.4472 & 0 & 0.2236 \\ -0.4472 & 0.5 & -0.2236 & 0 \\ 0 & -0.2236 & 0.5 & -0.4472 \\ 0.2236 & 0 & -0.4472 & 0.5 \end{bmatrix}$$

Therefore:

b34 $= \Psi^{T}H^{1(3,4)}\Psi = \varepsilon_3 \Psi_3{}^{T}\Psi_3 + \varepsilon_4 \Psi_4{}^{T}\Psi_4$

And $C_{jk} = (\varepsilon_3 b_3 + \varepsilon_4 b_4)/(E_i - E_j) = \dfrac{\varepsilon_3 \Psi_3{}^{o2} + \varepsilon_4 \Psi_4{}^{o2}}{(E_i - E_j)}$

With $\varepsilon_3 = \varepsilon_4 = 1$ for this problem.

To generalize, suppose:

$$H^1 = \begin{vmatrix} 0 & & & & & & & \\ & 0 & & & & & & \\ & & \varepsilon_3 B_{33} & & & & & \\ & & & 0 & & & & \\ & & & & \varepsilon_5 B_{55} & & & \\ & & & & & 0 & & \\ & & & & & & \cdot & \\ & & & & & & & \cdot \\ & & & & & & & \varepsilon_j B_{jj} \\ & & & & & & & & 0 \\ & & & & & & & & & \cdot \\ & & & & & & & & & & \varepsilon_n B_{nn} \end{vmatrix}$$

Then $b_{jk} = \varepsilon_3 b_3 + \varepsilon_5 b_5 + \varepsilon_j b_j + \varepsilon_n b_n = \boldsymbol{\varepsilon_3 \Psi_3^{\ T} \Psi_3 + \varepsilon_5 \Psi_5^{\ T} \Psi_5 + \varepsilon_j \Psi_j^{\ T} \Psi_j + \varepsilon_n \Psi_n^{\ T} \Psi_n} =$

$$= \varepsilon_j \sum_{j=1}^{n} b_j = \sum_{j=1}^{n} \varepsilon_j \Psi_j^{\circ 2} = \varepsilon_j \sum_{j=1}^{n} \Psi_j^{\ T} \Psi_j$$

And

$$C_{jk} = \Psi^T H^{1(3,5,j,n)} \Psi / (E^0_{\ i} - E^0_{\ j}) = \frac{\varepsilon_j \sum\limits_{j=1}^{n} \Psi_j^{\ T} \Psi_j}{((E_i^-)_{nn}^{\ T} \circ [1]_{nn}) + I_{nn}}$$

And the working computer program for this problem in perturbations is: (NOTE: we have written the perturbation computation in a single line of command code.)

$$\boldsymbol{\Psi^1_{\ 1} = ((\gamma_{jj})^{-1})^{1/2} [10EEX(LOG(\textstyle\sum_j \varepsilon_j \Psi_j^{\ T} \Psi_j - E_{jj}) + [I] - LOG((E_j \circ [I] - (E_j \circ [I])^T + [I])]\Psi^0_{\ 1}}$$

Lets look at the perturbation of butadiene to the molecule C=N-C-O

$$E_O = A + B \qquad \varepsilon_O = 1$$
$$E_N = A + 1/2B \qquad \varepsilon_N = .5$$

Let's use MathCad to determine the energies:

MULTIPLE PERTURBATIONS

$$A := \begin{bmatrix} X & 1 & 0 & 0 \\ 1 & X+.5 & 1 & 0 \\ 0 & 1 & X & 1 \\ 0 & 0 & 1 & X+1 \end{bmatrix} \begin{bmatrix} 1 \\ -2.5 \\ -2.5 \\ 1.5 \\ 1 \end{bmatrix} \quad \text{polyroots}(A) = \begin{bmatrix} -2 \\ -1.1617 \\ 0.32104 \\ 1.34067 \end{bmatrix}$$

ENERGIES FOR THE MO
WAVE FUNCTION

$x^4 + 1.5x^3 - 2.5x^2 - 2.5x + 1$ Symbolic determinant of matrix for C=N-C=O.

To compute the b_{jk} matrix:

$$b_{24} = \varepsilon_2 \Psi_2{}^T \Psi_2 + \varepsilon_4 \Psi_4{}^T \Psi_4 =$$

$$.5\left(\begin{vmatrix} .6015 \\ .3717 \\ -.3717 \\ -.6015 \end{vmatrix} [.6015 \ .3717 \ -.3717 \ -.6015]\right) + \begin{vmatrix} .3717 \\ -.6015 \\ .6015 \\ -.3717 \end{vmatrix} [.3717 \ -.6015 \ .6015 \ -.3717]$$

$\Psi2 := (.6015 \ .3717 \ -.3717 \ -.6015)$ $\Psi4 := (.3717 \ -.6015 \ .6015 \ -.3717)$

$$b24 := .5 \cdot \Psi2^T \cdot \Psi2 + \Psi4^T \cdot \Psi4 \qquad b24 = \begin{bmatrix} 0.31906 & -0.11179 & 0.11179 & -0.31906 \\ -0.11179 & 0.43088 & -0.43088 & 0.11179 \\ 0.11179 & -0.43088 & 0.43088 & -0.11179 \\ -0.31906 & 0.11179 & -0.11179 & 0.31906 \end{bmatrix}$$

$E_1 = 1.618 + .3191 = 1.9371$
$E_2 = .618 + .43088 = 1.04888$
$E_3 = -.618 + .43088 = -0.18712$
$E_4 = -1.618 + .31906 = -1.29894$

$$\overrightarrow{B24A := (I4 \cdot b24) \cdot FOUR1} \qquad B24A = \begin{bmatrix} 0.31906 \\ 0.43088 \\ 0.43088 \\ 0.31906 \end{bmatrix}$$

$$\gamma 24 := \left(B24A^{-1}\right)^{\overrightarrow{\frac{1}{2}}}$$

$$\gamma 24 = \begin{bmatrix} 1.77036 \\ 1.52342 \\ 1.52342 \\ 1.77036 \end{bmatrix}$$

$$\Delta E := \left[\begin{bmatrix} 0 & 1.618 & 1.618 & 1.618 \\ .618 & 0 & .618 & .618 \\ -.618 & -.618 & 0 & -.618 \\ -1.618 & -1.618 & -1.618 & 0 \end{bmatrix} - \begin{bmatrix} 0 & .618 & -.618 & -1.618 \\ 1.618 & 0 & -.618 & -1.618 \\ 1.618 & .618 & 0 & -1.618 \\ 1.618 & .618 & -.618 & 0 \end{bmatrix} \right] + \begin{bmatrix} 1 & 0 & 0 & 0 \\ 0 & 1 & 0 & 0 \\ 0 & 0 & 1 & 0 \\ 0 & 0 & 0 & 1 \end{bmatrix}$$

$$\Delta E = \begin{bmatrix} 1 & 1 & 2.236 & 3.236 \\ -1 & 1 & 1.236 & 2.236 \\ -2.236 & -1.236 & 1 & 1 \\ -3.236 & -2.236 & -1 & 1 \end{bmatrix}$$

$$\text{LOG}\Delta E := \overrightarrow{\log(\Delta E)}$$

$$\text{LOG}\Delta E = \begin{bmatrix} 0 & 0 & 0.34947 & 0.51001 \\ 1.36438i & 0 & 0.09202 & 0.34947 \\ 0.34947+ 1.36438i & 0.09202+ 1.36438i & 0 & 0 \\ 0.51001+ 1.36438i & 0.34947+ 1.36438i & 1.36438i & 0 \end{bmatrix}$$

$$\text{LOGb24} := \overrightarrow{\log(b24)}$$

$$\text{LOGb24} = \begin{bmatrix} -0.49612 & -0.9516+ 1.36438i & -0.9516 & -0.49612+ 1.36438i \\ -0.9516+ 1.36438i & -0.36564 & -0.36564+ 1.36438i & -0.9516 \\ -0.9516 & -0.36564+ 1.36438i & -0.36564 & -0.9516+ 1.36438i \\ -0.49612+ 1.36438i & -0.9516 & -0.9516+ 1.36438i & -0.49612 \end{bmatrix}$$

$$\text{LOGCjk} := \text{LOGb24} - \text{LOG}\Delta E$$

$$\text{LOGCjk} = \begin{bmatrix} -0.49612 & -0.9516+ 1.36438i & -1.30107 & -1.00613+ 1.36438i \\ -0.9516 & -0.36564 & -0.45766+ 1.36438i & -1.30107 \\ -1.30107- 1.36438i & -0.45766 & -0.36564 & -0.9516+ 1.36438i \\ -1.00613 & -1.30107- 1.36438i & -0.9516 & -0.49612 \end{bmatrix}$$

$$Cjk = \begin{bmatrix} 0.31906 & -0.11179 & 0.04999 & -0.0986 \\ 0.11179 & 0.43088 & -0.34861 & 0.04999 \\ -0.04999 & 0.34861 & 0.43088 & -0.11179 \\ 0.0986 & -0.04999 & 0.11179 & 0.31906 \end{bmatrix}$$

$\Psi\text{perturbed} = \text{diag}(\gamma 24) \cdot Cjk \cdot \Psi$

$$\Psi\text{perturbed} = \begin{bmatrix} 0.07927 & 0.33829 & 0.27543 & 0.44712 \\ 0.167 & 0.49802 & 0.10166 & -0.67929 \\ 0.62267 & 0.01004 & -0.58964 & 0.11038 \\ 0.34064 & -0.34123 & 0.40409 & 0.0272 \end{bmatrix}$$

The energy diagrams look like:

$\Psi_1 =$

$\Psi_2 =$

PERCENT ERROR

$$E := \begin{bmatrix} 2 \\ 1.167 \\ -.32104 \\ -1.34067 \end{bmatrix} \qquad \overrightarrow{INVE} := E^{-1} \qquad ECALC := \begin{bmatrix} 1.9371 \\ 1.0488 \\ -.18712 \\ -1.29894 \end{bmatrix}$$

$E1 := E - ECALC$

$PCTERROR := (\overrightarrow{INVE \cdot E1}) \cdot 100$
$\qquad PCTERROR = \begin{bmatrix} 3.145 \\ 10.12853 \\ 41.71443 \\ 3.11262 \end{bmatrix}$

SECOND ORDER CORRECTION

$b2 := Cjk \cdot b24$

$E2 := (\overrightarrow{I4 \cdot b2}) \cdot FOUR1$
$\qquad E2 = \begin{bmatrix} 0.15134 \\ 0.32896 \\ 0.04236 \\ 0.05226 \end{bmatrix}$

$E2CORR := ECALC + E2$
$\qquad E2CORR = \begin{bmatrix} 2.08844 \\ 1.37776 \\ -0.14476 \\ -1.24668 \end{bmatrix}$

$E2C := E - E2CORR$

$PCTERRORE2C := (\overrightarrow{INVE \cdot E2C}) \cdot 100$
$\qquad PCTERRORE2C = \begin{bmatrix} -4.42225 \\ -18.06018 \\ 54.90826 \\ 7.01039 \end{bmatrix}$

SOME OF THESE ROOTS ARE NOT VERY CLOSE, BUT REMEMBER, WE
PERTURBED 50% OF THE BUTADIENE MOLECULE, WHICH IS A VERY LARGE
CHANGE IN THE ENERGY. PERTURBATION THEORY MUST BE BASED UPON
SMALL CHANGES FOR IT TO WORK THE BEST (1% - 2% MAX).

BIBLIOGRAPHY

Goodrich, "A Primer of Quantum Chemistry", John Wiley & Sons, New York, 1972

Turner, "Methods in MO Theory", Prentice-Hall, N.J. 1974

Roberts, "MO Calculations", W.A. Benjamin Co., N.Y. 1962

Flurry, "MO Theories of Bonding in Organic Molecules", Marcel Dekker Inc., N.Y. 1968

Dewar, "MO Theory of Organic Chemistry", McGraw-Hill, N.Y. 1969

Liberles, "Introduction to MO Theory", Holt, Rinehart and Winston, N.Y., 1966

STATISTICS

Yesterday (8-26-96) I received a response from the Indiana Journal of Mathematics. They rejected my paper as it was written up to this point without even reading it (the paper clip had never been removed from the manuscript). On the phone earlier with them they were at a loss to explain why it was rejected.

They told me that everything worked, nothing was wrong with it, that they at the journal had never seen anything like it before, but it could only work for the examples included in my paper, but was impossible to work for anything else, the Half-multiplier Operator was a fluke. I did not have the heart to tell them that I had already begun working on Statistics. They made me so mad I decided that rather than solve a few problems, that I would rewrite the entire textbook, if only to find where this operator did **not** work in Statistics. That is why this section is so long. With every problem, I learned something new about the way this new operator worked. I did rewrite the whole book called the Computational Handbook of Statistics, 2nd edition, by James L. Bruning and B. L. Kintz. As a note, this book is a mish-mash of styles. I started to write it on October 22, 1996 (believe it or not (inside joke), it was at about 11:53 PM) when I saw the connection between cross multiplying a matrix and matrix multiplication as we now know it. I do not explain the Statistics, you must purchase their book if you wish to follow what we are doing in this section.

Lets go back to the Unified Field equation: $c\Sigma \left([A]^T_{ij} \circ [B]_{jk}\right) = c\ C_{jk}$ Letting $[B]_{jK} = [I]$, the equation reduces to:

$$c\Sigma A_{ij} = c\ C_{ij}$$

where c is a constant that defines what the field equation will represent. In the case of Statistics, $c = 1/N$. So the Field equation for a simple Statistical system becomes:

$$(1/N)\ \Sigma(A_{ij}) = (1/N)\ C_{ij}$$

But to make it Statistical, we must square the above expression so that it becomes:

$$(1/N)\ \Sigma A_{ij})^2 = (1/N)\ CC^T$$

Which is our basic Statistical equation.

The Σ operators are:

$$^{C}\Sigma = [1]_{1,i}\ A_{ij} = C_{j,1}$$

$$^{R}\Sigma = A_{ij}\ [1]_{j,1} = C_{i,1}$$

$$^{M}\Sigma = A_{ij}$$

This is the Statistics for simple systems, for Analysis of Variance, $[B] \neq [I]$.

$$(1/N)\ \Sigma([A]^{T}_{ij}\ o\ B_{jk}) = (1/N)C_{jk}$$

And the Statistics Become:

$$(1/N)\ \Sigma([A]^{T}_{ij}\ o\ [B]_{jk})^{2} = (1/N)\ CC^{T}\_CORRFACT$$

but we must sum the data in the Matrix A, and in Statistics we need only multiply the Matrices, not Half-Multiply them, so the final two Statistical Field Equations become

$$(1/N)\ \Sigma([1]_{1i}[A]_{ij}\ [B]_{jk})^{2} = (1/N)\ CC^{T}\_CORRFACT$$
(Between Subjects term)

and

$$(1/N)\ \Sigma([A]_{ij}\ [B]_{jk}\ [1]_{k1})^{2} = (1/N)\ CC^{T}\_CORRFACT$$
(Within Subjects term)

To make it a little clearer this is essentially

$$([1][A][B])^{2}\ Between\ Subjects$$

and

$$([A][B][1]^{2}\ Within\ Subjects$$

They are non-commutative. In some cases $[B] = [I]$, which reduces the computations.

The Σ operators are:

Where A represents the data taken and B = Database Matrix.

$$^{1/2}\Sigma([A]^T_{ij} \ o \ [B]_{jk}) = C_{j,k} \qquad \text{(Half-multiplier mode)}$$

$$^{C}\Sigma([A]^T_{ij} \ o \ [B]_{jk}) = C_{i,k} \qquad \text{(regular matrix multiplication)}$$

$$^{R}\Sigma([A]^T_{ij} \ o \ [B]_{jk}) = B_{jk} \ [1]_{j,1} \ o \ [A]^T_{ij}$$

$$^{M}\Sigma([A]^T_{ij} \ o \ [B]_{jk}) = [A]^T_{ij} \ [1]_{i,1} \ o \ B_{jk}$$

Note: these equations hold for Physics, Astrophysics and Accounting & Inventories also.

EXAMPLES: THE MEAN

The Statistics equation for the mean is:

$$x = (1/N) \sum (X)$$

(Sorry, can't find x bar anywhere)

The matric equivalent is: $(1/N) {}^c\Sigma A_{ij} =$

$$(1/N) \; [1 \; 1 \; 1 \; . \; . \; . \; 1]_{1,j} \begin{bmatrix} D \\ A \\ T \\ A \end{bmatrix}_{j,1}$$

EXAMPLE: We have 5 people whose height in inches has been recorded. What is the average height:

$$(1/5) \; [1 \; 1 \; 1 \; 1 \; 1] \begin{vmatrix} 60 \\ 53 \\ 57 \\ 52 \\ 58 \end{vmatrix} = 280/5 = 56 \text{ Inches}$$

STANDARD DEVIATION

**Computational Handbook of Statistics,
2nd Edition, section 1.2, pg. 4-6**

The Statistics equation for standard deviation is given by:

$$\text{STANDARDDEVIATION} := \left(\frac{\Sigma x^2 - \Sigma(X)^2}{N - 1} \right)^{\frac{1}{2}} \square$$

The matric equivalent is given by:

$$\text{STDDEV} := \left[\left(\frac{1}{N - 1} \right) \cdot (D \quad A \quad T \quad A) \cdot \begin{pmatrix} D & 1 \\ A & 1 \\ T & 1 \\ A & 1 \end{pmatrix} \cdot \begin{pmatrix} 1 \\ -\dfrac{b}{N} \end{pmatrix} \right]^{\frac{1}{2}} \square$$

Where Σx^2 = [DATA] $\begin{vmatrix} D \\ A \\ T \\ A \end{vmatrix}$ = a

We combine the two into one matrix operation. ↑

And $\Sigma(x)$ = [DATA] $\begin{vmatrix} 1 \\ 1 \\ 1 \\ 1 \end{vmatrix}$ = b

When we've calculated b, we plug it in the matrix, first dividing by N and then multiply. We then take the square root of the resulting single number.

EXAMPLE: Find the standard deviation (s.d.) for the following list of heights measured from twenty 12 year old students.

VARIABLES:

$$A_{20,1} := \begin{bmatrix} 64 \\ 48 \\ 55 \\ 68 \\ 72 \\ 59 \\ 57 \\ 61 \\ 63 \\ 60 \\ 60 \\ 43 \\ 67 \\ 70 \\ 65 \\ 55 \\ 56 \\ 64 \\ 61 \\ 60 \end{bmatrix} \quad TWENTY1 := \begin{bmatrix} 1 \\ 1 \\ 1 \\ 1 \\ 1 \\ 1 \\ 1 \\ 1 \\ 1 \\ 1 \\ 1 \\ 1 \\ 1 \\ 1 \\ 1 \\ 1 \\ 1 \\ 1 \\ 1 \\ 1 \end{bmatrix} \quad X1 = \begin{bmatrix} 64 & 1 \\ 48 & 1 \\ 55 & 1 \\ 68 & 1 \\ 72 & 1 \\ 59 & 1 \\ 57 & 1 \\ 61 & 1 \\ 63 & 1 \\ 60 & 1 \\ 60 & 1 \\ 43 & 1 \\ 67 & 1 \\ 70 & 1 \\ 65 & 1 \\ 55 & 1 \\ 56 & 1 \\ 64 & 1 \\ 61 & 1 \\ 60 & 1 \end{bmatrix} \quad X1 := \text{augment}\left(A_{20,1}, TWENTY1\right)$$

$$STDDEV := \left[\left(\frac{1}{N-1}\right) \cdot (D \quad A \quad T \quad A) \cdot \begin{pmatrix} D & 1 \\ A & 1 \\ T & 1 \\ A & 1 \end{pmatrix} \cdot \begin{pmatrix} 1 \\ -\dfrac{b}{N} \end{pmatrix} \right]^{\frac{1}{2}} \square$$

$$\xrightarrow{\hspace{5cm}}$$

$$STDDEV := \left[A_{20,1}{}^{T} \cdot X1 \cdot \begin{pmatrix} 1 \\ -\dfrac{1208}{20} \end{pmatrix} \right]^{\frac{1}{2}} \square$$

Mathcad cannot solve this as is, so we have to break the
problem up in two parts, first solve the problem, then take the
square root.

$$\text{STDEV1} := \left[\left(\frac{1}{20 - 1} \right) \cdot \left[A_{20,1}{}^T \cdot X1 \cdot \left(\frac{1}{-\frac{1208}{20}} \right) \right] \right]$$

STDEV1 = 48.989

Putting the arrow over the second expression allows us to take
the square root of a 1x1 Matrix, which Mathcad won't allow us
to do.

$$\text{STDEV} := \overrightarrow{\text{STDEV1}^{\frac{1}{2}}}$$

STDEV = 6.999

s.d. = $((1/19)[64\ 48\ 55\ 68\ 72\ 59\ 57\ 61\ 63\ 60\ 60\ 43\ 67\ 70\ 65\ 55\ 56\ 64\ 61\ 60]$ $\begin{vmatrix} 64 & 1 \\ 48 & 1 \\ 55 & 1 \\ \cdot \\ \cdot \\ \cdot \\ 60 & 1 \end{vmatrix}$$)^2$ $\begin{vmatrix} 1 \\ -b/20 \end{vmatrix}_{20,2}$

$1/19\ ([73,894\ 1208]$ $\begin{vmatrix} 1 \\ -1208/20 \end{vmatrix})^{1/2}$ $= (1/19)[73,894\ -72,963] = (931/19)^{1/2} = (49)^{1/2} = 7$

STANDARD ERROR OF THE MEAN: s.d/(# of terms)$^{1/2}$ = 7/(20)$^{1/2}$ =
7/4.47 = 1.57.

To do this totally by computer, proceed as follows:

$(1/(N-1)\ ([DATA]^2 - (1/N)[DATA][1])^2)^{1/2}$

TWENTY1

$((1/19)\ ([DATA]$ $\begin{vmatrix} D \\ A \\ T \\ A \end{vmatrix}_{20,1}$ $- 1/20[DATA]$ $\begin{vmatrix} 1 \\ \cdot \\ \cdot \\ 1 \end{vmatrix}_{20,1}))^{1/2}$

HP-48G PROGRAM:

```
RCL DATA
↑
TRN
SWAP
x
RCL DATA
RCL TWENTY1
x
RCL N = 20
/
-
RCL N-1 = 19
/
SQRT
STO SD 6.999
```

t-TEST FOR A DIFFERENCE BETWEEN A SAMPLE MEAN & POPULATION MEAN

Computational Handbook of Statistics,
2nd Edition, Section 1.5, pg. 8-10

The Statistical equation is given by:

$$t := \frac{x_{AV} - \mu}{\left[\dfrac{\sum x^2 - \left(\dfrac{\sum (x)^2}{N} \right)}{N \cdot (N-1)} \right]^{\frac{1}{2}}}$$

The matric equation is given by:

$$t := \frac{\dfrac{b}{N} - \mu}{\left[\left[\dfrac{1}{N \cdot (N-1)} \right] \cdot (D \quad A \quad T \quad A) \cdot \begin{pmatrix} D & 1 \\ A & 1 \\ T & 1 \\ A & 1 \end{pmatrix} \cdot \begin{pmatrix} 1 \\ -b \\ N \end{pmatrix} \right]^{\frac{1}{2}\square}}$$

Using the data from the preceding problem, μ is obtained from tables as the average height of 12 year olds in the whole USA = 58". We have already calculated

$$t := \frac{\dfrac{1208}{20} - 58}{\left[\left(\dfrac{1}{20 \cdot 19} \right) \cdot \left[(73894 \quad 1208) \cdot \begin{pmatrix} 1 \\ -\dfrac{1208}{20} \end{pmatrix} \right] \right]^{\frac{1}{2}\square}}$$

Again Mathcad can't handle the math, so we must break it into simpler parts.

$$t1 := \left(\frac{1208}{20} - 58\right) \cdot \left[\left(\frac{1}{20 \cdot 19}\right) \cdot \left[\begin{array}{cc}(73894 & 1208)\end{array} \cdot \begin{pmatrix} 1 \\ -\dfrac{1208}{20} \end{pmatrix}\right]\right]^{-1}$$

$$t2 := \left[\left(\frac{1}{20 \cdot 19}\right) \cdot \left[\begin{array}{cc}(73894 & 1208)\end{array} \cdot \begin{pmatrix} 1 \\ -\dfrac{1208}{20} \end{pmatrix}\right]\right]^{-1}$$

$t2 = 0.408$

$$t3 := t2^{\overrightarrow{\frac{1}{2}}}$$

$t3 = 0.639$

$$t := \left(\frac{1208}{20} - 58\right) \cdot t3$$

We multiply by t3 because we have already taken it's inverse, so we multiply to divide.

$t = 1.533$

To do this totally by computer or calculator, proceed as follows:

$$1/20 \ [DATA]_{1,20} \ [1]_{20,1} \ -\mu$$

$$\left(\frac{1}{20(20-1)} \ [DATA] \left| \begin{array}{c} D \\ A \\ T \\ A \end{array} \right| -1/20([DATA] \left| \begin{array}{c} 1 \\ . \\ . \\ 1 \end{array} \right|^2 \right)^{1/2}$$

PROGRAM:

```
RCL DATA
RCL TWENTY1
x
STO SUM
RCL N = 20
/
RCL μ = 58
```

```
−
RCL DATA
↑
TRN
SWAP
x
RCL SUM
SQ
RCL N = 20
/
−
RCL N(N−1) = 380
/
SQRT
/
STO t-TEST2     1.53
```

t-TEST FOR THE DIFFERENCE BETWEEN TWO INDEPENDENT MEANS

(COMPUTATIONAL HANDBOOK OF STATISTICS,
SECOND EDITION SEC. 1.6, PG. 10-13)

The Statistical equation is given by:

$$\frac{\text{Mean}_1 - \text{Mean}_2}{\left(\left|\frac{\Sigma X_1{}^2 - 1/N_1(\Sigma X_1)^2 + \Sigma X_2{}^2 - 1/N_2(\Sigma X_2)^2}{(N_1 + N_2) - 2}\right| (1/N_1 + 1/N_2)\right)^{1/2}}$$

The matrix form of this equation is:

$$\frac{b_1/N_1 - b_2/N_2}{\left|\frac{(1/N_1 - 1/N_2)([a_1\ b_1]\left|\begin{matrix}1\\-b_1/N_1\end{matrix}\right| + [a_2\ b_2]\left|\begin{matrix}1\\-b_2/N_2\end{matrix}\right|}{N_1 + N_2 - 2}\right|^{1/2}}$$

Where [DATA 1] $\begin{vmatrix} D & 1 \\ A & 1 \\ T & . \\ A & . \\ 1 & 1 \end{vmatrix} = [a_1\ b_1]$ And [DATA 2] $\begin{vmatrix} D & 1 \\ A & 1 \\ T & . \\ A & . \\ 2 & 1 \end{vmatrix} = [a_2\ b_2].$

EXAMPLE: A principal wishes to determine whether there is a significant difference between the IQ scores of boys and girls in a particular grade. The score on the Wechsler Intelligence Scale for Children is recorded for each student.

GROUP 1 (BOYS)	GROUP 2 (GIRLS)
107	109
96	94
88	127
131	76
109	115
84	121
79	87
105	92
108	91
92	98
96	104
101	96
0	110
0	108

VARIABLES:

$$\text{DATA} := \begin{bmatrix} 107 & 109 \\ 96 & 94 \\ 88 & 127 \\ 131 & 76 \\ 109 & 115 \\ 84 & 121 \\ 79 & 87 \\ 105 & 92 \\ 108 & 91 \\ 92 & 98 \\ 96 & 104 \\ 101 & 96 \\ 0 & 110 \\ 0 & 108 \end{bmatrix}$$

$$\text{DATA2} := \text{DATA} \cdot \begin{pmatrix} 0 \\ 1 \end{pmatrix}$$

$$\text{DATA1} := \text{DATA} \cdot \begin{pmatrix} 1 \\ 0 \end{pmatrix}$$

$$
DATA1 = \begin{bmatrix} 107 \\ 96 \\ 88 \\ 131 \\ 109 \\ 84 \\ 79 \\ 105 \\ 108 \\ 92 \\ 96 \\ 101 \\ 0 \\ 0 \end{bmatrix} \quad DATA2 = \begin{bmatrix} 109 \\ 94 \\ 127 \\ 76 \\ 115 \\ 121 \\ 87 \\ 92 \\ 91 \\ 98 \\ 104 \\ 96 \\ 110 \\ 108 \end{bmatrix} \quad DATA11 = \begin{bmatrix} 107 & 1 \\ 96 & 1 \\ 88 & 1 \\ 131 & 1 \\ 109 & 1 \\ 84 & 1 \\ 79 & 1 \\ 105 & 1 \\ 108 & 1 \\ 92 & 1 \\ 96 & 1 \\ 101 & 1 \\ 0 & 1 \\ 0 & 1 \end{bmatrix} \quad DATA21 = \begin{bmatrix} 109 & 1 \\ 94 & 1 \\ 127 & 1 \\ 76 & 1 \\ 115 & 1 \\ 121 & 1 \\ 87 & 1 \\ 92 & 1 \\ 91 & 1 \\ 98 & 1 \\ 104 & 1 \\ 96 & 1 \\ 110 & 1 \\ 108 & 1 \end{bmatrix}
$$

$ONE14 := (1 \quad 1 \quad 1 \quad 1 \quad 1 \quad 1 \quad 1 \quad 1 \quad 1 \quad 1 \quad 1 \quad 1 \quad 1 \quad 1)$

$DATA11 := augment\left(DATA1, ONE14^{T}\right)$

$DATA21 := augment\left(DATA2, ONE14^{T}\right)$

we are going to compute this completely from the Matric equation given below. To see how I derived this equation, please go to the next section (Sect. 1.7) for the derivation.

To calculate this problem totally on a computer and/or calculator:

$$
t := \frac{\left(\dfrac{1}{12}\right) \cdot (ONE14 \cdot DATA1) - \left(\dfrac{1}{14}\right) \cdot (ONE14 \cdot DATA2)}{\left[\left(\dfrac{1}{N_1} + \dfrac{1}{N_2}\right)\left[\left(\dfrac{1}{N_1 + N_2 - 2}\right)\cdot\left[DATA1^{T}\cdot DATA11 \cdot \begin{pmatrix} 1 \\ \dfrac{-b_1}{N_1} \end{pmatrix} + DATA2^{T}\cdot DATA21 \cdot \begin{pmatrix} 1 \\ \dfrac{-b_2}{N_2} \end{pmatrix}\right]\right]\right]^{\frac{1}{2}}}
$$

$$
t := \frac{\left(\dfrac{1}{12}\right) \cdot (ONE14 \cdot DATA1) - \left(\dfrac{1}{14}\right) \cdot (ONE14 \cdot DATA2)}{\left[\left(\dfrac{1}{12} + \dfrac{1}{14}\right)\left[\left(\dfrac{1}{12 + 14 - 2}\right)\cdot\left[DATA1^{T}\cdot DATA11 \cdot \begin{pmatrix} 1 \\ \dfrac{-1196}{12} \end{pmatrix} + DATA2^{T}\cdot DATA21 \cdot \begin{pmatrix} 1 \\ \dfrac{-1428}{14} \end{pmatrix}\right]\right]\right]^{\frac{1}{2}}}
$$

$$\text{DENOMINATOR1} := \left[\left(\frac{1}{12} + \frac{1}{14}\right) \cdot \left[\left(\frac{1}{12 + 14 - 2}\right) \cdot \left[\text{DATA1}^T \cdot \text{DATA11} \cdot \begin{pmatrix} 1 \\ -1196 \\ 12 \end{pmatrix} + \text{DATA2}^T \cdot \text{DATA21} \cdot \begin{pmatrix} 1 \\ -1428 \\ 14 \end{pmatrix}\right]\right]\right]$$

Of this big equation,

DENOMINATOR1 = 30.067

We can't take the square root of a matrix, so we use the arrow
function of mathcad which evaluates matrices element by element,
thus we can take the square root as follows:

$$\text{DENOMINATOR} := \overrightarrow{\text{DENOMINATOR1}^{\frac{1}{2}}}$$

$$30.066799^{\frac{1}{2}} = 5.483$$

DENOMINATOR = 5.483

Now we compute the numerator, Mathcad can do this expression
in one step, so we have:

$$\text{NUMERATOR} := \left(\frac{1}{12}\right) \cdot (\text{ONE14} \cdot \text{DATA1}) - \left(\frac{1}{14}\right) \cdot (\text{ONE14} \cdot \text{DATA2})$$

NUMERATOR = ⁻2.333

$$t := \frac{\text{NUMERATOR}}{\text{DENOMINATOR}}{}_{\square}$$

Whoops! Can't divide Matrices, we must multiply by the inverse:

$$t := \text{NUMERATOR} \cdot \text{DENOMINATOR}^{-1}$$

t = ⁻0.426

Since the t-test is the absolute value, t = .426

This program is for using another method.

HP-48G PROGRAM:

RCL FOURTEEN1
RCL A
x   [1196 1428]
STO SUM1
RCL A
↑
TRN
SWAP
x
MTH,MATR,NXT,→DIAG
2
↑
DIAG→
RCL TWO1 | 1 |
                   | 1 |

x
STO SUM2 | 121,318 |
RCL SUM2 | 148,202 |
RCL SUM1
TRN
↑
2
↑
DIAG→
SWAP
x | 1/12 |
RCL N = | 1/14 |
2
↑
MTH,MATR,NXT,DIAG→
STO NDIAG
RCL NDIAG

SWAP
                   | 119,201 |
                   | 145,656 |

x
−
STO SUM3 | 2116.667 |
                   | 2546.000 |

RCL N$_1$
INV
RCL N$_2$
INV
+
x   30
<<SQRT>>
TEACH
VAR,EXAM,PRGS, APLY = 5.48
STO SUM4
RCL SUM4
RCL SUM1
TRN
RCL NDIAG
SWAP
x
RCL MIN1   [1 -1]
SWAP
x   use absolute value if
SWAP   negative.
/
STO t   0.43

Note: This is long and complex
compared to the Analysis of Variance
due to all the little nit-picking
math steps that need to be done.

```
RCL TWO1
TRN
SWAP
x              4662.667
RCL N₁ =       12
RCL N₂ =       14
+
2
-
/              194.277
```

STEP 17: To determine if the t value is significant, the degrees of freedom df = $(N_1 + N_2) - 2 = 24$. The t value which is significant at the .05 level for df = 24 is 2.064. But we obtained t = .43, therefore it is concluded there is no significant difference between the IQ scores of boys and girls in this particular class.

MATHCAD +6 PROGRAM FOR t-TEST FOR DIFFERENCE BETWEEN TWO INDEPENDENT MEANS.

$$ONE14 := (1\ 1\ 1\ 1\ 1\ 1\ 1\ 1\ 1\ 1\ 1\ 1\ 1\ 1)$$

$$MINUS1 := \begin{pmatrix} 1 \\ -1 \end{pmatrix}$$

$$TWELVE1 := \begin{bmatrix} 1 \\ 1 \\ 1 \\ 1 \\ 1 \\ 1 \\ 1 \\ 1 \\ 1 \\ 1 \\ 1 \\ 1 \end{bmatrix} \quad TWO1 := \begin{pmatrix} 1 \\ 1 \end{pmatrix} \quad A := \begin{bmatrix} 107 & 109 \\ 96 & 94 \\ 88 & 127 \\ 131 & 76 \\ 109 & 115 \\ 84 & 121 \\ 79 & 87 \\ 105 & 92 \\ 108 & 91 \\ 92 & 98 \\ 96 & 104 \\ 101 & 96 \\ 0 & 110 \\ 0 & 108 \end{bmatrix}$$

$N1 := 12$

$N2 := 14$

$$N := \begin{bmatrix} \dfrac{1}{12} \\ \dfrac{1}{14} \end{bmatrix}$$

The above equalities are all we need to solve this problem.

$DATA2 := (109 \ 94 \ 127 \ 76 \ 115 \ 121 \ 87 \ 92 \ 91 \ 98 \ 104 \ 96 \ 110 \ 108)$

$DATA1 := (107 \ 96 \ 88 \ 131 \ 109 \ 84 \ 79 \ 105 \ 108 \ 92 \ 96 \ 101 \ 0 \ 0)$

$t1 := ONE14 \, A \cdot diag(N)$ Sum the columns of A and multiply to the diagonal of N

$t1 = (\ 99.667 \quad 102 \ )$

$t1A := t1 \cdot MINUS1$ Add the two values obtained in t1

$t1A = -2.333$

$$t2 := \left(\frac{1}{N1} + \frac{1}{N2} \right) \cdot \left[DATA1 \cdot DATA1^T - \left(\frac{1}{N1} \right) \cdot \left(DATA1 \cdot ONE14^T \right)^2 + DATA2 \cdot DATA2^T - \frac{1}{N2} \cdot \left(DATA2 \cdot ONE14^T \right)^2 \right]$$

This is $(\Sigma X_1^2 - 1/N_1 (\Sigma X_1)^2 + \Sigma X_2^2 - 1/N_2 (\Sigma X_2)^2 \ (1/N_1 + 1/N_2))^{1/2}$ in matrix form.

$t2 = 721.603$

$t3 := N1 + N2 - 2$

$$t4 := \frac{t2}{t3}$$

$t4 = 30.067$

$$\frac{721.603}{24} = 30.067$$

$$t4A := t4^{\frac{\overrightarrow{1}}{2}}$$

t4A = 5.483

$$t := \frac{\overrightarrow{t1A}}{t4A}$$

t = −0.426

ALL WE NEED FOR t IS THE ABSOLUTE VALUE, t = 0.426.

t-TEST FOR RELATED MEASURES

(COMPUTATIONAL HANDBOOK OF STATISTICS,
2ND EDITION SECTION 1.7, PG. 13-16

EXAMPLE: An experimenter is interested in determining the effects of a special education program on the intelligence test scores of underprivileged children. The first step is to match several pairs of students on the basis of their Wechsler IQ test scores. One student from each pair is randomly assigned to either the special training group or the control group that receives no special treatment. After 6 weeks of training, an alternate form of the Wechsler test is given to the students in both groups to determine the effects of the program. The score on this second test is recorded for each student.

STATISTICS FORMULA

$$t = \frac{\overline{X} - \overline{Y}}{(\Sigma D^2 - 1/N(\Sigma D)^2)/N(N-1)}$$

MATRIX METHOD

STEP 1: Table the data:

VARIABLES:

GROUP 1   GROUP 2
(NO SPECIAL (SPECIAL
TREATMENT)   TREATMENT)

$$A_{14,2} := \begin{bmatrix} 89 & 94 \\ 86 & 94 \\ 96 & 101 \\ 100 & 105 \\ 94 & 100 \\ 86 & 84 \\ 81 & 81 \\ 93 & 96 \\ 87 & 90 \\ 89 & 88 \\ 110 & 115 \\ 95 & 100 \\ 107 & 110 \\ 96 & 102 \end{bmatrix} \qquad MINPLUS1 := \begin{pmatrix} -1 \\ 1 \end{pmatrix} \qquad PLUSMIN1 := \begin{pmatrix} 1 \\ -1 \end{pmatrix}$$

$$ONE14 := (1 \ 1 \ 1 \ 1 \ 1 \ 1 \ 1 \ 1 \ 1 \ 1 \ 1 \ 1 \ 1 \ 1)$$

$$FOURTEEN1 := ONE14^T$$

EQUATION:

$$t := \cfrac{ \dfrac{\left[\left(\dfrac{1}{14}\right) \cdot \left(ONE14 \cdot A_{14,2}\right) \right] \cdot MINPLUS1 }{ \left[\left(\dfrac{1}{N-1}\right) \cdot \left[DIFF^T \cdot DIFF - \left(\dfrac{1}{N}\right) \cdot (DIFF \cdot FOURTEEN1)^2 \right] \right]^{\frac{1}{2}} } }{ N^{\frac{1}{2}} }$$

STEP 2: Obtain the difference between each pair of scores:

$$DIFF := A_{14,2} \cdot PLUSMIN1$$

$$
\text{DIFF} :=
\begin{bmatrix}
89 & 94 \\
86 & 94 \\
96 & 101 \\
100 & 105 \\
94 & 100 \\
86 & 84 \\
81 & 81 \\
93 & 96 \\
87 & 90 \\
89 & 88 \\
110 & 115 \\
95 & 100 \\
107 & 110 \\
96 & 102
\end{bmatrix}
\cdot
\begin{pmatrix}
1 \\
-1
\end{pmatrix}
\qquad
\text{DIFF} =
\begin{bmatrix}
^-5 \\
^-8 \\
^-5 \\
^-5 \\
^-6 \\
2 \\
0 \\
^-3 \\
^-3 \\
1 \\
^-5 \\
^-5 \\
^-3 \\
^-6
\end{bmatrix}
$$

HP-48G PROGRAM:

```
RCL A
RCL MINUS1
x
STO DIFF
```

STEP 3:   Square the difference scores and sum:

$\text{GRNDSUMSQ} := \text{DIFF}^{T} \cdot \text{DIFF}$

$$\text{GRNDSUMSQ} := (-5 \quad -8 \quad -5 \quad -5 \quad -6 \quad 2 \quad 0 \quad -3 \quad -3 \quad 1 \quad -5 \quad -5 \quad -3 \quad -6) \cdot \begin{bmatrix} -5 \\ -8 \\ -5 \\ -5 \\ -6 \\ 2 \\ 0 \\ -3 \\ -3 \\ 1 \\ -5 \\ -5 \\ -3 \\ -6 \end{bmatrix}$$

GRNDSUMSQ = 293

HP-48G PROGRAM:

```
RCL DIFF
↑
TRN
SWAP
x
STO GNDSUMSQ
```

STEP 4: Obtain the algebraic sum of the difference scores, square it and divide by N = 14.

PROGRAM: GRANDSUM = (1/N) ([diff]$_{1,14}$[1]$_{14,1}$)^2

$$\text{GRANDSUM} := \left(\frac{1}{N}\right) \cdot (\text{DIFF} \cdot \text{FOURTEEN1})^2 \,_\square$$

$$\text{GRANDSUM} := \left(\frac{1}{14}\right) \cdot \left[(-5 \;\; -8 \;\; -5 \;\; -5 \;\; -6 \;\; 2 \;\; 0 \;\; -3 \;\; -3 \;\; 1 \;\; -5 \;\; -5 \;\; -3 \;\; -6) \cdot \begin{bmatrix} 1 \\ 1 \\ 1 \\ 1 \\ 1 \\ 1 \\ 1 \\ 1 \\ 1 \\ 1 \\ 1 \\ 1 \\ 1 \\ 1 \end{bmatrix} \right]^2$$

$$\text{GRANDSUM} := \left(\frac{1}{14}\right) \cdot (-51)^2$$

$$\text{GRANDSUM} = 185.786$$

HP-48G PROGRAM:

```
RCL ONE14
RCL DIFF
x
<<SQ>>
TEACH
VAR,EXAM,PRG,APLY
RCL N = 14
/
STO GRNDSUM    186
```

STEP 5: Subtract step 4 from step 3. $(\Sigma x^2 - (\Sigma x)^2) = 293 - 186 = \mathbf{107}$.

$$\text{SS5} := \text{DIFF}^T \cdot \text{DIFF} - \left(\frac{1}{N}\right) \cdot (\text{DIFF} \cdot \text{FOURTEEN1})^2 \square$$

$$\text{SS5} := \text{GRNDSUMSQ} - \text{GRANDSUM}$$

$$\text{SS5} = 107.214$$

STEP 6:   Divide by N-1.   107/13 = **8.23**.

$$SS6 := \left(\frac{1}{N-1}\right) \cdot \left[DIFF^{T} \cdot DIFF - \left(\frac{1}{N}\right) \cdot (DIFF \cdot FOURTEEN1)^{2} \right] \square$$

$$SS6 := \frac{107.214}{14-1}$$

$$SS6 = 8.247$$

STEP 7:   Take the square root:

$$\sqrt{8.23} = 2.87$$

$$SS7 := \left[\left(\frac{1}{N-1}\right) \cdot \left[DIFF^{T} \cdot DIFF - \left(\frac{1}{N}\right) \cdot (DIFF \cdot FOURTEEN1)^{2} \right] \right]^{\frac{1}{2}} \square$$

$$8.247^{\frac{1}{2}} = 2.872$$

STEP 8:   Divide by $(N)^{1/2} = (14)^{1/2} = 3.74$.

$$SS8 := \frac{\left[\left(\frac{1}{N-1}\right) \cdot \left[DIFF^{T} \cdot DIFF - \left(\frac{1}{N}\right) \cdot (DIFF \cdot FOURTEEN1)^{2} \right] \right]^{\frac{1}{2}}}{N^{\frac{1}{2}}} \square$$

$$SS8 := \frac{2.872}{14^{\frac{1}{2}}}$$

$$SS8 = 0.768$$

$$2.87/3.74 = .767$$

STEP 9:   Obtain the mean score of each of the two groups:

PROGRAM: $(1/N)\,[1]_{1,14}[A]_{14,2}$

$$\text{MEAN} := \left(\frac{1}{N}\right) \cdot \left(\text{ONE14} \cdot A_{14,2}\right) \square$$

$$\text{MEAN} := \left(\frac{1}{14}\right) \cdot \left(\text{ONE14} \cdot A_{14,2}\right)$$

$$\text{MEAN} := \left(\frac{1}{14}\right) \cdot \left[(1\ \ 1\ \ 1\ \ 1\ \ 1\ \ 1\ \ 1\ \ 1\ \ 1\ \ 1\ \ 1\ \ 1\ \ 1\ \ 1) \cdot \begin{bmatrix} 89 & 94 \\ 86 & 94 \\ 96 & 101 \\ 100 & 105 \\ 94 & 100 \\ 86 & 84 \\ 81 & 81 \\ 93 & 96 \\ 87 & 90 \\ 89 & 88 \\ 110 & 115 \\ 95 & 100 \\ 107 & 110 \\ 96 & 102 \end{bmatrix} \right]$$

$$\text{MEAN} = (\ 93.5 \quad 97.143\ )$$

HP-48G PROGRAM:

```
RCL ONE14
RCL A
x
RCL N = 14
/
STO MEAN
```

STEP 10: Subtract the mean from group 1 from group 2:

PROGRAM: $\text{MEANDIFF} = [\text{MEAN}]_{1,2}[\text{MINUS1}]_{2,1}.$

$$[93.50 \quad 97.14] \begin{vmatrix} -1 \\ 1 \end{vmatrix} = 3.64 \text{ (use absolute value if difference is negative.}$$

MEANDIFF := MEAN·MINPLUS1

$$\text{MEANDIFF} := (93.5 \quad 97.143) \cdot \begin{pmatrix} -1 \\ 1 \end{pmatrix}$$

MEANDIFF = 3.643

STEP 11: Divide value obtained in step 10 by step 8.

$$t := \frac{\text{MEANDIFF}}{\text{SS8}} \square$$

$$t := \text{MEANDIFF} \cdot \text{SS8}^{-1}$$

t = 4.746

t = 3.64/.767 = **4.74.**

STEP 12: To determine significance:

Degrees of freedom df = N−1 = 13. From tables, the t value
that is significant at the .05 level for df = 13 is 2.16. But we
obtained t = 4.74, so it is concluded that the special training
program improved the IQ test scores.

The entire equation in one step (one step, that is, if modern
computers could handle the math) is given by:

$$t := \frac{\left[\left(\dfrac{1}{14}\right) \cdot \left(\text{ONE14} \cdot A_{14,2}\right)\right] \cdot \text{MINPLUS1}}{\dfrac{\left[\left(\dfrac{1}{N-1}\right) \cdot \left[\text{DIFF}^T \cdot \text{DIFF} - \left(\dfrac{1}{N}\right) \cdot (\text{DIFF} \cdot \text{FOURTEEN1})^2\right]\right]^{\frac{1}{2}}}{N^{\frac{1}{2}}}} \square$$

SANDLERS A TEST

An alternate and somewhat simpler technique for determining whether the difference between correlated samples is significant is by using the A test which is derived directly from the t-test.

STATISTICS FORMULA

$$A = \Sigma D^2 / (\Sigma D)^2 = 293/(-51)^2 = 293/2601 = .113$$
$$df = 13 \ (N-1)$$

From tables, for an A value to be significant at the .05 level, it must be equal to **or smaller than** 0.270. Since A = .113, it's concluded the difference is significant.

STEP 3: Square the difference scores and sum:

[-5 -8 -5 -5 -6 2 0 -3 -3 -1 -5 -5 -3 -6]

-5
-8
-5
-5
-6
2
0	= 293
-3
-3
1
-5
-5
-3
-6

HP-48G PROGRAM:

```
RCL DIFF
↑
TRN
SWAP
x
STO GNDSUMSQ
```

$$(1/14) \ \left([-5 \ -8 \ -5 \ -5 \ -6 \ 2 \ 0 \ -3 \ -3-1 \ -5 \ -5 \ -3 \ -6] \begin{bmatrix} 1 \\ 1 \\ 1 \\ 1 \\ 1 \\ 1 \\ 1 \\ 1 \\ 1 \\ 1 \\ 1 \\ 1 \\ 1 \\ 1 \end{bmatrix} \right)^2 = -51^2/14 = 186$$

PROGRAM:
$$\frac{(1/N) \, ([\text{diff}]^2{}_{1,14}}{(1/N) \, ([\text{diff}]_{1,14}[1]_{14,1})^2}$$

THE ANALYSIS OF VARIANCE

INTRODUCTION

Before we begin with the analysis of variance and other advanced statistical testing, I am going to try to explain a little about what I will be doing, especially since the method is new and there is no information on it anywhere in the world except here in this book. (And in my head).

First about the structure of the examples. I wrote the book up to section 2.9, Simple Latin Square when my word processor got electrocuted. I had all the information on floppy disk, but when I got use of a computer the matrices were all garbled and it would be simpler to re-type the information rather than try to rearrange the numbers. I got use of a computer and started at section 2.10, Latin Square Design: Complex. At this point I did not have MathCad +6 nor a HP-48G calculator. All I had was my HP-48 SX calculator to write the programs with. When I got to trend analysis, somewhere in there I got my HP-48G calculator and the MathCad program. When I got to the end of trend analysis, I began retyping the paper from page one. The book is presented as a progression of my exploration of the properties of this operator, sort of like a journal, except as I rewrote the earlier portions, I used the most recent properties I discovered and left out most of my earlier archaic methods. Then we get to section 2.10, which I am going to leave as is, though at the end of each section I will update the method using MathCad +6 and HP-48G programs. So don't be surprised if the math looks different after Latin Square: Simple ends. It will also help those who have the SX model to set up the statistical programs.

The math is basically simple. The basic function for the analysis of variance is

$$1/N(1 \times A \times B)^2 = C^2 \text{ Between Subjects}$$
$$1/N(A \times B \times 1)^2 = C^2 \text{ Within Subjects}$$

In matrix form, A is equal to the data we have taken, I call this the datastream matrix. B is the database matrix [DB]. This matrix is arbitrary, and with it you define how you manipulate the information in the datastream matrix [A]. C^2 is the solution

408

matrix. Usually, but not in all cases, it will be a single number. So basically all advanced statistics is based on A x B = C. The [1] matrix when it is in a row in front of A just sums the columns of [A] when multiplied. The number of 1's must equal the number of rows in [A]. If [1] is a column on the right side of the matrix [A], it sums the rows of [A] when multiplied. The number of 1's must equal the number of columns in [A]. i.e.

$$1\ 1\ 1] \begin{vmatrix} 1 & 2 & 3 \\ 4 & 5 & 6 \\ 7 & 8 & 9 \end{vmatrix} = [12\ 15\ 18] \quad \text{and} \quad \begin{vmatrix} 1 & 2 & 3 \\ 4 & 5 & 6 \\ 7 & 8 & 9 \end{vmatrix} \begin{vmatrix} 1 \\ 1 \\ 1 \end{vmatrix} = \begin{vmatrix} 6 \\ 15 \\ 24 \end{vmatrix}$$

SUMMING THE COLUMNS   SUMMING THE ROWS
(WITHIN SUBJECTS)   (BETWEEN SUBJECTS)

The third operator is found in Gaussian reduction. Had anyone tried to prove Gaussian Reduction, the half-multiplier operator would have been discovered over 200 years ago. It is simply this:

$$\begin{vmatrix} 2 \\ 3 \\ 4 \end{vmatrix} \circ \begin{vmatrix} 5 \\ 6 \\ 7 \end{vmatrix} = \begin{vmatrix} 10 \\ 18 \\ 28 \end{vmatrix} \quad \begin{matrix} (5 \times 2 = 10) \\ (6 \times 3 = 18) \\ (4 \times 7 = 28) \end{matrix}$$

But mathematicians have been told throughout the years that this is impossible, therefore, illegal to use. I have proved it in this paper. The translation I use to accomplish this half-multiplication is to diagonalize the operator and matrix multiply:

$$\begin{vmatrix} 2 & 0 & 0 \\ 0 & 3 & 0 \\ 0 & 0 & 4 \end{vmatrix} \begin{vmatrix} 5 \\ 6 \\ 7 \end{vmatrix} = \begin{vmatrix} 10 \\ 18 \\ 28 \end{vmatrix}$$

And now we get to the mysterious looking, but beautifully symmetric statistical database matrix. This is the heart and soul of statistics, physics, accounting and inventories and whatever other math this half-multiplier operator covers.

So far, I have covered [1][A] and [A][1] and [A] o [B], but the equation for the analysis of variance is:

$$1/N([1][A][DB])^2 = 1/NC^2 \text{ where } C^2 = CC^T$$

We multiply C^2 as CC^T to get a single number as an answer. C^TC will give us a NxN matrix, though the sum of its diagonal is equal to CC^T. The use and meaning of the statistical database is as simple as it is elegant. Once we have multiplied [1][A] we get a single row matrix. The statistical database tells us how to manipulate the data in this row matrix. For example, suppose we have the datastream matrix:

$$A] = \begin{vmatrix} 1 & 2 & 3 & 4 & 5 & 6 \\ 6 & 5 & 4 & 3 & 2 & 1 \\ 2 & 4 & 6 & 1 & 3 & 5 \end{vmatrix} \quad then \quad [1 \ 1 \ 1] \begin{vmatrix} 1 & 2 & 3 & 4 & 5 & 6 \\ 6 & 5 & 4 & 3 & 2 & 1 \\ 2 & 4 & 6 & 1 & 3 & 5 \end{vmatrix} = [9 \ 11 \ 13 \ 8 \ 10 \ 12]$$

$$\begin{aligned} 1+6+2&=9 \\ 2+5+4&=11 \\ 6+4+3&=13 \\ &etc. \end{aligned}$$

Now for accounting or statistical purposes, suppose we wish to add the first three numbers and ignore the last three, then ignore the first three numbers and add the last three numbers, and keep the sums separate. The database matrix that will accomplish this is:

$$\begin{vmatrix} 1 & 0 \\ 1 & 0 \\ 1 & 0 \\ 0 & 1 \\ 0 & 1 \\ 0 & 1 \end{vmatrix}$$

I call this [DB1]. In fact, I define this form as database 1 in all cases. Where the 1's in a column are not separated by a zero. Another form of database 1 is if we wish to add the first two numbers and ignore the rest, add the third and fourth numbers only and ignore the rest, and add the fifth and sixth numbers and ignore the rest. This database matrix is:

$$\begin{vmatrix} 1 & 0 & 0 \\ 1 & 0 & 0 \\ 0 & 1 & 0 \\ 0 & 1 & 0 \\ 0 & 0 & 1 \\ 0 & 0 & 1 \end{vmatrix}$$

Note again the ones in the columns are next to each other. To be consistent, since the above database matrix is called [DB1] I call this second matrix [DB11} to show it is the second database matrix of type one. i.e.

$$\begin{aligned} 9+11+13&=33 \\ 8+10+12&=30 \end{aligned} \qquad [9 \ 11 \ 13 \ 8 \ 10 \ 12] \begin{vmatrix} 1 & 0 \\ 1 & 0 \\ 1 & 0 \\ 0 & 1 \\ 0 & 1 \\ 0 & 1 \end{vmatrix} = [33 \ 30] \quad and$$

$$\begin{aligned} 9+11&=20 \\ 13+8&=21 \\ 10+12&=22 \end{aligned} \qquad [9 \ 11 \ 13 \ 8 \ 10 \ 12] \begin{vmatrix} 1 & 0 & 0 \\ 1 & 0 & 0 \\ 0 & 1 & 0 \\ 0 & 1 & 0 \\ 0 & 0 & 1 \\ 0 & 0 & 1 \end{vmatrix} = [20 \ 21 \ 22]$$

Note: for all you experimenters out there, we can use the database matrices to help set up our experiment properly. Lets look at databases 1 and 11. For each [DB1] the number of columns (we can have as many rows as we want) must be evenly divisible by 3. That is, we can have 6 columns, 9 columns, 12 columns, etc. For the [DB11] matrix the number of columns in [A] must be divisible by two. If we want to calculate both [DB1] and [DB11], the number of columns in [A] must be divisible by both 2 and 3. We can have 6 columns, 12 columns, 18 columns etc. for the additions to work out, but we can't have 9, 12, etc.

In the [DB2] type matrices, we add every other number, i.e. the first and third, the second and fourth, fifth and seventh, and the sixth and eighth numbers. We will end up with four sums with eight columns of data. The number of columns in [A] must therefore be divisible by four. So if we want to analyze with [DB1], [DB11] and [DB2], our common divisors are 2, 3 and 4. The only numbers that fit this category are 12, 24 etc. so in the experiment we must have 12 columns (again the number of rows (subjects) is arbitrary since we add them together, the number doesn't really count. The [DB2] matrix looks like:

$$DB2] = \begin{vmatrix} 1 & 0 & 0 & 0 \\ 0 & 1 & 0 & 0 \\ 1 & 0 & 0 & 0 \\ 0 & 1 & 0 & 0 \\ 0 & 0 & 1 & 0 \\ 0 & 0 & 0 & 1 \\ 0 & 0 & 1 & 0 \\ 0 & 0 & 0 & 1 \end{vmatrix}$$

Adding arbitrarily two more columns to [1] [A] example above (so we can do this problem) we get:

$$[9\ 11\ 13\ 8\ 10\ 12\ 6\ 14] \qquad 1\ 0\ 0\ 0 \qquad = [22\ 19\ 16\ 26]$$

$$\begin{vmatrix} 0 & 1 & 0 & 0 \\ 1 & 0 & 0 & 0 \\ 0 & 1 & 0 & 0 \\ 0 & 0 & 1 & 0 \\ 0 & 0 & 0 & 1 \\ 0 & 0 & 1 & 0 \\ 0 & 0 & 0 & 1 \end{vmatrix}$$

9 +13=22
11+8 =19
10+6 =16
12+14=26

In the [DB3] type matrices, we add every third number, i.e. first to the fourth, second to the fifth, etc. Note there are two zero's underneath every 1.

$$[DB3] = \begin{vmatrix} 1 & 0 & 0 \\ 0 & 1 & 0 \\ 0 & 0 & 1 \\ 1 & 0 & 0 \\ 0 & 1 & 0 \\ 0 & 0 & 1 \end{vmatrix} \quad \text{And } [1][A][DB3] = \quad [9 \; 11 \; 13 \; 8 \; 10 \; 12] \begin{vmatrix} 1 & 0 & 0 \\ 0 & 1 & 0 \\ 0 & 0 & 1 \\ 1 & 0 & 0 \\ 0 & 1 & 0 \\ 0 & 0 & 1 \end{vmatrix} \begin{array}{l} = 17 \; 21 \; 25] \\ \\ 9+8=17 \\ 11+10=21 \\ 13+12=25 \end{array}$$

Finally, we get to the calculation of N for each of these databases. N for each database equals the number of elements used to calculate C. It may or may not be the same as other databases. With the exception of the calculation of Σx^2 and $(\Sigma x)^2$, N is equal to the number of elements in [A] divided by the number of columns in the database matrix. That is, in the matrix [A], there are 18 elements, [DB1] has two columns, so $N_{[DB1]} = 18/2 = 9$.

$N_{[DB11]} = 18/2 = 9$. $N_{[DB3]} = 18/3 = 6$. It's that simple.

Once this first part is complete, we square the resulting row matrix. To get a single number as our solution, we multiply as $(1/N)CC^T$. We do **not** square the 1/N term. i.e. to get the variance for [DB1] we proceed as follows:

$1/N \; [1][A][DB1] = (1/9)[33 \; 30]$. And $1/N \; ([1][A][DB1])^2 = 1/9 \; [33 \; 30] \begin{vmatrix} 33 \\ 30 \end{vmatrix} =$

$(1/9) \; (33^2 + 30^2) = 1/9 \; (1089 + 900) = 1/9 \; (1989) = [221]$.

Note that if we multiplied as $1/N \; C^T C$ we would get: $1/9 \begin{bmatrix} 33 \\ 30 \end{bmatrix} [33 \; 30] =$

$(1/9) \begin{bmatrix} 1089 & 990 \\ 990 & 900 \end{bmatrix}$ and we would have to sum the diagonal and divide by $1/N = 1/9$ to get $(1/N)C^2$. Which is a lot more math steps to go through.

A few more items to cover and we can get to the math. When we need to determine the grand sum of squares (Σx^2), the easiest way I've found is to square the [A] matrix and sum the diagonal. In general I multiply to get the smallest NxN matrix, but when we get to F-tests for simple effects and trend analysis, we have to multiply to get the largest matrix possible because we need to keep track of each of the squares separately. Using the HP-48G calculator to get the diagonal matrix we must pull out the diagonal and make a row matrix, then reassemble it to a diagonal. In MathCad +6 I create an identity matrix and vectorize an onto mapping rather than the usual into mapping. This multiplies one element by one element rather than the normal row x column and summing. I'll not give an example here, you'll see it in the math soon enough.

As I began to write this paper, I started out by using a [b] post-multiplier. But I discovered if all the rows are equal, all we need to do is square the C matrix. The [b] seems to be archaic, but in reality it is very important, especially if you are running an experiment and one or more of the subjects has to drop out early and you do not want to discard the data. The N's then are not all the same and you must use the [b] form. I will put it in as many problems as I can remember.

Lets see how using the b method will work for the t-test for the difference between two independent means.

$$
\frac{b_1/N_1 - b_2/N_2}{(1/N_1 + 1/N_2)\left([a_1\ b_1]\begin{vmatrix}1\\-b_1/N_1\end{vmatrix} + [a_2\ b_2]\begin{vmatrix}1\\-b_2/N_2\end{vmatrix}\right)}
$$

$$
N_1 + N_2 - 2
$$

Where $[DATA\ X]\begin{vmatrix}D\ 1\\A\ .\\T\ .\\A\ .\\X\ 1\end{vmatrix} = [a_1\ b_1]$ and $[DATA\ Y]\begin{vmatrix}D\ 1\\A\ .\\T\ .\\A\ .\\Y\ 1\end{vmatrix}\begin{vmatrix}1\\-b2/N2\end{vmatrix} = [a_2\ b_2]$

$$
\frac{1196/12 - 1428/14}{((1/12 + 1/14)([121,318\quad 1196]\begin{vmatrix}1\\-1196/12\end{vmatrix} + [148,202\ 1428]\begin{vmatrix}1\\-1428/14\end{vmatrix}))/24\ ]^{1/2}} = .43
$$

As a final note, throughout this paper, I am going to refer to the sum [1][A] = SUM1. So whenever we encounter the term SUM1 we know we are talking about the datastream matrix [A] whose columns have been added. Therefore, there will always be room for SUM1 in this mathematics.

BASIC TRANSLATIONS FOR STATISTICS

$N\Sigma X^2 - (\Sigma X)^2 = [DATA] \begin{vmatrix} D & 1 \\ A & . \\ T & . \\ A & 1 \end{vmatrix} \begin{vmatrix} N_T \\ -b \end{vmatrix} = [DATA] \begin{vmatrix} D & 1 \\ A & . \\ T & . \\ A & 1 \end{vmatrix} \begin{vmatrix} 1 \\ -b/N_T \end{vmatrix} = N[DATA][DATA]^T - ([DATA] \begin{vmatrix} 1 \\ . \\ . \\ 1 \end{vmatrix}^2)$

$N\Sigma XY = N_T[DATAX] \begin{vmatrix} D \\ A \\ T \\ A \\ Y \end{vmatrix}$

$(\Sigma X)(\Sigma Y) = ([DATAX] \begin{bmatrix} 1 \\ . \\ . \\ 1 \end{bmatrix})([DATAY] \begin{bmatrix} 1 \\ . \\ . \\ 1 \end{bmatrix})$ Or $=$ Anti-Log$([1\ 1](Log \begin{bmatrix} DATAX \\ DATAY \end{bmatrix} \begin{bmatrix} 1 \\ . \\ . \\ 1 \end{bmatrix}))$

CROSS PRODUCT: $\Sigma XY - (1/N)\Sigma X\Sigma Y = [DATAX] \begin{vmatrix} D \\ A \\ T \\ A \\ Y \end{vmatrix} - (1/N)([DATAX] \begin{vmatrix} 1 \\ . \\ . \\ 1 \end{vmatrix})([DATAY] \begin{vmatrix} 1 \\ . \\ . \\ 1 \end{vmatrix})$

PEARSON PRODUCT-MOMENT CORRELATION

$r = \dfrac{N_P\Sigma XY - (\Sigma X)(\Sigma Y)}{([N\Sigma X^2 - (\Sigma X)^2][\ N\Sigma Y^2 - (\Sigma Y)^2])^{1/2}}$ Which can be translated to:

$(N_P[DATAX] \begin{vmatrix} D \\ A \\ T \\ A \\ Y \end{vmatrix} - ([DATAX] \begin{vmatrix} 1 \\ . \\ . \\ 1 \end{vmatrix})([DATAY] \begin{vmatrix} 1 \\ . \\ . \\ 1 \end{vmatrix})$

$[([DATAX] \begin{vmatrix} D & 1 \\ A & . \\ T & . \\ A & . \\ X & 1 \end{vmatrix} \begin{vmatrix} 1 \\ -b_x/N_P \end{vmatrix})([DATAY] \begin{vmatrix} D & 1 \\ A & . \\ T & . \\ A & . \\ Y & 1 \end{vmatrix} \begin{vmatrix} 1 \\ -b_y/N_P \end{vmatrix})]^{1/2}$

Or calculating without the b's:

$N_P[DATAX][DATAY]^T -$ Anti-Log$([1\ 1]$ LOG $\begin{vmatrix} DATAX \\ DATAY \end{vmatrix} \begin{vmatrix} 1 \\ . \\ . \\ 1 \end{vmatrix})$

$N_P[DATAX][DATAX]^T - ([DATAX] \begin{vmatrix} 1 \\ . \\ . \\ 1 \end{vmatrix})^2 (N_P[DATAY][DATAY]^T - ([DATAY] \begin{vmatrix} 1 \\ . \\ . \\ 1 \end{vmatrix})^2]$

PROBLEM 2.1: ANALYSIS OF VARIANCE, COMPLETELY RANDOMIZED DESIGN

Let me remind you, I am not a Statistician, what I do in the following study of Statistics has evolved from problems solved in the "Computational Handbook of Statistics" I do not know enough to even invent an example and do not want to plagiarize, so there is no example, just the math in all of the following Statistics. The numbers and size of the Matrices are different where possible. You will need to buy the book 'Computational Handbook of Statistics' for examples, tables and the methods used for computation that are consolidated in this section.

STEP1:   Consolidate the data into a Matrix:

$$DATA := \begin{pmatrix} 9 & 5 & 15 & 5 \\ 6 & 9 & 14 & 13 \\ 11 & 5 & 19 & 15 \\ 7 & 6 & 15 & 17 \\ 10 & 5 & 4 & 11 \\ 5 & 11 & 10 & 12 \\ 7 & 11 & 15 & 17 \\ 8 & 6 & 12 & 13 \\ 9 & 9 & 18 & 14 \\ 10 & 8 & 7 & 12 \\ 4 & 6 & 13 & 7 \\ 18 & 17 & 16 & 19 \end{pmatrix}$$

VARIABLES:

$$ONE12 := (1 \quad 1 \quad 1 \quad 1 \quad 1 \quad 1 \quad 1 \quad 1 \quad 1 \quad 1 \quad 1 \quad 1) \qquad FOUR1 := \begin{pmatrix} 1 \\ 1 \\ 1 \\ 1 \end{pmatrix}$$

STEP 2:   Add the scores in each Group. Each Group in this example consists of one column. This gives us SUM1, the most important value in all of Statistics.

$$\text{SUM1} := \left(1_{1,12}\right) \cdot \left(A_{12,4}\right)^{\blacksquare}$$

SUM1 := ONE12 · DATA

SUM1 = (104　98　158　155)

STEP 3:　Square each number in the Matrix and sum the values.

If you believe modern Mathematics that 3x2 is equal to six only in certain circumstances but is not equal to six in element by element multiplication of Matrices, then close this book and go back to the Random Variable Theory of Statistics. But if you are willing to take a chance that 3x2 = 6 in all cases . . . then read on.

$$\text{GRANDSUMSQ} := \left(1_{1,12}\right) \cdot \overline{\left[\left(A_{12,4}\right) \cdot \left(A_{12,4}\right)\right]} \cdot \left(1_{4,1}\right)^{\blacksquare}$$

$$\text{GRANDSUMSQ} := \text{ONE12} \cdot \overline{\left(\text{DATA} \cdot \text{DATA}\right)} \cdot \text{FOUR1}$$

$$\text{GRANDSUMSQ} := (1\ 1\ 1\ 1\ 1\ 1\ 1\ 1\ 1\ 1\ 1\ 1) \cdot \overrightarrow{\begin{pmatrix} 9 & 5 & 15 & 5 \\ 6 & 9 & 14 & 13 \\ 11 & 5 & 19 & 15 \\ 7 & 6 & 15 & 17 \\ 10 & 5 & 4 & 11 \\ 5 & 11 & 10 & 12 \\ 7 & 11 & 15 & 17 \\ 8 & 6 & 12 & 13 \\ 9 & 9 & 18 & 14 \\ 10 & 8 & 7 & 12 \\ 4 & 6 & 13 & 7 \\ 18 & 17 & 16 & 19 \end{pmatrix} \cdot \begin{pmatrix} 9 & 5 & 15 & 5 \\ 6 & 9 & 14 & 13 \\ 11 & 5 & 19 & 15 \\ 7 & 6 & 15 & 17 \\ 10 & 5 & 4 & 11 \\ 5 & 11 & 10 & 12 \\ 7 & 11 & 15 & 17 \\ 8 & 6 & 12 & 13 \\ 9 & 9 & 18 & 14 \\ 10 & 8 & 7 & 12 \\ 4 & 6 & 13 & 7 \\ 18 & 17 & 16 & 19 \end{pmatrix}} \cdot \begin{pmatrix} 1 \\ 1 \\ 1 \\ 1 \end{pmatrix}$$

GRANDSUMSQ = 6457

STEP 4:　Compute the Grand Sum of the Matrix.

$$\text{GRANDSUM} := \left(1_{1,12}\right) \cdot \left(A_{12,4}\right) \cdot \left(1_{4,1}\right)^{\blacksquare}$$

GRANDSUM := ONE12 · DATA · FOUR1

$$\text{GRANDSUM} := (1 \quad 1 \quad 1 \quad 1 \quad 1 \quad 1 \quad 1 \quad 1 \quad 1 \quad 1 \quad 1 \quad 1) \cdot \left(\begin{pmatrix} 9 & 5 & 15 & 5 \\ 6 & 9 & 14 & 13 \\ 11 & 5 & 19 & 15 \\ 7 & 6 & 15 & 17 \\ 10 & 5 & 4 & 11 \\ 5 & 11 & 10 & 12 \\ 7 & 11 & 15 & 17 \\ 8 & 6 & 12 & 13 \\ 9 & 9 & 18 & 14 \\ 10 & 8 & 7 & 12 \\ 4 & 6 & 13 & 7 \\ 18 & 17 & 16 & 19 \end{pmatrix} \cdot \begin{pmatrix} 1 \\ 1 \\ 1 \\ 1 \end{pmatrix} \right)$$

GRANDSUM = 515

STEP 5: Compute the Correction Factor (CORRFACT) by squaring the Grand Sum and dividing by the number of elements in the Matrix. m = 12 x 6 = 72

Nt := 48   m := 48

$$\text{CORRFACT} := \left(\frac{1}{Nt} \right) \cdot \left[(1_{1,12}) \cdot (A_{12,4}) \cdot (1_{4,1}) \right] \cdot \left[(1_{1,12}) \cdot (A_{12,4}) \cdot (1_{4,1}) \right]^T \cdot (1_{4,1})^{\blacksquare}$$

$$\text{CORRFACT} := \left(\frac{1}{Nt} \right) \cdot (\text{SUM1} \cdot \text{FOUR1})^2$$

CORRFACT = 5525.521

STEP 6: Find the total Sum of Squares SSt by subtracting GRANDSUMSQ - CORRFACT.

$$\text{SSt} := (1_{1,12}) \cdot \overrightarrow{\left[(A_{12,4}) \cdot (A_{12,4}) \right]} \cdot (1_{4,1}) - \left[\left(\frac{1}{Nt} \right) \cdot \left[(1_{1,12}) \cdot (A_{12,4}) \cdot (1_{4,1}) \right] \cdot \left[(1_{1,12}) \cdot (A_{12,4}) \cdot (1_{4,1}) \right]^T \cdot (1_{4,1}) \right]^{\blacksquare}$$

SSt := GRANDSUMSQ - CORRFACT

SSt = 931.479

STEP 7: Square the sum of each of the groups and divide by the number of elements in each group. Nb = Nt/6 columns Nb = 72/6 = 12

Nb := 12

$$SSb := \left(\frac{1}{Nb}\right) \cdot \left[(1_{1,12}) (A_{12,4}) \right] \left[(1_{1,12}) (A_{12,4}) \right]^{T} - \left[\left(\frac{1}{Nt}\right) \cdot \left[(1_{1,12}) (A_{12,4}) (1_{4,1}) \right] \left[(1_{1,12}) (A_{12,4}) (1_{4,1}) \right]^{T} \cdot (1_{4,1}) \right]$$

$$SSb := \left(\frac{1}{Nb}\right) \cdot (SUM1) \cdot SUM1^{T} - CORRFACT \qquad\qquad \left(\frac{1}{Nb}\right) \cdot (SUM1) \cdot SUM1^{T} = 5784.083$$

SSb = 258.563

STEP 7b: Suppose all the values in Nb are not equal ie some of the values in the data Matrix = zero. This is a real world condition and may be computed by:

SUM1 = (104   98   158   155)

N1 := 12   N2 := 12   N3 := 12   N4 := 12   N5 := 12   N6 := 12

$$SSb1 := \left[\overrightarrow{ \left(\frac{1}{N1} \quad \frac{1}{N2} \quad \frac{1}{N3} \quad \frac{1}{N4}\right) \cdot (104 \quad 98 \quad 158 \quad 155) } \right] \cdot \begin{pmatrix} 104 \\ 98 \\ 158 \\ 155 \end{pmatrix} - CORRFACT$$

SSb1 = 258.563

STEP 8: Computation of the Within Subjects Sum of Squares

SSw := SSt - SSb

$$SSw := \left[(1_{1,12}) \overrightarrow{(A_{12,4}) (A_{12,4})} \cdot ((1_{4,1})) - \left[\left(\frac{1}{Nt}\right) \cdot (1_{1,12}) (A_{12,4}) (1_{4,1}) \left[(1_{1,12}) (A_{12,4}) (1_{4,1}) \right]^{T} \cdot (1_{4,1}) \right] \right]$$

$$\left(\frac{1}{Nb}\right) \cdot \left[(1_{1,12}) (A_{12,4}) \right] \left[(1_{1,12}) (A_{12,4}) \right]^{T} - \left[\left(\frac{1}{Nt}\right) \cdot (1_{1,12}) (A_{12,4}) (1_{4,1}) \left[(1_{1,12}) (A_{12,4}) (1_{4,1}) \right]^{T} \cdot (1_{4,1}) \right]$$

$$SSw := ONE12 \cdot \overrightarrow{((DATA \cdot DATA))} \cdot FOUR1 - \left[\left(\frac{1}{Nt}\right) \cdot (SUM1 \cdot FOUR1)^{2} \right] - \left[\left(\frac{1}{Nb}\right) \cdot (SUM1) \cdot SUM1^{T} - CORRFACT \right]$$

SSw = 672.917

STEP 9: We now find the degrees of freedom df. These are needed because the F in the F-test is a ratio of the mean squares SS/df. Where m = # of elements in the Matrix and G = # groups (in this case the # of columns in the Matrix).

$$m := 48 \qquad G := 2$$

$$df := \begin{bmatrix} m - 1 \\ G - 1 \\ (m-1) - (G-1) \end{bmatrix} \qquad df = \begin{pmatrix} 47 \\ 1 \\ 46 \end{pmatrix}$$

STEP 10: Calculation of F.

$$SS := \begin{pmatrix} SSt \\ SSb \\ SSw \end{pmatrix} \qquad ERROR := \frac{SSw^{\blacksquare}}{dfSSw} \qquad ERROR := \frac{672.917}{44}$$

$$SS = \begin{pmatrix} 931.479 \\ 258.563 \\ 672.917 \end{pmatrix}$$

$$F := \overrightarrow{\left(SS \cdot df^{-1} \cdot ERROR^{-1}\right)}$$

$$F = \begin{pmatrix} 1.296 \\ 16.907 \\ 0.957 \end{pmatrix}$$

We have a F of 16.907 with df of 1 and 46, therefore the probability is:

$$p < .001$$

This F value with a df of 3 and 44 would occur by chance less than once in 1000 times. It is concluded that the experimental conditions do affect the results of the tests.

PROBLEM 2.1 USING NESTED ARRAYS

$$SUM1a := (\begin{matrix} 104 & 98 & 158 & 155 \end{matrix})$$

$$SUM1a := ONE12 \cdot DATA^{\blacksquare}$$

What we need to do here is multiply SUM1 by DBALL. In this case the CF, GS and GSQ are computed from other equations so we have to compute them separately and add them to the Matrix when it is necessary. DBALL is the compilation of all the DB Matrices used to solve the problems that have the form [1]x{A} x[B] for Between Subjects analysis or of the form [A]x[B]x[1] for Within Subjects analysis

$$
DBa := \begin{pmatrix} 0 & 0 & 1 & 0 & 0 & 0 \\ 0 & 0 & 0 & 1 & 0 & 0 \\ 0 & 0 & 0 & 0 & 1 & 0 \\ 0 & 0 & 0 & 0 & 0 & 1 \end{pmatrix}
\qquad N := (\,1 \quad 48 \quad 12 \quad 12 \quad 12 \quad 12\,)
$$

$SUM1a \cdot DBa = (\,0 \quad 0 \quad 104 \quad 98 \quad 158 \quad 155\,)$

Adding the Grand Sum to the Matrix we get

$SUM2a := (\,0 \quad 515 \quad 0 \quad 0 \quad 0 \quad 0\,)$

$SUM2 := SUM1 \cdot DBa + SUM2a$

$SUM2 = (\,0 \quad 515 \quad 104 \quad 98 \quad 158 \quad 155\,)$

Now we need to square SUM2

$$
\overrightarrow{\left[(SUM1 \cdot DBa + SUM2a) \cdot N^{-1} \right]} = \blacksquare
$$

Mathcad cannot handle this so we must simplify

$SUM2 := SUM1 \cdot DBa + SUM2a$

$$
SUM3 := \overrightarrow{\left(SUM2 \cdot SUM2 \cdot N^{-1} \right)} \qquad\qquad \overrightarrow{\left[(SUM1 \cdot DBa + SUM2a) \cdot N^{-1} \right]} + SUM3a
$$

$SUM3 = (\,0 \quad 5525.521 \quad 901.333 \quad 800.333 \quad 2080.333 \quad 2002.083\,)$

Here we have squared the values computed for SUM2.

$$
\overrightarrow{\left(SUM2 \cdot SUM2 \cdot N^{-1} \right)} = (\,0 \quad 5525.521 \quad 901.333 \quad 800.333 \quad 2080.333 \quad 2002.083\,)
$$

Now we need to add the Grand Sum of Squares

SUM3 = (0   5525.521   901.333   800.333   2080.333   2002.083)

GSQ := (6457   0   0   0   0   0)

$$\overrightarrow{\left(\text{SUM2}\cdot\text{SUM2}\cdot\text{N}^{-1}\right)} + \text{GSQ}$$

or

$$\overrightarrow{\left(\text{SUM2}\cdot\text{SUM2}\cdot\text{N}^{-1}\right)} + \text{GSQ} = (\ 6457\quad 5525.521\quad 901.333\quad 800.333\quad 2080.333\quad 2002.083\ )$$

The basic Matrix is complete, now we must sum the elements using DB. Elements 1 and 2 we leave alone, but we must sum elements 3, 4, 5 and 6. The DB Matrix for this computation is as follows:

$$\text{SUM4} := \overrightarrow{\left(\text{SUM2}\cdot\text{SUM2}\cdot\text{N}^{-1}\right)} + \text{GSQ}$$

SUM4 = (6457   5525.521   901.333   800.333   2080.333   2002.083)

Now we sum the last 4 elements in SUM4 using DB:

$$\text{SUM5} := \left(\overrightarrow{\left(\text{SUM2}\cdot\text{SUM2}\cdot\text{N}^{-1}\right)} + \text{GSQ}\right)\cdot\text{DB}
\qquad
\text{DB} := \begin{pmatrix} 1 & 0 & 0 \\ 0 & 1 & 0 \\ 0 & 0 & 1 \\ 0 & 0 & 1 \\ 0 & 0 & 1 \\ 0 & 0 & 1 \end{pmatrix}$$

SUM5 = (6457   5525.521   5784.083)

$$\begin{pmatrix} \text{GSQ} & \text{CF} & \text{SSb} \\ 6457 & 5525.521 & 5784.083 \end{pmatrix}$$

Now we subtract the various Correction Factors using DB2.

$$\text{DB2} := \begin{pmatrix} 1 & 0 & 1 \\ -1 & -1 & 0 \\ 0 & 1 & -1 \end{pmatrix}$$

DB2 means we do the following:

$$\begin{pmatrix} GSQ - CF & SSb - CF & GSQ - SSb \\ 1 & 0 & 1 \\ -1 & -1 & 0 \\ 0 & 1 & -1 \end{pmatrix}$$

In the first column we subtract the Correction Factor from GRANDSUMSQ. In the second column, we subtract the Correction Factor from SSb. In the third column we subtract SSb from the Grand Sum of Squares. Doing this we get:

$$SS := \left[\overrightarrow{\left(\overline{\left(SUM2 \cdot SUM2 \cdot N^{-1} \right)} + GSQ \right) \cdot DB \cdot DB2} \right]$$

$$SS = (\, 931.479 \quad 258.563 \quad 672.917 \,)$$

$$m := 48 \qquad G := 2$$

$$df := \begin{bmatrix} m - 1 \\ G - 1 \\ (m - 1) - (G - 1) \end{bmatrix} \qquad df = \begin{pmatrix} 47 \\ 1 \\ 46 \end{pmatrix}$$

$$ERRORTERM := \left[\overrightarrow{(\, 931.479 \quad 258.563 \quad 672.917 \,) \cdot (\, 0 \quad 0 \quad 1 \,) \cdot \left(df^{-1} \right)^T} \right]$$

$$ERRORTERM = (\, 0 \quad 0 \quad 14.629 \,)$$

$$ERROR := (\, 14.294 \quad 14.294 \quad 14.294 \,)$$

$$ERROR := (\, 15.294 \quad 15.294 \quad 15.294 \,)^T \qquad df = \begin{pmatrix} 47 \\ 1 \\ 46 \end{pmatrix}$$

$$SS := \left[\overrightarrow{\left(\overline{\left(SUM2 \cdot SUM2 \cdot N^{-1} \right)} + GSQ \right) \cdot DB \cdot DB2} \right] \qquad ERROR = \begin{pmatrix} 15.294 \\ 15.294 \\ 15.294 \end{pmatrix}$$

$$F := \left(\overrightarrow{SS^T \cdot df^{-1} \cdot ERROR^{-1}} \right)$$

$$F = \begin{pmatrix} 1.296 \\ 16.906 \\ 0.956 \end{pmatrix}$$

Which matches the F value computed above.

The final SS equation (the equation expressed in a single line)
should look like:

$$SS := \left[\left(\overrightarrow{\left(SUM2 \cdot SUM2 \cdot N^{-1} \right)} + GSQ \right) \cdot DB \cdot DB2 \right]$$

$$SS = (\; 931.479 \quad 258.563 \quad 672.917 \;)$$

The final evaluation for F should be (if MathCad could compute
this far):

$$\overrightarrow{\left[\left[\left(\overrightarrow{(SUM1 \cdot DB + SUM2a) \cdot N^{-1}} \right] + SUM3a \right) \cdot DB \cdot DB2 \right] \cdot df^{-1} \cdot ERROR^{-1} \right]} = \blacksquare$$

But again MathCad cannot solve it without substitutions. This
is the final computer program written in a single line for this
problem. Remember, this Half-Multiplier Operator, when we have
the final mathematical solution, we then automatically have the
computer program for all problems of this type, we just change
the dimensions of the arrays.

FACTORIAL DESIGN: TWO FACTORS

SECTION 2.2, COMPUTATIONAL HANDBOOK OF STATISTICS,
SECOND EDITION, PG. 27-31.

STATISTICAL EQUATIONS

ROWS

$$cm\sum_{i=1}^{r}(\overline{x}_{i..}-\overline{x})^2 \qquad L_R=r\sum_{i=1}^{r}R_i^2-T^2 \qquad r-1 \qquad M_R \qquad M_R/M_W \qquad M_R/M_I \qquad M_R/M_I$$

COLUMNS

$$rm\sum_{j=1}^{c}(\overline{x}_{.j.}-\overline{x})^2 \qquad L_C=c\sum_{j=1}^{c}C_j^2-T^2 \qquad c-1 \qquad M_C \qquad M_C/M_W \qquad M_C/M_I \qquad M_C/M_W$$

INTERACTION

$$m\sum_{i=1}^{r}\sum_{j=1}^{c}(\overline{x}_{ij}-\overline{x}_{i..}-\overline{x}_{.j.}+\overline{x})^2 \qquad L_I=rc\sum_{i=1}^{r}\sum_{j=1}^{c}T_{ij}^2 \qquad (r-1)(c-1) \qquad M_I \qquad M_I/M_W \qquad M_I/M_W \qquad M_I/M_W$$

$$-r\sum_{i=1}^{r}R_i^2-c\sum_{j=1}^{c}C_j^2+T^2$$

WITHIN CELLS

$$\sum_{i=1}^{r}\sum_{j=1}^{c}\sum_{m=1}^{m}(x_{ijk}-\overline{x}_{ij})^2 \qquad L_W=n\sum_{i=1}^{r}\sum_{j=1}^{c}\sum_{m=1}^{m}x^2_{ijk} \qquad rc(m-1) \qquad M_W$$

$$-rc\sum_{i=1j=1}^{r}\sum^{c}T^2_{ij}$$

TOTAL

$$\sum_{i=1}^{r}\sum_{j=1}^{c}\sum_{m=1}^{m}(x_{ijk}-\overline{x})^2 \qquad L_T=n\sum_{i=1}^{r}\sum_{j=1}^{c}\sum_{m=1}^{m}x^2_{ijk}-T^2 \qquad n-1$$

SOURCE: STATISTICS WITH APPLICATIONS TO THE BIOLOGICAL AND HEALTH SCIENCES, REMINGTON & SCHORK, PRENTICE/HALL PG. 298.

MATRIX METHOD

ARCHAIC METHOD: Used if # of elements in each column are not equal.

and N_T = total # of data elements

$$(1\ \ 1\ \ 1\ \ .\ \ .\ \ 1)\cdot
\begin{vmatrix}
a_{11} & b_{12} & . & d_{1m} \\
a_{21} & b_{22} & . & d_{2m} \\
a_{31} & b_{32} & . & d_{3m} \\
. & . & . & . \\
. & . & . & . \\
a_{n1} & b_{n2} & . & d_{nm}
\end{vmatrix}
=(A\ \ B\ \ C\ \ .\ \ D)$$

$$SS_T := (A \quad B \quad C \quad . \quad D) \cdot \begin{bmatrix} \dfrac{A}{N_1} & 1 \\[2ex] \dfrac{B}{N_2} & 1 \\[2ex] \dfrac{C}{N_3} & 1 \\[1ex] . & . \\[1ex] \dfrac{D}{N_n} & 1 \end{bmatrix} \cdot \begin{pmatrix} 1 \\[2ex] -\dfrac{b}{N_T} \end{pmatrix} \square= \text{ For all N's not being equal}$$

$$SS_T := (A \quad B \quad C \quad . \quad D) \cdot \begin{vmatrix} A & 1 \\ B & 1 \\ C & 1 \\ . & . \\ D & 1 \end{vmatrix} \cdot \begin{pmatrix} 1 \\[2ex] -\dfrac{b}{N_T} \end{pmatrix} \square= \text{ For all N's being equal}$$

FIRST FACTOR EFFECT:

$$(A \quad B \quad C \quad D) \cdot \begin{pmatrix} 1 & 0 \\ 0 & 1 \\ 1 & 0 \\ 0 & 1 \end{pmatrix} \cdot \begin{pmatrix} \dfrac{A+C}{N_1 + N_3} \\[2ex] \dfrac{B+D}{N_2 + N_4} \end{pmatrix} \square$$

$$SS_{SHOCK} := \left[(A \quad B \quad C \quad D) \cdot \begin{pmatrix} 1 & 0 \\ 0 & 1 \\ 1 & 0 \\ 0 & 1 \end{pmatrix} \cdot \begin{pmatrix} \dfrac{A+C}{N_1 + N_3} \\[2ex] \dfrac{B+D}{N_2 + N_4} \end{pmatrix} \right] - \left[(A \quad B \quad C \quad D) \cdot \begin{pmatrix} 1 \\ 1 \\ 1 \\ 1 \end{pmatrix} \cdot \left(\dfrac{b}{N_T} \right) \right] \square - \text{ CORRECTION FACTOR}$$

UPGRADED:

$$SS_S := (SUM1 \cdot DB2)^2 - CORRFACT \; \square$$

$$SS_{SHOCK} := SS_S \; \square$$

$$SS_{SHOCK} := \left(\dfrac{1}{N_S} \right) \cdot \left[(A \quad B \quad C \quad D) \cdot \begin{pmatrix} 1 & 0 \\ 0 & 1 \\ 1 & 0 \\ 0 & 1 \end{pmatrix} \right]^2 - CORRFACT \; \square$$

PROGRAM: $1/N_s$ $([SUM1]_{1,4}[DB2]_{4,2})^2-$**CORRFACT**

SECOND FACTOR SHOCK: $SS_{LIST}=SS_L.$

$$SS_L := \left(\frac{1}{N_L}\right) \cdot (SUM1 \cdot DB1)^2 - \text{CORRFACT}_\square$$

$$SS_{LIST} := SS_{L\square}$$

$$SSL := \left[(A\ \ B\ \ C\ \ D) \cdot \begin{pmatrix} 1 & 0 \\ 1 & 0 \\ 0 & 1 \\ 0 & 1 \end{pmatrix} \begin{pmatrix} \dfrac{A+B}{N1+N2} \\ \dfrac{C+D}{N3+N4} \end{pmatrix} \right] - \left[(A\ \ B\ \ C\ \ D) \cdot \begin{pmatrix} 1 \\ 1 \\ 1 \\ 1 \end{pmatrix} \cdot \left(\dfrac{b}{Nt}\right) \right]^{\blacksquare}$$

Upgraded:

$$SS_L := \left(\frac{1}{N_L}\right) \cdot \left[(A\ \ B\ \ C\ \ D) \cdot \begin{pmatrix} 1 & 0 \\ 1 & 0 \\ 0 & 1 \\ 0 & 1 \end{pmatrix} \right]^2 - \text{CORRFACT}_\square$$

$$\frac{b^2}{N_T}$$

PROGRAM: $(1/N_L)$ $([SUM1]_{1,4}[DB1]_{4,2})^2-$**CORRFACT**

INTERACTIVE SHOCK: $SS_{SHOCK \times LIST} = SS_{SxL}.$

$$SS_L := \left(\frac{1}{N_L}\right) \cdot \left[(A \quad B \quad C \quad D) \cdot (A \quad B \quad C \quad D)^T \right] - CORRFACT - SS_L - SS_S \square$$

$$SS_{SHOCKxLIST} := SS_{SxL} \square$$

$$SS_L := (A \quad B \quad C \quad D) \cdot \begin{bmatrix} \dfrac{A}{N_1} \\[6pt] \dfrac{B}{N_2} \\[6pt] \dfrac{C}{N_3} \\[6pt] \dfrac{D}{N_4} \end{bmatrix} - CORRFACT - SS_L - SS_S \square$$

$$[A \ B \ C \ D] \begin{vmatrix} A/N_1 \\ B/N_2 \\ C/N_3 \\ D/N_4 \end{vmatrix} - b^2/N_T - SS_L - SS_S = SS_{SxL}$$

UPGRADE: $[SUM1][SUM1]^T - CORRFACT - SS_L - SS_S$

$$SS_{ERROR} = SS_T - SS_L - SS_S - SS_{SxL}$$

STEP 1:   Table the data as a Matrix:

$$A := \begin{pmatrix} 10 & 16 & 11 & 20 \\ 17 & 14 & 13 & 17 \\ 15 & 10 & 19 & 19 \\ 12 & 10 & 17 & 24 \\ 11 & 7 & 16 & 13 \\ 7 & 10 & 14 & 14 \end{pmatrix}$$

$ONE6 := (1 \ 1 \ 1 \ 1 \ 1 \ 1)$

$ONE4 := (1 \ 1 \ 1 \ 1)$

$SIX1 := ONE6^T$

$FOUR1 := ONE4^T$

$$DB1 := \begin{pmatrix} 1 & 0 \\ 1 & 0 \\ 0 & 1 \\ 0 & 1 \end{pmatrix}$$

$$DB2 := \begin{pmatrix} 1 & 0 \\ 0 & 1 \\ 1 & 0 \\ 0 & 1 \end{pmatrix}$$

$$DB := \begin{pmatrix} 1 \\ 1 \\ 1 \\ 1 \end{pmatrix}$$

$$DB0 := \begin{pmatrix} 0 \\ 0 \\ 0 \\ 0 \end{pmatrix}$$

STEP 2:   Add the scores in each Group: SUM1

$$\text{SUM1} := \left(1_{1,6}\right)\left(A_{6,4}\right)^{\blacksquare}$$

$$\text{SUM1} := \text{ONE6\,A}$$

$$\text{SUM1} = (72 \quad 67 \quad 90 \quad 107)$$

STEP 3:   Square each term in the Matrix and sum. GRANDSUMSQ

$$\text{GRANDSUMSQ} := \left(1_{1,6}\right)\overrightarrow{\left[\left(A_{6,4}\right)\left(A_{6,4}\right)\right]} \cdot \left(1_{4,1}\right)^{\blacksquare}$$

$$\text{GRANDSUMSQ} := \text{ONE6}\,\overrightarrow{[(A) \cdot (A)]} \cdot \text{FOUR1}$$

$$\text{GRANDSUMSQ} := (1 \quad 1 \quad 1 \quad 1 \quad 1 \quad 1) \cdot \overrightarrow{\left[\begin{pmatrix} 10 & 16 & 11 & 20 \\ 17 & 14 & 13 & 17 \\ 15 & 10 & 19 & 19 \\ 12 & 10 & 17 & 24 \\ 11 & 7 & 16 & 13 \\ 7 & 10 & 14 & 14 \end{pmatrix} \cdot \begin{pmatrix} 10 & 16 & 11 & 20 \\ 17 & 14 & 13 & 17 \\ 15 & 10 & 19 & 19 \\ 12 & 10 & 17 & 24 \\ 11 & 7 & 16 & 13 \\ 7 & 10 & 14 & 14 \end{pmatrix}\right]} \cdot \begin{pmatrix} 1 \\ 1 \\ 1 \\ 1 \end{pmatrix}$$

$$\text{GRANDSUMSQ} = 5112$$

STEP 4:   Add the elements in the entire Matrix, sum, square and divide by the total number of elements in the Matrix. ie find the Correction Factor CORRFACT. Nt = 6 X 4 = 24

$$\text{Nt} := 24$$

$$\text{CORRFACT} := \left(\frac{1}{\text{Nt}}\right) \cdot \left[\left(1_{1,6}\right)\left(A_{6,6}\right)\left(1_{6,1}\right)\right]^{2^{\blacksquare}}$$

$$\text{CORRFACT} := \left(\frac{1}{\text{Nt}}\right) \cdot (\text{ONE6\,A} \cdot \text{FOUR1})^2$$

$$\text{CORRFACT} := \left(\frac{1}{\text{Nt}}\right) \cdot (\text{ONE6 A} \cdot \text{FOUR1}) \cdot (\text{ONE6 A} \cdot \text{FOUR1})$$

$$\text{CORRFACT} = 4704$$

STEP 5: Computation of the Total Sum of Squares SSt: Nt = total number of elements in Data Matrix. In this case Nt = 6 x 4 = 24

$$\text{Nt} := 24$$

$\text{SSt} := \text{GRANDSUMSQ} - \text{CORRFACT}$	$\text{GRANDSUMSQ} = 5112$
$\text{SSt} = 408$	$\text{CORRFACT} = 4704$

$$\text{SSt} := \left(1_{1,6}\right)\overline{\left[\left(A_{6,4}\right)\left(A_{6,4}\right)\right]}\cdot\left(1_{4,1}\right) - \left(\frac{1}{\text{Nt}}\right)\cdot\left[\left(1_{1,6}\right)\left(A_{6,4}\right)\left(1_{4,1}\right)\right]^{2^{\blacksquare}}$$

$$\text{SSta} := \text{ONE6}\,\overline{[(A)\cdot(A)]}\cdot\text{FOUR1} - \left(\frac{1}{\text{Nt}}\right)\cdot(\text{ONE6 A}\cdot\text{FOUR1})^{2}$$

$\text{ONE6}\,\overline{[(A)\cdot(A)]}\cdot\text{FOUR1} = 5112$	$\left(\frac{1}{\text{Nt}}\right)\cdot(\text{ONE6 A}\cdot\text{FOUR1})^{2} = 4704$

$$\text{SSta} = 408$$

STEP 6: Computation of the effects of the first factor vs. the second factor, SSs. Ns=Nt/#columns in DB2. 36/3 = 12

$$\text{Ns} := 12$$

$$\text{SSSa} := \left(\frac{1}{\text{Ns}}\right)\cdot\left[\left(1_{1,6}\right)\left(A_{6,4}\right)\text{DB2}_{4,2}\right]^{2} - \left(\frac{1}{\text{Nt}}\right)\cdot\left[\left(1_{1,6}\right)\left(A_{6,4}\right)\left(1_{4,1}\right)\right]^{2^{\blacksquare}}$$

$$\text{SSS} := \left(\frac{1}{\text{Ns}}\right)\cdot(\text{SUM1}\cdot\text{DB2})\cdot(\text{SUM1}\cdot\text{DB2})^{T} - \text{CORRFACT}$$

SSS = 6

$$DB2 = \begin{pmatrix} 1 & 0 \\ 0 & 1 \\ 1 & 0 \\ 0 & 1 \end{pmatrix}$$

$$\left(\frac{1}{12}\right) \cdot \left[(72 \; 67 \; 90 \; 107) \cdot \begin{pmatrix} 1 & 0 \\ 0 & 1 \\ 1 & 0 \\ 0 & 1 \end{pmatrix}\right] \cdot (SUM1 \cdot DB2)^T - CORRFACT = 6$$

STEP 7:   Computation of the effects of the second factor (the overall effects of the second two test conditions) disregarding the SSs dimension. SSlist = SSL. NL = Nt/# columns in DB1. NL = 36/3 = 12.

NL := 12

$$SSLa := \left(\frac{1}{NL}\right) \cdot (SUM1 \cdot DB1) \cdot \left((SUM1 \cdot DB1)^T\right) - CORRFACT$$

CORRFACT = 4704

SUM1·DB1 = (139 197)

The Math looks like:

$$\left(\frac{1}{12}\right) \cdot \left[(72 \; 67 \; 90 \; 107) \cdot \begin{pmatrix} 1 & 0 \\ 1 & 0 \\ 0 & 1 \\ 0 & 1 \end{pmatrix}\right] \left[(72 \; 67 \; 90 \; 107) \cdot \begin{pmatrix} 1 & 0 \\ 1 & 0 \\ 0 & 1 \\ 0 & 1 \end{pmatrix}\right]^T - CORRFACT$$

Which becomes:

$$\left(\frac{1}{12}\right) \cdot \left[(139 \; 197) \cdot \begin{pmatrix} 139 \\ 197 \end{pmatrix}\right] - 4704 = 140.167$$

SSL := 4844.167 − 4704

SSL = 140.167

NOTE: This step is exactly the same as STEP 6 above except we change DB1 to DB2 and SSs to SSL.

STEP 8: Computation of the interactive effects of the first and second experimental conditions. SSsxl. Nsxl = Nt/#elements in SUM1 Nsxl = 36/6= 6

Nsxl := 6

SUM1 = (72 67 90 107)   SSS = 6   CORRFACT = 4704   SSL = 140.167

$$\text{SSSxL} := \left(\frac{1}{\text{Nsxl}}\right) \cdot \left(\text{SUM1} \cdot \text{SUM1}^{\text{T}}\right) - \text{CORRFACT} - \text{SSS} - \text{SSL}$$

The Math looks like:

$$\text{SSSxL} := \left(\frac{1}{6}\right) \cdot \left[(72 \quad 67 \quad 90 \quad 107) \cdot \begin{pmatrix} 72 \\ 67 \\ 90 \\ 107 \end{pmatrix}\right] - 4704 - 6 - 140.167$$

$$\left(\frac{1}{\text{Nsxl}}\right) \cdot \left(\text{SUM1} \cdot \text{SUM1}^{\text{T}}\right) = 4870.333$$

SSSxL := 4870.333 − 4704 − 6 − 140.163

SSSxL = 20.17

STEP 9: Computation of the Error Term Sum of Squares SSerror.

SSt = 408   SSS = 6   SSL = 140.167   SSSxL = 20.17

SSERROR := SSt − SSS − SSL − SSSxL

$$SSERROR := 408 - 6 - 140.163 - 20.17$$

$$SSERROR = 241.667$$

$$\begin{pmatrix} SSt \\ SSS \\ SSL \\ SSSxL \\ SSERROR \end{pmatrix} = \begin{pmatrix} 408 \\ 6 \\ 140.167 \\ 20.17 \\ 241.667 \end{pmatrix}$$

STEP 10: All computations are done, now we must compute the degrees of Freedom for the system df.

$$m := 24 \quad Ex := 2 \quad Cond := 2$$

$$df := \begin{bmatrix} m - 1 \\ Ex - 1 \\ Cond - 1 \\ (Ex - 1) \cdot (Cond - 1) \\ (m - 1) - (Ex - 1) - (Cond - 1) - (Ex - 1) \cdot (Cond - 1) \end{bmatrix}$$

$$SS := \begin{pmatrix} SSt \\ SSS \\ SSL \\ SSSxL \\ SSERROR \end{pmatrix} \quad SS := \begin{pmatrix} SSt \\ SSS \\ SSL \\ SSSxL \\ SSERROR \end{pmatrix} \quad SS = \begin{pmatrix} 408 \\ 6 \\ 140.167 \\ 20.17 \\ 241.667 \end{pmatrix} \quad df = \begin{pmatrix} 23 \\ 1 \\ 1 \\ 1 \\ 20 \end{pmatrix}$$

$$\begin{pmatrix} 5112 - 4704 \\ 4710 - 4704 \\ 4844.167 - 4704 \\ 4870.333 - 4704 - 6 - 140.167 \\ 408 - 6 - 140.167 - 20.166 \end{pmatrix} = \begin{pmatrix} 408 \\ 6 \\ 140.167 \\ 20.166 \\ 241.667 \end{pmatrix} \quad F = \begin{pmatrix} 1.468 \\ 0.497 \\ 11.6 \\ 1.669 \\ 1 \end{pmatrix}$$

The Error Term is computed as:

$$ERR := \frac{241.667}{20}$$

$$ERR = 12.083$$

Thus the Error Matrix is defined as:

$$\text{ERROR} := \begin{pmatrix} 12.083 \\ 12.083 \\ 12.083 \\ 12.083 \\ 12.083 \end{pmatrix}$$

The final values to divide SS by the mean squares and compute the F ratio are:

$$\text{SS} := \begin{pmatrix} 408 \\ 6 \\ 140.167 \\ 20.166 \\ 241.667 \end{pmatrix} \quad \text{df} = \begin{pmatrix} 23 \\ 1 \\ 1 \\ 1 \\ 20 \end{pmatrix} \quad \text{ERROR} = \begin{pmatrix} 12.083 \\ 12.083 \\ 12.083 \\ 12.083 \\ 12.083 \end{pmatrix}$$

The F-ratio is given by:

$$F := \overrightarrow{\left(\text{SS} \cdot \text{df}^{-1} \cdot \text{ERROR}^{-1} \right)}$$

The Math looks like:

$$F := \overrightarrow{\left[\begin{pmatrix} 408 \\ 6 \\ 140.167 \\ 20.166 \\ 241.667 \end{pmatrix} \cdot \begin{pmatrix} 23 \\ 1 \\ 1 \\ 1 \\ 20 \end{pmatrix}^{-1} \cdot \begin{pmatrix} 12.083 \\ 12.083 \\ 12.083 \\ 12.083 \\ 12.083 \end{pmatrix}^{-1} \right]}$$

$$F = \begin{pmatrix} 1.468 \\ 0.497 \\ 11.6 \\ 1.669 \\ 1 \end{pmatrix} \quad \text{df} = \begin{pmatrix} 23 \\ 1 \\ 1 \\ 1 \\ 20 \end{pmatrix}$$

SSs   for df's 1/1 with F = .497 p< ns

SSl   for df's 1/1 with F = 11.6 p< .05

SSsxl for df's 4/27 with F = 1.669 p< ns

We conclude the effect of factor 1 had no affect on performance; the effect of SSL was significant and would be expected to occur in less than one in .05 times and that factor one compared to factor 2 is not significant.

PROBLEM USING A SINGLE EQUATION
USING NESTED ARRAYS

First we collect all the DB Matrices used in computing the problem.

$$ONE6 := (1 \ 1 \ 1 \ 1 \ 1 \ 1)$$

$$SIX1 := ONE6^T$$

$$DB1 = \begin{pmatrix} 1 & 0 \\ 1 & 0 \\ 0 & 1 \\ 0 & 1 \end{pmatrix} \quad DB2 = \begin{pmatrix} 1 & 0 \\ 0 & 1 \\ 1 & 0 \\ 0 & 1 \end{pmatrix} \quad DB = \begin{pmatrix} 1 \\ 1 \\ 1 \\ 1 \end{pmatrix} \quad DB0 = \begin{pmatrix} 0 \\ 0 \\ 0 \\ 0 \end{pmatrix}$$

$$DB4 := \begin{pmatrix} 1 & 0 & 0 & 0 \\ 0 & 1 & 0 & 0 \\ 0 & 0 & 1 & 0 \\ 0 & 0 & 0 & 1 \end{pmatrix} \qquad SUM1 := (72 \ 67 \ 90 \ 107)$$

Now we put them into a Nested array: DBALLNA := (DB DB0 DB1 DB2 DB4)

$$DBALLNA = \left[\begin{pmatrix} 1 \\ 1 \\ 1 \\ 1 \end{pmatrix} \begin{pmatrix} 0 \\ 0 \\ 0 \\ 0 \end{pmatrix} \begin{pmatrix} 1 & 0 \\ 1 & 0 \\ 0 & 1 \\ 0 & 1 \end{pmatrix} \begin{pmatrix} 1 & 0 \\ 0 & 1 \\ 1 & 0 \\ 0 & 1 \end{pmatrix} \begin{pmatrix} 1 & 0 & 0 & 0 \\ 0 & 1 & 0 & 0 \\ 0 & 0 & 1 & 0 \\ 0 & 0 & 0 & 1 \end{pmatrix} \right]$$

Partitioning the Nested Array we get:

$$DBALL := \begin{pmatrix} 1 & 0 & 1 & 0 & 1 & 0 & 1 & 0 & 0 & 0 \\ 1 & 0 & 0 & 1 & 1 & 0 & 0 & 1 & 0 & 0 \\ 1 & 0 & 1 & 0 & 0 & 1 & 0 & 0 & 1 & 0 \\ 1 & 0 & 0 & 1 & 0 & 1 & 0 & 0 & 0 & 1 \end{pmatrix}$$

Now we multiply by SUM!

$\text{SUM1·DBALL} = (336 \quad 0 \quad 162 \quad 174 \quad 139 \quad 197 \quad 72 \quad 67 \quad 90 \quad 107)$

Now, we know MathCad cannot handle the math so we'll substitute as follows:

$\text{SUM1a} := \text{SUM1·DBALL}$

We might as well divide the values by N at the same time. We will now square each value in SUM1A and divide by the corresponding N values.

$$N := \overrightarrow{(24 \quad 1 \quad 12 \quad 12 \quad 12 \quad 12 \quad 6 \quad 6 \quad 6 \quad 6)}^{-1}$$

The arrow over N and SS represents element by element multiplication

$$SS := \overrightarrow{(\text{SUM1a·SUM1a·N})}$$

$SS = (4704 \quad 0 \quad 2187 \quad 2523 \quad 1610.083 \quad 3234.083 \quad 864 \quad 748.167 \quad 1350 \quad 1908.167)$

Now we must post multiply to sum the sums of the different DB Matrices, the first one is one by four, the second is zero, the 2 next are 2x4 so we sum the next 4 numbers and the last Matrix is 4x4 so we sum the last 4 elements. The DB Sum Matrix looks like:

$$SS^T = \begin{pmatrix} 4704 \\ 0 \\ 2187 \\ 2523 \\ 1610.083 \\ 3234.083 \\ 864 \\ 748.167 \\ 1350 \\ 1908.167 \end{pmatrix} \qquad DBSUM := \begin{pmatrix} 1 & 0 & 0 & 0 & 0 \\ 0 & 0 & 0 & 0 & 0 \\ 0 & 0 & 1 & 0 & 0 \\ 0 & 0 & 1 & 0 & 0 \\ 0 & 0 & 0 & 1 & 0 \\ 0 & 0 & 0 & 1 & 0 \\ 0 & 0 & 0 & 0 & 1 \\ 0 & 0 & 0 & 0 & 1 \\ 0 & 0 & 0 & 0 & 1 \\ 0 & 0 & 0 & 0 & 1 \end{pmatrix}$$

Remember, Post-multiplier Matrices multiply and add the rows of the pre-multiplier. (This seems to be the opposite of what should seem logical, but it works).

$$SS \cdot DBSUM = (4704 \quad 0 \quad 4710 \quad 4844.167 \quad 4870.333)$$

Or the single equation becomes:

$$SSa := \overline{(SUM1a \cdot SUM1a \cdot N)} \, DBSUM$$

$$SSa = (4704 \quad 0 \quad 4710 \quad 4844.167 \quad 4870.333)$$

Now we are at the point where we can add the Grand Sum of Squares to the solution Matrix.

NOTE: We must add this in by hand as this equation is not of the same form as all the rest and must be calculated separately.

$$GRANDSUMSQ = 5112 \qquad GSQ := GRANDSUMSQ$$

$$GSQ := (0 \quad 5112 \quad 0 \quad 0 \quad 0)$$

$$SSb := \overline{(SUM1a \cdot SUM1a \cdot N)} \, DBSUM + GSQ$$

$$SSb = (4704 \quad 5112 \quad 4710 \quad 4844.167 \quad 4870.333)$$

Now we subtract the Correction Factor from each term:

$$CORRFACT1 := \begin{pmatrix} -1 & -1 & -1 & -1 & 0 \\ 1 & 0 & 0 & 0 & 0 \\ 0 & 1 & 0 & 0 & 0 \\ 0 & 0 & 1 & 0 & 0 \\ 0 & 0 & 0 & 1 & 0 \end{pmatrix}$$

$$SScf := \left(\overline{(SUM1a \cdot SUM1a \cdot N)} \, DBSUM + GSQ \right) CORRFACT1$$

$$SScf = (408 \quad 6 \quad 140.167 \quad 166.333 \quad 0)$$

$$\left(\overrightarrow{(\text{SUM1a}\cdot\text{SUM1a}\cdot\text{N})\,\text{DBSUM} + \text{GSQ}}\right)\text{CORRFACT1} = (408 \quad 6 \quad 140.167 \quad 166.333 \quad 0)$$

This DB Matrix subtracts all the Correction Factors other than the first CORRFACT which we have already done. In this example, this occurs in the 4th column, the error term is computed in the 5th column. The DBSS Matrix looks like

$$\text{DBSS} := \begin{pmatrix} 1 & 0 & 0 & 0 & 0 \\ 0 & 1 & 0 & -1 & 0 \\ 0 & 0 & 1 & -1 & 0 \\ 0 & 0 & 0 & 1 & 0 \\ 0 & 0 & 0 & 0 & 0 \end{pmatrix}$$

The pure diagonal values are the numbers we don't want to change.

$$\text{SSd} := \left(\overrightarrow{(\text{SUM1a}\cdot\text{SUM1a}\cdot\text{N})\,\text{DBSUM} + \text{GSQ}}\right)\text{CORRFACT1}\cdot\text{DBSS}$$

$$\text{SSd} = (408 \quad 6 \quad 140.167 \quad 20.167 \quad 0)$$

$$\text{SSd} := \left(\overrightarrow{(\text{SUM1a}\cdot\text{SUM1a}\cdot\text{N})\,\text{DBSUM} + \text{GSQ}}\right)\text{CORRFACT1}\cdot\text{DBSS}$$

Now we compute the Error Term:

$$\text{DBERR} := \begin{pmatrix} 1 & 0 & 0 & 0 & 1 \\ 0 & 1 & 0 & 0 & -1 \\ 0 & 0 & 1 & 0 & -1 \\ 0 & 0 & 0 & 1 & -1 \\ 0 & 0 & 0 & 0 & 0 \end{pmatrix}$$

$$\text{SSe} := \left(\overrightarrow{(\text{SUM1a}\cdot\text{SUM1a}\cdot\text{N})\,\text{DBSUM} + \text{GSQ}}\right)\text{CORRFACT1}\cdot\text{DBSS}\cdot\text{DBERR}$$

$$\text{SSe} = (408 \quad 6 \quad 140.167 \quad 20.167 \quad 241.667)$$

Our final Sum of Squares Matrix is computed to be:

$$
SSe^T = \begin{pmatrix} 408 \\ 6 \\ 140.167 \\ 20.167 \\ 241.667 \end{pmatrix}
\qquad
df = \begin{pmatrix} 23 \\ 1 \\ 1 \\ 1 \\ 20 \end{pmatrix}
\qquad
ERROR = \begin{pmatrix} 12.083 \\ 12.083 \\ 12.083 \\ 12.083 \\ 12.083 \end{pmatrix}
$$

We are now ready to compute the F-ratio:

$$
F := \overrightarrow{\left(SSe^T \cdot df^{-1} \cdot ERROR^{-1} \right)}
$$

Or:

$$
F := \overrightarrow{\left[\begin{pmatrix} 408 \\ 6 \\ 140.167 \\ 20.167 \\ 241.667 \end{pmatrix} \cdot \begin{pmatrix} 23 \\ 1 \\ 1 \\ 1 \\ 20 \end{pmatrix}^{-1} \cdot \begin{pmatrix} 12.083 \\ 12.083 \\ 12.083 \\ 12.083 \\ 12.083 \end{pmatrix}^{-1} \right]}
$$

$$
F1 = \begin{pmatrix} 1.468 \\ 0.497 \\ 11.6 \\ 1.669 \\ 1 \end{pmatrix}
$$

Another way to compute F is using diagonal Matrices. ie

$$
\text{diag}\left(df^{-1} \right) = \begin{pmatrix} 0.043 & 0 & 0 & 0 & 0 \\ 0 & 1 & 0 & 0 & 0 \\ 0 & 0 & 1 & 0 & 0 \\ 0 & 0 & 0 & 1 & 0 \\ 0 & 0 & 0 & 0 & 0.05 \end{pmatrix}
$$

$$
\mathrm{diag}\left(SSe^{T}\right) = \begin{pmatrix} 408 & 0 & 0 & 0 & 0 \\ 0 & 6 & 0 & 0 & 0 \\ 0 & 0 & 140.167 & 0 & 0 \\ 0 & 0 & 0 & 20.167 & 0 \\ 0 & 0 & 0 & 0 & 241.667 \end{pmatrix}
$$

$$
\mathrm{diag}\left(ERROR^{-1}\right) = \begin{pmatrix} 0.083 & 0 & 0 & 0 & 0 \\ 0 & 0.083 & 0 & 0 & 0 \\ 0 & 0 & 0.083 & 0 & 0 \\ 0 & 0 & 0 & 0.083 & 0 \\ 0 & 0 & 0 & 0 & 0.083 \end{pmatrix}
$$

The elements we are diagonalizing must be column Matrices, not row Matrices.

$$
F1 := \mathrm{diag}\left(SSe^{T}\right)\mathrm{diag}\left(df^{-1}\right)\mathrm{diag}\left(ERROR^{-1}\right)\begin{pmatrix}1\\1\\1\\1\\1\end{pmatrix} \qquad F1 = \begin{pmatrix}1.468\\0.497\\11.6\\1.669\\1\end{pmatrix}
$$

Or this looks like Mathematically:

$$
F1 := \begin{pmatrix} 408 & 0 & 0 & 0 & 0 \\ 0 & 6 & 0 & 0 & 0 \\ 0 & 0 & 140.167 & 0 & 0 \\ 0 & 0 & 0 & 20.167 & 0 \\ 0 & 0 & 0 & 0 & 241.667 \end{pmatrix} \cdot \begin{pmatrix} 0.043 & 0 & 0 & 0 & 0 \\ 0 & 1 & 0 & 0 & 0 \\ 0 & 0 & 1 & 0 & 0 \\ 0 & 0 & 0 & 1 & 0 \\ 0 & 0 & 0 & 0 & 0.05 \end{pmatrix} \cdot \begin{pmatrix} 0.083 & 0 & 0 & 0 & 0 \\ 0 & 0.083 & 0 & 0 & 0 \\ 0 & 0 & 0.083 & 0 & 0 \\ 0 & 0 & 0 & 0.083 & 0 \\ 0 & 0 & 0 & 0 & 0.083 \end{pmatrix} \cdot \begin{pmatrix}1\\1\\1\\1\\1\end{pmatrix}
$$

Funny, in the equation form the F's are equal, but in the graphical form they differ, must be due to rounding errors.

$$
F1 = \begin{pmatrix}1.456\\0.498\\11.634\\1.674\\1.003\end{pmatrix} \qquad F = \begin{pmatrix}1.468\\0.497\\11.6\\1.669\\1\end{pmatrix}
$$

PROBLEM 2.3: THREE FACTOR ANALYSIS OF VARIANCE

Again, I do not know enough about Statistics to make an example problem, so we'll just do the math. I have no access to an University Library so I cannot find the traditional Statistical Equations. We will solve the math without knowing what they are. A is the data collected, the rest are the Database Matrices. This example I have chosen to solve the equations 4 different ways (all complete solutions). So this long, but hopefully worth it.

$$A_{6,8} := \begin{pmatrix} 14 & 25 & 12 & 22 & 17 & 14 & 19 & 35 \\ 9 & 15 & 18 & 25 & 8 & 19 & 18 & 30 \\ 10 & 18 & 10 & 26 & 5 & 15 & 10 & 39 \\ 5 & 17 & 15 & 32 & 4 & 20 & 16 & 41 \\ 7 & 16 & 19 & 36 & 8 & 13 & 23 & 35 \\ 6 & 13 & 18 & 29 & 15 & 19 & 20 & 33 \end{pmatrix}$$

$$ONE6 := (1 \ 1 \ 1 \ 1 \ 1 \ 1)$$

$$SIX1 := ONE6^T$$

$$EIGHT1_{8,1} := \begin{pmatrix} 1 \\ 1 \\ 1 \\ 1 \\ 1 \\ 1 \\ 1 \\ 1 \end{pmatrix}$$

$$FIVE1 := \begin{pmatrix} 1 \\ 1 \\ 1 \\ 1 \\ 1 \end{pmatrix} \qquad THREE1 := \begin{pmatrix} 1 \\ 1 \\ 1 \end{pmatrix} \qquad ONE5 := FIVE1^T$$

$$DB11_{8,2} := \begin{pmatrix} 1 & 0 \\ 1 & 0 \\ 0 & 1 \\ 0 & 1 \\ 1 & 0 \\ 1 & 0 \\ 0 & 1 \\ 0 & 1 \end{pmatrix} \quad DB11b_{8,3} := \begin{pmatrix} 1 & 0 & 1 \\ 1 & 0 & 1 \\ 0 & 1 & 1 \\ 0 & 1 & 1 \\ 1 & 0 & 1 \\ 1 & 0 & 1 \\ 0 & 1 & 1 \\ 0 & 1 & 1 \end{pmatrix} \quad DB12_{8,4} := \begin{pmatrix} 1 & 0 & 0 & 0 \\ 1 & 0 & 0 & 0 \\ 0 & 1 & 0 & 0 \\ 0 & 1 & 0 & 0 \\ 0 & 0 & 1 & 0 \\ 0 & 0 & 1 & 0 \\ 0 & 0 & 0 & 1 \\ 0 & 0 & 0 & 1 \end{pmatrix} \quad DB12b_{8,5} := \begin{pmatrix} 1 & 0 & 0 & 0 & 1 \\ 1 & 0 & 0 & 0 & 1 \\ 0 & 1 & 0 & 0 & 1 \\ 0 & 1 & 0 & 0 & 1 \\ 0 & 0 & 1 & 0 & 1 \\ 0 & 0 & 1 & 0 & 1 \\ 0 & 0 & 0 & 1 & 1 \\ 0 & 0 & 0 & 1 & 1 \end{pmatrix}$$

$$DB1_{8,2} := \begin{pmatrix} 1 & 0 \\ 1 & 0 \\ 1 & 0 \\ 1 & 0 \\ 0 & 1 \\ 0 & 1 \\ 0 & 1 \\ 0 & 1 \end{pmatrix} \quad DB2_{8,2} := \begin{pmatrix} 1 & 0 \\ 0 & 1 \\ 1 & 0 \\ 0 & 1 \\ 1 & 0 \\ 0 & 1 \\ 1 & 0 \\ 0 & 1 \end{pmatrix} \quad DB2b_{8,3} := \begin{pmatrix} 1 & 0 & 1 \\ 0 & 1 & 1 \\ 1 & 0 & 1 \\ 0 & 1 & 1 \\ 1 & 0 & 1 \\ 0 & 1 & 1 \\ 1 & 0 & 1 \\ 0 & 1 & 1 \end{pmatrix} \quad DB1b_{8,3} := \begin{pmatrix} 1 & 0 & 1 \\ 1 & 0 & 1 \\ 1 & 0 & 1 \\ 1 & 0 & 1 \\ 0 & 1 & 1 \\ 0 & 1 & 1 \\ 0 & 1 & 1 \\ 0 & 1 & 1 \end{pmatrix}$$

$$SUM1 := (1 \ 1 \ 1 \ 1 \ 1 \ 1) \cdot \begin{pmatrix} 14 & 25 & 12 & 22 & 17 & 14 & 19 & 35 \\ 9 & 15 & 18 & 25 & 8 & 19 & 18 & 30 \\ 10 & 18 & 10 & 26 & 5 & 15 & 10 & 39 \\ 5 & 17 & 15 & 32 & 4 & 20 & 16 & 41 \\ 7 & 16 & 19 & 36 & 8 & 13 & 23 & 35 \\ 6 & 13 & 18 & 29 & 15 & 19 & 20 & 33 \end{pmatrix}$$

$$SUM1 = (51 \ 104 \ 92 \ 170 \ 57 \ 100 \ 106 \ 213)$$

Again, the term SUM1 seems to be a universal value in Statistics being the product of [1]x[A], the value we obtain before multiplying to the Database Matrices (at least in the evaluation of between-subject evaluations). It acts as a constant in the evaluations.

$$GRANDSUMSQ := ONE6 \cdot \overrightarrow{\left(A_{6,8} \cdot A_{6,8} \right)} EIGHT1_{8,1}$$

$$GRANDSUMSQ = 20805$$

METHOD 2:

With this method, we square the data Matrix, remove the diagonal and sum the elements of the diagonal to compute the GRANDSUMSQ of the data. Though this is a little morze complex than the first method, we must realize the first method is considered illegal by the standards of modern math even though it does give the correct answer.

$$GRANDSUMSQ := ONE6 \cdot diag \left(A_{6,8} \cdot A_{6,8}^T \right)$$

$$\left(A_{6,8} \cdot A_{6,8}{}^T\right) = \begin{pmatrix} 3520 & 3061 & 3132 & 3466 & 3498 & 3319 \\ 3061 & 2904 & 2865 & 3300 & 3320 & 3129 \\ 3132 & 2865 & 3071 & 3417 & 3314 & 3075 \\ 3466 & 3300 & 3417 & 3916 & 3839 & 3562 \\ 3498 & 3320 & 3314 & 3839 & 3949 & 3618 \\ 3319 & 3129 & 3075 & 3562 & 3618 & 3445 \end{pmatrix} \qquad \text{diag}\left(\left(A_{6,8} \cdot A_{6,8}{}^T\right)\right) = \begin{pmatrix} 3520 \\ 2904 \\ 3071 \\ 3916 \\ 3949 \\ 3445 \end{pmatrix}$$

$$\text{ONE6} \cdot \text{diag}\left(\left(A_{6,8} \cdot A_{6,8}{}^T\right)\right) = 20805$$

GRANDSUMSQ = 20805

STEP 3:   Computation of the Grand Sum of the Matrix $A_{6,8}$

$$\text{GRANDSUM} := \left(1_{1,6}\right) A_{6,8} \cdot \left(1_{8,1}\right)$$

$$\text{GRANDSUM} := \text{ONE6} \cdot A_{6,8} \cdot \text{EIGHT1}_{8,1}$$

Lets remember, pre-multiplying by [1] adds the columns ([1] x[A]), post-multiplying by [1] sums the rows ([A]x[1]). This is the accounting equation.

$$(1 \ 1 \ 1 \ 1 \ 1 \ 1) \cdot \begin{pmatrix} 14 & 25 & 12 & 22 & 17 & 14 & 19 & 35 \\ 9 & 15 & 18 & 25 & 8 & 19 & 18 & 30 \\ 10 & 18 & 10 & 26 & 5 & 15 & 10 & 39 \\ 5 & 17 & 15 & 32 & 4 & 20 & 16 & 41 \\ 7 & 16 & 19 & 36 & 8 & 13 & 23 & 35 \\ 6 & 13 & 18 & 29 & 15 & 19 & 20 & 33 \end{pmatrix} \cdot \begin{pmatrix} 1 \\ 1 \\ 1 \\ 1 \\ 1 \\ 1 \\ 1 \\ 1 \end{pmatrix} = 893$$

GRANDSUM = 893

STEP 4:   Computation of Total Sum of Squares SSt. Nt = number of elements in the Data Matrix Nt = 6x8=48.

Nt := 48

$$\text{CORRFACT} := \left(\frac{1}{\text{Nt1}}\right)\left[\left(1_{1,6}\right)\left(A_{6,8}\right)\left(1_{8,1}\right)\right]^2$$

$$\text{CORRFACT} := \left(\frac{1}{Nt}\right)\left[\left(\text{SUM1}\cdot\text{EIGHT1}_{8,1}\right)\left(\text{SUM1}\cdot\text{EIGHT1}_{8,1}\right)\right]$$

$$\text{SSt} := \text{ONE6}\cdot\overrightarrow{\left(A_{6,8}\cdot A_{6,8}\right)}\text{EIGHT1}_{8,1} - \left(\frac{1}{Nt1}\right)\left[\left(1_{1,6}\right)\left(A_{6,8}\right)\left(1_{8,1}\right)\right]^{2}$$

$$\text{SSt} := \text{ONE6}\cdot\overrightarrow{\left(A_{6,8}\cdot A_{6,8}\right)}\text{EIGHT1}_{8,1} - \left(\frac{1}{Nt}\right)\left[\left(\text{ONE6}\cdot A_{6,8}\cdot\text{EIGHT1}_{8,1}\right)\left(\text{ONE6}\cdot A_{6,8}\cdot\text{EIGHT1}_{8,1}\right)\right]$$

Or   CORRFACT = 16613.52

$$\text{SSt} := \text{ONE6}\cdot\overrightarrow{\left(A_{6,8}\cdot A_{6,8}\right)}\text{EIGHT1}_{8,1} - \left(\frac{1}{Nt}\right)\left[\left(\text{SUM1}\cdot\text{EIGHT1}_{8,1}\right)\left(\text{SUM1}\cdot\text{EIGHT1}_{8,1}\right)\right]$$

GRANDSUMSQ = 20805

SSt = 4191.48

CORRFACT = 16613.52

METHOD 2:

The b method. The b method is used when a value in the Data Matrix A is missing and we have to find a different average than the rest of the values.

$$b := \text{SUM1}\cdot\text{EIGHT1}_{8,1}$$

$$b = 893$$

$$\text{SStb} := \text{GRANDSUMSQ} - b\cdot\left(\frac{b}{Nt}\right)$$

$$\text{SStb} = 4191.48$$

METHOD 3:

The b method. This is the real world model to use in case the N's are not all the same. Like one subject drops out of the study early but it is felt their values were valid and not to be thrown out just to simplify the study. In the case of SSt this is not necessary, but is included for study.

$$\text{SStb1} := \text{GRANDSUMSQ} - \left(\overrightarrow{\left[\left(\frac{1}{48}\ \frac{1}{48}\ \frac{1}{48}\ \frac{1}{48}\ \frac{1}{48}\ \frac{1}{48}\ \frac{1}{48}\ \frac{1}{48}\right)\cdot\text{SUM1}\right]}\cdot\text{EIGHT1}_{8,1}\cdot\left(\text{SUM1}\cdot\text{EIGHT1}_{8,1}\right)\right)$$

SStb1 = 4191.48

STEP 6: Computation of SS1, the computation of the first experimental conditions (the First Factor Effect. N1 = Nt/#columns in DB1. N1 = 48/2 = 24

$N1 := \dfrac{48}{2}$ $b := SUM1 \cdot EIGHT1_{8,1}$

$SUM1 \cdot DB1_{8,2} = (417 \quad 476)$

$B := SUM1 \cdot DB1_{8,2}$

$SS1 := \left[SUM1 \cdot DB1_{8,2} \cdot \left(\dfrac{b}{N1} \right) - \left(SUM1 \cdot EIGHT1_{8,1} \right) \left(\dfrac{b}{Nt} \right) \right]$

$b1a := SUM1 \cdot DB1_{8,2} \cdot \begin{pmatrix} 1 \\ 0 \end{pmatrix}$

$SUM1 = (51 \quad 104 \quad 92 \quad 170 \quad 57 \quad 100 \quad 106 \quad 213)$

$b2a := SUM1 \cdot DB1_{8,2} \cdot \begin{pmatrix} 0 \\ 1 \end{pmatrix}$

$$SS1 := SUM1 \cdot \begin{pmatrix} 1 & 0 \\ 1 & 0 \\ 1 & 0 \\ 1 & 0 \\ 0 & 1 \\ 0 & 1 \\ 0 & 1 \\ 0 & 1 \end{pmatrix} \cdot \begin{pmatrix} \dfrac{b1a}{N1} \\ \dfrac{b2a}{N1} \end{pmatrix} - SUM1 \cdot \begin{pmatrix} 1 \\ 1 \\ 1 \\ 1 \\ 1 \\ 1 \\ 1 \\ 1 \end{pmatrix} \cdot \left(\dfrac{b}{Nt} \right)$$

SS1 = 72.52

NOTE: Suppose we write the equation as follows. Written in this manner we compute the Corection Factor automatically.

$$DB1b_{8,3} = \begin{pmatrix} 1 & 0 & 1 \\ 1 & 0 & 1 \\ 1 & 0 & 1 \\ 1 & 0 & 1 \\ 0 & 1 & 1 \\ 0 & 1 & 1 \\ 0 & 1 & 1 \\ 0 & 1 & 1 \end{pmatrix}$$

$$\text{SUM1} \cdot \text{DB1b}_{8,3} \cdot \begin{pmatrix} \dfrac{b1a}{Ns} \\ \dfrac{b2a}{Ns} \\ \dfrac{-b}{Nt} \end{pmatrix} \qquad\qquad \text{SS1} := \text{SUM1} \cdot \text{DB1b}_{8,3} \cdot \begin{pmatrix} \dfrac{b1a}{N1} \\ \dfrac{b2a}{N1} \\ \dfrac{-b}{Nt} \end{pmatrix}$$

$$\text{SS1} = 72.52$$

$$\text{SS1} := (51 \quad 104 \quad 92 \quad 170 \quad 57 \quad 100 \quad 106 \quad 213) \cdot \begin{pmatrix} 1 & 0 & 1 \\ 1 & 0 & 1 \\ 1 & 0 & 1 \\ 1 & 0 & 1 \\ 0 & 1 & 1 \\ 0 & 1 & 1 \\ 0 & 1 & 1 \\ 0 & 1 & 1 \end{pmatrix} \cdot \begin{pmatrix} \dfrac{b1a}{N1} \\ \dfrac{b2a}{N1} \\ \dfrac{-b}{Nt} \end{pmatrix}$$

$$\text{SS1} = 72.52$$

METHOD 3:

This method uses the properties of the Half-Multiplier Operator, symbolized by the operator o.

$$\text{SS1b1} := \left[\left(\text{SUM1} \cdot \text{DB1b}_{8,3}\right)^{o2} \cdot \text{diag}\!\left(\dfrac{1}{Ni}\right) \right] \cdot \text{THREE1}^{\blacksquare} \qquad Ni := \begin{pmatrix} 24 \\ 24 \\ -48 \end{pmatrix} \qquad \text{SUM1b1} := \left(\text{SUM1} \cdot \text{DB1b}_{8,3}\right)$$

$$\text{SUM1b1} = (417 \quad 476 \quad 893) \qquad\qquad \text{diag}\!\left(\dfrac{1}{Ni}\right) = \begin{pmatrix} 0.04 & 0 & 0 \\ 0 & 0.04 & 0 \\ 0 & 0 & -0.02 \end{pmatrix}$$

$$\text{SS1b1} := \left(\overrightarrow{(\text{SUM1b1} \cdot \text{SUM1b1})} \cdot \text{diag}\!\left(\dfrac{1}{Ni}\right) \cdot \text{THREE1} \right)$$

Or

$$\text{SS1b1} := \left[\overrightarrow{(\text{SUM1b1} \cdot \text{SUM1b1})} \cdot \begin{pmatrix} 0.04 & 0 & 0 \\ 0 & 0.04 & 0 \\ 0 & 0 & -0.02 \end{pmatrix} \cdot \text{THREE1} \right]$$

Kind of crowded here, but the solution is

SS1b1 = 69.62

Again, the Correction Factor is automatically subtracted in the Mathematics.

METHOD 4:

And now for the method which I use through most of this book.

$$SS1 := \left(\frac{1}{N1}\right)\left[\left(1_{1,6}\right)A_{6,8}\cdot DB1_{8,2}\right]^2 - \left(\frac{1}{Nt}\right)\left[\left(1_{1,6}\right)A_{6,8}\cdot \left(1_{8,1}\right)\right]^2 \blacksquare$$

$$SS1 := \left(\frac{1}{N1}\right)\cdot\left(ONE6\cdot A_{6,8}\cdot DB1_{8,2}\right)^2 - \left(\frac{1}{Nt}\right)\cdot GRANDSUM^2$$

$$SS1 := \left(\frac{1}{N1}\right)\left[\left(SUM1\cdot DB1_{8,2}\right)\left(SUM1\cdot DB1_{8,2}\right)^T\right] - \left(\frac{1}{Nt}\right)\cdot\left(ONE6\cdot A_{6,8}\cdot EIGHT1_{8,1}\right)\left(ONE6\cdot A_{6,8}\cdot EIGHT1_{8,1}\right)$$

SS1 = 72.52

b Method:

I have to split the b method equation to get it on a single page

$$SS1b := \left(\left[\left(\overrightarrow{\frac{1}{24}\ \frac{1}{24}\ \frac{1}{24}\ \frac{1}{24}\ \frac{1}{24}\ \frac{1}{24}\ \frac{1}{24}\ \frac{1}{24}}\right)\cdot SUM1\right]\cdot DB1_{8,2}\right)\left(SUM1\cdot DB1_{8,2}\right)^T - \blacksquare$$

$$\left[\left(\overrightarrow{\frac{1}{48}\ \frac{1}{48}\ \frac{1}{48}\ \frac{1}{48}\ \frac{1}{48}\ \frac{1}{48}\ \frac{1}{48}\ \frac{1}{48}}\right)\cdot SUM1\right]\cdot EIGHT1_{8,1}\cdot\left(ONE6\cdot A_{6,8}\cdot EIGHT1_{8,1}\right)$$

$$SS1b := \left(\left[\left(\overrightarrow{\frac{1}{24}\ \frac{1}{24}\ \frac{1}{24}\ \frac{1}{24}\ \frac{1}{24}\ \frac{1}{24}\ \frac{1}{24}\ \frac{1}{24}}\right)\cdot SUM1\right]\cdot DB1_{8,2}\right)\left(SUM1\cdot DB1_{8,2}\right)^T - \left[\left(\overrightarrow{\frac{1}{48}\ \frac{1}{48}\ \frac{1}{48}\ \frac{1}{48}\ \frac{1}{48}\ \frac{1}{48}\ \frac{1}{48}\ \frac{1}{48}}\right)\cdot SUM1\right]\cdot EIGHT1_{8,1}\cdot\left(ONE6\cdot A_{6,8}\cdot EIGHT1_{8,1}\right)$$

The solution is

SS1b = 72.52

STEP 7: Computation of the effects of the second factor or experimental condition, SS_2. N2 = Nt /#Columns in DB11. N_2 = 48/2 = 24

$N_2 := \dfrac{48}{2}$ **NOTE: all the Math is the same as in Step 6 except we Multiply by DB11 instead of DB1.**

b:= SUM1·EIGHT1$_{8,1}$

$$SS2 := \left[SUM1 \cdot DB11_{8,2}\left(\frac{b}{N_2}\right) - \left(SUM1 \cdot EIGHT1_{8,1}\right)\left(\frac{b}{Nt}\right) \right]$$

SUM1 = (51 104 92 170 57 100 106 213)

SUM1·DB11$_{8,2}$ = (312 581)

B := SUM1·DB11$_{8,2}$

b1a:= SUM1·DB11$_{8,2}$$\begin{pmatrix}1\\0\end{pmatrix}$

b2a:= SUM1·DB11$_{8,2}$$\begin{pmatrix}0\\1\end{pmatrix}$

$$SS_2 := SUM1 \cdot \begin{pmatrix} 1 & 0 \\ 1 & 0 \\ 0 & 1 \\ 0 & 1 \\ 1 & 0 \\ 1 & 0 \\ 0 & 1 \\ 0 & 1 \end{pmatrix} \cdot \begin{pmatrix} \dfrac{b1a}{N_2} \\ \dfrac{b2a}{N_2} \end{pmatrix} - SUM1 \cdot \begin{pmatrix} 1 \\ 1 \\ 1 \\ 1 \\ 1 \\ 1 \\ 1 \\ 1 \end{pmatrix} \cdot \left(\frac{b}{Nt}\right)$$

SS_2 = 1507.52

NOTE: Suppose we write the equation as follows. Written in this manner we compute the Correction Factor automatically.

$$SUM1 \cdot DB11b_{8,3} \cdot \begin{pmatrix} \dfrac{b1a}{N_2} \\ \dfrac{b2a}{N_2} \\ \dfrac{-b}{Nt} \end{pmatrix}$$

$$SS2 := SUM1 \cdot DB11b_{8,3} \cdot \begin{pmatrix} \dfrac{b1a}{N_2} \\ \dfrac{b2a}{N_2} \\ \dfrac{-b}{Nt} \end{pmatrix}$$

SS2 = 1507.52

The math looks like:

$$SS2 := SUM1 \cdot \begin{pmatrix} 1 & 0 & 1 \\ 1 & 0 & 1 \\ 0 & 1 & 1 \\ 0 & 1 & 1 \\ 1 & 0 & 1 \\ 1 & 0 & 1 \\ 0 & 1 & 1 \\ 0 & 1 & 1 \end{pmatrix} \cdot \begin{pmatrix} \dfrac{b1a}{N_2} \\ \dfrac{b2a}{N_2} \\ \dfrac{-b}{Nt} \end{pmatrix}$$

$SS2 = 1507.52$

METHOD 3:

$$SS2b1 := \left[\left(SUM1 \cdot DB11b_{8,3} \right)^{o2} \cdot diag\left(\frac{1}{Ni} \right) \right] \cdot THREE1 \quad\blacksquare \qquad\qquad Ni := \begin{pmatrix} 24 \\ 24 \\ -48 \end{pmatrix}$$

$$SUM1b1 := \left(SUM1 \cdot DB11b_{8,3} \right) \qquad\qquad SUM1b1 = (312 \quad 581 \quad 893)$$

$$SS2b1 := \left(\overrightarrow{(SUM1b1 \cdot SUM1b1)} \cdot diag\left(\frac{1}{Ni} \right) \cdot THREE1 \right)$$

$SS2b1 = 1507.52$

$$SS2b1 := \overrightarrow{(SUM1b1 \cdot SUM1b1)} \cdot \begin{pmatrix} 0.04 & 0 & 0 \\ 0 & 0.04 & 0 \\ 0 & 0 & -0.02 \end{pmatrix} \cdot THREE1$$

$SS2b1 = 1447.22$

Again, the Correction Factor is automatically subtracted in the Mathematics.

METHOD 4:

And now for the method which I use through most of this book.

$$SS2 := \left(\frac{1}{Ns} \right) \left[\left(1_{1,6} \right) A_{6,8} \cdot DB11_{8,2} \right]^2 - \left(\frac{1}{Nt} \right) \left[\left(1_{1,6} \right) A_{6,8} \cdot \left(1_{8,1} \right) \right]^2 \quad\blacksquare$$

$$SS_2 := \left(\frac{1}{N_2}\right) \cdot \left(ONE6 \cdot A_{6,8} \cdot DB11_{8,2}\right)^2 - \left(\frac{1}{Nt}\right) \cdot GRANDSUM^2$$

$$SS_2 := \left(\frac{1}{N_2}\right)\left[\left(SUM1 \cdot DB11_{8,2}\right)\left(SUM1 \cdot DB11_{8,2}\right)^T\right] - \left(\frac{1}{Nt}\right) \cdot \left(ONE6 \cdot A_{6,8} \cdot EIGHT1_{8,1}\right)\left(ONE6 \cdot A_{6,8} \cdot EIGHT1_{8,1}\right)$$

$$SS_2 = 1507.52$$

B METHOD:

$$SS_2 := \left(\left[\left[\left(\frac{1}{24}\ \frac{1}{24}\ \frac{1}{24}\ \frac{1}{24}\ \frac{1}{24}\ \frac{1}{24}\ \frac{1}{24}\ \frac{1}{24}\right) \cdot SUM1\right] \cdot DB11_{8,2}\right)\left(SUM1 \cdot DB11_{8,2}\right)^T - \left[\left[\left(\frac{1}{48}\ \frac{1}{48}\ \frac{1}{48}\ \frac{1}{48}\ \frac{1}{48}\ \frac{1}{48}\ \frac{1}{48}\ \frac{1}{48}\right) \cdot SUM1\right] \cdot EIGHT1_{8,1}\left(ONE6 \cdot A_{6,8} \cdot EIGHT1_{8,1}\right)\right]$$

Splitting the equation so we can get it on a single page we have:

$$SS2 := \left(\left[\left[\overrightarrow{\left(\frac{1}{24}\ \frac{1}{24}\ \frac{1}{24}\ \frac{1}{24}\ \frac{1}{24}\ \frac{1}{24}\ \frac{1}{24}\ \frac{1}{24}\right) \cdot SUM1}\right] \cdot DB11_{8,2}\right) \cdot \left(SUM1 \cdot DB11_{8,2}\right)^T - \blacksquare$$

$$\left[\left[\overrightarrow{\left(\frac{1}{48}\ \frac{1}{48}\ \frac{1}{48}\ \frac{1}{48}\ \frac{1}{48}\ \frac{1}{48}\ \frac{1}{48}\ \frac{1}{48}\right) \cdot SUM1}\right] \cdot EIGHT1_{8,1}\left(ONE6 \cdot A_{6,8} \cdot EIGHT1_{8,1}\right)\right]$$

$$SS_2 = 1507.52$$

So far, the final SS Matrix looks like:

$$SS := \begin{pmatrix} SSt \\ SS1 \\ SS_2 \\ 0 \\ 0 \\ 0 \\ 0 \\ 0 \\ 0 \end{pmatrix} \qquad SS = \begin{pmatrix} 4191.48 \\ 72.52 \\ 1507.52 \\ 0 \\ 0 \\ 0 \\ 0 \\ 0 \\ 0 \end{pmatrix}$$

STEP 8: Computation of the effects of the third experimental factor SS3. N3 = Nt/#columns in DB2 = 48/2 = 24.

$N3 := 12$

$$DB2_{8,2} = \begin{pmatrix} 1 & 0 \\ 0 & 1 \\ 1 & 0 \\ 0 & 1 \\ 1 & 0 \\ 0 & 1 \\ 1 & 0 \\ 0 & 1 \end{pmatrix} \qquad\qquad DB2b_{8,3} = \begin{pmatrix} 1 & 0 & 1 \\ 0 & 1 & 1 \\ 1 & 0 & 1 \\ 0 & 1 & 1 \\ 1 & 0 & 1 \\ 0 & 1 & 1 \\ 1 & 0 & 1 \\ 0 & 1 & 1 \end{pmatrix}$$

NOTE: all the Math is the same as in Step 6 except we Multiply by DB2 instead of DB1.

$N_3 := \dfrac{48}{2}$ $\qquad\qquad\qquad\qquad\qquad\qquad$ $SUM1 \cdot DB2_{8,2} = (306 \quad 587)$

$b := SUM1 \cdot EIGHT1_{8,1}$ $\qquad\qquad\qquad\qquad$ $B := SUM1 \cdot DB2_{8,2}$

$SS3 := \left[SUM1 \cdot DB2_{8,2}\left(\dfrac{b}{N_3}\right) - \left(SUM1 \cdot EIGHT1_{8,1}\right)\left(\dfrac{b}{Nt}\right) \right]$ $\qquad$ $b1a := SUM1 \cdot DB2_{8,2} \cdot \begin{pmatrix} 1 \\ 0 \end{pmatrix}$

$SUM1 = (51 \quad 104 \quad 92 \quad 170 \quad 57 \quad 100 \quad 106 \quad 213)$ $\qquad$ $b2a := SUM1 \cdot DB2_{8,2} \cdot \begin{pmatrix} 0 \\ 1 \end{pmatrix}$

$$SS_3 := SUM1 \cdot \begin{pmatrix} 1 & 0 \\ 0 & 1 \\ 1 & 0 \\ 0 & 1 \\ 1 & 0 \\ 0 & 1 \\ 1 & 0 \\ 0 & 1 \end{pmatrix} \cdot \begin{pmatrix} \dfrac{b1a}{N_3} \\ \dfrac{b2a}{N_3} \end{pmatrix} - SUM1 \cdot \begin{pmatrix} 1 \\ 1 \\ 1 \\ 1 \\ 1 \\ 1 \\ 1 \\ 1 \end{pmatrix} \cdot \left(\dfrac{b}{Nt}\right)$$

$SS_3 = 1645.02$

NOTE: Suppose we write the equation as follows. Written in this manner we compute the Correction Factor automatically.

$$SS3 := SUM1 \cdot \begin{pmatrix} 1 & 0 & 1 \\ 0 & 1 & 1 \\ 1 & 0 & 1 \\ 0 & 1 & 1 \\ 1 & 0 & 1 \\ 0 & 1 & 1 \\ 1 & 0 & 1 \\ 0 & 1 & 1 \end{pmatrix} \cdot \begin{pmatrix} \dfrac{b1a}{N_3} \\ \dfrac{b2a}{N_3} \\ \dfrac{-b}{Nt} \end{pmatrix} \qquad DB2b_{8,3} = \begin{pmatrix} 1 & 0 & 1 \\ 0 & 1 & 1 \\ 1 & 0 & 1 \\ 0 & 1 & 1 \\ 1 & 0 & 1 \\ 0 & 1 & 1 \\ 1 & 0 & 1 \\ 0 & 1 & 1 \end{pmatrix}$$

$SS3 = 1645.02$

METHOD 3:

$$SS3b1 := \left[\left(SUM1 \cdot DB2b_{8,3} \right)^{\triangleright 2} \cdot diag\left(\frac{1}{Ni} \right) \right] \cdot THREE1 \blacksquare \qquad Ni := \begin{pmatrix} 24 \\ 24 \\ -48 \end{pmatrix}$$

$$SUM1b1 := \left(SUM1 \cdot DB2b_{8,3} \right) \qquad\qquad SUM1b1 = (306 \quad 587 \quad 893)$$

$$SS3b1 := \left(\overrightarrow{(SUM1b1 \cdot SUM1b1)} \cdot diag\left(\frac{1}{Ni} \right) \cdot THREE1 \right)$$

$SS3b1 = 1645.02$

Again, the Correction Factor is automatically subtracted in the Mathematics.

METHOD 4:

And now for the method which I use through most of this book.

$$SS3 := \left(\frac{1}{Ns} \right) \left[\left(1_{1,6} \right) A_{6,8} \cdot DB2_{8,2} \right]^2 - \left(\frac{1}{Nt} \right) \left[\left(1_{1,6} \right) A_{6,8} \cdot \left(1_{8,1} \right) \right]^2 \blacksquare$$

$$SS_3 := \left(\frac{1}{N_2} \right) \cdot \left(ONE6 \cdot A_{6,8} \cdot DB2_{8,2} \right)^2 - \left(\frac{1}{Nt} \right) \cdot GRANDSUM^2$$

$$SS_3 := \left(\frac{1}{N_3} \right) \left[\left(SUM1 \cdot DB2_{8,2} \right) \left(SUM1 \cdot DB2_{8,2} \right)^T \right] - \left(\frac{1}{Nt} \right) \cdot \left(ONE6 \cdot A_{6,8} \cdot EIGHT1_{8,1} \right) \left(ONE6 \cdot A_{6,8} \cdot EIGHT1_{8,1} \right)$$

$SS_3 = 1645.02$

B METHOD:

$$SS_3 := \left(\left[\left(\frac{1}{24} \ \frac{1}{24} \ \frac{1}{24} \ \frac{1}{24} \ \frac{1}{24} \ \frac{1}{24} \ \frac{1}{24} \ \frac{1}{24} \right) \cdot SUM1 \right] \cdot DB2_{8,2} \right) \cdot \left(SUM1 \cdot DB2_{8,2} \right)^T - \left[\left[\left(\frac{1}{48} \ \frac{1}{48} \ \frac{1}{48} \ \frac{1}{48} \ \frac{1}{48} \ \frac{1}{48} \ \frac{1}{48} \ \frac{1}{48} \right) \cdot SUM1 \right] \cdot EIGHT1_{8,1} \cdot \left(ONE6 \cdot A_{6,8} \cdot EIGHT1_{8,1} \right) \right]$$

Splitting the equation so we can get it on a single page we get:

$$SS3 := \left(\left[\left(\frac{1}{24} \ \frac{1}{24} \ \frac{1}{24} \ \frac{1}{24} \ \frac{1}{24} \ \frac{1}{24} \ \frac{1}{24} \ \frac{1}{24} \right) \cdot SUM1 \right] \cdot DB2_{8,2} \right) \cdot \left(SUM1 \cdot DB2_{8,2} \right)^T - \blacksquare$$

$$\left[\left[\left(\frac{1}{48} \ \frac{1}{48} \ \frac{1}{48} \ \frac{1}{48} \ \frac{1}{48} \ \frac{1}{48} \ \frac{1}{48} \ \frac{1}{48} \right) \cdot SUM1 \right] \cdot EIGHT1_{8,1} \cdot \left(ONE6 \cdot A_{6,8} \cdot EIGHT1_{8,1} \right) \right]$$

$$SS_3 = 1645.02$$

The final SS Matrix now looks like

$$SS := \begin{pmatrix} SSt \\ SS1 \\ SS_2 \\ SS3 \\ 0 \\ 0 \\ 0 \\ 0 \\ 0 \end{pmatrix} \qquad SS = \begin{pmatrix} 4191.48 \\ 72.52 \\ 1507.52 \\ 1645.02 \\ 0 \\ 0 \\ 0 \\ 0 \\ 0 \end{pmatrix}$$

STEP 9: Computation of the interactive effects of the First and Second experimental conditions SS1x2. Nsx2 = Nt/#columns in DB12 N1x2 = 48/4=12

$$N1x2 := \frac{48}{4}$$

$$N3 := 12$$

$$DB12_{8,4} = \begin{pmatrix} 1 & 0 & 0 & 0 \\ 1 & 0 & 0 & 0 \\ 0 & 1 & 0 & 0 \\ 0 & 1 & 0 & 0 \\ 0 & 0 & 1 & 0 \\ 0 & 0 & 1 & 0 \\ 0 & 0 & 0 & 1 \\ 0 & 0 & 0 & 1 \end{pmatrix} \qquad DB12b_{8,5} = \begin{pmatrix} 1 & 0 & 0 & 0 & 1 \\ 1 & 0 & 0 & 0 & 1 \\ 0 & 1 & 0 & 0 & 1 \\ 0 & 1 & 0 & 0 & 1 \\ 0 & 0 & 1 & 0 & 1 \\ 0 & 0 & 1 & 0 & 1 \\ 0 & 0 & 0 & 1 & 1 \\ 0 & 0 & 0 & 1 & 1 \end{pmatrix}$$

NOTE: all the Math is the same as in Step 6 except we Multiply by DB12 instead of DB1, and of course the DB Matrices are larger.

$$SS1x2 := \left(1_{1,6}\right)A_{(6,8)} \cdot DB12_{8,4} \left(\frac{b}{N1x2}\right)_{4,1} - \left[\left(\!\left(1_{1,6}\right)\!\right)A_{6,8} \cdot \left(1_{8,1}\right)\left(\frac{b}{Nt}\right)_{1,1}\right] - SS1 - SS2 \qquad \blacksquare$$

$$b1 := 155$$

$$b2 := 262$$

$$SUM1 \cdot DB12_{8,4} = (155 \quad 262 \quad 157 \quad 319)$$

$$b3 := 157$$

$$b4 := 319$$

$$SS1x2 := SUM1 \cdot \begin{pmatrix} 1 & 0 & 0 & 0 \\ 1 & 0 & 0 & 0 \\ 0 & 1 & 0 & 0 \\ 0 & 1 & 0 & 0 \\ 0 & 0 & 1 & 0 \\ 0 & 0 & 1 & 0 \\ 0 & 0 & 0 & 1 \\ 0 & 0 & 0 & 1 \end{pmatrix} \begin{pmatrix} \dfrac{b1}{N1x2} \\ \dfrac{b2}{N1x2} \\ \dfrac{b3}{N1x2} \\ \dfrac{b4}{N1x2} \end{pmatrix} - CORRFACT - SS1 - SS2$$

$$SS1x2 = 63.02$$

METHOD 2: Another b method

NOTE: Suppose we write the equation as follows. Written in this manner we compute the Correction Factor automatically.

$$DB12_{8,4} = \begin{pmatrix} 1 & 0 & 0 & 0 \\ 1 & 0 & 0 & 0 \\ 0 & 1 & 0 & 0 \\ 0 & 1 & 0 & 0 \\ 0 & 0 & 1 & 0 \\ 0 & 0 & 1 & 0 \\ 0 & 0 & 0 & 1 \\ 0 & 0 & 0 & 1 \end{pmatrix} \qquad DB12b_{8,5} = \begin{pmatrix} 1 & 0 & 0 & 0 & 1 \\ 1 & 0 & 0 & 0 & 1 \\ 0 & 1 & 0 & 0 & 1 \\ 0 & 1 & 0 & 0 & 1 \\ 0 & 0 & 1 & 0 & 1 \\ 0 & 0 & 1 & 0 & 1 \\ 0 & 0 & 0 & 1 & 1 \\ 0 & 0 & 0 & 1 & 1 \end{pmatrix}$$

$$SS1x2 := SUM1 \cdot \begin{pmatrix} 1 & 0 & 0 & 0 & 1 \\ 1 & 0 & 0 & 0 & 1 \\ 0 & 1 & 0 & 0 & 1 \\ 0 & 1 & 0 & 0 & 1 \\ 0 & 0 & 1 & 0 & 1 \\ 0 & 0 & 1 & 0 & 1 \\ 0 & 0 & 0 & 1 & 1 \\ 0 & 0 & 0 & 1 & 1 \end{pmatrix} \begin{pmatrix} \dfrac{b1}{N1x2} \\ \dfrac{b2}{N1x2} \\ \dfrac{b3}{N1x2} \\ \dfrac{b4}{N1x2} \\ \dfrac{-b}{Nt} \end{pmatrix} - SS1 - SS2$$

$$SS1x2 = 63.02$$

METHOD 3:

$$SUM1b := \left(SUM1 \cdot DB12b_{8,5} \right)$$

$$SSsx2 := \left[\left(\overrightarrow{(SUM1b \cdot SUM1b)} \cdot diag\left(\frac{1}{Ni} \right) \right) \cdot FIVE1 - SS1 - SS2 \right]$$

$$Ni := \begin{pmatrix} 12 \\ 12 \\ 12 \\ 12 \\ -48 \end{pmatrix}$$

$$SSsx2 := \left[\left[\overrightarrow{\left(SUM1 \cdot DB12b_{8,5} \right)\left(SUM1 \cdot DB12b_{8,5} \right)} \right] \begin{pmatrix} 12 & 0 & 0 & 0 & 0 \\ 0 & 12 & 0 & 0 & 0 \\ 0 & 0 & 12 & 0 & 0 \\ 0 & 0 & 0 & 12 & 0 \\ 0 & 0 & 0 & 0 & -48 \end{pmatrix} \right] \cdot FIVE1 - SSs - SS2 \right]^{\blacksquare}$$

$$SSsx2 = 63.02$$

METHOD 4: $N_3 := 12$

And now for the method which I use through most of this book.

$$SS3 := \left(\frac{1}{N3}\right)\left[\left(1_{1,6}\right)A_{6,8}\cdot DB2_{8,2}\right]^2 - \left(\frac{1}{Nt}\right)\left[\left(1_{1,6}\right)A_{6,8}\cdot\left(1_{8,1}\right)\right]^2$$

$$SS_3 := \left(\frac{1}{12}\right)\cdot\left(ONE6\cdot A_{6,8}\cdot DB12_{8,4}\right)^2 - \left(\frac{1}{Nt}\right)\cdot GRANDSUM^2$$

$$SS_3 := \left(\frac{1}{12}\right)\left[\left(SUM1\cdot DB12_{8,4}\right)\left(SUM1\cdot DB12_{8,4}\right)^T\right] - \left(\frac{1}{Nt}\right)\cdot\left(ONE6\cdot A_{6,8}\cdot EIGHT1_{8,1}\right)\left(ONE6\cdot A_{6,8}\cdot EIGHT1_{8,1}\right) - SS1 - SS2$$

$$SS_3 = 63.02$$

B METHOD:

$$SS_3 := \left(\left[\left(\frac{1}{12}\ \frac{1}{12}\ \frac{1}{12}\ \frac{1}{12}\ \frac{1}{12}\ \frac{1}{12}\ \frac{1}{12}\ \frac{1}{12}\right)\cdot SUM1\right]\cdot DB12_{8,4}\right)\left(SUM1\cdot DB12_{8,4}\right)^T - \left[\left(\frac{1}{48}\ \frac{1}{48}\ \frac{1}{48}\ \frac{1}{48}\ \frac{1}{48}\ \frac{1}{48}\ \frac{1}{48}\ \frac{1}{48}\right)\cdot SUM1\right]\cdot EIGHT1_{8,1}\left(ONE6\cdot A_{6,8}\cdot EIGHT1_{8,1}\right) - SS1 - SS2$$

Again SS3 will not fit on a single page, so we must separate the equation so we can see it all.

$$SS3 := \left(\left[\overrightarrow{\left(\frac{1}{12}\ \frac{1}{12}\ \frac{1}{12}\ \frac{1}{12}\ \frac{1}{12}\ \frac{1}{12}\ \frac{1}{12}\ \frac{1}{12}\right)\cdot SUM1}\right]\cdot DB12_{8,4}\right)\cdot\left(SUM1\cdot DB12_{8,4}\right)^T - $$

$$\left[\overrightarrow{\left(\frac{1}{48}\ \frac{1}{48}\ \frac{1}{48}\ \frac{1}{48}\ \frac{1}{48}\ \frac{1}{48}\ \frac{1}{48}\ \frac{1}{48}\right)\cdot SUM1}\right]\cdot EIGHT1_{8,1}\left(ONE6\cdot A_{6,8}\cdot EIGHT1_{8,1}\right) - SS1 - SS2$$

$$SS_3 = 63.02$$

The final SS Matrix now looks like

$$SS := \begin{pmatrix} SSt \\ SS1 \\ SS2 \\ SS3 \\ SSsx2 \\ 0 \\ 0 \\ 0 \\ 0 \end{pmatrix} \qquad SS = \begin{pmatrix} 4191.48 \\ 72.52 \\ 1507.52 \\ 63.02 \\ 0 \\ 0 \\ 0 \\ 0 \\ 0 \end{pmatrix}$$

STEP 10: Computation of the interactive effects of the first and third experimental factors SS1x3. N1x3 = Nt/# columns in DB21. N1x3 = 48/4 = 12

NOTE: All the math is the same as in step 9 except we multiply by DB21 instead of DB12 and subtract SSs and SS3

METHOD 1:

$$N1x3 := \frac{48}{4} \qquad\qquad\qquad N3 := 12$$

$$DB21_{8,4} = \begin{pmatrix} 1 & 0 & 0 & 0 \\ 0 & 1 & 0 & 0 \\ 1 & 0 & 0 & 0 \\ 0 & 1 & 0 & 0 \\ 0 & 0 & 1 & 0 \\ 0 & 0 & 0 & 1 \\ 0 & 0 & 1 & 0 \\ 0 & 0 & 0 & 1 \end{pmatrix} \qquad\qquad DB21b_{8,5} = \begin{pmatrix} 1 & 0 & 0 & 0 & 1 \\ 0 & 1 & 0 & 0 & 1 \\ 1 & 0 & 0 & 0 & 1 \\ 0 & 1 & 0 & 0 & 1 \\ 0 & 0 & 1 & 0 & 1 \\ 0 & 0 & 0 & 1 & 1 \\ 0 & 0 & 1 & 0 & 1 \\ 0 & 0 & 0 & 1 & 1 \end{pmatrix}$$

NOTE: all the Math is the same as in Step 6 except we Multiply by DB21 instead of DB12.

$$N1x3 := \frac{48}{4}$$

$$SS1x3 := \left(1_{1,6}\right)A_{(6,8)} \cdot DB12_{8,4} \cdot \left(\frac{b}{N1x3}\right)_{4,1} - \left[\left(\left(1_{1,6}\right)A_{6,8} \cdot \left(1_{8,1}\right)\right)\left(\frac{b}{Nt}\right)_{1,1}\right] - SSs - SS3 \quad \blacksquare$$

$$SUM1 \cdot DB21_{8,4} = (143 \quad 274 \quad 163 \quad 313)$$

$$b1 := 143$$

$$b2 := 274$$

$$b3 := 163$$

$$b4 := 313$$

$$SS1x3 := SUM1 \cdot \begin{pmatrix} 1 & 0 & 0 & 0 \\ 0 & 1 & 0 & 0 \\ 1 & 0 & 0 & 0 \\ 0 & 1 & 0 & 0 \\ 0 & 0 & 1 & 0 \\ 0 & 0 & 0 & 1 \\ 0 & 0 & 1 & 0 \\ 0 & 0 & 0 & 1 \end{pmatrix} \begin{pmatrix} \dfrac{b1}{N1x3} \\ \dfrac{b2}{N1x3} \\ \dfrac{b3}{N1x3} \\ \dfrac{b4}{N1x3} \end{pmatrix} - CORRFACT - SS1 - SS3$$

$$SS1x3 = 7.52$$

METHOD 2: Another b method

NOTE: Suppose we write the equation as follows. Written in this manner we compute the Correction Factor automatically.

$$DB21_{8,4} = \begin{pmatrix} 1 & 0 & 0 & 0 \\ 0 & 1 & 0 & 0 \\ 1 & 0 & 0 & 0 \\ 0 & 1 & 0 & 0 \\ 0 & 0 & 1 & 0 \\ 0 & 0 & 0 & 1 \\ 0 & 0 & 1 & 0 \\ 0 & 0 & 0 & 1 \end{pmatrix} \qquad DB21b_{8,5} = \begin{pmatrix} 1 & 0 & 0 & 0 & 1 \\ 0 & 1 & 0 & 0 & 1 \\ 1 & 0 & 0 & 0 & 1 \\ 0 & 1 & 0 & 0 & 1 \\ 0 & 0 & 1 & 0 & 1 \\ 0 & 0 & 0 & 1 & 1 \\ 0 & 0 & 1 & 0 & 1 \\ 0 & 0 & 0 & 1 & 1 \end{pmatrix}$$

$$SS1x3 := SUM1 \cdot \begin{pmatrix} 1 & 0 & 0 & 0 & 1 \\ 0 & 1 & 0 & 0 & 1 \\ 1 & 0 & 0 & 0 & 1 \\ 0 & 1 & 0 & 0 & 1 \\ 0 & 0 & 1 & 0 & 1 \\ 0 & 0 & 0 & 1 & 1 \\ 0 & 0 & 1 & 0 & 1 \\ 0 & 0 & 0 & 1 & 1 \end{pmatrix} \begin{pmatrix} \dfrac{b1}{N1x3} \\ \dfrac{b2}{N1x3} \\ \dfrac{b3}{N1x3} \\ \dfrac{b4}{N1x3} \\ \dfrac{-b}{Nt} \end{pmatrix} - SS1 - SS3$$

$$SS1x3 = 7.52$$

METHOD 3: Another b method

$$Ni := \begin{pmatrix} 12 \\ 12 \\ 12 \\ 12 \\ -48 \end{pmatrix} \qquad diag\left(\frac{1}{Ni}\right) = \begin{pmatrix} 0.08 & 0 & 0 & 0 & 0 \\ 0 & 0.08 & 0 & 0 & 0 \\ 0 & 0 & 0.08 & 0 & 0 \\ 0 & 0 & 0 & 0.08 & 0 \\ 0 & 0 & 0 & 0 & -0.02 \end{pmatrix}$$

$$SS1x3 := \left[\left(\overrightarrow{\left[\left(SUM1 \cdot DB21b_{8,5}\right)\left(SUM1 \cdot DB21b_{8,5}\right)\right] \cdot diag\left(\frac{1}{Ni}\right)}\right) \cdot FIVE1 - SS1 - SS3\right]^{\blacksquare}$$

$$SUM1b := \left(SUM1 \cdot DB21b_{8,5}\right)$$

$$SS1x3 := \left[\left(\overrightarrow{(SUM1b \cdot SUM1b) \cdot diag\left(\frac{1}{Ni}\right)}\right) \cdot FIVE1 - SS1 - SS3\right]$$

$$SS1x3 := \left[\left[\overrightarrow{\left(SUM1 \cdot DB21b_{8,5}\right)\left(SUM1 \cdot DB21b_{8,5}\right)} \cdot \begin{pmatrix} 0.08 & 0 & 0 & 0 & 0 \\ 0 & 0.08 & 0 & 0 & 0 \\ 0 & 0 & 0.08 & 0 & 0 \\ 0 & 0 & 0 & 0.08 & 0 \\ 0 & 0 & 0 & 0 & -0.02 \end{pmatrix}\right] \cdot FIVE1 - SS1 - SS3\right]^{\blacksquare}$$

$$SS1x3 = 7.52$$

Lets see what this looks like:

$$SS1x3 := \left[\left[\overrightarrow{\left\|SUM1 \cdot \begin{pmatrix} 1&0&0&0&1 \\ 0&1&0&0&1 \\ 1&0&0&0&1 \\ 0&1&0&0&1 \\ 0&0&1&0&1 \\ 0&0&0&1&1 \\ 0&0&1&0&1 \\ 0&0&0&1&1 \end{pmatrix}\right\| \cdot \left\|SUM1 \cdot \begin{pmatrix} 1&0&0&0&1 \\ 0&1&0&0&1 \\ 1&0&0&0&1 \\ 0&1&0&0&1 \\ 0&0&1&0&1 \\ 0&0&0&1&1 \\ 0&0&1&0&1 \\ 0&0&0&1&1 \end{pmatrix}\right\|} \cdot \begin{pmatrix} 12&0&0&0&0 \\ 0&12&0&0&0 \\ 0&0&12&0&0 \\ 0&0&0&12&0 \\ 0&0&0&0&-48 \end{pmatrix}^{-1} \begin{pmatrix} 1 \\ 1 \\ 1 \\ 1 \\ 1 \end{pmatrix}\right] - SS1 - SS3\right]^{\blacksquare}$$

$$SS1x3 = 7.52$$

$$\text{SUM1b} := (51 \quad 104 \quad 92 \quad 170 \quad 57 \quad 100 \quad 106 \quad 213) \cdot \begin{pmatrix} 1 & 0 & 0 & 0 & 1 \\ 0 & 1 & 0 & 0 & 1 \\ 1 & 0 & 0 & 0 & 1 \\ 0 & 1 & 0 & 0 & 1 \\ 0 & 0 & 1 & 0 & 1 \\ 0 & 0 & 0 & 1 & 1 \\ 0 & 0 & 1 & 0 & 1 \\ 0 & 0 & 0 & 1 & 1 \end{pmatrix}$$

$$\text{SS1x3} := \overrightarrow{[(\text{SUM1b}) \cdot (\text{SUM1b})]} \cdot \begin{pmatrix} 12 & 0 & 0 & 0 & 0 \\ 0 & 12 & 0 & 0 & 0 \\ 0 & 0 & 12 & 0 & 0 \\ 0 & 0 & 0 & 12 & 0 \\ 0 & 0 & 0 & 0 & -48 \end{pmatrix}^{-1} \begin{pmatrix} 1 \\ 1 \\ 1 \\ 1 \\ 1 \end{pmatrix} - \text{SS1} - \text{SS3}$$

$$\text{SS1x3} = 7.52$$

METHOD 4 :

Now for the theoretical way where all N's are the same order.

$$\text{SS1x3} := \left(\frac{1}{N_{1,3}}\right) \overrightarrow{\left[\left(1_{1,6}\right)\left(A_{6,8}\right)\left(\text{DB21}_{8,4}\right)\right]\left[\left(1_{1,6}\right)\left(A_{6,8}\right)\left(\text{DB21}_{8,4}\right)\right]} - \left(\frac{1}{\text{Nt}}\right) \cdot \text{SUM1} \cdot \text{EIGHT1}_{8,1} - \text{SS1} - \text{SS3} \quad \blacksquare$$

$$\text{SS1x3a} := \left(\frac{1}{12}\right) \overrightarrow{\left[\text{ONE6} \cdot \left(A_{6,8}\right)\left(\text{DB21}_{8,4}\right)\right]\left[(\text{ONE6}) \cdot \left(A_{6,8}\right)\left(\text{DB21}_{8,4}\right)\right]} - \left(\frac{1}{\text{Nt}}\right) \cdot \text{SUM1} \cdot \text{EIGHT1}_{8,1} - \text{SS1} - \text{SS3} \quad \blacksquare$$

$$\text{SUM1b} := \text{SUM1} \cdot \text{DB21}_{8,4}$$

$$\text{SS1x3a} := \left(\frac{1}{12}\right) \cdot \text{SUM1b} \cdot \text{SUM1b}^T - \text{CORRFACT} - \text{SS1} - \text{SS3}$$

$$\text{SS1x3a} = 7.52$$

$$\text{SS} := \begin{pmatrix} \text{SSt} \\ \text{SS1} \\ \text{SS2} \\ \text{SS3} \\ \text{SS1x2} \\ \text{SS1x3} \\ 0 \\ 0 \\ 0 \end{pmatrix} \qquad \text{SS} = \begin{pmatrix} 4191.479 \\ 72.521 \\ 1507.521 \\ 1645.021 \\ 63.021 \\ 7.521 \\ 0 \\ 0 \\ 0 \end{pmatrix}$$

STEP 11: Computation of the interactive effects of the second and third experimental conditions (factors) SS2x3. N2x3 = Nt/#columns in DB4 = 48/4 = 12.

$$DB4_{8,4} = \begin{pmatrix} 1 & 0 & 0 & 0 \\ 0 & 1 & 0 & 0 \\ 0 & 0 & 1 & 0 \\ 0 & 0 & 0 & 1 \\ 1 & 0 & 0 & 0 \\ 0 & 1 & 0 & 0 \\ 0 & 0 & 1 & 0 \\ 0 & 0 & 0 & 1 \end{pmatrix} \qquad DB4b_{8,5} = \begin{pmatrix} 1 & 0 & 0 & 0 & 1 \\ 0 & 1 & 0 & 0 & 1 \\ 0 & 0 & 1 & 0 & 1 \\ 0 & 0 & 0 & 1 & 1 \\ 1 & 0 & 0 & 0 & 1 \\ 0 & 1 & 0 & 0 & 1 \\ 0 & 0 & 1 & 0 & 1 \\ 0 & 0 & 0 & 1 & 1 \end{pmatrix} \qquad N2x3 := 12$$

NOTE: This is getting redundant but all the math is the same as in STEP 9 & 10 except we multiply bt DB4 instead of DB12 or DB21 and subtract CORRFACT, SS2 AND SS3. This redundancy is fascinating, for it means we might be able to do all the math all at the same time. We will explore this possibility as soon as we solve the computations for the problem.

METHOD 1:

$N1x3 := \dfrac{48}{4}$ NOTE: all the Math is the same as in Step 6 except we Multiply by DB4 instead of DB12.

$$SS1x3 := \left(1_{1,6}\right)A_{(6,8)} \cdot DB4_{8,4} \left(\frac{b}{N2x3}\right)_{4,1} - \left[\left(\left(1_{1,6}\right)\right)A_{6,8} \cdot \left(1_{8,1}\right)\left(\frac{b}{Nt}\right)_{1,1}\right] - SS2 - SS3 \quad \blacksquare$$

$SUM1 \cdot DB4_{8,4} = (108 \quad 204 \quad 198 \quad 383)$ b1 := 108

                                                                        b2 := 204

                                                                        b3 := 198

                                                                        b4 := 383

$$SS2x3 := SUM1 \cdot \begin{pmatrix} 1 & 0 & 0 & 0 \\ 0 & 1 & 0 & 0 \\ 0 & 0 & 1 & 0 \\ 0 & 0 & 0 & 1 \\ 1 & 0 & 0 & 0 \\ 0 & 1 & 0 & 0 \\ 0 & 0 & 1 & 0 \\ 0 & 0 & 0 & 1 \end{pmatrix} \cdot \begin{pmatrix} \dfrac{b1}{N2x3} \\ \dfrac{b2}{N2x3} \\ \dfrac{b3}{N2x3} \\ \dfrac{b4}{N2x3} \end{pmatrix} - CORRFACT - SS2 - SS3$$

$SS2x3 = 165.02$

METHOD 2:

$N2x3 := \dfrac{48}{4}$ NOTE: all the Math is the same as in Step 6 except we Multiply by DB4b instead of DB12b.

$$SS1x3 := \left(1_{1,6}\right)A_{(6,8)} \cdot DB4_{8,5}\left(\frac{b}{N2x3}\right)_{5,1} - \left[\left(\left(1_{1,6}\right)\right)A_{6,8}\cdot\left(1_{8,1}\right)\left(\frac{b}{Nt}\right)_{1,1}\right] - SS2 - SS3 \quad \blacksquare$$

$SUM1 \cdot DB4b_{8,5} = (108 \quad 204 \quad 198 \quad 383 \quad 893)$ $b1 := 108$

$b2 := 204$

$b3 := 198$

$b4 := 383$

$$SS2x3 := SUM1 \cdot \begin{pmatrix} 1 & 0 & 0 & 0 & 1 \\ 0 & 1 & 0 & 0 & 1 \\ 0 & 0 & 1 & 0 & 1 \\ 0 & 0 & 0 & 1 & 1 \\ 1 & 0 & 0 & 0 & 1 \\ 0 & 1 & 0 & 0 & 1 \\ 0 & 0 & 1 & 0 & 1 \\ 0 & 0 & 0 & 1 & 1 \end{pmatrix} \cdot \begin{pmatrix} \dfrac{b1}{N2x3} \\ \dfrac{b2}{N2x3} \\ \dfrac{b3}{N2x3} \\ \dfrac{b4}{N2x3} \\ \dfrac{-b}{Nt} \end{pmatrix} - SS2 - SS3$$

$SS2x3 = 165.02$

METHOD 3:

NOTE: Suppose we write the above equation as: $SS2x3 = [SUM1]_{1,8}[DB4b]_{8,5}[b/Ni] - SS2 - SS3$

$$\frac{b}{N_i} := \begin{pmatrix} \dfrac{b_1}{N2x3} \\[2ex] \dfrac{2}{N2x3} \\[2ex] \dfrac{b_3}{N2x3} \\[2ex] \dfrac{b_4}{N2x3} \end{pmatrix}^{\blacksquare}$$

SS2 = 1507.52

SS3 = 1645.02

$$SS2x3 := SUM1 \cdot DB4b_{8,5} \cdot diag\left(\frac{b}{Ni}\right) - SS_2 - SS_{\blacksquare}$$

SS2x3 = 165.02

METHOD 4:

Now for the theoretical way where all N's are the same order.

$$SS2x3 := \left(\frac{1}{N_{2,3}}\right)\overline{\left[\left(1_{1,6}\right)\left(A_{6,8}\right)\left(DB4_{8,4}\right)\right]\left[\left(1_{1,6}\right)\left(A_{6,8}\right)\left(DB4_{8,4}\right)\right]} - \left(\frac{1}{Nt}\right)\cdot SUM1 \cdot EIGHT1_{8,1} - SS2 - SS3^{\blacksquare}$$

$$SS2x3a := \left(\frac{1}{12}\right)\overline{\left[ONE6\cdot\left(A_{6,8}\right)\left(DB4_{8,4}\right)\right]\left[\left(ONE6\right)\cdot\left(A_{6,8}\right)\left(DB4_{8,4}\right)\right]} - \left(\frac{1}{Nt}\right)\cdot SUM1 \cdot EIGHT1_{8,1} - SS2 - SS3^{\blacksquare}$$

$$SUM1b := SUM1 \cdot DB4_{8,4}$$

$$SS2x3a := \left(\frac{1}{12}\right)\cdot SUM1b \cdot SUM1b^T - CORRFACT - SS2 - SS3$$

SS2x3a = 165.02

We are nearing the end of our computations, the final SS Matrix now becomes:

$$SS := \begin{pmatrix} SSt \\ SS1 \\ SS2 \\ SS3 \\ SS1x2 \\ SS1x3 \\ SS2x3 \\ 0 \\ 0 \end{pmatrix} \qquad SS = \begin{pmatrix} 4191.479 \\ 72.521 \\ 1507.521 \\ 1645.021 \\ 63.021 \\ 7.521 \\ 165.021 \\ 0 \\ 0 \end{pmatrix}$$

STEP 12: Computation of the interactive effects of the 1st, 2nd and
3rd experimental factors SS1x2x3. N1x2x3 = Nt/#elements in SUM1.

This computation is simple, we just square SUM1 and subtract
every Correction Factor found.

$\text{N1x2x3} := 6$

METHOD 1:

$$\text{SS1x2x3} := \text{SUM1} \cdot (\text{SUM1}, b)^T \cdot \left(\frac{b}{\text{Ni}}\right) - \text{SS1} - \text{SS2} - \text{SS3} - \text{SS1x2} - \text{SS1x3} - \text{SS2x3} \quad\blacksquare$$

$$\text{SS1x2x3b} := \text{SUM1} \cdot \begin{pmatrix} 51 & 1 \\ 104 & 1 \\ 92 & 1 \\ 170 & 1 \\ 57 & 1 \\ 100 & 1 \\ 106 & 1 \\ 213 & 1 \end{pmatrix} \qquad\qquad \text{SUM1} = (51 \quad 104 \quad 92 \quad 170 \quad 57 \quad 100 \quad 106 \quad 213)$$

$$\text{SS1x2x3} := \text{SUM1} \cdot \begin{pmatrix} 51 & 1 \\ 104 & 1 \\ 92 & 1 \\ 170 & 1 \\ 57 & 1 \\ 100 & 1 \\ 106 & 1 \\ 213 & 1 \end{pmatrix} \cdot \begin{pmatrix} \dfrac{1}{6} \\ \dfrac{-893}{48} \end{pmatrix} - \text{SS1} - \text{SS2} - \text{SS3} - \text{SS1x2} - \text{SS1x3} - \text{SS2x3}$$

$\text{SS1x2x3} = 31.69$

METHOD 2:

$$\text{SS1x2x3} := \left(\frac{1}{\text{N1x2x3}}\right) \cdot \left(\text{SUM1} \cdot \text{SUM1}^T\right) - \text{CORRFACT} - \text{SS1} - \text{SS2} - \text{SS3} - \text{SS1x2} - \text{SS1x3} - \text{SS2x3}$$

$\text{SS1x2x3} = 31.69$

STEP 13: Computation of the Error term Sum of Squares. SSerror.

SSerror := SSt − SS1 − SS2 − SS3 − SS1x2 − SS1x3 − SS2x3 − SS1x2x3

SSerror = 699.17

STEP 14: All computations are done we must now compute the df's for each of the omponents.

m := 48   E1 := 2   E2 := 2   E3 := 2   dfSS1 := E1 − 1   dfSSt := m − 1

                                              dfSS2 := E2 − 1   dfSS3 := E3 − 1

dfSS1x2 := dfSS1·dfSS2   dfSS1x2x3 := dfSS1·dfSS2·dfSS3

dfSS1x3 := dfSS1·dfSS3   dfSS2x3 := dfSS2·dfSS3

$$
df := \begin{pmatrix} m - 1 \\ E1 - 1 \\ E2 - 1 \\ E3 - 1 \\ dfSS1 \cdot dfSS2 \\ dfSS1 \cdot dfSS3 \\ dfSS2 \cdot dfSS3 \\ dfSS1 \cdot dfSS2 \cdot dfSS3 \\ dfSSt - dfSS1 - dfSS2 - dfSS3 - dfSS1x2 - dfSS1x3 - dfSS2x3 - dfSS1x2x3 \end{pmatrix}
\qquad
df = \begin{pmatrix} 47 \\ 1 \\ 1 \\ 1 \\ 1 \\ 1 \\ 1 \\ 1 \\ 40 \end{pmatrix}
$$

SSerror is the last value in the final SS Matrix.

SSerror = 699.17

STEP 15: Computation of F

$$F := (SS) \cdot odf^{-1} o \cdot ERROR^{-1} \blacksquare$$

$$SS := \begin{pmatrix} SSt \\ SS1 \\ SS2 \\ SS3 \\ SS1x2 \\ SS1x3 \\ SS2x3 \\ SS1x2x3 \\ SSerror \end{pmatrix}$$

$$SS = \begin{pmatrix} 4191.48 \\ 72.52 \\ 1507.52 \\ 1645.02 \\ 63.02 \\ 7.52 \\ 165.02 \\ 31.69 \\ 699.17 \end{pmatrix} \quad df^{-1} = \begin{pmatrix} 0.02 \\ 1 \\ 1 \\ 1 \\ 1 \\ 1 \\ 1 \\ 1 \\ 0.03 \end{pmatrix} \quad ERROR := \begin{pmatrix} \dfrac{699.167}{40} \\ \dfrac{699.167}{40} \\ 17.479 \\ 17.479 \\ 17.479 \\ 17.479 \\ 17.479 \\ 17.479 \\ 17.479 \end{pmatrix} \quad ERROR^{-1} = \begin{pmatrix} 0.06 \\ 0.06 \\ 0.06 \\ 0.06 \\ 0.06 \\ 0.06 \\ 0.06 \\ 0.06 \\ 0.06 \end{pmatrix}$$

$$F := \overrightarrow{\left(SS \cdot df^{-1} \cdot ERROR^{-1} \right)}$$

$$\begin{pmatrix} SSt \\ SS1 \\ SS2 \\ SS3 \\ SS1x2 \\ SS1x3 \\ SS2x3 \\ SS1x2x3 \\ SSerror \end{pmatrix} \quad F = \begin{pmatrix} 5.1 \\ 4.15 \\ 86.25 \\ 94.11 \\ 3.61 \\ 0.43 \\ 9.44 \\ 1.81 \\ 1 \end{pmatrix} \quad p := \begin{pmatrix} 0 \\ 0 \\ .001 \\ .001 \\ ns \\ ns \\ .001 \\ ns \\ ns \end{pmatrix}$$

SS2 < .001

SS3 < .001

SS2x3 < .001

Therefore it is concluded that experimental condition 2 and experimental condition 3 affect the rate of errors and the effects of experimental condition 2 with condition 3 are interactive by factors 1000 times greater than by chance.

PRESENTATION OF THIS 3-VARIABLE ANALYSIS OF VARIANCE USING NESTED ARRAYS.

This is a three variable Analysis of Variance Statistic. The data is in Matrix A Note we need at least 4 columns of data to solve 3 variables. The rest are the variables used to solve the equation. The purpose is to solve the corrected Sum of Squares and the F value in one equation.

$$A := \begin{pmatrix} 14 & 25 & 12 & 22 & 17 & 14 & 19 & 35 \\ 9 & 15 & 18 & 25 & 8 & 19 & 18 & 30 \\ 10 & 18 & 10 & 26 & 5 & 15 & 10 & 39 \\ 5 & 17 & 15 & 32 & 4 & 20 & 16 & 41 \\ 7 & 16 & 19 & 36 & 8 & 13 & 23 & 35 \\ 6 & 13 & 18 & 29 & 15 & 19 & 20 & 33 \end{pmatrix} \quad EIGHT1 := \begin{pmatrix} 1 \\ 1 \\ 1 \\ 1 \\ 1 \\ 1 \\ 1 \\ 1 \end{pmatrix} \quad df := \begin{pmatrix} 47 \\ 1 \\ 1 \\ 1 \\ 1 \\ 1 \\ 1 \\ 1 \\ 40 \end{pmatrix} \quad ERROR := \begin{pmatrix} \dfrac{40}{692} \\ .058 \\ .058 \\ .058 \\ .058 \\ .058 \\ .058 \\ .058 \\ .058 \end{pmatrix}$$

$$ONE6 := (1 \quad 1 \quad 1 \quad 1 \quad 1 \quad 1)$$

$$N := (48 \quad 1 \quad 24 \quad 24 \quad 24 \quad 24 \quad 24 \quad 12 \quad 12 \quad 12 \quad 12 \quad 12 \quad 12 \quad 12 \quad 12 \quad 12 \quad 12 \quad 6 \quad 6 \quad 6 \quad 6 \quad 6 \quad 6 \quad 6)$$

$$SUM1 := ONE6 \cdot A$$

$$SUM1 = (51 \quad 104 \quad 92 \quad 170 \quad 57 \quad 100 \quad 106 \quad 213)$$

Here we sum the columns in A.

$$SUM1 := (1 \quad 1 \quad 1 \quad 1 \quad 1 \quad 1) \cdot \begin{bmatrix} \begin{pmatrix} 14 & 25 & 12 & 22 & 17 & 14 & 19 & 35 \\ 9 & 15 & 18 & 25 & 8 & 19 & 18 & 30 \\ 10 & 18 & 10 & 26 & 5 & 15 & 10 & 39 \\ 5 & 17 & 15 & 32 & 4 & 20 & 16 & 41 \\ 7 & 16 & 19 & 36 & 8 & 13 & 23 & 35 \\ 6 & 13 & 18 & 29 & 15 & 19 & 20 & 33 \end{pmatrix} \end{bmatrix}$$

Next we to compute the Grand Sum of Squares

$$GRANDSUMSQ := ONE6 \cdot \overrightarrow{(A \cdot A)} \cdot EIGHT1$$

$$\text{GRANDSUMSQ} := (1\ 1\ 1\ 1\ 1\ 1) \cdot \left[\begin{pmatrix} 14 & 25 & 12 & 22 & 17 & 14 & 19 & 35 \\ 9 & 15 & 18 & 25 & 8 & 19 & 18 & 30 \\ 10 & 18 & 10 & 26 & 5 & 15 & 10 & 39 \\ 5 & 17 & 15 & 32 & 4 & 20 & 16 & 41 \\ 7 & 16 & 19 & 36 & 8 & 13 & 23 & 35 \\ 6 & 13 & 18 & 29 & 15 & 19 & 20 & 33 \end{pmatrix} \begin{pmatrix} 14 & 25 & 12 & 22 & 17 & 14 & 19 & 35 \\ 9 & 15 & 18 & 25 & 8 & 19 & 18 & 30 \\ 10 & 18 & 10 & 26 & 5 & 15 & 10 & 39 \\ 5 & 17 & 15 & 32 & 4 & 20 & 16 & 41 \\ 7 & 16 & 19 & 36 & 8 & 13 & 23 & 35 \\ 6 & 13 & 18 & 29 & 15 & 19 & 20 & 33 \end{pmatrix}\right] \begin{pmatrix} 1 \\ 1 \\ 1 \\ 1 \\ 1 \\ 1 \\ 1 \\ 1 \end{pmatrix}$$

GRANDSUMSQ = 20805

Lets go ahead and put this value into Matrix form so it will be ready for computation when it is needed. Since we will be putting it in the SStotal Matrix we will put it in:

SStotal1 := (0 20805 0 0 0 0 0 0)

Now remember when I pointed out that all the computations were all the same except for changing the DB Matrices and subtracting the Correction Factor(s)? In DBALL we put all the individual Data Bases into a single Nested Array in the order we computed them above. Since there is no such thing as using Nested Arrays mathematically, we will just partition the Matrix and treat it as if it were a Nested Array.

That is, with the DB Matrices below we get the Nested Array

$$A1 := \begin{pmatrix} 1 \\ 1 \\ 1 \\ 1 \\ 1 \\ 1 \\ 1 \\ 1 \end{pmatrix} \quad B := \begin{pmatrix} 0 \\ 0 \\ 0 \\ 0 \\ 0 \\ 0 \\ 0 \\ 0 \end{pmatrix} \quad C := \begin{pmatrix} 1 & 0 \\ 1 & 0 \\ 1 & 0 \\ 1 & 0 \\ 0 & 1 \\ 0 & 1 \\ 0 & 1 \\ 0 & 1 \end{pmatrix} \quad D := \begin{pmatrix} 1 & 0 \\ 1 & 0 \\ 0 & 1 \\ 0 & 1 \\ 1 & 0 \\ 1 & 0 \\ 0 & 1 \\ 0 & 1 \end{pmatrix} \quad E := \begin{pmatrix} 1 & 0 \\ 0 & 1 \\ 1 & 0 \\ 0 & 1 \\ 1 & 0 \\ 0 & 1 \\ 1 & 0 \\ 0 & 1 \end{pmatrix} \quad G := \begin{pmatrix} 1 & 0 & 0 & 0 \\ 1 & 0 & 0 & 0 \\ 0 & 1 & 0 & 0 \\ 0 & 1 & 0 & 0 \\ 0 & 0 & 1 & 0 \\ 0 & 0 & 1 & 0 \\ 0 & 0 & 0 & 1 \\ 0 & 0 & 0 & 1 \end{pmatrix}$$

$$H := \begin{pmatrix} 1 & 0 & 0 & 0 \\ 0 & 1 & 0 & 0 \\ 1 & 0 & 0 & 0 \\ 0 & 1 & 0 & 0 \\ 0 & 0 & 1 & 0 \\ 0 & 0 & 0 & 1 \\ 0 & 0 & 1 & 0 \\ 0 & 0 & 0 & 1 \end{pmatrix} \quad I := \begin{pmatrix} 1 & 0 & 0 & 0 \\ 0 & 1 & 0 & 0 \\ 0 & 0 & 1 & 0 \\ 0 & 0 & 0 & 1 \\ 1 & 0 & 0 & 00 \\ 0 & 1 & 0 & 0 \\ 0 & 0 & 1 & 0 \\ 0 & 0 & 0 & 1 \end{pmatrix} \quad J := \begin{pmatrix} 1 & 0 & 0 & 0 & 0 & 0 & 0 & 0 \\ 0 & 1 & 0 & 0 & 0 & 0 & 0 & 0 \\ 0 & 0 & 1 & 0 & 0 & 0 & 0 & 0 \\ 0 & 0 & 0 & 1 & 0 & 0 & 0 & 0 \\ 0 & 0 & 0 & 0 & 1 & 0 & 0 & 0 \\ 0 & 0 & 0 & 0 & 0 & 1 & 0 & 0 \\ 0 & 0 & 0 & 0 & 0 & 0 & 1 & 0 \\ 0 & 0 & 0 & 0 & 0 & 0 & 0 & 1 \end{pmatrix}$$

$NA := (A1 \;\; B \;\; C \;\; D \;\; E \;\; G \;\; H \;\; I \;\; J)$

$$NA = \left[\begin{pmatrix}1\\1\\1\\1\\1\\1\\1\\1\end{pmatrix} \begin{pmatrix}0\\0\\0\\0\\0\\0\\0\\0\end{pmatrix} \begin{pmatrix}1&0\\1&0\\1&0\\1&0\\0&1\\0&1\\0&1\\0&1\end{pmatrix} \begin{pmatrix}1&0\\1&0\\0&1\\0&1\\1&0\\1&0\\0&1\\0&1\end{pmatrix} \begin{pmatrix}1&0\\0&1\\1&0\\0&1\\1&0\\0&1\\1&0\\0&1\end{pmatrix} \begin{pmatrix}1&0&0&0\\1&0&0&0\\0&1&0&0\\0&1&0&0\\0&0&1&0\\0&0&1&0\\0&0&0&1\\0&0&0&1\end{pmatrix} \begin{pmatrix}1&0&0&0\\0&1&0&0\\1&0&0&0\\0&1&0&0\\0&0&1&0\\0&0&0&1\\0&0&1&0\\0&0&0&1\end{pmatrix} \begin{pmatrix}1&0&0&0\\0&1&0&0\\0&0&1&0\\0&0&0&1\\1&0&0&0\\0&1&0&0\\0&0&1&0\\0&0&0&1\end{pmatrix} \begin{pmatrix}1&0&0&0&0&0&0&0\\0&1&0&0&0&0&0&0\\0&0&1&0&0&0&0&0\\0&0&0&1&0&0&0&0\\0&0&0&0&1&0&0&0\\0&0&0&0&0&1&0&0\\0&0&0&0&0&0&1&0\\0&0&0&0&0&0&0&1\end{pmatrix} \right]$$

When we partition this out we get the 9 Matric DBALL Nested Array Matrix. To adjust this Matrix so we can use it (because the Mathematics does not yet exist) we partition the Matrix so it becomes:

NOTE: There are 9 Nested Arrays, so there will be 9 values in the final F Matrix.

$$DBALL := \begin{pmatrix}
1&0&1&0&1&0&1&0&1&0&0&0&1&0&0&0&1&0&0&0&1&0&0&0&0&0&0&0\\
1&0&1&0&1&0&0&1&1&0&0&0&0&1&0&0&0&1&0&0&0&1&0&0&0&0&0&0\\
1&0&1&0&0&1&1&0&0&1&0&0&1&0&0&0&0&0&1&0&0&0&1&0&0&0&0&0\\
1&0&1&0&0&1&0&1&0&1&0&0&0&1&0&0&0&0&0&1&0&0&0&1&0&0&0&0\\
1&0&0&1&1&0&1&0&0&0&1&0&0&0&1&0&1&0&0&0&0&0&0&0&1&0&0&0\\
1&0&0&1&1&0&0&1&0&0&1&0&0&0&0&1&0&1&0&0&0&0&0&0&0&1&0&0\\
1&0&0&1&0&1&1&0&0&0&0&1&0&0&1&0&0&0&1&0&0&0&0&0&0&0&1&0\\
1&0&0&1&0&1&0&1&0&0&0&1&0&0&0&1&0&0&0&1&0&0&0&0&0&0&0&1
\end{pmatrix}$$

Now we multiply SUM1 by all it's DB Matrices:

$SS1 := SUM1 \cdot DBALL$

$$SS1 := SUM1 \cdot \begin{pmatrix}
1&0&1&0&1&0&1&0&1&0&0&0&1&0&0&0&1&0&0&0&1&0&0&0&0&0&0&0\\
1&0&1&0&1&0&0&1&1&0&0&0&0&1&0&0&0&1&0&0&0&1&0&0&0&0&0&0\\
1&0&1&0&0&1&1&0&0&1&0&0&1&0&0&0&0&0&1&0&0&0&1&0&0&0&0&0\\
1&0&1&0&0&1&0&1&0&1&0&0&0&1&0&0&0&0&0&1&0&0&0&1&0&0&0&0\\
1&0&0&1&1&0&1&0&0&0&1&0&0&0&1&0&1&0&0&0&0&0&0&0&1&0&0&0\\
1&0&0&1&1&0&0&1&0&0&1&0&0&0&0&1&0&1&0&0&0&0&0&0&0&1&0&0\\
1&0&0&1&0&1&1&0&0&0&0&1&0&0&1&0&0&0&1&0&0&0&0&0&0&0&1&0\\
1&0&0&1&0&1&0&1&0&0&0&1&0&0&0&1&0&0&0&1&0&0&0&0&0&0&0&1
\end{pmatrix}$$

$SS1 = (893\;\;0\;\;417\;\;476\;\;312\;\;581\;\;306\;\;587\;\;155\;\;262\;\;157\;\;319\;\;143\;\;274\;\;163\;\;313\;\;108\;\;204\;\;198\;\;383\;\;51\;\;104\;\;92\;\;170\;\;57\;\;100\;\;106\;\;213)$

There is not room for SS1 so we will transpose it so it will
fit on a single page.

$$
SS1^T = \begin{pmatrix}
893 \\
0 \\
417 \\
476 \\
312 \\
581 \\
306 \\
587 \\
155 \\
262 \\
157 \\
319 \\
143 \\
274 \\
163 \\
313 \\
108 \\
204 \\
198 \\
383 \\
51 \\
104 \\
92 \\
170 \\
57 \\
100 \\
106 \\
213
\end{pmatrix}
$$

Note the last 8 values equal SUM1.

Now lets square each value in SS1 and divide by N.

$$N := (48 \;\; 1 \;\; 24 \;\; 24 \;\; 24 \;\; 24 \;\; 24 \;\; 24 \;\; 12 \;\; 12 \;\; 12 \;\; 12 \;\; 12 \;\; 12 \;\; 12 \;\; 12 \;\; 12 \;\; 12 \;\; 12 \;\; 12 \;\; 6 \;\; 6 \;\; 6 \;\; 6 \;\; 6 \;\; 6 \;\; 6 \;\; 6)$$

$$SS2 := \overrightarrow{\left[(SS1 \cdot SS1) \cdot N^{-1} \right]}$$

SS2 = (16613.52 0 7245.38 9440.67 4056 14065.04 3901.5 14357.04 2002.08 5720.33 2054.08 8480.08 1704.08 6256.33 2214.08 8164.08 972 3468 3267 12224.08 433.5 1802.67 1410.67 4816.67 541.5 1666.67 1872.67 7561.5)

Whoa, this result goes off the page so we'll transpose it so we can see the answers.

$$
SS2^T = \begin{pmatrix}
16613.52 \\
0 \\
7245.38 \\
9440.67 \\
4056 \\
14065.04 \\
3901.5 \\
14357.04 \\
2002.08 \\
5720.33 \\
2054.08 \\
8480.08 \\
1704.08 \\
6256.33 \\
2214.08 \\
8164.08 \\
972 \\
3468 \\
3267 \\
12224.08 \\
433.5 \\
1802.67 \\
1410.67 \\
4816.67 \\
541.5 \\
1666.67 \\
1872.67 \\
7561.5
\end{pmatrix}
$$

The inner, or core parenthetical expression is computed, now we will compute the uncorrected SS values. First we need to sum the values for the individual Nested Arrays. The first and second numbers remain unchanged. Then we add the 3 two column Matrices and the 3 four column Matrices and the final 8 values in the 8x8 matrix. We do this all at the same time as follows:

$$
\text{DBSUM} := \begin{pmatrix}
1 & 0 & 0 & 0 & 0 & 0 & 0 & 0 & 0 \\
0 & 0 & 0 & 0 & 0 & 0 & 0 & 0 & 0 \\
0 & 0 & 1 & 0 & 0 & 0 & 0 & 0 & 0 \\
0 & 0 & 1 & 0 & 0 & 0 & 0 & 0 & 0 \\
0 & 0 & 0 & 1 & 0 & 0 & 0 & 0 & 0 \\
0 & 0 & 0 & 1 & 0 & 0 & 0 & 0 & 0 \\
0 & 0 & 0 & 0 & 1 & 0 & 0 & 0 & 0 \\
0 & 0 & 0 & 0 & 1 & 0 & 0 & 0 & 0 \\
0 & 0 & 0 & 0 & 0 & 1 & 0 & 0 & 0 \\
0 & 0 & 0 & 0 & 0 & 1 & 0 & 0 & 0 \\
0 & 0 & 0 & 0 & 0 & 1 & 0 & 0 & 0 \\
0 & 0 & 0 & 0 & 0 & 1 & 0 & 0 & 0 \\
0 & 0 & 0 & 0 & 0 & 0 & 1 & 0 & 0 \\
0 & 0 & 0 & 0 & 0 & 0 & 1 & 0 & 0 \\
0 & 0 & 0 & 0 & 0 & 0 & 1 & 0 & 0 \\
0 & 0 & 0 & 0 & 0 & 0 & 1 & 0 & 0 \\
0 & 0 & 0 & 0 & 0 & 0 & 0 & 1 & 0 \\
0 & 0 & 0 & 0 & 0 & 0 & 0 & 1 & 0 \\
0 & 0 & 0 & 0 & 0 & 0 & 0 & 1 & 0 \\
0 & 0 & 0 & 0 & 0 & 0 & 0 & 1 & 0 \\
0 & 0 & 0 & 0 & 0 & 0 & 0 & 0 & 1 \\
0 & 0 & 0 & 0 & 0 & 0 & 0 & 0 & 1 \\
0 & 0 & 0 & 0 & 0 & 0 & 0 & 0 & 1 \\
0 & 0 & 0 & 0 & 0 & 0 & 0 & 0 & 1 \\
0 & 0 & 0 & 0 & 0 & 0 & 0 & 0 & 1 \\
0 & 0 & 0 & 0 & 0 & 0 & 0 & 0 & 1 \\
0 & 0 & 0 & 0 & 0 & 0 & 0 & 0 & 1 \\
0 & 0 & 0 & 0 & 0 & 0 & 0 & 0 & 1
\end{pmatrix}
$$

SSUNCORRECTED := SS2 ·DBSUM

SSUNCORRECTED = (16613.52　0　16686.04　18121.04　18258.54　18256.58　18338.58　19931.08　20105.83)

To see where we are at lets condense these steps:

$$
\left[\overrightarrow{(\text{SS1} \cdot \text{SS1}) \cdot N^{-1}} \right] \cdot \text{DBSUM} = (\, 16613.52 \quad 0 \quad 16686.04 \quad 18121.04 \quad 18258.54 \quad 18256.58 \quad 18338.58 \quad 19931.08 \quad 20105.83 \,)
$$

Now lets add the Grand Sum of Squares since all the other needed
values are squared:

$\overline{\left[(SS1 \cdot SS1) \cdot N^{-1}\right]} \cdot DBSUM + SStotal1 = (16613.52 \quad 20805 \quad 16686.04 \quad 18121.04 \quad 18258.54 \quad 18256.58 \quad 18338.58 \quad 19931.08 \quad 20105.83)$

$$SStotal := \overline{\left[(SS1 \cdot SS1) \cdot N^{-1}\right]} \cdot DBSUM + SStotal1$$

I have transposed SStotal so we can see the entire solution.

$$SStotal^{T} = \begin{pmatrix} 16613.52 \\ 20805 \\ 16686.04 \\ 18121.04 \\ 18258.54 \\ 18256.58 \\ 18338.58 \\ 19931.08 \\ 20105.83 \end{pmatrix}$$

Now we have the Matrix that holds all the seed values for finishing the computation of the F values.

The next step is to subtract the Correction Factor 16613.52 from all of these values. This is accomplished by multiplying SStotal to the following Matrix: NOTE: These Correction Factor Matrices are valid for all solutions of this 3-factor Statistic, no matter the size of the Data Matrix A.

$$DBCORRFACT := \begin{pmatrix} -1 & -1 & -1 & -1 & -1 & -1 & -1 & -1 & 0 \\ 1 & 0 & 0 & 0 & 0 & 0 & 0 & 0 & 0 \\ 0 & 1 & 0 & 0 & 0 & 0 & 0 & 0 & 0 \\ 0 & 0 & 1 & 0 & 0 & 0 & 0 & 0 & 0 \\ 0 & 0 & 0 & 1 & 0 & 0 & 0 & 0 & 0 \\ 0 & 0 & 0 & 0 & 1 & 0 & 0 & 0 & 0 \\ 0 & 0 & 0 & 0 & 0 & 1 & 0 & 0 & 0 \\ 0 & 0 & 0 & 0 & 0 & 0 & 1 & 0 & 0 \\ 0 & 0 & 0 & 0 & 0 & 0 & 0 & 1 & 0 \end{pmatrix}$$

$$\text{SStotal} \cdot \begin{pmatrix} -1 & -1 & -1 & -1 & -1 & -1 & -1 & -1 & 0 \\ 1 & 0 & 0 & 0 & 0 & 0 & 0 & 0 & 0 \\ 0 & 1 & 0 & 0 & 0 & 0 & 0 & 0 & 0 \\ 0 & 0 & 1 & 0 & 0 & 0 & 0 & 0 & 0 \\ 0 & 0 & 0 & 1 & 0 & 0 & 0 & 0 & 0 \\ 0 & 0 & 0 & 0 & 1 & 0 & 0 & 0 & 0 \\ 0 & 0 & 0 & 0 & 0 & 1 & 0 & 0 & 0 \\ 0 & 0 & 0 & 0 & 0 & 0 & 1 & 0 & 0 \\ 0 & 0 & 0 & 0 & 0 & 0 & 0 & 1 & 0 \end{pmatrix} = (4191.48 \quad 72.52 \quad 1507.52 \quad 1645.02 \quad 1643.06 \quad 1725.06 \quad 3317.56 \quad 3492.31 \quad 0)$$

$$\left(\overrightarrow{\left[(\text{SS1} \cdot \text{SS1}) \cdot \text{N}^{-1} \right]} \cdot \text{DBSUM} + \text{SStotal1} \right) \cdot \text{DBCORRFACT} = (4191.48 \quad 72.52 \quad 1507.52 \quad 1645.02 \quad 1643.06 \quad 1725.06 \quad 3317.56 \quad 3492.31 \quad 0)$$

Transposing so we can get the total solution on a single page
we get:

$$\left[\left(\overrightarrow{\left[(\text{SS1} \cdot \text{SS1}) \cdot \text{N}^{-1} \right]} \cdot \text{DBSUM} + \text{SStotal1} \right) \cdot \text{DBCORRFACT} \right]^{T} = \begin{pmatrix} 4191.48 \\ 72.52 \\ 1507.52 \\ 1645.02 \\ 1643.06 \\ 1725.06 \\ 3317.56 \\ 3492.31 \\ 0 \end{pmatrix}$$

Now we have subtracted the Correction Factor from each of the
terms leaving a zero space at the end for the error value which
will be computed after all the other values are completed.

The next step is to subtract the other various Correction Factors
from their SS values.

DB12 represents the values for SS1x2, SS1x3 AND SS2x3. DB11
represents the completed computations SS1, SS2 And SS3 and SSall
which do not need to be changed.

$$DB11 := \begin{pmatrix} 1 & 0 & 0 & 0 & 0 & 0 & 0 & 0 & 0 \\ 0 & 1 & 0 & 0 & 0 & 0 & 0 & 0 & 0 \\ 0 & 0 & 1 & 0 & 0 & 0 & 0 & 0 & 0 \\ 0 & 0 & 0 & 1 & 0 & 0 & 0 & 0 & 0 \\ 0 & 0 & 0 & 0 & 0 & 0 & 0 & 0 & 0 \\ 0 & 0 & 0 & 0 & 0 & 0 & 0 & 0 & 0 \\ 0 & 0 & 0 & 0 & 0 & 0 & 0 & 0 & 0 \\ 0 & 0 & 0 & 0 & 0 & 0 & 0 & 1 & 0 \\ 0 & 0 & 0 & 0 & 0 & 0 & 0 & 0 & 0 \end{pmatrix} \qquad DB12 := \begin{pmatrix} 0 & 0 & 0 & 0 & 0 & 0 & 0 & 0 & 0 \\ 0 & 0 & 0 & 0 & -1 & -1 & 0 & 0 & 0 \\ 0 & 0 & 0 & 0 & -1 & 0 & -1 & 0 & 0 \\ 0 & 0 & 0 & 0 & 0 & -1 & -1 & 0 & 0 \\ 0 & 0 & 0 & 0 & 1 & 0 & 0 & 0 & 0 \\ 0 & 0 & 0 & 0 & 0 & 1 & 0 & 0 & 0 \\ 0 & 0 & 0 & 0 & 0 & 0 & 1 & 0 & 0 \\ 0 & 0 & 0 & 0 & 0 & 0 & 0 & 0 & 0 \\ 0 & 0 & 0 & 0 & 0 & 0 & 0 & 0 & 0 \end{pmatrix}$$

Hmmmmm . . . no numbers here overlap, so perhaps it is possible to add the two Matrices and compute the values of the two separate matrices in one operation instead of two operations.

$$DB11 + DB12 = \begin{pmatrix} 1 & 0 & 0 & 0 & 0 & 0 & 0 & 0 & 0 \\ 0 & 1 & 0 & 0 & -1 & -1 & 0 & 0 & 0 \\ 0 & 0 & 1 & 0 & -1 & 0 & -1 & 0 & 0 \\ 0 & 0 & 0 & 1 & 0 & -1 & -1 & 0 & 0 \\ 0 & 0 & 0 & 0 & 1 & 0 & 0 & 0 & 0 \\ 0 & 0 & 0 & 0 & 0 & 1 & 0 & 0 & 0 \\ 0 & 0 & 0 & 0 & 0 & 0 & 1 & 0 & 0 \\ 0 & 0 & 0 & 0 & 0 & 0 & 0 & 1 & 0 \\ 0 & 0 & 0 & 0 & 0 & 0 & 0 & 0 & 0 \end{pmatrix}$$

$$\left(\overrightarrow{\left[(SS1 \cdot SS1) \cdot N^{-1}\right] \cdot DBSUM + SStotal1}\right) \cdot DBCORRFACT \cdot (DB11 + DB12) = (4191.48 \quad 72.52 \quad 1507.52 \quad 1645.02 \quad 63.02 \quad 7.52 \quad 165.02 \quad 3492.31 \quad 0)$$

Or to get the solution on a single page we write:

$$\left(\overrightarrow{\left[(SS1 \cdot SS1) \cdot N^{-1}\right] \cdot DBSUM + SStotal1}\right) \cdot DBCORRFACT \cdot (DB11 + DB12)^{T} = \begin{pmatrix} 4191.48 \\ 72.52 \\ 1507.52 \\ 1645.02 \\ 63.02 \\ 7.52 \\ 165.02 \\ 3492.31 \\ 0 \end{pmatrix}$$

Now we multiply by DB13 to get SS1x2x3

$$DB13 := \begin{pmatrix} 1 & 0 & 0 & 0 & 0 & 0 & 0 & 0 & 0 \\ 0 & 1 & 0 & 0 & 0 & 0 & 0 & -1 & 0 \\ 0 & 0 & 1 & 0 & 0 & 0 & 0 & -1 & 0 \\ 0 & 0 & 0 & 1 & 0 & 0 & 0 & -1 & 0 \\ 0 & 0 & 0 & 0 & 1 & 0 & 0 & -1 & 0 \\ 0 & 0 & 0 & 0 & 0 & 1 & 0 & -1 & 0 \\ 0 & 0 & 0 & 0 & 0 & 0 & 1 & -1 & 0 \\ 0 & 0 & 0 & 0 & 0 & 0 & 0 & 1 & 0 \\ 0 & 0 & 0 & 0 & 0 & 0 & 0 & 0 & 0 \end{pmatrix} \qquad DB14 := \begin{pmatrix} 1 & 0 & 0 & 0 & 0 & 0 & 0 & 0 & 1 \\ 0 & 1 & 0 & 0 & 0 & 0 & 0 & 0 & -1 \\ 0 & 0 & 1 & 0 & 0 & 0 & 0 & 0 & -1 \\ 0 & 0 & 0 & 1 & 0 & 0 & 0 & 0 & -1 \\ 0 & 0 & 0 & 0 & 1 & 0 & 0 & 0 & -1 \\ 0 & 0 & 0 & 0 & 0 & 1 & 0 & 0 & -1 \\ 0 & 0 & 0 & 0 & 0 & 0 & 1 & 0 & -1 \\ 0 & 0 & 0 & 0 & 0 & 0 & 0 & 1 & -1 \\ 0 & 0 & 0 & 0 & 0 & 0 & 0 & 0 & 0 \end{pmatrix}$$

$$\left(\overrightarrow{\left[\overrightarrow{(SS1 \cdot SS1) \cdot N^{-1}} \right] \cdot DBSUM + SStotal1} \right) \cdot DBCORRFACT \cdot (DB11 + DB12) \cdot DB13 = (4191.48 \quad 72.52 \quad 1507.52 \quad 1645.02 \quad 63.02 \quad 7.52 \quad 165.02 \quad 31.69 \quad 0)$$

Or, to get the solution to fit on the page, we transpose and get:

$$\left(\overrightarrow{\left[\overrightarrow{(SS1 \cdot SS1) \cdot N^{-1}} \right] \cdot DBSUM + SStotal1} \right) \cdot DBCORRFACT \cdot (DB11 + DB12) \cdot DB13 \quad \overset{T}{=} \begin{pmatrix} 4191.48 \\ 72.52 \\ 1507.52 \\ 1645.02 \\ 63.02 \\ 7.52 \\ 165.02 \\ 31.69 \\ 0 \end{pmatrix}$$

And finally we multiply by DB14 to compute the error term which
will fit in the last column.

$$SS := \left(\overrightarrow{\left[\overrightarrow{(SS1 \cdot SS1) \cdot N^{-1}} \right] \cdot DBSUM + SStotal1} \right) \cdot DBCORRFACT \cdot (DB11 + DB12) \cdot DB13 \cdot DB14$$

$$\left(\overrightarrow{\left[\overrightarrow{(SS1 \cdot SS1) \cdot N^{-1}} \right] \cdot DBSUM + SStotal1} \right) \cdot DBCORRFACT \cdot (DB11 + DB12) \cdot DB13 \cdot DB14 \quad \overset{T}{=} \begin{pmatrix} 4191.48 \\ 72.52 \\ 1507.52 \\ 1645.02 \\ 63.02 \\ 7.52 \\ 165.02 \\ 31.69 \\ 699.17 \end{pmatrix}$$

From the problem above we computed df and ERROR as

$m := 48$ $E1 := 2$ $E2 := 2$ $E3 := 2$

$dfSS1 := E1 - 1$

$dfSSt := m - 1$

$dfSS2 := E2 - 1$ $dfSS3 := E3 - 1$

$dfSS1x2 := dfSS1 \cdot dfSS2$

$dfSS1x3 := dfSS1 \cdot dfSS3$

$dfSS2x3 := dfSS2 \cdot dfSS3$

$dfSS1x2x3 := dfSS1 \cdot dfSS2 \cdot dfSS3$

$$df := \begin{pmatrix} m - 1 \\ E1 - 1 \\ E2 - 1 \\ E3 - 1 \\ dfSS1 \cdot dfSS2 \\ dfSS1 \cdot dfSS3 \\ dfSS2 \cdot dfSS3 \\ dfSS1 \cdot dfSS2 \cdot dfSS3 \\ dfSSt - dfSS1 - dfSS2 - dfSS3 - dfSS1x2 - dfSS1x3 - dfSS2x3 - dfSS1x2x3 \end{pmatrix}$$

$$df := \begin{pmatrix} 47 \\ 1 \\ 1 \\ 1 \\ 1 \\ 1 \\ 1 \\ 1 \\ 40 \end{pmatrix} \qquad ERROR^{-1} := \begin{pmatrix} \dfrac{40}{699} \\ .057 \\ .057 \\ .057 \\ .057 \\ .057 \\ .057 \\ .057 \\ .057 \end{pmatrix}^{\blacksquare}$$

The next computation just puts the output as a Column Matrix to help compute F, although F can be computed as a Row Matrix if we so wish.

$$SS := \left[\left[\overrightarrow{(SS1 \cdot SS1) \cdot N^{-1}} \cdot DBSUM + SStotal1\right) \cdot DBCORRFACT \cdot (DB11 + DB12) \cdot DB13 \cdot DB14\right]^T$$

$$F := \overrightarrow{\left(SS \cdot df^{-1} \cdot ERROR\right)}$$

Or
$$F := \begin{pmatrix} 4191.479 \\ 72.521 \\ 1507.521 \\ 1645.021 \\ 63.021 \\ 7.521 \\ 165.021 \\ 31.687 \\ 699.167 \end{pmatrix} \cdot \begin{pmatrix} 0.021 \\ 1 \\ 1 \\ 1 \\ 1 \\ 1 \\ 1 \\ 1 \\ 0.025 \end{pmatrix} \cdot \overrightarrow{\begin{pmatrix} 0.057 \\ 0.057 \\ 0.057 \\ 0.057 \\ 0.057 \\ 0.057 \\ 0.057 \\ 0.057 \\ 0.057 \end{pmatrix}}$$

$$F = \begin{pmatrix} 5.02 \\ 4.13 \\ 85.93 \\ 93.77 \\ 3.59 \\ 0.43 \\ 9.41 \\ 1.81 \\ 1 \end{pmatrix} \qquad p := \begin{pmatrix} 0 \\ 0 \\ .001 \\ .001 \\ ns \\ ns \\ .001 \\ ns \\ ns \end{pmatrix}$$

probabilities

SS2x3 < .001

SS3 < .001

SS2 < .001

Therefore it is concluded that experimental condition 2 and experimental condition 3 affect the rate of errors and the effects of experimental condition 2 with condition 3 are interactive by factors 1000 times greater than by chance. SS2x3, SS1 and SS1x2x3 are non-significant (ns).

TREATMENT BY LEVELS DESIGN

In this method, we have 2 variables, Levels (first experimental condition). This is like high intelligence, normal and low intelligence being the different levels being treated by experimental condition 2. We see if there is significance for differing intelligence, by what they were trained in and is there any interactive effect between the two.

The data obtained in the experiment is given in Matrix A. The other values are the constants we will need to solve the Statistic. I am trying a different way to solve the Statistic All at Once. This step by step example will have 3 groups, but in the All at Once computation, I will solve for four groups so we can see how the Data Base Matrices change.

$$
A := \begin{pmatrix}
77 & 96 & 88 & 72 & 79 & 71 & 94 & 94 & 71 \\
78 & 94 & 73 & 75 & 89 & 85 & 84 & 81 & 85 \\
96 & 93 & 97 & 99 & 95 & 93 & 87 & 82 & 82 \\
89 & 81 & 90 & 85 & 83 & 72 & 70 & 71 & 68 \\
89 & 88 & 100 & 82 & 88 & 91 & 77 & 70 & 67 \\
84 & 87 & 70 & 90 & 88 & 76 & 76 & 68 & 62
\end{pmatrix}
$$

CONSTANTS:

$$
NINE1 := \begin{pmatrix}
1 \\
1 \\
1 \\
1 \\
1 \\
1 \\
1 \\
1 \\
1
\end{pmatrix}
$$

ONE6 := $(1\ 1\ 1\ 1\ 1\ 1)$

$$DB1 := \begin{pmatrix} 1 & 0 & 0 \\ 1 & 0 & 0 \\ 1 & 0 & 0 \\ 0 & 1 & 0 \\ 0 & 1 & 0 \\ 0 & 1 & 0 \\ 0 & 0 & 1 \\ 0 & 0 & 1 \\ 0 & 0 & 1 \end{pmatrix} \quad DB3 := \begin{pmatrix} 1 & 0 & 0 \\ 0 & 1 & 0 \\ 0 & 0 & 1 \\ 1 & 0 & 0 \\ 0 & 1 & 0 \\ 0 & 0 & 1 \\ 1 & 0 & 0 \\ 0 & 1 & 0 \\ 0 & 0 & 1 \end{pmatrix} \quad ZERO9 := \begin{pmatrix} 0 \\ 0 \\ 0 \\ 0 \\ 0 \\ 0 \\ 0 \\ 0 \\ 0 \end{pmatrix} \quad NINE9 := \begin{pmatrix} 1 & 0 & 0 & 0 & 0 & 0 & 0 & 0 & 0 \\ 0 & 1 & 0 & 0 & 0 & 0 & 0 & 0 & 0 \\ 0 & 0 & 1 & 0 & 0 & 0 & 0 & 0 & 0 \\ 0 & 0 & 0 & 1 & 0 & 0 & 0 & 0 & 0 \\ 0 & 0 & 0 & 0 & 1 & 0 & 0 & 0 & 0 \\ 0 & 0 & 0 & 0 & 0 & 1 & 0 & 0 & 0 \\ 0 & 0 & 0 & 0 & 0 & 0 & 1 & 0 & 0 \\ 0 & 0 & 0 & 0 & 0 & 0 & 0 & 1 & 0 \\ 0 & 0 & 0 & 0 & 0 & 0 & 0 & 0 & 1 \end{pmatrix}$$

STEP 2: Sum the scores in each group: SUM1. Then add the values
in SUM1 to get the Grand Sum and find the Correction Factor.
N = # elements in the Matrix N = 6 x 9 = 54.

SUM1 := $1_{1,6} \cdot A_{6,9}\ \blacksquare$ Nt := 54   NA := (NINE1 ZERO9 DB1 DB3 NINE9)

SUM1 := ONE6 A

$$SUM1 := (1\ \ 1\ \ 1\ \ 1\ \ 1\ \ 1) \cdot \begin{pmatrix} 77 & 96 & 88 & 72 & 79 & 71 & 94 & 94 & 71 \\ 78 & 94 & 73 & 75 & 89 & 85 & 84 & 81 & 85 \\ 96 & 93 & 97 & 99 & 95 & 93 & 87 & 82 & 82 \\ 89 & 81 & 90 & 85 & 83 & 72 & 70 & 71 & 68 \\ 89 & 88 & 100 & 82 & 88 & 91 & 77 & 70 & 67 \\ 84 & 87 & 70 & 90 & 88 & 76 & 76 & 68 & 62 \end{pmatrix}$$

SUM1 = $(513\ \ 539\ \ 518\ \ 503\ \ 522\ \ 488\ \ 488\ \ 466\ \ 435)$

Now we compute the Grand Sum:

GRANDSUM := ONE6 A · NINE1

$$\text{GRANDSUM} := (1\ 1\ 1\ 1\ 1\ 1) \cdot \begin{pmatrix} 77 & 96 & 88 & 72 & 79 & 71 & 94 & 94 & 71 \\ 78 & 94 & 73 & 75 & 89 & 85 & 84 & 81 & 85 \\ 96 & 93 & 97 & 99 & 95 & 93 & 87 & 82 & 82 \\ 89 & 81 & 90 & 85 & 83 & 72 & 70 & 71 & 68 \\ 89 & 88 & 100 & 82 & 88 & 91 & 77 & 70 & 67 \\ 84 & 87 & 70 & 90 & 88 & 76 & 76 & 68 & 62 \end{pmatrix} \cdot \begin{pmatrix} 1 \\ 1 \\ 1 \\ 1 \\ 1 \\ 1 \\ 1 \\ 1 \\ 1 \end{pmatrix}$$

GRANDSUM = 4472

Now we compute the Correction Factor:

$$\text{CORRFACT} := \left(\frac{1}{\text{Nt}}\right) \cdot \text{GRANDSUM}^2$$

CORRFACT = 370347.9

STEP 3: Now we find the Grand Sum of Squares GSQ by squaring each number in the Matrix and summing the values.

$\text{PROGRAM} := \blacksquare[1]_{1,6}([A]_{6,9} \ddot{A} [A]_{6,9})\,[1]_{9,1}$

$\text{GSQ} := \text{ONE6}\left(\overrightarrow{A \cdot A}\right)\text{NINE1}$

$\text{GSQ} := \text{ONE6}\left(\overrightarrow{A \cdot A}\right)\text{NINE1}$

$$\text{GSQ} := (1\ 1\ 1\ 1\ 1\ 1) \cdot \left[\begin{pmatrix} 77 & 96 & 88 & 72 & 79 & 71 & 94 & 94 & 71 \\ 78 & 94 & 73 & 75 & 89 & 85 & 84 & 81 & 85 \\ 96 & 93 & 97 & 99 & 95 & 93 & 87 & 82 & 82 \\ 89 & 81 & 90 & 85 & 83 & 72 & 70 & 71 & 68 \\ 89 & 88 & 100 & 82 & 88 & 91 & 77 & 70 & 67 \\ 84 & 87 & 70 & 90 & 88 & 76 & 76 & 68 & 62 \end{pmatrix} \cdot \overrightarrow{\begin{pmatrix} 77 & 96 & 88 & 72 & 79 & 71 & 94 & 94 & 71 \\ 78 & 94 & 73 & 75 & 89 & 85 & 84 & 81 & 85 \\ 96 & 93 & 97 & 99 & 95 & 93 & 87 & 82 & 82 \\ 89 & 81 & 90 & 85 & 83 & 72 & 70 & 71 & 68 \\ 89 & 88 & 100 & 82 & 88 & 91 & 77 & 70 & 67 \\ 84 & 87 & 70 & 90 & 88 & 76 & 76 & 68 & 62 \end{pmatrix}} \right] \cdot \begin{pmatrix} 1 \\ 1 \\ 1 \\ 1 \\ 1 \\ 1 \\ 1 \\ 1 \\ 1 \end{pmatrix}$$

GSQ = 375262

STEP 4: Computation of the Total Sum of Squares SSt. Nt = 54.

SSt := GSQ – CORRFACT

$$SSt := ONE6\overrightarrow{\left(A \cdot A\right)}NINE1 - \left(\frac{1}{Nt}\right) \cdot GRANDSUM^2$$

SSt = 4914.1

STEP 5: Computation of the effect of Levels (the overall
performance of subjects from 3 different environmental, social
or experimental levels) disregarding the second experimental
condition. N1 = Nt/ # columns in DB1 = 54/3 = 18.

N1 := 18

PROGRAM: SS1 = (1/N1)x([SUM1]$_{1,9}$ x [DB1]$_{9,3}$)^2 - CORRFACT

$$SS1 := \left(\frac{1}{N1}\right) \cdot \left[(SUM1 \cdot DB1) \cdot (SUM1 \cdot DB1)^T\right] - CORRFACT$$

Of course, when we square the computation below, the square must
be in the form of C*C transpose. But the math formula won't fit
on a single page. But we can try it the following way solving
for SS1a to give us an idea of the multiplication involved.

$$SS1a := \left(\frac{1}{N1}\right) \cdot \left[(SUM1 \cdot DB1) \cdot \left[(513 \quad 539 \quad 518 \quad 503 \quad 522 \quad 488 \quad 488 \quad 466 \quad 435) \cdot \begin{pmatrix} 1 & 0 & 0 \\ 1 & 0 & 0 \\ 1 & 0 & 0 \\ 0 & 1 & 0 \\ 0 & 1 & 0 \\ 0 & 1 & 0 \\ 0 & 0 & 1 \\ 0 & 0 & 1 \\ 0 & 0 & 1 \end{pmatrix}^T\right]\right] - CORRFACT$$

SS1 = 951.6

SS1a = 951.6

They match.

STEP 6:   Computation of the treatment effects (the experimental conditions on each of the Levels tested) disregarding the Levels dimension. SS2. N2 = Nt/ # columns in DB3. N2 = 18.

$N3 := 18$

$$\text{PROGRAM} := \left(\frac{1}{N3}\right) \cdot \left[\left[(\text{SUM1} \cdot \text{DB3}) \cdot (\text{SUM1} \cdot \text{DB3})^{T}\right] - \text{CORRFACT}\right]$$

$$\text{SS2} := \left(\frac{1}{N3}\right) \cdot \left[\left[(\text{SUM1} \cdot \text{DB3}) \cdot (\text{SUM1} \cdot \text{DB3})^{T}\right] - \text{CORRFACT}\right]$$

$$\text{SS2a} := \left(\frac{1}{N3}\right) \cdot \left[(\text{SUM1} \cdot \text{DB3}) \cdot \left[(513\ 539\ 518\ 503\ 522\ 488\ 488\ 466\ 435) \cdot \begin{pmatrix} 1 & 0 & 0 \\ 0 & 1 & 0 \\ 0 & 0 & 1 \\ 1 & 0 & 0 \\ 0 & 1 & 0 \\ 0 & 0 & 1 \\ 1 & 0 & 0 \\ 0 & 1 & 0 \\ 0 & 0 & 1 \end{pmatrix}^{T}\right]\right] - \text{CORRFACT}$$

$$\text{SS2} := \left(\frac{1}{18}\right) \cdot \left[(\text{SUM1} \cdot \text{DB3}) \cdot (\text{SUM1} \cdot \text{DB3})^{T}\right] - \text{CORRFACT}$$

SS2a = 220.3

SS2 = 220.3

STEP 7:   Computation of the interactive effects of the two experimental conditions SS1x2. N1x2 = Nt/# elements in SUM1 = 54/9 = 6.

$N1x2 := 6$

PROGRAM: SS1x2 = (1/6)*([SUM1]$_{1,9}$ x [SUM1]$_{1,9}^{T}$)-CORRFACT-SS1-SS2

$$SS1x2 := \left(\frac{1}{6}\right) \cdot \left[(SUM1)(SUM1)^{T}\right] - CORRFACT - SS1 - SS2$$

$$SS1x2 = 176.3$$

$$SS1x2a := \left(\frac{1}{6}\right) \cdot \left[(513\ 539\ 518\ 503\ 522\ 488\ 488\ 466\ 435) \cdot \begin{pmatrix} 513 \\ 539 \\ 518 \\ 503 \\ 522 \\ 488 \\ 488 \\ 466 \\ 435 \end{pmatrix}\right] - CORRFACT - SS1 - SS2$$

$$SS1x2a = 176.3$$

STEP 8: Computation of the ERROR term.

$$SSerror := SSt - SS1 - SS2 - SS1x2$$

$$SSerror = 3566$$

$$SS := \begin{pmatrix} SSt \\ SS1 \\ SS2 \\ SS1x2 \\ SSerror \end{pmatrix} \qquad SS = \begin{pmatrix} 4914.1 \\ 951.6 \\ 220.3 \\ 176.3 \\ 3566 \end{pmatrix}$$

$$m := 54 \qquad ExpCon1 := 3 \qquad ExpCon2 := 3$$

$$df := \begin{bmatrix} m - 1 \\ ExpCon1 - 1 \\ ExpCon2 - 1 \\ (ExpCon1 - 1) \cdot (ExpCon2 - 1) \\ (m - 1) - (ExpCon1 - 1) - (ExpCon2 - 1) - (ExpCon1 - 1) \cdot (ExpCon2 - 1) \end{bmatrix}$$

$$df = \begin{pmatrix} 53 \\ 2 \\ 2 \\ 4 \\ 45 \end{pmatrix}$$

$$\text{ERROR1} := \frac{\text{SSerror}^{\blacksquare}}{\text{dferror}}$$

$$\text{ERROR1} := \frac{3566}{45}$$

$$\text{ERROR} := \begin{pmatrix} \dfrac{3566}{45} \\ 79.2 \\ 79.2 \\ 79.2 \\ 79.2 \end{pmatrix}$$

$$F := \overrightarrow{\left(SS \cdot df^{-1} \cdot \text{ERROR}^{-1}\right)}$$

$$F = \begin{pmatrix} 1.2 \\ 6 \\ 1.4 \\ 0.6 \\ 1 \end{pmatrix} \qquad df = \begin{pmatrix} 53 \\ 2 \\ 2 \\ 4 \\ 45 \end{pmatrix} \qquad p := \begin{pmatrix} 0 \\ .01 \\ ns \\ ns \\ -\blacksquare \end{pmatrix}$$

INTERPRETATION: Although there is significance in the First Experimental condition of different levels that would happen by chance 1 in 100 times, the treatment (second experimental condition) of these levels have no significance and there is no significant interaction between conditions one and two.

ALL IN ONE EQUATION:

$$A1 := \begin{pmatrix} 74 & 94 & 87 & 70 & 76 & 70 & 92 & 91 & 70 & 74 & 94 & 87 \\ 79 & 96 & 76 & 78 & 91 & 86 & 86 & 84 & 86 & 79 & 96 & 76 \\ 94 & 92 & 94 & 96 & 94 & 91 & 84 & 80 & 81 & 94 & 92 & 94 \\ 90 & 84 & 92 & 86 & 86 & 74 & 71 & 73 & 71 & 90 & 84 & 92 \\ 88 & 85 & 98 & 80 & 87 & 88 & 74 & 69 & 65 & 88 & 85 & 98 \\ 87 & 89 & 71 & 92 & 91 & 77 & 78 & 71 & 63 & 87 & 89 & 71 \end{pmatrix}$$

Since we changed the size of the Matrix, we must change the size of the DB Matrices.

$$NINE9 := \begin{pmatrix} 1 & 0 & 0 & 0 & 0 & 0 & 0 & 0 & 0 \\ 0 & 1 & 0 & 0 & 0 & 0 & 0 & 0 & 0 \\ 0 & 0 & 1 & 0 & 0 & 0 & 0 & 0 & 0 \\ 0 & 0 & 0 & 1 & 0 & 0 & 0 & 0 & 0 \\ 0 & 0 & 0 & 0 & 1 & 0 & 0 & 0 & 0 \\ 0 & 0 & 0 & 0 & 0 & 1 & 0 & 0 & 0 \\ 0 & 0 & 0 & 0 & 0 & 0 & 1 & 0 & 0 \\ 0 & 0 & 0 & 0 & 0 & 0 & 0 & 1 & 0 \\ 0 & 0 & 0 & 0 & 0 & 0 & 0 & 0 & 1 \end{pmatrix}$$

$$DB1a := \begin{pmatrix} 1 & 0 & 0 & 0 \\ 1 & 0 & 0 & 0 \\ 1 & 0 & 0 & 0 \\ 0 & 1 & 0 & 0 \\ 0 & 1 & 0 & 0 \\ 0 & 1 & 0 & 0 \\ 0 & 0 & 1 & 0 \\ 0 & 0 & 1 & 0 \\ 0 & 0 & 1 & 0 \\ 0 & 0 & 0 & 1 \\ 0 & 0 & 0 & 1 \\ 0 & 0 & 0 & 1 \end{pmatrix}$$

$$DB3a := \begin{pmatrix} 1 & 0 & 0 & 0 \\ 0 & 1 & 0 & 0 \\ 0 & 0 & 1 & 0 \\ 0 & 0 & 0 & 1 \\ 1 & 0 & 0 & 0 \\ 0 & 1 & 0 & 0 \\ 0 & 0 & 1 & 0 \\ 0 & 0 & 0 & 1 \\ 1 & 0 & 0 & 0 \\ 0 & 1 & 0 & 0 \\ 0 & 0 & 1 & 0 \\ 0 & 0 & 0 & 1 \end{pmatrix}$$

$$TWELVE1 := \begin{pmatrix} 1 \\ 1 \\ 1 \\ 1 \\ 1 \\ 1 \\ 1 \\ 1 \\ 1 \\ 1 \\ 1 \\ 1 \end{pmatrix}$$

$$ZERO12 := \begin{pmatrix} 0 \\ 0 \\ 0 \\ 0 \\ 0 \\ 0 \\ 0 \\ 0 \\ 0 \\ 0 \\ 0 \\ 0 \end{pmatrix}$$

$$TWELVE12 := \begin{pmatrix} 1 & 0 & 0 & 0 & 0 & 0 & 0 & 0 & 0 & 0 & 0 & 0 \\ 0 & 1 & 0 & 0 & 0 & 0 & 0 & 0 & 0 & 0 & 0 & 0 \\ 0 & 0 & 1 & 0 & 0 & 0 & 0 & 0 & 0 & 0 & 0 & 0 \\ 0 & 0 & 0 & 1 & 0 & 0 & 0 & 0 & 0 & 0 & 0 & 0 \\ 0 & 0 & 0 & 0 & 1 & 0 & 0 & 0 & 0 & 0 & 0 & 0 \\ 0 & 0 & 0 & 0 & 0 & 1 & 0 & 0 & 0 & 0 & 0 & 0 \\ 0 & 0 & 0 & 0 & 0 & 0 & 1 & 0 & 0 & 0 & 0 & 0 \\ 0 & 0 & 0 & 0 & 0 & 0 & 0 & 1 & 0 & 0 & 0 & 0 \\ 0 & 0 & 0 & 0 & 0 & 0 & 0 & 0 & 1 & 0 & 0 & 0 \\ 0 & 0 & 0 & 0 & 0 & 0 & 0 & 0 & 0 & 1 & 0 & 0 \\ 0 & 0 & 0 & 0 & 0 & 0 & 0 & 0 & 0 & 0 & 1 & 0 \\ 0 & 0 & 0 & 0 & 0 & 0 & 0 & 0 & 0 & 0 & 0 & 1 \end{pmatrix}$$

SUM1 := ONE6a

SUM1 = (513 539 518 503 522 488 488 466 435 512 540 518)

NA := (TWELVE1 ZERO12 DB1a DB3a TWELVE12)

$$NA = \begin{bmatrix} \begin{pmatrix} 1 \\ 1 \\ 1 \\ 1 \\ 1 \\ 1 \\ 1 \\ 1 \\ 1 \\ 1 \\ 1 \\ 1 \end{pmatrix} \begin{pmatrix} 0 \\ 0 \\ 0 \\ 0 \\ 0 \\ 0 \\ 0 \\ 0 \\ 0 \\ 0 \\ 0 \\ 0 \end{pmatrix} \begin{pmatrix} 1 & 0 & 0 & 0 \\ 1 & 0 & 0 & 0 \\ 1 & 0 & 0 & 0 \\ 0 & 1 & 0 & 0 \\ 0 & 1 & 0 & 0 \\ 0 & 1 & 0 & 0 \\ 0 & 0 & 1 & 0 \\ 0 & 0 & 1 & 0 \\ 0 & 0 & 1 & 0 \\ 0 & 0 & 0 & 1 \\ 0 & 0 & 0 & 1 \\ 0 & 0 & 0 & 1 \end{pmatrix} \begin{pmatrix} 1 & 0 & 0 & 0 \\ 0 & 1 & 0 & 0 \\ 0 & 0 & 1 & 0 \\ 0 & 0 & 0 & 1 \\ 1 & 0 & 0 & 0 \\ 0 & 1 & 0 & 0 \\ 0 & 0 & 1 & 0 \\ 0 & 0 & 0 & 1 \\ 1 & 0 & 0 & 0 \\ 0 & 1 & 0 & 0 \\ 0 & 0 & 1 & 0 \\ 0 & 0 & 0 & 1 \end{pmatrix} \begin{pmatrix} 1 & 0 & 0 & 0 & 0 & 0 & 0 & 0 & 0 & 0 & 0 & 0 \\ 0 & 1 & 0 & 0 & 0 & 0 & 0 & 0 & 0 & 0 & 0 & 0 \\ 0 & 0 & 1 & 0 & 0 & 0 & 0 & 0 & 0 & 0 & 0 & 0 \\ 0 & 0 & 0 & 1 & 0 & 0 & 0 & 0 & 0 & 0 & 0 & 0 \\ 0 & 0 & 0 & 0 & 1 & 0 & 0 & 0 & 0 & 0 & 0 & 0 \\ 0 & 0 & 0 & 0 & 0 & 1 & 0 & 0 & 0 & 0 & 0 & 0 \\ 0 & 0 & 0 & 0 & 0 & 0 & 1 & 0 & 0 & 0 & 0 & 0 \\ 0 & 0 & 0 & 0 & 0 & 0 & 0 & 1 & 0 & 0 & 0 & 0 \\ 0 & 0 & 0 & 0 & 0 & 0 & 0 & 0 & 1 & 0 & 0 & 0 \\ 0 & 0 & 0 & 0 & 0 & 0 & 0 & 0 & 0 & 1 & 0 & 0 \\ 0 & 0 & 0 & 0 & 0 & 0 & 0 & 0 & 0 & 0 & 1 & 0 \\ 0 & 0 & 0 & 0 & 0 & 0 & 0 & 0 & 0 & 0 & 0 & 1 \end{pmatrix} \end{bmatrix}$$

Partitioning the Nested Array we get:

DBALL := augment(TWELVE1, ZERO12, DB1a, DB3a, TWELVE12)

$$DBALL = \begin{pmatrix} 1 & 0 & 1 & 0 & 0 & 0 & 1 & 0 & 0 & 0 & 1 & 0 & 0 & 0 & 0 & 0 & 0 & 0 & 0 & 0 & 0 & 0 \\ 1 & 0 & 1 & 0 & 0 & 0 & 0 & 1 & 0 & 0 & 0 & 1 & 0 & 0 & 0 & 0 & 0 & 0 & 0 & 0 & 0 & 0 \\ 1 & 0 & 1 & 0 & 0 & 0 & 0 & 0 & 1 & 0 & 0 & 0 & 1 & 0 & 0 & 0 & 0 & 0 & 0 & 0 & 0 & 0 \\ 1 & 0 & 0 & 1 & 0 & 0 & 0 & 0 & 1 & 0 & 0 & 0 & 1 & 0 & 0 & 0 & 0 & 0 & 0 & 0 & 0 & 0 \\ 1 & 0 & 0 & 1 & 0 & 0 & 1 & 0 & 0 & 0 & 0 & 0 & 0 & 1 & 0 & 0 & 0 & 0 & 0 & 0 & 0 & 0 \\ 1 & 0 & 0 & 1 & 0 & 0 & 0 & 1 & 0 & 0 & 0 & 0 & 0 & 1 & 0 & 0 & 0 & 0 & 0 & 0 & 0 & 0 \\ 1 & 0 & 0 & 0 & 1 & 0 & 0 & 0 & 1 & 0 & 0 & 0 & 0 & 0 & 1 & 0 & 0 & 0 & 0 & 0 & 0 & 0 \\ 1 & 0 & 0 & 0 & 1 & 0 & 0 & 0 & 0 & 1 & 0 & 0 & 0 & 0 & 0 & 1 & 0 & 0 & 0 & 0 & 0 & 0 \\ 1 & 0 & 0 & 0 & 1 & 0 & 1 & 0 & 0 & 0 & 0 & 0 & 0 & 0 & 0 & 0 & 1 & 0 & 0 & 0 & 0 & 0 \\ 1 & 0 & 0 & 0 & 0 & 1 & 0 & 1 & 0 & 0 & 0 & 0 & 0 & 0 & 0 & 0 & 0 & 1 & 0 & 0 & 0 & 0 \\ 1 & 0 & 0 & 0 & 0 & 1 & 0 & 0 & 1 & 0 & 0 & 0 & 0 & 0 & 0 & 0 & 0 & 0 & 0 & 1 & 0 \\ 1 & 0 & 0 & 0 & 0 & 1 & 0 & 0 & 0 & 1 & 0 & 0 & 0 & 0 & 0 & 0 & 0 & 0 & 0 & 0 & 0 & 1 \end{pmatrix}$$

SUM1 = (513 539 518 503 522 488 488 466 435 512 540 518)

STEP 2:2: MULTIPLY DBALL BY SUM1

SUM1·DBALL = (6042 0 1570 1513 1389 1570 1470 1539 1546 1487 513 539 518 503 522 488 488 466 435 512 540 518)

$$N := (72 \ 1 \ 18 \ 18 \ 18 \ 18 \ 18 \ 18 \ 18 \ 18 \ 6 \ 6 \ 6 \ 6 \ 6 \ 6 \ 6 \ 6 \ 6 \ 6 \ 6 \ 6)$$

$$SS3a := SUM1 \cdot DBALl$$

$$SS3 := \overrightarrow{\left(SS3a \cdot SS3a \cdot N^{-1}\right)}$$

SS3 = (507024.5 0 136938.9 127176.1 107184.5 136938.9 120050 131584.5 132784.2 122842.7 43861.5 48420.2 44720.7 42168.2 45414 39690.7 39690.7 36192.7 31537.5 43690.7 48600 44720.7)

We put DBSUM, SUM1 x DBALL and SS3 as transposes on the next
page so we can see all the solutions on a single page.

$$DBSUM := \begin{pmatrix} 1 & 0 & 0 & 0 & 0 \\ 0 & 0 & 0 & 0 & 0 \\ 0 & 0 & 1 & 0 & 0 \\ 0 & 0 & 1 & 0 & 0 \\ 0 & 0 & 1 & 0 & 0 \\ 0 & 0 & 1 & 0 & 0 \\ 0 & 0 & 0 & 1 & 0 \\ 0 & 0 & 0 & 1 & 0 \\ 0 & 0 & 0 & 1 & 0 \\ 0 & 0 & 0 & 1 & 0 \\ 0 & 0 & 0 & 0 & 1 \\ 0 & 0 & 0 & 0 & 1 \\ 0 & 0 & 0 & 0 & 1 \\ 0 & 0 & 0 & 0 & 1 \\ 0 & 0 & 0 & 0 & 1 \\ 0 & 0 & 0 & 0 & 1 \\ 0 & 0 & 0 & 0 & 1 \\ 0 & 0 & 0 & 0 & 1 \\ 0 & 0 & 0 & 0 & 1 \\ 0 & 0 & 0 & 0 & 1 \\ 0 & 0 & 0 & 0 & 1 \\ 0 & 0 & 0 & 0 & 1 \end{pmatrix} \quad (SUM1 \cdot DBALL)^T = \begin{pmatrix} 6042 \\ 0 \\ 1570 \\ 1513 \\ 1389 \\ 1570 \\ 1470 \\ 1539 \\ 1546 \\ 1487 \\ 513 \\ 539 \\ 518 \\ 503 \\ 522 \\ 488 \\ 488 \\ 466 \\ 435 \\ 512 \\ 540 \\ 518 \end{pmatrix} \quad SS3^T = \begin{pmatrix} 507024.5 \\ 0 \\ 136938.9 \\ 127176.1 \\ 107184.5 \\ 136938.9 \\ 120050 \\ 131584.5 \\ 132784.2 \\ 122842.7 \\ 43861.5 \\ 48420.2 \\ 44720.7 \\ 42168.2 \\ 45414 \\ 39690.7 \\ 39690.7 \\ 36192.7 \\ 31537.5 \\ 43690.7 \\ 48600 \\ 44720.7 \end{pmatrix}$$

Using DBSUM we sum the elements in SS3 together as per the the
number of columns in their respective DB Matrices.

$$SS4 := \overrightarrow{\left(SS3a \cdot SS3a \cdot N^{-1}\right)DBSUM}$$

$$SS4 = (507024.5 \ \ 0 \ \ 508238.3 \ \ 507261.4 \ \ 508707.3)$$

Now is the time to add GSQ to the SS4 Matrix:

$$GSQa := ONE6\overrightarrow{\left((a \cdot a)\right)}TWELVE$$

$$GSQa = 513272$$

$$GSQ := (0 \ \ 512750 \ \ 0 \ \ 0 \ \ 0)$$

$$SS5 := \overrightarrow{\left(SS3a \cdot SS3a \cdot N^{-1}\right)DBSUM} + GSQ$$

$$SS5 = (507024.5 \ \ 512750 \ \ 508238.3 \ \ 507261.4 \ \ 508707.3)$$

Now its time to subtract the Correction Factor(s)

$$CF := \begin{pmatrix} -1 & -1 & -1 & -1 & 0 \\ 1 & 0 & 0 & 0 & 0 \\ 0 & 1 & 0 & 0 & 0 \\ 0 & 0 & 1 & 0 & 0 \\ 0 & 0 & 0 & 1 & 0 \end{pmatrix} \quad SS7a := \begin{pmatrix} 1 & 0 & 0 & 0 & 0 \\ 0 & 1 & 0 & -1 & 0 \\ 0 & 0 & 1 & -1 & 0 \\ 0 & 0 & 0 & 1 & 0 \\ 0 & 0 & 0 & 0 & 0 \end{pmatrix} \quad SS8a := \begin{pmatrix} 1 & 0 & 0 & 0 & 1 \\ 0 & 1 & 0 & 0 & -1 \\ 0 & 0 & 1 & 0 & -1 \\ 0 & 0 & 0 & 1 & -1 \\ 0 & 0 & 0 & 0 & 0 \end{pmatrix}$$

CF computes the CORRFACT subtraction, SS7a computes the subtraction of SS1 and SS2 from SS1x2 and SS8a computes the Error term.

Multiplying by CF subtracts the CORRFACT from each term

$$SS6 := \left(\overrightarrow{\left(SS3a \cdot SS3a \cdot N^{-1}\right)DBSUM} + GSQ\right) \cdot CF$$

$$SS6 = (5725.5 \ \ 1213.8 \ \ 236.9 \ \ 1682.8 \ \ 0)$$

Multiplying by SS7a we subtract SS1 and SS2 from SS1x2

$$SS7 := \left(\overrightarrow{\left(SS3a \cdot SS3a \cdot N^{-1} \right)}DBSUM + GSQ \right) \cdot CF \cdot \begin{pmatrix} 1 & 0 & 0 & 0 & 0 \\ 0 & 1 & 0 & -1 & 0 \\ 0 & 0 & 1 & -1 & 0 \\ 0 & 0 & 0 & 1 & 0 \\ 0 & 0 & 0 & 0 & 0 \end{pmatrix}$$

$$SS7 := \left(\overrightarrow{\left(SS3a \cdot SS3a \cdot N^{-1} \right)}DBSUM + GSQ \right) \cdot CF \cdot SS7a$$

$$\left(\overrightarrow{\left(SS3a \cdot SS3a \cdot N^{-1} \right)}DBSUM + GSQ \right) \cdot CF \cdot \begin{pmatrix} 1 & 0 & 0 & 0 & 0 \\ 0 & 1 & 0 & -1 & 0 \\ 0 & 0 & 1 & -1 & 0 \\ 0 & 0 & 0 & 1 & 0 \\ 0 & 0 & 0 & 0 & 0 \end{pmatrix} = (5725.5 \quad 1213.8 \quad 236.9 \quad 232.1 \quad 0)$$

Now we compute the ERROR term which will be in the last term
in the SS8 multiplication.

$$SS8 := \left(\overrightarrow{\left(SS3a \cdot SS3a \cdot N^{-1} \right)}DBSUM + GSQ \right) \cdot CF \cdot SS7a \cdot SS8a$$

$$SS8 = (5725.5 \quad 1213.8 \quad 236.9 \quad 232.1 \quad 4042.7)$$

$$SS8^T = \begin{pmatrix} 5725.5 \\ 1213.8 \\ 236.9 \\ 232.1 \\ 4042.7 \end{pmatrix}$$

$$SS := \begin{pmatrix} SSt \\ SS1 \\ SS2 \\ SS1x2 \\ SSerror \end{pmatrix} \qquad SS8^T = \begin{pmatrix} 5725.5 \\ 1213.8 \\ 236.9 \\ 232.1 \\ 4042.7 \end{pmatrix}$$

$$m := 72 \qquad\qquad ExpCon1 := 4 \qquad\qquad ExpCon2 := 4$$

CLINTON L. HOLT

$$df := \begin{bmatrix} m - 1 \\ \text{ExpCon1} - 1 \\ \text{ExpCon2} - 1 \\ (\text{ExpCon1} - 1) \cdot (\text{ExpCon2} - 1) \\ (m - 1) - (\text{ExpCon1} - 1) - (\text{ExpCon2} - 1) - (\text{ExpCon1} - 1) \cdot (\text{ExpCon2} - 1) \end{bmatrix}$$

TREATMENTS-BY-SUBJECTS, OR REPEATED MEASURE DESIGN

PROB. 2.5, PG. 44, Computational Handbook of Statistics, 2ND Ed. Bruning & Kintz, Scott Forseman & Co.

$$df = \begin{pmatrix} 71 \\ 3 \\ 3 \\ 9 \\ 56 \end{pmatrix} \qquad ERROR1 := \frac{SSerror^{\blacksquare}}{dferror} \qquad ERROR1 := \frac{4042.7}{56} \qquad ERROR := \begin{pmatrix} \dfrac{4042.7}{56} \\ 72.2 \\ 72.2 \\ 72.2 \\ 72.2 \end{pmatrix}$$

$$ERROR1 = 72.2$$

$$F := \overrightarrow{\left(SS \cdot df^{-1} \cdot ERROR^{-1} \right)}$$

$$F = \begin{pmatrix} 1 \\ 4.4 \\ 1 \\ 0.3 \\ 0.9 \end{pmatrix} \qquad df = \begin{pmatrix} 71 \\ 3 \\ 3 \\ 9 \\ 56 \end{pmatrix} \qquad p := \begin{pmatrix} 0 \\ .01 \\ ns \\ ns \\ 0 \end{pmatrix}^{\blacksquare}$$

1: COMPUTE SUM1, THEN PARTITION ALL THE DATA BASE MATRICES INTO A SINGLE MATRIX DBALL.
2: MULTIPLY DBALL BY SUM1
3: NOW WE SQUARE THE ELEMENTS OF SUM1 x DBALL AND DIVIDE BY N
4: NOW WE SUM ALL THE SQUARED TERMS IN RELATION TO THE NUMBER OF COLUMNS IN THEIR RESPECTIVE
5: ADD THE GRAND SUM OF SQUARES TO SS4
6: SUBTRACT THE CORRECTION FACTOR
7: NOW WE SUBTRACT THE INTERACTIVE TERMS SS1x2 SS1
8: NOW WE SUBTRACT TO FIND THE ERROR TERM

Again, I have no example, please refer to the 'Computational Handbook of Statistics' for a representative problem. Here we just solve the Math. In this type of Statistic, we use the same subjects, give them the same tests but over a period of time and compare if there is a statistically significant difference in the scores. Therefore each column represents the same subject but a new score over the years.

Data and constants: SINGLE EQUATION DATA

$$A := \begin{pmatrix} 99 & 104 & 116 & 109 \\ 106 & 103 & 106 & 113 \\ 129 & 132 & 130 & 139 \\ 73 & 77 & 87 & 80 \\ 84 & 82 & 89 & 86 \\ 89 & 91 & 86 & 92 \\ 99 & 98 & 104 & 104 \\ 112 & 118 & 108 & 120 \\ 121 & 123 & 131 & 123 \\ 96 & 94 & 101 & 99 \\ 81 & 82 & 79 & 86 \\ 121 & 127 & 133 & 135 \\ 134 & 134 & 133 & 139 \\ 103 & 115 & 116 & 116 \\ 115 & 109 & 118 & 117 \\ 95 & 89 & 92 & 97 \\ 87 & 93 & 89 & 95 \\ 92 & 100 & 97 & 101 \\ 88 & 97 & 97 & 98 \\ 95 & 102 & 93 & 104 \end{pmatrix}$$

$$a := \begin{pmatrix} 99 & 104 & 116 & 109 \\ 106 & 103 & 106 & 113 \\ 129 & 132 & 130 & 139 \\ 73 & 77 & 87 & 80 \\ 84 & 82 & 89 & 86 \\ 89 & 91 & 86 & 92 \\ 99 & 98 & 104 & 104 \\ 112 & 118 & 108 & 120 \\ 121 & 123 & 131 & 123 \\ 96 & 94 & 101 & 99 \\ 81 & 82 & 79 & 86 \\ 121 & 127 & 133 & 135 \\ 134 & 134 & 133 & 139 \\ 103 & 115 & 116 & 116 \\ 115 & 109 & 118 & 117 \\ 95 & 89 & 92 & 97 \\ 87 & 93 & 89 & 95 \\ 92 & 100 & 97 & 101 \\ 88 & 97 & 97 & 98 \\ 95 & 102 & 93 & 104 \end{pmatrix}$$

ONE20 := (1 1 1 1 1 1 1 1 1 1 1 1 1 1 1 1 1 1 1 1)

one20 := (1 1 1 1 1 1 1 1 1 1 1 1 1 1 1 1 1 1 1 1)

$$FOUR1 := \begin{pmatrix} 1 \\ 1 \\ 1 \\ 1 \end{pmatrix}$$

ONE4 := FOUR1^T

First we will compute SUM1. Here we just sum the columns of A.

PROGRAM = $[1]_{1,20} \times [A]_{20,4}$

SUM1 := ONE20 A

$$\text{SUM1} := (1\ 1\ 1\ 1\ 1\ 1\ 1\ 1\ 1\ 1\ 1\ 1\ 1\ 1\ 1\ 1\ 1\ 1\ 1\ 1) \cdot \begin{pmatrix} 99 & 104 & 116 & 109 \\ 106 & 103 & 106 & 113 \\ 129 & 132 & 130 & 139 \\ 73 & 77 & 87 & 80 \\ 84 & 82 & 89 & 86 \\ 89 & 91 & 86 & 92 \\ 99 & 98 & 104 & 104 \\ 112 & 118 & 108 & 120 \\ 121 & 123 & 131 & 123 \\ 96 & 94 & 101 & 99 \\ 81 & 82 & 79 & 86 \\ 121 & 127 & 133 & 135 \\ 134 & 134 & 133 & 139 \\ 103 & 115 & 116 & 116 \\ 115 & 109 & 118 & 117 \\ 95 & 89 & 92 & 97 \\ 87 & 93 & 89 & 95 \\ 92 & 100 & 97 & 101 \\ 88 & 97 & 97 & 98 \\ 95 & 102 & 93 & 104 \end{pmatrix}$$

SUM1 = (2019 2070 2105 2153)

Now we compute the Grand Sum of A:

PROGRAM: GRANDSUM = $[1]_{1,20}$ **x** $[A]_{20,4}$ **x** $[1]_{4,1}$

GRANDSUM := ONE20 A · FOUR1

GRANDSUM = 8347

Now we'll compute the Grand Sum of Squares:

$$\text{GSQ} := \text{ONE20} \overparen{(A \cdot A)} \text{FOUR1}$$

$$
\text{GSQ} := \text{ONE20} \left[\begin{pmatrix} 99 & 104 & 116 & 109 \\ 106 & 103 & 106 & 113 \\ 129 & 132 & 130 & 139 \\ 73 & 77 & 87 & 80 \\ 84 & 82 & 89 & 86 \\ 89 & 91 & 86 & 92 \\ 99 & 98 & 104 & 104 \\ 112 & 118 & 108 & 120 \\ 121 & 123 & 131 & 123 \\ 96 & 94 & 101 & 99 \\ 81 & 82 & 79 & 86 \\ 121 & 127 & 133 & 135 \\ 134 & 134 & 133 & 139 \\ 103 & 115 & 116 & 116 \\ 115 & 109 & 118 & 117 \\ 95 & 89 & 92 & 97 \\ 87 & 93 & 89 & 95 \\ 92 & 100 & 97 & 101 \\ 88 & 97 & 97 & 98 \\ 95 & 102 & 93 & 104 \end{pmatrix} \cdot \begin{pmatrix} 99 & 104 & 116 & 109 \\ 106 & 103 & 106 & 113 \\ 129 & 132 & 130 & 139 \\ 73 & 77 & 87 & 80 \\ 84 & 82 & 89 & 86 \\ 89 & 91 & 86 & 92 \\ 99 & 98 & 104 & 104 \\ 112 & 118 & 108 & 120 \\ 121 & 123 & 131 & 123 \\ 96 & 94 & 101 & 99 \\ 81 & 82 & 79 & 86 \\ 121 & 127 & 133 & 135 \\ 134 & 134 & 133 & 139 \\ 103 & 115 & 116 & 116 \\ 115 & 109 & 118 & 117 \\ 95 & 89 & 92 & 97 \\ 87 & 93 & 89 & 95 \\ 92 & 100 & 97 & 101 \\ 88 & 97 & 97 & 98 \\ 95 & 102 & 93 & 104 \end{pmatrix} \cdot \begin{pmatrix} 1 \\ 1 \\ 1 \\ 1 \end{pmatrix} \right]
$$

GSQ = 893269

STEP 5: Computation of the Correction Factor. Nt = # elements in A. Nt = 20x4 = 80

Nt := 80

$$
\text{CORRFACT} := \left(\frac{1}{\text{Nt}} \right) \cdot (\text{ONE20} \, \text{A} \cdot \text{FOUR1})^2
$$

CORRFACT = 870905.113

STEP 6: Here we will compute the Total Sum of Squares, SSt.

$$SSt := GSQ - CORRFACT$$

$$SSt = 22363.887$$

STEPS 7, 8 AND 9: Hey, this is our first Within Subjects Computation ([A] x [1] instead of [1] x [A]). Here we are summing the rows instead of the columns. N in this example is different from the other computations. N1 = # elements/column in A.

$$N1 := 4$$

PROGRAM: $(1/N1)X ([A]_{20,4} \times [1]_{4,1})^2$

$$SS1 := \left(\frac{1}{N1}\right) \cdot (A \cdot FOUR1)^2 - CORRFACT$$

$$SS1a := A \cdot FOUR1$$

$$SS1 := \left(\frac{1}{4}\right) \cdot \left(SS1a^T \cdot SS1a\right) - CORRFACT$$

$$SS1 = 21114.137$$

$$SSt = 22363.887$$

I want to get the math all on the same page so we skip a few lines and now we compute the Within Subjects computation:

$$\left[\frac{1}{4}\cdot\begin{pmatrix} 99 & 104 & 116 & 109 \\ 106 & 103 & 106 & 113 \\ 129 & 132 & 130 & 139 \\ 73 & 77 & 87 & 80 \\ 84 & 82 & 89 & 86 \\ 89 & 91 & 86 & 92 \\ 99 & 98 & 104 & 104 \\ 112 & 118 & 108 & 120 \\ 121 & 123 & 131 & 123 \\ 96 & 94 & 101 & 99 \\ 81 & 82 & 79 & 86 \\ 121 & 127 & 133 & 135 \\ 134 & 134 & 133 & 139 \\ 103 & 115 & 116 & 116 \\ 115 & 109 & 118 & 117 \\ 95 & 89 & 92 & 97 \\ 87 & 93 & 89 & 95 \\ 92 & 100 & 97 & 101 \\ 88 & 97 & 97 & 98 \\ 95 & 102 & 93 & 104 \end{pmatrix}\cdot\begin{pmatrix}1\\1\\1\\1\end{pmatrix}\right]^{2}$$

Or $\left(\dfrac{1}{4}\right)\cdot A\cdot FOUR1 =$

$$\begin{pmatrix} 107 \\ 107 \\ 132.5 \\ 79.25 \\ 85.25 \\ 89.5 \\ 101.25 \\ 114.5 \\ 124.5 \\ 97.5 \\ 82 \\ 129 \\ 135 \\ 112.5 \\ 114.75 \\ 93.25 \\ 91 \\ 97.5 \\ 95 \\ 98.5 \end{pmatrix}$$

WHERE SS1=(1/4)(AxFOUR1)^{T} x (AxFOUR1))

$$SS1 := \left(\frac{1}{4}\right)\cdot\left(SS1a^{T}\cdot SS1a\right) - CORRFACT$$

SS1 = 21114.137

STEP 11 & 12: Computation of the effects of the Treatments.
These, in previous computations are the computations of the
interactive effects. But here we have only SS1 so there are
no interactive effects. We'll call this SS2. N2 = # scores on
which each sum in SUM1 is based. N2= 20.

N2 := 20

PROGRAM: SS2 = (1/N2)x([SUM1]$_{1,4}$x([SUM1]$_{1,4}$^{T}-CORRFACT

Note: This is the same equation as SS1 except we have SUM1xSUM1^T instead of SUM1^{T}xSUM1. Remember, these Matrices are not necessarily commutative, so the answers will be different.

$$SS2 := \left(\frac{1}{20}\right) \cdot \left[(SUM1) \cdot SUM1^T \right] - CORRFACT$$

$$SS2a := \left(\frac{1}{20}\right) \cdot \left[(2016 \quad 2065 \quad 2095 \quad 2150) \cdot \begin{pmatrix} 2019 \\ 2070 \\ 2105 \\ 2153 \end{pmatrix} \right] - CORRFACT$$

SS2a = −1716.162

SS2 = 479.637

STEP 13: Computation of the Error term Sum of Squares SSerror.

SSerror := SSt − SS1 − SS2

SSerror = 770.113

$$SS := \begin{pmatrix} SSt \\ SS1 \\ SS2 \\ SSerror \end{pmatrix} \qquad SS = \begin{pmatrix} 22363.887 \\ 21114.137 \\ 479.637 \\ 770.113 \end{pmatrix}$$

STEP 14: All computations are complete, however, since the F-ratio is a ratio of the mean squares, these mean square values must be computed. To do this the degrees of Freedom (df) must first be determined.

m := 80   S := 20   T := 4

$$df := \begin{bmatrix} m - 1 \\ S - 1 \\ T - 1 \\ (m - 1) - (S - 1) - (T - 1) \end{bmatrix} \qquad SS = \begin{pmatrix} 22363.887 \\ 21114.137 \\ 479.637 \\ 770.113 \end{pmatrix}$$

$$df = \begin{pmatrix} 79 \\ 19 \\ 3 \\ 57 \end{pmatrix} \qquad ERR := \frac{770.113}{57} \qquad ERROR := \begin{pmatrix} \dfrac{770.113}{57} \\ 13.511 \\ 13.511 \\ 13.511 \end{pmatrix}$$

$$F := \overrightarrow{\left(SS \cdot df^{-1} \cdot ERROR^{-1} \right)}$$

$$F = \begin{pmatrix} 20.953 \\ 82.249 \\ 11.833 \\ 1 \end{pmatrix}$$

$ERR = 13.511$

$SUM1 = (2019 \quad 2070 \quad 2105 \quad 2153)$

$GRANDSUM = 8347$

$GSQ = 893269$

$SSt = 22363.887$

$SS1 = 21114.137$

$SS2 = 479.637$

$SSerror = 770.113$

PROBLEM SOLVED "ALMOST" ALL AT ONCE

In trying to get a single equation for this Statistic, we run into a problem. Let's look at the Statistical Field Equations.

Between Subjects equation = $([1]x[A]x[B])^2$ which all of our equations have been following so far. So we can treat all the mathematics in the same way. But now we are also concerned with the Within Subjects equation = $([A]x[B]x[1])^2$. These equations are non-commutative, so, at least at this present time with my limitations on the understanding of Statistics, we have to compute them as two separate operations

NOTE: Here, a = A, Mathcad kept getting confused with all the previous equations containing A.

SUM1 := ONE20a

SUM1 = (2019 2070 2105 2153)

What we need to do here is multiply SUM1 by DBALL. In this case the CF, GS and GSQ are computed from other equations so we have to compute them separately and add them to the Matrix when it is necessary. DBALL is the compilation of all the DB Matrices used to solve the problems that have the form [1]x{A] x[B] for Between Subjects analysis or of the form [A]x[B]x[1] for Within Subjects analysis

Note: Here we have 2 zero's in the summing Matrix, the one on the left is for the Grand Sum of Squares, the one to it's right is for the Within Subjects Sum of Squares SSw which is a different form of math.

$$
dballA := \begin{pmatrix} 1 & 0 & 0 & 1 & 0 & 0 & 0 \\ 1 & 0 & 0 & 0 & 1 & 0 & 0 \\ 1 & 0 & 0 & 0 & 0 & 1 & 0 \\ 1 & 0 & 0 & 0 & 0 & 0 & 1 \end{pmatrix}
$$

sum1 := one20·a

sum1 = (2019 2070 2105 2153)

First we combine all the Data Base Matrices together and multiply by SUM1:

sum1·dballA = (8347 0 0 2019 2070 2105 2153)

$$(2019\ 2070\ 2105\ 2153)\cdot\begin{pmatrix} 1 & 0 & 0 & 1 & 0 & 0 & 0 \\ 1 & 0 & 0 & 0 & 1 & 0 & 0 \\ 1 & 0 & 0 & 0 & 0 & 1 & 0 \\ 1 & 0 & 0 & 0 & 0 & 0 & 1 \end{pmatrix} = (8347\ 0\ 0\ 2019\ 2070\ 2105\ 2153)$$

The N values we divide by are given by: $N := (80\ 1\ 4\ 20\ 20\ 20\ 20)$

Since we cannot handle the math as it is, we will let:

$$sum2A := sum1\cdot dballA$$

Then multiplying element by element we get the squares of the terms.

$$\overrightarrow{\left(sum2A\cdot sum2A\cdot N^{-1} \right)} = (870905.113\ 0\ 0\ 203818.05\ 214245\ 221551.25\ 231770.45)$$

sum2cf := (0 870905.113 0 0 0 0 0)

Where the   CORRFACT := 870905.11

Here we add the Grand Sum of Squares and the Within Subjects terms which cannot be computed using this all at once method. But first let's compute the SSw (Within Subjects term).

This is the Within Subjects equation we will use. Note that it is of the form $([A]x[1])^2$

$$SSw := \left(\frac{1}{4} \right)\cdot(A\cdot FOUR1)^2 - CORRFACT$$

$$SS1w := A\cdot FOUR1$$

To get this to work we must transpose SS1w, since it is a column Matrix and we need to start the multiplication with a row Matrix to get a single number value.

$$SS1 := \left(\frac{1}{4}\right) \cdot \left(SS1w^T \cdot SS1w\right) - CORRFACT$$

$$SS1 = 21114.137$$

$$SS1w^T = (428 \ \ 428 \ \ 530 \ \ 317 \ \ 341 \ \ 358 \ \ 405 \ \ 458 \ \ 498 \ \ 390 \ \ 328 \ \ 516 \ \ 540 \ \ 450 \ \ 459 \ \ 373 \ \ 364 \ \ 390 \ \ 380 \ \ 394)$$

Now we will square each term in SS1 and divide by the corresponding N values

$$N1 := 4$$

$$\overrightarrow{\left(SS1w^T \cdot SS1w^T \cdot N1^{-1}\right)} = \blacksquare$$

This will not fit on a single page, so we'll transpose the solution so we can see the solution.

$$\left(\overrightarrow{\left(SS1w^T \cdot SS1w^T \cdot N1^{-1}\right)}\right)^T = \begin{pmatrix} 45796 \\ 45796 \\ 70225 \\ 25122.25 \\ 29070.25 \\ 32041 \\ 41006.25 \\ 52441 \\ 62001 \\ 38025 \\ 26896 \\ 66564 \\ 72900 \\ 50625 \\ 52670.25 \\ 34782.25 \\ 33124 \\ 38025 \\ 36100 \\ 38809 \end{pmatrix}$$

Now we will sum all the values into a single number. This is SSw.

$$\overrightarrow{\left(SS1w^T \cdot SS1w^T \cdot N1^{-1}\right)ONE20^T} = 892019.25 \quad SSw := \overrightarrow{\left(SS1w^T \cdot SS1w^T \cdot N1^{-1}\right)ONE20^T}$$

$$SSw = 892019.25 \qquad\qquad GSQ = 893269$$

Which is the answer we want. Now we add the Grand Sum of Squares and SSw to the Matrix. First we must make them into Matrices so we can add them together.

$$SSwm := (0 \ \ 0 \ \ 892019.25 \ 0 \ \ 0 \ \ 0 \ \ 0)$$

$$GSQm := (1 \ \ 893269 \ 0 \ \ 0 \ \ 0 \ \ 0 \ \ 0)$$

$$\overrightarrow{\left(sum2A \cdot sum2A \cdot N^{-1}\right)} + GSQm + SSwm = (870906.113 \ \ 893269 \ \ 892019.25 \ \ 203818.05 \ \ 214245 \ \ 221551.25 \ \ 231770.45)$$

Some of these values fit the page when we save the math to Word, but some don't. Just to be safe we'll put the result as a transpose below.

$$\left(\overrightarrow{\left(sum2A \cdot sum2A \cdot N^{-1}\right)} + GSQm + SSwm\right)^T = \begin{pmatrix} 870906.113 \\ 893269 \\ 892019.25 \\ 203818.05 \\ 214245 \\ 221551.25 \\ 231770.45 \end{pmatrix}$$

The basic Matrix is complete, now we must sum the elements using DBSUM. Elements 1, 2 and 3 we leave alone, but we must sum elements 4, 5, 6 and 7. The DB Matrix for this computation is as follows:

$$\text{DBSUM1} := \begin{pmatrix} 1 & 0 & 0 & 0 \\ 0 & 1 & 0 & 0 \\ 0 & 0 & 1 & 0 \\ 0 & 0 & 0 & 1 \\ 0 & 0 & 0 & 1 \\ 0 & 0 & 0 & 1 \\ 0 & 0 & 0 & 1 \end{pmatrix}$$

$$\left(\overrightarrow{\left(\text{sum2A} \cdot \text{sum2A} \cdot \text{N}^{-1} \right)} + \text{GSQm} + \text{SSwm} \right) \cdot \text{DBSUM1} = (870906.113 \ \ 893269 \ \ 892019.25 \ \ 871384.75)$$

The math looks like:

$$\left(\overrightarrow{\left(\text{sum2A} \cdot \text{sum2A} \cdot \text{N}^{-1} \right)} + \text{GSQm} + \text{SSwm} \right) \cdot \begin{pmatrix} 1 & 0 & 0 & 0 \\ 0 & 1 & 0 & 0 \\ 0 & 0 & 1 & 0 \\ 0 & 0 & 0 & 1 \\ 0 & 0 & 0 & 1 \\ 0 & 0 & 0 & 1 \\ 0 & 0 & 0 & 1 \end{pmatrix} = (870906.113 \ \ 893269 \ \ 892019.25 \ \ 871384.75)$$

Now we subtract the Correction Factor CF from each of the computed terms. We subtract by multiplying by CF.

$$\text{CF} := \begin{pmatrix} -1 & -1 & -1 & 0 \\ 1 & 0 & 0 & 0 \\ 0 & 1 & 0 & 0 \\ 0 & 0 & 1 & 0 \end{pmatrix}$$

$$\left(\overrightarrow{\left(\text{sum2A} \cdot \text{sum2A} \cdot \text{N}^{-1} \right)} + \text{GSQm} + \text{SSwm} \right) \cdot \text{DBSUM1} \cdot \text{CF} = (22362.887 \ \ 21113.137 \ \ 478.637 \ \ 0)$$

The math looks like:

$$\left(\overrightarrow{\left(sum2A \cdot sum2A \cdot N^{-1}\right)} + GSQm + SSwm\right) \cdot \begin{pmatrix} 1 & 0 & 0 & 0 \\ 0 & 1 & 0 & 0 \\ 0 & 0 & 1 & 0 \\ 0 & 0 & 0 & 1 \\ 0 & 0 & 0 & 1 \\ 0 & 0 & 0 & 1 \\ 0 & 0 & 0 & 1 \end{pmatrix} \cdot \begin{pmatrix} -1 & -1 & -1 & 0 \\ 1 & 0 & 0 & 0 \\ 0 & 1 & 0 & 0 \\ 0 & 0 & 1 & 0 \end{pmatrix} = (22362.887 \ 21113.137 \ 478.637 \ 0)$$

And finally we compute the Error term. The Error Matrix for this Statistic is given by:

$$\text{Error} := \begin{pmatrix} 1 & 0 & 0 & 1 \\ 0 & 1 & 0 & -1 \\ 0 & 0 & 1 & -1 \\ 0 & 0 & 0 & -1 \end{pmatrix}$$

$$\left(\overrightarrow{\left(sum2A \cdot sum2A \cdot N^{-1}\right)} + GSQm + SSwm\right) \cdot DBSUM1 \cdot CF \cdot Error = (22362.887 \ 21113.137 \ 478.637 \ 771.113)$$

$$SS := \left(\overrightarrow{\left(sum2A \cdot sum2A \cdot N^{-1}\right)} + GSQm + SSwm\right) \cdot DBSUM1 \cdot CF \cdot Error$$

$$SS = (22362.887 \ 21113.137 \ 478.637 \ 771.113)$$

$$\text{Error Term} = \frac{771.113}{57} = 13.528$$

See above to get the df values.

$$df := \begin{pmatrix} 79 \\ 19 \\ 3 \\ 57 \end{pmatrix} \quad ERROR := \begin{pmatrix} 13.528 \\ 13.528 \\ 13.528 \\ 13.528 \end{pmatrix} \quad SS^T = \begin{pmatrix} 22362.887 \\ 21113.137 \\ 478.637 \\ 771.113 \end{pmatrix} \quad df^{-1} = \begin{pmatrix} 0.013 \\ 0.053 \\ 0.333 \\ 0.018 \end{pmatrix}$$

The F values are given by:

$$F := \overrightarrow{\left(SS^T \cdot df^{-1} \cdot ERROR^{-1}\right)}$$

This is the single equation   This is the F-value
       F-value   previously computed

$$F = \begin{pmatrix} 20.925 \\ 82.142 \\ 11.794 \\ 1 \end{pmatrix} \qquad \begin{pmatrix} 20.953 \\ 82.249 \\ 11.833 \\ 1 \end{pmatrix}$$

The values differ in that in the long method we rounded the
interim results while we left the results alone in the single
equation method.

TREATMENT-BY-TREATMENT-SUBJECTS OR REPEATED MEASURES, 2-LEVEL DESIGN.

PROB. 2.6, PG. 48, Computational Handbook of Statistics, 2nd Ed., Bruning & Kintz, Scott Forseman & Co.

Tests where the same subjects in a 2-factor experiment are tested under all the different conditions.

COMMENT: This is the last AOV problem in this book in which I will compute the solution as a single equation, therefore, so you may follow the procedures better, I have left the values as presented in the Computational Handbook of Statistics unchanged.

VARIABLES:

$$\text{ONE10} := (1\ 1\ 1\ 1\ 1\ 1\ 1\ 1\ 1\ 1) \qquad \text{ONE2} := (1\ 1) \qquad \text{TWO1} := \begin{pmatrix} 1 \\ 1 \end{pmatrix}$$

$$A := \begin{pmatrix} 83 & 100 & 75 & 40 \\ 100 & 83 & 71 & 41 \\ 100 & 100 & 50 & 43 \\ 89 & 92 & 74 & 47 \\ 100 & 90 & 83 & 62 \\ 100 & 100 & 100 & 70 \\ 92 & 92 & 72 & 44 \\ 100 & 89 & 83 & 62 \\ 50 & 54 & 33 & 38 \\ 100 & 100 & 72 & 50 \end{pmatrix} \quad \text{DB1} := \begin{pmatrix} 1 & 0 \\ 1 & 0 \\ 0 & 1 \\ 0 & 1 \end{pmatrix} \quad \text{DB2} := \begin{pmatrix} 1 & 0 \\ 0 & 1 \\ 1 & 0 \\ 0 & 1 \end{pmatrix} \quad \text{FOUR1} := \begin{pmatrix} 1 \\ 1 \\ 1 \\ 1 \end{pmatrix}$$

$$\text{DBALL} := \begin{pmatrix} 1 & 0 & 0 & 1 & 0 & 1 & 0 & 1 & 0 & 1 & 0 & 0 & 0 \\ 1 & 0 & 0 & 1 & 0 & 0 & 1 & 0 & 1 & 0 & 1 & 0 & 0 \\ 1 & 0 & 0 & 0 & 1 & 1 & 0 & 0 & 0 & 0 & 1 & 0 \\ 1 & 0 & 0 & 0 & 1 & 0 & 1 & 0 & 0 & 0 & 1 \end{pmatrix}$$

STEP 1:   Computation of SUM1, Grand Sum, Correction Factor and
the Grand Sum of Squares.

SUM1 := ONE10 A

$$
\text{SUM1} := (1 \ \ 1 \ \ 1 \ \ 1 \ \ 1 \ \ 1 \ \ 1 \ \ 1 \ \ 1 \ \ 1) \cdot
\begin{pmatrix}
83 & 100 & 75 & 40 \\
100 & 83 & 71 & 41 \\
100 & 100 & 50 & 43 \\
89 & 92 & 74 & 47 \\
100 & 90 & 83 & 62 \\
100 & 100 & 100 & 70 \\
92 & 92 & 72 & 44 \\
100 & 89 & 83 & 62 \\
50 & 54 & 33 & 38 \\
100 & 100 & 72 & 50
\end{pmatrix}
$$

SUM1 = (914 900 713 497)

#2: GRANDSUM:

GRANDSUM := ONE10 A·FOUR1

GRANDSUM = 3024

#3: CORRECTION FACTOR   Nt := 40

$$
\text{CORRFACT} := \left(\frac{1}{\text{Nt}} \right) \cdot \text{GRANDSUM}^2
$$

$$
\text{CORRFACT} := \left(\frac{1}{40} \right) \cdot 3024^2
$$

CORRFACT = 228614.4

#4: GRANDSUMSQUARE:

$$
\text{GSQ} := \text{ONE10} \overleftrightarrow{(A \cdot A)} \text{FOUR1}
$$

GSQ = 248212

METHODS USING LOGARITHMS. First we square A, remove the diagonal, take it's anti-log and sum the values together.

Let $ASQ := A \cdot A^T$ then

$$\overrightarrow{\log(ASQ)} = \begin{pmatrix} 4.3823 & 4.3723 & 4.376 & 4.3805 & 4.4151 & 4.4564 & 4.3801 & 4.4134 & 4.1318 & 4.4099 \\ 4.3723 & 4.3731 & 4.3732 & 4.3751 & 4.4134 & 4.4513 & 4.3757 & 4.412 & 4.1266 & 4.4059 \\ 4.376 & 4.3732 & 4.3865 & 4.377 & 4.4119 & 4.4473 & 4.3783 & 4.4102 & 4.1362 & 4.4108 \\ 4.3805 & 4.3751 & 4.377 & 4.3815 & 4.4189 & 4.4592 & 4.3811 & 4.4174 & 4.135 & 4.4112 \\ 4.4151 & 4.4134 & 4.4119 & 4.4189 & 4.4599 & 4.5002 & 4.418 & 4.4585 & 4.1748 & 4.4483 \\ 4.4564 & 4.4513 & 4.4473 & 4.4592 & 4.5002 & 4.5428 & 4.4576 & 4.4989 & 4.2138 & 4.4871 \\ 4.3801 & 4.3757 & 4.3783 & 4.3811 & 4.418 & 4.4576 & 4.3811 & 4.4165 & 4.134 & 4.4114 \\ 4.4134 & 4.412 & 4.4102 & 4.4174 & 4.4585 & 4.4989 & 4.4165 & 4.4572 & 4.1732 & 4.4468 \\ 4.1318 & 4.1266 & 4.1362 & 4.135 & 4.1748 & 4.2138 & 4.134 & 4.1732 & 3.9003 & 4.1666 \\ 4.4099 & 4.4059 & 4.4108 & 4.4112 & 4.4483 & 4.4871 & 4.4114 & 4.4468 & 4.1666 & 4.4422 \end{pmatrix}$$

Now we remove the diagonal, take the anti-log to get back into the realm of numbers and sum the elements.

$$GSQa := ONE10 \cdot 10^{\overrightarrow{diag\left(\overrightarrow{\log(ASQ)}\right)}}$$

$$10^{\overrightarrow{diag\left(\overrightarrow{\log(ASQ)}\right)}} = \begin{pmatrix} 24114 \\ 23611 \\ 24349 \\ 24070 \\ 28833 \\ 34900 \\ 24048 \\ 28654 \\ 7949 \\ 27684 \end{pmatrix}$$

$GSQa = 248212$ This is the diagonal now we sum its elements.

STEP 6: Computation of the Total Sum of Squares SSt.

$SSt := GSQ - CORRFACT$

$SSt = 19597.6$

$$\text{SSta} := \text{ONE10}\overrightarrow{(A \cdot A)}\text{FOUR1} - \left(\frac{1}{Nt}\right) \cdot (\text{ONE10 A} \cdot \text{FOUR1})^2$$

$$\text{SSta} = 19597.6$$

$$\text{CORRFACT} = 228614.4$$

STEPS 7, 8, 9 AND 10: Do the Within Subjects analysis, summing each row instead of column. N = # of scores that were added to get the sum.

$$N1 := 4$$

PROGRAM: $(1/N1)$ x $([A]_{10,4} \times [1]_{4,1})^2$ - **CORRFACT**

$$\text{SSw} := \left(\frac{1}{N1}\right) \cdot (A \cdot \text{FOUR1})^T \cdot (A \cdot \text{FOUR1}) - \text{CORRFACT}$$

The Math is:

$$\text{SSw} := \left(\frac{1}{4}\right) \cdot \left[(298\ \ 295\ \ 293\ \ 302\ \ 335\ \ 370\ \ 300\ \ 334\ \ 175\ \ 322) \cdot \begin{pmatrix} 298 \\ 295 \\ 293 \\ 302 \\ 335 \\ 370 \\ 300 \\ 334 \\ 175 \\ 322 \end{pmatrix} \right] - \text{CORRFACT}$$

$$\text{SSw} = 5853.6$$

STEP 11: Computation of the Sum of Squares for the First Factor effects SS1. N1 = Nt/ # columns in DB1. 40/2 = 20

N1 := 20

PROGRAM: SS1 = (1/N1) x ([SUM1]$_{1,4}$ x [DB1]$_{4,2}$)^2 - CORRFACT

The equation is given by:

$$SS1 := \left(\frac{1}{N1}\right) \cdot (SUM1 \cdot DB1) \cdot (SUM1 \cdot DB1)^T - CORRFACT$$

And the Math looks like:

$$SS1 := \left(\frac{1}{N1}\right) \cdot \left[(914 \quad 900 \quad 713 \quad 497) \cdot \begin{pmatrix} 1 & 0 \\ 1 & 0 \\ 0 & 1 \\ 0 & 1 \end{pmatrix} \right] \cdot \left[(914 \quad 900 \quad 713 \quad 497) \cdot \begin{pmatrix} 1 & 0 \\ 1 & 0 \\ 0 & 1 \\ 0 & 1 \end{pmatrix} \right]^T - CORRFACT$$

$$SS1 := \left(\frac{1}{N1}\right) \cdot \left[((1814 \quad 1210)) \cdot \begin{pmatrix} 1814 \\ 1210 \end{pmatrix} \right] - CORRFACT$$

$$SS1 = 9120.4$$

STEP 12: Computation of the effects of the second experimental factor SS2. N2 = Nt / #columns in DB2. N2 = 40/2 = 20.

PROGRAM: SS2 = (1/N1) x ([SUM1]$_{1,4}$ x [DB2]$_{4,2}$)^2 - CORRFACT

N2 := 20

$$SS2 := \left(\frac{1}{N2}\right) \cdot (SUM1 \cdot DB2) \cdot (SUM1 \cdot DB2)^T - CORRFACT$$

$$SS2 = 1322.5$$

Or the Math looks like:

$$SS2 := \left(\frac{1}{N2}\right) \cdot \left[(914 \quad 900 \quad 713 \quad 497) \cdot \begin{pmatrix} 1 & 0 \\ 0 & 1 \\ 1 & 0 \\ 0 & 1 \end{pmatrix} \right] \cdot \left[(914 \quad 900 \quad 713 \quad 497) \cdot \begin{pmatrix} 1 & 0 \\ 0 & 1 \\ 1 & 0 \\ 0 & 1 \end{pmatrix} \right]^T - CORRFACT$$

The next computation equals:

$$SS2 := \left(\frac{1}{N2}\right) \cdot \left[(1627 \quad 1397) \cdot \begin{pmatrix} 1627 \\ 1397 \end{pmatrix} \right] - CORRFACT$$

$SS2 = 1322.5$

STEP 13: Computation of the interactive effects of the first and second experimental factors. SS1x2. N1x2 = Nt / # elements in SUM1. N1X2 = 40/4 = 10.

$N1x2 := 10$

PROGRAM: SS1x2 = (1/N1x2) x ([SUM1]$_{1,4}$ x [SUM1]$^T_{4,1}$) - CORRFACT-SS1-SS2

The equation is given by:

$$SS1x2 := \left(\frac{1}{N1x2}\right) \cdot \left(SUM1 \cdot SUM1^T\right) - CORRFACT - SS1 - SS2$$

$SS1x2 = 1020.1$

The Math looks like:

$$SS1x2 := \left(\frac{1}{10}\right) \cdot \left[(914 \quad 900 \quad 713 \quad 497) \cdot \begin{pmatrix} 914 \\ 900 \\ 713 \\ 497 \end{pmatrix} \right] - CORRFACT - SS1 - SS2$$

$$\left(\frac{1}{N1x2}\right) \cdot \left(SUM1 \cdot SUM1^T\right) = 240077.4$$

$CORRFACT = 228614.4$

Now we have three Error terms to compute.

STEP 14: Computation of the Error term of the Within Subjects factor. SSerr1.

NOTE: This is a within Subjects Multiplication of the forn ([A] x[1])^2. (Instead of the normal between subjects computation ([1]x[A])^2.)

$$\text{SSerr1} := \left(\frac{1}{\text{Nerr1}}\right) \cdot \left[\text{ONE2}(A \cdot DB1)^2 \cdot \text{TWO1}\right] - \text{CORRFACT} - \text{SS1} - \text{SSw} \quad \blacksquare$$

Again, this is not solvable in MathCad as is, so we make the following substitution:

$$\text{SUMwe} := A \cdot DB1$$

$$\text{SUMwe} = \begin{pmatrix} 183 & 115 \\ 183 & 112 \\ 200 & 93 \\ 181 & 121 \\ 190 & 145 \\ 200 & 170 \\ 184 & 116 \\ 189 & 145 \\ 104 & 71 \\ 200 & 122 \end{pmatrix}$$

Then we have:

$$SS1 = 9120.4$$

$$\text{ONE10}(\overrightarrow{\text{SUMwe} \cdot \text{SUMwe}}) \cdot \text{TWO1} = 489622$$

$$SSw := 5853.6$$

$$\text{CORRFACT} = 228614.4$$

And dividing by Nerr1 we get:

$$\left(\frac{1}{2}\right) \cdot \left(\text{ONE10}(\overrightarrow{\text{SUMwe} \cdot \text{SUMwe}}) \cdot \text{TWO1}\right) = 244811$$

$$\text{SSerr1} := \left(\frac{1}{2}\right) \cdot \left(\text{ONE10}\overrightarrow{(\text{SUMwe} \cdot \text{SUMwe}) \cdot \text{TWO1}}\right) - \text{CORRFACT} - \text{SSw} - \text{SS1}$$

$\text{SSerr1} = 1222.6$ Or looking at the math we have

$$\left(\frac{1}{2}\right) \cdot (1\ 1\ 1\ 1\ 1\ 1\ 1\ 1\ 1\ 1) \cdot \overrightarrow{\begin{bmatrix} \begin{pmatrix} 183 & 115 \\ 183 & 112 \\ 200 & 93 \\ 181 & 121 \\ 190 & 145 \\ 200 & 170 \\ 184 & 116 \\ 189 & 145 \\ 104 & 71 \\ 200 & 122 \end{pmatrix} \cdot \begin{pmatrix} 183 & 115 \\ 183 & 112 \\ 200 & 93 \\ 181 & 121 \\ 190 & 145 \\ 200 & 170 \\ 184 & 116 \\ 189 & 145 \\ 104 & 71 \\ 200 & 122 \end{pmatrix} \end{bmatrix}} \cdot \begin{pmatrix} 1 \\ 1 \end{pmatrix} - \text{CORRFACT} - \text{SSw} - \text{SS1} = 1222.6$$

I'm thinking this might be a clearer substitution variable.

$$\text{AxDB1} := \text{A} \cdot \text{DB1}$$

$$\left(\frac{1}{2}\right) \cdot \left[\left(\text{ONE10}\overrightarrow{[(\text{AxDB1}) \cdot (\text{AxDB1})]}\right)\text{TWO1}\right] - \text{CORRFACT} - \text{SS2} - \text{SSw} = 9020.5$$

STEP 15: Computation of the Error term of the Within Subjects
factor. SSerrw.

$$\text{Nerr2} := 2$$

$$\text{SSerr2} := \left(\frac{1}{\text{Nerr2}}\right) \cdot \left[\text{ONE2}(\text{A} \cdot \text{DB2})^2 \cdot \text{TWO1}\right] - \text{CORRFACT} - \text{SS1} - \text{SSw} \quad \blacksquare$$

We again have to substitute, I wrote MathCad about this problem
5 years ago, but guess they do not believe that the Matrix
Operators are as important as the Differential Operator as a
Non-Commutative Operator. In fact, from MathCad 8 to MathCad
13, their attention on Matrices has deteriorated; I can do less
with them now than I could with MathCad 6, which came out only
on floppy disk. We must make the following substitution:

$$\text{AxDB2} := \text{A} \cdot \text{DB2}$$

$$\text{ONE10} \overrightarrow{(\text{AxDB2 AxDB2})} \cdot \text{TWO1} = 472632$$

Dividing by Nerr2 we get:

$$\left(\frac{1}{2}\right) \cdot \left(\text{ONE10} \overrightarrow{(\text{AxDB2 AxDB2})} \cdot \text{TWO1}\right) = 236316$$

$$\text{SSerr2} := \left(\frac{1}{2}\right) \cdot \left(\text{ONE10} \overrightarrow{(\text{AxDB2 AxDB2})} \cdot \text{TWO1}\right) - \text{CORRFACT} - \text{SS2} - \text{SSw}$$

$$\text{SSerr2} = 525.5$$

This is what the math looks like.

$$\left(\frac{1}{2}\right) \cdot (1\ 1\ 1\ 1\ 1\ 1\ 1\ 1\ 1\ 1) \cdot \left[\begin{pmatrix} 158 & 140 \\ 171 & 124 \\ 150 & 143 \\ 163 & 139 \\ 183 & 152 \\ 200 & 170 \\ 164 & 136 \\ 183 & 151 \\ 83 & 92 \\ 172 & 150 \end{pmatrix} \cdot \overrightarrow{\begin{pmatrix} 158 & 140 \\ 171 & 124 \\ 150 & 143 \\ 163 & 139 \\ 183 & 152 \\ 200 & 170 \\ 164 & 136 \\ 183 & 151 \\ 83 & 92 \\ 172 & 150 \end{pmatrix}} \cdot \begin{pmatrix} 1 \\ 1 \end{pmatrix} \right] - 228614.4 - 1322.5 - 5853.6 = 525.5$$

STEP 16: Computation of the Error term for the interaction of experimental conditions SS1x2.

$$\text{SSerr1x2} := \text{SSt} - \text{SSw} - \text{SS1} - \text{SS2} - \text{SS1x2} - \text{SSerr1} - \text{SSerr2}$$

$$\text{SSerr1x2} = 532.9$$

$$\text{SSt} = 19597.6$$
$$\text{SSw} = 5853.6$$
$$\text{SS1} = 9120.4$$
$$\text{SS2} = 1322.5$$
$$\text{SS1x2} = 1020.1$$
$$\text{SSerr1} = 1222.6$$
$$\text{SSerr2} = 525.5$$

STEP 17: All computations based on the data are complete, now we must find the degrees of freedom (df) for each of the components. NOTE: These df values work for all problems of this category.

$m := 40$ $s := 10$ $EXPCOND1 := 2$ $EXPCOND2 := 2$

$dfSSt := m - 1$

$dfSSw := s - 1$

$dfSS1 := EXPCOND1 - 1$

$dfSSerr1 := dfSSw \cdot dfSS1$

$dfSS2 := EXPCOND2 - 1$

$dfSSerr2 := dfSSw \cdot dfSS2$

$dfSS1x2 := dfSS1 \cdot dfSS2$

$dfSSerr1x2 := dfSSw \cdot dfSS1 \cdot dfSS2$

$$df := \begin{bmatrix} m - 1 \\ s - 1 \\ EXPCOND1 - 1 \\ (s - 1) \cdot (EXPCOND1 - 1) \\ EXPCOND2 - 1 \\ (s - 1) \cdot (EXPCOND2 - 1) \\ (EXPCOND1 - 1) \cdot (EXPCOND2 - 1) \\ (s - 1) \cdot [(EXPCOND1 - 1) \cdot (EXPCOND2 - 1)] \end{bmatrix}$$

$$df = \begin{pmatrix} 39 \\ 9 \\ 1 \\ 9 \\ 1 \\ 9 \\ 1 \\ 9 \end{pmatrix} \qquad \begin{pmatrix} SSt \\ SSw \\ SS1 \\ SS2 \\ SS1x2 \\ SSerr1 \\ SSerr2 \\ SSerr1x2 \end{pmatrix} = \begin{pmatrix} 19597.6 \\ 5853.6 \\ 9120.4 \\ 1322.5 \\ 1020.1 \\ 1222.6 \\ 525.5 \\ 532.9 \end{pmatrix}$$

Computing for the mean square we get:

$$SSremix := \begin{pmatrix} SSt \\ SSw \\ SS1 \\ SSerr1 \\ SS2 \\ SSerr2 \\ SS1x2 \\ SSerr1x2 \end{pmatrix} \begin{pmatrix} 39 \\ 9 \\ 1 \\ 9 \\ 1 \\ 9 \\ 1 \\ 9 \end{pmatrix} \begin{pmatrix} 19597.6 \\ 5853.6 \\ 9120.4 \\ 1322.5 \\ 1020.1 \\ 1222.6 \\ 525.5 \\ 532.9 \end{pmatrix}$$

$\dfrac{1322.5}{9} = 146.9444$

$\dfrac{1222.6}{9} = 135.8444$

$\dfrac{525.5}{9} = 58.3889$

$\dfrac{532.9}{9} = 59.2111$

The ERROR Matrix is thus given by:

$$
\text{ERROR} := \begin{pmatrix} 135.8 \\ 135.8 \\ 135.8 \\ 135.8 \\ 58.4 \\ 58.4 \\ 59.2 \\ 59.2 \end{pmatrix}
$$

And the F-ratio is computed as: $F := \overrightarrow{\left(\text{SSremixdf}^{-1} \cdot \text{ERROR}^{-1}\right)}$

Looking in Appendix E of the Computational Handbook of Statistics we compute the probabilities:

$$
F = \begin{pmatrix} 3.7003 \\ 4.7894 \\ 67.1605 \\ 1.0003 \\ 22.6455 \\ 0.9998 \\ 17.2314 \\ 1.0002 \end{pmatrix} \qquad p := \begin{pmatrix} 0 \\ 0 \\ .001 \\ .005 \\ .005 \\ 0 \\ 0 \\ 0 \end{pmatrix}
$$

For Experimental Condition 1, the time between had a significant effect.

For Experimental Condition 2, the more advanced the conditions the more significantly the effects were affected.

The effects between Experimental Condition 1 and Experimental Condition 2 interacted to a significant degree.

ALL AT ONCE

$$N := (40 \ \ 1 \ \ 1 \ \ 20 \ \ 20 \ \ 20 \ \ 20 \ \ 10 \ \ 10 \ \ 10 \ \ 10) \qquad DBALL = \begin{pmatrix} 1 & 0 & 0 & 1 & 0 & 1 & 0 & 1 & 0 & 0 & 0 \\ 1 & 0 & 0 & 1 & 0 & 0 & 1 & 0 & 1 & 0 & 0 \\ 1 & 0 & 0 & 0 & 1 & 1 & 0 & 0 & 0 & 1 & 0 \\ 1 & 0 & 0 & 0 & 1 & 0 & 1 & 0 & 0 & 0 & 1 \end{pmatrix}$$

In SUNDBALL we just combine all the Database Matrices into one single Nested Array.

SUM1xDBALL:= SUM1·DBALL

$$B := \begin{pmatrix} 1 \\ 1 \\ 1 \\ 1 \end{pmatrix} \quad C := \begin{pmatrix} 0 \\ 0 \\ 0 \\ 0 \end{pmatrix} \quad D := \begin{pmatrix} 0 \\ 0 \\ 0 \\ 0 \end{pmatrix} \quad E := \begin{pmatrix} 1 & 0 \\ 1 & 0 \\ 0 & 1 \\ 0 & 1 \end{pmatrix} \quad F := \begin{pmatrix} 1 & 0 \\ 0 & 1 \\ 1 & 0 \\ 0 & 1 \end{pmatrix} \quad G := \begin{pmatrix} 1 & 0 & 0 & 0 \\ 0 & 1 & 0 & 0 \\ 0 & 0 & 1 & 0 \\ 0 & 0 & 0 & 1 \end{pmatrix}$$

DBALLa := (B C D E F G)

$$DBALLa = \left[\begin{pmatrix} 1 \\ 1 \\ 1 \\ 1 \end{pmatrix} \begin{pmatrix} 0 \\ 0 \\ 0 \\ 0 \end{pmatrix} \begin{pmatrix} 0 \\ 0 \\ 0 \\ 0 \end{pmatrix} \begin{pmatrix} 1 & 0 \\ 1 & 0 \\ 0 & 1 \\ 0 & 1 \end{pmatrix} \begin{pmatrix} 1 & 0 \\ 0 & 1 \\ 1 & 0 \\ 0 & 1 \end{pmatrix} \begin{pmatrix} 1 & 0 & 0 & 0 \\ 0 & 1 & 0 & 0 \\ 0 & 0 & 1 & 0 \\ 0 & 0 & 0 & 1 \end{pmatrix} \right]$$

Modern Mathematics cannot handle Nested Arrays, so we partition the Matrix into DBALL and just 'remember' where the brackets go as we compute the following solutions. (please refer to this Nested Array as we use DBSUM to sum the individual Sums of Squares further on.)

SUM1·DBALL = (3024 0 0 1814 1210 1627 1397 914 900 713 497)

$$\overline{\left(\text{SUM1xDBALL} \cdot \text{SUM1xDBALL} \cdot N^{-1} \right)} = (228614.4 \ \ 0 \ \ 0 \ \ 164529.8 \ \ 73205 \ \ 132356.45 \ \ 97580.45 \ \ 83539.6 \ \ 81000 \ \ 50836.9 \ \ 24700.9)$$

To fit on a single page, we transpose to get:

$$\left(\overrightarrow{\left(\text{SUM1xDBALLSUM1xDBALLN}^{-1}\right)}\right)^{T} = \begin{pmatrix} 228614.4 \\ 0 \\ 0 \\ 164529.8 \\ 73205 \\ 132356.45 \\ 97580.45 \\ 83539.6 \\ 81000 \\ 50836.9 \\ 24700.9 \end{pmatrix}$$

Here we add the Grand Sum of Squares and the Within Subjects terms which cannot be computed using this all at once method. But first let's compute the SSw (Within Subjects term).

This is the Within Subjects equation we will use. Note that it is of the form $([A]x[1])^2$

$$SSw := \left(\frac{1}{4}\right) \cdot (A \cdot FOUR1)^2 \cdot ONE10^{T}{}^{\blacksquare}$$

$$SS1w := A \cdot FOUR1$$

To get this to work we must transpose SS1w, since it is a column Matrix and we need to start the multiplication with a row Matrix to get a single number value.

$$SS1w^{T} = (298 \quad 295 \quad 293 \quad 302 \quad 335 \quad 370 \quad 300 \quad 334 \quad 175 \quad 322)$$

$$SS1 := \left(\frac{1}{4}\right) \cdot \left(SS1w^{T} \cdot SS1w\right) - CORRFACT$$

$$SS1 = 5853.6$$

Now we will square each term in SS1 and divide by the
corresponding N values.

$$\left(\frac{1}{4}\right) \cdot \left(\overrightarrow{\left(SS1w^T \cdot SS1w^T\right)}\right) = (22201\ \ 21756.25\ \ 21462.25\ \ 22801\ \ 28056.25\ \ 34225\ \ 22500\ \ 27889\ \ 7656.25\ \ 25921)$$

Now we will sum all the values into a single number. This is SSw.

$$SSw := \left(\frac{1}{4}\right) \cdot \left(\overrightarrow{\left(SS1w^T \cdot SS1w^T\right)}\right) \cdot \overrightarrow{ONE10}^T$$

$$SSw = 234468$$

$$GSQ = 248212$$

We will put these in by hand. We can possibly compute both at
the same time with proper placement in the 1x11 Matrix. This
is a homework problem.

$$SSadd := (0\ \ 248212\ \ 234468\ \ 0\ \ 0\ \ 0\ \ 0\ \ 0\ \ 0\ \ 0\ \ 0)$$

$$\overrightarrow{\left(SUM1xDBALL \cdot SUM1xDBALL \cdot N^{-1}\right)} = (228614.4\ \ 0\ \ 0\ \ 164529.8\ \ 73205\ \ 132356.45\ \ 97580.45\ \ 83539.6\ \ 81000\ \ 50836.9\ \ 24700.9)$$

The Post-Multiplier Nested Array for DBSUM looks like:

$$B := (1\ \ 0\ \ 0\ \ 0\ \ 0\ \ 0\ \ 0\ \ 0) \quad C := (0\ \ 1\ \ 0\ \ 0\ \ 0\ \ 0\ \ 0\ \ 0) \quad D := (0\ \ 0\ \ 1\ \ 0\ \ 0\ \ 0\ \ 0\ \ 0)$$

$$E := \begin{pmatrix} 0 & 0 & 0 & 1 & 0 & 0 & 0 & 0 \\ 0 & 0 & 0 & 1 & 0 & 0 & 0 & 0 \end{pmatrix} \quad F := \begin{pmatrix} 0 & 0 & 0 & 0 & 1 & 0 & 0 & 0 \\ 0 & 0 & 0 & 0 & 1 & 0 & 0 & 0 \end{pmatrix} \quad G := \begin{pmatrix} 0 & 0 & 0 & 0 & 0 & 1 & 0 & 0 \\ 0 & 0 & 0 & 0 & 0 & 1 & 0 & 0 \\ 0 & 0 & 0 & 0 & 0 & 1 & 0 & 0 \\ 0 & 0 & 0 & 0 & 0 & 1 & 0 & 0 \end{pmatrix}$$

$$\text{SSadda} := \begin{pmatrix} B \\ C \\ D \\ E \\ F \\ G \end{pmatrix} \qquad \text{SSadda} = \begin{bmatrix} (1\ 0\ 0\ 0\ 0\ 0\ 0\ 0) \\ (0\ 1\ 0\ 0\ 0\ 0\ 0\ 0) \\ (0\ 0\ 1\ 0\ 0\ 0\ 0\ 0) \\ \begin{pmatrix} 0\ 0\ 0\ 1\ 0\ 0\ 0\ 0 \\ 0\ 0\ 0\ 1\ 0\ 0\ 0\ 0 \end{pmatrix} \\ \begin{pmatrix} 0\ 0\ 0\ 0\ 1\ 0\ 0\ 0 \\ 0\ 0\ 0\ 0\ 1\ 0\ 0\ 0 \end{pmatrix} \\ \begin{pmatrix} 0\ 0\ 0\ 0\ 0\ 1\ 0\ 0 \\ 0\ 0\ 0\ 0\ 0\ 1\ 0\ 0 \\ 0\ 0\ 0\ 0\ 0\ 1\ 0\ 0 \\ 0\ 0\ 0\ 0\ 0\ 1\ 0\ 0 \end{pmatrix} \end{bmatrix}$$

But again, Computer Programs cannot handle Nested Arrays, so we partition it to the following single Matrix, 'remembering' we are dealing with Nested Arrays, not single Matrices. Plus we are adding SSadd. So DBSUM becomes:

$$\text{DBSUM} := \begin{pmatrix} 1\ 0\ 0\ 0\ 0\ 0\ 0\ 0 \\ 0\ 1\ 0\ 0\ 0\ 0\ 0\ 0 \\ 0\ 0\ 1\ 0\ 0\ 0\ 0\ 0 \\ 0\ 0\ 0\ 1\ 0\ 0\ 0\ 0 \\ 0\ 0\ 0\ 1\ 0\ 0\ 0\ 0 \\ 0\ 0\ 0\ 0\ 1\ 0\ 0\ 0 \\ 0\ 0\ 0\ 0\ 1\ 0\ 0\ 0 \\ 0\ 0\ 0\ 0\ 0\ 1\ 0\ 0 \\ 0\ 0\ 0\ 0\ 0\ 1\ 0\ 0 \\ 0\ 0\ 0\ 0\ 0\ 1\ 0\ 0 \\ 0\ 0\ 0\ 0\ 0\ 1\ 0\ 0 \end{pmatrix} \qquad \left(\overrightarrow{\text{SUM1xDBALLSUM1xDBALLN}^{-1}} \right)^{T} = \begin{pmatrix} 228614.4 \\ 0 \\ 0 \\ 164529.8 \\ 73205 \\ 132356.45 \\ 97580.45 \\ 83539.6 \\ 81000 \\ 50836.9 \\ 24700.9 \end{pmatrix}$$

$$\text{SSadd} := (0\ \ 248212\ \ 234468\ \ 0\ \ 0\ \ 0\ \ 0\ \ 0\ \ 0\ \ 0\ \ 0)$$

$$\left(\overrightarrow{\text{SUM1xDBALLSUM1xDBALLN}^{-1}} + \text{SSadd} \right) \cdot \text{DBSUM} = (228614.4\ \ 248212\ \ 234468\ \ 237734.8\ \ 229936.9\ \ 240077.4\ \ 0\ \ 0)$$

Transposing to get it all on a single page we get:

$$\left[\left(\overrightarrow{\left(\text{SUM1xDBALL·SUM1xDBALL·N}^{-1}\right)} + \text{SSadd}\right)\cdot\text{DBSUM}\right]^{T} = \begin{pmatrix} 228614.4 \\ 248212 \\ 234468 \\ 237734.8 \\ 229936.9 \\ 240077.4 \\ 0 \\ 0 \end{pmatrix}$$

Let's compute the error1 and error2 terms separately from the above values. The DBALL for the error terms is given by

$$\text{DBALLerr} := \begin{pmatrix} 0 & 0 & 0 & 0 & 0 & 0 & 0 & 1 & 0 & 1 & 0 & 0 \\ 0 & 0 & 0 & 0 & 0 & 0 & 0 & 1 & 0 & 0 & 1 & 0 \\ 0 & 0 & 0 & 0 & 0 & 0 & 0 & 0 & 1 & 1 & 0 & 0 \\ 0 & 0 & 0 & 0 & 0 & 0 & 0 & 0 & 1 & 0 & 1 & 0 \end{pmatrix}$$

Multiplying A by this Matrix we get

$$\text{A}\cdot\text{DBALLerr} = \begin{pmatrix} 0 & 0 & 0 & 0 & 0 & 0 & 0 & 183 & 115 & 158 & 140 & 0 \\ 0 & 0 & 0 & 0 & 0 & 0 & 0 & 183 & 112 & 171 & 124 & 0 \\ 0 & 0 & 0 & 0 & 0 & 0 & 0 & 200 & 93 & 150 & 143 & 0 \\ 0 & 0 & 0 & 0 & 0 & 0 & 0 & 181 & 121 & 163 & 139 & 0 \\ 0 & 0 & 0 & 0 & 0 & 0 & 0 & 190 & 145 & 183 & 152 & 0 \\ 0 & 0 & 0 & 0 & 0 & 0 & 0 & 200 & 170 & 200 & 170 & 0 \\ 0 & 0 & 0 & 0 & 0 & 0 & 0 & 184 & 116 & 164 & 136 & 0 \\ 0 & 0 & 0 & 0 & 0 & 0 & 0 & 189 & 145 & 183 & 151 & 0 \\ 0 & 0 & 0 & 0 & 0 & 0 & 0 & 104 & 71 & 83 & 92 & 0 \\ 0 & 0 & 0 & 0 & 0 & 0 & 0 & 200 & 122 & 172 & 150 & 0 \end{pmatrix}$$

We make the following substitution.

AxDBALLerr := A·DBALLerr

The equation now becomes:

$$\left(\frac{1}{2}\right) \cdot \left(\text{ONE10}\overrightarrow{(\text{AxDBALLerr AxDBALLerr})}\right) = (0\ \ 0\ \ 0\ \ 0\ \ 0\ \ 0\ \ 0\ \ 168116\ \ 76695\ \ 136810.5\ \ 99505.5\ \ 0)$$

Now we must sum the first 2 values and the last two values together. The DBSUM Matrix is given by (remember, this is also a Nested Array).

$$\text{SUMALLerr} := \begin{pmatrix} 0 & 0 & 0 & 0 & 0 & 0 & 0 & 0 \\ 0 & 0 & 0 & 0 & 0 & 0 & 0 & 0 \\ 0 & 0 & 0 & 0 & 0 & 0 & 0 & 0 \\ 0 & 0 & 0 & 0 & 0 & 0 & 0 & 0 \\ 0 & 0 & 0 & 0 & 0 & 0 & 0 & 0 \\ 0 & 0 & 0 & 0 & 0 & 0 & 0 & 0 \\ 0 & 0 & 0 & 0 & 0 & 0 & 0 & 0 \\ 0 & 0 & 0 & 0 & 0 & 0 & 1 & 0 \\ 0 & 0 & 0 & 0 & 0 & 0 & 1 & 0 \\ 0 & 0 & 0 & 0 & 0 & 0 & 0 & 1 \\ 0 & 0 & 0 & 0 & 0 & 0 & 0 & 1 \\ 0 & 0 & 0 & 0 & 0 & 0 & 0 & 0 \end{pmatrix}$$

Multiplying we get:

$$\left(\frac{1}{2}\right) \cdot \left(\text{ONE10}\overrightarrow{(\text{AxDBALLerr AxDBALLerr})}\right)\text{SUMALLerr} = (0\ \ 0\ \ 0\ \ 0\ \ 0\ \ 0\ \ 244811\ \ 236316)$$

$$\text{SSadderr} := \left(\frac{1}{2}\right) \cdot \left(\text{ONE10}\overrightarrow{(\text{AxDBALLerr AxDBALLerr})}\right)\text{SUMALLerr}$$

The Correction Factor Matrix is given by:

$$CF := \begin{pmatrix} -1 & -1 & -1 & -1 & -1 & -1 & -1 \\ 1 & 0 & 0 & 0 & 0 & 0 & 0 \\ 0 & 1 & 0 & 0 & 0 & 0 & 0 \\ 0 & 0 & 1 & 0 & 0 & 0 & 0 \\ 0 & 0 & 0 & 1 & 0 & 0 & 0 \\ 0 & 0 & 0 & 0 & 1 & 0 & 0 \\ 0 & 0 & 0 & 0 & 0 & 1 & 0 \\ 0 & 0 & 0 & 0 & 0 & 0 & 1 \end{pmatrix}$$

Here we subtract the Correction Factor from the raw Sum of Squares values.

$$\left[\left(\overrightarrow{\left(\text{SUM1xDBALLSUM1xDBALLN}^{-1}\right)} + \text{SSadd}\right) \cdot \text{DBSUM}\right] + \text{SSadderr} = (228614.4\ 248212\ 234468\ 237734.8\ 229936.9\ 240077.4\ 244811\ 236316)$$

Since this won't fit on a page, we will transpose the result so we can see it all.

$$\left[\left[\left(\overrightarrow{\left(\text{SUM1xDBALLSUM1xDBALLN}^{-1}\right)} + \text{SSadd}\right) \cdot \text{DBSUM}\right] + \text{SSadderr}\right]^T = \begin{pmatrix} 228614.4 \\ 248212 \\ 234468 \\ 237734.8 \\ 229936.9 \\ 240077.4 \\ 244811 \\ 236316 \end{pmatrix}$$

We subtract the correction Factor from all the terms.

$$\left[\left[\left(\overrightarrow{\left(\text{SUM1xDBALLSUM1xDBALLN}^{-1}\right)} + \text{SSadd}\right) \cdot \text{DBSUM}\right] + \text{SSadderr}\right] \cdot \text{CF} = (19597.6\ 5853.6\ 9120.4\ 1322.5\ 11463\ 16196.6\ 7701.6)$$

The transpose of this equation is:

$$\left[\left[\left(\overrightarrow{\text{SUM1xDBALLSUM1xDBALLN}}^{-1}\right) + \text{SSadd}\right) \cdot \text{DBSUM}\right] + \text{SSadderr}\right] \cdot \text{CF} \right]^T = \begin{pmatrix} 19597.6 \\ 5853.6 \\ 9120.4 \\ 1322.5 \\ 11463 \\ 16196.6 \\ 7701.6 \end{pmatrix}$$

We now will compose the Matrix which subtracts the secondary Correction Factors SS1, SS2 and SS1x2. We will call this Matrix SS12.

$$\text{SS12} := \begin{pmatrix} 1 & 0 & 0 & 0 & 0 & 0 & 0 & 0 \\ 0 & 1 & 0 & 0 & 0 & -1 & -1 & 0 \\ 0 & 0 & 1 & 0 & -1 & -1 & 0 & 0 \\ 0 & 0 & 0 & 1 & -1 & 0 & -1 & 0 \\ 0 & 0 & 0 & 0 & 1 & 0 & 0 & 0 \\ 0 & 0 & 0 & 0 & 0 & 1 & 0 & 0 \\ 0 & 0 & 0 & 0 & 0 & 0 & 1 & 0 \end{pmatrix}$$

The Multiplication now becomes:

$$\left[\left[\left(\overrightarrow{\text{SUM1xDBALLSUM1xDBALLN}}^{-1}\right) + \text{SSadd}\right) \cdot \text{DBSUM}\right] + \text{SSadderr}\right] \cdot \text{CF} \cdot \text{SS12} = (19597.6 \ 5853.6 \ 9120.4 \ 1322.5 \ 1020.1 \ 1222.6 \ 525.5 \ 0)$$

To fit on a single page, we again transpose the equation.

$$\left[\left[\left(\overrightarrow{\text{SUM1xDBALLSUM1xDBALLN}}^{-1}\right) + \text{SSadd}\right) \cdot \text{DBSUM}\right] + \text{SSadderr}\right] \cdot \text{CF} \cdot \text{SS12} \right]^T = \begin{pmatrix} 19597.6 \\ 5853.6 \\ 9120.4 \\ 1322.5 \\ 1020.1 \\ 1222.6 \\ 525.5 \\ 0 \end{pmatrix}$$

Now we solve for the final total Error for the system. The Matrix for this is given by SStotalerr

$$\text{SStotalerr} := \begin{pmatrix} 1 & 0 & 0 & 0 & 0 & 0 & 0 & 1 \\ 0 & 1 & 0 & 0 & 0 & 0 & 0 & -1 \\ 0 & 0 & 1 & 0 & 0 & 0 & 0 & -1 \\ 0 & 0 & 0 & 1 & 0 & 0 & 0 & -1 \\ 0 & 0 & 0 & 0 & 1 & 0 & 0 & -1 \\ 0 & 0 & 0 & 0 & 0 & 1 & 0 & -1 \\ 0 & 0 & 0 & 0 & 0 & 0 & 1 & -1 \\ 0 & 0 & 0 & 0 & 0 & 0 & 0 & -1 \end{pmatrix}$$

$$\left[\left[\left[\left(\overrightarrow{\text{SUM1xDBALL}\cdot\text{SUM1xDBALL}\cdot N^{-1}}\right) + \text{SSadd}\right)\cdot\text{DBSUM}\right] + \text{SSadderr}\right]\cdot\text{CF}\cdot\text{SS12}\cdot\text{SStotalerr} = (\,19597.6\ \ 5853.6\ \ 9120.4\ \ 1322.5\ \ 1020.1\ \ 1222.6\ \ 525.5\ \ 532.9\,)$$

$$\left[\left[\left[\left(\overrightarrow{\text{SUM1xDBALL}\cdot\text{SUM1xDBALL}\cdot N^{-1}}\right) + \text{SSadd}\right)\cdot\text{DBSUM}\right] + \text{SSadderr}\right]\cdot\text{CF}\cdot\text{SS12}\cdot\text{SStotalerr}\right]^{T} = \begin{pmatrix} 19597.6 \\ 5853.6 \\ 9120.4 \\ 1322.5 \\ 1020.1 \\ 1222.6 \\ 525.5 \\ 532.9 \end{pmatrix}$$

We have solved this Statistic in a single programming step, (a
single command line of programming, as opposed to many pages
of regular computer programming code).

PROBLEM 2.7: TWO-FACTOR MIXED DESIGN, REPEATED MEASURES ON ONE FACTOR

COMPUTATIONAL HANDBOOKS OF STATISTICS, 2ND ED. SECTION 2.7, PG. 55-61

STEP 1: First we table the data as a Matrix, then we will define the [1] operators and other variables:

$$
DATA1 := \begin{pmatrix}
5 & 5 & 6 & 4 & 5 & 8 & 4 & 6 & 10 & 3 & 4 & 7 \\
2 & 3 & 3 & 1 & 3 & 4 & 2 & 5 & 2 & 2 & 5 & 5 \\
2 & 4 & 5 & 4 & 6 & 8 & 5 & 8 & 9 & 3 & 5 & 4 \\
1 & 1 & 2 & 1 & 3 & 5 & 1 & 6 & 10 & 2 & 3 & 4 \\
3 & 4 & 2 & 5 & 8 & 3 & 2 & 4 & 4 & 3 & 5 & 2 \\
4 & 5 & 4 & 3 & 2 & 3 & 6 & 5 & 6 & 7 & 6 & 7 \\
4 & 4 & 5 & 3 & 4 & 5 & 5 & 4 & 3 & 1 & 2 & 3 \\
1 & 3 & 5 & 7 & 9 & 11 & 2 & 4 & 6 & 5 & 4 & 3 \\
5 & 6 & 6 & 2 & 2 & 4 & 3 & 3 & 1 & 1 & 6 & 6 \\
6 & 5 & 3 & 2 & 1 & 1 & 3 & 2 & 5 & 5 & 13 & 2 \\
3 & 2 & 1 & 6 & 7 & 2 & 2 & 3 & 6 & 7 & 8 & 9
\end{pmatrix}
$$

VARIABLES:

$ONE11 := (\ 1\quad 1\quad 1\quad 1\quad 1\quad 1\quad 1\quad 1\quad 1\quad 1\quad 1\ )$

$THREE1 := \begin{pmatrix} 1 \\ 1 \\ 1 \end{pmatrix}$

$TWELVE1 := \begin{pmatrix} 1 \\ 1 \\ 1 \\ 1 \\ 1 \\ 1 \\ 1 \\ 1 \\ 1 \\ 1 \\ 1 \\ 1 \end{pmatrix}$

$ONE12 := TWELVE1^{T}$

$$\text{DB1a} := \begin{pmatrix} 1 & 0 & 0 & 0 \\ 1 & 0 & 0 & 0 \\ 1 & 0 & 0 & 0 \\ 0 & 1 & 0 & 0 \\ 0 & 1 & 0 & 0 \\ 0 & 1 & 0 & 0 \\ 0 & 0 & 1 & 0 \\ 0 & 0 & 1 & 0 \\ 0 & 0 & 1 & 0 \\ 0 & 0 & 0 & 1 \\ 0 & 0 & 0 & 1 \\ 0 & 0 & 0 & 1 \end{pmatrix} \qquad \text{DB2a} := \begin{pmatrix} 1 & 0 & 0 & 0 \\ 0 & 1 & 0 & 0 \\ 0 & 0 & 1 & 0 \\ 0 & 0 & 0 & 1 \\ 1 & 0 & 0 & 0 \\ 0 & 1 & 0 & 0 \\ 0 & 0 & 1 & 0 \\ 0 & 0 & 0 & 1 \\ 1 & 0 & 0 & 0 \\ 0 & 1 & 0 & 0 \\ 0 & 0 & 1 & 0 \\ 0 & 0 & 0 & 1 \end{pmatrix}$$

$\text{ONE11} \cdot \text{DATA1} \cdot \text{DB2a} = (\ 148 \quad 135 \quad 138 \quad 140\ )$

STEP 2: Add the scores in each Group, we call this value SUM1.

$$\text{SUM1a} := \left[\left(1_{1,\,11} \right) \cdot \left(A_{11,\,12} \right) \right]_{\blacksquare}^{\blacksquare}$$

$\text{SUM1a} := \text{ONE11} \cdot \text{DATA1}$

$\text{SUM1a} = (\ 36 \quad 42 \quad 42 \quad 38 \quad 50 \quad 54 \quad 35 \quad 50 \quad 62 \quad 39 \quad 61 \quad 52\ )$

STEP 3: Add the sums of each line in each Group.

$$\text{GROUPSUMS} := \left(1_{1,\,11} \right) \cdot \left(A_{11,\,12} \right) \cdot \text{DB1a}_{12,\,4}^{\blacksquare}$$

$\text{GROUPSUMS} := \text{ONE11} \cdot \text{DATA1} \cdot \text{DB1a}$

$\text{GROUPSUMS} = (\ 120 \quad 142 \quad 147 \quad 152\ )$

STEP 4: Now we will add the scores for each individual person (subject) in the Group:

$$\text{SUBJECTSUMS1} := \left(A_{11,\,12} \cdot DB1_{12,\,4}\right)^{\blacksquare}$$

$$\text{SUBJECTSUMS1} = \begin{pmatrix} 16 & 17 & 20 & 14 \\ 8 & 8 & 9 & 12 \\ 11 & 18 & 22 & 12 \\ 4 & 9 & 17 & 9 \\ 9 & 16 & 10 & 10 \\ 13 & 8 & 17 & 20 \\ 13 & 12 & 12 & 6 \\ 9 & 27 & 12 & 12 \\ 17 & 8 & 7 & 13 \\ 14 & 4 & 10 & 20 \\ 6 & 15 & 11 & 24 \end{pmatrix}$$

$$\text{SUBJECTSUMS1} := \text{DATA1} \cdot \text{DB1a}$$

STEP 5: Now we Square each term in the DATA and sum them together. This can be done 2 ways, one easy, one a little more complicated. First the complicated:

We square the data matrix, remove the diagonal and sum it together.

$$D := \text{DATA1} \cdot \text{DATA1}^T$$

$$\text{DATA1} \cdot \text{DATA1}^T = \begin{pmatrix} 417 & 213 & 385 & 267 & 246 & 323 & 248 & 355 & 250 & 249 & 308 \\ 213 & 135 & 204 & 132 & 138 & 184 & 133 & 185 & 155 & 156 & 180 \\ 385 & 204 & 381 & 258 & 244 & 299 & 234 & 352 & 221 & 233 & 286 \\ 267 & 132 & 258 & 207 & 150 & 197 & 137 & 223 & 126 & 149 & 198 \\ 246 & 138 & 244 & 150 & 201 & 201 & 157 & 250 & 156 & 183 & 230 \\ 323 & 184 & 299 & 197 & 201 & 310 & 196 & 259 & 220 & 257 & 287 \\ 248 & 133 & 234 & 137 & 157 & 196 & 171 & 219 & 169 & 149 & 171 \\ 355 & 185 & 352 & 223 & 250 & 259 & 219 & 392 & 200 & 197 & 287 \\ 250 & 155 & 221 & 126 & 156 & 220 & 169 & 200 & 213 & 203 & 197 \\ 249 & 156 & 233 & 149 & 183 & 257 & 149 & 197 & 203 & 312 & 251 \\ 308 & 180 & 286 & 198 & 230 & 287 & 171 & 287 & 197 & 251 & 346 \end{pmatrix}$$

$$\text{GRANDSUMSQ1} := \left(1_{1,\,11}\right) \cdot \text{diag}\left[\left(A_{11,\,12}\right)\left(\left(A_{11,\,12}\right)^T\right)\right]^{\blacksquare}$$

$$\text{GRANDSUMSQ1} := \text{ONE11} \cdot \text{diag}\left(\text{DATA1} \cdot \text{DATA1}^T\right)$$

$$\text{diag}(D) = \begin{pmatrix} 417 \\ 135 \\ 381 \\ 207 \\ 201 \\ 310 \\ 171 \\ 392 \\ 213 \\ 312 \\ 346 \end{pmatrix}$$

GRANDSUMSQ1 = 3085

$$\text{GRANDSUMSQ2} := \left(1_{1,\,11}\right)\cdot\left(\overrightarrow{\left[\left(A_{11,\,12}\right)\cdot\left(A_{11,\,12}\right)\right]}\right)\cdot\left(1_{12,\,1}\right)^{\blacksquare}$$

$$\text{GRANDSUMSQ2} := \text{ONE11}\cdot\overrightarrow{[\,(\text{DATA1})\cdot(\text{DATA1})\,]}\cdot\text{TWELVE1}$$

$$(1\ 1\ 1\ 1\ 1\ 1\ 1\ 1\ 1\ 1\ 1)\left[\begin{pmatrix} 5 & 5 & 6 & 4 & 5 & 8 & 4 & 6 & 10 & 3 & 4 & 7 \\ 2 & 3 & 3 & 1 & 3 & 4 & 2 & 5 & 2 & 2 & 5 & 5 \\ 2 & 4 & 5 & 4 & 6 & 8 & 5 & 8 & 9 & 3 & 5 & 4 \\ 1 & 1 & 2 & 1 & 3 & 5 & 1 & 6 & 10 & 2 & 3 & 4 \\ 3 & 4 & 2 & 5 & 8 & 3 & 2 & 4 & 4 & 3 & 5 & 2 \\ 4 & 5 & 4 & 3 & 2 & 3 & 6 & 5 & 6 & 7 & 6 & 7 \\ 4 & 4 & 5 & 3 & 4 & 5 & 5 & 4 & 3 & 1 & 2 & 3 \\ 1 & 3 & 5 & 7 & 9 & 11 & 2 & 4 & 6 & 5 & 4 & 3 \\ 5 & 6 & 6 & 2 & 2 & 4 & 3 & 3 & 1 & 1 & 6 & 6 \\ 6 & 5 & 3 & 2 & 1 & 1 & 3 & 2 & 5 & 5 & 13 & 2 \\ 3 & 2 & 1 & 6 & 7 & 2 & 2 & 3 & 6 & 7 & 8 & 9 \end{pmatrix}\cdot\begin{pmatrix} 5 & 5 & 6 & 4 & 5 & 8 & 4 & 6 & 10 & 3 & 4 & 7 \\ 2 & 3 & 3 & 1 & 3 & 4 & 2 & 5 & 2 & 2 & 5 & 5 \\ 2 & 4 & 5 & 4 & 6 & 8 & 5 & 8 & 9 & 3 & 5 & 4 \\ 1 & 1 & 2 & 1 & 3 & 5 & 1 & 6 & 10 & 2 & 3 & 4 \\ 3 & 4 & 2 & 5 & 8 & 3 & 2 & 4 & 4 & 3 & 5 & 2 \\ 4 & 5 & 4 & 3 & 2 & 3 & 6 & 5 & 6 & 7 & 6 & 7 \\ 4 & 4 & 5 & 3 & 4 & 5 & 5 & 4 & 3 & 1 & 2 & 3 \\ 1 & 3 & 5 & 7 & 9 & 11 & 2 & 4 & 6 & 5 & 4 & 3 \\ 5 & 6 & 6 & 2 & 2 & 4 & 3 & 3 & 1 & 1 & 6 & 6 \\ 6 & 5 & 3 & 2 & 1 & 1 & 3 & 2 & 5 & 5 & 13 & 2 \\ 3 & 2 & 1 & 6 & 7 & 2 & 2 & 3 & 6 & 7 & 8 & 9 \end{pmatrix}\right]\cdot\begin{pmatrix} 1 \\ 1 \\ 1 \\ 1 \\ 1 \\ 1 \\ 1 \\ 1 \\ 1 \\ 1 \\ 1 \\ 1 \end{pmatrix} = 3085$$

GRANDSUMSQ2 = 3085

GRANDSUMSQ2 = 3085

STEP 6:   Sum each score in the data for the Grand Sum:

$$\text{GRANDSUM1} := \left(1_{1,\,11}\right)\cdot\left(A_{11,\,12}\right)\cdot\left(1_{12,\,1}\right)^{\blacksquare}$$

$$\text{GRANDSUM1} := \text{ONE11}\cdot\text{DATA1}\cdot\text{TWELVE1}$$

GRANDSUM1 = 561

And the Math looks like:

$$
\text{GRANDSUM1} := \begin{pmatrix} 1 & 1 & 1 & 1 & 1 & 1 & 1 & 1 & 1 & 1 & 1 \end{pmatrix} \cdot \begin{pmatrix} 5 & 5 & 6 & 4 & 5 & 8 & 4 & 6 & 10 & 3 & 4 & 7 \\ 2 & 3 & 3 & 1 & 3 & 4 & 2 & 5 & 2 & 2 & 5 & 5 \\ 2 & 4 & 5 & 4 & 6 & 8 & 5 & 8 & 9 & 3 & 5 & 4 \\ 1 & 1 & 2 & 1 & 3 & 5 & 1 & 6 & 10 & 2 & 3 & 4 \\ 3 & 4 & 2 & 5 & 8 & 3 & 2 & 4 & 4 & 3 & 5 & 2 \\ 4 & 5 & 4 & 3 & 2 & 3 & 6 & 5 & 6 & 7 & 6 & 7 \\ 4 & 4 & 5 & 3 & 4 & 5 & 5 & 4 & 3 & 1 & 2 & 3 \\ 1 & 3 & 5 & 7 & 9 & 11 & 2 & 4 & 6 & 5 & 4 & 3 \\ 5 & 6 & 6 & 2 & 2 & 4 & 3 & 3 & 1 & 1 & 6 & 6 \\ 6 & 5 & 3 & 2 & 1 & 1 & 3 & 2 & 5 & 5 & 13 & 2 \\ 3 & 2 & 1 & 6 & 7 & 2 & 2 & 3 & 6 & 7 & 8 & 9 \end{pmatrix} \cdot \begin{pmatrix} 1 \\ 1 \\ 1 \\ 1 \\ 1 \\ 1 \\ 1 \\ 1 \\ 1 \\ 1 \\ 1 \\ 1 \end{pmatrix}
$$

$\text{GRANDSUM1} = 561$

STEP 7: Now we will compute the Correction Factor CORRFACT: Square the Grand Sum and divide by the total number (#) of measures in the table. Nt = 11*12=132

$$\text{CORRFACT1} := \left(\frac{1}{\text{Nt1}} \right) \cdot \left[\left(\vec{1}_{1,\,11} \right) \left(\overset{A}{}_{11,\,9} \right) \left(\vec{1}_{9,\,1} \right) \right]^{2^{\blacksquare}} \qquad\qquad \text{Nt1} := 132$$

$$\text{CORRFACT1} := \left(\frac{1}{\text{Nt1}} \right) \cdot \left(\text{ONE11} \cdot \text{DATA1} \cdot \text{TWELVE1} \right)^2$$

$$\text{CORRFACT1} := \left(\frac{1}{\text{Nt1}} \right) \cdot \text{GRANDSUM1}^2$$

$$\text{CORRFACT1} := \left(\frac{1}{132} \right) \cdot 561^2$$

$\text{CORRFACT1} = 2384.25$

STEP 8: Computation of the Total Sum of Squares SST1: Subtract the Correction Factor from the Grand Sum of Squares.

$$\text{SST1} := \left(\vec{1}_{1,\,11} \right) \cdot \left(\overrightarrow{\left[\overset{A}{}_{11,\,12} \right) \left(\overset{A}{}_{11,\,12}} \right] \right) \cdot \left(\vec{1}_{12,\,1} \right) - \left(\frac{1}{\text{Nt1}} \right) \cdot \left[\left(\vec{1}_{1,\,11} \right) \left(\overset{A}{}_{11,\,9} \right) \left(\vec{1}_{9,\,1} \right) \right]^{2^{\blacksquare}}$$

$$SST1 := ONE11 \cdot \overrightarrow{[(DATA1) \cdot (DATA1)]} \cdot TWELVE1 - \left(\frac{1}{Nt1} \right) \cdot (ONE11 \cdot DATA1 \cdot TWELVE1)^2$$

$$SST1 := GRANDSUMSQ1 - CORRFACT1$$

$$SST1 = 700.75$$

BETWEEN SUBJECTS EFFECTS

STEP 9: Computation of the Between Subjects Sum of Squares SSb. Nb = # of trials given each subject. N=3

$$SSb1 := \left(\frac{1}{Nb} \right) \cdot \overrightarrow{\left[\overline{\left(A_{11, 12} \cdot DB1_{12, 4} \right) \cdot \left(A_{11, 12} \cdot DB1_{12, 4} \right)} \right] \cdot \left(1_{4, 1} \right)} - \left(\frac{1}{Nt1} \right) \cdot \left[\left(1_{1, 11} \right) \cdot \left(A_{11, 9} \right) \cdot \left(1_{9, 1} \right) \right]^2 \blacksquare$$

Now the math for SSb looks complicated, let's simplify it if we can.

$$SSb := \left(\frac{1}{Nb} \right) [ONE11 \, [\, \overline{\left[(DATA1 \cdot DB1a) \cdot (DATA1 \cdot DB1a) \right] \cdot FOUR1} \,) - CORRFACT1$$

Mathcad will not let us solve it in this particular order, so we will have to pre-compute DATA*DB1. Note: We have to be very careful with the placing of parenthesis.

$$SUM1B := DATA1 \cdot DB1a \qquad\qquad Nb1 := 3 \qquad\qquad FOUR1 := \begin{pmatrix} 1 \\ 1 \\ 1 \\ 1 \end{pmatrix}$$

So now we have a solvable computation:

$$SSb1 := \left(\frac{1}{Nb1} \right) \cdot \left[ONE11 \cdot \overrightarrow{\left([\, (SUM1B) \cdot (SUM1B) \,] \right)} \right] \cdot FOUR1 - CORRFACT1$$

$$SSb1 = 390.75$$

Lets check and see how this works: first we compute SUM1b, then square it element by element.

$$SUM1B = \begin{pmatrix} 16 & 17 & 20 & 14 \\ 8 & 8 & 9 & 12 \\ 11 & 18 & 22 & 12 \\ 4 & 9 & 17 & 9 \\ 9 & 16 & 10 & 10 \\ 13 & 8 & 17 & 20 \\ 13 & 12 & 12 & 6 \\ 9 & 27 & 12 & 12 \\ 17 & 8 & 7 & 13 \\ 14 & 4 & 10 & 20 \\ 6 & 15 & 11 & 24 \end{pmatrix} \qquad \overrightarrow{(SUM1B \cdot SUM1B)} = \begin{pmatrix} 256 & 289 & 400 & 196 \\ 64 & 64 & 81 & 144 \\ 121 & 324 & 484 & 144 \\ 16 & 81 & 289 & 81 \\ 81 & 256 & 100 & 100 \\ 169 & 64 & 289 & 400 \\ 169 & 144 & 144 & 36 \\ 81 & 729 & 144 & 144 \\ 289 & 64 & 49 & 169 \\ 196 & 16 & 100 & 400 \\ 36 & 225 & 121 & 576 \end{pmatrix}$$

Now we sum the columns, divide by Nb and add the rows to get 2775,

$$ONE11 \cdot \overrightarrow{(SUM1B \cdot SUM1B)} = (\,1478 \quad 2256 \quad 2201 \quad 2390\,)$$

$$\left(\frac{1}{3}\right) \cdot \left(ONE11 \cdot \overrightarrow{(SUM1B \cdot SUM1B)}\right) \cdot \begin{pmatrix} 1 \\ 1 \\ 1 \\ 1 \end{pmatrix} = 2775$$

Subtracting the Correction Factor we get 391.

$$\left(\frac{1}{3}\right) \cdot \left(ONE11 \cdot \overrightarrow{(SUM1B \cdot SUM1B)}\right) \cdot \begin{pmatrix} 1 \\ 1 \\ 1 \\ 1 \end{pmatrix} - CORRFACT1 = 390.75$$

STEP 10: Computation of the effects of the conditions on overall performance (the meaningfulness effects) SSc. Nc= Nt/#columns in DB! 132/4=33.

$Nc := 33$

$$SSc1 := \left(\frac{1}{Nc}\right) \cdot \left[\left(1_{1,\,11}\right) \cdot A_{11,\,12} \cdot DB1_{12,\,4}\right]^2 - \left(\frac{1}{Nt}\right) \cdot \left[\left(1_{1,\,11}\right) \cdot \left(A_{11,\,12}\right) \cdot \left(1_{12,\,1}\right)\right]^{2\blacksquare}$$

$$SSc1 := \left(\frac{1}{Nc}\right) \cdot \left(ONE11 \cdot DATA1 \cdot DB1a\right)^2 - CORRFACT1$$

Again, Mathcad cannot do what we wish here, so we have to rewrite the equation.

$$\text{SSc1} := \left(\frac{1}{Nc}\right) \cdot \left[(\text{ONE11} \cdot \text{DATA1} \cdot \text{DB1a}) \cdot (\text{ONE11} \cdot \text{DATA1} \cdot \text{DB1a})^T \right] - \text{CORRFACT1}$$

$$\text{SSc1} = 18.083$$

STEP 11: Computation of Between Subjects Error term SSerrb.

$$\text{SSerrb1} := \text{SSb1} - \text{SSc1}$$

$$\text{SSerrb1} = 372.667$$

The Program for this computation is:

$$\text{SSerrb1} := \left[\left(\frac{1}{Nb1}\right)\left[\overrightarrow{\left(\left(\!\left(_{11,\,12}\cdot^{DB1}_{12,\,4}\right)\right)\!\left(_{11,\,12}\cdot^{DB1}_{12,\,4}\right)}\right)\!\left(_{4,\,1}\right) - \left(\frac{1}{Nt1}\right)\left[\left(_{1,\,11}\right)\left(_{11,\,9}\right)\left(_{9,\,1}\right)\right]^2\right] - \left[\left(\frac{1}{Nc}\right)\left[\left(_{1,\,11}\right)^A_{11,\,12}\cdot^{DB1}_{12,\,4}\right]^2 - \left(\frac{1}{Nt1}\right)\left[\left(_{1,\,11}\right)\left(_{11,\,12}\right)\left(_{12,\,1}\right)\right]^2\right]$$

The actual equation looks like:

$$\text{SSerrb1} := \left(\frac{1}{Nb1}\right)\cdot\left[\text{ONE11}\cdot\overrightarrow{\left[(\text{SUM1B})\cdot(\text{SUM1B})\right]}\right]\cdot\text{FOUR1} - \text{CORRFACT1} - \left[\left(\frac{1}{Nc}\right)\left[(\text{ONE11}\cdot\text{DATA1}\cdot\text{DB1a})\cdot(\text{ONE11}\cdot\text{DATA1}\cdot\text{DB1a})^T\right] - \text{CORRFACT1}\right]$$

The solution is:

$$\text{SSerrb1} = 372.667$$

STEP 12: Computation of Within Subjects Sum of Squares SSw

$$\text{SSw1} := \text{SST1} - \text{SSb1}$$

The program looks like:

$$\text{SSw1} := \left[\left(_{1,\,11}\right)\left(\overrightarrow{\left[\left(_{11,\,12}\right)\left(_{11,\,12}\right)\right]}\right)\left(_{12,\,1}\right) - \left(\frac{1}{Nt1}\right)\left[\left(_{1,\,11}\right)\left(_{11,\,9}\right)\left(_{9,\,1}\right)\right]^2\right] - \left[\left(\frac{1}{Nb}\right)\cdot\left[\overrightarrow{\left(\!\left(_{11,\,12}\cdot^{DB1}_{12,\,4}\right)\right)\!\left(_{11,\,12}\cdot^{DB1}_{12,\,4}\right)}\right]\left(_{4,\,1}\right) - \left(\frac{1}{Nt1}\right)\left[\left(_{1,\,11}\right)\left(_{11,\,9}\right)\left(_{9,\,1}\right)\right]^2\right]$$

The equation looks like:

$$\text{SSw1} := \left[\text{ONE11}\cdot\overrightarrow{\left[(\text{DATA1})\cdot(\text{DATA1})\right]}\cdot\text{TWELVE1} - \left(\frac{1}{Nt1}\right)\cdot(\text{ONE11}\cdot\text{DATA1}\cdot\text{TWELVE1})^2 - \left[\left(\frac{1}{Nb1}\right)\cdot\left[\text{ONE11}\cdot\overrightarrow{\left[(\text{SUM1B})\cdot(\text{SUM1B})\right]}\right]\cdot\text{FOUR1} - \text{CORRFACT1}\right]\right]$$

And the solution is:

SSw1 = 310

STEP 13: Computation for Sum of Squares for Trials SStr. Ntr = Nt/#columns in DB#3. Ntr = 132/4 = 33

Ntr := 33

$$\text{SStr1} := \left(\frac{1}{\text{Ntr}}\right) \cdot \left[\left(1_{1,11}\right) \cdot \left(A_{11,9}\right) \cdot \left(DB2_{9,3}\right)\right]^2 - \text{CORRFACT} \quad \blacksquare$$

$$\text{SStr1} := \left(\frac{1}{\text{Ntr}}\right) \cdot \left[(\text{ONE11} \cdot \text{DATA1} \cdot \text{DB2a}) \cdot (\text{ONE11} \cdot \text{DATA1} \cdot \text{DB2a})^T\right] - \text{CORRFACT1}$$

SUM1a·DB2a = (148   135   138   140)

$$\left(\frac{\text{SUM1a} \cdot \text{DB2a} \cdot (\text{SUM1a} \cdot \text{DB2a})^T}{33}\right) - \text{CORRFACT1} = 2.811$$

SStr1 := 3

STEP 14: Computation of the Sum of Squares of SUM1a: SStrxc. Ntrxc1= Nt/# elements in SUM1a. Ntrxc1 = 132/12 = 11

Ntrxc1 := 11

We already know that MathCad cannot handle the following equation, so we will make The substitution now. Let

SUM1a := ONE11·DATA1

Then

$$\text{SStrxc1} := \left(\frac{1}{\text{Ntrxc1}}\right) \cdot \left(\text{SUM1a} \cdot \text{SUM1a}^T\right) - \text{CORRFACT1} - \text{SSc1} - \text{SStr1}$$

SStrxc1 = 67.303

STEP 15: Computation of Within Subjects Error term SSerrw1.

SSerrw1 := SSw1 − SStr1 − SStrxc1

SSerrw1 = 239.697

The computations are complete, now we will find the Degrees of Freedom for this Statistic. The computation for df holds for all computations of this type no matter how large the study.

$$
SS1 := \begin{pmatrix} SST1 \\ SSb1 \\ SSc1 \\ SSerrb1 \\ SSw1 \\ SStr1 \\ SStrxc1 \\ SSerrw1 \end{pmatrix}
\qquad
SS1 = \begin{pmatrix} 700.75 \\ 390.75 \\ 18.083 \\ 372.667 \\ 310 \\ 3 \\ 67.303 \\ 239.697 \end{pmatrix}
$$

STEP 16: Computation of Degrees of Freedom

m1 := 132   s1 := 33   G1 := 4   Tr1 := 4

$$
df1 := \begin{bmatrix} m1 - 1 \\ s1 - 1 \\ G1 - 1 \\ (s1 - 1) - (G1 - 1) \\ (m1 - 1) - (s1 - 1) \\ Tr1 - 1 \\ (Tr1 - 1) \cdot (G1 - 1) \\ (m1 - 1) - (s1 - 1) - (Tr1 - 1) - (Tr1 - 1) \cdot (G1 - 1) \end{bmatrix}
$$

$$df1 = \begin{pmatrix} 131 \\ 32 \\ 3 \\ 29 \\ 99 \\ 3 \\ 9 \\ 87 \end{pmatrix} \quad SS1 = \begin{pmatrix} 700.75 \\ 390.75 \\ 18.083 \\ 372.667 \\ 310 \\ 3 \\ 67.303 \\ 239.697 \end{pmatrix} \quad df1^{-1} = \begin{pmatrix} 0.008 \\ 0.031 \\ 0.333 \\ 0.034 \\ 0.01 \\ 0.333 \\ 0.111 \\ 0.011 \end{pmatrix} \quad ERROR1 := \begin{pmatrix} \dfrac{373}{29} \\ \dfrac{373}{29} \\ \dfrac{373}{29} \\ \dfrac{373}{29} \\ \dfrac{240}{87} \\ \dfrac{240}{87} \\ \dfrac{240}{87} \\ \dfrac{240}{87} \end{pmatrix}$$

The final computation is given by:

$$F1 := \overrightarrow{\left(SS1 \cdot df1^{-1} \cdot ERROR1^{-1} \right)}$$

$$F1 = \begin{pmatrix} 0.416 \\ 0.949 \\ 0.469 \\ 0.999 \\ 1.135 \\ 0.363 \\ 2.711 \\ 0.999 \end{pmatrix} \quad h := \begin{pmatrix} 0 \\ 0 \\ 1 \\ 0 \\ 0 \\ 1 \\ 1 \\ 0 \end{pmatrix}$$

$$F2 := \overrightarrow{(F1 \cdot h)}$$

This template Matrix h just gets rid of the values we don't need.

$$
df1 = \begin{pmatrix} 131 \\ 32 \\ 3 \\ 29 \\ 99 \\ 3 \\ 9 \\ 87 \end{pmatrix} \qquad
F2 = \begin{pmatrix} 0 \\ 0 \\ 0.469 \\ 0 \\ 0 \\ 0.363 \\ 2.711 \\ 0 \end{pmatrix} \qquad
p := \begin{pmatrix} 0 \\ 0 \\ .20 \\ 0 \\ 0 \\ .20 \\ .005 \\ 0 \end{pmatrix}
$$

HOMEWORK: Reduce this problem and the next four Analysis of Variance problems to a single equation.

THREE-FACTOR MIXED DESIGN;
REPEATED MEASURES ON ONE FACTOR

Based on Prob. 2.8, Computational Handbook of Statistics,
pg. 62. Please refer to book for problem description.

**This is a combination of the Factorial Design and the Treatments
by Subjects Design.**

$$A := \begin{pmatrix} 3 & 5 & 6 & 8 & 4 & 7 & 8 & 11 & 2 & 3 & 3 & 4 & 3 & 7 & 9 & 13 \\ 4 & 5 & 7 & 9 & 1 & 3 & 8 & 12 & 3 & 4 & 6 & 7 & 3 & 8 & 12 & 16 \\ 6 & 9 & 11 & 14 & 8 & 11 & 13 & 14 & 6 & 7 & 7 & 9 & 4 & 8 & 13 & 19 \\ 5 & 6 & 8 & 11 & 7 & 10 & 13 & 17 & 7 & 8 & 9 & 10 & 7 & 10 & 13 & 19 \\ 2 & 4 & 7 & 8 & 2 & 7 & 11 & 13 & 1 & 2 & 2 & 3 & 4 & 9 & 13 & 17 \\ 4 & 4 & 5 & 7 & 2 & 4 & 8 & 11 & 6 & 6 & 7 & 6 & 5 & 7 & 9 & 14 \\ 8 & 9 & 10 & 13 & 7 & 8 & 9 & 12 & 7 & 9 & 9 & 9 & 6 & 12 & 18 & 21 \end{pmatrix}$$

$$DB1 := \begin{pmatrix} 1 & 0 \\ 1 & 0 \\ 0 & 1 \\ 0 & 1 \end{pmatrix} \qquad DB2 := \begin{pmatrix} 1 & 0 \\ 0 & 1 \\ 1 & 0 \\ 0 & 1 \end{pmatrix}$$

$$ONE7 := (1 \ 1 \ 1 \ 1 \ 1 \ 1 \ 1)$$

$$FOUR1 := \begin{pmatrix} 1 \\ 1 \\ 1 \\ 1 \end{pmatrix}$$

$$DB3 := \begin{pmatrix} 1 & 0 & 0 & 0 \\ 1 & 0 & 0 & 0 \\ 1 & 0 & 0 & 0 \\ 1 & 0 & 0 & 0 \\ 0 & 1 & 0 & 0 \\ 0 & 1 & 0 & 0 \\ 0 & 1 & 0 & 0 \\ 0 & 1 & 0 & 0 \\ 0 & 0 & 1 & 0 \\ 0 & 0 & 1 & 0 \\ 0 & 0 & 1 & 0 \\ 0 & 0 & 1 & 0 \\ 0 & 0 & 0 & 1 \\ 0 & 0 & 0 & 1 \\ 0 & 0 & 0 & 1 \\ 0 & 0 & 0 & 1 \end{pmatrix} \qquad DB4 := \begin{pmatrix} 1 & 0 & 0 & 0 \\ 0 & 1 & 0 & 0 \\ 0 & 0 & 1 & 0 \\ 0 & 0 & 0 & 1 \\ 1 & 0 & 0 & 0 \\ 0 & 1 & 0 & 0 \\ 0 & 0 & 1 & 0 \\ 0 & 0 & 0 & 1 \\ 1 & 0 & 0 & 0 \\ 0 & 1 & 0 & 0 \\ 0 & 0 & 1 & 0 \\ 0 & 0 & 0 & 1 \\ 1 & 0 & 0 & 0 \\ 0 & 1 & 0 & 0 \\ 0 & 0 & 1 & 0 \\ 0 & 0 & 0 & 1 \end{pmatrix} \qquad SIXTEEN1 := \begin{pmatrix} 1 \\ 1 \\ 1 \\ 1 \\ 1 \\ 1 \\ 1 \\ 1 \\ 1 \\ 1 \\ 1 \\ 1 \\ 1 \\ 1 \\ 1 \\ 1 \end{pmatrix}$$

$$DB41 := \begin{pmatrix} 1 & 0 & 0 & 0 & 0 & 0 & 0 & 0 \\ 0 & 1 & 0 & 0 & 0 & 0 & 0 & 0 \\ 0 & 0 & 1 & 0 & 0 & 0 & 0 & 0 \\ 0 & 0 & 0 & 1 & 0 & 0 & 0 & 0 \\ 1 & 0 & 0 & 0 & 0 & 0 & 0 & 0 \\ 0 & 1 & 0 & 0 & 0 & 0 & 0 & 0 \\ 0 & 0 & 1 & 0 & 0 & 0 & 0 & 0 \\ 0 & 0 & 0 & 1 & 0 & 0 & 0 & 0 \\ 0 & 0 & 0 & 0 & 1 & 0 & 0 & 0 \\ 0 & 0 & 0 & 0 & 0 & 1 & 0 & 0 \\ 0 & 0 & 0 & 0 & 0 & 0 & 1 & 0 \\ 0 & 0 & 0 & 0 & 0 & 0 & 0 & 1 \\ 0 & 0 & 0 & 0 & 1 & 0 & 0 & 0 \\ 0 & 0 & 0 & 0 & 0 & 1 & 0 & 0 \\ 0 & 0 & 0 & 0 & 0 & 0 & 1 & 0 \\ 0 & 0 & 0 & 0 & 0 & 0 & 0 & 1 \end{pmatrix} \qquad DB8 := \begin{pmatrix} 1 & 0 & 0 & 0 & 0 & 0 & 0 & 0 \\ 0 & 1 & 0 & 0 & 0 & 0 & 0 & 0 \\ 0 & 0 & 1 & 0 & 0 & 0 & 0 & 0 \\ 0 & 0 & 0 & 1 & 0 & 0 & 0 & 0 \\ 0 & 0 & 0 & 0 & 1 & 0 & 0 & 0 \\ 0 & 0 & 0 & 0 & 0 & 1 & 0 & 0 \\ 0 & 0 & 0 & 0 & 0 & 0 & 1 & 0 \\ 0 & 0 & 0 & 0 & 0 & 0 & 0 & 1 \\ 1 & 0 & 0 & 0 & 0 & 0 & 0 & 0 \\ 0 & 1 & 0 & 0 & 0 & 0 & 0 & 0 \\ 0 & 0 & 1 & 0 & 0 & 0 & 0 & 0 \\ 0 & 0 & 0 & 1 & 0 & 0 & 0 & 0 \\ 0 & 0 & 0 & 0 & 1 & 0 & 0 & 0 \\ 0 & 0 & 0 & 0 & 0 & 1 & 0 & 0 \\ 0 & 0 & 0 & 0 & 0 & 0 & 1 & 0 \\ 0 & 0 & 0 & 0 & 0 & 0 & 0 & 1 \end{pmatrix}$$

$$df := \begin{pmatrix} 111 \\ 27 \\ 1 \\ 1 \\ 1 \\ 24 \\ 84 \\ 3 \\ 3 \\ 3 \\ 3 \\ 72 \end{pmatrix} \qquad SS := \begin{pmatrix} 1656.429 \\ 575.429 \\ 4.321 \\ 289.286 \\ 78.893 \\ 209.929 \\ 1081.000 \\ 783.500 \\ 1.036 \\ 167.214 \\ 49.036 \\ 81.250 \end{pmatrix}$$

$$ERROR := \begin{pmatrix} 8.455 \\ 8.455 \\ 8.455 \\ 8.455 \\ 8.455 \\ 8.455 \\ 1.128 \\ 1.128 \\ 1.128 \\ 1.128 \\ 1.128 \\ 1.128 \end{pmatrix}$$

STEP 2: Add the scores in each group to obtain SUM1.

PROGRAM: $SUM1 = [1]_{1,7} \, [A]_{7,16}$

$SUM1 := ONE7 \cdot A$

$$SUM1 := (1 \ 1 \ 1 \ 1 \ 1 \ 1 \ 1) \cdot \begin{pmatrix} 3 & 5 & 6 & 8 & 4 & 7 & 8 & 11 & 2 & 3 & 3 & 4 & 3 & 7 & 9 & 13 \\ 4 & 5 & 7 & 9 & 1 & 3 & 8 & 12 & 3 & 4 & 6 & 7 & 3 & 8 & 12 & 16 \\ 6 & 9 & 11 & 14 & 8 & 11 & 13 & 14 & 6 & 7 & 7 & 9 & 4 & 8 & 13 & 19 \\ 5 & 6 & 8 & 11 & 7 & 10 & 13 & 17 & 7 & 8 & 9 & 10 & 7 & 10 & 13 & 19 \\ 2 & 4 & 7 & 8 & 2 & 7 & 11 & 13 & 1 & 2 & 2 & 3 & 4 & 9 & 13 & 17 \\ 4 & 4 & 5 & 7 & 2 & 4 & 8 & 11 & 6 & 6 & 7 & 6 & 5 & 7 & 9 & 14 \\ 8 & 9 & 10 & 13 & 7 & 8 & 9 & 12 & 7 & 9 & 9 & 9 & 6 & 12 & 18 & 21 \end{pmatrix}$$

$SUM1 = (32 \ \ 42 \ \ 54 \ \ 70 \ \ 31 \ \ 50 \ \ 70 \ \ 90 \ \ 32 \ \ 39 \ \ 43 \ \ 48 \ \ 32 \ \ 61 \ \ 87 \ \ 119)$

STEP 3: Obtain the sum for each group by addig the sums of the
idividual trials.

PROGRAM GROUPSUM: $[1]_{1,7}[A]_{7,17}[DB1]_{16,4}$

$GROUPSUM := ONE7 \cdot A \cdot DB3$

$GROUPSUM = (198 \ \ 241 \ \ 162 \ \ 299)$

STEPS 4 & 5: Add the scores for each subject in each group.

PROGRAM: SUBJECTSCORES = $[A]_{7,16}[DB3]_{16,4}$

SUBJECTSCORES := A·DB3

$$\text{SUBJECTSCORES} := \begin{pmatrix} 3 & 5 & 6 & 8 & 4 & 7 & 8 & 11 & 2 & 3 & 3 & 4 & 3 & 7 & 9 & 13 \\ 4 & 5 & 7 & 9 & 1 & 3 & 8 & 12 & 3 & 4 & 6 & 7 & 3 & 8 & 12 & 16 \\ 6 & 9 & 11 & 14 & 8 & 11 & 13 & 14 & 6 & 7 & 7 & 9 & 4 & 8 & 13 & 19 \\ 5 & 6 & 8 & 11 & 7 & 10 & 13 & 17 & 7 & 8 & 9 & 10 & 7 & 10 & 13 & 19 \\ 2 & 4 & 7 & 8 & 2 & 7 & 11 & 13 & 1 & 2 & 2 & 3 & 4 & 9 & 13 & 17 \\ 4 & 4 & 5 & 7 & 2 & 4 & 8 & 11 & 6 & 6 & 7 & 6 & 5 & 7 & 9 & 14 \\ 8 & 9 & 10 & 13 & 7 & 8 & 9 & 12 & 7 & 9 & 9 & 9 & 6 & 12 & 18 & 21 \end{pmatrix} \cdot \begin{pmatrix} 1 & 0 & 0 & 0 \\ 1 & 0 & 0 & 0 \\ 1 & 0 & 0 & 0 \\ 1 & 0 & 0 & 0 \\ 0 & 1 & 0 & 0 \\ 0 & 1 & 0 & 0 \\ 0 & 1 & 0 & 0 \\ 0 & 1 & 0 & 0 \\ 0 & 0 & 1 & 0 \\ 0 & 0 & 1 & 0 \\ 0 & 0 & 1 & 0 \\ 0 & 0 & 1 & 0 \\ 0 & 0 & 0 & 1 \\ 0 & 0 & 0 & 1 \\ 0 & 0 & 0 & 1 \\ 0 & 0 & 0 & 1 \end{pmatrix}$$

$$\text{SUBJECTSCORES} = \begin{pmatrix} 22 & 30 & 12 & 32 \\ 25 & 24 & 20 & 39 \\ 40 & 46 & 29 & 44 \\ 30 & 47 & 34 & 49 \\ 21 & 33 & 8 & 43 \\ 20 & 25 & 25 & 35 \\ 40 & 36 & 34 & 57 \end{pmatrix}$$

STEP 6: Square each score in the entire table ad sum to get the Grand Sum of the squared numbers.

PROGRAM: $[1]_{1,7} * ([A1]_{7,16} \otimes [A1]_{7,16}) * [1]_{16,1}$

GRANDSUMSQ := ONE7·$\overrightarrow{(A \cdot A)}$·SIXTEEN1

This problem is too large to display on a single page so we will display it this way.

$$\text{GRANDSUMSQ} := (1 \ 1 \ 1 \ 1 \ 1 \ 1 \ 1) \cdot \overrightarrow{(A \cdot A)} \cdot \begin{pmatrix} 1 \\ 1 \\ 1 \\ 1 \\ 1 \\ 1 \\ 1 \\ 1 \\ 1 \\ 1 \\ 1 \\ 1 \\ 1 \\ 1 \\ 1 \\ 1 \end{pmatrix}$$

GRANDSUMSQ = 9194

STEP 7 & 8: Computation of the Grand Sum and the Correction Factor.

PROGRAM: $[1]_{1,7} \ [A]_{7,16} [1]_{16,1}$

GS := ONE7·A·SIXTEEN1

$$\text{GS} := (1 \ 1 \ 1 \ 1 \ 1 \ 1 \ 1) \cdot \begin{pmatrix} 3 & 5 & 6 & 8 & 4 & 7 & 8 & 11 & 2 & 3 & 3 & 4 & 3 & 7 & 9 & 13 \\ 4 & 5 & 7 & 9 & 1 & 3 & 8 & 12 & 3 & 4 & 6 & 7 & 3 & 8 & 12 & 16 \\ 6 & 9 & 11 & 14 & 8 & 11 & 13 & 14 & 6 & 7 & 7 & 9 & 4 & 8 & 13 & 19 \\ 5 & 6 & 8 & 11 & 7 & 10 & 13 & 17 & 7 & 8 & 9 & 10 & 7 & 10 & 13 & 19 \\ 2 & 4 & 7 & 8 & 2 & 7 & 11 & 13 & 1 & 2 & 2 & 3 & 4 & 9 & 13 & 17 \\ 4 & 4 & 5 & 7 & 2 & 4 & 8 & 11 & 6 & 6 & 7 & 6 & 5 & 7 & 9 & 14 \\ 8 & 9 & 10 & 13 & 7 & 8 & 9 & 12 & 7 & 9 & 9 & 9 & 6 & 12 & 18 & 21 \end{pmatrix} \cdot \begin{pmatrix} 1 \\ 1 \\ 1 \\ 1 \\ 1 \\ 1 \\ 1 \\ 1 \\ 1 \\ 1 \\ 1 \\ 1 \\ 1 \\ 1 \\ 1 \\ 1 \end{pmatrix}$$

GS = 900

We find the Correction Factor by squaring the Grand Sum and dividing by the number of elements in the Data Matrix A

$Nt := 16 \cdot 7$

$Nt = 112$

$CORRFACT := \left(\dfrac{1}{Nt}\right) \cdot GS^2$

$CORRFACT = 7232.143$

STEP 9: Computation of the Total Sum of Squares SSt. To accomplish this we subtract the Correction Factor from the Grand Sum of Squares.

$SSt := GRANDSUMSQ - CORRFACT$

$SSt = 1961.857$

$ONE7 \cdot \overrightarrow{(A \cdot A)} \cdot SIXTEEN1 - \left(\dfrac{1}{Nt}\right) \cdot GS^2$

$SSt := ONE7 \cdot \overrightarrow{(A \cdot A)} \cdot SIXTEEN1 - \left(\dfrac{1}{Nt}\right) \cdot GS^2$

$SSt = 1961.857$

STEP 10: Computation of the Between Subjects Sum of Squares SSb. We square the Matrix computed in Step 4 and sum.

$Nb := 28$

PROGRAM: $SSb = (1/Nb) * ([1]_{1,7} * ([A]_{7,16} * DB3)^2 * [1]_{4,1} - CORRFACT$

$SSb1 := \left(\dfrac{1}{Nb}\right) \left[ONE7 \cdot \overrightarrow{[(A \cdot DB3) \cdot (A \cdot DB3)]} \right] \cdot FOUR1 - CORRFACT$

Again, MathCad cannot handle this computation, so we make the following substitution

AxDB3 := A·DB3

Now we have

$$\text{SSb} := \left(\frac{1}{\text{Nb}}\right)\left[\text{ONE7}\cdot\overrightarrow{[(\text{AxDB3})\cdot(\text{AxDB3})]}\right]\cdot\text{FOUR1} - \text{CORRFACT}$$

$$\text{SSb} := \left(\frac{1}{4}\right)\cdot(1\ 1\ 1\ 1\ 1\ 1\ 1)\cdot\overrightarrow{\left[\begin{pmatrix} 22 & 30 & 12 & 32 \\ 25 & 24 & 20 & 39 \\ 40 & 46 & 29 & 44 \\ 30 & 47 & 34 & 49 \\ 21 & 33 & 8 & 43 \\ 20 & 25 & 25 & 35 \\ 40 & 36 & 34 & 57 \end{pmatrix}\cdot\begin{pmatrix} 22 & 30 & 12 & 32 \\ 25 & 24 & 20 & 39 \\ 40 & 46 & 29 & 44 \\ 30 & 47 & 34 & 49 \\ 21 & 33 & 8 & 43 \\ 20 & 25 & 25 & 35 \\ 40 & 36 & 34 & 57 \end{pmatrix}\right]}\cdot\begin{pmatrix} 1 \\ 1 \\ 1 \\ 1 \end{pmatrix} - \text{CORRFACT}$$

SSb = 880.857

STEP 11: Computation of the First Experimental Condition on the overall performance, SS1. N1 = Nt/ number of columns in DB2. 112/2 = 56

N1 := 56

PROGRAM: $\text{SS1} = (1/\text{N1})*([1]_{1,7}[A]_{7,16}[\text{DB3}]_{16,4}[\text{DB1}]_{4,2})^2 - \text{CORRFACT}$

NOTE: Here we come across the first expanded definition of our Statistical Field Equation. Instead of $([1][A][\text{DB}])^2$ we have a new equation of the form $([1][A][\text{DB1}][\text{DB2}])^2$. We will look at the theoretical aspects of this in the 3-variable Analysis of Variance coming up in problem 2-11.

$$\text{SS1} := \left(\frac{1}{\text{N1}}\right)\cdot(\text{ONE7}\cdot\text{A}\cdot\text{DB3}\cdot\text{DB1})^2 - \text{CORRFACT}$$

$$\text{SS1} := \left(\frac{1}{\text{N1}}\right)\cdot(\text{ONE7}\cdot\text{A}\cdot\text{DB3}\cdot\text{DB1})\cdot(\text{ONE7}\cdot\text{A}\cdot\text{DB3}\cdot\text{DB1})^T - \text{CORRFACT}$$

This equation is too big to fit on a page, so I'll shorten it, remembering $c^2 = cc^T$.

$$\frac{1}{56} \cdot \left[\left(32\ 42\ 54\ 70\ 31\ 50\ 70\ 90\ 32\ 39\ 43\ 48\ 32\ 61\ 87\ 119 \right) \cdot \begin{pmatrix} 1 & 0 & 0 & 0 \\ 1 & 0 & 0 & 0 \\ 1 & 0 & 0 & 0 \\ 1 & 0 & 0 & 0 \\ 0 & 1 & 0 & 0 \\ 0 & 1 & 0 & 0 \\ 0 & 1 & 0 & 0 \\ 0 & 1 & 0 & 0 \\ 0 & 0 & 1 & 0 \\ 0 & 0 & 1 & 0 \\ 0 & 0 & 1 & 0 \\ 0 & 0 & 1 & 0 \\ 0 & 0 & 0 & 1 \\ 0 & 0 & 0 & 1 \\ 0 & 0 & 0 & 1 \\ 0 & 0 & 0 & 1 \end{pmatrix} \cdot \begin{pmatrix} 1 & 0 \\ 1 & 0 \\ 0 & 1 \\ 0 & 1 \end{pmatrix} \right]^2 - 7232.1$$

SS1 = 4.321

STEP 12: Computation of the effects of the Second Experimental
Condition on the overall performance. SS2. N2 = Nt/# columns
in DB2. Note this is exactly the same problem as above except
we substitute DB2 for DB1.

N2 := 56

PROGRAM: $\mathrm{SS2} = (1/\mathrm{N2}) \ast \left([1]_{1,7} [A]_{7,16} [DB3]_{16,4} [DB2]_{4,2} \right)^2 - \mathrm{CORRFACT}$

$\mathrm{SS2} := \left(\dfrac{1}{\mathrm{N2}} \right) \cdot \left(\mathrm{ONE7 \cdot A \cdot DB3 \cdot DB2} \right)^2 - \mathrm{CORRFACT}$

$\mathrm{SS2} := \left(\dfrac{1}{\mathrm{N2}} \right) \cdot \left(\mathrm{ONE7 \cdot A \cdot DB3 \cdot DB2} \right) \cdot \left(\mathrm{ONE7 \cdot A \cdot DB3 \cdot DB2} \right)^{\mathrm{T}} - \mathrm{CORRFACT}$

SS2 = 289.286

Simplifying we get

$\mathrm{ONE7 \cdot A \cdot DB3 \cdot DB2} = \left(360\ \ 540 \right) \qquad \left(360\ \ 540 \right)^{\mathrm{T}} = \begin{pmatrix} 360 \\ 540 \end{pmatrix}$

$$\left(\frac{1}{56}\right)\left[(360 \quad 540)\cdot\binom{360}{540}\right] - \text{CORRFACT} = 289.286$$

Or by expanding so we can fit this on a single page we get:

$$\frac{1}{56}\cdot\left[(32 \ 42 \ 54 \ 70 \ 31 \ 50 \ 70 \ 90 \ 32 \ 39 \ 43 \ 48 \ 32 \ 61 \ 87 \ 119)\cdot\begin{pmatrix}1&0&0&0\\1&0&0&0\\1&0&0&0\\1&0&0&0\\0&1&0&0\\0&1&0&0\\0&1&0&0\\0&1&0&0\\0&0&1&0\\0&0&1&0\\0&0&1&0\\0&0&1&0\\0&0&0&1\\0&0&0&1\\0&0&0&1\\0&0&0&1\end{pmatrix}\cdot\begin{pmatrix}1&0\\0&1\\1&0\\0&1\end{pmatrix}\right]^{2} - 7232.1$$

SS2 = 289.286

STEP 13: Computation of the interactive effects of Experimental Condition 1 and Experimental Condition 2. SS1x2. N1x2 = Nt/#Columns in DB1. N1x2 = 112/4= 28

N1x2 := 28

PROGRAM: SS1x2 = (1/N1x2)*([1]$_{1,7}$[A]$_{7,16}$[DB3]$_{16,4}$)^{2}-CORRFACT-SS1-SS2

$$\text{SS1x2} := \left(\frac{1}{\text{N1x2}}\right)\left[(\text{SUM1}\cdot\text{DB3})\cdot(\text{SUM1}\cdot\text{DB3})^{T}\right] - \text{CORRFACT} - \text{SS1} - \text{SS2}$$

SS1x2 = 78.893

STEP 14: Computation of the Between Subjects Error term: SSerrb.

SSerb := SSb − SS1 − SS2 − SS1x2 $880.857 − 4 − 290 − 79 = 507.857$

SSerb := $880.857 − 4 − 290 − 79$

SSerb = 507.857

STEP 15: Computation of the within Subjects Error term: SSw.

SSw := SSt − SSb   SSt = 1961.857

SSw := $1961.857 − 880.857$ SSb = 880.857

SSw = 1081

STEP 16: Computation of the Sum of Squares for Trials where we sum the corresponding values in each group together: SS3. N3 = Nt /# of columns in DB4. N3 = 112/4=28.

N3 := 28

PROGRAM: SS3 = $(1/N3) \times ([1]_{1,7}[A]_{7,16}[DB4]_{16,4})^2$ −CORRFACT

$$SS3 := \left(\frac{1}{N3}\right) \cdot (SUM1 \cdot DB4) \cdot (SUM1 \cdot DB4)^T − CORRFACT$$

SS3 = 783.5

Or to see the computation and trying to keep it to a single page, the equation/program looks like:

$$SS3 := \left(\frac{1}{28}\right) \cdot \left[(32\ \ 42\ \ 54\ \ 70\ \ 31\ \ 50\ \ 70\ \ 90\ \ 32\ \ 39\ \ 43\ \ 48\ \ 32\ \ 61\ \ 87\ \ 119) \cdot \begin{pmatrix} 1 & 0 & 0 & 0 \\ 0 & 1 & 0 & 0 \\ 0 & 0 & 1 & 0 \\ 0 & 0 & 0 & 1 \\ 1 & 0 & 0 & 0 \\ 0 & 1 & 0 & 0 \\ 0 & 0 & 1 & 0 \\ 0 & 0 & 0 & 1 \\ 1 & 0 & 0 & 0 \\ 0 & 1 & 0 & 0 \\ 0 & 0 & 1 & 0 \\ 0 & 0 & 0 & 1 \\ 1 & 0 & 0 & 0 \\ 0 & 1 & 0 & 0 \\ 0 & 0 & 1 & 0 \\ 0 & 0 & 0 & 1 \end{pmatrix} \cdot (SUM1 \cdot DB4)^T \right] - 7232.143$$

SS3 = 783.5

STEP 17: Computation of the Sum of Squares for Trials by First Experimental Condition Interaction: SS1x3. N1x3 = Nt /# of columns in DB4. N1x3 = 112/8=14.

N1x3 := 14

PROGRAM: $SS1x3 = (1/N1x3) x ([1]_{1,7}[A]_{7,16}[DB41]_{16,4})^2 - CORRFACT$

$$SS1x3 := \left(\frac{1}{N1x3}\right)\left[(SUM1 \cdot DB41) \cdot (SUM1 \cdot DB41)^T\right] - CORRFACT - SS1 - SS3$$

SS1x3 = 1.036   SS1 = 4.321   SS3 = 783.5

The math looks like:

$$\left(\frac{1}{14}\right)\cdot(32\ 42\ 54\ 70\ 31\ 50\ 70\ 90\ 32\ 39\ 43\ 48\ 32\ 61\ 87\ 119)\cdot\begin{pmatrix}1&0&0&0&0&0&0&0\\0&1&0&0&0&0&0&0\\0&0&1&0&0&0&0&0\\0&0&0&1&0&0&0&0\\1&0&0&0&0&0&0&0\\0&1&0&0&0&0&0&0\\0&0&1&0&0&0&0&0\\0&0&0&1&0&0&0&0\\0&0&0&0&1&0&0&0\\0&0&0&0&0&1&0&0\\0&0&0&0&0&0&1&0\\0&0&0&0&0&0&0&1\\0&0&0&0&1&0&0&0\\0&0&0&0&0&1&0&0\\0&0&0&0&0&0&1&0\\0&0&0&0&0&0&0&1\end{pmatrix}\cdot(\text{SUM1}\cdot\text{DB41})^{T}-7232.143-4.321-783.5=1.036$$

Which reduces to:

$$\left(\frac{1}{14}\right)\cdot(63\ 92\ 124\ 160\ 64\ 100\ 130\ 167)\cdot\begin{pmatrix}63\\92\\124\\160\\64\\100\\130\\167\end{pmatrix}-7232.143-4.321-783.5=1.036$$

STEP 18: Computation of the Sum of Squares for Trials by Second
Experimental Condition interaction: SS2x3. N2x3 = Nt /# of
columns in DB8. N2x3 = 112/8=14.

N2x3 := 14

PROGRAM: $SS2x3 = (1/N2x3)x\ ([1]_{1,7}[A]_{7,16}[DB8]_{16,8})^{2}$
$-CORRFACT-SS2-SS3$

$$SS2x3 := \left(\frac{1}{N2x3}\right)\left[(\text{SUM1}\cdot\text{DB8})\cdot(\text{SUM1}\cdot\text{DB8})^{T}\right] - CORRFACT - SS2 - SS3$$

$SS2x3 = 167.214$

The math looks like: $SS2 = 289.286$ $SS3 = 783.5$

$$\left(\frac{1}{14}\right)\cdot(32 \;\; 42 \;\; 54 \;\; 70 \;\; 31 \;\; 50 \;\; 70 \;\; 90 \;\; 32 \;\; 39 \;\; 43 \;\; 48 \;\; 32 \;\; 61 \;\; 87 \;\; 119)\cdot\begin{pmatrix} 1 & 0 & 0 & 0 & 0 & 0 & 0 & 0 \\ 0 & 1 & 0 & 0 & 0 & 0 & 0 & 0 \\ 0 & 0 & 1 & 0 & 0 & 0 & 0 & 0 \\ 0 & 0 & 0 & 1 & 0 & 0 & 0 & 0 \\ 0 & 0 & 0 & 0 & 1 & 0 & 0 & 0 \\ 0 & 0 & 0 & 0 & 0 & 1 & 0 & 0 \\ 0 & 0 & 0 & 0 & 0 & 0 & 1 & 0 \\ 0 & 0 & 0 & 0 & 0 & 0 & 0 & 1 \\ 1 & 0 & 0 & 0 & 0 & 0 & 0 & 0 \\ 0 & 1 & 0 & 0 & 0 & 0 & 0 & 0 \\ 0 & 0 & 1 & 0 & 0 & 0 & 0 & 0 \\ 0 & 0 & 0 & 1 & 0 & 0 & 0 & 0 \\ 0 & 0 & 0 & 0 & 1 & 0 & 0 & 0 \\ 0 & 0 & 0 & 0 & 0 & 1 & 0 & 0 \\ 0 & 0 & 0 & 0 & 0 & 0 & 1 & 0 \\ 0 & 0 & 0 & 0 & 0 & 0 & 0 & 1 \end{pmatrix}\cdot(\text{SUM1}\cdot\text{DB8})^{\mathrm{T}} - 7232.143 - 289.286 - 783.5 = 167.214$$

Which reduces to:

$$\left(\frac{1}{14}\right)\cdot(64 \;\; 81 \;\; 97 \;\; 118 \;\; 63 \;\; 111 \;\; 157 \;\; 209)\cdot\begin{pmatrix} 64 \\ 81 \\ 97 \\ 118 \\ 63 \\ 111 \\ 157 \\ 209 \end{pmatrix} - 7232.143 - 289.286 - 783.5 = 167.214$$

SS2x3 = 167.214

STEP 19: Computation of the Sum of Squares for trials by First and Second Experimental Conditions Interaction. SS1x2x3. N1x2x3 = Nt / # Elements in A. Nt = 112/16 = 7.

N1x2x3 := 7

PROGRAM: $(1/\text{N1X2X3})\times([1]_{1,7}[A]_{7,16})\times([1]_{1,7}[A]_{7,16})^{\mathrm{T}}-\text{CORRFACT}-\text{SS1}-\text{SS2}-\text{SSb}-\text{SS1x2}-\text{SS1x3}-\text{SS2x3}$

$$\text{SS1x2x3} := \left(\frac{1}{\text{N1x2x3}}\right)\left[(\text{SUM1})\cdot(\text{SUM1})^{\mathrm{T}}\right] - \text{CORRFACT} - \text{SS1} - \text{SS2} - \text{SS3} - \text{SS1x2} - \text{SS1x3} - \text{SS2x3}$$

SS1 = 4.321	SS2 = 289.286	SS3 = 783.5	SSb = 880.857
SS2x3 = 167.214	SS1x3 = 1.036	SS1x2 = 78.893	CORRFACT = 7232.143

$$SS1x2x3 := \left(\frac{1}{7}\right) \cdot (SUM1) \cdot \left[\left(\left(\begin{array}{c} 32 \\ 42 \\ 54 \\ 70 \\ 31 \\ 50 \\ 70 \\ 90 \\ 32 \\ 39 \\ 43 \\ 48 \\ 32 \\ 61 \\ 87 \\ 119 \end{array} \right) \right) \right] - 7232.143 - 4.321 - 289.286 - 783.5 - 78.893 - 1.036 - 167.214$$

$SS1x2x3 = 49.036$

STEP 20: Computation of the Within Subjects Error Term SSerw.

$SSerw := SSw - SS3 - SS1x3 - SS1x2x3 - SS2x3$

$SSerw = 80.214$

$SSw = 1081 \qquad SS3 = 783.5 \qquad SS1x3 = 1.036 \qquad SS2x3 = 167.214 \qquad SS1x2x3 = 49.036$

$SSerw := 1081 - 783.5 - 1.036 - 167.214 - 49.036$

$SSerw = 80.214$

STEP 21: Computation of the degrees of Freedom for the Data in the study.

$m := 112 \qquad tr := 4 \qquad s := 28 \qquad FB := 2 \qquad G := 2$

m = # elements in Data Matrix, G = # Groups in study, G = # Experimental Conditions 1, FB = # Experimental Conditions 2, tr = # trials = 4, s = # subjects = 28.

$$\text{dfSSt} := m - 1$$
$$\text{dfSSb} := s - 1$$
$$\text{dfSS1} := G - 1$$
$$\text{dfSS2} := FB - 1$$
$$\text{dfSS1x2} := \text{dfSS1} \cdot \text{dfSS2}$$
$$\text{dfSSerb} := \text{dfSSb} - \text{dfSS2} - \text{dfSS1} - \text{dfSS1x2}$$
$$\text{dfSSw} := \text{dfSSt} - \text{dfSSb}$$
$$\text{dfSS3} := tr - 1$$
$$\text{dfSS1x3} := \text{dfSS1} \cdot \text{dfSS3}$$
$$\text{dfSS2x3} := \text{dfSS2} \cdot \text{dfSS3}$$
$$\text{dfSS1x2x3} := \text{dfSS1} \cdot \text{dfSS2} \cdot \text{dfSS3}$$
$$\text{dfSSerrw} := \text{dfSSw} - \text{dfSS3} - \text{dfSS1x3} - \text{dfSS2x3} - \text{dfSS1x2x3}$$

$$
\begin{pmatrix}
\text{SSt} \\
\text{SSb} \\
\text{SS1} \\
\text{SS2} \\
\text{SS1x2} \\
\text{SSerb} \\
\text{SSw} \\
\text{SS3} \\
\text{SS1x3} \\
\text{SS2x3} \\
\text{SS1x2x3} \\
\text{SSerw}
\end{pmatrix}
=
\begin{pmatrix}
1961.857 \\
880.857 \\
4.321 \\
289.286 \\
78.893 \\
507.857 \\
1081 \\
783.5 \\
1.036 \\
167.214 \\
49.036 \\
80.214
\end{pmatrix}
$$

$$
\text{df} :=
\begin{pmatrix}
\text{dfSSt} \\
\text{dfSSb} \\
\text{dfSS1} \\
\text{dfSS2} \\
\text{dfSS1x2} \\
\text{dfSSerrb} \\
\text{dfSSw} \\
\text{dfSS3} \\
\text{dfSS1x3} \\
\text{dfSS2x3} \\
\text{dfSS1x2x3} \\
\text{dfSSerrw}
\end{pmatrix}
\qquad
\text{df} :=
\begin{pmatrix}
m - 1 \\
s - 1 \\
G - 1 \\
FB - 1 \\
\text{dfSS1} \cdot \text{dfSS2} \\
\text{dfSSb} - \text{dfSS2} - \text{dfSS1} - \text{dfSS1x2} \\
\text{dfSSt} - \text{dfSSb} \\
tr - 1 \\
\text{dfSS1} \cdot \text{dfSS3} \\
\text{dfSS2} \cdot \text{dfSS3} \\
\text{dfSS1} \cdot \text{dfSS2} \cdot \text{dfSS3} \\
\text{dfSSw} - \text{dfSS3} - \text{dfSS1x3} - \text{dfSS2x3} - \text{dfSS1x2x3}
\end{pmatrix}
\qquad
\text{df} =
\begin{pmatrix}
111 \\
27 \\
1 \\
1 \\
1 \\
24 \\
84 \\
3 \\
3 \\
3 \\
3 \\
72
\end{pmatrix}
$$

STEP 22: Computation of the Within and Between Error Matrix and the F values for this system.

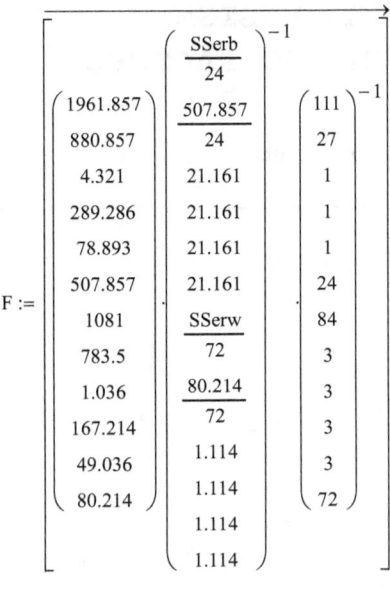

$$F := \begin{bmatrix} \begin{pmatrix} 1961.857 \\ 880.857 \\ 4.321 \\ 289.286 \\ 78.893 \\ 507.857 \\ 1081 \\ 783.5 \\ 1.036 \\ 167.214 \\ 49.036 \\ 80.214 \end{pmatrix} \cdot \begin{pmatrix} \dfrac{SSerb}{24} \\ \dfrac{507.857}{24} \\ 21.161 \\ 21.161 \\ 21.161 \\ 21.161 \\ \dfrac{SSerw}{72} \\ \dfrac{80.214}{72} \\ 1.114 \\ 1.114 \\ 1.114 \\ 1.114 \end{pmatrix}^{-1} \cdot \begin{pmatrix} 111 \\ 27 \\ 1 \\ 1 \\ 1 \\ 24 \\ 84 \\ 3 \\ 3 \\ 3 \\ 3 \\ 72 \end{pmatrix}^{-1} \end{bmatrix}$$

$$F = \begin{pmatrix} 0.835 \\ 1.542 \\ 0.204 \\ 13.671 \\ 3.728 \\ 1 \\ 11.551 \\ 234.423 \\ 0.31 \\ 50.034 \\ 14.673 \\ 1 \end{pmatrix} \qquad p := \begin{pmatrix} ns \\ ns \\ ns \\ .001 \\ .001 \\ ns \\ ns \\ .001 \\ ns \\ .001 \\ .001 \\ ns \end{pmatrix}$$

HOMEWORK: You've seen how this is done, reduce this problem to a single equation.

THREE FACTOR MIXED DESIGN;
REPEATED MEASURES ON TWO FACTORS

Prob. 2.9 Pg.73, Computational Handbook of Statistics, 2'd ed.

STEP 1:   Here we define the Data Matrix A and all constants and all sub-Statistical Data Base Matrices.

VARIABLES:

$ONE2 := (\ 1 \quad 1\ )$

$ONE4 := (\ 1 \quad 1 \quad 1 \quad 1\ )$

$FOUR1 := ONE4^T$

$ONE10 := (\ 1 \quad 1 \quad 1 \quad 1 \quad 1 \quad 1 \quad 1 \quad 1 \quad 1 \quad 1\ )$

$TEN1 := ONE10^T$

$ONE8 := (\ 1 \quad 1 \quad 1 \quad 1 \quad 1 \quad 1 \quad 1 \quad 1\ )$

$EIGHT1 := ONE8^T$

$TWO1 := ONE2^T$

$ONE16 := (\ 1 \quad 1 \quad 1 \quad 1 \quad 1 \quad 1 \quad 1 \quad 1 \quad 1 \quad 1 \quad 1 \quad 1 \quad 1 \quad 1 \quad 1 \quad 1\ )$

$$A := \begin{pmatrix}
24 & 29 & 38 & 46 & 5 & 12 & 23 & 27 & 18 & 20 & 29 & 33 & 14 & 17 & 20 & 23 \\
20 & 30 & 43 & 48 & 7 & 12 & 16 & 17 & 8 & 11 & 19 & 26 & 17 & 16 & 21 & 24 \\
26 & 25 & 46 & 49 & 11 & 16 & 22 & 22 & 23 & 24 & 29 & 27 & 11 & 19 & 17 & 14 \\
18 & 19 & 37 & 36 & 3 & 2 & 7 & 6 & 17 & 25 & 38 & 21 & 15 & 15 & 14 & 20 \\
12 & 16 & 23 & 31 & 2 & 5 & 10 & 18 & 9 & 14 & 13 & 14 & 3 & 1 & 4 & 7 \\
14 & 13 & 25 & 26 & 3 & 7 & 15 & 22 & 11 & 19 & 19 & 29 & 1 & 8 & 7 & 11 \\
13 & 19 & 30 & 36 & 7 & 14 & 25 & 29 & 20 & 22 & 25 & 30 & 10 & 13 & 12 & 18 \\
17 & 19 & 29 & 32 & 3 & 9 & 19 & 28 & 18 & 19 & 19 & 24 & 15 & 19 & 23 & 22 \\
23 & 31 & 29 & 32 & 5 & 7 & 19 & 17 & 15 & 19 & 28 & 26 & 7 & 12 & 17 & 22 \\
17 & 19 & 30 & 37 & 8 & 22 & 35 & 38 & 9 & 11 & 15 & 16 & 7 & 9 & 14 & 14
\end{pmatrix}$$

$$
DB11 := \begin{pmatrix} 1 & 0 \\ 1 & 0 \\ 1 & 0 \\ 1 & 0 \\ 0 & 1 \\ 0 & 1 \\ 0 & 1 \\ 0 & 1 \\ 1 & 0 \\ 1 & 0 \\ 1 & 0 \\ 1 & 0 \\ 0 & 1 \\ 0 & 1 \\ 0 & 1 \\ 0 & 1 \end{pmatrix}
\qquad
SIXTEEN1 := \begin{pmatrix} 1 \\ 1 \\ 1 \\ 1 \\ 1 \\ 1 \\ 1 \\ 1 \\ 1 \\ 1 \\ 1 \\ 1 \\ 1 \\ 1 \\ 1 \\ 1 \end{pmatrix}
\qquad
DB1 := \begin{pmatrix} 1 & 0 \\ 1 & 0 \\ 1 & 0 \\ 1 & 0 \\ 1 & 0 \\ 1 & 0 \\ 1 & 0 \\ 1 & 0 \\ 0 & 1 \\ 0 & 1 \\ 0 & 1 \\ 0 & 1 \\ 0 & 1 \\ 0 & 1 \\ 0 & 1 \\ 0 & 1 \end{pmatrix}
\qquad
DB4 := \begin{pmatrix} 1 & 0 & 0 & 0 \\ 0 & 1 & 0 & 0 \\ 0 & 0 & 1 & 0 \\ 0 & 0 & 0 & 1 \\ 1 & 0 & 0 & 0 \\ 0 & 1 & 0 & 0 \\ 0 & 0 & 1 & 0 \\ 0 & 0 & 0 & 1 \\ 1 & 0 & 0 & 0 \\ 0 & 1 & 0 & 0 \\ 0 & 0 & 1 & 0 \\ 0 & 0 & 0 & 1 \\ 1 & 0 & 0 & 0 \\ 0 & 1 & 0 & 0 \\ 0 & 0 & 1 & 0 \\ 0 & 0 & 0 & 1 \end{pmatrix}
$$

$$
DB12 := \begin{pmatrix} 1 & 0 & 0 & 0 \\ 1 & 0 & 0 & 0 \\ 1 & 0 & 0 & 0 \\ 1 & 0 & 0 & 0 \\ 0 & 1 & 0 & 0 \\ 0 & 1 & 0 & 0 \\ 0 & 1 & 0 & 0 \\ 0 & 1 & 0 & 0 \\ 0 & 0 & 1 & 0 \\ 0 & 0 & 1 & 0 \\ 0 & 0 & 1 & 0 \\ 0 & 0 & 1 & 0 \\ 0 & 0 & 0 & 1 \\ 0 & 0 & 0 & 1 \\ 0 & 0 & 0 & 1 \\ 0 & 0 & 0 & 1 \end{pmatrix}
\qquad
DB8 := \begin{pmatrix} 1 & 0 & 0 & 0 & 0 & 0 & 0 & 0 \\ 0 & 1 & 0 & 0 & 0 & 0 & 0 & 0 \\ 0 & 0 & 1 & 0 & 0 & 0 & 0 & 0 \\ 0 & 0 & 0 & 1 & 0 & 0 & 0 & 0 \\ 0 & 0 & 0 & 0 & 1 & 0 & 0 & 0 \\ 0 & 0 & 0 & 0 & 0 & 1 & 0 & 0 \\ 0 & 0 & 0 & 0 & 0 & 0 & 1 & 0 \\ 0 & 0 & 0 & 0 & 0 & 0 & 0 & 1 \\ 1 & 0 & 0 & 0 & 0 & 0 & 0 & 0 \\ 0 & 1 & 0 & 0 & 0 & 0 & 0 & 0 \\ 0 & 0 & 1 & 0 & 0 & 0 & 0 & 0 \\ 0 & 0 & 0 & 1 & 0 & 0 & 0 & 0 \\ 0 & 0 & 0 & 0 & 1 & 0 & 0 & 0 \\ 0 & 0 & 0 & 0 & 0 & 1 & 0 & 0 \\ 0 & 0 & 0 & 0 & 0 & 0 & 1 & 0 \\ 0 & 0 & 0 & 0 & 0 & 0 & 0 & 1 \end{pmatrix}
\qquad
DB41 := \begin{pmatrix} 1 & 0 & 0 & 0 & 0 & 0 & 0 & 0 \\ 0 & 1 & 0 & 0 & 0 & 0 & 0 & 0 \\ 0 & 0 & 1 & 0 & 0 & 0 & 0 & 0 \\ 0 & 0 & 0 & 1 & 0 & 0 & 0 & 0 \\ 1 & 0 & 0 & 0 & 0 & 0 & 0 & 0 \\ 0 & 1 & 0 & 0 & 0 & 0 & 0 & 0 \\ 0 & 0 & 1 & 0 & 0 & 0 & 0 & 0 \\ 0 & 0 & 0 & 1 & 0 & 0 & 0 & 0 \\ 0 & 0 & 0 & 0 & 1 & 0 & 0 & 0 \\ 0 & 0 & 0 & 0 & 0 & 1 & 0 & 0 \\ 0 & 0 & 0 & 0 & 0 & 0 & 1 & 0 \\ 0 & 0 & 0 & 0 & 0 & 0 & 0 & 1 \\ 0 & 0 & 0 & 0 & 1 & 0 & 0 & 0 \\ 0 & 0 & 0 & 0 & 0 & 1 & 0 & 0 \\ 0 & 0 & 0 & 0 & 0 & 0 & 1 & 0 \\ 0 & 0 & 0 & 0 & 0 & 0 & 0 & 1 \end{pmatrix}
$$

STEP 2: Sum the Columns of the Data Matrix A. SUM1.

PROGRAM: $SUM1 = [1]_{1,10}[A]_{10,16}$

SUM1 := ONE10 · A

$$\text{SUM1} := (1\ \ 1\ \ 1\ \ 1\ \ 1\ \ 1\ \ 1\ \ 1\ \ 1\ \ 1) \cdot \begin{pmatrix} 24 & 29 & 38 & 46 & 5 & 12 & 23 & 27 & 18 & 20 & 29 & 33 & 14 & 17 & 20 & 23 \\ 20 & 30 & 43 & 48 & 7 & 12 & 16 & 17 & 8 & 11 & 19 & 26 & 17 & 16 & 21 & 24 \\ 26 & 25 & 46 & 49 & 11 & 16 & 22 & 22 & 23 & 24 & 29 & 27 & 11 & 19 & 17 & 14 \\ 18 & 19 & 37 & 36 & 3 & 2 & 7 & 6 & 17 & 25 & 38 & 21 & 15 & 15 & 14 & 20 \\ 12 & 16 & 23 & 31 & 2 & 5 & 10 & 18 & 9 & 14 & 13 & 14 & 3 & 1 & 4 & 7 \\ 14 & 13 & 25 & 26 & 3 & 7 & 15 & 22 & 11 & 19 & 19 & 29 & 1 & 8 & 7 & 11 \\ 13 & 19 & 30 & 36 & 7 & 14 & 25 & 29 & 20 & 22 & 25 & 30 & 10 & 13 & 12 & 18 \\ 17 & 19 & 29 & 32 & 3 & 9 & 19 & 28 & 18 & 19 & 19 & 24 & 15 & 19 & 23 & 22 \\ 23 & 31 & 29 & 32 & 5 & 7 & 19 & 17 & 15 & 19 & 28 & 26 & 7 & 12 & 17 & 22 \\ 17 & 19 & 30 & 37 & 8 & 22 & 35 & 38 & 9 & 11 & 15 & 16 & 7 & 9 & 14 & 14 \end{pmatrix}$$

$$\text{SUM1} = (184 \quad 220 \quad 330 \quad 373 \quad 54 \quad 106 \quad 191 \quad 224 \quad 148 \quad 184 \quad 234 \quad 246 \quad 100 \quad 129 \quad 149 \quad 175)$$

STEP 3:   Square each element (term) in Data Matrix A and Sum. Grand Sum Square = GRANDSUMSQ.

PROGRAM: $\text{GRANDSUMSQ} = [1]_{1,10}\,([A]_{10,16} \otimes [A]_{10,16})\,[1]_{16,1}$

$\text{GRANDSUMSQ} := \text{ONE10} \cdot \overrightarrow{(A \cdot A)} \cdot \text{SIXTEEN1}$

$$(1\ \ 1\ \ 1\ \ 1\ \ 1\ \ 1\ \ 1\ \ 1\ \ 1\ \ 1) \cdot \overrightarrow{A \cdot \begin{pmatrix} 24 & 29 & 38 & 46 & 5 & 12 & 23 & 27 & 18 & 20 & 29 & 33 & 14 & 17 & 20 & 23 \\ 20 & 30 & 43 & 48 & 7 & 12 & 16 & 17 & 8 & 11 & 19 & 26 & 17 & 16 & 21 & 24 \\ 26 & 25 & 46 & 49 & 11 & 16 & 22 & 22 & 23 & 24 & 29 & 27 & 11 & 19 & 17 & 14 \\ 18 & 19 & 37 & 36 & 3 & 2 & 7 & 6 & 17 & 25 & 38 & 21 & 15 & 15 & 14 & 20 \\ 12 & 16 & 23 & 31 & 2 & 5 & 10 & 18 & 9 & 14 & 13 & 14 & 3 & 1 & 4 & 7 \\ 14 & 13 & 25 & 26 & 3 & 7 & 15 & 22 & 11 & 19 & 19 & 29 & 1 & 8 & 7 & 11 \\ 13 & 19 & 30 & 36 & 7 & 14 & 25 & 29 & 20 & 22 & 25 & 30 & 10 & 13 & 12 & 18 \\ 17 & 19 & 29 & 32 & 3 & 9 & 19 & 28 & 18 & 19 & 19 & 24 & 15 & 19 & 23 & 22 \\ 23 & 31 & 29 & 32 & 5 & 7 & 19 & 17 & 15 & 19 & 28 & 26 & 7 & 12 & 17 & 22 \\ 17 & 19 & 30 & 37 & 8 & 22 & 35 & 38 & 9 & 11 & 15 & 16 & 7 & 9 & 14 & 14 \end{pmatrix}} \cdot \begin{pmatrix} 1 \\ 1 \\ 1 \\ 1 \\ 1 \\ 1 \\ 1 \\ 1 \\ 1 \\ 1 \\ 1 \\ 1 \\ 1 \\ 1 \\ 1 \\ 1 \end{pmatrix} = 73943$$

GRANDSUMSQ = 73943

Here is an alternative method of computing GRANDSUMSQ. We square A using $A^{T}A$, remove the diagonal and sum the diagonal. ie

$$
\mathrm{diag}\left(A^T \cdot A\right) = \begin{pmatrix} 3592 \\ 5196 \\ 11414 \\ 14467 \\ 364 \\ 1432 \\ 4215 \\ 5704 \\ 2438 \\ 3606 \\ 6012 \\ 6380 \\ 1264 \\ 1951 \\ 2549 \\ 3359 \end{pmatrix}
\qquad
\mathrm{ONE16} \cdot \mathrm{diag}\left(A^T \cdot A\right) = 73943
$$

STEP 4:   Computation of the Grand Sum GS and the Correction
Factor CORRFACT. Nt = Row x Column in Data Matrix A. Nt =10 x
16 = 160.

GRANDSUM := ONE10·A·SIXTEEN1

Nt := 160

ONE10 = (1   1   1   1   1   1   1   1   1   1)

$$
(1\ 1\ 1\ 1\ 1\ 1\ 1\ 1\ 1\ 1) \cdot
\begin{pmatrix}
24 & 29 & 38 & 46 & 5 & 12 & 23 & 27 & 18 & 20 & 29 & 33 & 14 & 17 & 20 & 23 \\
20 & 30 & 43 & 48 & 7 & 12 & 16 & 17 & 8 & 11 & 19 & 26 & 17 & 16 & 21 & 24 \\
26 & 25 & 46 & 49 & 11 & 16 & 22 & 22 & 23 & 24 & 29 & 27 & 11 & 19 & 17 & 14 \\
18 & 19 & 37 & 36 & 3 & 2 & 7 & 6 & 17 & 25 & 38 & 21 & 15 & 15 & 14 & 20 \\
12 & 16 & 23 & 31 & 2 & 5 & 10 & 18 & 9 & 14 & 13 & 14 & 3 & 1 & 4 & 7 \\
14 & 13 & 25 & 26 & 3 & 7 & 15 & 22 & 11 & 19 & 19 & 29 & 1 & 8 & 7 & 11 \\
13 & 19 & 30 & 36 & 7 & 14 & 25 & 29 & 20 & 22 & 25 & 30 & 10 & 13 & 12 & 18 \\
17 & 19 & 29 & 32 & 3 & 9 & 19 & 28 & 18 & 19 & 19 & 24 & 15 & 19 & 23 & 22 \\
23 & 31 & 29 & 32 & 5 & 7 & 19 & 17 & 15 & 19 & 28 & 26 & 7 & 12 & 17 & 22 \\
17 & 19 & 30 & 37 & 8 & 22 & 35 & 38 & 9 & 11 & 15 & 16 & 7 & 9 & 14 & 14
\end{pmatrix}
\cdot
\begin{pmatrix} 1 \\ 1 \\ 1 \\ 1 \\ 1 \\ 1 \\ 1 \\ 1 \\ 1 \\ 1 \\ 1 \\ 1 \\ 1 \\ 1 \\ 1 \\ 1 \end{pmatrix}
= 3047
$$

GRANDSUM = 3047

$$\text{CORRFACT} := \frac{1}{\text{Nt}} \cdot \text{GRANDSUM}^2$$

$$\text{CORRFACT} := \left(\frac{1}{160}\right) \cdot 3047^2$$

CORRFACT = 58026.306

STEP 5: Computation of the Total Sum of Squares SSt.

SSt := GRANDSUMSQ − CORRFACT

There's no room to expand the Matrices into equations, so we'll just leave the computation as is.

$$\text{SSt} := \text{ONE10} \cdot \overrightarrow{(A \cdot A)} \cdot \text{SIXTEEN1} - \frac{1}{\text{Nt}} \cdot \text{GRANDSUM}^2$$

SSt = 15916.694

STEP 6: Computation of the effects of the 2 Experimental Conditions on the Total Performance. ie the effects of the first 8 columns vs the 2nd 8 columns. Therefore we sum the first 8 numbers in SUM1 and the 2nd 8 numbers, keeping them separate. Where are 2 columns in DB1, so N1 = 160/2 = 80 We proceed as follows.

PROGRAM: SS1= $(1/\text{N1}) * [1]_{1,10} [A]_{10,16} [\text{DB1}]_{16,2})^2$−CORRFACT

N1 := 80

$$\text{SS1} := \left(\frac{1}{\text{N1}}\right) \cdot (\text{SUM1} \cdot \text{DB1})^2 - \text{CORRFACT}$$

$$\text{SS1} := \left(\frac{1}{\text{N1}}\right) \cdot (\text{SUM1} \cdot \text{DB1}) \cdot (\text{SUM1} \cdot \text{DB1})^T - \text{CORRFACT}$$

SS1 = 628.056

Expanding we get:

$$\text{SUM1A} := (\ 184 \quad 220 \quad 330 \quad 373 \quad 54 \quad 106 \quad 191 \quad 224 \quad 148 \quad 184 \quad 234 \quad 246 \quad 100 \quad 129 \quad 149 \quad 175\ ) \cdot \begin{pmatrix} 1 & 0 \\ 1 & 0 \\ 1 & 0 \\ 1 & 0 \\ 1 & 0 \\ 1 & 0 \\ 1 & 0 \\ 1 & 0 \\ 0 & 1 \\ 0 & 1 \\ 0 & 1 \\ 0 & 1 \\ 0 & 1 \\ 0 & 1 \\ 0 & 1 \\ 0 & 1 \end{pmatrix}$$

$$\text{SUM1A} = (\ 1682 \quad 1365\ ) \qquad\qquad \text{SUM1A}^T = \begin{pmatrix} 1682 \\ 1365 \end{pmatrix}$$

$$\text{SS1} := \left(\frac{1}{80}\right) \cdot \left[\ (\ 1682 \quad 1365\ ) \cdot \begin{pmatrix} 1682 \\ 1365 \end{pmatrix}\right] - \text{CORRFACT}$$

$$\text{SS1} = 628.056$$

STEP 6: Computation of the Between Subjects Sum of Squares.
SSb. Nb = Nt/# elements in SUM2. Nb = 160/20 = 8

$$\text{SUM2} := \text{A} \cdot \text{DB1} \qquad \text{Nb} := 8$$

$$\text{SSb} := \left(\frac{1}{\text{Nb}}\right) \cdot \overrightarrow{[\ (\text{A} \cdot \text{DB1}) \cdot (\text{A} \cdot \text{DB1})\ ]}^{\blacksquare} - \text{CORRFACT}$$

MathCad cannot handle the math, se we define

$$\text{ADB1} := \text{A} \cdot \text{DB1}$$

Then we have:

$$\text{SSb} := \left(\frac{1}{\text{Nb}}\right) \cdot \text{ONE10}\ \overrightarrow{[\ (\text{ADB1}) \cdot (\text{ADB1})\ ]}\ \text{TWO1} - \text{CORRFACT}$$

$$ADB1 = \begin{pmatrix} 204 & 174 \\ 193 & 142 \\ 217 & 164 \\ 128 & 165 \\ 117 & 65 \\ 125 & 105 \\ 173 & 150 \\ 156 & 159 \\ 163 & 146 \\ 206 & 95 \end{pmatrix}$$

$$SSb := \left(\frac{1}{Nb}\right) \cdot ONE10 \left[\left(\begin{pmatrix} 204 & 174 \\ 193 & 142 \\ 217 & 164 \\ 128 & 165 \\ 117 & 65 \\ 125 & 105 \\ 173 & 150 \\ 156 & 159 \\ 163 & 146 \\ 206 & 95 \end{pmatrix} \right) \cdot \left(\begin{pmatrix} 204 & 174 \\ 193 & 142 \\ 217 & 164 \\ 128 & 165 \\ 117 & 65 \\ 125 & 105 \\ 173 & 150 \\ 156 & 159 \\ 163 & 146 \\ 206 & 95 \end{pmatrix} \right) \right] TWO1 - CORRFACT$$

SSb = 3580.569

SSerb := SSb − SS1

SSb = 3580.569   SS1 = 628.056

SSerb := 3580.569 − 628.056

SSerb = 2952.513

WITHIN SUBJECTS EFFECTS

STEP 8: Computation of the Within Subjects Sum of Squares. SSw.

SSw := SSt − SSb

SSw = 12336.125

STEP 9: Computation of the Sum of Squares for the First Within Subject Factors (2nd Experimental Conditions). SS2. N2 = # columns in DB11. N2 = 160/2 = 80

N2 := 80

PROGRAM: $(1/N2) \times ([1]_{1,10}[A]_{10,16}[DB11]_{16,2})^2 - \text{CORRFACT}$

$$SS2 := \left(\frac{1}{N2}\right) \cdot \left[(SUM1 \cdot DB11) \cdot (SUM1 \cdot DB11)^T\right] - \text{CORRFACT} \qquad \text{CORRFACT} = 58026.306$$

SS2 = 3910.506

$$(184 \quad 220 \quad 330 \quad 373 \quad 54 \quad 106 \quad 191 \quad 224 \quad 148 \quad 184 \quad 234 \quad 246 \quad 100 \quad 129 \quad 149 \quad 175) \cdot \begin{pmatrix} 1 & 0 \\ 1 & 0 \\ 1 & 0 \\ 1 & 0 \\ 0 & 1 \\ 0 & 1 \\ 0 & 1 \\ 0 & 1 \\ 1 & 0 \\ 1 & 0 \\ 1 & 0 \\ 1 & 0 \\ 0 & 1 \\ 0 & 1 \\ 0 & 1 \\ 0 & 1 \end{pmatrix} = (1919 \quad 1128)$$

$SUM1 \cdot DB11 = (1919 \quad 1128)$

$$SS2 := \left(\frac{1}{80}\right) \cdot \left[(1919 \quad 1128) \cdot \begin{pmatrix} 1919 \\ 1128 \end{pmatrix}\right] - \text{CORRFACT}$$

SS2 = 3910.506

STEP 10: Computation of the Sum of Squares for the Second Within Subject Factors (effects of the different Trials). SS3. N3 = # columns in DB4. N3 = 160/4 = 40

N3 := 40

PROGRAM: $(1/N3) \times ([1]_{1,10}[A]_{10,16}[DB4]_{16,4})^2 - \text{CORRFACT}$

$$SS3 := \left(\frac{1}{N3}\right) \cdot \left[(SUM1 \cdot DB4) \cdot (SUM1 \cdot DB4)^T\right] - CORRFACT$$

$$SS3 = 4425.119$$

$$SUM1DB4 := SUM1 \cdot DB4$$

$$(184 \ 220 \ 330 \ 373 \ 54 \ 106 \ 191 \ 224 \ 148 \ 184 \ 234 \ 246 \ 100 \ 129 \ 149 \ 175) \cdot \begin{pmatrix} 1 & 0 & 0 & 0 \\ 0 & 1 & 0 & 0 \\ 0 & 0 & 1 & 0 \\ 0 & 0 & 0 & 1 \\ 1 & 0 & 0 & 0 \\ 0 & 1 & 0 & 0 \\ 0 & 0 & 1 & 0 \\ 0 & 0 & 0 & 1 \\ 1 & 0 & 0 & 0 \\ 0 & 1 & 0 & 0 \\ 0 & 0 & 1 & 0 \\ 0 & 0 & 0 & 1 \\ 1 & 0 & 0 & 0 \\ 0 & 1 & 0 & 0 \\ 0 & 0 & 1 & 0 \\ 0 & 0 & 0 & 1 \end{pmatrix} = (486 \ 639 \ 904 \ 1018)$$

$$SUM1DB4 = (486 \ \ 639 \ \ 904 \ \ 1018)$$

$$SS3 := \left(\frac{1}{N3}\right) \cdot \left[(SUM1DB4) \cdot (SUM1DB4)^T\right] - CORRFACT$$

$$\left(\frac{1}{40}\right) \cdot \left[(486 \ \ 639 \ \ 904 \ \ 1018) \cdot \begin{pmatrix} 486 \\ 639 \\ 904 \\ 1018 \end{pmatrix}\right] - CORRFACT = 4425.119$$

$$SS3 = 4425.119$$

STEP 11: Computation of the Sum of Squares for the Interaction of the First and Second Experimental Conditions. SS1x2. N1x2 = # columns in DB12. N1x2 = 160/4 = 40

$$N1x2 := 40$$

PROGRAM: $(1/N1x2) \ \times \ ([1]_{1,10}[A]_{10,16}[DB12]_{16,4})^2 - CORRFACT$

$$SS1x2 := \left(\frac{1}{N1x2}\right) \cdot \left[(SUM1 \cdot DB12) \cdot (SUM1 \cdot DB12)^T\right] - CORRFACT - SS1 - SS2$$

SS1x2 = 465.806

$$(184 \quad 220 \quad 330 \quad 373 \quad 54 \quad 106 \quad 191 \quad 224 \quad 148 \quad 184 \quad 234 \quad 246 \quad 100 \quad 129 \quad 149 \quad 175) \cdot \begin{pmatrix} 1 & 0 & 0 & 0 \\ 1 & 0 & 0 & 0 \\ 1 & 0 & 0 & 0 \\ 1 & 0 & 0 & 0 \\ 0 & 1 & 0 & 0 \\ 0 & 1 & 0 & 0 \\ 0 & 1 & 0 & 0 \\ 0 & 1 & 0 & 0 \\ 0 & 0 & 1 & 0 \\ 0 & 0 & 1 & 0 \\ 0 & 0 & 1 & 0 \\ 0 & 0 & 1 & 0 \\ 0 & 0 & 0 & 1 \\ 0 & 0 & 0 & 1 \\ 0 & 0 & 0 & 1 \\ 0 & 0 & 0 & 1 \end{pmatrix} = (1107 \quad 575 \quad 812 \quad 553)$$

SUM1DB12 := SUM1·DB12

$$\text{SS1x2} := \left(\frac{1}{\text{N1x2}}\right) \cdot \left[(\text{SUM1DB12}) \cdot (\text{SUM1DB12})^T \right] - \text{CORRFACT} - \text{SS1} - \text{SS2}$$

$$\left(\frac{1}{40}\right) \cdot \left[(1107 \quad 575 \quad 812 \quad 553) \cdot \begin{pmatrix} 1107 \\ 575 \\ 812 \\ 553 \end{pmatrix} \right] - \text{CORRFACT} - \text{SS1} - \text{SS2} = 465.806$$

SS1 = 628.056   SS2 = 3910.506   CORRFACT = 58026.306

$$\left(\frac{1}{\text{N1x2}}\right) \cdot \left[(\text{SUM1·DB12}) \cdot (\text{SUM1·DB12})^T \right] - 58026.306 - 628.056 - 3910.506 = 465.807$$

SS1x2 = 465.806

STEP 11: Computation of the Sum of Squares for the Interaction
of the 2 Experimental Groups. SS1x3. N1x3 = # columns in DB41.
N1x3 = 160/8 = 20

PROGRAM: $(1/\text{N1x3}) \times ([1]_{1,10} [A]_{10,16} [\text{DB41}]_{16,8})^2 - \text{CORRFACT} - \text{SS1} - \text{SS3}$

N1x3 := 20

$$\text{SS1x3} := \left(\frac{1}{\text{N1x3}}\right) \cdot \left[(\text{SUM1·DB41}) \cdot (\text{SUM1·DB41})^T \right] - \text{CORRFACT} - \text{SS1} - \text{SS3}$$

SS1x3 = 629.169

$$
SUM1DB41 := \begin{pmatrix} 184 & 220 & 330 & 373 & 54 & 106 & 191 & 224 & 148 & 184 & 234 & 246 & 100 & 129 & 149 & 175 \end{pmatrix} \cdot \begin{pmatrix} 1 & 0 & 0 & 0 & 0 & 0 & 0 & 0 \\ 0 & 1 & 0 & 0 & 0 & 0 & 0 & 0 \\ 0 & 0 & 1 & 0 & 0 & 0 & 0 & 0 \\ 0 & 0 & 0 & 1 & 0 & 0 & 0 & 0 \\ 1 & 0 & 0 & 0 & 0 & 0 & 0 & 0 \\ 0 & 1 & 0 & 0 & 0 & 0 & 0 & 0 \\ 0 & 0 & 1 & 0 & 0 & 0 & 0 & 0 \\ 0 & 0 & 0 & 1 & 0 & 0 & 0 & 0 \\ 0 & 0 & 0 & 0 & 1 & 0 & 0 & 0 \\ 0 & 0 & 0 & 0 & 0 & 1 & 0 & 0 \\ 0 & 0 & 0 & 0 & 0 & 0 & 1 & 0 \\ 0 & 0 & 0 & 0 & 0 & 0 & 0 & 1 \\ 0 & 0 & 0 & 0 & 1 & 0 & 0 & 0 \\ 0 & 0 & 0 & 0 & 0 & 1 & 0 & 0 \\ 0 & 0 & 0 & 0 & 0 & 0 & 1 & 0 \\ 0 & 0 & 0 & 0 & 0 & 0 & 0 & 1 \end{pmatrix}
$$

$$SUM1 \cdot DB41 = \begin{pmatrix} 238 & 326 & 521 & 597 & 248 & 313 & 383 & 421 \end{pmatrix}$$

$$SS1x3 := \left(\frac{1}{N1x3} \right) \cdot \left[(SUM1 \cdot DB41) \cdot (SUM1 \cdot DB41)^T \right] - CORRFACT - SS1 - SS3$$

SS1 = 628.056

SS3 = 4425.119

$$
\left(\frac{1}{20} \right) \cdot \left[\begin{pmatrix} 238 & 326 & 521 & 597 & 248 & 313 & 383 & 421 \end{pmatrix} \cdot \begin{pmatrix} 238 \\ 326 \\ 521 \\ 597 \\ 248 \\ 313 \\ 383 \\ 421 \end{pmatrix} \right] - CORRFACT - 628.056 - 4425.119 = 629.169
$$

SS1x3 = 629.169

STEP 12: Computation of the Sum of Squares for the Interaction of the First and Second Experimental Factors. SS2x3. N2x3 = # columns in DB8. N2x3 = 160/8 = 20

N2x3 := 20

PROGRAM: $1/N2x3) \times ([1]_{1,10}[A]_{10,16}[DB8]_{16,8})^2 - CORRFACT - SS2 - SS3$

$$\text{SS2x3} := \left(\frac{1}{\text{N2x3}}\right) \cdot \left[(\text{SUM1} \cdot \text{DB8}) \cdot (\text{SUM1} \cdot \text{DB8})^T\right] - \text{CORRFACT} - \text{SS2} - \text{SS3}$$

$$\text{SS2x3} = 60.019$$

$$\text{SUM1DB8} := (\begin{matrix} 184 & 220 & 330 & 373 & 54 & 106 & 191 & 224 & 148 & 184 & 234 & 246 & 100 & 129 & 149 & 175 \end{matrix}) \cdot \begin{pmatrix} 1 & 0 & 0 & 0 & 0 & 0 & 0 & 0 \\ 0 & 1 & 0 & 0 & 0 & 0 & 0 & 0 \\ 0 & 0 & 1 & 0 & 0 & 0 & 0 & 0 \\ 0 & 0 & 0 & 1 & 0 & 0 & 0 & 0 \\ 0 & 0 & 0 & 0 & 1 & 0 & 0 & 0 \\ 0 & 0 & 0 & 0 & 0 & 1 & 0 & 0 \\ 0 & 0 & 0 & 0 & 0 & 0 & 1 & 0 \\ 0 & 0 & 0 & 0 & 0 & 0 & 0 & 1 \\ 1 & 0 & 0 & 0 & 0 & 0 & 0 & 0 \\ 0 & 1 & 0 & 0 & 0 & 0 & 0 & 0 \\ 0 & 0 & 1 & 0 & 0 & 0 & 0 & 0 \\ 0 & 0 & 0 & 1 & 0 & 0 & 0 & 0 \\ 0 & 0 & 0 & 0 & 1 & 0 & 0 & 0 \\ 0 & 0 & 0 & 0 & 0 & 1 & 0 & 0 \\ 0 & 0 & 0 & 0 & 0 & 0 & 1 & 0 \\ 0 & 0 & 0 & 0 & 0 & 0 & 0 & 1 \end{pmatrix}$$

$$\text{SUM1} \cdot \text{DB8} = (\begin{matrix} 332 & 404 & 564 & 619 & 154 & 235 & 340 & 399 \end{matrix})$$

$$\text{SS2x3} := \left(\frac{1}{\text{N2x3}}\right) \cdot \left[(\text{SUM1} \cdot \text{DB8}) \cdot (\text{SUM1} \cdot \text{DB8})^T\right] - \text{CORRFACT} - \text{SS2} - \text{SS3}$$

$$\text{SS2} = 3910.506$$

$$\text{SS3} = 4425.119$$

$$\left(\frac{1}{20}\right) \cdot (\begin{matrix} 332 & 404 & 564 & 619 & 154 & 235 & 340 & 399 \end{matrix}) \cdot \begin{pmatrix} 332 \\ 404 \\ 564 \\ 619 \\ 154 \\ 235 \\ 340 \\ 399 \end{pmatrix} - \text{CORRFACT} - 3910.506 - 4425.119 = 60.019$$

$$\text{SS2x3} = 60.019$$

STEP 13: Computation of the Interaction Between all Experimental Groups and Experimental Conditions. SS1x2x3. N1x2x3 = Nt/# elements in SUM1 = 160/16 = 10

$$\text{N1x2x3} := 10$$

$$SS1x2x3 := \left(\frac{1}{N1x2x3}\right) \cdot \left(SUM1 \cdot SUM1^T\right) - CORRFACT - SS1 - SS2 - SS3 - SS1x2 - SS1x3 - SS2x3$$

$$SS1x2x3 = 14.319$$

$$SS1x2x3 := \left(\frac{1}{10}\right) \cdot \left[SUM1 \cdot \begin{pmatrix} 184 \\ 220 \\ 330 \\ 373 \\ 54 \\ 106 \\ 191 \\ 224 \\ 148 \\ 184 \\ 234 \\ 246 \\ 100 \\ 129 \\ 149 \\ 175 \end{pmatrix} \right] - 58026.306 - 628.056 - 3910.506 - 4425.119 - 465.806 - 629.169 - 60.019$$

CORRFACT = 58026.306   SS1x2 = 465.806

SS1 = 628.056   SS1x3 = 629.169

SS2 = 3910.506   SS2x3 = 60.019

SS3 = 4425.119   SS1x2x3 = 14.319

STEP 14: Computation of the Total Error for Within Subjects. SSerw

SSw = 12336.125

$$SSerw := SSw - SS2 - SS3 - SS1x2 - SS1x3 - SS2x3 - SS1x2x3$$

SSerw = 2831.187

STEP 15: Computation of the Error Term for the SS1 and SS2 interaction.

$$Ner1 := \frac{160}{40}$$

SSer1 = (1/4) x (ONE10*(A1*DB12)$^{\otimes 2}$*FOUR1) - CORRFACT-SSb-SS2-SS1x2

NOTE: The term $[A]_{10,16}[DB12]_{16,8})^2$ is an element by element multiplication, so I will note this by using ⊠ as the element by element multiplicative symbol.

$$SSer1 := \left(\frac{1}{Ner1}\right) \cdot \left[ONE10 \cdot \overrightarrow{(A \cdot DB12 \boxtimes A \cdot DB12)} \cdot FOUR1\right] - CORRFACTa - SSb - SS2a - SS1x2a \quad \blacksquare$$

MathCad cannot solve this type of equation, so we make the following substitution:

ADB12 := A·DB12

The Error 1 equation now becomes:

$$SSer1 := \left(\frac{1}{Ner1}\right) \cdot \left(ONE10 \cdot \overrightarrow{(ADB12 \boxtimes ADB12)} \cdot FOUR1\right) - CORRFACT - SSb - SS2 - SS1x2$$

$$
ADB12 :=
\begin{pmatrix}
24 & 29 & 38 & 46 & 5 & 12 & 23 & 27 & 18 & 20 & 29 & 33 & 14 & 17 & 20 & 23 \\
20 & 30 & 43 & 48 & 7 & 12 & 16 & 17 & 8 & 11 & 19 & 26 & 17 & 16 & 21 & 24 \\
26 & 25 & 46 & 49 & 11 & 16 & 22 & 22 & 23 & 24 & 29 & 27 & 11 & 19 & 17 & 14 \\
18 & 19 & 37 & 36 & 3 & 2 & 7 & 6 & 17 & 25 & 38 & 21 & 15 & 15 & 14 & 20 \\
12 & 16 & 23 & 31 & 2 & 5 & 10 & 18 & 9 & 14 & 13 & 14 & 3 & 1 & 4 & 7 \\
14 & 13 & 25 & 26 & 3 & 7 & 15 & 22 & 11 & 19 & 19 & 29 & 1 & 8 & 7 & 11 \\
13 & 19 & 30 & 36 & 7 & 14 & 25 & 29 & 20 & 22 & 25 & 30 & 10 & 13 & 12 & 18 \\
17 & 19 & 29 & 32 & 3 & 9 & 19 & 28 & 18 & 19 & 19 & 24 & 15 & 19 & 23 & 22 \\
23 & 31 & 29 & 32 & 5 & 7 & 19 & 17 & 15 & 19 & 28 & 26 & 7 & 12 & 17 & 22 \\
17 & 19 & 30 & 37 & 8 & 22 & 35 & 38 & 9 & 11 & 15 & 16 & 7 & 9 & 14 & 14
\end{pmatrix}
\cdot
\begin{pmatrix}
1 & 0 & 0 & 0 \\
1 & 0 & 0 & 0 \\
1 & 0 & 0 & 0 \\
1 & 0 & 0 & 0 \\
0 & 1 & 0 & 0 \\
0 & 1 & 0 & 0 \\
0 & 1 & 0 & 0 \\
0 & 1 & 0 & 0 \\
0 & 0 & 1 & 0 \\
0 & 0 & 1 & 0 \\
0 & 0 & 1 & 0 \\
0 & 0 & 1 & 0 \\
0 & 0 & 0 & 1 \\
0 & 0 & 0 & 1 \\
0 & 0 & 0 & 1 \\
0 & 0 & 0 & 1
\end{pmatrix}
$$

$$A \cdot DB12 = \begin{pmatrix} 137 & 67 & 100 & 74 \\ 141 & 52 & 64 & 78 \\ 146 & 71 & 103 & 61 \\ 110 & 18 & 101 & 64 \\ 82 & 35 & 50 & 15 \\ 78 & 47 & 78 & 27 \\ 98 & 75 & 97 & 53 \\ 97 & 59 & 80 & 79 \\ 115 & 48 & 88 & 58 \\ 103 & 103 & 51 & 44 \end{pmatrix}$$

$$\left(\frac{1}{Ner1}\right) \cdot \left(ONE10 \cdot \overrightarrow{(ADB12 \cdot ADB12)} \cdot FOUR1\right) = 67514.25$$

SS2 = 3910.506   SS1x2 = 465.806

SSb = 3580.569   CORRFACT = 58026.306

$$\left(\frac{1}{Ner1}\right) \cdot \left(ONE10 \cdot \overrightarrow{(ADB12 \cdot ADB12)} \cdot FOUR1\right) - CORRFACT - SSb - SS2 - SS1x2 = 1531.063$$

$$SSer1 := \left(\frac{1}{Ner1}\right) \cdot \left(ONE10 \cdot \overrightarrow{(ADB12 \cdot ADB12)} \cdot FOUR1\right) - CORRFACT - SSb - SS2 - SS1x2$$

67514.25 − 58026.306 − 3910.506 − 465.806 − 3580.569 = 1531.063

SSer1 = 1531.063

STEP 16: Computation of the Error Term for the SS3 and SS1x3 interaction. Ner2 = 2 (the # of scores to obtain each sum)

Ner2 := 2

PROGRAM:

SSer2 = (1/Ner2) x ([1]$_{1,10}$([A]$_{10,16}$[DB41]$_{16,8}$)^2 [1]$_{8,1}$)− CORRFACT−SSb−SS3−SS1x3

NOTE: The term [A]$_{10,16}$[db41]$_{16,8}$)^2 is an element by element multiplication, so I will note this by using $\otimes$ as the element by element multiplicative symbol.

$$\text{SSer2} = (1/2) \times (\text{ONE10} * (A*DB41)^{\otimes 2} * \text{EIGHT1}) - \text{CORRFACT} - \text{SSb} - \text{SS3} - \text{SS1x3}$$

$$\text{SSer2a} := \left(\frac{1}{\text{Ner2}}\right) \cdot \left(\text{ONE10} \cdot \overrightarrow{(A \cdot DB41 \cdot A1 \cdot DB41)} \cdot \text{EIGHT1}\right) - \text{CORRFACT} - \text{SSb} - \text{SS3} - \text{SS1x3}$$

Again, MathCad cannot compute the above expression, so we will substitute the following expression:

$$\text{ADB41} := A \cdot DB41$$

So now we've got:

$$\text{SSer2} := \left(\frac{1}{\text{Ner2}}\right) \cdot \left(\text{ONE10} \cdot \overrightarrow{(\text{ADB41} \cdot \text{ADB41})} \cdot \text{EIGHT1}\right) - \text{CORRFACT} - \text{SSb} - \text{SS3} - \text{SS1x3}$$

$$\text{SSer2} = 603.338$$

$$\begin{pmatrix} 24 & 29 & 38 & 46 & 5 & 12 & 23 & 27 & 18 & 20 & 29 & 33 & 14 & 17 & 20 & 23 \\ 20 & 30 & 43 & 48 & 7 & 12 & 16 & 17 & 8 & 11 & 19 & 26 & 17 & 16 & 21 & 24 \\ 26 & 25 & 46 & 49 & 11 & 16 & 22 & 22 & 23 & 24 & 29 & 27 & 11 & 19 & 17 & 14 \\ 18 & 19 & 37 & 36 & 3 & 2 & 7 & 6 & 17 & 25 & 38 & 21 & 15 & 15 & 14 & 20 \\ 12 & 16 & 23 & 31 & 2 & 5 & 10 & 18 & 9 & 14 & 13 & 14 & 3 & 1 & 4 & 7 \\ 14 & 13 & 25 & 26 & 3 & 7 & 15 & 22 & 11 & 19 & 19 & 29 & 1 & 8 & 7 & 11 \\ 13 & 19 & 30 & 36 & 7 & 14 & 25 & 29 & 20 & 22 & 25 & 30 & 10 & 13 & 12 & 18 \\ 17 & 19 & 29 & 32 & 3 & 9 & 19 & 28 & 18 & 19 & 19 & 24 & 15 & 19 & 23 & 22 \\ 23 & 31 & 29 & 32 & 5 & 7 & 19 & 17 & 15 & 19 & 28 & 26 & 7 & 12 & 17 & 22 \\ 17 & 19 & 30 & 37 & 8 & 22 & 35 & 38 & 9 & 11 & 15 & 16 & 7 & 9 & 14 & 14 \end{pmatrix} \cdot \begin{pmatrix} 1 & 0 & 0 & 0 & 0 & 0 & 0 & 0 \\ 0 & 1 & 0 & 0 & 0 & 0 & 0 & 0 \\ 0 & 0 & 1 & 0 & 0 & 0 & 0 & 0 \\ 0 & 0 & 0 & 1 & 0 & 0 & 0 & 0 \\ 1 & 0 & 0 & 0 & 0 & 0 & 0 & 0 \\ 0 & 1 & 0 & 0 & 0 & 0 & 0 & 0 \\ 0 & 0 & 1 & 0 & 0 & 0 & 0 & 0 \\ 0 & 0 & 0 & 1 & 0 & 0 & 0 & 0 \\ 0 & 0 & 0 & 0 & 1 & 0 & 0 & 0 \\ 0 & 0 & 0 & 0 & 0 & 1 & 0 & 0 \\ 0 & 0 & 0 & 0 & 0 & 0 & 1 & 0 \\ 0 & 0 & 0 & 0 & 0 & 0 & 0 & 1 \\ 0 & 0 & 0 & 0 & 1 & 0 & 0 & 0 \\ 0 & 0 & 0 & 0 & 0 & 1 & 0 & 0 \\ 0 & 0 & 0 & 0 & 0 & 0 & 1 & 0 \\ 0 & 0 & 0 & 0 & 0 & 0 & 0 & 1 \end{pmatrix}$$

$$\text{ADB41} = \begin{pmatrix} 29 & 41 & 61 & 73 & 32 & 37 & 49 & 56 \\ 27 & 42 & 59 & 65 & 25 & 27 & 40 & 50 \\ 37 & 41 & 68 & 71 & 34 & 43 & 46 & 41 \\ 21 & 21 & 44 & 42 & 32 & 40 & 52 & 41 \\ 14 & 21 & 33 & 49 & 12 & 15 & 17 & 21 \\ 17 & 20 & 40 & 48 & 12 & 27 & 26 & 40 \\ 20 & 33 & 55 & 65 & 30 & 35 & 37 & 48 \\ 20 & 28 & 48 & 60 & 33 & 38 & 42 & 46 \\ 28 & 38 & 48 & 49 & 22 & 31 & 45 & 48 \\ 25 & 41 & 65 & 75 & 16 & 20 & 29 & 30 \end{pmatrix}$$

CLINTON L. HOLT

Lets put numbers in for this problem.

$$\left(\frac{1}{Ner2}\right) \cdot \left(\overrightarrow{ONE10 \cdot (ADB41 \cdot ADB41)} \cdot EIGHT1 \right) = 67264.5$$

SSb = 3580.569   SS3 = 4425.119

CORRFACT = 58026.306   SS1x3 = 629.169

$$SSer2 := \left(\frac{1}{Ner2}\right) \cdot \left(\overrightarrow{ONE10 \cdot (ADB41 \cdot ADB41)} \cdot EIGHT1 \right) - 58026.306 - 3580.569 - 4425.119 - 629.169$$

SSer2 = 603.337

STEP 17: Computation of the df and Error Term for the ss2x3 and SS1x2x3 interaction.

SSer3 := SSerw − SSer1 − SSer2

SSerw = 2831.187   SSer1 = 1531.063   SSer2 = 603.337

SSer3 = 696.788

For the Degrees of Freedom: m = # elements in A; G = # Groups in the study; s = # subjects in each Group, T = # trials, R = # experimental conditioned and S = s/G.

$S := \dfrac{s}{G}$ m := 160   s := 20   G := 2 $S := \dfrac{s}{G}$ R := 2   T := 4   S := 10

We compute the Degrees of Freedom using the following program w

$$df := \begin{bmatrix} m-1 \\ s-1 \\ G-1 \\ (s-1)-1 \\ (m-1)-(s-1) \\ R-1 \\ T-1 \\ (G-1)\cdot(R-1) \\ (G-1)\cdot(T-1) \\ (R-1)\cdot(T-1) \\ (G-1)\cdot(R-1)\cdot(T-1) \\ (m-1)-(s-1)-(R-1)-(T-1)-(G-1)\cdot(R-1)-(G-1)\cdot(T-1)-(R-1)\cdot(T-1)-(G-1)\cdot(R-1)\cdot(T-1) \\ (R-1)\cdot(s-1)\cdot G \\ (T-1)\cdot(s-1)\cdot G \\ (R-1)\cdot(T-1)\cdot(s-1)\cdot G \end{bmatrix}$$

These df values are good for all df's of this type of problem, no matter how many groups, trials or subjects.

$$x := \begin{pmatrix} \text{dfSSt} \\ \text{dfSSb} \\ \text{dfSS1} \\ \text{dfSSerb} \\ \text{dfSSw} \\ \text{dfSS2} \\ \text{dfSS3} \\ \text{dfSS1x2} \\ \text{dfSS1x3} \\ \text{dfSS2x3} \\ \text{dfSS1x2x3} \\ \text{dfSSerw} \\ \text{dfSSer1} \\ \text{dfSSer2} \\ \text{dfSSer3} \end{pmatrix} \blacksquare$$

ERRO computes the Error Terms, the Matrix to the right gives the order of the differing error terms and their placements.

$$
SS := \begin{pmatrix}
14753.6 \\
3580.569 \\
628.056 \\
2952.513 \\
13412.844 \\
3910.506 \\
4425.119 \\
465.806 \\
629.169 \\
60.019 \\
14.319 \\
3907.906 \\
1531.063 \\
603.338 \\
1773.506
\end{pmatrix}
\qquad
df = \begin{pmatrix}
159 \\
19 \\
1 \\
18 \\
140 \\
1 \\
3 \\
1 \\
3 \\
3 \\
3 \\
126 \\
18 \\
54 \\
54
\end{pmatrix}
\qquad
ERRO := \begin{pmatrix}
\dfrac{2952.513}{18} \\
164.029 \\
164.029 \\
164.029 \\
1 \\
\dfrac{1531.063}{18} \\
\dfrac{603.338}{54} \\
85.059 \\
11.173 \\
\dfrac{552.775}{54} \\
10.237 \\
1 \\
85.059 \\
11.173 \\
10.237
\end{pmatrix}
\qquad
\begin{pmatrix}
b \\
b \\
b \\
b \\
w \\
1 \\
2 \\
1 \\
2 \\
3 \\
3 \\
w \\
1 \\
2 \\
3
\end{pmatrix}
$$

$$ F := \overrightarrow{\left(SS \cdot df^{-1} \cdot ERRO^{-1} \right)} $$

This is the final F value   These are the probabilities.
for the new problem.

$$
F = \begin{pmatrix}
0.566 \\
1.149 \\
3.829 \\
1 \\
95.806 \\
45.974 \\
132.019 \\
5.476 \\
18.771 \\
1.954 \\
0.466 \\
31.015 \\
1 \\
1 \\
3.208
\end{pmatrix}
\quad
Fcorrected := \begin{pmatrix}
0 \\
0 \\
3.829 \\
0 \\
0 \\
45.95 \\
132.019 \\
5.473 \\
18.771 \\
1.954 \\
.466 \\
0 \\
0 \\
0 \\
0
\end{pmatrix}
\quad
p := \begin{pmatrix}
0 \\
0 \\
.001 \\
0 \\
0 \\
.001 \\
.005 \\
.001 \\
.0025 \\
.ns \\
ns \\
0 \\
0 \\
0 \\
0
\end{pmatrix}
\quad
df = \begin{pmatrix}
159 \\
19 \\
1 \\
18 \\
140 \\
1 \\
3 \\
1 \\
3 \\
3 \\
3 \\
126 \\
18 \\
54 \\
54
\end{pmatrix}
$$

Looking at the Probability tables we get:

$$
p := \begin{pmatrix} 0 \\ 0 \\ .001 \\ 0 \\ 0 \\ .001 \\ .005 \\ .001 \\ .0025 \\ ns \\ ns \\ 0 \\ 0 \\ 0 \\ 0 \end{pmatrix}
$$

The overall performance of the Groups was different

The effects of the First Experimental Condition were significant

The Subjects performance changed across Trials

There was a significant 1x3 interaction

There was a significant 1x2 interaction

HOMEWORK: You've seen how this is done, reduce this problem to a single equation.

LATIN SQUARE DESIGN: SIMPLE

STEP 1:   Here we define the Data Matrix A and all constants and all sub-Statistical Data Base Matrices.

$$A := \begin{pmatrix} 5 & 7 & 8 & 9 & 7 & 3 & 9 & 10 & 10 \\ 5 & 7 & 8 & 8 & 3 & 5 & 7 & 8 & 11 \\ 6 & 9 & 11 & 12 & 9 & 10 & 11 & 10 & 11 \\ 2 & 4 & 5 & 5 & 5 & 5 & 8 & 10 & 10 \\ 5 & 6 & 10 & 15 & 9 & 18 & 8 & 9 & 10 \\ 4 & 5 & 5 & 8 & 4 & 5 & 9 & 7 & 12 \\ 8 & 20 & 11 & 9 & 6 & 9 & 8 & 5 & 9 \end{pmatrix}$$

VARIABLES:

$\text{ONE7} := \begin{pmatrix} 1 & 1 & 1 & 1 & 1 & 1 & 1 \end{pmatrix}$

$\text{ONE3} := \begin{pmatrix} 1 & 1 & 1 \end{pmatrix}$

$\text{THREE1} := \text{ONE3}^{T}$

$$\text{NINE1} := \begin{pmatrix} 1 \\ 1 \\ 1 \\ 1 \\ 1 \\ 1 \\ 1 \\ 1 \\ 1 \end{pmatrix}$$

$$\text{DB3} := \begin{pmatrix} 1 & 0 & 0 \\ 0 & 1 & 0 \\ 0 & 0 & 1 \\ 1 & 0 & 0 \\ 0 & 1 & 0 \\ 0 & 0 & 1 \\ 1 & 0 & 0 \\ 0 & 1 & 0 \\ 0 & 0 & 1 \end{pmatrix} \quad \text{DBL} := \begin{pmatrix} 1 & 0 & 0 \\ 0 & 1 & 0 \\ 0 & 0 & 1 \\ 0 & 0 & 1 \\ 1 & 0 & 0 \\ 0 & 1 & 0 \\ 0 & 1 & 0 \\ 0 & 0 & 1 \\ 1 & 0 & 0 \end{pmatrix} \quad \text{DB1} := \begin{pmatrix} 1 & 0 & 0 \\ 1 & 0 & 0 \\ 1 & 0 & 0 \\ 0 & 1 & 0 \\ 0 & 1 & 0 \\ 0 & 1 & 0 \\ 0 & 0 & 1 \\ 0 & 0 & 1 \\ 0 & 0 & 1 \end{pmatrix} \quad \text{df} := \begin{pmatrix} 62 \\ 20 \\ 2 \\ 18 \\ 42 \\ 2 \\ 2 \\ 2 \\ 36 \end{pmatrix}$$

STEP 2:   Sum the columns of A to obtain SUM1:

PROGRAM: $\text{SUM1} = [1]_{1,7}[A]_{7,9}$

SUM1 := ONE7 · A

$$(1 \quad 1 \quad 1 \quad 1 \quad 1 \quad 1 \quad 1) \cdot \begin{pmatrix} 5 & 7 & 8 & 9 & 7 & 3 & 9 & 10 & 10 \\ 5 & 7 & 8 & 8 & 3 & 5 & 7 & 8 & 11 \\ 6 & 9 & 11 & 12 & 9 & 10 & 11 & 10 & 11 \\ 2 & 4 & 5 & 5 & 5 & 5 & 8 & 10 & 10 \\ 5 & 6 & 10 & 15 & 9 & 18 & 8 & 9 & 10 \\ 4 & 5 & 5 & 8 & 4 & 5 & 9 & 7 & 12 \\ 8 & 20 & 11 & 9 & 6 & 9 & 8 & 5 & 9 \end{pmatrix} = (35 \quad 58 \quad 58 \quad 66 \quad 43 \quad 55 \quad 60 \quad 59 \quad 73)$$

SUM1 = (35   58   58   66   43   55   60   59   73)

STEP 3:   Now we obtain the Sum for Each Group. There are 3 Groups:

GROUPSUM = [SUM1]$_{1,9}$ [DB1]$_{9,3}$ = [1]$_{1,7}$ [A]$_{7,9}$ [DB1]$_{9,3}$

GROUPSUM := SUM1 · DB1

NOTE: Here we sum the 1st 3 numbers in SUM1, the 2nd and 3rd 3 numbers. ie 35+58+58=151.

$$(35 \quad 58 \quad 58 \quad 66 \quad 43 \quad 55 \quad 60 \quad 59 \quad 73) \cdot \begin{pmatrix} 1 & 0 & 0 \\ 1 & 0 & 0 \\ 1 & 0 & 0 \\ 0 & 1 & 0 \\ 0 & 1 & 0 \\ 0 & 1 & 0 \\ 0 & 0 & 1 \\ 0 & 0 & 1 \\ 0 & 0 & 1 \end{pmatrix} = (151 \quad 164 \quad 192)$$

GROUPSUM = (151   164   192)

STEP 4:   Now we obtain the Sum for each Subject in each Group.

SUBJECTSUM = [A]$_{7,9}$[DB1]$_{9,3}$

NOTE: This is a Between Subjects computation.

SUBJECTSUM := A · DB1

NOTE: Here we sum the first 3 numbers in row 1, the second three numbers and the 3rd three numbers, 5+7+8=20, 9+7+3=19, etc.

$$
\begin{pmatrix}
5 & 7 & 8 & 9 & 7 & 3 & 9 & 10 & 10 \\
5 & 7 & 8 & 8 & 3 & 5 & 7 & 8 & 11 \\
6 & 9 & 11 & 12 & 9 & 10 & 11 & 10 & 11 \\
2 & 4 & 5 & 5 & 5 & 5 & 8 & 10 & 10 \\
5 & 6 & 10 & 15 & 9 & 18 & 8 & 9 & 10 \\
4 & 5 & 5 & 8 & 4 & 5 & 9 & 7 & 12 \\
8 & 20 & 11 & 9 & 6 & 9 & 8 & 5 & 9
\end{pmatrix}
\cdot
\begin{pmatrix}
1 & 0 & 0 \\
1 & 0 & 0 \\
1 & 0 & 0 \\
0 & 1 & 0 \\
0 & 1 & 0 \\
0 & 1 & 0 \\
0 & 0 & 1 \\
0 & 0 & 1 \\
0 & 0 & 1
\end{pmatrix}
=
\begin{pmatrix}
20 & 19 & 29 \\
20 & 16 & 26 \\
26 & 31 & 32 \\
11 & 15 & 28 \\
21 & 42 & 27 \\
14 & 17 & 28 \\
39 & 24 & 22
\end{pmatrix}
$$

$$
\mathrm{SUBJECTSUM} =
\begin{pmatrix}
20 & 19 & 29 \\
20 & 16 & 26 \\
26 & 31 & 32 \\
11 & 15 & 28 \\
21 & 42 & 27 \\
14 & 17 & 28 \\
39 & 24 & 22
\end{pmatrix}
$$

STEP 5: Square each Score in the table to find the Grand Sum of Squares GSQ.

PROGRAM: $[\mathbf{1}]_{1,7}([\mathbf{A}]_{7,9} \otimes [\mathbf{A}]_{7,9})[\mathbf{1}]_{9,1}$

$\mathrm{GSQ} := \mathrm{ONE7} \cdot \overrightarrow{\left(A \cdot A \right)} \cdot \mathrm{NINE1}$

Here is the Math, but we must compress ONE7 to allow the problem to fit a single page.

$$
\mathrm{ONE7} \cdot
\left[
\begin{pmatrix}
5 & 7 & 8 & 9 & 7 & 3 & 9 & 10 & 10 \\
5 & 7 & 8 & 8 & 3 & 5 & 7 & 8 & 11 \\
6 & 9 & 11 & 12 & 9 & 10 & 11 & 10 & 11 \\
2 & 4 & 5 & 5 & 5 & 5 & 8 & 10 & 10 \\
5 & 6 & 10 & 15 & 9 & 18 & 8 & 9 & 10 \\
4 & 5 & 5 & 8 & 4 & 5 & 9 & 7 & 12 \\
8 & 20 & 11 & 9 & 6 & 9 & 8 & 5 & 9
\end{pmatrix}
\cdot
\overrightarrow{
\begin{pmatrix}
5 & 7 & 8 & 9 & 7 & 3 & 9 & 10 & 10 \\
5 & 7 & 8 & 8 & 3 & 5 & 7 & 8 & 11 \\
6 & 9 & 11 & 12 & 9 & 10 & 11 & 10 & 11 \\
2 & 4 & 5 & 5 & 5 & 5 & 8 & 10 & 10 \\
5 & 6 & 10 & 15 & 9 & 18 & 8 & 9 & 10 \\
4 & 5 & 5 & 8 & 4 & 5 & 9 & 7 & 12 \\
8 & 20 & 11 & 9 & 6 & 9 & 8 & 5 & 9
\end{pmatrix}
}
\cdot
\begin{pmatrix}
1 \\ 1 \\ 1 \\ 1 \\ 1 \\ 1 \\ 1 \\ 1 \\ 1
\end{pmatrix}
\right] = 4751
$$

GSQ = 4751

STEP 6: Computation of the Grand Sum (GS) and the Correction Factor (CORRFACT).

PROGRAM: GRANDSUM = $[1]_{1,7}[A]_{7,9}[1]_{9,1}$

GS := ONE7 ·A ·NINE1

$$
(1\ \ 1\ \ 1\ \ 1\ \ 1\ \ 1\ \ 1) \cdot
\begin{pmatrix}
5 & 7 & 8 & 9 & 7 & 3 & 9 & 10 & 10 \\
5 & 7 & 8 & 8 & 3 & 5 & 7 & 8 & 11 \\
6 & 9 & 11 & 12 & 9 & 10 & 11 & 10 & 11 \\
2 & 4 & 5 & 5 & 5 & 5 & 8 & 10 & 10 \\
5 & 6 & 10 & 15 & 9 & 18 & 8 & 9 & 10 \\
4 & 5 & 5 & 8 & 4 & 5 & 9 & 7 & 12 \\
8 & 20 & 11 & 9 & 6 & 9 & 8 & 5 & 9
\end{pmatrix}
\cdot
\begin{pmatrix} 1 \\ 1 \\ 1 \\ 1 \\ 1 \\ 1 \\ 1 \\ 1 \\ 1 \end{pmatrix}
= 507
$$

GS = 507

CORRFACT = $(1/N_t)\,([1]_{1,7}\ [A]_{7,9}\ [1]_{9,1}\,)^2$

The Correction Factor is obtained by squaring the Grand Sum and dividing by the total number of elements in the Data Matrix A

Nt := 9 ·7   Nt = number of elements in Data Matrix A = 9x7=63

$$
\text{CORRFACT} := \left(\frac{1}{Nt}\right) \cdot (\text{ONE7} \cdot A \cdot \text{NINE1})^2
$$

$$
\left(\frac{1}{63}\right) \cdot
\left[
(1\ \ 1\ \ 1\ \ 1\ \ 1\ \ 1\ \ 1) \cdot
\begin{pmatrix}
5 & 7 & 8 & 9 & 7 & 3 & 9 & 10 & 10 \\
5 & 7 & 8 & 8 & 3 & 5 & 7 & 8 & 11 \\
6 & 9 & 11 & 12 & 9 & 10 & 11 & 10 & 11 \\
2 & 4 & 5 & 5 & 5 & 5 & 8 & 10 & 10 \\
5 & 6 & 10 & 15 & 9 & 18 & 8 & 9 & 10 \\
4 & 5 & 5 & 8 & 4 & 5 & 9 & 7 & 12 \\
8 & 20 & 11 & 9 & 6 & 9 & 8 & 5 & 9
\end{pmatrix}
\cdot
\begin{pmatrix} 1 \\ 1 \\ 1 \\ 1 \\ 1 \\ 1 \\ 1 \\ 1 \\ 1 \end{pmatrix}
\right]^2
= 4080.143
$$

CORRFACT = 4080.143

STEP 7:   Computation of the Total Sum of Squares SSt.

SSt := GSQ − CORRFACT

SSt = 670.857

Or the program looks like:

$$SSt := ONE7 \cdot \overrightarrow{(A \cdot A)} \cdot NINE1 - \left(\frac{1}{Nt}\right) \cdot (ONE7 \cdot A \cdot NINE1)^2$$

SSt = 670.857

STEP 8:   Computation of the Between Subjects Sum of Squares SSb. In STEP 4 we computed the scores for each Subject in each Group (SUBJECTSUM) which is a row (rather than a column) computation

SUBJECTSUM := A · DB1

Nb = # Latin orders, in this case = 3

Nb := 3

$$SS_b := \left[\left[\frac{1}{(N_b)}\right] \cdot \overrightarrow{[ONE7 \cdot (A \cdot DB1) \cdot (A \cdot DB1) \cdot THREE1]} - CORRFACT\right.$$

MathCad cannot compute this equation, but let A*DB1 = SUBJECTSUM, therefore

$$SSb := \left(\frac{1}{Nb}\right) \cdot \left[ONE7 \cdot \overrightarrow{[(SUBJECTSUM) \cdot (SUBJECTSUM)] \cdot THREE1}\right] - CORRFACT$$

SSb = 422.857

$$
\text{SUBJECTSUM} = \begin{pmatrix} 20 & 19 & 29 \\ 20 & 16 & 26 \\ 26 & 31 & 32 \\ 11 & 15 & 28 \\ 21 & 42 & 27 \\ 14 & 17 & 28 \\ 39 & 24 & 22 \end{pmatrix}
$$

And the math looks like: $\qquad \text{THREE1} = \begin{pmatrix} 1 \\ 1 \\ 1 \end{pmatrix}$

$$
\left(\frac{1}{3}\right) \cdot \left(1\ 1\ 1\ 1\ 1\ 1\ 1\right) \cdot \left[\begin{pmatrix} 20 & 19 & 29 \\ 20 & 16 & 26 \\ 26 & 31 & 32 \\ 11 & 15 & 28 \\ 21 & 42 & 27 \\ 14 & 17 & 28 \\ 39 & 24 & 22 \end{pmatrix} \cdot \begin{pmatrix} 20 & 19 & 29 \\ 20 & 16 & 26 \\ 26 & 31 & 32 \\ 11 & 15 & 28 \\ 21 & 42 & 27 \\ 14 & 17 & 28 \\ 39 & 24 & 22 \end{pmatrix} \cdot \begin{pmatrix} 1 \\ 1 \\ 1 \end{pmatrix} \right] - \text{CORRFACT} = 422.857
$$

STEP 9: Computation of the Effect of Groups, SSg. (analysis of the effects of the Order of Presentation of Experimental Condition 1. Ng = Nt/#columns in DB1 = 3.

$$
\text{Ng} := \frac{63}{3}
$$

PROGRAM: $1/N_G[1]_{1,3}([A]_{7,9}\ [DB1]_{9,3})^2 = 1/N_G(\text{SUM1 DB1})^2 - \text{CORRFACT} = \text{SSG}$

$$
\text{SSg} := \left(\frac{1}{\text{Ng}}\right) \cdot (\text{SUM1} \cdot \text{DB1})^2 - \text{CORRFACT}
$$

$$
\text{SSg} := \left(\frac{1}{\text{Ng}}\right) \cdot (\text{SUM1} \cdot \text{DB1}) \cdot (\text{SUM1} \cdot \text{DB1})^T - \text{CORRFACT}
$$

$$
\text{SUM1DB1} := \text{SUM1} \cdot \text{DB1}
$$

$$\text{SUM1DB1} := (\; 35 \quad 58 \quad 58 \quad 66 \quad 43 \quad 55 \quad 60 \quad 59 \quad 73 \;) \cdot \begin{pmatrix} 1 & 0 & 0 \\ 1 & 0 & 0 \\ 1 & 0 & 0 \\ 0 & 1 & 0 \\ 0 & 1 & 0 \\ 0 & 1 & 0 \\ 0 & 0 & 1 \\ 0 & 0 & 1 \\ 0 & 0 & 1 \end{pmatrix}$$

$$\text{SUM1DB1} = (\; 151 \quad 164 \quad 192 \;)$$

The final evaluation becomes:

$$\left(\frac{1}{21}\right) \cdot \left[(\; 151 \quad 164 \quad 192 \;) \cdot \begin{pmatrix} 151 \\ 164 \\ 192 \end{pmatrix} \right] - 4080.143 = 41.809$$

$\text{SSg} = 41.81$

STEP 10: Computation of the Between Subjects Error Term, SSerb.

$\text{SSerb} := \text{SSb} - \text{SSg}$

$\text{SSerb} = 381.048$

STEP 11: Computation of the Within Subjects Sum of Squares, SSw.

$\text{SSw} := \text{SSt} - \text{SSb}$

$\text{SSw} = 248$

STEP 12: Computation of the 2nd Experimental Factor, SS2. N2 = Nt/#columns in DB3.

PROGRAM: 1/N2) x ($[\text{SUM1}]_{1,9}$ $[\text{DB3}]_{9,3}$) x ($[\text{SUM1}]_{1,9}$ $[\text{DB3}]_{9,3}$)^T – CORRFACT

$$N2 := \frac{63}{3}$$

$$SS2 := \left(\frac{1}{N2}\right) \cdot (SUM1 \cdot DB3)^2 - CORRFACT$$

$$SS2 := \left(\frac{1}{N2}\right) \cdot (SUM1 \cdot DB3) \cdot (SUM1 \cdot DB3)^T - CORRFACT$$

$$\left(\frac{1}{N2}\right) \cdot (SUM1 \cdot DB3) \cdot (SUM1 \cdot DB3)^T - CORRFACT = 20.667$$

$$SS2 := 20.667$$

$$SUM1DB1 := (\begin{matrix} 35 & 58 & 58 & 66 & 43 & 55 & 60 & 59 & 73 \end{matrix}) \cdot \begin{pmatrix} 1 & 0 & 0 \\ 0 & 1 & 0 \\ 0 & 0 & 1 \\ 1 & 0 & 0 \\ 0 & 1 & 0 \\ 0 & 0 & 1 \\ 1 & 0 & 0 \\ 0 & 1 & 0 \\ 0 & 0 & 1 \end{pmatrix}$$

$$CORRFACT = 4080.143$$

The final evaluation becomes:

$$\left(\frac{1}{21}\right) \cdot \left[(\begin{matrix} 161 & 160 & 186 \end{matrix}) \cdot \begin{pmatrix} 161 \\ 160 \\ 186 \end{pmatrix} \right] - 4080.143 = 20.667$$

$$SS2 := 20.667$$

STEP 13: Computation of the Order of Effects. In the Latin
Square design, we change the order of presentation from first
to second to third to 2nd to 1st to 3rd, or 3rd to 1st to 2nd,
etc. Here we take the random order and return it to it's original
order, ie 2nd to 3rd to 1st is returned mathematically to 1st
to 2nd to 3rd. These are set up in the experiment and are not
set values. SSL. NL = Nt/#columns in DBL.

$$NL := \frac{63}{3}$$

PROGRAM: SSL $= 1/N_L[1]_{1,3}([A]_{7,9} \ [DBL]_{9,3})^2 = 1/21(SUM1 \ DBL)^2 - CORRFACT$

$$SSL := \left(\frac{1}{NL}\right) \cdot (SUM1 \cdot DBL)^2 - CORRFACT$$

$$SSL := \left(\frac{1}{NL}\right) \cdot (SUM1 \cdot DBL) \cdot \left((SUM1 \cdot DBL)^T\right) - CORRFACT$$

$$SUM1 = (\ 35 \quad 58 \quad 58 \quad 66 \quad 43 \quad 55 \quad 60 \quad 59 \quad 73\ )$$

$$(\ 35 \quad 58 \quad 58 \quad 66 \quad 43 \quad 55 \quad 60 \quad 59 \quad 73\ ) \cdot \begin{pmatrix} 1 & 0 & 0 \\ 0 & 1 & 0 \\ 0 & 0 & 1 \\ 0 & 0 & 1 \\ 1 & 0 & 0 \\ 0 & 1 & 0 \\ 0 & 1 & 0 \\ 0 & 0 & 1 \\ 1 & 0 & 0 \end{pmatrix} = (\ 151 \quad 173 \quad 183\ )$$

$$SSL := \left(\frac{1}{21}\right) \cdot (\ 151 \quad 173 \quad 183\ ) \cdot \begin{pmatrix} 151 \\ 173 \\ 183 \end{pmatrix} - CORRFACT$$

$$SSL = 25.524$$

STEP 14: Computation of the effects of a particular Experimental Condition being presented 1st, 2nd or 3rd. SS2xL. N2xL = Nt/# elements in SUM1.

$$N2xL := \frac{63}{9}$$

PROGRAM: $(1/N2xL) \ [1]_{1,7}([A]_{7,9})^2 - CORRFACT - SSg - SS2 - SSL$

$$SS2xL := \left(\frac{1}{N2xL}\right) \cdot \left(SUM1 \cdot SUM1^T\right) - CORRFACT - SSg - SS2 - SSL$$

The math looks like:

$$
SS2xL := \left(\frac{1}{7}\right) \cdot \left[(35\ \ 58\ \ 58\ \ 66\ \ 43\ \ 55\ \ 60\ \ 59\ \ 73) \cdot \begin{pmatrix} 35 \\ 58 \\ 58 \\ 66 \\ 43 \\ 55 \\ 60 \\ 59 \\ 73 \end{pmatrix} \right] - CORRFACT - SSg - SS2 - SSL
$$

$$
SS2xL := \left(\frac{1}{7}\right) \cdot 29593 - 4080.143 - 41.81 - 20.667 - 25.524
$$

$$
SS2xL = 59.427
$$

STEP 15: Computation of the Within Subjects Error Term SSerw.

This is just the difference of several terms we have already computed. ie

$SSerw := SSw - SS2 - SSL - SS2xL$

$SSerw = 142.382$

$SSw - SS2 - SSL - SS2xL = 142.382$

$SSw = 248$ $SSL = 25.524$ $SS2 = 20.667$ $SS2xL = 59.427$

$SSerw := 248 - 20.667 - 25.524 - 59.427$

$SSerw = 142.382$

STEP 16: Computation of the Degrees of Freedom for the system.

$m := 63$ $s := 21$ $G := 3$ $t := 3$ $L := 3$

m = # elements in Data Matrix A = 63
s = # Subjects in each part of study s = 21
G = # Groups, in this study G = 3
t = # experimental treatments t = 3
L = # Latin orders (# of tests scrambled) L = 3

This df is good for all 2 variable Latin Square Studies.

$$
\text{df} := \begin{bmatrix} m-1 \\ s-1 \\ G-1 \\ (s-1)-(G-1) \\ (m-1)-(s-1) \\ t-1 \\ L-1 \\ (t-2)\cdot(L-1) \\ [(m-1)-(s-1)]-(t-1)-(L-1)-(t-2)\cdot(L-1) \end{bmatrix}
\qquad
\text{df} = \begin{pmatrix} 62 \\ 20 \\ 2 \\ 18 \\ 42 \\ 2 \\ 2 \\ 2 \\ 36 \end{pmatrix}
$$

The Error Matrix is computed as the first 4 values in the Error Matrix are SSb divided by it's Degrees of Freedom, while the last 5 values are the value of SSerw divided by it's df value.

$$
\text{ERR} = \begin{pmatrix} 21.169 \\ 21.169 \\ 21.169 \\ 21.169 \\ 3.955 \\ 3.955 \\ 3.955 \\ 3.955 \\ 3.955 \end{pmatrix}
$$

$$
\text{ERR} := \begin{pmatrix} \dfrac{381.048}{18} \\ \dfrac{381.048}{18} \\ \dfrac{381.048}{18} \\ \dfrac{381.048}{18} \\ \dfrac{142.382}{36} \\ \dfrac{142.382}{36} \\ \dfrac{142.382}{36} \\ \dfrac{142.382}{36} \\ \dfrac{142.382}{36} \end{pmatrix}
\qquad
\text{SS} := \begin{pmatrix} \text{SSt} \\ \text{SSb} \\ \text{SSg} \\ \text{SSerb} \\ \text{SSw} \\ \text{SS2} \\ \text{SSL} \\ \text{SS2xL} \\ \text{SSerw} \end{pmatrix}
\qquad
\text{SS} = \begin{pmatrix} 670.857 \\ 422.857 \\ 41.81 \\ 381.048 \\ 248 \\ 20.667 \\ 25.524 \\ 59.427 \\ 142.382 \end{pmatrix}
\qquad
\text{df} = \begin{pmatrix} 62 \\ 20 \\ 2 \\ 18 \\ 42 \\ 2 \\ 2 \\ 2 \\ 36 \end{pmatrix}
$$

SS definition defines the different Sums of Squares we have computed. There is no math value, just for us to see what we have computed.

$$
\text{SSdefinition} := \begin{pmatrix} \text{SSt} & 670.857 \\ \text{SSb} & 422.857 \\ \text{SSg} & 41.81 \\ \text{SSerb} & 381.048 \\ \text{SSw} & 248 \\ \text{SS2} & 20.667 \\ \text{SSL} & 25.524 \\ \text{SS2xL} & 59.427 \\ \text{SSerw} & 142.382 \end{pmatrix}
$$

OK, now the F value is computed from dividing SS by df and dividing by the Error term,

$$F := \overrightarrow{\left(SS \cdot df^{-1} \cdot ERR^{-1}\right)} \qquad df^{-1} = \begin{pmatrix} 0.016 \\ 0.05 \\ 0.5 \\ 0.056 \\ 0.024 \\ 0.5 \\ 0.5 \\ 0.5 \\ 0.028 \end{pmatrix} \qquad ERR^{-1} = \begin{pmatrix} 0.047 \\ 0.047 \\ 0.047 \\ 0.047 \\ 0.253 \\ 0.253 \\ 0.253 \\ 0.253 \\ 0.253 \end{pmatrix} \qquad SS = \begin{pmatrix} 670.857 \\ 422.857 \\ 41.81 \\ 381.048 \\ 248 \\ 20.667 \\ 25.524 \\ 59.427 \\ 142.382 \end{pmatrix}$$

$$F := \overrightarrow{\left[\begin{pmatrix} 670.857 \\ 422.857 \\ 41.81 \\ 381.048 \\ 248 \\ 20.667 \\ 25.524 \\ 59.427 \\ 142.382 \end{pmatrix} \cdot \begin{pmatrix} 0.016 \\ 0.05 \\ 0.5 \\ 0.056 \\ 0.024 \\ 0.5 \\ 0.5 \\ 0.5 \\ 0.028 \end{pmatrix} \cdot \begin{pmatrix} 0.047 \\ 0.047 \\ 0.047 \\ 0.047 \\ 0.253 \\ 0.253 \\ 0.253 \\ 0.253 \\ 0.253 \end{pmatrix} \right]}$$

$$F = \begin{pmatrix} 0.504 \\ 0.994 \\ 0.983 \\ 1.003 \\ 1.506 \\ 2.614 \\ 3.229 \\ 7.518 \\ 1.009 \end{pmatrix} \qquad p := \begin{pmatrix} 0 \\ 0 \\ .983 \\ 0 \\ 0 \\ 2.61 \\ 3.23 \\ 7.51 \\ 0 \end{pmatrix} \qquad df = \begin{pmatrix} 62 \\ 20 \\ 2 \\ 18 \\ 42 \\ 2 \\ 2 \\ 2 \\ 36 \end{pmatrix} \qquad pf := \begin{pmatrix} 0 \\ 0 \\ ns \\ 0 \\ 0 \\ ns \\ ns \\ .005 \\ ns \end{pmatrix} \blacksquare$$

Therefore the interaction of SS2xL are interactive, all other values are non significant.

HOMEWORK: You've seen how this is done, reduce this problem to a single equation.

LATIN SQUARE DESIGN: COMPLEX

COMPUTATIONAL Handbook of statistics, Pg. 93-109

NOTE: Before, the Statistical Field Equation was represented by $1/N([1]*[A]*[B])^2 - CF$. In this example, the Equation expands to $1/N([1]*[A]*[B]*[C])^2 - CF$. Therefore, it follows that the complete Statistical Field Equation follows the form:

$$1/N([1]_{ab}*[A]_{bc}*[B]_{cd}*[C]_{de}\cdots[I]_{ij}*[J]_{jk}*\cdots[M]_{mn}*[N]_{no})^2 - \text{CORRFACT}$$

$$A := \begin{pmatrix}
21 & 17 & 10 & 18 & 18 & 10 & 20 & 10 & 12 & 27 & 17 & 8 & 15 & 18 & 9 & 20 & 11 & 11 \\
15 & 16 & 8 & 17 & 19 & 10 & 15 & 9 & 14 & 24 & 17 & 7 & 16 & 19 & 7 & 16 & 10 & 12 \\
19 & 11 & 12 & 13 & 16 & 9 & 14 & 7 & 15 & 22 & 12 & 9 & 12 & 17 & 7 & 15 & 8 & 7 \\
15 & 11 & 9 & 10 & 14 & 9 & 13 & 11 & 9 & 18 & 10 & 5 & 12 & 15 & 8 & 12 & 11 & 9 \\
17 & 8 & 11 & 12 & 12 & 11 & 15 & 9 & 14 & 23 & 9 & 7 & 14 & 12 & 10 & 14 & 9 & 8 \\
18 & 13 & 9 & 14 & 15 & 11 & 17 & 15 & 12 & 26 & 11 & 9 & 12 & 10 & 9 & 16 & 10 & 7
\end{pmatrix}$$

VARIABLES:

$$DB1 := \begin{pmatrix}
1 & 0 & 0 & 0 & 0 & 0 \\
1 & 0 & 0 & 0 & 0 & 0 \\
1 & 0 & 0 & 0 & 0 & 0 \\
0 & 1 & 0 & 0 & 0 & 0 \\
0 & 1 & 0 & 0 & 0 & 0 \\
0 & 1 & 0 & 0 & 0 & 0 \\
0 & 0 & 1 & 0 & 0 & 0 \\
0 & 0 & 1 & 0 & 0 & 0 \\
0 & 0 & 1 & 0 & 0 & 0 \\
0 & 0 & 0 & 1 & 0 & 0 \\
0 & 0 & 0 & 1 & 0 & 0 \\
0 & 0 & 0 & 1 & 0 & 0 \\
0 & 0 & 0 & 0 & 1 & 0 \\
0 & 0 & 0 & 0 & 1 & 0 \\
0 & 0 & 0 & 0 & 1 & 0 \\
0 & 0 & 0 & 0 & 0 & 1 \\
0 & 0 & 0 & 0 & 0 & 1 \\
0 & 0 & 0 & 0 & 0 & 1
\end{pmatrix}$$

$ONE10 := (1\ 1\ 1\ 1\ 1\ 1\ 1\ 1\ 1\ 1)$

$TEN1 := ONE10^T$

$ONE6 := (1\ 1\ 1\ 1\ 1\ 1)$

$SIX1 := ONE6^T$

$$DB2 := \begin{pmatrix}
1 & 0 \\
1 & 0 \\
1 & 0 \\
0 & 1 \\
0 & 1 \\
0 & 1
\end{pmatrix}$$

$$DB3 := \begin{pmatrix}
1 & 0 & 0 \\
0 & 1 & 0 \\
0 & 0 & 1 \\
1 & 0 & 0 \\
0 & 1 & 0 \\
0 & 0 & 1
\end{pmatrix}$$

$$
\mathbf{DB4} := \begin{pmatrix}
1 & 0 & 0 \\
0 & 1 & 0 \\
0 & 0 & 1 \\
1 & 0 & 0 \\
0 & 1 & 0 \\
0 & 0 & 1 \\
1 & 0 & 0 \\
0 & 1 & 0 \\
0 & 0 & 1 \\
1 & 0 & 0 \\
0 & 1 & 0 \\
0 & 0 & 1 \\
1 & 0 & 0 \\
0 & 1 & 0 \\
0 & 0 & 1 \\
1 & 0 & 0 \\
0 & 1 & 0 \\
0 & 0 & 1
\end{pmatrix}
\qquad
\mathbf{DB5} := \begin{pmatrix}
1 & 0 & 0 \\
0 & 1 & 0 \\
0 & 0 & 1 \\
0 & 0 & 1 \\
1 & 0 & 0 \\
0 & 1 & 0 \\
0 & 1 & 0 \\
0 & 0 & 1 \\
1 & 0 & 0 \\
1 & 0 & 0 \\
0 & 1 & 0 \\
0 & 0 & 1 \\
0 & 0 & 1 \\
1 & 0 & 0 \\
0 & 1 & 0 \\
0 & 1 & 0 \\
0 & 0 & 1 \\
1 & 0 & 0
\end{pmatrix}
\qquad
\mathbf{DB6} := \begin{pmatrix}
1 & 0 & 0 & 0 & 0 & 0 & 0 & 0 & 0 \\
0 & 1 & 0 & 0 & 0 & 0 & 0 & 0 & 0 \\
0 & 0 & 1 & 0 & 0 & 0 & 0 & 0 & 0 \\
0 & 0 & 0 & 1 & 0 & 0 & 0 & 0 & 0 \\
0 & 0 & 0 & 0 & 1 & 0 & 0 & 0 & 0 \\
0 & 0 & 0 & 0 & 0 & 1 & 0 & 0 & 0 \\
0 & 0 & 0 & 0 & 0 & 0 & 1 & 0 & 0 \\
0 & 0 & 0 & 0 & 0 & 0 & 0 & 1 & 0 \\
0 & 0 & 0 & 0 & 0 & 0 & 0 & 0 & 1 \\
1 & 0 & 0 & 0 & 0 & 0 & 0 & 0 & 0 \\
0 & 1 & 0 & 0 & 0 & 0 & 0 & 0 & 0 \\
0 & 0 & 1 & 0 & 0 & 0 & 0 & 0 & 0 \\
0 & 0 & 0 & 1 & 0 & 0 & 0 & 0 & 0 \\
0 & 0 & 0 & 0 & 1 & 0 & 0 & 0 & 0 \\
0 & 0 & 0 & 0 & 0 & 1 & 0 & 0 & 0 \\
0 & 0 & 0 & 0 & 0 & 0 & 1 & 0 & 0 \\
0 & 0 & 0 & 0 & 0 & 0 & 0 & 1 & 0 \\
0 & 0 & 0 & 0 & 0 & 0 & 0 & 0 & 1
\end{pmatrix}
$$

$$
\mathbf{DB7} := \begin{pmatrix}
1 & 0 & 0 & 0 & 0 & 0 \\
0 & 1 & 0 & 0 & 0 & 0 \\
0 & 0 & 1 & 0 & 0 & 0 \\
1 & 0 & 0 & 0 & 0 & 0 \\
0 & 1 & 0 & 0 & 0 & 0 \\
0 & 0 & 1 & 0 & 0 & 0 \\
1 & 0 & 0 & 0 & 0 & 0 \\
0 & 1 & 0 & 0 & 0 & 0 \\
0 & 0 & 1 & 0 & 0 & 0 \\
0 & 0 & 0 & 1 & 0 & 0 \\
0 & 0 & 0 & 0 & 1 & 0 \\
0 & 0 & 0 & 0 & 0 & 1 \\
0 & 0 & 0 & 1 & 0 & 0 \\
0 & 0 & 0 & 0 & 1 & 0 \\
0 & 0 & 0 & 0 & 0 & 1 \\
0 & 0 & 0 & 1 & 0 & 0 \\
0 & 0 & 0 & 0 & 1 & 0 \\
0 & 0 & 0 & 0 & 0 & 1
\end{pmatrix}
\qquad
\mathbf{DBL} := \begin{pmatrix}
1 & 0 & 0 & 0 & 0 & 0 \\
0 & 1 & 0 & 0 & 0 & 0 \\
0 & 0 & 1 & 0 & 0 & 0 \\
0 & 0 & 1 & 0 & 0 & 0 \\
1 & 0 & 0 & 0 & 0 & 0 \\
0 & 1 & 0 & 0 & 0 & 0 \\
0 & 1 & 0 & 0 & 0 & 0 \\
0 & 0 & 1 & 0 & 0 & 0 \\
1 & 0 & 0 & 0 & 0 & 0 \\
0 & 0 & 0 & 1 & 0 & 0 \\
0 & 0 & 0 & 0 & 1 & 0 \\
0 & 0 & 0 & 0 & 0 & 1 \\
0 & 0 & 0 & 1 & 0 & 0 \\
0 & 0 & 0 & 0 & 1 & 0 \\
0 & 0 & 0 & 0 & 1 & 0 \\
0 & 0 & 0 & 0 & 0 & 1 \\
0 & 0 & 0 & 1 & 0 & 0
\end{pmatrix}
\qquad
\mathbf{EIGHTEEN1} := \begin{pmatrix}
1 \\ 1 \\ 1 \\ 1 \\ 1 \\ 1 \\ 1 \\ 1 \\ 1 \\ 1 \\ 1 \\ 1 \\ 1 \\ 1 \\ 1 \\ 1 \\ 1 \\ 1
\end{pmatrix}
\qquad
\mathbf{ONE18} := \mathbf{EIGHTEEN1}^{\mathbf{T}}
$$

STEP 2:   Sum the columns of A

$\mathbf{SUM1} := \mathbf{ONE6} \cdot \mathbf{A}$

$$\mathbf{SUM1} := (1 \quad 1 \quad 1 \quad 1 \quad 1 \quad 1) \cdot \begin{pmatrix} 21 & 17 & 10 & 18 & 18 & 10 & 20 & 10 & 12 & 27 & 17 & 8 & 15 & 18 & 9 & 20 & 11 & 11 \\ 15 & 16 & 8 & 17 & 19 & 10 & 15 & 9 & 14 & 24 & 17 & 7 & 16 & 19 & 7 & 16 & 10 & 12 \\ 19 & 11 & 12 & 13 & 16 & 9 & 14 & 7 & 15 & 22 & 12 & 9 & 12 & 17 & 7 & 15 & 8 & 7 \\ 15 & 11 & 9 & 10 & 14 & 9 & 13 & 11 & 9 & 18 & 10 & 5 & 12 & 15 & 8 & 12 & 11 & 9 \\ 17 & 8 & 11 & 12 & 12 & 11 & 15 & 9 & 14 & 23 & 9 & 7 & 14 & 12 & 10 & 14 & 9 & 8 \\ 18 & 13 & 9 & 14 & 15 & 11 & 17 & 15 & 12 & 26 & 11 & 9 & 12 & 10 & 9 & 16 & 10 & 7 \end{pmatrix}$$

$$\mathbf{SUM1} = (105 \quad 76 \quad 59 \quad 84 \quad 94 \quad 60 \quad 94 \quad 61 \quad 76 \quad 140 \quad 76 \quad 45 \quad 81 \quad 91 \quad 50 \quad 93 \quad 59 \quad 54)$$

STEP 3: Sum each of the Groups separately

$\mathbf{SUM1} \cdot \mathbf{DB1} = (240 \quad 238 \quad 231 \quad 261 \quad 222 \quad 206)$

The math is too big to fit a single page, so we won't expand SUM1

$$\mathbf{SUM1} \cdot \begin{pmatrix} 1 & 0 & 0 & 0 & 0 & 0 \\ 1 & 0 & 0 & 0 & 0 & 0 \\ 1 & 0 & 0 & 0 & 0 & 0 \\ 0 & 1 & 0 & 0 & 0 & 0 \\ 0 & 1 & 0 & 0 & 0 & 0 \\ 0 & 1 & 0 & 0 & 0 & 0 \\ 0 & 0 & 1 & 0 & 0 & 0 \\ 0 & 0 & 1 & 0 & 0 & 0 \\ 0 & 0 & 1 & 0 & 0 & 0 \\ 0 & 0 & 0 & 1 & 0 & 0 \\ 0 & 0 & 0 & 1 & 0 & 0 \\ 0 & 0 & 0 & 1 & 0 & 0 \\ 0 & 0 & 0 & 0 & 1 & 0 \\ 0 & 0 & 0 & 0 & 1 & 0 \\ 0 & 0 & 0 & 0 & 1 & 0 \\ 0 & 0 & 0 & 0 & 0 & 1 \\ 0 & 0 & 0 & 0 & 0 & 1 \\ 0 & 0 & 0 & 0 & 0 & 1 \end{pmatrix} = (240 \quad 238 \quad 231 \quad 261 \quad 222 \quad 206)$$

STEP 4:   Add the Scores for each Subject in each Group

$$
\mathbf{A \cdot DB1} =
\begin{pmatrix}
48 & 46 & 42 & 52 & 42 & 42 \\
39 & 46 & 38 & 48 & 42 & 38 \\
42 & 38 & 36 & 43 & 36 & 30 \\
35 & 33 & 33 & 33 & 35 & 32 \\
36 & 35 & 38 & 39 & 36 & 31 \\
40 & 40 & 44 & 46 & 31 & 33
\end{pmatrix}
$$

And the math is:

$$
\begin{pmatrix}
21 & 17 & 10 & 18 & 18 & 10 & 20 & 10 & 12 & 27 & 17 & 8 & 15 & 18 & 9 & 20 & 11 & 11 \\
15 & 16 & 8 & 17 & 19 & 10 & 15 & 9 & 14 & 24 & 17 & 7 & 16 & 19 & 7 & 16 & 10 & 12 \\
19 & 11 & 12 & 13 & 16 & 9 & 14 & 7 & 15 & 22 & 12 & 9 & 12 & 17 & 7 & 15 & 8 & 7 \\
15 & 11 & 9 & 10 & 14 & 9 & 13 & 11 & 9 & 18 & 10 & 5 & 12 & 15 & 8 & 12 & 11 & 9 \\
17 & 8 & 11 & 12 & 12 & 11 & 15 & 9 & 14 & 23 & 9 & 7 & 14 & 12 & 10 & 14 & 9 & 8 \\
18 & 13 & 9 & 14 & 15 & 11 & 17 & 15 & 12 & 26 & 11 & 9 & 12 & 10 & 9 & 16 & 10 & 7
\end{pmatrix}
\cdot
\begin{pmatrix}
1 & 0 & 0 & 0 & 0 & 0 \\
1 & 0 & 0 & 0 & 0 & 0 \\
1 & 0 & 0 & 0 & 0 & 0 \\
0 & 1 & 0 & 0 & 0 & 0 \\
0 & 1 & 0 & 0 & 0 & 0 \\
0 & 1 & 0 & 0 & 0 & 0 \\
0 & 0 & 1 & 0 & 0 & 0 \\
0 & 0 & 1 & 0 & 0 & 0 \\
0 & 0 & 1 & 0 & 0 & 0 \\
0 & 0 & 0 & 1 & 0 & 0 \\
0 & 0 & 0 & 1 & 0 & 0 \\
0 & 0 & 0 & 1 & 0 & 0 \\
0 & 0 & 0 & 1 & 0 & 0 \\
0 & 0 & 0 & 1 & 0 & 0 \\
0 & 0 & 0 & 1 & 0 & 0 \\
0 & 0 & 0 & 0 & 1 & 0 \\
0 & 0 & 0 & 0 & 1 & 0 \\
0 & 0 & 0 & 0 & 1 & 0 \\
0 & 0 & 0 & 0 & 0 & 1 \\
0 & 0 & 0 & 0 & 0 & 1 \\
0 & 0 & 0 & 0 & 0 & 1
\end{pmatrix}
=
\begin{pmatrix}
48 & 46 & 42 & 52 & 42 & 42 \\
39 & 46 & 38 & 48 & 42 & 38 \\
42 & 38 & 36 & 43 & 36 & 30 \\
35 & 33 & 33 & 33 & 35 & 32 \\
36 & 35 & 38 & 39 & 36 & 31 \\
40 & 40 & 44 & 46 & 31 & 33
\end{pmatrix}
$$

STEP 5:   Square each term in the Data Matrix A and sum to get the Grand Sum of Squares GSQ

$$
\mathbf{GSQ} := \mathbf{ONE6} \cdot \left(\overrightarrow{\mathbf{A \cdot A}} \right) \cdot \mathbf{EIGHTEEN1}
$$

$\mathbf{GSQ} = 20194$

The GSQ is too large to fit on a single page so we'll show no math here.

STEP 6:   Computation for the Grand Sum of the Data GS.

$$
\mathbf{GS} := \mathbf{SUM1} \cdot \mathbf{EIGHTEEN1}
$$

$\mathbf{GS} = 1398$

$$GS := (105 \quad 76 \quad 59 \quad 84 \quad 94 \quad 60 \quad 94 \quad 61 \quad 76 \quad 140 \quad 76 \quad 45 \quad 81 \quad 91 \quad 50 \quad 93 \quad 59 \quad 54) \cdot \begin{pmatrix} 1 \\ 1 \\ 1 \\ 1 \\ 1 \\ 1 \\ 1 \\ 1 \\ 1 \\ 1 \\ 1 \\ 1 \\ 1 \\ 1 \\ 1 \\ 1 \\ 1 \\ 1 \end{pmatrix}$$

$GS = 1398$

STEP 7: Computation of the Correction Factor CF. Nt = # elements in Data Matrix A. Nt=18x6 = 108

$Nt := 108$

$$CF := \left(\frac{1}{Nt} \right) \cdot GS^2$$

The Math is:

$$\left(\frac{1}{108} \right) \cdot 1398^2 = 18096.333$$

$CF = 18096.333$

STEP 8: Computation of the Total Sum of Squares SSt

$SSt := GSQ - CF$

SSt = 2097.667

STEP 9: Computation of the Between Subjects Sum of Squares SSb. Nb = # Groups summed to get A*DB1. in this case Nb = 3.

Nb := 3

$$\text{SS}_b := \left[\left(\frac{1}{N_b} \right) \cdot \overrightarrow{[\text{ONE6} \cdot (\text{A} \cdot \text{DB1}) \cdot (\text{A} \cdot \text{DB1}) \cdot \text{SIX1}]} - \text{CF} \right] \blacksquare$$

MathCad will not compute this, so we make the following substitution:

ADB1 := **A** · **DB1**   then we get:

$$\text{SSb} := \left(\frac{1}{\text{Nb}} \right) \cdot \left[\text{ONE6} \cdot \overrightarrow{(\text{ADB1} \cdot \text{ADB1})} \cdot \text{SIX1} \right] - \text{CF}$$

SSb = 351.667

$$\text{SSb} := \left(\frac{1}{3} \right) \cdot \left[\text{ONE6} \cdot \overrightarrow{ \left(\begin{pmatrix} 48 & 46 & 42 & 52 & 42 & 42 \\ 39 & 46 & 38 & 48 & 42 & 38 \\ 42 & 38 & 36 & 43 & 36 & 30 \\ 35 & 33 & 33 & 33 & 35 & 32 \\ 36 & 35 & 38 & 39 & 36 & 31 \\ 40 & 40 & 44 & 46 & 31 & 33 \end{pmatrix} \cdot \begin{pmatrix} 48 & 46 & 42 & 52 & 42 & 42 \\ 42 & 49 & 41 & 51 & 45 & 41 \\ 36 & 32 & 30 & 37 & 30 & 24 \\ 32 & 30 & 30 & 30 & 32 & 29 \\ 42 & 41 & 44 & 45 & 42 & 37 \\ 40 & 40 & 44 & 46 & 31 & 33 \end{pmatrix} \right) } \cdot \text{SIX1} \right] - \text{CF}$$

SSb = 381.667

STEP 10: Computation of the Pure Between Group Effects SS1. N1 = Nt/#columns in DB2. 108/2=54

N1 := 54

$$\text{SS1} := \left(\frac{1}{\text{N1}} \right) \cdot (\text{SUM1} \cdot \text{DB1} \cdot \text{DB2}) \cdot (\text{SUM1} \cdot \text{DB1} \cdot \text{DB2})^{\text{T}} - \text{CF} \qquad (\text{SUM1} \cdot \text{DB1} \cdot \text{DB2})^{\text{T}} = \begin{pmatrix} 709 \\ 689 \end{pmatrix}$$

SS1 = 3.704

For the Math, SUM1 is too long to fit on a page, so we'll write
it this way:

$$
\left(\frac{1}{54}\right) \cdot \text{SUM1} \cdot \left[\begin{pmatrix} 1&0&0&0&0&0 \\ 1&0&0&0&0&0 \\ 1&0&0&0&0&0 \\ 0&1&0&0&0&0 \\ 0&1&0&0&0&0 \\ 0&1&0&0&0&0 \\ 0&0&1&0&0&0 \\ 0&0&1&0&0&0 \\ 0&0&1&0&0&0 \\ 0&0&0&1&0&0 \\ 0&0&0&1&0&0 \\ 0&0&0&1&0&0 \\ 0&0&0&0&1&0 \\ 0&0&0&0&1&0 \\ 0&0&0&0&1&0 \\ 0&0&0&0&0&1 \\ 0&0&0&0&0&1 \\ 0&0&0&0&0&1 \end{pmatrix} \cdot \begin{pmatrix} 1&0 \\ 1&0 \\ 1&0 \\ 0&1 \\ 0&1 \\ 0&1 \end{pmatrix} \right] \cdot \left[\text{SUM1} \cdot \left(\begin{pmatrix} 1&0&0&0&0&0 \\ 1&0&0&0&0&0 \\ 1&0&0&0&0&0 \\ 0&1&0&0&0&0 \\ 0&1&0&0&0&0 \\ 0&1&0&0&0&0 \\ 0&0&1&0&0&0 \\ 0&0&1&0&0&0 \\ 0&0&1&0&0&0 \\ 0&0&0&1&0&0 \\ 0&0&0&1&0&0 \\ 0&0&0&1&0&0 \\ 0&0&0&0&1&0 \\ 0&0&0&0&1&0 \\ 0&0&0&0&1&0 \\ 0&0&0&0&0&1 \\ 0&0&0&0&0&1 \\ 0&0&0&0&0&1 \end{pmatrix} \cdot \begin{pmatrix} 1&0 \\ 1&0 \\ 1&0 \\ 0&1 \\ 0&1 \\ 0&1 \end{pmatrix} \right)^T \right] - \text{CF} = 3.704
$$

STEP 11: Computation of the First Experimental Condition, SS2
N2 = Nt/# columns in DB3 N2 = 108/3 = 36.

$N2 := 36$

$$
\text{SS2} := \left(\frac{1}{N2}\right) \cdot (\text{SUM1} \cdot \text{DB1} \cdot \text{DB3}) \cdot \left((\text{SUM1} \cdot \text{DB1} \cdot \text{DB3})^T\right) - \text{CF}
$$

$\text{SS2} = 58.389$

$$\left(\frac{1}{36}\right)\cdot \mathbf{SUM1}\cdot\begin{bmatrix}\begin{pmatrix}1&0&0&0&0&0\\1&0&0&0&0&0\\1&0&0&0&0&0\\0&1&0&0&0&0\\0&1&0&0&0&0\\0&1&0&0&0&0\\0&0&1&0&0&0\\0&0&1&0&0&0\\0&0&1&0&0&0\\0&0&0&1&0&0\\0&0&0&1&0&0\\0&0&0&1&0&0\\0&0&0&0&1&0\\0&0&0&0&1&0\\0&0&0&0&1&0\\0&0&0&0&0&1\\0&0&0&0&0&1\\0&0&0&0&0&1\end{pmatrix}\cdot\begin{pmatrix}1&0&0\\0&1&0\\0&0&1\\1&0&0\\0&1&0\\0&0&1\end{pmatrix}\cdot \mathbf{SUM1}\cdot\begin{bmatrix}\begin{pmatrix}1&0&0&0&0&0\\1&0&0&0&0&0\\1&0&0&0&0&0\\0&1&0&0&0&0\\0&1&0&0&0&0\\0&1&0&0&0&0\\0&0&1&0&0&0\\0&0&1&0&0&0\\0&0&1&0&0&0\\0&0&0&1&0&0\\0&0&0&1&0&0\\0&0&0&1&0&0\\0&0&0&0&1&0\\0&0&0&0&1&0\\0&0&0&0&1&0\\0&0&0&0&0&1\\0&0&0&0&0&1\\0&0&0&0&0&1\end{pmatrix}\cdot\begin{pmatrix}1&0&0\\0&1&0\\0&0&1\\1&0&0\\0&1&0\\0&0&1\end{pmatrix}\end{bmatrix}^{\mathbf{T}}\end{bmatrix}-\mathbf{CF}=58.389$$

$\mathbf{SS2} = 58.389$

STEP 12: Computation of the interaction of the First 2 Experimental Conditions SS1x2. N1x2 = Nt/#columns in DB1 = 108/6=18

$$\mathbf{N1x2} := \frac{108}{6}$$

$$\mathbf{SS1x2} := \left(\frac{1}{\mathbf{N1x2}}\right)\cdot(\mathbf{SUM1}\cdot\mathbf{DB1})\cdot(\mathbf{SUM1}\cdot\mathbf{DB1})^{\mathbf{T}} - \mathbf{CF} - \mathbf{SS1} - \mathbf{SS2}$$

And the Math for this becomes:

$$SS1x2 := \left(\frac{1}{18}\right) \cdot \left[SUM1 \cdot \begin{pmatrix} 1 & 0 & 0 & 0 & 0 & 0 \\ 1 & 0 & 0 & 0 & 0 & 0 \\ 1 & 0 & 0 & 0 & 0 & 0 \\ 0 & 1 & 0 & 0 & 0 & 0 \\ 0 & 1 & 0 & 0 & 0 & 0 \\ 0 & 1 & 0 & 0 & 0 & 0 \\ 0 & 0 & 1 & 0 & 0 & 0 \\ 0 & 0 & 1 & 0 & 0 & 0 \\ 0 & 0 & 1 & 0 & 0 & 0 \\ 0 & 0 & 0 & 1 & 0 & 0 \\ 0 & 0 & 0 & 1 & 0 & 0 \\ 0 & 0 & 0 & 1 & 0 & 0 \\ 0 & 0 & 0 & 0 & 1 & 0 \\ 0 & 0 & 0 & 0 & 1 & 0 \\ 0 & 0 & 0 & 0 & 1 & 0 \\ 0 & 0 & 0 & 0 & 0 & 1 \\ 0 & 0 & 0 & 0 & 0 & 1 \\ 0 & 0 & 0 & 0 & 0 & 1 \end{pmatrix} \cdot SUM1 \cdot \begin{pmatrix} 1 & 0 & 0 & 0 & 0 & 0 \\ 1 & 0 & 0 & 0 & 0 & 0 \\ 1 & 0 & 0 & 0 & 0 & 0 \\ 0 & 1 & 0 & 0 & 0 & 0 \\ 0 & 1 & 0 & 0 & 0 & 0 \\ 0 & 1 & 0 & 0 & 0 & 0 \\ 0 & 0 & 1 & 0 & 0 & 0 \\ 0 & 0 & 1 & 0 & 0 & 0 \\ 0 & 0 & 1 & 0 & 0 & 0 \\ 0 & 0 & 0 & 1 & 0 & 0 \\ 0 & 0 & 0 & 1 & 0 & 0 \\ 0 & 0 & 0 & 1 & 0 & 0 \\ 0 & 0 & 0 & 0 & 1 & 0 \\ 0 & 0 & 0 & 0 & 1 & 0 \\ 0 & 0 & 0 & 0 & 1 & 0 \\ 0 & 0 & 0 & 0 & 0 & 1 \\ 0 & 0 & 0 & 0 & 0 & 1 \\ 0 & 0 & 0 & 0 & 0 & 1 \end{pmatrix}^{T} \right] - CF - SS1 - SS2$$

$SS1x2 = 33.019$

STEP 13: Computation of the Between Subjects Error Term SSerb

$SSerb := SSb - SS1 - SS2 - SS1x2$

$SSerb = 286.556$

STEP 14: Computation of the Within Subjects Sum of Squares SSw

$SSw := SSt - SSb$

$SSw = 1716$

STEP 15: Computation of the effects of the 2nd experimental Factor, SS4. N4= Nt/# columns in DB4. N4 = 108/3

$N2 := \frac{108}{3}$ $N2 = 36$

SUM1 ·DB4 = (597   457   344)

$$\mathbf{SS4} := \left(\frac{1}{N2}\right) \cdot (\mathbf{SUM1 \cdot DB4}) \cdot (\mathbf{SUM1 \cdot DB4})^{T} - \mathbf{CF}$$

SS4 = 892.389

$$\mathbf{SS4} := \left(\frac{1}{36}\right) \cdot \mathbf{SUM1} \cdot \begin{bmatrix} \begin{pmatrix} 1 & 0 & 0 \\ 0 & 1 & 0 \\ 0 & 0 & 1 \\ 1 & 0 & 0 \\ 0 & 1 & 0 \\ 0 & 0 & 1 \\ 1 & 0 & 0 \\ 0 & 1 & 0 \\ 0 & 0 & 1 \\ 1 & 0 & 0 \\ 0 & 1 & 0 \\ 0 & 0 & 1 \\ 1 & 0 & 0 \\ 0 & 1 & 0 \\ 0 & 0 & 1 \\ 1 & 0 & 0 \\ 0 & 1 & 0 \\ 0 & 0 & 1 \end{pmatrix} \end{bmatrix} \cdot \mathbf{SUM1} \cdot \begin{bmatrix} \begin{pmatrix} 1 & 0 & 0 \\ 0 & 1 & 0 \\ 0 & 0 & 1 \\ 1 & 0 & 0 \\ 0 & 1 & 0 \\ 0 & 0 & 1 \\ 1 & 0 & 0 \\ 0 & 1 & 0 \\ 0 & 0 & 1 \\ 1 & 0 & 0 \\ 0 & 1 & 0 \\ 0 & 0 & 1 \\ 1 & 0 & 0 \\ 0 & 1 & 0 \\ 0 & 0 & 1 \\ 1 & 0 & 0 \\ 0 & 1 & 0 \\ 0 & 0 & 1 \end{pmatrix} \end{bmatrix}^{T} - \mathbf{CF}$$

Which reduces to

$$\left(\frac{1}{36}\right) \cdot ((597 \quad 457 \quad 344)) \cdot \begin{pmatrix} 597 \\ 457 \\ 344 \end{pmatrix} - \mathbf{CF} = 892.389$$

SS4 = 892.389

STEP 16: Computation of the effects of the 3rd experimental Factor, SS5. N5= Nt/# columns in DB5. N5 = 108/3.

NOTE: DB5 is one of the 2 de-scrambling Matrices to re-put the Experiments in proper order.

$$\mathbf{N5} := \frac{108}{3}$$

$$\mathbf{SUM1 \cdot DB5} = (\;560 \quad 449 \quad 389\;)$$

$$\mathbf{SS5} := \left(\frac{1}{\mathbf{N5}}\right) \cdot (\mathbf{SUM1 \cdot DB5}) \cdot (\mathbf{SUM1 \cdot DB5})^{\mathbf{T}} - \mathbf{CF}$$

$$\mathbf{SS5} = 418.167$$

And the Math is:

$$\left(\frac{1}{36}\right) \cdot \mathbf{SUM1} \cdot \begin{pmatrix} 1 & 0 & 0 \\ 0 & 1 & 0 \\ 0 & 0 & 1 \\ 0 & 0 & 1 \\ 1 & 0 & 0 \\ 0 & 1 & 0 \\ 0 & 1 & 0 \\ 0 & 0 & 1 \\ 1 & 0 & 0 \\ 1 & 0 & 0 \\ 0 & 1 & 0 \\ 0 & 0 & 1 \\ 0 & 0 & 1 \\ 1 & 0 & 0 \\ 0 & 1 & 0 \\ 0 & 1 & 0 \\ 0 & 0 & 1 \\ 1 & 0 & 0 \end{pmatrix} \cdot \mathbf{SUM1} \cdot \begin{pmatrix} 1 & 0 & 0 \\ 0 & 1 & 0 \\ 0 & 0 & 1 \\ 0 & 0 & 1 \\ 1 & 0 & 0 \\ 0 & 1 & 0 \\ 0 & 1 & 0 \\ 0 & 0 & 1 \\ 1 & 0 & 0 \\ 1 & 0 & 0 \\ 0 & 1 & 0 \\ 0 & 0 & 1 \\ 0 & 0 & 1 \\ 1 & 0 & 0 \\ 0 & 1 & 0 \\ 0 & 1 & 0 \\ 0 & 0 & 1 \\ 1 & 0 & 0 \end{pmatrix}^{\mathbf{T}} - \mathbf{CF}$$

Which simplifies to:

$$\mathbf{SS5} := \left(\frac{1}{36}\right) \cdot (\;560 \quad 449 \quad 389\;) \cdot \begin{pmatrix} 560 \\ 449 \\ 389 \end{pmatrix} - \mathbf{CF}$$

$$\mathbf{SS5} = 418.167$$

STEP 17: Computation of the Order of Effects of the 3 Experimental Factors, SS6. N5= Nt/# columns in DB6. N6 = 108/9.

NOTE: By the logic of all the Labeling I have done before, The Sum of Squares should be SS2x4x5

$$\mathbf{N6} := \frac{108}{9} \qquad \mathbf{N6} = 12$$

$$\mathbf{SUM1 \cdot DB6} = (\,245 \quad 152 \quad 104 \quad 165 \quad 185 \quad 110 \quad 187 \quad 120 \quad 130\,)$$

$$\left(\frac{1}{\mathbf{N6}}\right) \cdot (\mathbf{SUM1 \cdot DB6}) \cdot (\mathbf{SUM1 \cdot DB6})^{\mathbf{T}} - \mathbf{CF} - \mathbf{SS5} - \mathbf{SS4} - \mathbf{SS2} = 15.056$$

$$\mathbf{SS6} := \left(\frac{1}{12}\right) \cdot \mathbf{SUM1} \cdot \begin{bmatrix} 1 & 0 & 0 & 0 & 0 & 0 & 0 & 0 & 0 \\ 0 & 1 & 0 & 0 & 0 & 0 & 0 & 0 & 0 \\ 0 & 0 & 1 & 0 & 0 & 0 & 0 & 0 & 0 \\ 0 & 0 & 0 & 1 & 0 & 0 & 0 & 0 & 0 \\ 0 & 0 & 0 & 0 & 1 & 0 & 0 & 0 & 0 \\ 0 & 0 & 0 & 0 & 0 & 1 & 0 & 0 & 0 \\ 0 & 0 & 0 & 0 & 0 & 0 & 1 & 0 & 0 \\ 0 & 0 & 0 & 0 & 0 & 0 & 0 & 1 & 0 \\ 0 & 0 & 0 & 0 & 0 & 0 & 0 & 0 & 1 \\ 1 & 0 & 0 & 0 & 0 & 0 & 0 & 0 & 0 \\ 0 & 1 & 0 & 0 & 0 & 0 & 0 & 0 & 0 \\ 0 & 0 & 1 & 0 & 0 & 0 & 0 & 0 & 0 \\ 0 & 0 & 0 & 1 & 0 & 0 & 0 & 0 & 0 \\ 0 & 0 & 0 & 0 & 1 & 0 & 0 & 0 & 0 \\ 0 & 0 & 0 & 0 & 0 & 1 & 0 & 0 & 0 \\ 0 & 0 & 0 & 0 & 0 & 0 & 1 & 0 & 0 \\ 0 & 0 & 0 & 0 & 0 & 0 & 0 & 1 & 0 \\ 0 & 0 & 0 & 0 & 0 & 0 & 0 & 0 & 1 \end{bmatrix} \cdot \mathbf{SUM1} \cdot \begin{bmatrix} 1 & 0 & 0 & 0 & 0 & 0 & 0 & 0 & 0 \\ 0 & 1 & 0 & 0 & 0 & 0 & 0 & 0 & 0 \\ 0 & 0 & 1 & 0 & 0 & 0 & 0 & 0 & 0 \\ 0 & 0 & 0 & 1 & 0 & 0 & 0 & 0 & 0 \\ 0 & 0 & 0 & 0 & 1 & 0 & 0 & 0 & 0 \\ 0 & 0 & 0 & 0 & 0 & 1 & 0 & 0 & 0 \\ 0 & 0 & 0 & 0 & 0 & 0 & 1 & 0 & 0 \\ 0 & 0 & 0 & 0 & 0 & 0 & 0 & 1 & 0 \\ 0 & 0 & 0 & 0 & 0 & 0 & 0 & 0 & 1 \\ 1 & 0 & 0 & 0 & 0 & 0 & 0 & 0 & 0 \\ 0 & 1 & 0 & 0 & 0 & 0 & 0 & 0 & 0 \\ 0 & 0 & 1 & 0 & 0 & 0 & 0 & 0 & 0 \\ 0 & 0 & 0 & 1 & 0 & 0 & 0 & 0 & 0 \\ 0 & 0 & 0 & 0 & 1 & 0 & 0 & 0 & 0 \\ 0 & 0 & 0 & 0 & 0 & 1 & 0 & 0 & 0 \\ 0 & 0 & 0 & 0 & 0 & 0 & 1 & 0 & 0 \\ 0 & 0 & 0 & 0 & 0 & 0 & 0 & 1 & 0 \\ 0 & 0 & 0 & 0 & 0 & 0 & 0 & 0 & 1 \end{bmatrix}^{\mathbf{T}} - \mathbf{CF} - \mathbf{SS5} - \mathbf{SS4} - \mathbf{SS2}$$

Which reduces to:

$$\mathbf{SS6} := \left(\frac{1}{12}\right) \cdot (\,(\,245 \quad 152 \quad 104 \quad 165 \quad 185 \quad 110 \quad 187 \quad 120 \quad 130\,)\,) \cdot \begin{pmatrix} 245 \\ 152 \\ 104 \\ 165 \\ 185 \\ 110 \\ 187 \\ 120 \\ 130 \end{pmatrix} - \mathbf{CF} - \mathbf{SS5} - \mathbf{SS4} - \mathbf{SS2}$$

$$\mathbf{SS6} = 15.056$$

STEP 18: Computation of the Treatments by Group Interaction, SS7. N7= Nt/# columns in DB7. N7 = 108/6.

$$\mathbf{N7} := \frac{108}{6} \qquad \mathbf{N7} = 18$$

$$\mathbf{SUM1 \cdot DB7} = (283 \quad 231 \quad 195 \quad 314 \quad 226 \quad 149)$$

$$\left(\frac{1}{\mathbf{N7}}\right) \cdot (\mathbf{SUM1 \cdot DB7}) \cdot (\mathbf{SUM1 \cdot DB7})^{\mathbf{T}} - \mathbf{CF} - \mathbf{SS1} - \mathbf{SS4} = 82.463$$

$$\mathbf{SS7} := \left(\frac{1}{\mathbf{N7}}\right) \cdot (\mathbf{SUM1 \cdot DB7}) \cdot (\mathbf{SUM1 \cdot DB7})^{\mathbf{T}} - \mathbf{CF} - \mathbf{SS1} - \mathbf{SS4}$$

$$\mathbf{SS7} := \left(\frac{1}{18}\right) \cdot \mathbf{SUM1} \cdot \begin{pmatrix} 1 & 0 & 0 & 0 & 0 & 0 \\ 0 & 1 & 0 & 0 & 0 & 0 \\ 0 & 0 & 1 & 0 & 0 & 0 \\ 1 & 0 & 0 & 0 & 0 & 0 \\ 0 & 1 & 0 & 0 & 0 & 0 \\ 0 & 0 & 1 & 0 & 0 & 0 \\ 1 & 0 & 0 & 0 & 0 & 0 \\ 0 & 1 & 0 & 0 & 0 & 0 \\ 0 & 0 & 1 & 0 & 0 & 0 \\ 0 & 0 & 0 & 1 & 0 & 0 \\ 0 & 0 & 0 & 0 & 1 & 0 \\ 0 & 0 & 0 & 0 & 0 & 1 \\ 0 & 0 & 0 & 1 & 0 & 0 \\ 0 & 0 & 0 & 0 & 1 & 0 \\ 0 & 0 & 0 & 0 & 0 & 1 \\ 0 & 0 & 0 & 1 & 0 & 0 \\ 0 & 0 & 0 & 0 & 1 & 0 \\ 0 & 0 & 0 & 0 & 0 & 1 \end{pmatrix} \cdot \mathbf{SUM1} \cdot \begin{pmatrix} 1 & 0 & 0 & 0 & 0 & 0 \\ 0 & 1 & 0 & 0 & 0 & 0 \\ 0 & 0 & 1 & 0 & 0 & 0 \\ 1 & 0 & 0 & 0 & 0 & 0 \\ 0 & 1 & 0 & 0 & 0 & 0 \\ 0 & 0 & 1 & 0 & 0 & 0 \\ 1 & 0 & 0 & 0 & 0 & 0 \\ 0 & 1 & 0 & 0 & 0 & 0 \\ 0 & 0 & 1 & 0 & 0 & 0 \\ 0 & 0 & 0 & 1 & 0 & 0 \\ 0 & 0 & 0 & 0 & 1 & 0 \\ 0 & 0 & 0 & 0 & 0 & 1 \\ 0 & 0 & 0 & 1 & 0 & 0 \\ 0 & 0 & 0 & 0 & 1 & 0 \\ 0 & 0 & 0 & 0 & 0 & 1 \\ 0 & 0 & 0 & 1 & 0 & 0 \\ 0 & 0 & 0 & 0 & 1 & 0 \\ 0 & 0 & 0 & 0 & 0 & 1 \end{pmatrix}^{\mathbf{T}} - \mathbf{CF} - \mathbf{SS1} - \mathbf{SS4}$$

Which reduces to:

$$\left(\frac{1}{18}\right) \cdot (283 \quad 231 \quad 195 \quad 314 \quad 226 \quad 149) \cdot \begin{pmatrix} 283 \\ 231 \\ 195 \\ 314 \\ 226 \\ 149 \end{pmatrix} - \mathbf{CF} - \mathbf{SS1} - \mathbf{SS4} = 82.463$$

$$\mathbf{SS7} = 82.463$$

STEP 19: Computation of the Latin Order by Group Interaction, SSlatin. Nt/# columns in DBL. Nlatin = 108/6.

NOTE: This is the second de-scrambling Matrix used to put all tests in original order.

SUM1·DBL = (275   230   204   285   219   185)

Nlatin := 18

$$\textbf{SSlatin} := \left(\frac{1}{\textbf{Nlatin}} \right) \cdot (\textbf{SUM1·DBL}) \cdot (\textbf{SUM1·DBL})^{\textbf{T}} - \textbf{CF} - \textbf{SS1} - \textbf{SS5}$$

SSlatin = 12.463

$$\textbf{SSlatin} := \left(\frac{1}{\textbf{Nlatin}} \right) \cdot \textbf{SUM1} \cdot \begin{pmatrix} 1&0&0&0&0&0 \\ 0&1&0&0&0&0 \\ 0&0&1&0&0&0 \\ 0&0&1&0&0&0 \\ 1&0&0&0&0&0 \\ 0&1&0&0&0&0 \\ 0&1&0&0&0&0 \\ 0&0&1&0&0&0 \\ 1&0&0&0&0&0 \\ 0&0&0&1&0&0 \\ 0&0&0&0&1&0 \\ 0&0&0&0&0&1 \\ 0&0&0&0&0&1 \\ 0&0&0&1&0&0 \\ 0&0&0&0&1&0 \\ 0&0&0&0&1&0 \\ 0&0&0&0&0&1 \\ 0&0&0&1&0&0 \end{pmatrix} \cdot \textbf{SUM1} \cdot \begin{pmatrix} 1&0&0&0&0&0 \\ 0&1&0&0&0&0 \\ 0&0&1&0&0&0 \\ 0&0&1&0&0&0 \\ 1&0&0&0&0&0 \\ 0&1&0&0&0&0 \\ 0&1&0&0&0&0 \\ 0&0&1&0&0&0 \\ 1&0&0&0&0&0 \\ 0&0&0&1&0&0 \\ 0&0&0&0&1&0 \\ 0&0&0&0&0&1 \\ 0&0&0&0&0&1 \\ 0&0&0&1&0&0 \\ 0&0&0&0&1&0 \\ 0&0&0&0&1&0 \\ 0&0&0&0&0&1 \\ 0&0&0&1&0&0 \end{pmatrix}^{\textbf{T}} - \textbf{CF} - \textbf{SS1} - \textbf{SS5}$$

Which reduces to:

$$\textbf{SSlatin} := \left(\frac{1}{18} \right) \cdot (275 \quad 230 \quad 204 \quad 285 \quad 219 \quad 185) \cdot \begin{pmatrix} 275 \\ 230 \\ 204 \\ 285 \\ 219 \\ 185 \end{pmatrix} - \textbf{CF} - \textbf{SS1} - \textbf{SS5}$$

SSlatin = 12.463

STEP 20: Computation of Within Subjects 1st, 2nd and 3rd Experimental Condition interaction, **SS1x2x3x4x5x6x7xL**.

$$\text{N1x2x3x4x5x6x7xL} := \frac{108}{18}$$

$$\text{N1x2x3x4x5x6x7xL} = 6$$

$$\text{SS1x2x3x4x5x6x7xL} := \left(\frac{1}{6}\right) \cdot \left(\text{SUM1} \cdot \text{SUM1}^T\right) - \text{CF} - \text{SS1} - \text{SS2} - \text{SS1x2} - \text{SS4} - \text{SS5} - \text{SS6} - \text{SS7} - \text{SSlatin}$$

$$\text{SS1x2x3x4x5x6x7xL} = 37.352$$

The Math is too long to get on a single page, so we write it as:

$$\left(\frac{1}{6}\right) \cdot \left[\text{SUM1} \cdot \begin{pmatrix} 105 \\ 76 \\ 59 \\ 84 \\ 94 \\ 60 \\ 94 \\ 61 \\ 76 \\ 140 \\ 76 \\ 45 \\ 81 \\ 91 \\ 50 \\ 93 \\ 59 \\ 54 \end{pmatrix}\right] - \text{CF} - \text{SS1} - \text{SS2} - \text{SS1x2} - \text{SS4} - \text{SS5} - \text{SS6} - \text{SS7} - \text{SSlatin} = 37.352$$

Which reduces to:

$$\text{SS1x2x3x4x5x6x7xL} := \left(\frac{1}{6}\right) \cdot (117896) - \text{CF} - \text{SS1} - \text{SS2} - \text{SS1x2} - \text{SS4} - \text{SS5} - \text{SS6} - \text{SS7} - \text{SSlatin}$$

$$\text{SS1x2x3x4x5x6x7xL} = 37.352$$

STEP 21: Computation of the Within Subjects Error Term SSerw

$$\text{SSerw} := \text{SSw} - \text{SS4} - \text{SS5} - \text{SS6} - \text{SS7} - \text{SSlatin} - \text{SS1x2x3x4x5x6x7xL}$$

$$\text{SSerw} = 258.111$$

Step 22: Computation of the Degrees of Freedom for this
Statistic. m = # elements in Data Matrix A, s = # subjects in
study, G = # Groups, L = # of experimental conditions, t = # of
Treatment Groups, O = # of Latin Orders. This df list is good
for all Latin Square Complex computations, we just change the
values of m, s, G, L, t and O as needed by the study.

$m := 108$ $s := 36$ $G := 2$ $L := 3$ $t := 3$ $O := 3$

The Degrees of Freedom are computed by the following Matrix.

$$df := \begin{bmatrix} m-1 \\ s-1 \\ G-1 \\ L-1 \\ (t-1)\cdot(G-1) \\ (s-1)-(G-1)-(L-1)-(t-1)\cdot(G-1) \\ (m-1)-(s-1) \\ t-1 \\ O-1 \\ (t-1)\cdot(O-2) \\ (t-1)\cdot(G-1) \\ (O-1)\cdot(G-1) \\ (t-1)\cdot(O-2)\cdot(G-1) \\ [(m-1)-(s-1)]-(t-1)-(O-1)-[(t-1)\cdot(O-2)]-[(t-1)\cdot(G-1)]-[(O-1)\cdot(G-1)]-(t-1)\cdot(O-2)\cdot(G-1) \end{bmatrix}$$

The Error Matrix is formed by SSerb divided by its Degree of
Freedom and fills the first 6 places in the Matrix, then by
SSerw divided by it's Degree of Freedom filling the rest of
the ERROR Matrix.

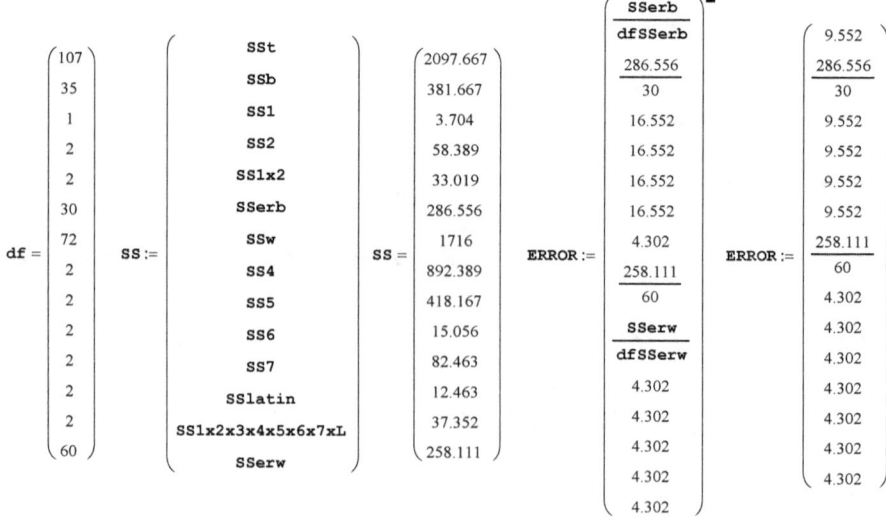

$$F := \overrightarrow{\left(SS \cdot df^{-1} \cdot ERROR^{-1}\right)}$$

$$df = \begin{pmatrix} 107 \\ 35 \\ 1 \\ 2 \\ 2 \\ 30 \\ 72 \\ 2 \\ 2 \\ 2 \\ 2 \\ 2 \\ 2 \\ 60 \end{pmatrix} \qquad F = \begin{pmatrix} 2.052 \\ 1.142 \\ 0.388 \\ 3.056 \\ 1.728 \\ 1 \\ 5.54 \\ 103.718 \\ 48.601 \\ 1.75 \\ 9.584 \\ 1.449 \\ 4.341 \\ 1 \end{pmatrix} \qquad p := \begin{pmatrix} 0 \\ 0 \\ ns \\ ns \\ ns \\ 0 \\ 0 \\ \blacksquare < .001 \\ \blacksquare < .025 \\ ns \\ \blacksquare < .001 \\ ns \\ \blacksquare < .021 \\ 0 \end{pmatrix}$$

The effects of the first experimental condition are significant to the one in a thousand level.

The effects of the order of presentation were significant 25 times out of a thousand

The First and third Experimental conditions (order of presentation) were significant.

Effects of the 3 experimental conditions interacted.

PS: Here I label the df values:

$$\begin{bmatrix} dfsst & m-1 \\ dfssb & s-1 \\ dfss1 & G-1 \\ dfss2 & L-1 \\ dfss1x2 & (t-1) \cdot (G-1) \\ dfsserb & (s-1) - (G-1) - (L-1) - (t-1) \cdot (G-1) \\ dfssw & (m-1) - (s-1) \\ dfss4 & (t-1) \\ dfss5 & (O-1) \\ dfss6 & (t-1) \cdot (O-2) \\ dfss7 & (t-1) \cdot (G-1) \\ dfsslatin & [(O-1) \cdot (G-1)] \\ dfss1x2x3x4x5x6x7xL & (t-1) \cdot (O-2) \cdot (G-1) \\ dfsserw & [(m-1) - (s-1)] - (t-1) - (O-1) - [(t-1) \cdot (O-2)] - [(t-1) \cdot (G-1)] - [(O-1) \cdot (G-1)] - (t-1) \cdot (O-2) \cdot (G-1) \end{bmatrix}$$

USE OF ORTHOGONAL COMPONENTS
IN TESTS FOR TREND COMPUTATION
OF THE LINEAR TERM

Example from Prob. 2.7

$$A := \begin{pmatrix} 5 & 5 & 6 & 4 & 5 & 8 & 4 & 6 & 10 \\ 2 & 3 & 3 & 1 & 3 & 4 & 2 & 5 & 2 \\ 2 & 4 & 5 & 4 & 6 & 8 & 5 & 8 & 9 \\ 1 & 1 & 2 & 1 & 3 & 5 & 1 & 6 & 10 \\ 3 & 4 & 2 & 5 & 8 & 3 & 2 & 4 & 4 \\ 4 & 5 & 4 & 3 & 2 & 3 & 6 & 5 & 6 \\ 4 & 4 & 5 & 3 & 4 & 5 & 5 & 4 & 3 \\ 1 & 3 & 5 & 7 & 9 & 11 & 2 & 4 & 6 \\ 5 & 6 & 6 & 2 & 2 & 4 & 3 & 3 & 1 \\ 6 & 5 & 3 & 2 & 1 & 1 & 3 & 2 & 5 \\ 3 & 2 & 1 & 6 & 7 & 2 & 2 & 3 & 6 \end{pmatrix} \qquad a := \begin{pmatrix} 3 & 3 & 4 & 2 & 3 & 6 & 2 & 4 & 8 \\ 3 & 4 & 4 & 1 & 4 & 5 & 3 & 6 & 3 \\ 1 & 3 & 4 & 3 & 6 & 7 & 4 & 7 & 7 \\ 1 & 2 & 3 & 2 & 4 & 6 & 1 & 7 & 4 \\ 2 & 3 & 5 & 1 & 2 & 3 & 3 & 7 & 12 \\ 4 & 5 & 6 & 4 & 5 & 5 & 1 & 4 & 2 \\ 4 & 5 & 6 & 1 & 3 & 4 & 1 & 5 & 3 \\ 1 & 4 & 5 & 1 & 2 & 4 & 2 & 5 & 6 \\ 1 & 4 & 5 & 2 & 3 & 4 & 3 & 6 & 6 \\ 2 & 3 & 3 & 3 & 5 & 6 & 1 & 5 & 12 \\ 2 & 2 & 4 & 4 & 6 & 7 & 4 & 9 & 7 \end{pmatrix} \qquad DBLN := \begin{pmatrix} -1 & 0 & 0 \\ 0 & 0 & 0 \\ 1 & 0 & 0 \\ 0 & -1 & 0 \\ 0 & 0 & 0 \\ 0 & 1 & 0 \\ 0 & 0 & -1 \\ 0 & 0 & 0 \\ 0 & 0 & 1 \end{pmatrix}$$

$$DB11 := \begin{pmatrix} 1 & 0 \\ 1 & 0 \\ 1 & 0 \\ 0 & 1 \\ 0 & 1 \\ 0 & 1 \end{pmatrix} \qquad ONE11 := (1\ 1\ 1\ 1\ 1\ 1\ 1\ 1\ 1\ 1\ 1) \qquad NINE1 := \begin{pmatrix} 1 \\ 1 \\ 1 \\ 1 \\ 1 \\ 1 \\ 1 \\ 1 \\ 1 \end{pmatrix} \qquad ONE9 := NINE1^T$$

$$DBLNQD := \begin{pmatrix} -1 & 0 & 0 & 1 & 0 & 0 \\ 0 & 0 & 0 & -2 & 0 & 0 \\ 1 & 0 & 0 & 1 & 0 & 0 \\ 0 & -1 & 0 & 0 & 1 & 0 \\ 0 & 0 & 0 & 0 & -2 & 0 \\ 0 & 1 & 0 & 0 & 1 & 0 \\ 0 & 0 & -1 & 0 & 0 & 1 \\ 0 & 0 & 0 & 0 & 0 & -2 \\ 0 & 0 & 1 & 0 & 0 & 1 \end{pmatrix} \qquad DB1 := \begin{pmatrix} 1 & 0 & 0 \\ 1 & 0 & 0 \\ 1 & 0 & 0 \\ 0 & 1 & 0 \\ 0 & 1 & 0 \\ 0 & 1 & 0 \\ 0 & 0 & 1 \\ 0 & 0 & 1 \\ 0 & 0 & 1 \end{pmatrix} \qquad DB2 := \begin{pmatrix} 1 & 0 & 0 \\ 0 & 1 & 0 \\ 0 & 0 & 1 \\ 1 & 0 & 0 \\ 0 & 1 & 0 \\ 0 & 0 & 1 \\ 1 & 0 & 0 \\ 0 & 1 & 0 \\ 0 & 0 & 1 \end{pmatrix}$$

$$ONE6 := (1\ 1\ 1\ 1\ 1\ 1)$$

$$DBLQ := (-1\ 0\ 1\ 1\ -2\ 1)$$

$$\textbf{THREE1} := \begin{pmatrix} 1 \\ 1 \\ 1 \end{pmatrix}$$

STEPS 1-3: We table the data as Matrix A then Multiply by both
Linear and Quadratic components in DBLNQD.

$$\textbf{A·DBLN} = \begin{pmatrix} 1 & 4 & 6 \\ 1 & 3 & 0 \\ 3 & 4 & 4 \\ 1 & 4 & 9 \\ -1 & -2 & 2 \\ 0 & 0 & 0 \\ 1 & 2 & -2 \\ 4 & 4 & 4 \\ 1 & 2 & -2 \\ -3 & -1 & 2 \\ -2 & -4 & 4 \end{pmatrix}$$

$$\begin{pmatrix} 5 & 5 & 6 & 4 & 5 & 8 & 4 & 6 & 10 \\ 2 & 3 & 3 & 1 & 3 & 4 & 2 & 5 & 2 \\ 2 & 4 & 5 & 4 & 6 & 8 & 5 & 8 & 9 \\ 1 & 1 & 2 & 1 & 3 & 5 & 1 & 6 & 10 \\ 3 & 4 & 2 & 5 & 8 & 3 & 2 & 4 & 4 \\ 4 & 5 & 4 & 3 & 2 & 3 & 6 & 5 & 6 \\ 4 & 4 & 5 & 3 & 4 & 5 & 5 & 4 & 3 \\ 1 & 3 & 5 & 7 & 9 & 11 & 2 & 4 & 6 \\ 5 & 6 & 6 & 2 & 2 & 4 & 3 & 3 & 1 \\ 6 & 5 & 3 & 2 & 1 & 1 & 3 & 2 & 5 \\ 3 & 2 & 1 & 6 & 7 & 2 & 2 & 3 & 6 \end{pmatrix} \cdot \begin{pmatrix} -1 & 0 & 0 \\ 0 & 0 & 0 \\ 1 & 0 & 0 \\ 0 & -1 & 0 \\ 0 & 0 & 0 \\ 0 & 1 & 0 \\ 0 & 0 & -1 \\ 0 & 0 & 0 \\ 0 & 0 & 1 \end{pmatrix} = \begin{pmatrix} 1 & 4 & 6 \\ 1 & 3 & 0 \\ 3 & 4 & 4 \\ 1 & 4 & 9 \\ -1 & -2 & 2 \\ 0 & 0 & 0 \\ 1 & 2 & -2 \\ 4 & 4 & 4 \\ 1 & 2 & -2 \\ -3 & -1 & 2 \\ -2 & -4 & 4 \end{pmatrix}$$

STEP 4: Add the Subject Sums to obtain the Algebraic Sum for
each Group.

PROGRAM: $1]_{1,11} [A]_{11,9} [DBLN]_{9,3}$

SUMLN := **ONE11 ·A·DBLN**

SUMLN = (6 16 27)

And the math looks like:

$$(1\ 1\ 1\ 1\ 1\ 1\ 1\ 1\ 1\ 1\ 1) \cdot \begin{pmatrix} 5 & 5 & 6 & 4 & 5 & 8 & 4 & 6 & 10 \\ 2 & 3 & 3 & 1 & 3 & 4 & 2 & 5 & 2 \\ 2 & 4 & 5 & 4 & 6 & 8 & 5 & 8 & 9 \\ 1 & 1 & 2 & 1 & 3 & 5 & 1 & 6 & 10 \\ 3 & 4 & 2 & 5 & 8 & 3 & 2 & 4 & 4 \\ 4 & 5 & 4 & 3 & 2 & 3 & 6 & 5 & 6 \\ 4 & 4 & 5 & 3 & 4 & 5 & 5 & 4 & 3 \\ 1 & 3 & 5 & 7 & 9 & 11 & 2 & 4 & 6 \\ 5 & 6 & 6 & 2 & 2 & 4 & 3 & 3 & 1 \\ 6 & 5 & 3 & 2 & 1 & 1 & 3 & 2 & 5 \\ 3 & 2 & 1 & 6 & 7 & 2 & 2 & 3 & 6 \end{pmatrix} \cdot \begin{pmatrix} -1 & 0 & 0 \\ 0 & 0 & 0 \\ 1 & 0 & 0 \\ 0 & -1 & 0 \\ 0 & 0 & 0 \\ 0 & 1 & 0 \\ 0 & 0 & -1 \\ 0 & 0 & 0 \\ 0 & 0 & 1 \end{pmatrix} = (6\ \ 16\ \ 27)$$

STEP 5:   Square each of the Subject Sums obtained in STEP 3 and add them together keeping the Linear and Quadratic components separate.

$$\text{SUBSUM} := \text{ONE11} \cdot [\overrightarrow{(A \cdot \text{DBLN}) \cdot (A \cdot \text{DBLN})}] \cdot \text{THREE1} \blacksquare$$

$$\text{ADBLN} := A \cdot \text{DBLN}$$

$$\text{SUBSUM} := \text{ONE11} \cdot [\overrightarrow{(\text{ADBLN}) \cdot (\text{ADBLN})}] \cdot \text{THREE1}$$

$$\text{SUBSUM} = 327$$

The math looks like:

$$\text{ONE11} \cdot \left[\overrightarrow{\left[\begin{pmatrix} 1 & 4 & 6 \\ 1 & 3 & 0 \\ 3 & 4 & 4 \\ 1 & 4 & 9 \\ -1 & -2 & 2 \\ 0 & 0 & 0 \\ 1 & 2 & -2 \\ 4 & 4 & 4 \\ 1 & 2 & -2 \\ -3 & -1 & 2 \\ -2 & -4 & 4 \end{pmatrix} \cdot \begin{pmatrix} 1 & 4 & 6 \\ 1 & 3 & 0 \\ 3 & 4 & 4 \\ 1 & 4 & 9 \\ -1 & -2 & 2 \\ 0 & 0 & 0 \\ 1 & 2 & -2 \\ 4 & 4 & 4 \\ 1 & 2 & -2 \\ -3 & -1 & 2 \\ -2 & -4 & 4 \end{pmatrix}\right]}\right] \cdot \begin{pmatrix} 1 \\ 1 \\ 1 \end{pmatrix} = 327$$

STEP 6: Add the Group Sums to get the overall Algebraic Sum
(from STEP 4)

ALGSUM := **ONE11 ·(A·DBLN·THREE1)**

ALGSUM = 49

$$
(1\ 1\ 1\ 1\ 1\ 1\ 1\ 1\ 1\ 1\ 1) \cdot \left[
\begin{pmatrix}
5 & 5 & 6 & 4 & 5 & 8 & 4 & 6 & 10 \\
2 & 3 & 3 & 1 & 3 & 4 & 2 & 5 & 2 \\
2 & 4 & 5 & 4 & 6 & 8 & 5 & 8 & 9 \\
1 & 1 & 2 & 1 & 3 & 5 & 1 & 6 & 10 \\
3 & 4 & 2 & 5 & 8 & 3 & 2 & 4 & 4 \\
4 & 5 & 4 & 3 & 2 & 3 & 6 & 5 & 6 \\
4 & 4 & 5 & 3 & 4 & 5 & 5 & 4 & 3 \\
1 & 3 & 5 & 7 & 9 & 11 & 2 & 4 & 6 \\
5 & 6 & 6 & 2 & 2 & 4 & 3 & 3 & 1 \\
6 & 5 & 3 & 2 & 1 & 1 & 3 & 2 & 5 \\
3 & 2 & 1 & 6 & 7 & 2 & 2 & 3 & 6
\end{pmatrix}
\cdot
\begin{pmatrix}
-1 & 0 & 0 \\
0 & 0 & 0 \\
1 & 0 & 0 \\
0 & -1 & 0 \\
0 & 0 & 0 \\
0 & 1 & 0 \\
0 & 0 & -1 \\
0 & 0 & 0 \\
0 & 0 & 1
\end{pmatrix}
\cdot
\begin{pmatrix}
1 \\
1 \\
1
\end{pmatrix}
\right] = 49
$$

STEP 7: Square the Linear component and add the squared values.
The squaring and summing are done at the same time under Matrix
Multiplication.

$$
\text{LINSQ} := (1\ \ 0\ \ -1) \cdot
\begin{pmatrix}
1 \\
0 \\
-1
\end{pmatrix}
$$

LINSQ = 2

STEP 8: Computation of the Within Subjects Linear component. In
terms of already solved computations we have: NOTE: It must be
a problem in converting from Windows XP to Vista, but MathCad
keeps pringing variables in red and I cannot get rid of the
Color, since the book is in black and white, the red variables
will appear as a lighter grey. Please bear with me on this as
I can find no way to remedy this.

$$
\text{SSwL} := \left(\overrightarrow{\text{ONE11} \cdot [(\text{A·DBLN}) \cdot (\text{A·DBLN})] \cdot \text{THREE1}} \right) \text{LINSQ}^{-1}
$$

Again, MathCad cannot compute this type of expression, so we must make the following substitution:

ADBLN := A·DBLN

$$\mathbf{SSwL} := \left(\overrightarrow{\mathbf{ONE11} \cdot [(\mathbf{ADBLN}) \cdot (\mathbf{ADBLN})]} \cdot \mathbf{THREE1}\right) \mathbf{LINSQ}^{-1}$$

SSwL = 163.5

The math looks like:

$$\mathbf{SSwL} := (1\ 1\ 1\ 1\ 1\ 1\ 1\ 1\ 1\ 1\ 1) \cdot \left[\left(\left(\begin{array}{ccc} 1 & 4 & 6 \\ 1 & 3 & 0 \\ 3 & 4 & 4 \\ 1 & 4 & 9 \\ -1 & -2 & 2 \\ 0 & 0 & 0 \\ 1 & 2 & -2 \\ 4 & 4 & 4 \\ 1 & 2 & -2 \\ -3 & -1 & 2 \\ -2 & -4 & 4 \end{array} \right) \overrightarrow{\left(\begin{array}{ccc} 1 & 4 & 6 \\ 1 & 3 & 0 \\ 3 & 4 & 4 \\ 1 & 4 & 9 \\ -1 & -2 & 2 \\ 0 & 0 & 0 \\ 1 & 2 & -2 \\ 4 & 4 & 4 \\ 1 & 2 & -2 \\ -3 & -1 & 2 \\ -2 & -4 & 4 \end{array} \right)} \right) \cdot \left(\begin{array}{c} 1 \\ 1 \\ 1 \end{array} \right) \right] \cdot \left[(1\ 0\ -1) \cdot \left(\begin{array}{c} 1 \\ 0 \\ -1 \end{array} \right) \right]^{-1}$$

SSwL = 163.5

STEP 9: Computation of the Trials effect for the Linear and Quadratic components.

NtrL := 33

$$\mathbf{SStrL} := \left(\frac{1}{\mathbf{NtrL}} \right) \left[\overrightarrow{(\mathbf{ALGSUM}) \cdot (\mathbf{ALGSUM}) \cdot \mathbf{LINSQ}^{-1}} \right]$$

SStrL = 36.379

The Math looks like

$$\left(\frac{1}{33} \right) \left[\overrightarrow{(49) \cdot (49) \cdot 2^{-1}} \right] = 36.379$$

STEP 10: Computation of the Trials by Groups effect for the
Linear component. NtxgL = # Subjects in a single Trial = 11.

NtxgL := 11

SUMLN = (6 16 27)

The expression is given by:

$$\textbf{SStxgL} := \left(\frac{1}{\textbf{NtxgL}} \right) \left[(\textbf{SUMLN}) \cdot \left(\textbf{diag} \left((\textbf{SUMLN})^{\textbf{T}} \right) \cdot \textbf{THREE1} \right) \cdot \textbf{LINSQ}^{-1} \right] - \textbf{SStrL}$$

SStxgL = 10.03

The math looks like:

$$\left(\frac{1}{11} \right) \cdot \left[((6 \quad 16 \quad 27)) \cdot \left[\begin{pmatrix} 6 & 0 & 0 \\ 0 & 16 & 0 \\ 0 & 0 & 27 \end{pmatrix} \cdot \begin{pmatrix} 1 \\ 1 \\ 1 \end{pmatrix} \right] \cdot \left[(1 \quad 0 \quad -1) \cdot \begin{pmatrix} 1 \\ 0 \\ -1 \end{pmatrix} \right]^{-1} \right] - 36.379 = 10.03$$

STEP 11: Computation of the Error term for the Linear component.

SSer2L := **SSwL** − **SStrL** − **SStxgL**

SSer2L = 117.091

STEP 12: Computation of the Degrees of Freedom df and the Error Term.

s = # Subjects in each Group, tr always = 1, G = # Groups in
study and error = s-tr-(G-1)

s := 33   **tr** := 1   **G** := 3

$$\textbf{df} := \begin{bmatrix} \textbf{s} \\ \textbf{tr} \\ \textbf{G} - 1 \\ \textbf{s} - \textbf{tr} - (\textbf{G} - 1) \end{bmatrix} \quad \textbf{df} = \begin{pmatrix} 33 \\ 1 \\ 2 \\ 30 \end{pmatrix} \quad \textbf{msL} := \frac{117.091}{30} \quad \textbf{SS} := \begin{pmatrix} \textbf{SSwL} \\ \textbf{SStrL} \\ \textbf{SStxgL} \\ \textbf{SSer2L} \end{pmatrix} \quad \textbf{ErrL} := \begin{pmatrix} \dfrac{117.091}{30} \\ 3.903 \\ 3.903 \\ 3.903 \end{pmatrix}$$

Now we will get all the terms ready for the final computation of F and compile all the Sums of Squares computed for the Trends.

$$\begin{pmatrix} \mathbf{SSwL} \\ \mathbf{SStrL} \\ \mathbf{SStxgL} \\ \mathbf{SSer2L} \end{pmatrix} = \begin{pmatrix} 163.5 \\ 36.379 \\ 10.03 \\ 117.091 \end{pmatrix} \qquad \mathbf{df} := \begin{pmatrix} 33 \\ 1 \\ 2 \\ 30 \end{pmatrix} \qquad \mathbf{FL} := \overrightarrow{\left(\mathbf{SS} \cdot \mathbf{df}^{-1} \cdot \mathbf{ErrL}^{-1}\right)}$$

Or in Half-Multiplier Mode we have:

$$\mathbf{FL} := \mathbf{SS} \cdot_0 \cdot \mathbf{df}^{-1} \cdot_0 \cdot \mathbf{ErrL}^{-1}$$

$$\mathbf{FL} := \overrightarrow{\left[\begin{pmatrix} 163.5 \\ 36.379 \\ 10.03 \\ 117.091 \end{pmatrix} \cdot \begin{pmatrix} 33 \\ 1 \\ 2 \\ 30 \end{pmatrix}^{-1} \cdot \begin{pmatrix} \dfrac{117.091}{30} \\ 3.903 \\ 3.903 \\ 3.903 \\ 3.903 \end{pmatrix}^{-1} \right]}$$

$$\mathbf{FL} = \begin{pmatrix} 1.269 \\ 9.321 \\ 1.285 \\ 1 \end{pmatrix} \qquad \mathbf{ss} := \begin{pmatrix} 198.667 \\ 37.838 \\ 10.222 \\ 150.666 \end{pmatrix}$$

We cannot tell the Trend until we compute the Trends for the Quadratic term.

USE OF ORTHOGONAL COMPONENTS IN TESTS FOR TREND COMPUTATION OF BOTH LINEAR AND QUADRATIC TERMS AT THE SAME TIME

Example from Prob. 2.7

We have just computed the Linear component and now we will compute the Quadratic component. As you have by now seen, I am a lazy Mathematician, why do it the hard way if there is an easier way? Since we are working in Nested Arrays (using partitions until Programming catches up), why don't we try to compute both the Linear and the Quadratic components at the same time?

NOTE: This is the problem where I first saw the connection between Accounting and Inventory systems and Statistics (1-29-97). Also, analysis of Trend is quite amenable to operations by the Half-Multiplier Operator.

STEPS 1-3: We table the data as Matrix A then Multiply by both Linear and Quadratic components in DBLNQD.

$$A \cdot DBLNQD = \begin{pmatrix} 1 & 4 & 6 & 1 & 2 & 2 \\ 1 & 3 & 0 & -1 & -1 & -6 \\ 3 & 4 & 4 & -1 & 0 & -2 \\ 1 & 4 & 9 & 1 & 0 & -1 \\ -1 & -2 & 2 & -3 & -8 & -2 \\ 0 & 0 & 0 & -2 & 2 & 2 \\ 1 & 2 & -2 & 1 & 0 & 0 \\ 4 & 4 & 4 & 0 & 0 & 0 \\ 1 & 2 & -2 & -1 & 2 & -2 \\ -3 & -1 & 2 & -1 & 1 & 4 \\ -2 & -4 & 4 & 0 & -6 & 2 \end{pmatrix}$$

The math looks like:

$$
\begin{pmatrix}
5 & 5 & 6 & 4 & 5 & 8 & 4 & 6 & 10 \\
2 & 3 & 3 & 1 & 3 & 4 & 2 & 5 & 2 \\
2 & 4 & 5 & 4 & 6 & 8 & 5 & 8 & 9 \\
1 & 1 & 2 & 1 & 3 & 5 & 1 & 6 & 10 \\
3 & 4 & 2 & 5 & 8 & 3 & 2 & 4 & 4 \\
4 & 5 & 4 & 3 & 2 & 3 & 6 & 5 & 6 \\
4 & 4 & 5 & 3 & 4 & 5 & 5 & 4 & 3 \\
1 & 3 & 5 & 7 & 9 & 11 & 2 & 4 & 6 \\
5 & 6 & 6 & 2 & 2 & 4 & 3 & 3 & 1 \\
6 & 5 & 3 & 2 & 1 & 1 & 3 & 2 & 5 \\
3 & 2 & 1 & 6 & 7 & 2 & 2 & 3 & 6
\end{pmatrix}
\begin{pmatrix}
-1 & 0 & 0 & 1 & 0 & 0 \\
0 & 0 & 0 & -2 & 0 & 0 \\
1 & 0 & 0 & 1 & 0 & 0 \\
0 & -1 & 0 & 0 & 1 & 0 \\
0 & 0 & 0 & 0 & -2 & 0 \\
0 & 1 & 0 & 0 & 1 & 0 \\
0 & 0 & -1 & 0 & 0 & 1 \\
0 & 0 & 0 & 0 & 0 & -2 \\
0 & 0 & 1 & 0 & 0 & 1
\end{pmatrix}
=
\begin{pmatrix}
1 & 4 & 6 & 1 & 2 & 2 \\
1 & 3 & 0 & -1 & -1 & -6 \\
3 & 4 & 4 & -1 & 0 & -2 \\
1 & 4 & 9 & 1 & 0 & -1 \\
-1 & -2 & 2 & -3 & -8 & -2 \\
0 & 0 & 0 & -2 & 2 & 2 \\
1 & 2 & -2 & 1 & 0 & 0 \\
4 & 4 & 4 & 0 & 0 & 0 \\
1 & 2 & -2 & -1 & 2 & -2 \\
-3 & -1 & 2 & -1 & 1 & 4 \\
-2 & -4 & 4 & 0 & -6 & 2
\end{pmatrix}
\qquad
A \cdot DBLN =
\begin{pmatrix}
1 & 4 & 6 \\
1 & 3 & 0 \\
3 & 4 & 4 \\
1 & 4 & 9 \\
-1 & -2 & 2 \\
0 & 0 & 0 \\
1 & 2 & -2 \\
4 & 4 & 4 \\
1 & 2 & -2 \\
-3 & -1 & 2 \\
-2 & -4 & 4
\end{pmatrix}
$$

The first 3 rows = the 3 rows in the Linear example.

STEP 4:   Add the Subject Sums to obtain the Algebraic Sum for each Group.

PROGRAM: [1]$_{1,11}$[A]$_{11,9}$[DBLNQD]$_{9,6}$

SUMLNQD := ONE11·A·DBLNQD

SUMLNQD = (6 16 27 –6 –8 –3)   **ONE11** = (1 1 1 1 1 1 1 1 1 1 1)

$$
(1\ 1\ 1\ 1\ 1\ 1\ 1\ 1\ 1\ 1\ 1) \cdot \left[
\begin{pmatrix}
5 & 5 & 6 & 4 & 5 & 8 & 4 & 6 & 10 \\
2 & 3 & 3 & 1 & 3 & 4 & 2 & 5 & 2 \\
2 & 4 & 5 & 4 & 6 & 8 & 5 & 8 & 9 \\
1 & 1 & 2 & 1 & 3 & 5 & 1 & 6 & 10 \\
3 & 4 & 2 & 5 & 8 & 3 & 2 & 4 & 4 \\
4 & 5 & 4 & 3 & 2 & 3 & 6 & 5 & 6 \\
4 & 4 & 5 & 3 & 4 & 5 & 5 & 4 & 3 \\
1 & 3 & 5 & 7 & 9 & 11 & 2 & 4 & 6 \\
5 & 6 & 6 & 2 & 2 & 4 & 3 & 3 & 1 \\
6 & 5 & 3 & 2 & 1 & 1 & 3 & 2 & 5 \\
3 & 2 & 1 & 6 & 7 & 2 & 2 & 3 & 6
\end{pmatrix}
\begin{pmatrix}
-1 & 0 & 0 & 1 & 0 & 0 \\
0 & 0 & 0 & -2 & 0 & 0 \\
1 & 0 & 0 & 1 & 0 & 0 \\
0 & -1 & 0 & 0 & 1 & 0 \\
0 & 0 & 0 & 0 & -2 & 0 \\
0 & 1 & 0 & 0 & 1 & 0 \\
0 & 0 & -1 & 0 & 0 & 1 \\
0 & 0 & 0 & 0 & 0 & -2 \\
0 & 0 & 1 & 0 & 0 & 1
\end{pmatrix}
\right]
= (6\ 16\ 27\ -6\ -8\ -3)
$$

Check: The Linear component computed above =: **SUMLN** = (6 16 27)
Which checks.

STEP 5: Square each of the Subject Sums obtained in STEP 3 and add them together keeping the Linear and Quadratic components separate.

$[1]_{1,11} ([A]_{11,9} Ä [A]_{11,9}) [1]_{9,1}$

$$\text{ALGSUM} := \text{ONE11} \cdot [\overrightarrow{(A \cdot \text{DBLNQD}) \cdot (A \cdot \text{DBLNQD})}] \cdot \text{DB11} \quad \blacksquare$$

MathCad cannot solve the equation as is, so we make the following substitution:

ADBLNQD := A·DBLNQD

$$\text{ALGSUM} := \text{ONE11} \cdot [\overrightarrow{(\text{ADBLNQD}) \cdot (\text{ADBLNQD})}] \cdot \text{DB11}$$

ALGSUM = (327 211)

The math looks like:

$$\text{ONE11} \cdot \left[\left[\begin{pmatrix} 1 & 4 & 6 & 1 & 2 & 2 \\ 1 & 3 & 0 & -1 & -1 & -6 \\ 3 & 4 & 4 & -1 & 0 & -2 \\ 1 & 4 & 9 & 1 & 0 & -1 \\ -1 & -2 & 2 & -3 & -8 & -2 \\ 0 & 0 & 0 & -2 & 2 & 2 \\ 1 & 2 & -2 & 1 & 0 & 0 \\ 4 & 4 & 4 & 0 & 0 & 0 \\ 1 & 2 & -2 & -1 & 2 & -2 \\ -3 & -1 & 2 & -1 & 1 & 4 \\ -2 & -4 & 4 & 0 & -6 & 2 \end{pmatrix} \cdot \begin{pmatrix} 1 & 4 & 6 & 1 & 2 & 2 \\ 1 & 3 & 0 & -1 & -1 & -6 \\ 3 & 4 & 4 & -1 & 0 & -2 \\ 1 & 4 & 9 & 1 & 0 & -1 \\ -1 & -2 & 2 & -3 & -8 & -2 \\ 0 & 0 & 0 & -2 & 2 & 2 \\ 1 & 2 & -2 & 1 & 0 & 0 \\ 4 & 4 & 4 & 0 & 0 & 0 \\ 1 & 2 & -2 & -1 & 2 & -2 \\ -3 & -1 & 2 & -1 & 1 & 4 \\ -2 & -4 & 4 & 0 & -6 & 2 \end{pmatrix}\right]\right] \cdot \begin{pmatrix} 1 & 0 \\ 1 & 0 \\ 1 & 0 \\ 0 & 1 \\ 0 & 1 \\ 0 & 1 \end{pmatrix} = (327 \ 211)$$

CHECK: **SUBSUM := ONE11 · [$\overrightarrow{(\text{ADBLN}) \cdot (\text{ADBLN})}$] · THREE1**   **SUBSUM = 327**
Which checks.

STEP 6: Add the Group Sums to get the overall Algebraic Sum (from STEP 4)

PROGRAM: $[1]_{1,11} \ ([A]_{11,9} [\text{DBLNQD}]_{9,6}) \{\text{DB}\}_{6,2}$

ALGSUM := ONE11 ·(A·DBLNQD ·DB11)

ALGSUM = (49 −17)

The math looks like:

$$(1\ 1\ 1\ 1\ 1\ 1\ 1\ 1\ 1\ 1\ 1) \cdot \begin{pmatrix} 5 & 5 & 6 & 4 & 5 & 8 & 4 & 6 & 10 \\ 2 & 3 & 3 & 1 & 3 & 4 & 2 & 5 & 2 \\ 2 & 4 & 5 & 4 & 6 & 8 & 5 & 8 & 9 \\ 1 & 1 & 2 & 1 & 3 & 5 & 1 & 6 & 10 \\ 3 & 4 & 2 & 5 & 8 & 3 & 2 & 4 & 4 \\ 4 & 5 & 4 & 3 & 2 & 3 & 6 & 5 & 6 \\ 4 & 4 & 5 & 3 & 4 & 5 & 5 & 4 & 3 \\ 1 & 3 & 5 & 7 & 9 & 11 & 2 & 4 & 6 \\ 5 & 6 & 6 & 2 & 2 & 4 & 3 & 3 & 1 \\ 6 & 5 & 3 & 2 & 1 & 1 & 3 & 2 & 5 \\ 3 & 2 & 1 & 6 & 7 & 2 & 2 & 3 & 6 \end{pmatrix} \cdot \begin{pmatrix} -1 & 0 & 0 & 1 & 0 & 0 \\ 0 & 0 & 0 & -2 & 0 & 0 \\ 1 & 0 & 0 & 1 & 0 & 0 \\ 0 & -1 & 0 & 0 & 1 & 0 \\ 0 & 0 & 0 & 0 & -2 & 0 \\ 0 & 1 & 0 & 0 & 1 & 0 \\ 0 & 0 & -1 & 0 & 0 & 1 \\ 0 & 0 & 0 & 0 & 0 & -2 \\ 0 & 0 & 1 & 0 & 0 & 1 \end{pmatrix} \cdot \begin{pmatrix} 1 & 0 \\ 1 & 0 \\ 1 & 0 \\ 0 & 1 \\ 0 & 1 \\ 0 & 1 \end{pmatrix} = (49\ -17)$$

CHECK: From the Linear computation we get:

ALGSUM := **ONE11** ·(**A·DBLN·THREE1**)

ALGSUM = 49

which checks.

STEP 7: Square each of the Linear and Quadratic components and add the squared values.

First we write the Linear and Quadratic terms as a row Matrix.

DBLQ := (−1 0 1 1 −2 1)

In Half-Multiplier mode the equation becomes:

LNQDSQ := **ONE6**·(**DBLQ** ·◦·**DB11**)

Translating the equation into diagonals so we can Half-Multiply by proxy we get:

$$\textbf{LNQDSQ} := \textbf{ONE6} \cdot \left(\text{diag}\left(\left(\overleftarrow{\overrightarrow{\textbf{DBLQ} \cdot \textbf{DBLQ}}} \right)^{\!T} \right) \cdot \textbf{DB11} \right)$$

Where

$$\left(\overrightarrow{\mathbf{DBLQ \cdot DBLQ}}\right)^{\mathbf{T}} = \begin{pmatrix} 1 \\ 0 \\ 1 \\ 1 \\ 4 \\ 1 \end{pmatrix} \quad \text{and} \quad \mathbf{diag}\left(\left(\overrightarrow{\mathbf{DBLQ \cdot DBLQ}}\right)^{\mathbf{T}}\right) = \begin{pmatrix} 1 & 0 & 0 & 0 & 0 & 0 \\ 0 & 0 & 0 & 0 & 0 & 0 \\ 0 & 0 & 1 & 0 & 0 & 0 \\ 0 & 0 & 0 & 1 & 0 & 0 \\ 0 & 0 & 0 & 0 & 4 & 0 \\ 0 & 0 & 0 & 0 & 0 & 1 \end{pmatrix}$$

LNQDSQ $= (2 \quad 6)$

The math looks like

$$(1 \ 1 \ 1 \ 1 \ 1 \ 1) \cdot \mathbf{diag}\left[\left(\overrightarrow{\begin{pmatrix} -1 \\ 0 \\ 1 \\ 1 \\ -2 \\ 1 \end{pmatrix} \cdot \begin{pmatrix} -1 \\ 0 \\ 1 \\ 1 \\ -2 \\ 1 \end{pmatrix}} \right) \cdot \begin{pmatrix} 1 & 0 \\ 1 & 0 \\ 1 & 0 \\ 0 & 1 \\ 0 & 1 \\ 0 & 1 \end{pmatrix} \right] \qquad \mathbf{diag}\left(\overrightarrow{\begin{pmatrix} -1 \\ 0 \\ 1 \\ 1 \\ -2 \\ 1 \end{pmatrix} \cdot \begin{pmatrix} -1 \\ 0 \\ 1 \\ 1 \\ -2 \\ 1 \end{pmatrix}} \right) = \begin{pmatrix} 1 & 0 & 0 & 0 & 0 & 0 \\ 0 & 0 & 0 & 0 & 0 & 0 \\ 0 & 0 & 1 & 0 & 0 & 0 \\ 0 & 0 & 0 & 1 & 0 & 0 \\ 0 & 0 & 0 & 0 & 4 & 0 \\ 0 & 0 & 0 & 0 & 0 & 1 \end{pmatrix}$$

Which becomes

$$(1 \ 1 \ 1 \ 1 \ 1 \ 1) \cdot \left[\begin{pmatrix} 1 & 0 & 0 & 0 & 0 & 0 \\ 0 & 0 & 0 & 0 & 0 & 0 \\ 0 & 0 & 1 & 0 & 0 & 0 \\ 0 & 0 & 0 & 1 & 0 & 0 \\ 0 & 0 & 0 & 0 & 4 & 0 \\ 0 & 0 & 0 & 0 & 0 & 1 \end{pmatrix} \cdot \begin{pmatrix} 1 & 0 \\ 1 & 0 \\ 1 & 0 \\ 0 & 1 \\ 0 & 1 \\ 0 & 1 \end{pmatrix} \right] = (2 \quad 6)$$

STEP 8: Computation of the Within Subjects Linear and Quadratic components. In terms of already solved computations we have:

$$\mathbf{SSwLQ} := \left(\mathbf{ONE11} \cdot \overrightarrow{[(A \cdot DBLNQD) \cdot (A \cdot DBLNQD)]} \cdot \mathbf{DB11} \right) \left[\mathbf{ONE6} \cdot \left(\mathbf{diag}\left(\left(\overrightarrow{\mathbf{DBLQ \cdot DBLQ}} \right)^{\mathbf{T}} \right) \right) \cdot \mathbf{DB11} \right]^{-1}$$

Again, MathCad cannot compute this type of expression, so we must make the following substitution:

ADBLNQD $:=$ **A·DBLNQD**

$$\mathbf{SSwLQ} := \left(\overrightarrow{\mathbf{ONE11}\cdot[(\mathbf{ADBLNQD})\cdot(\mathbf{ADBLNQD})]\cdot\mathbf{DB11}}\right)\cdot\mathbf{diag}\left(\mathbf{LNQDSQ}^{\mathbf{T}}\right)^{-1}$$

$$\mathbf{SSwLQ} = (\,163.5 \quad 35.167\,)$$

The math looks like:

$$(1\ 1\ 1\ 1\ 1\ 1\ 1\ 1\ 1\ 1\ 1)\cdot\left[\left[\begin{pmatrix} 1 & 4 & 6 & 1 & 2 & 2 \\ 1 & 3 & 0 & -1 & -1 & -6 \\ 3 & 4 & 4 & -1 & 0 & -2 \\ 1 & 4 & 9 & 1 & 0 & -1 \\ -1 & -2 & 2 & -3 & -8 & -2 \\ 0 & 0 & 0 & -2 & 2 & 2 \\ 1 & 2 & -2 & 1 & 0 & 0 \\ 4 & 4 & 4 & 0 & 0 & 0 \\ 1 & 2 & -2 & -1 & 2 & -2 \\ -3 & -1 & 2 & -1 & 1 & 4 \\ -2 & -4 & 4 & 0 & -6 & 2 \end{pmatrix}\cdot\begin{pmatrix} 1 & 4 & 6 & 1 & 2 & 2 \\ 1 & 3 & 0 & -1 & -1 & -6 \\ 3 & 4 & 4 & -1 & 0 & -2 \\ 1 & 4 & 9 & 1 & 0 & -1 \\ -1 & -2 & 2 & -3 & -8 & -2 \\ 0 & 0 & 0 & -2 & 2 & 2 \\ 1 & 2 & -2 & 1 & 0 & 0 \\ 4 & 4 & 4 & 0 & 0 & 0 \\ 1 & 2 & -2 & -1 & 2 & -2 \\ -3 & -1 & 2 & -1 & 1 & 4 \\ -2 & -4 & 4 & 0 & -6 & 2 \end{pmatrix}\right)\cdot\begin{pmatrix} 1 & 0 \\ 1 & 0 \\ 1 & 0 \\ 0 & 1 \\ 0 & 1 \\ 0 & 1 \end{pmatrix}\right]\cdot\begin{pmatrix} 2 & 0 \\ 0 & 6 \end{pmatrix}^{-1} = (\,163.5 \quad 35.167\,)$$

STEP 9: Computation of the Trials effect for the Linear and Quadratic components.

$$\mathbf{NtrLQ} := 33 \qquad\qquad\qquad \mathbf{ALGSUM} = 49$$

$$\mathbf{SStrLQ} := \left(\frac{1}{\mathbf{NtrLQ}}\right)\left[\overrightarrow{(\mathbf{ALGSUM})\cdot(\mathbf{ALGSUM})\cdot\mathbf{LNQDSQ}^{-1}}\right]$$

$$\mathbf{SStrLQ} = (\,36.379 \quad 12.126\,)$$

The Math looks like

$$\left(\frac{1}{33}\right)\left[\overrightarrow{((49\ \ -17))\cdot((49\ \ -17))\cdot(2\ \ 6)^{-1}}\right] = (\,36.379 \quad 1.46\,)$$

To check the Linear component we have

$$\left(\frac{1}{\mathbf{NtrL}}\right)\left[\overrightarrow{(\mathbf{ALGSUM})\cdot(\mathbf{ALGSUM})\cdot\mathbf{LINSQ}^{-1}}\right] = 36.379 \qquad\qquad \text{Which checks.}$$

STEP 10: Computation of the Trials by Groups effect for the Linear and Quadratic components. NtxgLQ = # Subjects in a single Trial = 11.

SUMLNQD = (6 16 27 –6 –8 –3)

NtxgLQ := 11

The expression is given by:

$$\textbf{SStxgLQ} := \left(\frac{1}{\textbf{NtxgLQ}}\right)\left[(\textbf{SUMLNQD})\cdot\left(\textbf{diag}\left(\textbf{SUMLNQD}^T\right)\cdot\textbf{DB11}\right)\cdot\textbf{diag}\left(\textbf{LNQDSQ}^T\right)^{-1}\right] - \textbf{SStrLQ}$$

SStxgLQ = (10.03 –10.475)

The math looks like:

$$\left(\frac{1}{11}\right)\cdot\left[((6\ \ 16\ \ 27\ \ -6\ \ -8\ \ -3))\cdot\left[\begin{pmatrix}6 & 0 & 0 & 0 & 0 & 0 \\ 0 & 16 & 0 & 0 & 0 & 0 \\ 0 & 0 & 27 & 0 & 0 & 0 \\ 0 & 0 & 0 & -6 & 0 & 0 \\ 0 & 0 & 0 & 0 & -8 & 0 \\ 0 & 0 & 0 & 0 & 0 & -3\end{pmatrix}\begin{pmatrix}1 & 0 \\ 1 & 0 \\ 1 & 0 \\ 0 & 1 \\ 0 & 1 \\ 0 & 1\end{pmatrix}\right]\cdot\begin{pmatrix}2 & 0 \\ 0 & 6\end{pmatrix}^{-1}\right] - (36.379\ \ 1.46) = (10.03\ \ 0.192)$$

To check the previous value for the Linear component we have:

$$\textbf{SStxgL} := \left(\frac{1}{\textbf{NtxgL}}\right)\left[(\textbf{SUMLN})\cdot\left(\textbf{diag}\left(\textbf{SUMLN}^T\right)\cdot\textbf{THREE1}\right)\cdot\textbf{LINSQ}^{-1}\right] - \textbf{SStrL}$$

SStxgL = 10.03   Which checks.

STEP 11: Computation of the Error term for the Linear and Quadratic components.

SSer2 := **SSwLQ** – **SStrLQ** – **SStxgLQ**

SSer2 = (117.091 33.515)

To check the already computed Linear component, we have:

SSer2 := **SSwL** – **SStrL** – **SStxgL**

SSer2 = 117.091   Which checks.

Our Trend computations now look like:

$$\begin{pmatrix} \textbf{SSwLQ} \\ \textbf{SStrLQ} \\ \textbf{SStxgLQ} \\ \textbf{SSer2} \end{pmatrix} = \begin{bmatrix} (163.5 \quad 35.167) \\ (36.379 \quad 12.126) \\ (10.03 \quad -10.475) \\ 117.091 \end{bmatrix} \qquad \textbf{df} := \begin{pmatrix} 33 \\ 1 \\ 2 \\ 30 \end{pmatrix}$$

STEP 12: Computation of the Degrees of Freedom df and the Error Term.

s = # Subjects in each Group, tr always = 1, G = # Groups in study and error = s-tr-(G-1)

$\textbf{tr} := 1$ $\qquad\qquad$ $\textbf{G} := 3$

$$\textbf{df} := \begin{bmatrix} s \\ \textbf{tr} \\ \textbf{G} - 1 \\ s - \textbf{tr} - (\textbf{G} - 1) \end{bmatrix} \qquad \textbf{df} = \begin{pmatrix} 33 \\ 1 \\ 2 \\ 30 \end{pmatrix}$$

Now we will get all the terms ready for the final computation of F and compile all the Sums of Squares computed for the Trends.

$$\begin{pmatrix} \textbf{SSwLQ} \\ \textbf{SStrLQ} \\ \textbf{SStxgLQ} \\ \textbf{SSer2} \end{pmatrix} = \begin{bmatrix} (163.5 \quad 35.167) \\ (36.379 \quad 12.126) \\ (10.03 \quad -10.475) \\ 117.091 \end{bmatrix} \qquad \textbf{df} := \begin{pmatrix} 33 \\ 1 \\ 2 \\ 30 \end{pmatrix} \qquad \textbf{ss} := \begin{pmatrix} \textbf{SSwLQ} \\ \textbf{SStrLQ} \\ \textbf{SStxgLQ} \\ \textbf{SSer2} \end{pmatrix}$$

$$\textbf{SSLQ} := \begin{pmatrix} 163.5 & 35.167 \\ 36.379 & 1.46 \\ 10.03 & 0.192 \\ 117.091 & 33.515 \end{pmatrix}$$

The mean Squares are computed as the Linear SSer2 term divided by dferr2 and the Quadratic SSer2 term divided by dfer2.

$$\textbf{ms} := \begin{pmatrix} \dfrac{117.091}{30} \\ \dfrac{33.515}{30} \end{pmatrix}$$

$$\textbf{ms} = \begin{pmatrix} 3.903 \\ 1.117 \end{pmatrix}$$

$$F := diag(df)^{-1} \cdot SSLQ \cdot diag(ms)^{-1}$$

SSw

SStr
$$FL = \begin{pmatrix} 1.269 \\ 9.321 \\ 1.285 \\ 1 \end{pmatrix}$$
SStxg

SSer2

Interpretation: The Trials seem to trend in a straight line
since it is 8x as big as the Quadratic component. The Trials
x Group Interaction tends to be Linear since it is 14x larger
than the Quadratic term.

To compare the Linear computation against the Linear and
Quadratic computations, the Sums of Squares for the Linear only
were computed as:

(This is the solution as computed below, the answers check.)

$$ss := \begin{pmatrix} 198.667 \\ 37.838 \\ 10.222 \\ 150.666 \end{pmatrix}$$

Now to check if this is correct, we will re-do Prob. 2.7 with
the new Data Matrix A and check to make sure SSw, SStr, SStxg
and SSer2 are equal to SSLQ x ONE2. Now we have already done
this problem in Sect. 2.7 so we will not re-do the problem,
just go through the steps to get a quick computation. NOTE:
The Following Statistic is a 2-Factor Mixed Design Analysis
of Variance.

SUM1 := ONE11·A

SUM1 = (36 42 42 38 50 54 35 50 62)

GROUPSUMS := ONE11·A·DB1

GROUPSUMS = (120 142 147)

SUBJECTSUM := A·DB1

$$\mathbf{ONE11 \cdot A \cdot DB1} = (\,120 \quad 142 \quad 147\,) \qquad \mathbf{DBLNQD} := \begin{pmatrix} -1 & 0 & 0 & 1 & 0 & 0 \\ 0 & 0 & 0 & -2 & 0 & 0 \\ 1 & 0 & 0 & 1 & 0 & 0 \\ 0 & -1 & 0 & 0 & 1 & 0 \\ 0 & 0 & 0 & 0 & -2 & 0 \\ 0 & 1 & 0 & 0 & 1 & 0 \\ 0 & 0 & -1 & 0 & 0 & 1 \\ 0 & 0 & 0 & 0 & 0 & -2 \\ 0 & 0 & 1 & 0 & 0 & 1 \end{pmatrix} \qquad \mathbf{SUBJECTSUM} = \begin{pmatrix} 16 & 17 & 20 \\ 8 & 8 & 9 \\ 11 & 18 & 22 \\ 4 & 9 & 17 \\ 9 & 16 & 10 \\ 13 & 8 & 17 \\ 13 & 12 & 12 \\ 9 & 27 & 12 \\ 17 & 8 & 7 \\ 14 & 4 & 10 \\ 6 & 15 & 11 \end{pmatrix}$$

$$\mathbf{GSQ} := \mathbf{ONE11} \cdot \overrightarrow{[(A) \cdot (A)]} \cdot \mathbf{NINE1}$$

$$\mathbf{GSQ} = 2177$$

$$\mathbf{GS} := \mathbf{ONE11} \cdot A \cdot \mathbf{NINE1}$$

$$\mathbf{F} = \begin{pmatrix} 1.269 & 0.954 \\ 9.321 & 1.307 \\ 1.285 & 0.086 \\ 1 & 1 \end{pmatrix} \qquad \mathbf{SSLQ} \cdot \begin{pmatrix} 1 \\ 1 \end{pmatrix} = \begin{pmatrix} 198.667 \\ 37.839 \\ 10.222 \\ 150.606 \end{pmatrix} \qquad \mathbf{ADB1} := \begin{pmatrix} 16 & 17 & 20 \\ 8 & 8 & 9 \\ 11 & 18 & 22 \\ 4 & 9 & 17 \\ 9 & 16 & 10 \\ 13 & 8 & 17 \\ 13 & 12 & 12 \\ 9 & 27 & 12 \\ 17 & 8 & 7 \\ 14 & 4 & 10 \\ 6 & 15 & 11 \end{pmatrix}$$

$$\mathbf{GS} = 409$$

$$\mathbf{CF} := \left(\frac{1}{99}\right) \cdot \mathbf{GS}^2$$

$$\mathbf{CF} = 1689.707$$

$$\mathbf{SSt} := \mathbf{ONE11} \cdot \overrightarrow{[(A) \cdot (A)]} \cdot \mathbf{NINE1} - \left(\frac{1}{99}\right) \cdot \mathbf{GS}^2$$

$$\mathbf{SSt} = 487.293$$

$$SSb := \left(\frac{1}{Nb}\right) \cdot (ONE11) \cdot \overrightarrow{\left[(A \cdot DB1) \cdot (A \cdot DB1)\right] \cdot THREE1} \bigg) - CF \quad \blacksquare$$

$$Nb := 3$$

$$SSb := \left(\frac{1}{Nb}\right) \left[ONE11 \cdot \left(\overrightarrow{\left[(ADB1) \cdot (ADB1)\right] \cdot THREE1}\right)\right] - CF$$

$$SSb = 288.626$$

$$Nc := 33$$

$$SSc := \left(\frac{1}{Nc}\right) \left[(ONE11 \cdot A \cdot DB1) \cdot (ONE11 \cdot A \cdot DB1)^T\right] - CF$$

$$SSc = 12.505$$

$$SSerb := SSb - SSc$$

$$SSerb = 276.121$$

$$SSw := SSt - SSb$$

$$SSw = 198.667$$

$$Ntr := 33$$

$$SStr := \left(\frac{1}{Ntr}\right) \left[(ONE11 \cdot A \cdot DB2) \cdot (ONE11 \cdot A \cdot DB2)^T\right] - CF$$

$$SStr = 37.838$$

$$Ntrxg := 11$$

$$SStrxg := \left(\frac{1}{Ntrxg}\right) \cdot \left(SUM1 \cdot SUM1^T\right) - CF - SSc - SStr$$

$$SStrxg = 10.222$$

$$SSerrw := SSw - SStr - SStrxg$$

SSerrw = 150.606

The values check.

$$
SS := \begin{pmatrix} SSt \\ SSb \\ SSc \\ SSerb \\ SSw \\ SStr \\ SStrxg \\ SSerrw \end{pmatrix} \quad \begin{pmatrix} SSt \\ SSb \\ SSc \\ SSerb \\ SSw \\ SStr \\ SStrxg \\ SSerrw \end{pmatrix} = \begin{pmatrix} 487.293 \\ 288.626 \\ 12.505 \\ 276.121 \\ 198.667 \\ 37.838 \\ 10.222 \\ 150.606 \end{pmatrix} \qquad ss := \begin{pmatrix} 198.667 \\ 37.838 \\ 10.222 \\ 150.666 \end{pmatrix} \quad \begin{matrix} SSw \\ SStr \\ SStxg \\ SSer2 \end{matrix}
$$

EXAMPLE 3 ON TREND ANALYSIS: THREE FACTOR MIXED DESIGN

(Section 3.11, pg. 154-168).

STEP 1: Tabling of the data Matrix A: I'm going to attempt to solve for Linear, Quadratic and Cubic components at the same time.

$$A := \begin{pmatrix} 5 & 7 & 8 & 10 & 6 & 9 & 10 & 13 & 4 & 5 & 5 & 6 & 5 & 9 & 11 & 15 \\ 7 & 8 & 10 & 12 & 4 & 6 & 11 & 15 & 6 & 7 & 9 & 10 & 6 & 11 & 15 & 19 \\ 6 & 9 & 11 & 14 & 8 & 11 & 13 & 14 & 6 & 7 & 7 & 9 & 4 & 8 & 13 & 19 \\ 4 & 5 & 7 & 10 & 6 & 9 & 12 & 16 & 6 & 7 & 8 & 9 & 6 & 9 & 12 & 18 \\ 3 & 5 & 8 & 9 & 3 & 8 & 12 & 14 & 2 & 3 & 3 & 4 & 5 & 10 & 14 & 18 \\ 4 & 4 & 5 & 7 & 2 & 4 & 8 & 11 & 6 & 6 & 7 & 6 & 5 & 7 & 9 & 14 \\ 6 & 7 & 8 & 11 & 5 & 6 & 7 & 10 & 5 & 7 & 7 & 7 & 4 & 10 & 16 & 19 \end{pmatrix}$$

VARIABLES:

$$ONE7 := \begin{pmatrix} 1 & 1 & 1 & 1 & 1 & 1 & 1 \end{pmatrix}$$

$$ONE12 := \begin{pmatrix} 1 & 1 & 1 & 1 & 1 & 1 & 1 & 1 & 1 & 1 & 1 & 1 \end{pmatrix}$$

$$ONE6 := \begin{pmatrix} 1 & 1 & 1 & 1 & 1 & 1 \end{pmatrix}$$

$$LQC := \begin{pmatrix} -3 & -1 & 1 & 3 & 1 & -1 & -1 & 1 & -1 & 3 & -3 & 1 \end{pmatrix}$$

$$DB1 := \begin{pmatrix} 1 & 0 & 0 \\ 1 & 0 & 0 \\ 1 & 0 & 0 \\ 1 & 0 & 0 \\ 0 & 1 & 0 \\ 0 & 1 & 0 \\ 0 & 1 & 0 \\ 0 & 1 & 0 \\ 0 & 0 & 1 \\ 0 & 0 & 1 \\ 0 & 0 & 1 \\ 0 & 0 & 1 \end{pmatrix} \quad DB13 := \begin{pmatrix} 1 & 0 & 0 \\ 1 & 0 & 0 \\ 0 & 1 & 0 \\ 0 & 1 & 0 \\ 0 & 0 & 1 \\ 0 & 0 & 1 \end{pmatrix} \quad DB2 := \begin{pmatrix} 1 & 0 & 0 & 0 & 0 & 0 \\ 0 & 1 & 0 & 0 & 0 & 0 \\ 1 & 0 & 0 & 0 & 0 & 0 \\ 0 & 1 & 0 & 0 & 0 & 0 \\ 0 & 0 & 1 & 0 & 0 & 0 \\ 0 & 0 & 0 & 1 & 0 & 0 \\ 0 & 0 & 1 & 0 & 0 & 0 \\ 0 & 0 & 0 & 1 & 0 & 0 \\ 0 & 0 & 0 & 0 & 1 & 0 \\ 0 & 0 & 0 & 0 & 0 & 1 \\ 0 & 0 & 0 & 0 & 1 & 0 \\ 0 & 0 & 0 & 0 & 0 & 1 \end{pmatrix} \quad DB12 := \begin{pmatrix} 1 & 0 & 0 & 0 & 0 & 0 \\ 1 & 0 & 0 & 0 & 0 & 0 \\ 0 & 1 & 0 & 0 & 0 & 0 \\ 0 & 1 & 0 & 0 & 0 & 0 \\ 0 & 0 & 1 & 0 & 0 & 0 \\ 0 & 0 & 1 & 0 & 0 & 0 \\ 0 & 0 & 0 & 1 & 0 & 0 \\ 0 & 0 & 0 & 1 & 0 & 0 \\ 0 & 0 & 0 & 0 & 1 & 0 \\ 0 & 0 & 0 & 0 & 1 & 0 \\ 0 & 0 & 0 & 0 & 0 & 1 \\ 0 & 0 & 0 & 0 & 0 & 1 \end{pmatrix}$$

$$
\text{DBLQC} :=
\begin{pmatrix}
-3 & 0 & 0 & 0 & 1 & 0 & 0 & 0 & -1 & 0 & 0 & 0 \\
-1 & 0 & 0 & 0 & -1 & 0 & 0 & 0 & 3 & 0 & 0 & 0 \\
1 & 0 & 0 & 0 & -1 & 0 & 0 & 0 & -3 & 0 & 0 & 0 \\
3 & 0 & 0 & 0 & 1 & 0 & 0 & 0 & 1 & 0 & 0 & 0 \\
0 & -3 & 0 & 0 & 0 & 1 & 0 & 0 & 0 & -1 & 0 & 0 \\
0 & -1 & 0 & 0 & 0 & -1 & 0 & 0 & 0 & 3 & 0 & 0 \\
0 & 1 & 0 & 0 & 0 & -1 & 0 & 0 & 0 & -3 & 0 & 0 \\
0 & 3 & 0 & 0 & 0 & 1 & 0 & 0 & 0 & 1 & 0 & 0 \\
0 & 0 & -3 & 0 & 0 & 0 & 1 & 0 & 0 & 0 & -1 & 0 \\
0 & 0 & -1 & 0 & 0 & 0 & -1 & 0 & 0 & 0 & 3 & 0 \\
0 & 0 & 1 & 0 & 0 & 0 & -1 & 0 & 0 & 0 & -3 & 0 \\
0 & 0 & 3 & 0 & 0 & 0 & 1 & 0 & 0 & 0 & 1 & 0 \\
0 & 0 & 0 & -3 & 0 & 0 & 0 & 1 & 0 & 0 & 0 & -1 \\
0 & 0 & 0 & -1 & 0 & 0 & 0 & -1 & 0 & 0 & 0 & 3 \\
0 & 0 & 0 & 1 & 0 & 0 & 0 & -1 & 0 & 0 & 0 & -3 \\
0 & 0 & 0 & 3 & 0 & 0 & 0 & 1 & 0 & 0 & 0 & 1
\end{pmatrix}
$$

STEP 1: Computation of the Algebraic Sum of the Linear, Quadratic and Cubic components.

ALGSUM := A·DBLQC

$$
\text{ALGSUM} =
\begin{pmatrix}
16 & 22 & 6 & 32 & 0 & 0 & 0 & 0 & 2 & 4 & 2 & 4 \\
17 & 38 & 14 & 43 & 1 & 2 & 0 & -1 & -1 & -4 & -2 & 1 \\
26 & 20 & 9 & 50 & 0 & -2 & 1 & 2 & 2 & 0 & 3 & 0 \\
20 & 33 & 10 & 39 & 2 & 1 & 0 & 3 & 0 & 1 & 0 & 3 \\
21 & 37 & 6 & 43 & -1 & -3 & 0 & -1 & -3 & -1 & 2 & 1 \\
10 & 31 & 1 & 29 & 2 & 1 & -1 & 3 & 0 & -3 & -3 & 3 \\
16 & 16 & 6 & 51 & 2 & 2 & -2 & -3 & 2 & 2 & 2 & -3
\end{pmatrix}
$$

The Math looks like:

$$
\begin{pmatrix}
5 & 7 & 8 & 10 & 6 & 9 & 10 & 13 & 4 & 5 & 5 & 6 & 5 & 9 & 11 & 15 \\
7 & 8 & 10 & 12 & 4 & 6 & 11 & 15 & 6 & 7 & 9 & 10 & 6 & 11 & 15 & 19 \\
6 & 9 & 11 & 14 & 8 & 11 & 13 & 14 & 6 & 7 & 7 & 9 & 4 & 8 & 13 & 19 \\
4 & 5 & 7 & 10 & 6 & 9 & 12 & 16 & 6 & 7 & 8 & 9 & 6 & 9 & 12 & 18 \\
3 & 5 & 8 & 9 & 3 & 8 & 12 & 14 & 2 & 3 & 3 & 4 & 5 & 10 & 14 & 18 \\
4 & 4 & 5 & 7 & 2 & 4 & 8 & 11 & 6 & 6 & 7 & 6 & 5 & 7 & 9 & 14 \\
6 & 7 & 8 & 11 & 5 & 6 & 7 & 10 & 5 & 7 & 7 & 7 & 4 & 10 & 16 & 19
\end{pmatrix}
\cdot
\begin{pmatrix}
-3 & 0 & 0 & 0 & 1 & 0 & 0 & 0 & -1 & 0 & 0 & 0 \\
-1 & 0 & 0 & 0 & -1 & 0 & 0 & 0 & 3 & 0 & 0 & 0 \\
1 & 0 & 0 & 0 & -1 & 0 & 0 & 0 & -3 & 0 & 0 & 0 \\
3 & 0 & 0 & 0 & 1 & 0 & 0 & 0 & 1 & 0 & 0 & 0 \\
0 & -3 & 0 & 0 & 0 & 1 & 0 & 0 & 0 & -1 & 0 & 0 \\
0 & -1 & 0 & 0 & 0 & -1 & 0 & 0 & 0 & 3 & 0 & 0 \\
0 & 1 & 0 & 0 & 0 & -1 & 0 & 0 & 0 & -3 & 0 & 0 \\
0 & 3 & 0 & 0 & 0 & 1 & 0 & 0 & 0 & 1 & 0 & 0 \\
0 & 0 & -3 & 0 & 0 & 0 & 1 & 0 & 0 & 0 & -1 & 0 \\
0 & 0 & -1 & 0 & 0 & 0 & -1 & 0 & 0 & 0 & 3 & 0 \\
0 & 0 & 1 & 0 & 0 & 0 & -1 & 0 & 0 & 0 & -3 & 0 \\
0 & 0 & 3 & 0 & 0 & 0 & 1 & 0 & 0 & 0 & 1 & 0 \\
0 & 0 & 0 & -3 & 0 & 0 & 0 & 1 & 0 & 0 & 0 & -1 \\
0 & 0 & 0 & -1 & 0 & 0 & 0 & -1 & 0 & 0 & 0 & 3 \\
0 & 0 & 0 & 1 & 0 & 0 & 0 & -1 & 0 & 0 & 0 & -3 \\
0 & 0 & 0 & 3 & 0 & 0 & 0 & 1 & 0 & 0 & 0 & 1
\end{pmatrix}
=
\begin{pmatrix}
16 & 22 & 6 & 32 & 0 & 0 & 0 & 2 & 4 & 2 & 4 \\
17 & 38 & 14 & 43 & 1 & 2 & 0 & -1 & -1 & -4 & -2 & 1 \\
26 & 20 & 9 & 50 & 0 & -2 & 1 & 2 & 2 & 0 & 3 & 0 \\
20 & 33 & 10 & 39 & 2 & 1 & 0 & 3 & 0 & 1 & 0 & 3 \\
21 & 37 & 6 & 43 & -1 & -3 & 0 & -1 & -3 & -1 & 2 & 1 \\
10 & 31 & 1 & 29 & 2 & 1 & -1 & 3 & 0 & -3 & -3 & 3 \\
16 & 16 & 6 & 51 & 2 & 2 & -2 & -3 & 2 & 2 & 2 & -3
\end{pmatrix}
$$

STEP 4: Add the sums of the columns of ALGSUM.

GROUPSUM:= ONE7·A·DBLQC

$$\text{GROUPSUM} = \begin{pmatrix} 126 & 197 & 52 & 287 & 6 & 1 & -2 & 3 & 2 & -1 & 4 & 9 \end{pmatrix}$$

The Math looks like:

$$
\text{ONE7}\cdot
\begin{pmatrix}
5 & 7 & 8 & 10 & 6 & 9 & 10 & 13 & 4 & 5 & 5 & 6 & 5 & 9 & 11 & 15 \\
7 & 8 & 10 & 12 & 4 & 6 & 11 & 15 & 6 & 7 & 9 & 10 & 6 & 11 & 15 & 19 \\
6 & 9 & 11 & 14 & 8 & 11 & 13 & 14 & 6 & 7 & 7 & 9 & 4 & 8 & 13 & 19 \\
4 & 5 & 7 & 10 & 6 & 9 & 12 & 16 & 6 & 7 & 8 & 9 & 6 & 9 & 12 & 18 \\
3 & 5 & 8 & 9 & 3 & 8 & 12 & 14 & 2 & 3 & 3 & 4 & 5 & 10 & 14 & 18 \\
4 & 4 & 5 & 7 & 2 & 4 & 8 & 11 & 6 & 6 & 7 & 6 & 5 & 7 & 9 & 14 \\
6 & 7 & 8 & 11 & 5 & 6 & 7 & 10 & 5 & 7 & 7 & 7 & 4 & 10 & 16 & 19
\end{pmatrix}
\cdot
\begin{pmatrix}
-3 & 0 & 0 & 0 & 1 & 0 & 0 & 0 & -1 & 0 & 0 & 0 \\
-1 & 0 & 0 & 0 & -1 & 0 & 0 & 0 & 3 & 0 & 0 & 0 \\
1 & 0 & 0 & 0 & -1 & 0 & 0 & 0 & -3 & 0 & 0 & 0 \\
3 & 0 & 0 & 0 & 1 & 0 & 0 & 0 & 1 & 0 & 0 & 0 \\
0 & -3 & 0 & 0 & 0 & 1 & 0 & 0 & 0 & -1 & 0 & 0 \\
0 & -1 & 0 & 0 & 0 & -1 & 0 & 0 & 0 & 3 & 0 & 0 \\
0 & 1 & 0 & 0 & 0 & -1 & 0 & 0 & 0 & -3 & 0 & 0 \\
0 & 3 & 0 & 0 & 0 & 1 & 0 & 0 & 0 & 1 & 0 & 0 \\
0 & 0 & -3 & 0 & 0 & 0 & 1 & 0 & 0 & 0 & -1 & 0 \\
0 & 0 & -1 & 0 & 0 & 0 & -1 & 0 & 0 & 0 & 3 & 0 \\
0 & 0 & 1 & 0 & 0 & 0 & -1 & 0 & 0 & 0 & -3 & 0 \\
0 & 0 & 3 & 0 & 0 & 0 & 1 & 0 & 0 & 0 & 1 & 0 \\
0 & 0 & 0 & -3 & 0 & 0 & 0 & 1 & 0 & 0 & 0 & -1 \\
0 & 0 & 0 & -1 & 0 & 0 & 0 & -1 & 0 & 0 & 0 & 3 \\
0 & 0 & 0 & 1 & 0 & 0 & 0 & -1 & 0 & 0 & 0 & -3 \\
0 & 0 & 0 & 3 & 0 & 0 & 0 & 1 & 0 & 0 & 0 & 1
\end{pmatrix}
$$

STEP 5: Square each of the Subject Sums obtained in STEP 3.

$$
\text{ONE12}\left(\left(\text{diag}\left(\overrightarrow{\left(\text{ONE7}\cdot(\text{ALGSUM}\cdot\text{ALGSUM})\right)}\right)^{\mathrm{T}}\right)\cdot\text{DB1}\right) = \begin{pmatrix} 21092 & 76 & 148 \end{pmatrix}
$$

The Math looks like:

$$
\text{ONE12}\left[\text{diag}\left(\left(\text{ONE7}\cdot\overrightarrow{\begin{pmatrix} 16 & 22 & 6 & 32 & 0 & 0 & 0 & 0 & 2 & 4 & 2 & 4 \\ 17 & 38 & 14 & 43 & 1 & 2 & 0 & -1 & -1 & -4 & -2 & 1 \\ 26 & 20 & 9 & 50 & 0 & -2 & 1 & 2 & 2 & 0 & 3 & 0 \\ 20 & 33 & 10 & 39 & 2 & 1 & 0 & 3 & 0 & 1 & 0 & 3 \\ 21 & 37 & 6 & 43 & -1 & -3 & 0 & -1 & -3 & -1 & 2 & 1 \\ 10 & 31 & 1 & 29 & 2 & 1 & -1 & 3 & 0 & -3 & -3 & 3 \\ 16 & 16 & 6 & 51 & 2 & 2 & -2 & -3 & 2 & 2 & 2 & -3 \end{pmatrix}}^2\cdot\text{ALGSUM}\right)^{\text{T}}\right)\right]\cdot\begin{pmatrix} 1 & 0 & 0 \\ 1 & 0 & 0 \\ 1 & 0 & 0 \\ 1 & 0 & 0 \\ 0 & 1 & 0 \\ 0 & 1 & 0 \\ 0 & 1 & 0 \\ 0 & 1 & 0 \\ 0 & 0 & 1 \\ 0 & 0 & 1 \\ 0 & 0 & 1 \\ 0 & 0 & 1 \end{pmatrix}
$$

Or

$$
\text{diag}\left(\left(\text{ONE7}\cdot\overrightarrow{(\text{ALGSUM}\cdot\text{ALGSUM})}\right)^{\text{T}}\right) = \begin{pmatrix} 2418 & 0 & 0 & 0 & 0 & 0 & 0 & 0 & 0 & 0 & 0 & 0 \\ 0 & 6003 & 0 & 0 & 0 & 0 & 0 & 0 & 0 & 0 & 0 & 0 \\ 0 & 0 & 486 & 0 & 0 & 0 & 0 & 0 & 0 & 0 & 0 & 0 \\ 0 & 0 & 0 & 12185 & 0 & 0 & 0 & 0 & 0 & 0 & 0 & 0 \\ 0 & 0 & 0 & 0 & 14 & 0 & 0 & 0 & 0 & 0 & 0 & 0 \\ 0 & 0 & 0 & 0 & 0 & 23 & 0 & 0 & 0 & 0 & 0 & 0 \\ 0 & 0 & 0 & 0 & 0 & 0 & 6 & 0 & 0 & 0 & 0 & 0 \\ 0 & 0 & 0 & 0 & 0 & 0 & 0 & 33 & 0 & 0 & 0 & 0 \\ 0 & 0 & 0 & 0 & 0 & 0 & 0 & 0 & 22 & 0 & 0 & 0 \\ 0 & 0 & 0 & 0 & 0 & 0 & 0 & 0 & 0 & 47 & 0 & 0 \\ 0 & 0 & 0 & 0 & 0 & 0 & 0 & 0 & 0 & 0 & 34 & 0 \\ 0 & 0 & 0 & 0 & 0 & 0 & 0 & 0 & 0 & 0 & 0 & 45 \end{pmatrix}
$$

$$
\text{ONE12}\begin{pmatrix} 2418 & 0 & 0 & 0 & 0 & 0 & 0 & 0 & 0 & 0 & 0 & 0 \\ 0 & 6003 & 0 & 0 & 0 & 0 & 0 & 0 & 0 & 0 & 0 & 0 \\ 0 & 0 & 486 & 0 & 0 & 0 & 0 & 0 & 0 & 0 & 0 & 0 \\ 0 & 0 & 0 & 12185 & 0 & 0 & 0 & 0 & 0 & 0 & 0 & 0 \\ 0 & 0 & 0 & 0 & 14 & 0 & 0 & 0 & 0 & 0 & 0 & 0 \\ 0 & 0 & 0 & 0 & 0 & 23 & 0 & 0 & 0 & 0 & 0 & 0 \\ 0 & 0 & 0 & 0 & 0 & 0 & 6 & 0 & 0 & 0 & 0 & 0 \\ 0 & 0 & 0 & 0 & 0 & 0 & 0 & 33 & 0 & 0 & 0 & 0 \\ 0 & 0 & 0 & 0 & 0 & 0 & 0 & 0 & 22 & 0 & 0 & 0 \\ 0 & 0 & 0 & 0 & 0 & 0 & 0 & 0 & 0 & 47 & 0 & 0 \\ 0 & 0 & 0 & 0 & 0 & 0 & 0 & 0 & 0 & 0 & 34 & 0 \\ 0 & 0 & 0 & 0 & 0 & 0 & 0 & 0 & 0 & 0 & 0 & 45 \end{pmatrix}\cdot\begin{pmatrix} 1 & 0 & 0 \\ 1 & 0 & 0 \\ 1 & 0 & 0 \\ 1 & 0 & 0 \\ 0 & 1 & 0 \\ 0 & 1 & 0 \\ 0 & 1 & 0 \\ 0 & 1 & 0 \\ 0 & 0 & 1 \\ 0 & 0 & 1 \\ 0 & 0 & 1 \\ 0 & 0 & 1 \end{pmatrix} = \begin{pmatrix} 21092 & 76 & 148 \end{pmatrix}
$$

STEP 6:   Add the Group Sums STEP 4) to get the overall Grand Sum:

GRANDSUM := ONE7·ALGSUM·DB1

$$\text{GRANDSUM} = \begin{pmatrix} 662 & 8 & 14 \end{pmatrix}$$

And the Math looks like:

$$\text{ONE7} = \begin{pmatrix} 1 & 1 & 1 & 1 & 1 & 1 & 1 \end{pmatrix}$$

$$\begin{pmatrix} 1 & 1 & 1 & 1 & 1 & 1 & 1 \end{pmatrix} \cdot \begin{pmatrix} 16 & 22 & 6 & 32 & 0 & 0 & 0 & 0 & 2 & 4 & 2 & 4 \\ 17 & 38 & 14 & 43 & 1 & 2 & 0 & -1 & -1 & -4 & -2 & 1 \\ 26 & 20 & 9 & 50 & 0 & -2 & 1 & 2 & 2 & 0 & 3 & 0 \\ 20 & 33 & 10 & 39 & 2 & 1 & 0 & 3 & 0 & 1 & 0 & 3 \\ 21 & 37 & 6 & 43 & -3 & -3 & 0 & -1 & -3 & -1 & 2 & 1 \\ 10 & 31 & 1 & 29 & 2 & 1 & -1 & 3 & 0 & -3 & -3 & 3 \\ 16 & 16 & 6 & 51 & 2 & 2 & -2 & -3 & 2 & 2 & 2 & -3 \end{pmatrix} \cdot \begin{pmatrix} 1 & 0 & 0 \\ 1 & 0 & 0 \\ 1 & 0 & 0 \\ 1 & 0 & 0 \\ 0 & 1 & 0 \\ 0 & 1 & 0 \\ 0 & 1 & 0 \\ 0 & 1 & 0 \\ 0 & 0 & 1 \\ 0 & 0 & 1 \\ 0 & 0 & 1 \\ 0 & 0 & 1 \end{pmatrix} = \begin{pmatrix} 662 & 8 & 14 \end{pmatrix}$$

STEP 7:   Square each of the Linear, Quadratic and Cubic coefficients and sum the squared values. Remember, we must keep the squared totals of each of the coefficients separate, so again we must use the Half-Multiplier Operator.

PROGRAM: (ONE12) * (LCQ^T x LCQ^T o DB1) = [LCQ^T]^{o2} x [DB1]

or

$$\text{SUMLQC} := \text{LQC} \left(\text{diag} \left(\text{LQC}^T \right) \text{DB1} \right)$$

$$\text{SUMLQC} = \begin{pmatrix} 20 & 4 & 20 \end{pmatrix}$$

The Math looks like:

$$\mathrm{diag}\left(LQC^T\right) = \begin{pmatrix} -3 & 0 & 0 & 0 & 0 & 0 & 0 & 0 & 0 & 0 & 0 & 0 \\ 0 & -1 & 0 & 0 & 0 & 0 & 0 & 0 & 0 & 0 & 0 & 0 \\ 0 & 0 & 1 & 0 & 0 & 0 & 0 & 0 & 0 & 0 & 0 & 0 \\ 0 & 0 & 0 & 3 & 0 & 0 & 0 & 0 & 0 & 0 & 0 & 0 \\ 0 & 0 & 0 & 0 & 1 & 0 & 0 & 0 & 0 & 0 & 0 & 0 \\ 0 & 0 & 0 & 0 & 0 & -1 & 0 & 0 & 0 & 0 & 0 & 0 \\ 0 & 0 & 0 & 0 & 0 & 0 & -1 & 0 & 0 & 0 & 0 & 0 \\ 0 & 0 & 0 & 0 & 0 & 0 & 0 & 1 & 0 & 0 & 0 & 0 \\ 0 & 0 & 0 & 0 & 0 & 0 & 0 & 0 & -1 & 0 & 0 & 0 \\ 0 & 0 & 0 & 0 & 0 & 0 & 0 & 0 & 0 & 3 & 0 & 0 \\ 0 & 0 & 0 & 0 & 0 & 0 & 0 & 0 & 0 & 0 & -3 & 0 \\ 0 & 0 & 0 & 0 & 0 & 0 & 0 & 0 & 0 & 0 & 0 & 1 \end{pmatrix}$$

$$LQC = \begin{pmatrix} -3 & -1 & 1 & 3 & 1 & -1 & -1 & 1 & -1 & 3 & -3 & 1 \end{pmatrix}$$

$$\begin{pmatrix} -3 & -1 & 1 & 3 & 1 & -1 & -1 & 1 & -1 & 3 & -3 & 1 \end{pmatrix} \cdot \left[\begin{pmatrix} -3 & 0 & 0 & 0 & 0 & 0 & 0 & 0 & 0 & 0 & 0 & 0 \\ 0 & -1 & 0 & 0 & 0 & 0 & 0 & 0 & 0 & 0 & 0 & 0 \\ 0 & 0 & 1 & 0 & 0 & 0 & 0 & 0 & 0 & 0 & 0 & 0 \\ 0 & 0 & 0 & 3 & 0 & 0 & 0 & 0 & 0 & 0 & 0 & 0 \\ 0 & 0 & 0 & 0 & 1 & 0 & 0 & 0 & 0 & 0 & 0 & 0 \\ 0 & 0 & 0 & 0 & 0 & -1 & 0 & 0 & 0 & 0 & 0 & 0 \\ 0 & 0 & 0 & 0 & 0 & 0 & -1 & 0 & 0 & 0 & 0 & 0 \\ 0 & 0 & 0 & 0 & 0 & 0 & 0 & 1 & 0 & 0 & 0 & 0 \\ 0 & 0 & 0 & 0 & 0 & 0 & 0 & 0 & -1 & 0 & 0 & 0 \\ 0 & 0 & 0 & 0 & 0 & 0 & 0 & 0 & 0 & 3 & 0 & 0 \\ 0 & 0 & 0 & 0 & 0 & 0 & 0 & 0 & 0 & 0 & -3 & 0 \\ 0 & 0 & 0 & 0 & 0 & 0 & 0 & 0 & 0 & 0 & 0 & 1 \end{pmatrix} \cdot \begin{pmatrix} 1 & 0 & 0 \\ 1 & 0 & 0 \\ 1 & 0 & 0 \\ 1 & 0 & 0 \\ 0 & 1 & 0 \\ 0 & 1 & 0 \\ 0 & 1 & 0 \\ 0 & 1 & 0 \\ 0 & 0 & 1 \\ 0 & 0 & 1 \\ 0 & 0 & 1 \\ 0 & 0 & 1 \end{pmatrix}\right] = \begin{pmatrix} 20 & 4 & 20 \end{pmatrix}$$

Or this is LCQ*(LCQ o DB1) in Half-Multiplier notation. This would look like (in Half-Multiplier notation); We must perform the parenthesis first.

$$\left(-3 \;\; -1 \;\; 1 \;\; 3 \;\; 1 \;\; -1 \;\; -1 \;\; 1 \;\; -1 \;\; 3 \;\; -3 \;\; 1\right) \cdot \left[\begin{pmatrix} -3 \\ -1 \\ 1 \\ 3 \\ 1 \\ -1 \\ -1 \\ 1 \\ -1 \\ 3 \\ -3 \\ 1 \end{pmatrix} \cdot o \cdot \begin{pmatrix} 1 & 0 & 0 \\ 1 & 0 & 0 \\ 1 & 0 & 0 \\ 1 & 0 & 0 \\ 0 & 1 & 0 \\ 0 & 1 & 0 \\ 0 & 1 & 0 \\ 0 & 1 & 0 \\ 0 & 0 & 1 \\ 0 & 0 & 1 \\ 0 & 0 & 1 \\ 0 & 0 & 1 \end{pmatrix}\right] = \left(20 \;\; 4 \;\; 20\right)$$

Vastly simpler and with less memory.

STEP 8: Computation of the Within Subjects Linear, Quadratic
and Cubic components. Divide the overall Sum of the Squared
values (STEP 5) by the values of STEP 7.

SSwLQC:= SUMLQC

$$\mathrm{ONE12}\left(\left(\mathrm{diag}\left(\left(\mathrm{ONE7}\cdot\overrightarrow{(\mathrm{ALGSUM}\cdot\mathrm{ALGSUM})}\right)^{\mathrm{T}}\right)\cdot\mathrm{DB1}\right)\right)\cdot\left[\left((\mathrm{LQC}\;\cdot\left(\mathrm{diag}\left(\mathrm{LQC}^{\mathrm{T}}\right)\mathrm{DB1}\right)\right)\right]^{-1} =$$

Since

$$\mathrm{SUMLQC} := \mathrm{LQC}\cdot\left(\mathrm{diag}\left(\mathrm{LQC}^{\mathrm{T}}\right)\mathrm{DB1}\right)$$

MathCad cannot handle the equation, let's use this as a
substitution. We have:

NOTE: There are a lot of parentheses here, so be sure to keep
track of them. The first part of the equation is easy, the
second part we will derive:

$$\mathrm{SSwLQC} := \mathrm{ONE12}\left(\left(\mathrm{diag}\left(\left(\mathrm{ONE7}\cdot\overrightarrow{(\mathrm{ALGSUM}\cdot\mathrm{ALGSUM})}\right)^{\mathrm{T}}\right)\cdot\mathrm{DB1}\right)\right)\cdot\mathrm{SUMLQC}^{-1}$$

$$\mathrm{SUMLQC} = \left(20 \;\; 4 \;\; 20\right)$$

$$\text{SUMLQC} := \left[\text{LQC}\left(\text{diag}\left(\text{LQC}^\text{T}\right)\text{DB1}\right)\right]^\text{T}$$

$$\text{diag}\left(\left[\text{LQC}\left(\text{diag}\left(\text{LQC}^\text{T}\right)\text{DB1}\right)\right]^\text{T}\right) = \begin{pmatrix} 20 & 0 & 0 \\ 0 & 4 & 0 \\ 0 & 0 & 20 \end{pmatrix}$$

$$\left[\text{LQC}\left(\text{diag}\left(\text{LQC}^\text{T}\right)\text{DB1}\right)\right]^\text{T} = \begin{pmatrix} 20 \\ 4 \\ 20 \end{pmatrix}$$

$$\left(\text{diag}\left(\left[\text{LQC}\left(\text{diag}\left(\text{LQC}^\text{T}\right)\text{DB1}\right)\right]^\text{T}\right)^{-1}\right) = \begin{pmatrix} 0.05 & 0 & 0 \\ 0 & 0.25 & 0 \\ 0 & 0 & 0.05 \end{pmatrix}$$

Now we have the second part of the equation and it becomes:

$$\text{SSwLQC} := \text{ONE12}\left(\left(\text{diag}\left(\left(\text{ONE7}\cdot\overrightarrow{(\text{ALGSUM}\cdot\text{ALGSUM})}\right)^\text{T}\right)\cdot\text{DB1}\right)\right)\cdot\text{diag}\left(\left[\text{LQC}\left(\text{diag}\left(\text{LQC}^\text{T}\right)\text{DB1}\right)\right]^\text{T}\right)^{-1}$$

$$\text{SSwLQC} = \begin{pmatrix} 1054.6 & 19 & 7.4 \end{pmatrix}$$

Using the Half-Multiplier Operator we have:

$$\left[\text{ONE12}\left(\left(\text{diag}\left(\left(\text{ONE7}\cdot\overrightarrow{(\text{ALGSUM}\cdot\text{ALGSUM})}\right)\right)^\text{T}\right)\cdot\text{DB1}\right)\right]^\text{T} \cdot_\text{o} \cdot\left(\left[\text{LQC}\left(\text{diag}\left(\text{LQC}^\text{T}\right)\text{DB1}\right)\right]^\text{T}\right)^{-1}$$

$$\left[\text{ONE12}\left(\left(\text{diag}\left(\left(\text{ONE7}\cdot\overrightarrow{(\text{ALGSUM}\cdot\text{ALGSUM})}\right)\right)^\text{T}\right)\cdot\text{DB1}\right)\right]^\text{T} = \begin{pmatrix} 21092 \\ 76 \\ 148 \end{pmatrix}$$

$$\overrightarrow{\left[\begin{pmatrix} 21092 \\ 76 \\ 148 \end{pmatrix} \cdot \begin{pmatrix} 0.05 \\ 0.25 \\ 0.05 \end{pmatrix}\right]} = \begin{pmatrix} 1054.6 \\ 19 \\ 7.4 \end{pmatrix} \qquad \left(\left[\text{LQC}\left(\text{diag}\left(\text{LQC}^\text{T}\right)\text{DB1}\right)\right]^\text{T}\right)^{-1} = \begin{pmatrix} 0.05 \\ 0.25 \\ 0.05 \end{pmatrix}$$

STEP 9: Computation of the Repeated Measures (Trials) effect for all Trend components.

$$\text{NtrLCQ} := 28$$

In Half-Multiplier mode we have:

$$\text{SStrLCQ} := \left(\frac{1}{\text{NtrLCQ}}\right) \cdot \left(\text{GRANDSUM}^T\right)^{o2} \cdot o \cdot \left(\text{SUMLQC}^T\right)^{-1}$$

This is actually quite simple as we have already computed the values:

$$\text{SStrLCQ} := \left(\frac{1}{28}\right) \cdot \overrightarrow{\left[\begin{pmatrix}662\\8\\14\end{pmatrix} \cdot \begin{pmatrix}662\\8\\14\end{pmatrix} \cdot \begin{pmatrix}20\\4\\20\end{pmatrix}^{-1}\right]}$$

$$\text{SStrLCQ} = \begin{pmatrix}782.5786\\0.5714\\0.35\end{pmatrix}$$

The equation is:

$$\left(\frac{1}{28}\right) \cdot \overrightarrow{\left[\left(\overline{\text{ONE7·ALGSUM·DB1}^T}\right)\left(\text{ONE7·ALGSUM·DB1}^T\right)\right]^T \cdot \left(\text{||||}_{\text{LQC}}\left(\text{diag}\left([((\text{LQC}))]^T\right)\text{DB1}\right)\text{||||}^T\right)^{-1}} = \blacksquare$$

Again MathCad cannot compute this problem, so we make the following substitution:

ONE7ALGSUMDB1:= ONE7·ALGSUM·DB1

Then we have:

$$\left(\frac{1}{28}\right) \cdot \left[\left(\overline{\text{ONE7ALGSUMDB1}}\right)\left(\overline{\text{ONE7ALGSUMDB1}}\right)\right] \cdot \left(\text{||||}_{\text{LQC}}\left(\text{diag}\left([((\text{LQC}))]^T\right)\text{DB1}\right)\text{||||}^T\right)^{-1}$$

We need another substitution, in STEP 7, we computed the Sum of Squares of the Trend components as SUMLQC, let's use this.

$$\text{SUMLQC} := \text{LQC}\left(\text{diag}\left(\text{LQC}^T\right)\text{DB1}\right)$$

$$\left(\text{SUMLQC}^T\right)^{-1} = \begin{pmatrix}0.05\\0.25\\0.05\end{pmatrix}$$

$$\overrightarrow{\left[\left(\text{ONE7ALGSUMDB}^T\right)\left(\text{ONE7ALGSUMDB}^T\right)\left(\text{SUMLQC}^T\right)^{-1}\right]} = \begin{pmatrix} 21912.2 \\ 16 \\ 9.8 \end{pmatrix}$$

$$\text{SStrLCQ} := \left(\frac{1}{28}\right) \cdot \overrightarrow{\left[\left(\text{ONE7ALGSUMDB}^T\right)\left(\text{ONE7ALGSUMDB}^T\right)\left(\text{SUMLQC}^T\right)^{-1}\right]}$$

$$\text{SStrLCQ} = \begin{pmatrix} 782.5786 \\ 0.5714 \\ 0.35 \end{pmatrix}$$

This is the solution we want.

STEP 10: Computation of Trials by First Experimental Factor
effect for all Trend components.

$\text{Ntrx1} := 14$

The number of scores upon which each sum is based (in
Half-Multiplier notation) is:

$$\text{SStrx1LQC} := \left(\frac{1}{\text{Ntrx1}}\right) \cdot \left[\left[(\text{GROUPSUMDB12}) \cdot \text{diag}(\text{GROUPSUMDB12})^T \cdot \text{DB13}\right]^T \cdot o \cdot \left(\text{SUMLQC}^T\right)^{-1} - \text{SStrLQC}\right.$$

$\text{GROUPSUM} := \text{ONE7} \cdot A \cdot \text{DBLQC}$

Let's compute the first part of the equation:

$$\left(\frac{1}{\text{Ntrx1}}\right) \cdot \left[\left[(\text{GROUPSUMDB12}) \cdot \left(\text{diag}(\text{GROUPSUMDB12})^T\right) \text{DB13}\right]^T\right] = \blacksquare$$

$\text{GROUPSUMDB12} := \text{GROUPSUMDB12}$

The equation now becomes:

$$\left(\frac{1}{\text{Ntrx1}}\right) \cdot \left[\left[(\text{GROUPSUMDB12}) \cdot \left(\text{diag}\left(\text{GROUPSUMDB12}^T\right)\right)\text{DB13}\right]^T\right] = \begin{pmatrix} 15660.7143 \\ 3.5714 \\ 12.1429 \end{pmatrix}$$

We must diagonalize to partition for the Nested Array for the Half-Multiplication. We do this as follows: Note: we leave out the term (1/Ntrx1) until we know when to place it into the equation.

$$\left[(\text{GROUPSUMDB12})\left(\text{diag}\left(\text{GROUPSUMDB12}^T\right)\right)\text{DB13}\right]^T = \begin{pmatrix} 219250 \\ 50 \\ 170 \end{pmatrix}$$

or diagonalizing:

$$\text{diag}\left(\left[(\text{GROUPSUMDB12}\cdot\left(\text{diag}\left(\text{GROUPSUMDB12}^T\right)\right)\text{DB13}\right]^T\right) = \begin{pmatrix} 219250 & 0 & 0 \\ 0 & 50 & 0 \\ 0 & 0 & 170 \end{pmatrix}$$

Now we must diagonalize the inverse of the Sum of Squares of the Linear, Quadratic and Cubic terms. This is done by:

$$\text{diag}\left[\left(\text{SUMLQC}^T\right)^{-1}\right] = \begin{pmatrix} 0.05 & 0 & 0 \\ 0 & 0.25 & 0 \\ 0 & 0 & 0.05 \end{pmatrix}$$

We now connect the two terms and divide by the constant:

$$\text{SStrx1LQC} := \left(\frac{1}{\text{Ntrx1}}\right)\cdot\left[\text{diag}\left(\left[(\text{GROUPSUMDB12})\left(\text{diag}\left(\text{GROUPSUMDB12}^T\right)\right)\cdot\text{DB13}\right]^T\right)\cdot\text{diag}\left[\left(\text{SUMLQC}^T\right)^{-1}\right]\right]$$

$$\left(\frac{1}{\text{Ntrx1}}\right)\cdot\left[\text{diag}\left(\left[(\text{GROUPSUMDB12}\cdot\left(\text{diag}\left(\text{GROUPSUMDB12}^T\right)\right)\text{DB13}\right]^T\right)\cdot\text{diag}\left[\left(\text{SUMLQC}^T\right)^{-1}\right]\right] = \begin{pmatrix} L & Q & C \\ 783.0357 & 0 & 0 \\ 0 & 0.8929 & 0 \\ 0 & 0 & 0.6071 \end{pmatrix}$$

Finally we need to subtract the correction factor SStrLQC, but we must change its form to match the solution above.

$$\text{diag}(\text{SStrLCQ}) = \begin{pmatrix} 782.5786 & 0 & 0 \\ 0 & 0.5714 & 0 \\ 0 & 0 & 0.35 \end{pmatrix}$$

So the final form of the equation becomes:

$$\text{SStrx1LQC} := \left(\frac{1}{\text{Ntrx1}}\right)\cdot\left[\text{diag}\left(\left[(\text{GROUPSUMDB12}\cdot\left(\text{diag}\left(\text{GROUPSUMDB12}^T\right)\right)\text{DB13}\right]^T\right)\cdot\text{diag}\left[\left(\text{SUMLQC}^T\right)^{-1}\right]\right] - \text{diag}((\text{SStrLCQ}))$$

$$
\text{SStrx1LQC} = \begin{pmatrix} \overset{L}{0.4571} & \overset{Q}{0} & \overset{C}{0} \\ 0 & 0.3214 & 0 \\ 0 & 0 & 0.2571 \end{pmatrix} \qquad \text{ONE3} := \begin{pmatrix} 1 & 1 & 1 \end{pmatrix}
$$

Which are the values we want using Half-Multiplication in a single equation (or a single line of computer programming, if you wish). Remember; keep close watch on the parenthesis. One slip and the solutions are skewed. One misplacement and we'll still get a solution, but not correct.

STEP 11: Computation of Trials by Second Experimental Factor effect for all Trend components. Repeat STEP10) except use the Algebraic Sum of Subjects that were the same in terms of the Second Experimental Condition, disregarding the effects of the First Experimental Condition.

$\text{Ntrx2} := 14$

Now this is complicated, let's see if we can simplify it.

$$
\text{SStrx2LQC} := \left(\tfrac{1}{\text{Ntrx2}}\right) \cdot \left[\text{diag}\left[\left((\text{ONE7 ALGSUM DB2}) \cdot \left(\text{diag}(\text{ONE7 ALGSUM DB2})^T \right) \text{DB2} \right]^T \right) \cdot \text{diag}\left[\left(\text{SUMLQC}^T \right)^{-1} \right] \right] - \text{diag}\left(\text{SStrLQC}^T \right)
$$

The parts of this equation are:

$$
\text{ONE7·ALGSUM·DB2} = \begin{pmatrix} 178 & 484 & 4 & 4 & 6 & 8 \end{pmatrix}
$$

$$
\text{ALGSUM·DB2} = \begin{pmatrix} 22 & 54 & 0 & 0 & 4 & 8 \\ 31 & 81 & 1 & 1 & -3 & -3 \\ 35 & 70 & 1 & 0 & 5 & 0 \\ 30 & 72 & 2 & 4 & 0 & 4 \\ 27 & 80 & -1 & -4 & -1 & 0 \\ 11 & 60 & 1 & 4 & -3 & 0 \\ 22 & 67 & 0 & -1 & 4 & -1 \end{pmatrix}
$$

$$
\text{ONE7ALGSUMDB2} := \text{ONE7·ALGSUM·DB2}
$$

$$
(\text{ONE7·ALGSUM·DB2}) \cdot \left(\text{diag}\left(\text{ONE7ALGSUMDB2}^T \right) \text{DB13} \right) = \begin{pmatrix} 265940 & 32 & 100 \end{pmatrix}
$$

$$\left(\frac{1}{14}\right) \cdot \overline{\left[(\text{ONE7ALGSUMDB2}) \cdot \left(\text{diag}\left(\text{ONE7ALGSUMDB2}^{\text{T}}\right) \cdot \text{DB13}\right) \cdot \text{diag}\left[\left(\text{SUMLQC}^{\text{T}}\right)^{-1}\right]\right]} =,$$

The intermediate answer is:

$$\left(\frac{1}{14}\right) \cdot \left(\overline{\left[\left(265940 \quad 32 \quad 100\right) \cdot \left(20 \quad 4 \quad 20\right)^{-1}\right]}\right) = \left(949.7857 \quad 0.5714 \quad 0.3571\right)$$

The Correction Factor is given by:

$$(\text{SStrLCQ}) = \begin{pmatrix} 782.5786 \\ 0.5714 \\ 0.35 \end{pmatrix}$$

In short form, the solution is:

$$\left(\frac{1}{14}\right) \cdot \left(\overline{\left[\left(265940 \quad 32 \quad 100\right) \cdot \left(20 \quad 4 \quad 20\right)^{-1}\right]}\right) - \text{SStrLCQ}^{\text{T}} = \left(167.2071 \quad 0 \quad 0.0071\right)$$

This is where we want to be. Now we will get MathCad to solve this equation as a single equation.

Here we have the values we need to work with.

$$(\text{ONE7ALGSUMDB2}) \cdot \left(\text{diag}\left(\text{ONE7ALGSUMDB2}^{\text{T}}\right) \cdot \text{DB13}\right) = \left(265940 \quad 32 \quad 100\right)$$

$$\overline{\text{SUMLQC}^{-1}} = \left(0.05 \quad 0.25 \quad 0.05\right) \qquad\qquad \text{SStrLCQ}^{\text{T}} = \left(782.5786 \quad 0.5714 \quad 0.35\right)$$

We will solve this differently as Row Matrices. The first equation is:

$$\overline{(\text{ONE7ALGSUMDB2} \cdot \text{ONE7ALGSUMDB2})} \cdot \text{DB13} = \left(265940 \quad 32 \quad 100\right)$$

We will re-write it as:

$$\text{ONE7ALGSUMDB2ONE7ALGSUMDB2DB13} := \overline{(\text{ONE7ALGSUMDB2} \cdot \text{ONE7ALGSUMDB2})} \cdot \text{DB13}$$

Now we will divide by the Sum of Squares of the Linear, Quadratic and Cubic Components SUMLQC.

$$\overline{\left(\text{ONE7ALGSUMDB2ONE7ALGSUMDB2DB}\cdot\overrightarrow{\text{ISUMLQC}}^{-1}\right)} = \begin{pmatrix} 13297 & 8 & 5 \end{pmatrix}$$

Now we will divide by 14:

$$\frac{1}{14}\cdot\overline{\left(\text{ONE7ALGSUMDB2ONE7ALGSUMDB2DB}\cdot\overrightarrow{\text{ISUMLQC}}^{-1}\right)}$$

And now the equation in it's final form.

$$\text{SStrx2LQC} := \left(\frac{1}{14}\right)\cdot\overline{\left(\text{ONE7ALGSUMDB2ONE7ALGSUMDB2DB}\cdot\overrightarrow{\text{ISUMLQC}}^{-1}\right)} - \text{SStrLCQ}^{T}$$

$$\left(\frac{1}{14}\right)\cdot\overline{\left(\text{ONE7ALGSUMDB2ONE7ALGSUMDB2DB}\cdot\overrightarrow{\text{ISUMLQC}}^{-1}\right)} - \text{SStrLCQ}^{T} = \begin{pmatrix} 167.2071 & 0 & 0.0071 \end{pmatrix}$$

This gives us the solution we want.

The Math looks like:

CHK ALT

$$\left(\frac{1}{14}\right)\cdot\left[\begin{pmatrix} 178 & 484 & 4 & 4 & 6 & 8 \end{pmatrix}\cdot\left[\begin{pmatrix} 178 & 0 & 0 & 0 & 0 & 0 \\ 0 & 484 & 0 & 0 & 0 & 0 \\ 0 & 0 & 4 & 0 & 0 & 0 \\ 0 & 0 & 0 & 4 & 0 & 0 \\ 0 & 0 & 0 & 0 & 6 & 0 \\ 0 & 0 & 0 & 0 & 0 & 8 \end{pmatrix}\begin{pmatrix} 1 & 0 & 0 \\ 1 & 0 & 0 \\ 0 & 1 & 0 \\ 0 & 1 & 0 \\ 0 & 0 & 1 \\ 0 & 0 & 1 \end{pmatrix}\right]\cdot\begin{pmatrix} 0.05 & 0 & 0 \\ 0 & 0.25 & 0 \\ 0 & 0 & 0.05 \end{pmatrix}\right] = \begin{pmatrix} 949.7857 & 0.5714 & 0.3571 \end{pmatrix}$$

$$\left(\frac{1}{14}\right)\cdot\left[\begin{pmatrix} 178 & 484 & 4 & 4 & 6 & 8 \end{pmatrix}\cdot\left[\begin{pmatrix} 178 & 0 & 0 & 0 & 0 & 0 \\ 0 & 484 & 0 & 0 & 0 & 0 \\ 0 & 0 & 4 & 0 & 0 & 0 \\ 0 & 0 & 0 & 4 & 0 & 0 \\ 0 & 0 & 0 & 0 & 6 & 0 \\ 0 & 0 & 0 & 0 & 0 & 8 \end{pmatrix}\begin{pmatrix} 1 & 0 & 0 \\ 1 & 0 & 0 \\ 0 & 1 & 0 \\ 0 & 1 & 0 \\ 0 & 0 & 1 \\ 0 & 0 & 1 \end{pmatrix}\right]\begin{pmatrix} 0.05 & 0 & 0 \\ 0 & 0.25 & 0 \\ 0 & 0 & 0.05 \end{pmatrix}\right] - \text{SStrLCQ}^{T} = \begin{pmatrix} 167.2071 & 0 & 0.0071 \end{pmatrix}$$

STEP 12: Computation of the Repeated Measures (Trials) by the
First and Second Experimental conditions. SStrx1x2

$\text{Ntrx1x2} := 7$

$$\text{SStrx1x2} := \left(\frac{1}{\text{Ntrx1x2}}\right) \cdot \left[\text{GROUPSUM}\left(\text{diag}\left(\text{GROUPSUM}^T\right)\text{DB1}\right)\overrightarrow{\text{SUMLQC}^{-1}}\right]\blacksquare$$

$$\left(\text{GROUPSUM}\right) \cdot \left(\text{diag}\left(\text{GROUPSUM}^T\right)\text{DB1}\right) = \left(139758 \quad 50 \quad 102\right)$$

$$\left[\left(\frac{1}{\text{Ntrx1x2}}\right) \cdot \left[\left(\text{GROUPSUM}\right) \cdot \left(\text{diag}\left(\text{GROUPSUM}^T\right)\text{DB1}\right)\right] \cdot \overrightarrow{\text{SUMLQC}^{-1}}\right] = \blacksquare$$

MathCad just will not handle these equations at all, so
in re-defining the sub-equations I will try to keep things
recognizable.

To heck with it, we will just re-define the entire equation:

$$\text{GROUPSUMxdiagGROUPSUMtranxDB1} = \left(\text{GROUPSUM}\right) \cdot \left(\text{diag}\left(\text{GROUPSUM}^T\right)\text{DB1}\right)$$

This equation now equals:

$$\left[\left(\text{GROUPSUMxdiagGROUPSUMtranxDB1}\right)\overrightarrow{\text{SUMLQC}^{-1}}\right] = \left(6987.9 \quad 12.5 \quad 5.1\right)$$

And dividing by 7

$$\left(\frac{1}{7}\right) \cdot \left[\left(\text{GROUPSUMxdiagGROUPSUMtranxDB1}\right)\overrightarrow{\text{SUMLQC}^{-1}}\right] = \left(998.2714 \quad 1.7857 \quad 0.7286\right)$$

And finally we subtract the Correction Factors.

$$SStrx1x2:=\left(\frac{1}{Ntrx1x2}\right)\cdot\overrightarrow{\left[(GROUPSUMxdiagGROUPSUMtranxDB)SUMLQC^{-1}\right]}-SStrLCQ^T-SStrx2LQC$$

Where　　$SStrLCQ^T=\begin{pmatrix}782.5786 & 0.5714 & 0.35\end{pmatrix}$　　$SStrx2LQC=\begin{pmatrix}167.2071 & 0 & 0.0071\end{pmatrix}$

And the Interactive solution is:

$SStrx1x2=\begin{pmatrix}48.4857 & 1.2143 & 0.3714\end{pmatrix}$

The Math Looks like:

$GROUPSUM=\begin{pmatrix}126 & 197 & 52 & 287 & 6 & 1 & -2 & 3 & 2 & -1 & 4 & 9\end{pmatrix}$　　$\overrightarrow{SUMLQC^{-1}}=\begin{pmatrix}0.05 & 0.25 & 0.05\end{pmatrix}$

The Math is:

$$\left(\frac{1}{7}\right)\cdot\begin{bmatrix}\begin{pmatrix}126 & 197 & 52 & 287 & 6 & 1 & -2 & 3 & 2 & -1 & 4 & 9\end{pmatrix}\cdot\begin{pmatrix}126 & 0 & 0 & 0 & 0 & 0 & 0 & 0 & 0 & 0 & 0 & 0\\0 & 197 & 0 & 0 & 0 & 0 & 0 & 0 & 0 & 0 & 0 & 0\\0 & 0 & 52 & 0 & 0 & 0 & 0 & 0 & 0 & 0 & 0 & 0\\0 & 0 & 0 & 287 & 0 & 0 & 0 & 0 & 0 & 0 & 0 & 0\\0 & 0 & 0 & 0 & 6 & 0 & 0 & 0 & 0 & 0 & 0 & 0\\0 & 0 & 0 & 0 & 0 & 1 & 0 & 0 & 0 & 0 & 0 & 0\\0 & 0 & 0 & 0 & 0 & 0 & -2 & 0 & 0 & 0 & 0 & 0\\0 & 0 & 0 & 0 & 0 & 0 & 0 & 3 & 0 & 0 & 0 & 0\\0 & 0 & 0 & 0 & 0 & 0 & 0 & 0 & 2 & 0 & 0 & 0\\0 & 0 & 0 & 0 & 0 & 0 & 0 & 0 & 0 & -1 & 0 & 0\\0 & 0 & 0 & 0 & 0 & 0 & 0 & 0 & 0 & 0 & 4 & 0\\0 & 0 & 0 & 0 & 0 & 0 & 0 & 0 & 0 & 0 & 0 & 9\end{pmatrix}\begin{pmatrix}1 & 0 & 0\\1 & 0 & 0\\1 & 0 & 0\\1 & 0 & 0\\0 & 1 & 0\\0 & 1 & 0\\0 & 1 & 0\\0 & 1 & 0\\0 & 0 & 1\\0 & 0 & 1\\0 & 0 & 1\\0 & 0 & 1\end{pmatrix}\end{bmatrix}\cdot\begin{pmatrix}0.05 & 0.25 & 0.05\end{pmatrix}-SStrLCQ^T-SStrx2LQC$$

This equals　　　　$\overrightarrow{\left[\begin{pmatrix}19965.429 & 7.143 & 14.571\end{pmatrix}\cdot\begin{pmatrix}0.05 & 0.25 & 0.05\end{pmatrix}\right]}$

$\overrightarrow{\left[\begin{pmatrix}19965.429 & 7.143 & 14.571\end{pmatrix}\cdot\begin{pmatrix}0.05 & 0.25 & 0.05\end{pmatrix}\right]}-\begin{pmatrix}782.579 & 0.571 & 0.35\end{pmatrix}-\begin{pmatrix}167.207 & 0 & 0.007\end{pmatrix}=\begin{pmatrix}48.4855 & 1.2147 & 0.3716\end{pmatrix}$

STEP 13: Computation of the Linear, Quadratic and Cubic components error terms.

$SSerr:=SSwLQC-SStrLCQ^T-SStrx2LQC-SStrx1x2$　　$SStrx1x2=\begin{pmatrix}48.4857 & 1.2143 & 0.3714\end{pmatrix}$

$SStrx2LQC=\begin{pmatrix}167.2071 & 0 & 0.0071\end{pmatrix}$

$SSerr=\begin{pmatrix}56.3286 & 17.2143 & 6.6714\end{pmatrix}$　　　　$SStrLCQ^T=\begin{pmatrix}782.5786 & 0.5714 & 0.35\end{pmatrix}$

$SSwLQC=\begin{pmatrix}1054.6 & 19 & 7.4\end{pmatrix}$

STEP 14: Computation of the Degrees of Freedom df for the Linear,
Quadratic and Cubic components.

dfSSw = # Subjects in each Group
dfSStr = 1 (always)
dfSStrx1 = # of First Experimental Condition factors - 1 = 2-1
dfSStrx2 = # Second Experimental condition factors - 1 = 2-1
dfSStrx1x2 = dfSStrx1 x dfSStrx2 = 1x1=1
dfSSerr = dfSSw-dfSStr-dfSStrx1-dfSStrx2-dfSStrx1x2 = 28-1-1-1-1 = 24

Putting this in the df Matrix (NOTE: This general df Matrix is
good for all 3 Factor Trends of this type).

$$s := 28 \quad t := 1 \quad tx1 := 2 \quad tx2 := 2 \quad t1x2 := 1$$

$$df := \begin{bmatrix} s \\ t \\ tx1 - 1 \\ tx1 - 1 \\ t \cdot t \\ s - t - (tx1 - 1) - (tx2 - 1) - t \cdot t \end{bmatrix} \quad df = \begin{pmatrix} 28 \\ 1 \\ 1 \\ 1 \\ 1 \\ 24 \end{pmatrix}$$

STEP 15: Computation of the Mean Squares for the Linear,
Quadratic and Cubic components and the F ratio.

All the Trend Sums of Squares we have calculated are given in
the following Nested Array:

$$ss := \begin{pmatrix} SSwLQC \\ SStrLCQ^T \\ SStrx1LQCa \\ SStrx2LQC \\ SStrx1x2 \\ SSerr \end{pmatrix} \quad ss = \begin{bmatrix} \left(1054.6 \quad 19 \quad 7.4 \right) \\ \left(782.5786 \quad 0.5714 \quad 0.35 \right) \\ \left(0.4571 \quad 0.3214 \quad 0.2571 \right) \\ \left(167.2071 \quad 0 \quad 0.0071 \right) \\ \left(48.4857 \quad 1.2143 \quad 0.3714 \right) \\ \left(56.3286 \quad 17.2143 \quad 6.6714 \right) \end{bmatrix}$$

SStrx1LQCa:= ONE3 SStrx1LQC
SStrx1x2= $\left(48.4857 \quad 1.2143 \quad 0.3714 \right)$
SStrx2LQC= $\left(167.2071 \quad 0 \quad 0.0071 \right)$
SStrLCQ^T = $\left(782.5786 \quad 0.5714 \quad 0.35 \right)$
SSwLQC= $\left(1054.6 \quad 19 \quad 7.4 \right)$
SSerr = $\left(56.3286 \quad 17.2143 \quad 6.6714 \right)$

NOTE: We computed SStrx1LQC as a diagonal Matrix so we need to
change it to a Row Matrix for the final computation.

Since there are no computer programs in existence that can handle Nested Arrays, we must partition a single Matrix and act upon it as if it were Nested.

$$SS := \begin{pmatrix} 1054.6 & 19 & 7.4 \\ 782.5786 & .5714 & .35 \\ .4571 & .3214 & .2571 \\ 167.2071 & 0 & .0071 \\ 48.4857 & 1.2143 & .3714 \\ 56.3286 & 17.2143 & 6.6714 \end{pmatrix}$$

The Error term is given by:

$$\left(\frac{1}{24}\right) \cdot SSerr = \begin{pmatrix} 2.347 & 0.7173 & 0.278 \end{pmatrix}$$

But since we need this solution as a Column Matrix we make the following transformation:

$$ERROR := \left(\frac{1}{24}\right) \cdot SSerr^{T}$$

Now we need to Half-Multiply the three terms to get the F-Ratio. Since there is no such thing in programming to do what we wish to do, we must change the Column Matrices to Diagonal Matrices to accomplish the desired Product.

$$F := \left(\text{diag}\left(df^{-1}\right)\right) \cdot SS \cdot \left(\text{diag}\left(ERROR^{-1}\right)\right) \qquad THREE1 := ONE3^{T}$$

$$F = \begin{pmatrix} 16.0477 & 0.9461 & 0.9507 \\ 333.4345 & 0.7966 & 1.2591 \\ 0.1948 & 0.4481 & 0.9249 \\ 71.2422 & 0 & 0.0255 \\ 20.6584 & 1.693 & 1.3361 \\ 1 & 1 & 1 \end{pmatrix}$$

And the Math looks like:

$$\left(\text{diag}\left(df^{-1}\right)\right) = \begin{pmatrix} 0.0357 & 0 & 0 & 0 & 0 & 0 \\ 0 & 1 & 0 & 0 & 0 & 0 \\ 0 & 0 & 1 & 0 & 0 & 0 \\ 0 & 0 & 0 & 1 & 0 & 0 \\ 0 & 0 & 0 & 0 & 1 & 0 \\ 0 & 0 & 0 & 0 & 0 & 0.0417 \end{pmatrix} \qquad SS = \begin{pmatrix} 1054.6 & 19 & 7.4 \\ 782.5786 & 0.5714 & 0.35 \\ 0.4571 & 0.3214 & 0.2571 \\ 167.2071 & 0 & 0.0071 \\ 48.4857 & 1.2143 & 0.3714 \\ 56.3286 & 17.2143 & 6.6714 \end{pmatrix}$$

$$\text{diag}\left(ERROR^{-1}\right) = \begin{pmatrix} 0.4261 & 0 & 0 \\ 0 & 1.3942 & 0 \\ 0 & 0 & 3.5974 \end{pmatrix}$$

$$F := \begin{pmatrix} 0.0357 & 0 & 0 & 0 & 0 & 0 \\ 0 & 1 & 0 & 0 & 0 & 0 \\ 0 & 0 & 1 & 0 & 0 & 0 \\ 0 & 0 & 0 & 1 & 0 & 0 \\ 0 & 0 & 0 & 0 & 1 & 0 \\ 0 & 0 & 0 & 0 & 0 & 0.0417 \end{pmatrix} \cdot \begin{pmatrix} 1054.6 & 19 & 7.4 \\ 782.5786 & 0.5714 & 0.35 \\ 0.4571 & 0.3214 & 0.2571 \\ 167.2071 & 0 & 0.0071 \\ 48.4857 & 1.2143 & 0.3714 \\ 56.3286 & 17.2143 & 6.6714 \end{pmatrix} \cdot \begin{pmatrix} 0.4261 & 0 & 0 \\ 0 & 1.3942 & 0 \\ 0 & 0 & 3.5974 \end{pmatrix}$$

$$F = \begin{matrix} L & Q & C \end{matrix}$$
$$F = \begin{pmatrix} 16.0423 & 0.9457 & 0.9504 \\ 333.4567 & 0.7966 & 1.2591 \\ 0.1948 & 0.4481 & 0.9249 \\ 71.2469 & 0 & 0.0255 \\ 20.6598 & 1.693 & 1.3361 \\ 1.0009 & 1.0008 & 1.0008 \end{pmatrix}$$

Interpretation: The large numbers in the Linear Column mean the data tends to fall in a straight line rather than a Parabola (for Q) or a double humped curve (for C).

CHECK: If we sum the 3 components in the SS Matrix, they should equal the original Sum of Squares for the problem computed before we calculated the Trends.

$$SS \cdot THREE1 = \begin{pmatrix} 1081 \\ 783.5 \\ 1.0356 \\ 167.2142 \\ 50.0714 \\ 80.2143 \end{pmatrix} \qquad ORIGINAL := \begin{pmatrix} 1081 \\ 783.5 \\ 1.03 \\ 167.21 \\ 49.05 \\ 80.21 \end{pmatrix}$$

ONE14 := (1 1 1 1 1 1 1 1 1 1 1 1 1 1)

ONE14 := (1 1 1 1 1 1 1 1 1 1 1 1 1 1)   ONE2 := (1 1)   They both check.

TEST FOR DIFFERENCE BETWEEN VARIANCES OF TWO INDEPENDENT SAMPLES (TEST FOR HOMOGENEITY OF INDEPENDENT VARIANCES)

SECTION 3.1, COMPUTATIONAL HANDBOOK OF STATISTICS, PG 109-110.

There are two cases when this method is viable: When an experimental hypothesis is concerned with the variability of the samples; and when there is doubt concerning the requirement of equality of variances in a mean difference test.

EXAMPLE: Recall the data for the t-test for two independent means in section 1.6. Note the number of elements in the two lists are not equal, so we must use the b method of computing our statistics.

$$A := \begin{pmatrix} 107 & 109 \\ 96 & 94 \\ 88 & 127 \\ 131 & 76 \\ 109 & 115 \\ 84 & 121 \\ 79 & 87 \\ 105 & 92 \\ 108 & 91 \\ 92 & 98 \\ 96 & 104 \\ 101 & 96 \\ 0 & 110 \\ 0 & 108 \end{pmatrix}$$

$$\textbf{AVERAGE} := \mathbf{N} \cdot (\textbf{ONE14} \cdot \mathbf{A})^{\mathbf{T}}$$

$$\mathbf{N} := \begin{pmatrix} \dfrac{1}{12} & 0 \\ 0 & \dfrac{1}{14} \end{pmatrix}$$

$$\mathbf{N} \cdot (1196 \quad 1428)^{\mathbf{T}} = \begin{pmatrix} 99.667 \\ 102 \end{pmatrix}$$

$$\textbf{ONE14} \cdot \mathbf{A} \cdot \textbf{diag}(\textbf{AVERAGE}) = (119201.333 \quad 145656)$$

STEP 2: Sum the columns to get SUM1, find the Grand Sum (GS) and thge Correction Factor (CF).

N column 1:   **Nc1** := 12

N column 2:   **Nc2** := 14

GS := **ONE14** · **A**

GS = (1196   1428)

MEANS := **GS** · **N**

MEANS = (99.667   102)

The Math for the Grand Sum is:

$$
(1\ \ 1\ \ 1\ \ 1\ \ 1\ \ 1\ \ 1\ \ 1\ \ 1\ \ 1\ \ 1\ \ 1\ \ 1\ \ 1) \cdot \begin{pmatrix} 107 & 109 \\ 96 & 94 \\ 88 & 127 \\ 131 & 76 \\ 109 & 115 \\ 84 & 121 \\ 79 & 87 \\ 105 & 92 \\ 108 & 91 \\ 92 & 98 \\ 96 & 104 \\ 101 & 96 \\ 0 & 110 \\ 0 & 108 \end{pmatrix} \cdot \begin{pmatrix} 0.083 & 0 \\ 0 & 0.071 \end{pmatrix} = (\,99.268 \quad 101.388\,)
$$

These are the Means for the two groups of data.

$\mathbf{CF} := (\mathbf{MEANS} \cdot \overrightarrow{\mathbf{GS}})$

$\mathbf{CF} = (\,119201.333 \quad 145656\,)$

STEP 3:   Square each column in the Matrix separately and Sum:

$\mathbf{GSQsep} := \mathbf{ONE14} \cdot \left(\overrightarrow{\mathbf{A} \cdot \mathbf{A}}\right)$

$\mathbf{GSQsep} = (\,121318 \quad 148202\,)$

STEP 4:   Compute the Total Sum of Squares SSt.

$\mathbf{SSt} := \mathbf{GSQsep} - \mathbf{CF}$

$\mathbf{SSt} := \mathbf{ONE14} \cdot \left(\overleftrightarrow{\mathbf{A} \cdot \mathbf{A}}\right) - \overrightarrow{(\mathbf{MEANS} \cdot \mathbf{GS})}$

$\mathbf{SSt} = (\,2116.667 \quad 2546\,)$

STEP 4:   NOTE: All the above was actually STEP 1 for this
problem. Now we obtain the F ratio by placing the larger Variance
in SSt over the smaller Variance and dividing. The first element
is smallest, so we will do this in one step.

$$F := \dfrac{\overrightarrow{\log(\mathbf{SSt})} \cdot \begin{pmatrix} -1 \\ 1 \end{pmatrix}}{10}$$

$$\mathbf{F} = 1.203$$

If MathCad could handle it, the F Ratio would look like

$$F := \dfrac{\overrightarrow{\log\left[\mathrm{ONE14}\overrightarrow{(A \cdot A)} - (\mathrm{MEAN} \cdot \mathrm{GS})\right]} \cdot \begin{pmatrix} -1 \\ 1 \end{pmatrix}^{\blacksquare}}{10}$$

and we could write the Program in a single step.

STEP 5:   Computation of the Degrees of Freedom df:

df Variance 1 = 12-1 = 11 $\mathbf{N1} := 12$
df Variance 2 = 14-1 = 13 $\mathbf{N2} := 14$

$$\mathbf{df} := \begin{pmatrix} N1 - 1 \\ N2 - 1 \end{pmatrix}$$

STEP 6: The F Ratio has 2 Degrees of Freedom (11 and 13)
associated with it. From tabled F values in the Appendix,
with a df ratio of 13/11, the F Ratio needs a value > 7.63 to
be significant. Since 1.203 < 7.63, it is concluded that the
Variances are homogeneous. If the ratio would have been greater
than 7.63, the Ratios would be considered non-homogeneous.

TEST FOR DIFFERENCE BETWEEN VARIANCES OF TWO RELATED SAMPLES

(TEST FOR HOMOGENEITY OF RELATED VARIANCES)
SECTION 3.2, COMPUTATIONAL HANDBOOK OF STATISTICS, PG. 110-112.

There are two cases where it is desired to do a significance test of the difference between two related (correlated) variances: When an experimental hypothesis is concerned with the variability of the samples. When there is doubt concerning the requirement of equality of variances in a mean-difference test.

STEP 1: Table the Data as a Matrix, then perform the same 5 Steps we computed before calculating the F Ratio.

VARIABLES:

$$\text{MINPLUS1} := \begin{pmatrix} -1 \\ 1 \end{pmatrix} \qquad \text{TWO1} := \text{ONE2}^T$$

$$\text{ONE14} := (1 \quad 1 \quad 1 \quad 1 \quad 1 \quad 1 \quad 1 \quad 1 \quad 1 \quad 1 \quad 1 \quad 1 \quad 1 \quad 1) \qquad \text{FOURTEEN1} := \text{ONE14}^T$$

$$\text{ONE2} := (1 \quad 1) \qquad N := 14$$

$$
B := \begin{pmatrix}
89 & 94 \\
86 & 94 \\
96 & 101 \\
100 & 105 \\
94 & 100 \\
86 & 84 \\
81 & 81 \\
93 & 96 \\
87 & 90 \\
89 & 88 \\
110 & 115 \\
95 & 100 \\
107 & 110 \\
96 & 102
\end{pmatrix}
\quad
B^{\langle 1 \rangle} = \begin{pmatrix}
89 \\
86 \\
96 \\
100 \\
94 \\
86 \\
81 \\
93 \\
87 \\
89 \\
110 \\
95 \\
107 \\
96
\end{pmatrix}
\quad
B^{\langle 2 \rangle} = \begin{pmatrix}
94 \\
94 \\
101 \\
105 \\
100 \\
84 \\
81 \\
96 \\
90 \\
88 \\
115 \\
100 \\
110 \\
102
\end{pmatrix}
\quad
\text{DATAX1} := \begin{pmatrix}
89 & 1 \\
86 & 1 \\
96 & 1 \\
100 & 1 \\
94 & 1 \\
86 & 1 \\
81 & 1 \\
93 & 1 \\
87 & 1 \\
89 & 1 \\
110 & 1 \\
95 & 1 \\
107 & 1 \\
96 & 1
\end{pmatrix}
\quad
\text{DATAY1} := \begin{pmatrix}
94 & 1 \\
94 & 1 \\
101 & 1 \\
105 & 1 \\
100 & 1 \\
84 & 1 \\
81 & 1 \\
96 & 1 \\
90 & 1 \\
88 & 1 \\
115 & 1 \\
100 & 1 \\
110 & 1 \\
102 & 1
\end{pmatrix}
$$

STEP 2: Sum the Columns, find the Grand Sum GS and compute the Correction Factor.

SUM1 := **ONE14** ·**B**   **N** := 14

SUM1 = (1309   1360)

$$\text{MEAN} := \left(\frac{1}{N}\right) \cdot (\text{ONE14} \cdot \text{B})$$

$$\overrightarrow{\left(\frac{1}{N} \cdot \text{SUM1} \cdot \text{SUM1}\right)} = (\,122391.5 \quad 132114.286\,)$$

$$\text{MEAN} := \left(\frac{1}{N}\right) \cdot \text{SUM1}$$

MEAN = (93.5   97.143)

$$\text{CF} := \overrightarrow{(\text{MEAN} \cdot \text{SUM1})}$$

CF = (122391.5   132114.286)

STEP 3: Square each column in the Matrix separately and Sum to get the Grand Sum Squared.

$$\text{GSQ} := \text{ONE14} \cdot \overparen{\left(\overrightarrow{(\text{B} \cdot \text{B})}\right)}$$

GSQ = (123255   133304)

CF = (122391.5   132114.286)

STEP 4: Subtract the Correction Factor CF from GSQ

SSt := **GSQ** − **CF**

SSt = (863.5   1189.714)

$$
\mathbf{B} = \begin{pmatrix}
89 & 94 \\
86 & 94 \\
96 & 101 \\
100 & 105 \\
94 & 100 \\
86 & 84 \\
81 & 81 \\
93 & 96 \\
87 & 90 \\
89 & 88 \\
110 & 115 \\
95 & 100 \\
107 & 110 \\
96 & 102
\end{pmatrix}
$$

STEP 5: Ok, now we apply the Pearson Product-Moment Correlation to the data.

PEARSON PRODUCT–MOMENT CORRELATION

$$r = \frac{N_P \Sigma XY - (\Sigma X)(\Sigma Y)}{([N\Sigma X^2 - (\Sigma X)^2][N\Sigma Y^2 - (\Sigma Y)^2])^{1/2}}$$

Which can be translated to:

The Numerator r1 is given by:

$$\mathbf{r1} := (14) \cdot \left(\mathbf{B}^{\langle 1 \rangle^{\mathbf{T}}} \cdot \mathbf{B}^{\langle 2 \rangle}\right) - \left(\mathbf{B}^{\langle 1 \rangle^{\mathbf{T}}} \cdot \mathbf{FOURTEEN1}\right) \cdot \left(\mathbf{B}^{\langle 2 \rangle^{\mathbf{T}}} \cdot \mathbf{FOURTEEN1}\right)$$

$\mathbf{r1} = 13622$

$$\mathbf{ONE14} \cdot \mathbf{B} = (1309 \quad 1360) \qquad \mathbf{ONE0} := \begin{pmatrix} 1 \\ 0 \end{pmatrix} \qquad \mathbf{ZERO1} := \begin{pmatrix} 0 \\ 1 \end{pmatrix}$$

The Un-square-rooted Denominator r2 is given by:

$$\mathbf{r2} := \left(\mathbf{B}^{\langle 1 \rangle}\right)^{\mathbf{T}} \cdot \mathbf{DATAX1} \cdot \begin{bmatrix} \mathbf{N} \\ -(\mathbf{ONE14} \cdot \mathbf{B} \cdot \mathbf{ONE0}) \end{bmatrix} \cdot \left(\mathbf{B}^{\langle 2 \rangle}\right)^{\mathbf{T}} \cdot \mathbf{DATAY1} \cdot \begin{bmatrix} \mathbf{N} \\ -(\mathbf{ONE14} \cdot \mathbf{B} \cdot \mathbf{ZERO1}) \end{bmatrix}$$

Now we must take the Square Root of this equation.

$$\left[\left(\left(\mathbf{B}^{\langle 1 \rangle}\right)\right)^{\mathbf{T}} \cdot \mathbf{DATAX1} \cdot \begin{bmatrix} \mathbf{N} \\ -(\mathbf{ONE14} \cdot \mathbf{B} \cdot \mathbf{ONE0}) \end{bmatrix} \cdot \left(\left(\mathbf{B}^{\langle 2 \rangle}\right)\right)^{\mathbf{T}} \cdot \mathbf{DATAY1} \cdot \begin{bmatrix} \mathbf{N} \\ -(\mathbf{ONE14} \cdot \mathbf{B} \cdot \mathbf{ZERO1}) \end{bmatrix}\right]^{\frac{1}{2}} = 14189.94$$

or

$$\mathbf{r2} := \left[\left(\mathbf{B}^{\langle 1 \rangle}\right)^{\mathbf{T}} \cdot \mathbf{DATAX1} \cdot \begin{pmatrix} 14 \\ -1309 \end{pmatrix} \cdot \left(\mathbf{B}^{\langle 2 \rangle}\right)^{\mathbf{T}} \cdot \mathbf{DATAY1} \cdot \begin{pmatrix} 14 \\ -1360 \end{pmatrix}\right]^{\frac{1}{2}}$$

$\mathbf{r2} = 14189.94$

Now we take the ratio of r1/r2

$$\frac{\mathbf{r1}}{\mathbf{r2}} = 0.96$$

So the math for the Pearson Product–Moment Correlation looks like:

$$r := \frac{r1}{r2}$$

$$r := \frac{(14) \cdot \left(B^{\langle 1 \rangle^T} \cdot B^{\langle 2 \rangle}\right) - \left(B^{\langle 1 \rangle^T} \cdot \text{FOURTEEN1}\right) \cdot \left(B^{\langle 2 \rangle^T} \cdot \text{FOURTEEN1}\right)}{\left[\left(\left(B^{\langle 1 \rangle}\right)^T\right) \cdot \text{DATAX1} \cdot \left[\begin{matrix} N \\ -(\text{ONE14} \cdot B \cdot \text{ONE0}) \end{matrix}\right] \cdot \left(\left(B^{\langle 2 \rangle}\right)^T\right) \cdot \text{DATAY1} \cdot \left[\begin{matrix} N \\ -(\text{ONE14} \cdot B \cdot \text{ZERO1}) \end{matrix}\right]\right]^{\frac{1}{2}}}$$

$$r = 0.96$$

Hey! We have r in a single equation without even trying!!!!!!

$$\text{SSt} = (863.5 \quad 1189.714)$$

$$\text{VARIANCEDIFFERENCE} := \text{SSt} \cdot \text{MINPLUS1}$$

$$\text{VARIANCEDIFFERENCE} = 326.214$$

STEP 6: Multiply the two Variances together.

$$\text{MULT} := 10^{\overrightarrow{\log(\text{SSt}) \cdot \text{TWO1}}}$$

$$\text{MULT} = 1027318.286$$

STEP 7: Subtract 2 from the number of scores in one case (# elements in one column) and take the Square Root.

$$\text{Nc} := 14$$

$$s := (\text{Nc} - 2)^{\frac{1}{2}}$$

$$s := (14 - 2)^{\frac{1}{2}}$$

$$s = 3.464$$

STEP 8: Multiply STEP 5 by STEP 3.

$\text{SSt} \cdot \text{MINPLUS1} = 326.214$

$$\text{SS6} := (\text{SSt} \cdot \text{MINPLUS1}) \cdot (\text{Nc} - 2)^{\frac{1}{2}}$$

$\text{SS6} = 1130.039$

STEP 9: Find the t-Statistic by using the following equation, since these are single numbers, we do not need to treat them as Matrices, so to divide, we do not have to multiply by the inverse of the denominator.

$$t := \frac{(\text{Nc} - 2)^{\frac{1}{2}} \cdot (\text{SSt} \cdot \text{MINPLUS1})}{\left[4 \cdot (1 - r) \cdot 10^{\overrightarrow{\log(\text{SSt}) \cdot \text{TWO1}}} \right]^{\frac{1}{2}}}$$

$(\text{Nc} - 2)^{\frac{1}{2}} \cdot (\text{SSt} \cdot \text{MINPLUS1}) = 1130.039$

Where

$$\left[4 \cdot (1 - r) \cdot 10^{\overrightarrow{\log(\text{SSt}) \cdot \text{TWO1}}} \right]^{\frac{1}{2}} = 405.549$$

Which is equal to:

$$\frac{1130.039}{405.549} = 2.786$$

$t = 2.786$

The Degree of Freedom is equal to the number of elements in a single column minus 2.

$\text{df} := 14 - 2$

With a df of 12, we check in Appendix B in the 2nd Edition "Computational Handbook of Statistics" (can't reproduce it here, please buy book or) we have for a t = 2.786: a value that is larger than 2.786 is significant at a .01 level using a 2-tailed test.

t-TEST FOR DIFFERENCES AMONG SEVERAL MEANS

Since the example of section 2.1 Simple Randomized Design is used in section 3.4-3.9, I will import it here so we will not have to keep leafing back and forth through this book to see the original procedures.

$$C := \begin{pmatrix} 9 & 5 & 15 & 5 \\ 6 & 9 & 14 & 13 \\ 11 & 5 & 19 & 15 \\ 7 & 6 & 15 & 17 \\ 10 & 5 & 4 & 11 \\ 5 & 11 & 10 & 12 \\ 7 & 11 & 15 & 17 \\ 8 & 6 & 12 & 13 \\ 9 & 9 & 18 & 14 \\ 10 & 8 & 7 & 12 \\ 4 & 6 & 13 & 7 \\ 18 & 17 & 16 & 19 \end{pmatrix}$$

$N := 12 \cdot 4$ $\qquad$ $Nms := \dfrac{N}{4}$

$ONE12 := (1 \ \ 1 \ \ 1 \ \ 1 \ \ 1 \ \ 1 \ \ 1 \ \ 1 \ \ 1 \ \ 1 \ \ 1 \ \ 1)$

$$MEAN := \left(\frac{1}{Nms} \right) \cdot ONE12 \cdot C$$

$MEAN = (8.667 \quad 8.167 \quad 13.167 \quad 12.917)$

STEP 2: Compute the standard error of the difference among means using the formula:

N = # scores in each group. $\qquad$ $SE_{diff} = \left(2ms_{errw}/n \right)^{1/2}$

$Nms := 12$

$msErrw := \dfrac{SSw}{dfSSw}$ $\qquad$ $SSw := 672.917$ $\qquad$ $dfSSw := 44$ $\qquad$ $N := 12$

$msErrw = 15.294$

$$\text{SEdiff} := \left(2\frac{\text{msErrw}}{N}\right)^{\frac{1}{2}} \qquad\qquad \left(2\frac{\text{msErrw}}{N}\right)^{\frac{1}{2}} = 1.597$$

$$\text{SEdiff} := \left(2\frac{\frac{\text{SSw}}{\text{dfSSw}}}{N}\right)^{\frac{1}{2}}$$

$$\text{SEdiff} = 1.597$$

Suppose the N's are not all equivalent but are different values and close to each other. We can compute the Harmonic Mean as follows. Nt = # Elements in each Group, in this case it is = 4, ONENt is a Row Matrix of Nt elements all equal to 1, and the Column Matrix contains the inverse of the # Elements in each Group. ie, suppose

$$n1 := 16 \qquad n2 := 12 \qquad n3 := 8 \qquad n4 := 12 \qquad Nt := 4 \qquad \text{ONENt} := (1 \quad 1 \quad 1 \quad 1)$$

$$\nu := Nt \cdot \left[\text{ONENt} \cdot \begin{bmatrix} \frac{1}{n1} \\ \frac{1}{n2} \\ \frac{1}{n3} \\ \frac{1}{n4} \end{bmatrix}\right]^{-1} \qquad\qquad \nu = 11.294$$

Then we would have:

$$\text{SEdiffa} := \left(2\frac{\text{msErrw}}{\nu}\right)^{\frac{1}{2}}$$

$$\text{SEdiffa} = 1.646$$

If the n's are widely different, the Standard Error must be computed separately for each comparison. ie Suppose:

$n1 := 20$　　　　$n2 := 8$

$$SEdiffb := \left[msErrw \cdot \left(\frac{1}{20} + \frac{1}{8} \right) \right]^{\frac{1}{2}}$$

$SEdiffb = 1.636$

If there are 4 Groups, then we must compare n1, n2; n1,n3; n1,n4; n2,n3; n2,n4; n3,n4. 6 t-test comparisons.

The needed t-value is obtained from Appendix B in the 2nd Ed of the Computational Handbook of Statistics. With a df of 44, at the .05 level

$t := 2.02$

STEP 3: Multiply t by the Standard Error of the difference among the Means to obtain the Critical Difference.

$$Cdiff := t \cdot \left[\left(2 \frac{msErrw}{N} \right)^{\frac{1}{2}} \right]$$

$Cdiff = 3.225$

STEP 4: In all instances, if the difference between any 2 Means is larger than the Critical Difference, then the Means are considered to be significantly different. We shall subtract everything all at the same time, but we only need the half of the Matrix above (or below) the Diagonal.

$$ONE44 := \begin{pmatrix} 1 & 1 & 1 & 1 \\ 1 & 1 & 1 & 1 \\ 1 & 1 & 1 & 1 \\ 1 & 1 & 1 & 1 \end{pmatrix}$$

$Diff := diag \left(MEAN^T \right) \cdot ONE44 - \left(diag \left(MEAN^T \right) \cdot ONE44 \right)^T$　　　$MEAN = (8.667 \quad 8.167 \quad 13.167 \quad 12.917)$

$$\text{diag}\left(\text{MEAN}^T\right)\cdot\text{ONE}44 = \begin{pmatrix} 8.667 & 8.667 & 8.667 & 8.667 \\ 8.167 & 8.167 & 8.167 & 8.167 \\ 13.167 & 13.167 & 13.167 & 13.167 \\ 12.917 & 12.917 & 12.917 & 12.917 \end{pmatrix} \quad \left(\text{diag}\left(\text{MEAN}^T\right)\cdot\text{ONE}44\right)^T = \begin{pmatrix} 8.667 & 8.167 & 13.167 & 12.917 \\ 8.667 & 8.167 & 13.167 & 12.917 \\ 8.667 & 8.167 & 13.167 & 12.917 \\ 8.667 & 8.167 & 13.167 & 12.917 \end{pmatrix}$$

$$\text{Diff} = \begin{pmatrix} 0 & 0.5 & -4.5 & -4.25 \\ -0.5 & 0 & -5 & -4.75 \\ 4.5 & 5 & 0 & 0.25 \\ 4.25 & 4.75 & -0.25 & 0 \end{pmatrix}$$

The Math looks like:
$$\text{diag}\left(\text{MEAN}^T\right) = \begin{pmatrix} 8.667 & 0 & 0 & 0 \\ 0 & 8.167 & 0 & 0 \\ 0 & 0 & 13.167 & 0 \\ 0 & 0 & 0 & 12.917 \end{pmatrix}$$

$$\text{Diff} := \begin{pmatrix} 8.667 & 0 & 0 & 0 \\ 0 & 8.167 & 0 & 0 \\ 0 & 0 & 13.167 & 0 \\ 0 & 0 & 0 & 12.917 \end{pmatrix}\begin{pmatrix} 1 & 1 & 1 & 1 \\ 1 & 1 & 1 & 1 \\ 1 & 1 & 1 & 1 \\ 1 & 1 & 1 & 1 \end{pmatrix} - \left[\begin{pmatrix} 8.667 & 0 & 0 & 0 \\ 0 & 8.167 & 0 & 0 \\ 0 & 0 & 13.167 & 0 \\ 0 & 0 & 0 & 12.917 \end{pmatrix}\cdot\begin{pmatrix} 1 & 1 & 1 & 1 \\ 1 & 1 & 1 & 1 \\ 1 & 1 & 1 & 1 \\ 1 & 1 & 1 & 1 \end{pmatrix}\right]^T$$

$$\text{Diff} = \begin{pmatrix} 0 & 0.5 & -4.5 & -4.25 \\ -0.5 & 0 & -5 & -4.75 \\ 4.5 & 5 & 0 & 0.25 \\ 4.25 & 4.75 & -0.25 & 0 \end{pmatrix}$$

We look only at the Absolute Values, lets see what we have.

G1 - G2 =	.05	<3.225	NS
G1 - G3 =	4.5	>3.225	Significant
G1 - G4 =	4.25	>3.225	Significant
G2 - G3 =	5	>3.225	Significant
G2 - G4 =	4.75	>3.22	Significant
G3 - G4 =	0.25	<3.225	NS

PEARSON PRODUCT–MOMENT CORRELATION

Since this equation was solved earlier as a part of the t–Test in Prob. 3.2, I will just write it here directly from the Equation.

VARIABLES:

$$A := \begin{pmatrix} 40 & 2.2 \\ 56 & 3 \\ 45 & 3.1 \\ 47 & 2.4 \\ 52 & 2.7 \\ 63 & 3.8 \\ 59 & 3.3 \\ 27 & 1.4 \\ 34 & 1.9 \\ 37 & 2.4 \\ 46 & 3.3 \\ 44 & 2.5 \\ 42 & 2.3 \\ 37 & 2.4 \\ 46 & 3.2 \end{pmatrix} \quad A^{\langle 1 \rangle} = \begin{pmatrix} 40 \\ 56 \\ 45 \\ 47 \\ 52 \\ 63 \\ 59 \\ 27 \\ 34 \\ 37 \\ 46 \\ 44 \\ 42 \\ 37 \\ 46 \end{pmatrix} \quad A^{\langle 2 \rangle} = \begin{pmatrix} 2.2 \\ 3 \\ 3.1 \\ 2.4 \\ 2.7 \\ 3.8 \\ 3.3 \\ 1.4 \\ 1.9 \\ 2.4 \\ 3.3 \\ 2.5 \\ 2.3 \\ 2.4 \\ 3.2 \end{pmatrix} \quad DATAX := \begin{pmatrix} 40 & 1 \\ 56 & 1 \\ 45 & 1 \\ 47 & 1 \\ 52 & 1 \\ 63 & 1 \\ 59 & 1 \\ 27 & 1 \\ 34 & 1 \\ 37 & 1 \\ 46 & 1 \\ 44 & 1 \\ 42 & 1 \\ 37 & 1 \\ 46 & 1 \end{pmatrix} \quad DATAY := \begin{pmatrix} 2.2 & 1 \\ 3 & 1 \\ 3.1 & 1 \\ 2.4 & 1 \\ 2.7 & 1 \\ 3.8 & 1 \\ 3.3 & 1 \\ 1.4 & 1 \\ 1.9 & 1 \\ 2.4 & 1 \\ 3.3 & 1 \\ 2.5 & 1 \\ 2.3 & 1 \\ 2.4 & 1 \\ 3.2 & 1 \end{pmatrix}$$

$$FIFTEEN1 := (1\ 1\ 1\ 1\ 1\ 1\ 1\ 1\ 1\ 1\ 1\ 1\ 1\ 1\ 1)$$

$$FOURTEEN1 := (1\ 1\ 1\ 1\ 1\ 1\ 1\ 1\ 1\ 1\ 1\ 1\ 1\ 1)$$

$$ONE0 := \begin{pmatrix} 1 \\ 0 \end{pmatrix} \qquad ZERO1 := \begin{pmatrix} 0 \\ 1 \end{pmatrix} \qquad N := 15 \qquad MINPLUS := \begin{pmatrix} -1 \\ 1 \end{pmatrix}$$

$$r = \frac{N_P \Sigma XY - (\Sigma X)(\Sigma Y)}{([N\Sigma X^2 - (\Sigma X)^2][N\Sigma Y^2 - (\Sigma Y)^2])^{1/2}} \qquad \text{Which can be translated to:}$$

(First we write the Numerator☺

$$r1 := (N) \cdot \left(A^{\langle 1 \rangle^T} \cdot A^{\langle 2 \rangle} \right) - \left(A^{\langle 1 \rangle^T} \right) \cdot FIFTEEN1^T \cdot \left(A^{\langle 2 \rangle^T} \cdot FIFTEEN1^T \right)$$

$$(N) \cdot \left(A^{\langle 1 \rangle^T} \cdot A^{\langle 2 \rangle} \right) - \left(A^{\langle 1 \rangle^T} \right) \cdot \text{FIFTEEN}^T \cdot \left(A^{\langle 2 \rangle^T} \cdot \text{FIFTEEN}^T \right) = 1093.5$$

Now we write the Denominator.

$$r2 := \left[\left(\left(A^{\langle 1 \rangle} \right) \right)^T \cdot \text{DATAX} \cdot \left[\begin{matrix} N \\ -(\text{FIFTEEN} + A \cdot \text{ONE0}) \end{matrix} \right] \cdot \left(\left(A^{\langle 2 \rangle} \right) \right)^T \cdot \text{DATAY} \cdot \left[\begin{matrix} N \\ -(\text{FIFTEEN} + A \cdot \text{ZERO}) \end{matrix} \right] \right]^{\frac{1}{2}}$$

$$\left[\left(\left(A^{\langle 1 \rangle} \right) \right)^T \cdot \text{DATAX} \cdot \left[\begin{matrix} N \\ -(\text{FIFTEEN} + A \cdot \text{ONE0}) \end{matrix} \right] \cdot \left(\left(A^{\langle 2 \rangle} \right) \right)^T \cdot \text{DATAY} \cdot \left[\begin{matrix} N \\ -(\text{FIFTEEN} + A \cdot \text{ZERO}) \end{matrix} \right] \right]^{\frac{1}{2}} = 1265.223$$

The ratio is given by:

$$\frac{r1}{r2} = 0.864$$

And the final equation looks like:

$$r := \frac{(N) \cdot \left(A^{\langle 1 \rangle^T} \cdot A^{\langle 2 \rangle} \right) - \left(A^{\langle 1 \rangle^T} \right) \cdot \text{FIFTEEN}^T \cdot \left(A^{\langle 2 \rangle^T} \cdot \text{FIFTEEN}^T \right)}{\left[\left(\left(A^{\langle 1 \rangle} \right) \right)^T \cdot \text{DATAX} \cdot \left[\begin{matrix} N \\ -(\text{FIFTEEN} + A \cdot \text{ONE0}) \end{matrix} \right] \cdot \left(\left(A^{\langle 2 \rangle} \right) \right)^T \cdot \text{DATAY} \cdot \left[\begin{matrix} N \\ -(\text{FIFTEEN} + A \cdot \text{ZERO}) \end{matrix} \right] \right]^{\frac{1}{2}}}$$

$$r = 0.864$$

SIMPLE ANALYSIS OF COVARIANCE:
ONE TREATMENT VARIABLE

COMPUTATIONAL HANDBOOK OF STATISTICS, SECT. 4.9, PG. 192-197.

When an experimental control of an important variable (such as age or IQ) is impossible or impractical, statistical control may be achieved by use of a covariance analysis. Below is the method when there is only one treatment variable.

VARIABLES:

$$\textbf{ZERO1} := \begin{pmatrix} 0 \\ 1 \end{pmatrix} \qquad \textbf{ONE0} := \begin{pmatrix} 1 \\ 0 \end{pmatrix} \qquad \textbf{ONE6} := \begin{pmatrix} 1 & 1 & 1 & 1 & 1 & 1 \end{pmatrix}$$

$$\textbf{A} := \begin{pmatrix} 33 & 63 & 22 & 73 & 39 & 96 \\ 47 & 67 & 25 & 86 & 46 & 89 \\ 26 & 63 & 17 & 60 & 51 & 103 \\ 34 & 47 & 31 & 86 & 47 & 85 \\ 32 & 52 & 36 & 70 & 42 & 73 \\ 39 & 73 & 25 & 73 & 36 & 62 \end{pmatrix}$$

$$\textbf{DB2XY} := \begin{pmatrix} 1 & 0 \\ 0 & 1 \\ 1 & 0 \\ 0 & 1 \\ 1 & 0 \\ 0 & 1 \end{pmatrix} \qquad \textbf{DB2X} := \textbf{DB2XY}^{\langle 1 \rangle} \qquad \textbf{DB2Y} := \textbf{DB2XY}^{\langle 2 \rangle} \qquad \textbf{DB2X} = \begin{pmatrix} 1 \\ 0 \\ 1 \\ 0 \\ 1 \\ 0 \end{pmatrix} \qquad \textbf{DB2Y} = \begin{pmatrix} 0 \\ 1 \\ 0 \\ 1 \\ 0 \\ 1 \end{pmatrix}$$

$$\textbf{DBdiag2X} := \text{diag} (\textbf{DB2X}) \qquad \textbf{DBdiag2Y} := \text{diag} (\textbf{DB2Y})$$

$$\textbf{DBdiag2X} = \begin{pmatrix} 1 & 0 & 0 & 0 & 0 & 0 \\ 0 & 0 & 0 & 0 & 0 & 0 \\ 0 & 0 & 1 & 0 & 0 & 0 \\ 0 & 0 & 0 & 0 & 0 & 0 \\ 0 & 0 & 0 & 0 & 1 & 0 \\ 0 & 0 & 0 & 0 & 0 & 0 \end{pmatrix} \qquad \textbf{DBdiag2Y} = \begin{pmatrix} 0 & 0 & 0 & 0 & 0 & 0 \\ 0 & 1 & 0 & 0 & 0 & 0 \\ 0 & 0 & 0 & 0 & 0 & 0 \\ 0 & 0 & 0 & 1 & 0 & 0 \\ 0 & 0 & 0 & 0 & 0 & 0 \\ 0 & 0 & 0 & 0 & 0 & 1 \end{pmatrix}$$

$$DB1 := \begin{pmatrix} 1 & 0 & 0 \\ 1 & 0 & 0 \\ 0 & 1 & 0 \\ 0 & 1 & 0 \\ 0 & 0 & 1 \\ 0 & 0 & 1 \end{pmatrix} \qquad THREE1 := \begin{pmatrix} 1 \\ 1 \\ 1 \end{pmatrix} \qquad TWO1 := \begin{pmatrix} 1 \\ 1 \end{pmatrix}$$

STEP 2: Sum the columns to get SUM1

$$SUM1 := ONE6 \cdot A$$

$$SUM1 = (211 \quad 365 \quad 156 \quad 448 \quad 261 \quad 508)$$

STEP 3: Square all the numbers in the X column and sum. In method 1 we sum just the X component, in Method 2 we Sum the X and Y columns simultaneously.

METHOD 1:

$$XONLY := A \cdot DBdiag2X$$

Now to square XONLY and compute GRANDSUMX:

$$GRANDSUMSQX := ONE6 \cdot \left(\overrightarrow{(XONLY \cdot XONLY)} \right) \cdot ONE6^T$$

$$GRANDSUMSQX = 23462$$

$$XONLY := A \cdot DBdiag2X$$

$$XONLY = \begin{pmatrix} 33 & 0 & 22 & 0 & 39 & 0 \\ 47 & 0 & 25 & 0 & 46 & 0 \\ 26 & 0 & 17 & 0 & 51 & 0 \\ 34 & 0 & 31 & 0 & 47 & 0 \\ 32 & 0 & 36 & 0 & 42 & 0 \\ 39 & 0 & 25 & 0 & 36 & 0 \end{pmatrix}$$

The Math looks like:

$$
\text{XONLY} := \begin{pmatrix} 33 & 63 & 22 & 73 & 39 & 96 \\ 47 & 67 & 25 & 86 & 46 & 89 \\ 26 & 63 & 17 & 60 & 51 & 103 \\ 34 & 47 & 31 & 86 & 47 & 85 \\ 32 & 52 & 36 & 70 & 42 & 73 \\ 39 & 73 & 25 & 73 & 36 & 62 \end{pmatrix} \cdot \begin{pmatrix} 1 & 0 & 0 & 0 & 0 & 0 \\ 0 & 0 & 0 & 0 & 0 & 0 \\ 0 & 0 & 1 & 0 & 0 & 0 \\ 0 & 0 & 0 & 0 & 0 & 0 \\ 0 & 0 & 0 & 0 & 1 & 0 \\ 0 & 0 & 0 & 0 & 0 & 0 \end{pmatrix}
$$

$$
\text{GRANDSUMSQX} := (1\ \ 1\ \ 1\ \ 1\ \ 1\ \ 1) \cdot \left[\overrightarrow{\begin{pmatrix} 33 & 0 & 22 & 0 & 39 & 0 \\ 47 & 0 & 25 & 0 & 46 & 0 \\ 26 & 0 & 17 & 0 & 51 & 0 \\ 34 & 0 & 31 & 0 & 47 & 0 \\ 32 & 0 & 36 & 0 & 42 & 0 \\ 39 & 0 & 25 & 0 & 36 & 0 \end{pmatrix} \cdot \begin{pmatrix} 33 & 0 & 22 & 0 & 39 & 0 \\ 47 & 0 & 25 & 0 & 46 & 0 \\ 26 & 0 & 17 & 0 & 51 & 0 \\ 34 & 0 & 31 & 0 & 47 & 0 \\ 32 & 0 & 36 & 0 & 42 & 0 \\ 39 & 0 & 25 & 0 & 36 & 0 \end{pmatrix}} \right] \cdot \begin{pmatrix} 1 \\ 1 \\ 1 \\ 1 \\ 1 \\ 1 \end{pmatrix}
$$

GRANDSUMSQX = 23462

METHOD 2:

$$
\text{GRANDSUMSQXY} := \text{ONE6} \cdot \overrightarrow{(A \cdot A)} \cdot \text{DB2XY}
$$

GRANDSUMSQXY = (23462   100763)

This is much simpler, there are no intermediate values to compute. The Math looks like:

$$
\text{GRANDSUMSQXY} := (1\ \ 1\ \ 1\ \ 1\ \ 1\ \ 1) \cdot \left[\overrightarrow{\begin{pmatrix} 33 & 63 & 22 & 73 & 39 & 96 \\ 47 & 67 & 25 & 86 & 46 & 89 \\ 26 & 63 & 17 & 60 & 51 & 103 \\ 34 & 47 & 31 & 86 & 47 & 85 \\ 32 & 52 & 36 & 70 & 42 & 73 \\ 39 & 73 & 25 & 73 & 36 & 62 \end{pmatrix} \cdot \begin{pmatrix} 33 & 63 & 22 & 73 & 39 & 96 \\ 47 & 67 & 25 & 86 & 46 & 89 \\ 26 & 63 & 17 & 60 & 51 & 103 \\ 34 & 47 & 31 & 86 & 47 & 85 \\ 32 & 52 & 36 & 70 & 42 & 73 \\ 39 & 73 & 25 & 73 & 36 & 62 \end{pmatrix}} \right] \cdot \begin{pmatrix} 1 & 0 \\ 0 & 1 \\ 1 & 0 \\ 0 & 1 \\ 1 & 0 \\ 0 & 1 \end{pmatrix}
$$

$$\Sigma x^2 \quad \Sigma y^2$$

GRANDSUMSQXY = (23462   100763)

METHOD 3:

Here we use the properties of the Half-Multiplier Operator to solve the problem.

PROGRAM: $SS3 := ONE6 \cdot \left(diag\left(diag\left(A^T \cdot A\right)\right) \circ \cdot DB2XY\right)$

Since the o Operator does not exist in computer programs yet, we must Square the Data Matrix, remove the Diagonal, Multiply the Diagonal to DB2XY and Sum this solution. SS3 below gives the Program we need to accomplish this operation.

$$SS3 := ONE6 \cdot \left(diag\left(diag\left(A^T \cdot A\right)\right) \cdot DB2XY\right)$$

$SS3 = (\,23462 \quad 100763\,)$

Lets go through the Math step by step: First we square A:

$$A^T \cdot A = \begin{pmatrix} 7675 & 12975 & 5524 & 16022 & 9121 & 17673 \\ 12975 & 22669 & 9286 & 27152 & 15773 & 30817 \\ 5524 & 9286 & 4280 & 11787 & 6744 & 12901 \\ 16022 & 27152 & 11787 & 33950 & 19473 & 37788 \\ 9121 & 15773 & 6744 & 19473 & 11507 & 22384 \\ 17673 & 30817 & 12901 & 37788 & 22384 & 44144 \end{pmatrix}$$

Next we remove the Diagonal:

$$diag\left(A^T \cdot A\right) = \begin{pmatrix} 7675 \\ 22669 \\ 4280 \\ 33950 \\ 11507 \\ 44144 \end{pmatrix}$$

Then we Diagonalize it:

$$diag\left(diag\left(A^T \cdot A\right)\right) = \begin{pmatrix} 7675 & 0 & 0 & 0 & 0 & 0 \\ 0 & 22669 & 0 & 0 & 0 & 0 \\ 0 & 0 & 4280 & 0 & 0 & 0 \\ 0 & 0 & 0 & 33950 & 0 & 0 \\ 0 & 0 & 0 & 0 & 11507 & 0 \\ 0 & 0 & 0 & 0 & 0 & 44144 \end{pmatrix}$$

Now we multiply this value to DB2XY

$$
\mathbf{diag}\left(\mathbf{diag}\left(A^T \cdot A\right)\right) \cdot \mathbf{DB2XY} = \begin{pmatrix} 7675 & 0 \\ 0 & 22669 \\ 4280 & 0 \\ 0 & 33950 \\ 11507 & 0 \\ 0 & 44144 \end{pmatrix} \qquad \mathbf{DB2XY} = \begin{pmatrix} 1 & 0 \\ 0 & 1 \\ 1 & 0 \\ 0 & 1 \\ 1 & 0 \\ 0 & 1 \end{pmatrix}
$$

And finally we sum this Matrix to get:

$$
\mathbf{ONE6} \cdot \begin{pmatrix} 7675 & 0 \\ 0 & 22669 \\ 4280 & 0 \\ 0 & 33950 \\ 11507 & 0 \\ 0 & 44144 \end{pmatrix} = (23462 \quad 100763)
$$

STEP 4:   Sum the X and Y components from STEP 2.

SUMXY := **SUM1** · **DB2XY**

SUMXY = (628   1321)

Square SUMXY and divide by the # of X and Y values, in this case, there are 18 X values and 18 Y values.

Nx := 18

$$
\mathbf{SS5} := \left(\frac{1}{\mathbf{Nx}}\right) \cdot \overrightarrow{[\,(\,\mathbf{SUM1} \cdot \mathbf{DB2XY}\,) \cdot (\,\mathbf{SUM1} \cdot \mathbf{DB2XY}\,)\,]}\blacksquare
$$

MathCad cannot handle this equation, so we substitute:

SUM1DB2XY := **SUM1** · **DB2XY**

$$
\mathbf{SS5} := \left(\frac{1}{\mathbf{Nx}}\right) \cdot \overrightarrow{[\,(\,\mathbf{SUM1DB2XY}\,) \cdot (\,\mathbf{SUM1DB2XY}\,)\,]}
$$

SS5 = (21910.222   96946.722)

Now we subtract the above solution from that computed in STEP 3.

$$SS6 := ONE6 \cdot \overrightarrow{\left(A \cdot A \right)} \cdot DB2XY - \left(\frac{1}{Nx} \right) \cdot \overrightarrow{[(SUM1DB2XY) \cdot (SUM1DB2XY)]}$$

$SS3 - SS5 = (1551.778 \quad 3816.278)$

$SS6 = (1551.778 \quad 3816.278)$

$$SS7 := \left(\frac{1}{6} \right) \cdot \overrightarrow{\left(\overrightarrow{(SUM1 \cdot SUM1)} \cdot DB2XY \right)}$$

$$\overrightarrow{(SUM1 \cdot SUM1)} = (44521 \quad 133225 \quad 24336 \quad 200704 \quad 68121 \quad 258064) \qquad DB2XY = \begin{pmatrix} 1 & 0 \\ 0 & 1 \\ 1 & 0 \\ 0 & 1 \\ 1 & 0 \\ 0 & 1 \end{pmatrix}$$

$$\overrightarrow{(SUM1 \cdot SUM1)} \cdot DB2XY = (136978 \quad 591993)$$

$$\left(\frac{1}{6} \right) \cdot \overrightarrow{\left(\overrightarrow{(SUM1 \cdot SUM1)} \cdot DB2XY \right)} = (22829.667 \quad 98665.5)$$

$$SS8 := \left(\frac{1}{6} \right) \cdot \overrightarrow{\left(\overrightarrow{(SUM1 \cdot SUM1)} \cdot DB2XY \right)} - \left(\frac{1}{Nx} \right) \cdot \overrightarrow{[(SUM1DB2XY) \cdot (SUM1DB2XY)]}$$

$SS7 - SS5 = (919.444 \quad 1718.778)$

$SS8 = (919.444 \quad 1718.778)$

$SS9 := SS6 - SS8$

$$SS9 := ONE6 \cdot \overrightarrow{(A \cdot A)} \cdot DB2XY - \left(\frac{1}{Nx} \right) \cdot \overrightarrow{[(SUM1DB2XY) \cdot (SUM1DB2XY)]} - \left[\left(\frac{1}{6} \right) \cdot \overrightarrow{\left(\overrightarrow{(SUM1 \cdot SUM1)} \cdot DB2XY \right)} - \left(\frac{1}{Nx} \right) \cdot \overrightarrow{[(SUM1DB2XY) \cdot (SUM1DB2XY)]} \right]$$

$SS9 = (632.333 \quad 2097.5)$

STEP 17: Multiply each X score by it's paired Y score and Sum the Products. Because of the way we tabled the values, we are going to have to un-scramble the Matrix. We do this as follows:

$$\text{PRODUCTXY} := \text{ONE6} \cdot \left(10^{\overrightarrow{\log(A) \cdot DB1}}\right) \cdot \text{THREE1} \quad \blacksquare \quad \overrightarrow{\log(A)} = \begin{pmatrix} 1.519 & 1.799 & 1.342 & 1.863 & 1.591 & 1.982 \\ 1.672 & 1.826 & 1.398 & 1.934 & 1.663 & 1.949 \\ 1.415 & 1.799 & 1.23 & 1.778 & 1.708 & 2.013 \\ 1.531 & 1.672 & 1.491 & 1.934 & 1.672 & 1.929 \\ 1.505 & 1.716 & 1.556 & 1.845 & 1.623 & 1.863 \\ 1.591 & 1.863 & 1.398 & 1.863 & 1.556 & 1.792 \end{pmatrix}$$

Now we substitute:

$$\text{logADB1} := \overrightarrow{\log(A) \cdot DB1}$$

The Math looks like: here we separate the X-Y pairs.

$$\text{logADB1} := \begin{pmatrix} 1.519 & 1.799 & 1.342 & 1.863 & 1.591 & 1.982 \\ 1.672 & 1.826 & 1.398 & 1.934 & 1.663 & 1.949 \\ 1.415 & 1.799 & 1.23 & 1.778 & 1.708 & 2.013 \\ 1.531 & 1.672 & 1.491 & 1.934 & 1.672 & 1.929 \\ 1.505 & 1.716 & 1.556 & 1.845 & 1.623 & 1.863 \\ 1.591 & 1.863 & 1.398 & 1.863 & 1.556 & 1.792 \end{pmatrix} \cdot \begin{pmatrix} 1 & 0 & 0 \\ 1 & 0 & 0 \\ 0 & 1 & 0 \\ 0 & 1 & 0 \\ 0 & 0 & 1 \\ 0 & 0 & 1 \end{pmatrix}$$

Or we have, if MathCad could solve it:

$$\text{PRODUCTXY} := \text{ONE6} \cdot \left(10^{\overrightarrow{\left(\begin{pmatrix} 1.519 & 1.799 & 1.342 & 1.863 & 1.591 & 1.982 \\ 1.672 & 1.826 & 1.398 & 1.934 & 1.663 & 1.949 \\ 1.415 & 1.799 & 1.23 & 1.778 & 1.708 & 2.013 \\ 1.531 & 1.672 & 1.491 & 1.934 & 1.672 & 1.929 \\ 1.505 & 1.716 & 1.556 & 1.845 & 1.623 & 1.863 \\ 1.591 & 1.863 & 1.398 & 1.863 & 1.556 & 1.792 \end{pmatrix} \cdot \begin{pmatrix} 1 & 0 & 0 \\ 1 & 0 & 0 \\ 0 & 1 & 0 \\ 0 & 1 & 0 \\ 0 & 0 & 1 \\ 0 & 0 & 1 \end{pmatrix}\right)}\right) \cdot \text{THREE1} \quad \blacksquare$$

Breaking the equation into parts we have first:

$$\overrightarrow{\log(A) \cdot DB1} = \begin{pmatrix} 3.318 & 3.206 & 3.573 \\ 3.498 & 3.332 & 3.612 \\ 3.214 & 3.009 & 3.72 \\ 3.204 & 3.426 & 3.602 \\ 3.221 & 3.401 & 3.487 \\ 3.454 & 3.261 & 3.349 \end{pmatrix}$$

Now we compute the anti-log.

$$10^{\overrightarrow{\log ADB1}} = \begin{pmatrix} 2079.697 & 1603.245 & 3741.106 \\ 3147.748 & 2147.83 & 4092.607 \\ 1636.817 & 1018.591 & 5260.173 \\ 1595.879 & 2660.725 & 3990.249 \\ 1663.413 & 2517.677 & 3061.963 \\ 2844.461 & 1823.896 & 2228.435 \end{pmatrix}$$

Now we do the final Sum:

$$\textbf{PRODUCTXY} := \textbf{ONE6} \cdot \left(10^{\overrightarrow{\log ADB1}} \right) \cdot \textbf{THREE1}$$

$\textbf{PRODUCTXY} = 47114.512$

We want the Sum of Squares to mirror the Step we are using, so we have:

$\textbf{SS17} := \textbf{PRODUCTXY}$

$\textbf{SS17} = 47114.512$

NOTE: logarithms will play a big role in solving this Statistic.

STEP 18: Multiply the Sum of column X and column Y together and divide by the number of XY pairs. Np = 18

$\textbf{Np} := 18$

$$\textbf{SS18} := \left(\frac{1}{\textbf{Np}} \right) \cdot \left(10^{\overrightarrow{\log(\textbf{SUMXY})} \cdot \textbf{TWO1}} \right)$$

$\textbf{SS18} = 46088.222$

The Math looks like:

$$\textbf{SS18} := \left(\frac{1}{\textbf{Np}} \right) \cdot \left[10^{\overrightarrow{\log((628 \quad 1321))} \cdot \begin{pmatrix} 1 \\ 1 \end{pmatrix}} \right]$$

SS18 $= 46088.222$

STEP 19: Subtract SS18 from SS17.

SS17 $-$ **SS18** $= 1026.29$

$$\textbf{SS19} := \textbf{ONE6} \cdot \left(10^{\overrightarrow{\textbf{logADB1}}}\right) \cdot \textbf{THREE1} - \left(\frac{1}{\textbf{Np}}\right) \cdot \left[10^{\overrightarrow{\textbf{log}\,((\,628\quad 1321\,))}} \cdot \begin{pmatrix}1\\1\end{pmatrix}\right]$$

SS19 $= 1026.29$

STEP 20: Multiply each X sum by it's paired Y sum, sum them and divide by the number of pairs used to obtain this sum, NpXY = 6.

NpXY $:= 6$

This equation is given by:

$$\textbf{SS20} := \left(\frac{1}{\textbf{NpXY}}\right) \cdot \left(10^{\overrightarrow{\textbf{log}\,(\textbf{SUM1})\,\cdot\textbf{DB1}}}\right) \cdot \textbf{THREE1} \quad \blacksquare$$

To solve this equation, we make the following substitution:

$$\textbf{logSUM1xDB1} := \overrightarrow{\textbf{log}\,(\textbf{SUM1})\,\cdot\textbf{DB1}}$$

We now have:

$$\textbf{SS20} := \left(\frac{1}{\textbf{NpXY}}\right) \cdot \left(10^{\overrightarrow{\textbf{logSUM1xDB1}}}\right) \cdot \textbf{THREE1} \qquad \textbf{ONE6} \cdot \textbf{A} = (\,211 \quad 365 \quad 156 \quad 448 \quad 261 \quad 508\,)$$

SS20 $= 46581.833$ $\qquad\qquad$ **logSUM1xDB1** $= (\,4.887 \quad 4.844 \quad 5.123\,)$

The Math looks like:

$$SS20 := \left(\frac{1}{6}\right) \cdot \left[10^{\overrightarrow{\log((211 \quad 365 \quad 156 \quad 448 \quad 261 \quad 508)) \cdot \begin{pmatrix} 1 & 0 & 0 \\ 1 & 0 & 0 \\ 0 & 1 & 0 \\ 0 & 1 & 0 \\ 0 & 0 & 1 \\ 0 & 0 & 1 \end{pmatrix}}} \right] \cdot \begin{pmatrix} 1 \\ 1 \\ 1 \end{pmatrix}$$

Simplifying:

$$\overrightarrow{\log((211 \quad 365 \quad 156 \quad 448 \quad 261 \quad 508))} = (2.324 \quad 2.562 \quad 2.193 \quad 2.651 \quad 2.417 \quad 2.706)$$

And

$$\overrightarrow{\log((211 \quad 365 \quad 156 \quad 448 \quad 261 \quad 508))} \cdot \begin{pmatrix} 1 & 0 & 0 \\ 1 & 0 & 0 \\ 0 & 1 & 0 \\ 0 & 1 & 0 \\ 0 & 0 & 1 \\ 0 & 0 & 1 \end{pmatrix} = (4.887 \quad 4.844 \quad 5.123)$$

We have:

$$SS20 := \left(\frac{1}{6}\right) \cdot \left(\overrightarrow{10^{(4.887 \quad 4.844 \quad 5.123)}} \right) \cdot \begin{pmatrix} 1 \\ 1 \\ 1 \end{pmatrix}$$

$$SS20 = 46608.839$$

STEP 21: Subtract the result of STEP 18 from the result of STEP 20.

$$SS21 := \left(\frac{1}{NpXY}\right) \cdot \left(10^{\overrightarrow{\log\,(SUM1)\,\cdot DB1}}\right) \cdot THREE1 - \left(\frac{1}{Np}\right) \cdot \left(10^{\overrightarrow{\log\,(SUMXY)\,\cdot TWO1}}\right)^{\blacksquare}$$

With substitutions we have:

$$SS21 := \left(\frac{1}{NpXY}\right) \cdot \left(10^{\overrightarrow{\log SUM1xDB1}}\right) \cdot THREE1 - \left(\frac{1}{Np}\right) \cdot \left(10^{\overrightarrow{\log\,(SUMXY)\,\cdot TWO1}}\right)$$

$$SS21 = 493.611$$

STEP 22: Subtract the result of STEP 21 from the result of STEP 19.

$$SS22 := SS19 - SS21$$

$$SS22 := ONE6 \cdot \left(10^{\overrightarrow{\log ADB1}}\right) \cdot THREE1 - \left(\frac{1}{Np}\right) \cdot \left[10^{\overrightarrow{\log\,(\,(628 \quad 1321)\,)\,\cdot \binom{1}{1}}}\right] - \left[\left(\frac{1}{NpXY}\right) \cdot \left(10^{\overrightarrow{\log SUM1xDB1}}\right) \cdot THREE1 - \left(\frac{1}{Np}\right) \cdot \left(10^{\overrightarrow{\log\,(SUMXY)\,\cdot TWO1}}\right)\right]$$

$$SS22 = 532.679$$

STEP 23: Square SS19 and divide by SS6.

$$\frac{\left[ONE6 \cdot \left(10^{\overrightarrow{\log ADB1}}\right) \cdot THREE1 - \left(\frac{1}{Np}\right) \cdot \left(10^{\overrightarrow{\log\,(SUMXY)\,\cdot TWO1}}\right)\right]^2}{\left[ONE6 \cdot \left(\overrightarrow{(A \cdot A)}\right) \cdot DB2XY - \left(\frac{1}{Nx}\right) \cdot \overrightarrow{[\,(SUM1DB2XY)\,\cdot(SUM1DB2XY)\,]}\right] \cdot ONE0} = 678.751$$

STEP 24: Subtract SS23 from SS13

$$SS24 := \left[ONE6 \cdot \left(\overrightarrow{(A \cdot A)}\right) DB2XY - \left(\frac{1}{Nx}\right) \overrightarrow{[\,(SUM1DB2XY)\,\cdot(SUM1DB2XY)\,]}\right] \cdot ZERO1 - \frac{\left[ONE6 \cdot \left(10^{\overrightarrow{\log ADB1}}\right) \cdot THREE1 - \left(\frac{1}{Np}\right) \cdot \left(10^{\overrightarrow{\log\,(SUMXY)\,\cdot TWO1}}\right)\right]^2}{\left[ONE6 \cdot \left(\overrightarrow{(A \cdot A)}\right) DB2XY - \left(\frac{1}{Nx}\right) \cdot \overrightarrow{[\,(SUM1DB2XY)\,\cdot(SUM1DB2XY)\,]}\right] \cdot ONE0}$$

$$SS24 = 3137.527$$

STEP 25: Square SS22 and divide by SS9

$$SS25 := \frac{\left[\left[ONE6 \cdot \left(\overrightarrow{10^{logADB1}}\right) \cdot THREE1 - \left(\frac{1}{Np}\right) \cdot \left[\overrightarrow{10^{log((628\ \ 1321))} \cdot \binom{1}{1}}\right]\right] - \left[\left(\frac{1}{NpXY}\right) \cdot \left(\overrightarrow{10^{logSUM1xDB1}}\right) \cdot THREE1 - \left(\frac{1}{Np}\right) \cdot \left(\overrightarrow{10^{log(SUMXY)} \cdot TWO1}\right)\right]\right]^2}{\left[ONE6 \cdot \left(\overrightarrow{A \cdot A}\right) \cdot DB2XY - \left(\frac{1}{Nx}\right) \cdot \overrightarrow{[(SUM1DB2XY) \cdot (SUM1DB2XY)]} - \left[\left(\frac{1}{6}\right) \cdot \left(\overrightarrow{SUM1 \cdot SUM1} \cdot DB2XY\right) - \left(\frac{1}{Nx}\right) \cdot \overrightarrow{[(SUM1DB2XY) \cdot (SUM1DB2XY)]}\right]\right] \cdot ONE0}$$

SS25 = 448.73

STEP 26: Subtract SS25 from SS16

$$SS26 := \left[ONE6 \cdot \left(\overrightarrow{A \cdot A}\right) \cdot DB2XY - \left(\frac{1}{Nx}\right) \cdot \overrightarrow{[(SUM1DB2XY) \cdot (SUM1DB2XY)]} - \left[\left(\frac{1}{6}\right) \cdot \left(\overrightarrow{SUM1 \cdot SUM1} \cdot DB2XY\right) - \left(\frac{1}{Nx}\right) \cdot \overrightarrow{[(SUM1DB2XY) \cdot (SUM1DB2XY)]}\right]\right] \cdot ZERO1 - SS25$$

SS26 = 1648.77

STEP 27: Divide SS26 by the Degree of Freedom for the Within-Groups measure, where df = N-G-1 where N is the number of XY pairs (N = 18 for this example) and G = number of experimental Groups = 3 in this case.

$$N := 18 \qquad G := 3 \qquad \left(\frac{1}{N-G-1}\right) \cdot SS26 = 117.769$$

$$df := N - G - 1$$

$$SS27 := \left(\frac{1}{N-G-1}\right) \left[\left[ONE6 \cdot \left(\overrightarrow{A \cdot A}\right) \cdot DB2XY - \left(\frac{1}{Nx}\right) \cdot \overrightarrow{[(SUM1DB2XY) \cdot (SUM1DB2XY)]} - \left[\left(\frac{1}{6}\right) \cdot \left(\overrightarrow{SUM1 \cdot SUM1} \cdot DB2XY\right) - \left(\frac{1}{Nx}\right) \cdot \overrightarrow{[(SUM1DB2XY) \cdot (SUM1DB2XY)]}\right]\right] \cdot ZERO1 - SS25\right]$$

SS27 = 117.769

STEP 28: Subtract SS26 from SS24.

$$SS24 - SS26 = 1488.756$$

And the Math is:

$$SS28 := \left[ONE6 \cdot \left(\overrightarrow{A \cdot A}\right) \cdot DB2XY - \left(\frac{1}{Nx}\right) \cdot \overrightarrow{[(SUM1DB2XY) \cdot (SUM1DB2XY)]}\right] \cdot ZERO1 - \frac{\left[ONE6 \cdot \left(\overrightarrow{10^{logADB1}}\right) \cdot THREE1 - \left(\frac{1}{Np}\right) \cdot \left(\overrightarrow{10^{log(SUMXY)} \cdot TWO1}\right)\right]^2}{\left[ONE6 \cdot \left(\overrightarrow{A \cdot A}\right) \cdot DB2XY - \left(\frac{1}{Nx}\right) \cdot \overrightarrow{[(SUM1DB2XY) \cdot (SUM1DB2XY)]}\right] \cdot ONE0} - SS26$$

SS28 = 1488.756

STEP 29: Divide SS28 by the Degree of Freedom for the Between Subjects measure. This is always G-1. G in this case = 3 experimental Groups.

$$\mathbf{SS29} := \left(\frac{1}{G-1}\right) \cdot \mathbf{SS28}$$

$\mathbf{SS29} = 744.378$

Or the equation is:

$$\mathbf{SS29} := \left(\frac{1}{G-1}\right) \cdot \left[\left[\mathrm{ONE6} \cdot \left(\overrightarrow{A \cdot A}\right) \cdot \mathrm{DB2XY} - \left(\frac{1}{Nx}\right) \cdot \overrightarrow{[(\mathrm{SUM1DB2XY}) \cdot (\mathrm{SUM1DB2XY})]}\right] \cdot \mathrm{ZERO1} - \frac{\left[\mathrm{ONE6} \cdot \left(10^{\overrightarrow{\log\mathrm{ADB1}}}\right) \cdot \mathrm{THREE1} - \left(\frac{1}{Np}\right) \cdot \left(10^{\overrightarrow{\log(\mathrm{SUMXY})} \cdot \mathrm{TWO1}}\right)\right]^2}{\left[\mathrm{ONE6} \cdot \left(\overrightarrow{A \cdot A}\right) \cdot \mathrm{DB2XY} - \left(\frac{1}{Nx}\right) \cdot \overrightarrow{[(\mathrm{SUM1DB2XY}) \cdot (\mathrm{SUM1DB2XY})]}\right] \cdot \mathrm{ONE0}} - \mathrm{SS26}\right]$$

$\mathbf{SS29} = 744.378$

STEP 30: Computation of the F-Ratio.

$$F := \frac{\mathbf{SS29}}{\mathbf{SS27}}$$

$F = 6.321$

The Math has a One Variable Analysis of Covariance in a **single equation** for the computation of F. This equation is given by:

NOTE: For the sake of space we use SS25 and SS26 instead of the Math so it can fit on a single page.

$$F := \frac{\left(\frac{1}{G-1}\right) \cdot \left[\left[\mathrm{ONE6} \cdot \left(\overrightarrow{A \cdot A}\right) \cdot \mathrm{DB2XY} - \left(\frac{1}{Nx}\right) \cdot \overrightarrow{[(\mathrm{SUM1DB2XY}) \cdot (\mathrm{SUM1DB2XY})]}\right] \cdot \mathrm{ZERO1} - \frac{\left[\mathrm{ONE6} \cdot \left(10^{\overrightarrow{\log\mathrm{ADB1}}}\right) \cdot \mathrm{THREE1} - \left(\frac{1}{Np}\right) \cdot \left(10^{\overrightarrow{\log(\mathrm{SUMXY})} \cdot \mathrm{TWO1}}\right)\right]^2}{\left[\mathrm{ONE6} \cdot \left(\overrightarrow{A \cdot A}\right) \cdot \mathrm{DB2XY} - \left(\frac{1}{Nx}\right) \cdot \overrightarrow{[(\mathrm{SUM1DB2XY}) \cdot (\mathrm{SUM1DB2XY})]}\right] \cdot \mathrm{ONE0}} - \mathrm{SS26}\right]}{\left(\frac{1}{N-G-1}\right) \cdot \left[\left[\mathrm{ONE6} \cdot \left(\overrightarrow{A \cdot A}\right) \cdot \mathrm{DB2XY} - \left(\frac{1}{Nx}\right) \cdot \overrightarrow{[(\mathrm{SUM1DB2XY}) \cdot (\mathrm{SUM1DB2XY})]}\right] - \left[\left(\frac{1}{6}\right) \cdot \left(\overrightarrow{\mathrm{SUM1} \cdot \mathrm{SUM1}} \cdot \mathrm{DB2XY}\right) - \left(\frac{1}{Nx}\right) \cdot \overrightarrow{[(\mathrm{SUM1DB2XY}) \cdot (\mathrm{SUM1DB2XY})]}\right]\right] \cdot \mathrm{ZERO1} - \mathrm{SS25}}$$

Note: We have computed F in a single line.

$F = 6.321$

This F has a df of (G-1)/(N-G-1) which = 2/14 in this example.
From Appendix E in the Computational Handbook of Statistics,
2nd Ed., it is seen that a F-ratio greater than 6.51 would be
significant at the .01 level.

FACTORIAL ANALYSIS OF COVARIANCE:
TWO TREATMENT VARIABLES

COMPUTATIONAL HANDBOOK OF STATISTICS,

2ND EDITION, SECT. 4.10, PG. 198-209.

STEP 1: Table the data: I am going to table this data differently from the simple covariance problem in section 4.9 just to see if the other way I see is really simpler.

Now for the problem I am currently working on in Section 4.10 of the Computational Handbook of Statistics, 2nd Ed., PP. 199-200, steps 2-6, 16-20. (5-26-97)

$$a := \begin{pmatrix} 63 & 6 & 47 & 6 & 85 & 3 & 59 & 10 \\ 74 & 6 & 52 & 3 & 90 & 2 & 71 & 9 \\ 42 & 4 & 58 & 4 & 69 & 2 & 62 & 9 \\ 87 & 7 & 48 & 6 & 76 & 1 & 64 & 7 \\ 73 & 8 & 63 & 7 & 86 & 4 & 60 & 11 \end{pmatrix}$$

$$a^{\langle 1 \rangle} = \begin{pmatrix} 63 \\ 74 \\ 42 \\ 87 \\ 73 \end{pmatrix} \quad a^{\langle 2 \rangle} = \begin{pmatrix} 6 \\ 6 \\ 4 \\ 7 \\ 8 \end{pmatrix} \quad a^{\langle 3 \rangle} = \begin{pmatrix} 47 \\ 52 \\ 58 \\ 48 \\ 63 \end{pmatrix} \quad a^{\langle 4 \rangle} = \begin{pmatrix} 6 \\ 3 \\ 4 \\ 6 \\ 7 \end{pmatrix} \quad a^{\langle 5 \rangle} = \begin{pmatrix} 85 \\ 90 \\ 69 \\ 76 \\ 86 \end{pmatrix} \quad a^{\langle 6 \rangle} = \begin{pmatrix} 3 \\ 2 \\ 2 \\ 1 \\ 4 \end{pmatrix} \quad a^{\langle 7 \rangle} = \begin{pmatrix} 59 \\ 71 \\ 62 \\ 64 \\ 60 \end{pmatrix} \quad a^{\langle 8 \rangle} = \begin{pmatrix} 10 \\ 9 \\ 9 \\ 7 \\ 11 \end{pmatrix}$$

$$an := \begin{pmatrix} a^{\langle 1 \rangle} & a^{\langle 2 \rangle} & a^{\langle 3 \rangle} & a^{\langle 4 \rangle} & a^{\langle 5 \rangle} & a^{\langle 6 \rangle} & a^{\langle 7 \rangle} & a^{\langle 8 \rangle} \end{pmatrix}$$

This is the Nested Array, but we wish to separate it into a 2x20 Matrix pairing the XY values.

$$an = \begin{bmatrix} \begin{pmatrix} 63 \\ 74 \\ 42 \\ 87 \\ 73 \end{pmatrix} & \begin{pmatrix} 6 \\ 6 \\ 4 \\ 7 \\ 8 \end{pmatrix} & \begin{pmatrix} 47 \\ 52 \\ 58 \\ 48 \\ 63 \end{pmatrix} & \begin{pmatrix} 6 \\ 3 \\ 4 \\ 6 \\ 7 \end{pmatrix} & \begin{pmatrix} 85 \\ 90 \\ 69 \\ 76 \\ 86 \end{pmatrix} & \begin{pmatrix} 3 \\ 2 \\ 2 \\ 1 \\ 4 \end{pmatrix} & \begin{pmatrix} 59 \\ 71 \\ 62 \\ 64 \\ 60 \end{pmatrix} & \begin{pmatrix} 10 \\ 9 \\ 9 \\ 7 \\ 11 \end{pmatrix} \end{bmatrix}$$

When I am doing the proofs, I use the symbol + between the sub-matrices, this does not necessarily mean to add, but to connect. If we transpose as is, we get a 1x40 matrix corresponding Y pair, so I do the connection $[X]_1 + [y]_1 = [X \ Y]_1$ or $[X]_I +$

$[Y]_I = [X\ Y]_I$ and then transpose the resulting matrix to get:
Now the column matrix of A =

$$
XY := \begin{pmatrix}
63 & 6 \\
74 & 6 \\
42 & 4 \\
87 & 7 \\
73 & 8 \\
47 & 6 \\
52 & 3 \\
58 & 4 \\
48 & 6 \\
63 & 7 \\
85 & 3 \\
90 & 2 \\
69 & 2 \\
76 & 1 \\
86 & 4 \\
59 & 10 \\
71 & 9 \\
62 & 9 \\
64 & 7 \\
60 & 11
\end{pmatrix}
\quad
X1 := \begin{pmatrix}
63 & 1 \\
74 & 1 \\
42 & 1 \\
87 & 1 \\
73 & 1 \\
47 & 1 \\
52 & 1 \\
58 & 1 \\
48 & 1 \\
63 & 1 \\
85 & 1 \\
90 & 1 \\
69 & 1 \\
76 & 1 \\
86 & 1 \\
59 & 1 \\
71 & 1 \\
62 & 1 \\
64 & 1 \\
60 & 1
\end{pmatrix}
\quad
Y1 := \begin{pmatrix}
6 & 1 \\
6 & 1 \\
4 & 1 \\
7 & 1 \\
8 & 1 \\
6 & 1 \\
3 & 1 \\
4 & 1 \\
6 & 1 \\
7 & 1 \\
3 & 1 \\
2 & 1 \\
2 & 1 \\
1 & 1 \\
4 & 1 \\
10 & 1 \\
9 & 1 \\
9 & 1 \\
7 & 1 \\
11 & 1
\end{pmatrix}
\quad
X := \begin{pmatrix}
63 \\
74 \\
42 \\
87 \\
73 \\
47 \\
52 \\
58 \\
48 \\
63 \\
85 \\
90 \\
69 \\
76 \\
86 \\
59 \\
71 \\
62 \\
64 \\
60
\end{pmatrix}
\quad
a1 := \begin{pmatrix}
63 & 6 \\
74 & 6 \\
42 & 4 \\
87 & 7 \\
73 & 8 \\
47 & 6 \\
52 & 3 \\
58 & 4 \\
48 & 6 \\
63 & 7 \\
85 & 3 \\
90 & 2 \\
69 & 2 \\
76 & 1 \\
86 & 4 \\
59 & 10 \\
71 & 9 \\
62 & 9 \\
64 & 7 \\
60 & 11
\end{pmatrix}
\quad
Y := \begin{pmatrix}
6 \\
6 \\
4 \\
7 \\
8 \\
6 \\
3 \\
4 \\
6 \\
7 \\
3 \\
2 \\
2 \\
1 \\
4 \\
10 \\
9 \\
9 \\
7 \\
11
\end{pmatrix}
$$

$$
ONEYA := \begin{pmatrix} 0 \\ 1 \\ 0 \\ 1 \\ 0 \\ 1 \\ 0 \\ 1 \end{pmatrix}
\quad
FOUR1 := \begin{pmatrix} 1 \\ 1 \\ 1 \\ 1 \end{pmatrix}
\quad
ONEXA := \begin{pmatrix} 1 \\ 0 \\ 1 \\ 0 \\ 1 \\ 0 \\ 1 \\ 0 \end{pmatrix}
\quad
ONE0 := \begin{pmatrix} 1 \\ 0 \end{pmatrix}
\quad
ZERO1 := \begin{pmatrix} 0 \\ 1 \end{pmatrix}
\quad
DB12 := \begin{pmatrix} 1 & 0 \\ 1 & 0 \\ 0 & 1 \\ 0 & 1 \end{pmatrix}
$$

$$
DB1 := \begin{pmatrix}
1 & 1 & 1 & 1 & 1 & 0 & 0 & 0 & 0 & 0 & 0 & 0 & 0 & 0 & 0 & 0 & 0 & 0 & 0 & 0 \\
0 & 0 & 0 & 0 & 0 & 1 & 1 & 1 & 1 & 1 & 0 & 0 & 0 & 0 & 0 & 0 & 0 & 0 & 0 & 0 \\
0 & 0 & 0 & 0 & 0 & 0 & 0 & 0 & 0 & 0 & 1 & 1 & 1 & 1 & 1 & 0 & 0 & 0 & 0 & 0 \\
0 & 0.0 & 0 & 0 & 0 & 0 & 0 & 0 & 0 & 0 & 0 & 0 & 0 & 0 & 0 & 1 & 1 & 1 & 1 & 1
\end{pmatrix}
\qquad
DB11 := \begin{pmatrix}
1 & 0 & 0 & 0 \\
1 & 0 & 0 & 0 \\
0 & 1 & 0 & 0 \\
0 & 1 & 0 & 0 \\
0 & 0 & 1 & 0 \\
0 & 0 & 1 & 0 \\
0 & 0 & 0 & 1 \\
0 & 0 & 0 & 1
\end{pmatrix}
$$

$ONE20 := (1\ 1\ 1\ 1\ 1\ 1\ 1\ 1\ 1\ 1\ 1\ 1\ 1\ 1\ 1\ 1\ 1\ 1\ 1\ 1)$

$ONE8 := (1\ 1\ 1\ 1\ 1\ 1\ 1\ 1)$

$ONE5 := (1\ 1\ 1\ 1\ 1)$

$ONE4 := (1\ 1\ 1\ 1)$

$ONE2 := (1\ 1)$ $TWO1 := ONE2^T$ $X1Y1 := augment(X1, Y1)$ $DB2 := \begin{pmatrix} 1 & 0 & 1 & 0 \\ 0 & 1 & 0 & 1 \\ 1 & 1 & 0 & 0 \\ 0 & 0 & 1 & 1 \end{pmatrix}$

$X1Y1 := \begin{pmatrix} 63 & 1 & 6 & 1 \\ 74 & 1 & 6 & 1 \\ 42 & 1 & 4 & 1 \\ 87 & 1 & 7 & 1 \\ 73 & 1 & 8 & 1 \\ 47 & 1 & 6 & 1 \\ 52 & 1 & 3 & 1 \\ 58 & 1 & 4 & 1 \\ 48 & 1 & 6 & 1 \\ 63 & 1 & 7 & 1 \\ 85 & 1 & 3 & 1 \\ 90 & 1 & 2 & 1 \\ 69 & 1 & 2 & 1 \\ 76 & 1 & 1 & 1 \\ 86 & 1 & 4 & 1 \\ 59 & 1 & 10 & 1 \\ 71 & 1 & 9 & 1 \\ 62 & 1 & 9 & 1 \\ 64 & 1 & 7 & 1 \\ 60 & 1 & 11 & 1 \end{pmatrix}$

$diagONEYA := diag(ONEYA)$ $diagONEXA := diag(ONEXA)$

$diagONEYA := \begin{pmatrix} 0 & 0 & 0 & 0 & 0 & 0 & 0 & 0 \\ 0 & 1 & 0 & 0 & 0 & 0 & 0 & 0 \\ 0 & 0 & 0 & 0 & 0 & 0 & 0 & 0 \\ 0 & 0 & 0 & 1 & 0 & 0 & 0 & 0 \\ 0 & 0 & 0 & 0 & 0 & 0 & 0 & 0 \\ 0 & 0 & 0 & 0 & 0 & 1 & 0 & 0 \\ 0 & 0 & 0 & 0 & 0 & 0 & 0 & 0 \\ 0 & 0 & 0 & 0 & 0 & 0 & 0 & 1 \end{pmatrix}$ $diagONEXA = \begin{pmatrix} 1 & 0 & 0 & 0 & 0 & 0 & 0 & 0 \\ 0 & 0 & 0 & 0 & 0 & 0 & 0 & 0 \\ 0 & 0 & 1 & 0 & 0 & 0 & 0 & 0 \\ 0 & 0 & 0 & 0 & 0 & 0 & 0 & 0 \\ 0 & 0 & 0 & 0 & 1 & 0 & 0 & 0 \\ 0 & 0 & 0 & 0 & 0 & 0 & 0 & 0 \\ 0 & 0 & 0 & 0 & 0 & 0 & 1 & 0 \\ 0 & 0 & 0 & 0 & 0 & 0 & 0 & 0 \end{pmatrix}$

$DB42 := \begin{pmatrix} 0 & 0 & 0 & 0 \\ 1 & 0 & 1 & 0 \\ 0 & 0 & 0 & 0 \\ 0 & 1 & 1 & 0 \\ 0 & 0 & 0 & 0 \\ 1 & 0 & 0 & 1 \\ 0 & 0 & 0 & 0 \\ 0 & 1 & 0 & 1 \end{pmatrix}$ $DB4 := \begin{pmatrix} 1 & 0 & 1 & 0 \\ 0 & 0 & 0 & 0 \\ 0 & 1 & 1 & 0 \\ 0 & 0 & 0 & 0 \\ 1 & 0 & 0 & 1 \\ 0 & 0 & 0 & 0 \\ 0 & 1 & 0 & 1 \\ 0 & 0 & 0 & 0 \end{pmatrix}$ $DB22 := \begin{pmatrix} 1 & 0 & 1 & 0 \\ 0 & 1 & 0 & 1 \\ 1 & 1 & 0 & 0 \\ 0 & 0 & 1 & 1 \end{pmatrix}$

$DB23 := \begin{pmatrix} 1 & 0 \\ 0 & 1 \\ 0 & 0 \\ 0 & 0 \\ 1 & 0 \\ 0 & 1 \\ 0 & 0 \\ 0 & 0 \end{pmatrix}$

These are the Variables, now to begin the 2-Factor Analysis of Co-variance.

STEP 2: Add the Scores in each column.

We may compute separating the X and Y components.

sum1 := ONE5·a   SUM1a := DB1·al

sum1 = (339 31 268 26 406 12 316 46) $$DB1 \cdot al = \begin{pmatrix} 339 & 31 \\ 268 & 26 \\ 406 & 12 \\ 316 & 46 \end{pmatrix}$$

THE CONTROL VARIABLE

(Steps 3-15)

STEPS 3-4: Find the Sum of Squares for the X column.

sqsumX := $X^T \cdot X1$

sqsumX = (91977 1329)

$$sqsumX := (63\ \ 74\ \ 42\ \ 87\ \ 73\ \ 47\ \ 52\ \ 58\ \ 48\ \ 63\ \ 85\ \ 90\ \ 69\ \ 76\ \ 86\ \ 59\ \ 71\ \ 62\ \ 64\ \ 60) \cdot \begin{pmatrix} 63 & 1 \\ 74 & 1 \\ 42 & 1 \\ 87 & 1 \\ 73 & 1 \\ 47 & 1 \\ 52 & 1 \\ 58 & 1 \\ 48 & 1 \\ 63 & 1 \\ 85 & 1 \\ 90 & 1 \\ 69 & 1 \\ 76 & 1 \\ 86 & 1 \\ 59 & 1 \\ 71 & 1 \\ 62 & 1 \\ 64 & 1 \\ 60 & 1 \end{pmatrix}$$

Here a = Sum of Squares, b = Sum of Elements.

$$\qquad\qquad a \qquad b$$
sqsumX = (91977 1329)

SS4 := $X^T \cdot X1 \cdot ZERO1$ $\qquad\qquad$ SS3 := $X^T \cdot X1 \cdot ONE0$

SS4 := $X^T \cdot X1 \cdot \begin{pmatrix} 0 \\ 1 \end{pmatrix}$ $\qquad\qquad$ SS3 := $X^T \cdot X1 \cdot \begin{pmatrix} 1 \\ 0 \end{pmatrix}$

676

SS4 = 1329   SS3 = 91977

(5) Square the Sum of the X column and divide by 20 (# elements
    in a column of Matrix A)

(6) Subtract STEP 5 from STEP 3.

We will do this all in one operation. b = 1329

Using the b method:

$$SS6 := sqsumX \cdot \begin{pmatrix} 1 \\ \dfrac{-1329}{20} \end{pmatrix}$$

SS6 = 3664.95

CORRECTION FACTOR, 3 METHODS:

METHOD 1:

$$CF := \left(\frac{1}{20}\right) \cdot (sqsumX \cdot ZERO1)^2$$

CF = 88312.05

METHOD 2:   b:= 1329

$$CF := \overrightarrow{\left(sqsumX \cdot ZERO1^T\right)} \cdot \begin{pmatrix} 1 \\ \dfrac{b}{20} \end{pmatrix}$$ $sqsumX = (91977 \quad 1329)$

CF = 88312.05

The Math looks like:

$$ZERO1^T = (0 \quad 1)$$

$$\overrightarrow{[(91977 \quad 1329) \cdot (0 \quad 1)]} \cdot \begin{pmatrix} 1 \\ \dfrac{b}{20} \end{pmatrix} = 88312.05$$

$$b := X^T \cdot X1 \cdot \begin{pmatrix} 0 \\ 1 \end{pmatrix} \qquad\qquad Nt := 20$$

$$CF := b \frac{b}{Nt}$$

$$CF = 88312.05$$

STEP 7 and 8: STEP 7: Add the Sums of the X column of the 2 Groups, Square and divide by the number of elements upon which the Sum is based. Nc1 = 5 x 2 = 10. STEP 8: Of the X column of both of the 2nd Groups, Square and divide by the number of elements upon which the Sum is based. In this case, Nc2 = 5 x 2 = 10. STEP 9: Sum the two and Subtract the Correction Factor.

$$SS78 := \left(\frac{1}{10}\right)\left[\overrightarrow{(ONE5 \cdot a \cdot DB4)^2 \cdot DB11}\right]$$

$$DB4 := \begin{pmatrix} 1 & 0 & 1 & 0 \\ 0 & 0 & 0 & 0 \\ 0 & 1 & 1 & 0 \\ 0 & 0 & 0 & 0 \\ 1 & 0 & 0 & 1 \\ 0 & 0 & 0 & 0 \\ 0 & 1 & 0 & 1 \\ 0 & 0 & 0 & 0 \end{pmatrix} \qquad DB11 = \begin{pmatrix} 1 & 0 & 0 & 0 \\ 1 & 0 & 0 & 0 \\ 0 & 1 & 0 & 0 \\ 0 & 1 & 0 & 0 \\ 0 & 0 & 1 & 0 \\ 0 & 0 & 1 & 0 \\ 0 & 0 & 0 & 1 \\ 0 & 0 & 0 & 1 \end{pmatrix}$$

$$SS78 := \left(\frac{1}{10}\right) \cdot \overrightarrow{\left[(ONE5 \cdot a \cdot DB4) \cdot (ONE5 \cdot a \cdot DB4)\right] \cdot DB12}$$

$$DB12 = \begin{pmatrix} 1 & 0 \\ 1 & 0 \\ 0 & 1 \\ 0 & 1 \end{pmatrix}$$

$$ONE5 \cdot a \cdot DB4 = (745 \quad 584 \quad 607 \quad 722)$$

$$\overrightarrow{[(ONE5 \cdot a \cdot DB4) \cdot (ONE5 \cdot a \cdot DB4)]} = \blacksquare$$

$$ONE5xaxDB4 := (ONE5 \cdot a \cdot DB4)$$

$\overrightarrow{[(ONE5xaxDB4) \cdot ONE5xaxDB4]} = (\ 555025 \quad 341056 \quad 368449 \quad 521284\ )$

$SS78 := \left(\dfrac{1}{10}\right) \cdot \overrightarrow{\left([(ONE5xaxDB4) \cdot ONE5xaxDB4] \cdot DB12\right)}$

$SS78 = (\ 89608.1 \quad 88973.3\ )$

STEP 9 and 12:

Here we have computed two values at the same line which need for the correction factor to be subtracted from each. First we compute the appropriate correction factor, then subtrct ss912 := ss78 − CORRFACT2

$CF = 88312.05$

$CF2 := CF \cdot (1 \quad 1)$

$CF2 = (\ 88312.05 \quad 88312.05\ )$

$SS9and12 := SS78 - CF2$

$SS9and12 = (\ 1296.05 \quad 661.25\ )$

$SS9 := (\ 1296.05 \quad 661.25\ ) \cdot \begin{pmatrix} 1 \\ 0 \end{pmatrix}$
$\qquad\qquad$
$SS12 := (\ 1296.05 \quad 661.25\ ) \cdot \begin{pmatrix} 0 \\ 1 \end{pmatrix}$

$SS9 = 1296.05$

$SS12 = 661.25$

STEP 13-14: **(13)** Square each of the sums under the X column, divide by # of cases upon which each sum is based and sum (n = 5); **(14)** Subtract results from steps 5, 9 and 12 from ss13; **(15)** Subtract ss13 from ss3.

METHOD 1: PROGRAM: $(1/N_x)\ [1]_{1,8} ([SIM1A]_{1,8} diag[DB2X]_{8,8})^{\circ 2} -[SS9]_{1,1} - [SS12] - [SS3]_{1,1}\ (1/N_x)\ [1]_{1,4} ([SUM1B]_{4,2}[ONE0]_{2,1})^{\circ 2} - [SS9] - [SS12] - [CORRFACT]$

$Nx := 5$

$$\text{SS13and14} := \left(\frac{1}{Nx}\right)\overrightarrow{\left[\left(\overrightarrow{\text{sum1}\cdot(\text{ONEXA})^T}\right)\cdot\left(\overrightarrow{\text{sum1}\cdot(\text{ONEXA})^T}\right)\right]}\cdot\text{ONE8}^T - CF - SS9 - SS12$$

$$\text{ONEXA} := \begin{pmatrix} 1 \\ 0 \\ 1 \\ 0 \\ 1 \\ 0 \\ 1 \\ 0 \end{pmatrix}$$

$$\overrightarrow{\left(\text{sum1}\cdot(\text{ONEXA})^T\right)} = (\,339 \quad 0 \quad 268 \quad 0 \quad 406 \quad 0 \quad 316 \quad 0\,)$$

$$\text{SS13and14} = 18.05$$

$$\text{SS13and14T} := \left(\frac{1}{Nx}\right)\cdot\text{ONE8}\cdot\overrightarrow{\left[\left(\left(\overrightarrow{\text{sum1}\cdot(\text{ONEXA})^T}\right)^T\right)\cdot\left(\left(\overrightarrow{\text{sum1}\cdot(\text{ONEXA})^T}\right)^T\right)\right]} - CF - SS9 - SS12$$

$$\left(\frac{1}{5}\right)\cdot\overrightarrow{\left[\left(\overrightarrow{339 \quad 0 \quad 268 \quad 0 \quad 406 \quad 0 \quad 316 \quad 0}\right)\left(\overrightarrow{339 \quad 0 \quad 268 \quad 0 \quad 406 \quad 0 \quad 316 \quad 0}\right)\right]}\cdot\begin{pmatrix} 1 \\ 1 \\ 1 \\ 1 \\ 1 \\ 1 \\ 1 \\ 1 \end{pmatrix} - 88312.05 - 1296.05 - 661.25 = 18.05$$

Lets make this smaller to fit better on a page:

$$\left(\left(\overrightarrow{\text{sum1}\cdot(\text{ONEXA})^T}\right)^T\right) = \begin{pmatrix} 339 \\ 0 \\ 268 \\ 0 \\ 406 \\ 0 \\ 316 \\ 0 \end{pmatrix}$$

$$\text{SS13and14T} = 18.05$$

$$\text{SS13and14T} := \left(\frac{1}{5}\right) \cdot (1 \ 1 \ 1 \ 1 \ 1 \ 1 \ 1 \ 1) \cdot \overrightarrow{\left[\begin{pmatrix} 339 \\ 0 \\ 268 \\ 0 \\ 406 \\ 0 \\ 316 \\ 0 \end{pmatrix} \cdot \begin{pmatrix} 339 \\ 0 \\ 268 \\ 0 \\ 406 \\ 0 \\ 316 \\ 0 \end{pmatrix}\right]} - CF - SS9 - SS12$$

$$\text{SS13} := \text{ONE2} \cdot \text{SS13and14} \cdot \text{ONE0}$$

$$\text{SS14} := \text{ONE2} \cdot \text{SS13and14} \cdot \text{ZERO1}$$

$$\text{SS14} = 18.05 \qquad\qquad \text{SS13} = 18.05$$

METHOD 2:

$$\text{DB1}\cdot\text{a1} = \begin{pmatrix} 339 & 31 \\ 268 & 26 \\ 406 & 12 \\ 316 & 46 \end{pmatrix} \qquad \text{SUM1a}\cdot\text{ONE0} = \begin{pmatrix} 339 \\ 268 \\ 406 \\ 316 \end{pmatrix} \qquad \text{SUM1a} := \text{DB1}\cdot\text{a1}$$

$$\text{SS13and14Ta} := \left(\frac{1}{Nx}\right) \cdot \text{ONE4} \cdot \overrightarrow{[(\text{SUM1axONE0}) \cdot (\text{SUM1axONE0})]} - CF - SS9 - SS12$$

We substitute the following expression:

$$\text{SUM1axONE0} := \text{SUM1a}\cdot\text{ONE0}$$

$$\text{SS13and14Ta} := \left(\frac{1}{Nx}\right) \cdot \text{ONE4} \cdot \overrightarrow{[(\text{SUM1a}\cdot\text{ONE0}) \cdot (\text{SUM1a}\cdot\text{ONE0})]} - CF - SS9 - SS12 \quad\blacksquare$$

The Math looks like:

$$\text{SS13and14Ta} := \left(\frac{1}{5}\right) \cdot (1 \ 1 \ 1 \ 1) \cdot \overrightarrow{\left[\begin{pmatrix} 339 \\ 268 \\ 406 \\ 316 \end{pmatrix} \cdot \begin{pmatrix} 339 \\ 268 \\ 406 \\ 316 \end{pmatrix}\right]} - CF - SS9 - SS12$$

$$\text{SS13and14Ta} = 18.05$$

STEP 15: Subtract SS13 from SS3;

$$\text{SS15} := X^T \cdot X1 \cdot ONE0 - \left[\left(\frac{1}{Nx} \right) \cdot ONE4 \cdot \overrightarrow{[(SUM1axONE0) \cdot (SUM1axONE0)]} \right]$$

SS15 = 1689.6

X^T is too long for the Math to fit on a single page, so he Math becomes:

$$X^T \cdot \begin{pmatrix} 63 & 1 \\ 74 & 1 \\ 42 & 1 \\ 87 & 1 \\ 73 & 1 \\ 47 & 1 \\ 52 & 1 \\ 58 & 1 \\ 48 & 1 \\ 63 & 1 \\ 85 & 1 \\ 90 & 1 \\ 69 & 1 \\ 76 & 1 \\ 86 & 1 \\ 59 & 1 \\ 71 & 1 \\ 62 & 1 \\ 64 & 1 \\ 60 & 1 \end{pmatrix} \cdot \begin{pmatrix} 1 \\ 0 \end{pmatrix} - \left[\left(\frac{1}{5} \right) \cdot (1\ 1\ 1\ 1) \cdot \overrightarrow{\left[\left(\begin{pmatrix} 339 \\ 268 \\ 406 \\ 316 \end{pmatrix} \right) \cdot \left(\begin{pmatrix} 339 \\ 268 \\ 406 \\ 316 \end{pmatrix} \right) \right]} \right] = 1689.6$$

THE CRITERION VARIABLE (STEPS 16-28)

NOTE: The math is exactly the same as for the X column above, except we post-multiply by ZERO1 instead of ONE0.

STEP 16 and 17: Square all the values in the Y column and Sum the values in the Y column.

STEPS 16-19: Find the sum of squares for the Y column:

$$sqsumY := Y^T \cdot Y1$$

$$Y^T = (6 \ 6 \ 4 \ 7 \ 8 \ 6 \ 3 \ 4 \ 6 \ 7 \ 3 \ 2 \ 2 \ 1 \ 4 \ 10 \ 9 \ 9 \ 7 \ 11)$$

$$sqsumY = (813 \ \ 115)$$

$$sqsumY := (6 \ 6 \ 4 \ 7 \ 8 \ 6 \ 3 \ 4 \ 6 \ 7 \ 3 \ 2 \ 2 \ 1 \ 4 \ 10 \ 9 \ 9 \ 7 \ 11) \cdot \begin{pmatrix} 6 & 1 \\ 6 & 1 \\ 4 & 1 \\ 7 & 1 \\ 8 & 1 \\ 6 & 1 \\ 3 & 1 \\ 4 & 1 \\ 6 & 1 \\ 7 & 1 \\ 3 & 1 \\ 2 & 1 \\ 2 & 1 \\ 1 & 1 \\ 4 & 1 \\ 10 & 1 \\ 9 & 1 \\ 9 & 1 \\ 7 & 1 \\ 11 & 1 \end{pmatrix}$$

Here a = Sum of Squares, b = Sum of Elements.

$$sqsumY = (813 \ \ 115)$$

$$SS17 := Y^T \cdot Y1 \cdot \begin{pmatrix} 0 \\ 1 \end{pmatrix} \qquad\qquad Y^T \cdot Y1 \cdot \begin{pmatrix} 0 \\ 1 \end{pmatrix} = 115$$

$$SS17 = 115$$

$$SS16 = 813 \qquad\qquad b$$

$$\text{SS18} := \left(\frac{1}{\text{Ny}}\right) \cdot \left(Y^T \cdot Y1\right)^2$$

STEP 18: Square SS17 and divide by the number of measures needed to compute the Sum. Ny = 20. Computation of the Correction Factor for the Y components.

$$\text{SS18} := \left(\frac{1}{\text{Ny}}\right) \cdot \overrightarrow{\left[\left((Y)^T \cdot Y1\right)\left((Y)^T \cdot Y1\right)\right]}$$

We need the following substitution to compute the equation:

$$\text{YtranY1} := Y^T \cdot Y1$$

$$\text{SS18a} := \left(\frac{1}{\text{Ny}}\right) \cdot \overrightarrow{[(\text{YtranY1}) \cdot (\text{YtranY1})]}$$

$$\text{SS18a} = (\,33048.45 \quad 661.25\,)$$

$$\text{SS18} := \left(\frac{1}{\text{Ny}}\right) \cdot \overrightarrow{[(\text{YtranY1}) \cdot (\text{YtranY1})]} \cdot \text{ZERO1}$$

$$\text{SS18} = 661.25$$

$$\text{CFY} := \text{SS18}$$

$$\text{CFY} = 661.25$$

$$\text{CFY} := \left(\frac{1}{20}\right)\left[(Y)^T \cdot Y1 \cdot \begin{pmatrix} 0 \\ 1 \end{pmatrix}\left[Y^T \cdot Y1 \cdot \begin{pmatrix} 0 \\ 1 \end{pmatrix}\right]\right]$$

STEP 19: Subtract SS18 from SS16.

$$\text{SS16} - \text{SS18} = 151.75$$

$$\text{SS19} := Y^T \cdot Y1 \cdot \begin{pmatrix} 1 \\ 0 \end{pmatrix} - \left(\frac{1}{\text{Ny}}\right) \cdot \overrightarrow{[(\text{YtranY1}) \cdot (\text{YtranY1})]} \cdot \text{ZERO1}$$

$$\text{SS19} = 151.75$$

STEPS 20, 21 & 22; 23-25: (20) Add the sums of the Y column of the two Experiment 1 Groups, square and divide by # of cases on which the sum is based N_{class} = 10. **(21)** Add the sums of the Y column of the second two Experimental Groups, square and divide by # of cases on which the sum is based N_{conv} = 10. **(22)** Add the two & subtract the correction factor. (Will actually do all steps in one set of operations from 20-25)

The numbers in Y we add together are given in DB42.

$$DB42 := \begin{pmatrix} 0 & 0 & 0 & 0 \\ 1 & 0 & 1 & 0 \\ 0 & 0 & 0 & 0 \\ 0 & 1 & 1 & 0 \\ 0 & 0 & 0 & 0 \\ 1 & 0 & 0 & 1 \\ 0 & 0 & 0 & 0 \\ 0 & 1 & 0 & 1 \end{pmatrix} \qquad DB12 = \begin{pmatrix} 1 & 0 \\ 1 & 0 \\ 0 & 1 \\ 0 & 1 \end{pmatrix}$$

The basic equation is given by:

$$\frac{1}{10}\left[(ONES \cdot a \cdot DB42)^2 \cdot DB12\right] - CFY$$

This translates to:

$$\frac{1}{10} \cdot \left(\overrightarrow{[(ONES \cdot a \cdot DB42) \cdot (ONES \cdot a \cdot DB42)]} \cdot DB12\right) - CFY = \blacksquare$$

The substitution equation we need is given by:

$ONE5xaxDB42 := ONE5 \cdot a \cdot DB42$

and

$ONE5 \cdot a \cdot DB42$

But this is equal to:

$ONE5 \cdot a \cdot DB42 = (43 \quad 72 \quad 57 \quad 58)$

Squaring this equation we get:

$$\overrightarrow{[ONE5xaxDB42 \cdot (ONE5xaxDB42)]} = (\, 1849 \quad 5184 \quad 3249 \quad 3364 \,)$$

The final computation for SS22 and SS25 is given by:

$$SS2225 := \left(\frac{1}{10}\right) \cdot \left(\overrightarrow{[(ONE5xaxDB42) \cdot ONE5xaxDB42] \cdot DB12}\right) - CFY$$

$$SS2225 = (\, 42.05 \quad 0.05 \,)$$

Splitting this equation into its component parts we get:

$$SS22 := \left[\left(\frac{1}{10}\right) \cdot \left(\overrightarrow{[(ONE5xaxDB42) \cdot ONE5xaxDB42] \cdot DB12}\right) - CFY\right] \cdot ONE0$$

$$SS22 = 42.05$$

$$SS25 := \left[\left(\frac{1}{10}\right) \cdot \left(\overrightarrow{[(ONE5xaxDB42) \cdot ONE5xaxDB42] \cdot DB12}\right) - CFY\right] \cdot ZERO1$$

$$SS25 = 0.05$$

Method 2:

$$SUM1a = \begin{pmatrix} 339 & 31 \\ 268 & 26 \\ 406 & 12 \\ 316 & 46 \end{pmatrix}$$

$$ZERO1 = \begin{pmatrix} 0 \\ 1 \end{pmatrix}$$

$$SUM1AY := SUM1a \cdot ZERO1$$

$$SUM1AY = \begin{pmatrix} 31 \\ 26 \\ 12 \\ 46 \end{pmatrix}$$

Now we Sum the 1st and 3rd numbers, the 2nd and 4th, the 1st and 2nd and the 3rd and 4th numbers. This Matrix is smaller because we are dealing with 4 elements rather than 8 in Method 1.

$$\text{DB22} := \begin{pmatrix} 1 & 0 & 1 & 0 \\ 0 & 1 & 0 & 1 \\ 1 & 1 & 0 & 0 \\ 0 & 0 & 1 & 1 \end{pmatrix}$$

$$\text{DB22·SUM1AY} = \begin{pmatrix} 43 \\ 72 \\ 57 \\ 58 \end{pmatrix}$$

We proceed from here as we did above in Method 1.

STEPS 26-27: **(26)** Square each of the sums under the Y column, divide by # of cases upon which each sum is based and sum (N = 5); **(27)** Subtract results from steps SS18, SS22 and SS25 from SS26; **(28)** Subtract SS26 from SS16.

METHOD 1:

PROGRAM: (1/NY) [1]1,8([SIM1A]1,8diag[DB2Y]8,8)o2 − [SS22]1,1 −
[SS25] − [SS18]1,1(1/NY) [1]1,4([SUM1B]4,2[ZERO1]2,1)
o2 − SS22 − SS25 − CORRFACT

Ny := 5

$$\text{SS26} := \left(\frac{1}{\text{Ny}}\right)\cdot\left(\text{ONE8}\cdot\left(\overrightarrow{\overrightarrow{\left(\text{sum1}^T\cdot\text{ONEYA}\right)}\cdot\overrightarrow{\left(\text{sum1}^T\cdot\text{ONEYA}\right)}}\right)\right) \quad \overrightarrow{\left(\text{sum1}^T\cdot\text{ONEYA}\right)} = \begin{pmatrix} 0 \\ 31 \\ 0 \\ 26 \\ 0 \\ 12 \\ 0 \\ 46 \end{pmatrix} \quad \text{ONEYA} = \begin{pmatrix} 0 \\ 1 \\ 0 \\ 1 \\ 0 \\ 1 \\ 0 \\ 1 \end{pmatrix}$$

SS26 = 779.4

sum1 = (339 31 268 26 406 12 316 46)

$$SS27 := \left(\frac{1}{Ny}\right) \cdot \left(ONE8 \cdot \left(\overrightarrow{\left(\overrightarrow{sum1^T \cdot ONEYA}\right)\left(\overrightarrow{sum1^T \cdot ONEYA}\right)}\right)\right) - CFY - SS22 - SS25$$

$$SS27 = 76.05$$

METHOD 2:

$$SUM1a = \begin{pmatrix} 339 & 31 \\ 268 & 26 \\ 406 & 12 \\ 316 & 46 \end{pmatrix}$$

$$SS27a := \left(\frac{1}{Ny}\right) \cdot \overrightarrow{[(SUM1a \cdot ZERO1) \cdot (SUM1a \cdot ZERO1)]} \cdot ONE4^T - CFY - SS22 - SS25 \blacksquare$$

Substituting we get:

$$(SUM1axZERO1) := (SUM1a \cdot ZERO1)$$

$$ZERO1 = \begin{pmatrix} 0 \\ 1 \end{pmatrix} \qquad\qquad SUM1a \cdot ZERO1 = \begin{pmatrix} 31 \\ 26 \\ 12 \\ 46 \end{pmatrix}$$

$$SS27a := \left(\frac{1}{Ny}\right) \cdot \overrightarrow{[(SUM1axZERO1) \cdot (SUM1axZERO1)]} \cdot ONE4^T - CFY - SS22 - SS25$$

$$SS27a = 76.05$$

STEP 28: Subtract SS26 from SS16.

$$SS28 := SS16 - SS26$$

$$SS28 = 33.6$$

Where $\qquad SS16 := Y^T \cdot Y1 \cdot \begin{pmatrix} 1 \\ 0 \end{pmatrix} \qquad\qquad$ and

$$SS26 := \left(\frac{1}{Ny}\right) \cdot \left(ONE8 \cdot \left(\overrightarrow{\left(\overrightarrow{sum1^T \cdot ONEYA}\right)\left(\overrightarrow{sum1^T \cdot ONEYA}\right)}\right)\right)$$

The equation becomes:

$$SS28 := Y^T \cdot Y1 \cdot \begin{pmatrix} 1 \\ 0 \end{pmatrix} - \left(\frac{1}{Ny}\right) \cdot \left(\overrightarrow{ONE8 \cdot \left(\overrightarrow{\left(sum1^T \cdot ONEYA\right) \cdot \left(sum1^T \cdot ONEYA\right)}\right)}\right)$$

$SS28 = 33.6$

STEP 29: Multiply each of the X scores with it's paired Y score and Sum.

$SS29 := X^T \cdot Y$

$SS29 = 7470$

STEP 30: Multiply SS4 by SS17 and divide by N_{XY} # of XY pairs = 20).

**PROGRAM: SS30 = $(1/N_{XY})$ $([X]^T)_{20,1}[X1]_{20,2}[ZERO1]_{2,1})$ $([Y]^T_{20,1}[Y1]_{20,2}$
        $[ZERO1]_{2,1})$**

$Nxy := 20$

$SS30 := \left(\frac{1}{Nxy}\right) \cdot SS4 \cdot SS17$ $SS4 = 1329$ $SS17 = 115$ $SS30 = 7641.75$

$SS4 := X^T \cdot X1 \cdot \begin{pmatrix} 0 \\ 1 \end{pmatrix}$ $X^T \cdot X1 \cdot \begin{pmatrix} 0 \\ 1 \end{pmatrix} = 1329$

$SS17 := Y^T \cdot Y1 \cdot \begin{pmatrix} 0 \\ 1 \end{pmatrix}$ $Y^T \cdot Y1 \cdot \begin{pmatrix} 0 \\ 1 \end{pmatrix} = 115$

$SS30 := \left(\frac{1}{Nxy}\right)\left[X^T \cdot X1 \cdot \begin{pmatrix} 0 \\ 1 \end{pmatrix}\right]\left[Y^T \cdot Y1 \cdot \begin{pmatrix} 0 \\ 1 \end{pmatrix}\right]$

$SS30 = 7641.75$

STEP 31: Subtract SS30 from SS29

**PROGRAM: SS31 = $X^T Y$ - $(1/N_{XY})$ $([X]^T_{20,1}[X1]_{20,2}[ZERO1]_{2,1})$ $([Y]$
        $^T_{20,1}[Y1]_{20,2}[ZERO1]_{2,1})$**

$$SS31 := X^T \cdot Y - \left(\frac{1}{Nxy}\right)\left[X^T \cdot X1 \cdot \begin{pmatrix} 0 \\ 1 \end{pmatrix}\right]\left[Y^T \cdot Y1 \cdot \begin{pmatrix} 0 \\ 1 \end{pmatrix}\right]$$

$$SS31 := SS29 - SS30$$

$$SS31 = -171.75$$

STEP 32: Obtain the X sum and the Y sum for the two Experimental 1 Groups, multiply X by Y and divide by the # of scores in each group. $N_{EXP1} = 10$.

$$Nexp1 := 10$$

$$SS32 := \left(\frac{1}{Nexp1}\right)\left[10^{\overrightarrow{\log(\text{sum1} \cdot DB23) \cdot \begin{pmatrix} 1 \\ 1 \end{pmatrix}}}\right] \qquad DB23 := \begin{pmatrix} 1 & 0 \\ 0 & 1 \\ 0 & 0 \\ 0 & 0 \\ 1 & 0 \\ 0 & 1 \\ 0 & 0 \\ 0 & 0 \end{pmatrix}$$

$$\text{sum1} \cdot DB23 = (\,745 \quad 43\,)$$

We make the following substitution:

$$\text{sum1xDB23} := \text{sum1} \cdot DB23$$

We now have:

$$SS32 := \left(\frac{1}{Nexp1}\right) \cdot 10^{\overrightarrow{\log(\text{sum1xDB23}) \cdot TWO1}}$$

And

$$\overrightarrow{\log(\text{sum1xDB23}) \cdot TWO1} = 4.5056$$

$$SS32 = 3203.5$$

STEP 33: Obtain the X sum and the Y sum for the two Experiment 2 Groups multiply X by Y and divide by the # of scores in each group. $N_{exp2} = 10$.

PROGRAM: $[SS3233] = (1/N_{CLCON})(10EEX(LOG(DB2)_{2,4}[DB1]_{4,20}[A1]_{20,2}))$
$([1]_{2,1})$

$$SS32 := \left(\frac{1}{Nexp1}\right)\left[10^{\overrightarrow{\log(DB2 \cdot SUM1a) \cdot \binom{1}{1}}}\right]^{\blacksquare}$$

$$SUM1a = \begin{pmatrix} 339 & 31 \\ 268 & 26 \\ 406 & 12 \\ 316 & 46 \end{pmatrix}$$

We make the following substitution:

DB2xSUM1a := DB2·SUM1a

$$SS32 := \left(\frac{1}{Nexp1}\right)\left[10^{\overrightarrow{\log(DB2xSUM1a) \cdot \binom{1}{1}}}\right] \qquad \overrightarrow{\log(DB2xSUM1a) \cdot \binom{1}{1}} = \begin{pmatrix} 4.5056 \\ 4.6237 \\ 4.5391 \\ 4.622 \end{pmatrix}$$

$$SS32 = \begin{pmatrix} 3203.5 \\ 4204.8 \\ 3459.9 \\ 4187.6 \end{pmatrix} \qquad DB2 \cdot SUM1a = \begin{pmatrix} 745 & 43 \\ 584 & 72 \\ 607 & 57 \\ 722 & 58 \end{pmatrix} \qquad DB2 = \begin{pmatrix} 1 & 0 & 1 & 0 \\ 0 & 1 & 0 & 1 \\ 1 & 1 & 0 & 0 \\ 0 & 0 & 1 & 1 \end{pmatrix}$$

Here is something interesting, we have just saved ourselves a lot of steps computing the solutions of SS32, SS33, SS35 and SS36. Let's define,

ONE4 := (1 0 0 0)   THREE4 := (0 0 1 0)   SS32 := 2972   SS33 := 4074

TWO4 := (0 1 0 0)   FOUR4 := (0 0 0 1)   SS35 := 3327.5   SS36 := 3960

$$SS32 := ONE4\left[\left(\frac{1}{Nexp1}\right)\left[10^{\overrightarrow{\log(DB2xSUM1a) \cdot \binom{1}{1}}}\right]\right] \qquad SS33 := TWO4\left[\left(\frac{1}{Nexp1}\right)\left[10^{\overrightarrow{\log(DB2xSUM1a) \cdot \binom{1}{1}}}\right]\right]$$

$$SS35 := THREE4 \cdot \left[\left(\frac{1}{Nexp1} \right) \left[10^{\overrightarrow{\log(DB2xSUM1a)} \cdot \binom{1}{1}} \right] \right] \qquad SS36 := FOUR4 \cdot \left[\left(\frac{1}{Nexp1} \right) \left[10^{\overrightarrow{\log(DB2xSUM1a)} \cdot \binom{1}{1}} \right] \right]$$

SS32 = 3203.5

SS33 = 4204.8

SS35 = 3459.9

SS36 = 4187.6

STEP 34: Add SS32 + SS33 - SS30 $N_{CLCON} = 10$; $N_{XY} = 20$

PROGRAM: SS34 = (1/N_{CLCON}) [1]$_{1,2}$ (10EEX(LOG([DB2]$_{2,4}$([DB1]$_{4,20}$[A1]$_{20,2}$)))
    [[1]$_{2,1}$) - (1/N_{XY})([X]^T[X1][ZERO1])([Y]^T[Y1][ZERO1])

SS34 := SS32 + SS33 - SS30

SS34 = -233.45

$$ONE4 \cdot \left[\left(\frac{1}{Nexp1} \right) \left[10^{\overrightarrow{\log(DB2xSUM1a)} \cdot \binom{1}{1}} \right] \right] + TWO4 \cdot \left[\left(\frac{1}{Nexp1} \right) \left[10^{\overrightarrow{\log(DB2xSUM1a)} \cdot \binom{1}{1}} \right] \right] - \left(\frac{1}{Nxy} \right) \left[X^T \cdot X1 \binom{0}{1} \right] \left[Y^T \cdot Y1 \binom{0}{1} \right] = -233.45$$

STEP 35 & 36: Obtain the X sum and the Y sum for the two LAB groups, multiply X by Y and divide by the # of scores in each group. $N_{LAB} = 10$.

STEP 36: Obtain the x sum and the Y sum for the two non-LAB groups multiply X by Y and divide by the # of scores in each group. $N_{NL} = 10$.

$$SS35 := THREE4 \cdot \left[\left(\frac{1}{Nexp1} \right) \left[10^{\overrightarrow{\log(DB2xSUM1a)} \cdot \binom{1}{1}} \right] \right]$$

$$SS36 := FOUR4 \cdot \left[\left(\frac{1}{Nexp1} \right) \left[10^{\overrightarrow{\log(DB2xSUM1a)} \cdot \binom{1}{1}} \right] \right]$$

SS35 = 3459.9

SS36 = 4187.6

STEP 37: Add SS35+SS36-SS30

SS35 = 3459.9

SS36 = 4187.6

SS30 = 7641.75

SS37 := SS35 + SS36 − SS30

SS37 = 5.75

$$SS35 := THREE4 \cdot \left[\left(\frac{1}{Nexp1} \right) \left[10^{\overrightarrow{\log(DB2xSUM1a)} \cdot \binom{1}{1}} \right] \right]$$

$$SS36 := FOUR4 \cdot \left[\left(\frac{1}{Nexp1} \right) \left[10^{\overrightarrow{\log(DB2xSUM1a)} \cdot \binom{1}{1}} \right] \right]$$

The math for SS37 looks like:

$$THREE4 \left[\left(\frac{1}{Nexp1} \right) \left[10^{\overrightarrow{\log(DB2xSUM1a)} \cdot \binom{1}{1}} \right] \right] + FOUR4 \left[\left(\frac{1}{Nexp1} \right) \left[10^{\overrightarrow{\log(DB2xSUM1a)} \cdot \binom{1}{1}} \right] \right] - \left(\frac{1}{Nxy} \right) \left[X^T \cdot X1 \binom{0}{1} \right] \left[Y^T \cdot Y1 \binom{0}{1} \right] = 5.75$$

SS35 = 3459.9

SS36 = 4187.6

STEP 38: Multiply each X sum by its paired Y sum in SUM1B. Divide each product by the number of scores upon which each sum is based (N_{YXP} = 5) and add the quotients.

SS38 = $(1/N_{XYP})$ [1]1,4($^{\circ}$S [SUM1B])$_{4,2}$o[1]$_{1,4}$)

Nxyp := 5   ONE4 := (1 1 1 1)

$$SS38 \cdot \left(\frac{1}{\text{Nxyp}}\right) \left[\text{ONE4} \cdot \overrightarrow{\left(\text{[SUM1a} \cdot \text{ZERO1} \cdot (\text{SUM1a} \cdot \text{ONE0})]\right)}\right] \quad \text{SUM1a} \cdot \text{ZERO1} = \begin{pmatrix} 31 \\ 26 \\ 12 \\ 46 \end{pmatrix} \quad \text{SUM1a} \cdot \text{ONE0} = \begin{pmatrix} 339 \\ 268 \\ 406 \\ 316 \end{pmatrix}$$

In order to solve this equation, we make the following substitutions:

$$\left(\frac{1}{\text{Nxyp}}\right) \left[\text{ONE4} \cdot \overrightarrow{\left(\text{[SUM1axZERO1} \cdot (\text{SUM1axONE0})]\right)}\right] \quad \text{SUM1axONE0} := \text{SUM1a} \cdot \text{ONE0}$$

$$\overrightarrow{\left(\text{[SUM1axZERO1} \cdot (\text{SUM1axONE0})]\right)} = \begin{pmatrix} 10509 \\ 6968 \\ 4872 \\ 14536 \end{pmatrix}$$

$$\text{SUM1axZERO1} := \text{SUM1a} \cdot \text{ZERO1}$$

$$SS38 := \left(\frac{1}{\text{Nxyp}}\right) \left[\text{ONE4} \cdot \overrightarrow{\left(\text{[SUM1axZERO1} \cdot (\text{SUM1axONE0})]\right)}\right]$$

The Math looks like:

$$(1 \;\; 1 \;\; 1 \;\; 1) \cdot \left(\left[\overrightarrow{\left[\begin{pmatrix} 31 \\ 26 \\ 12 \\ 46 \end{pmatrix} \cdot \begin{pmatrix} 339 \\ 268 \\ 406 \\ 316 \end{pmatrix}\right]}\right]\right)$$

$$\left(\frac{1}{5}\right) \cdot \left[(1 \;\; 1 \;\; 1 \;\; 1) \cdot \left(\left[\overrightarrow{\left[\begin{pmatrix} 31 \\ 26 \\ 12 \\ 46 \end{pmatrix} \cdot \begin{pmatrix} 339 \\ 268 \\ 406 \\ 316 \end{pmatrix}\right]}\right]\right)\right] = 7377$$

$$SS38 = 7377$$

METHOD 2:

NOTE: Here is a way to get the same solution using pure half-multipliers.

$$\left(\frac{1}{N_{XYP}}\right) \cdot \left[\text{ONE4} \cdot \left[\Sigma^{\circ} \cdot ((\text{SUM1A} \cdot \text{oFOUR1}))\right]\right]_{\square}$$

$$\text{SUM1a} = \begin{pmatrix} 339 & 31 \\ 268 & 26 \\ 406 & 12 \\ 316 & 46 \end{pmatrix} \qquad \text{FOUR1} = \begin{pmatrix} 1 \\ 1 \\ 1 \\ 1 \end{pmatrix}$$

$$\begin{pmatrix} 339 & 31 \\ 268 & 26 \\ 406 & 12 \\ 316 & 46 \end{pmatrix} \cdot\!\circ\! \cdot \begin{pmatrix} 1 \\ 1 \\ 1 \\ 1 \end{pmatrix} := \left[\begin{pmatrix} 339 \\ 268 \\ 406 \\ 316 \end{pmatrix} \cdot\!\circ\! \cdot \begin{pmatrix} 31 \\ 26 \\ 12 \\ 46 \end{pmatrix}\right] \quad \overrightarrow{\left[\begin{pmatrix} 31 \\ 26 \\ 12 \\ 46 \end{pmatrix} \cdot \begin{pmatrix} 339 \\ 268 \\ 406 \\ 316 \end{pmatrix}\right]} = \begin{pmatrix} 10509 \\ 6968 \\ 4872 \\ 14536 \end{pmatrix} \quad \left(\frac{1}{5}\right) \cdot \left[(1\ \ 1\ \ 1\ \ 1) \cdot \begin{pmatrix} 10140 \\ 6675 \\ 4050 \\ 14175 \end{pmatrix}\right] = 7008$$

STEP 39: Subtract SS30, SS34 and SS37 from SS38.

$$SS39 := SS38 - SS30 - SS34 - SS37$$

$$SS39 = -37.05$$

$$SS39 := \left(\frac{1}{Nxyp}\right)\left[ONE4 \cdot \overrightarrow{[SUM1axZERO1 \cdot (SUM1axONE0)]}\right] - \left(\frac{1}{Nxy}\right)\left[X^T \cdot X1 \cdot \begin{pmatrix} 0 \\ 1 \end{pmatrix}\right]\left[Y^T \cdot Y1 \cdot \begin{pmatrix} 0 \\ 1 \end{pmatrix}\right] - ONE4 \cdot \left[\left(\frac{1}{Nexp1}\right)\left[10^{\overrightarrow{\log(DB2xSUM1a)}} \cdot \begin{pmatrix} 1 \\ 1 \end{pmatrix}\right]\right] + TWO4 \cdot \left[\left(\frac{1}{Nexp1}\right)\left[10^{\overrightarrow{\log(DB2xSUM1a)}} \cdot \begin{pmatrix} 1 \\ 1 \end{pmatrix}\right]\right] - \left(\frac{1}{Nxy}\right)\left[X^T \cdot X1 \cdot \begin{pmatrix} 0 \\ 1 \end{pmatrix}\right]\left[Y^T \cdot Y1 \cdot \begin{pmatrix} 0 \\ 1 \end{pmatrix}\right] - THREE4 \cdot \left[\left(\frac{1}{Nexp1}\right)\left[10^{\overrightarrow{\log(DB2xSUM1a)}} \cdot \begin{pmatrix} 1 \\ 1 \end{pmatrix}\right]\right] + FOUR4 \cdot \left[\left(\frac{1}{Nexp1}\right)\left[10^{\overrightarrow{\log(DB2xSUM1a)}} \cdot \begin{pmatrix} 1 \\ 1 \end{pmatrix}\right]\right] - \left(\frac{1}{Nxy}\right)\left[X^T \cdot X1 \cdot \begin{pmatrix} 0 \\ 1 \end{pmatrix}\right]\left[Y^T \cdot Y1 \cdot \begin{pmatrix} 0 \\ 1 \end{pmatrix}\right]$$

This equation is too long to fit even on 3 pages, so I've broken it down into 3 parts so we can see how complex this computation is.

$$SS39 := \left(\frac{1}{Nxyp}\right)\left[ONE4 \cdot \overrightarrow{[SUM1axZERO1 \cdot (SUM1axONE0)]}\right] - \left(\frac{1}{Nxy}\right)\left[X^T \cdot X1 \cdot \begin{pmatrix} 0 \\ 1 \end{pmatrix}\right]\left[Y^T \cdot Y1 \cdot \begin{pmatrix} 0 \\ 1 \end{pmatrix}\right] - $$

$$ONE4 \cdot \left[\left(\frac{1}{Nexp1}\right)\left[10^{\overrightarrow{\log(DB2xSUM1a)}} \cdot \begin{pmatrix} 1 \\ 1 \end{pmatrix}\right]\right] + TWO4 \cdot \left[\left(\frac{1}{Nexp1}\right)\left[10^{\overrightarrow{\log(DB2xSUM1a)}} \cdot \begin{pmatrix} 1 \\ 1 \end{pmatrix}\right]\right] - \left(\frac{1}{Nxy}\right)\left[X^T \cdot X1 \cdot \begin{pmatrix} 0 \\ 1 \end{pmatrix}\right]\left[Y^T \cdot Y1 \cdot \begin{pmatrix} 0 \\ 1 \end{pmatrix}\right] - $$

$$THREE4 \cdot \left[\left(\frac{1}{Nexp1}\right)\left[10^{\overrightarrow{\log(DB2xSUM1a)}} \cdot \begin{pmatrix} 1 \\ 1 \end{pmatrix}\right]\right] + FOUR4 \cdot \left[\left(\frac{1}{Nexp1}\right)\left[10^{\overrightarrow{\log(DB2xSUM1a)}} \cdot \begin{pmatrix} 1 \\ 1 \end{pmatrix}\right]\right] - \left(\frac{1}{Nxy}\right)\left[X^T \cdot X1 \cdot \begin{pmatrix} 0 \\ 1 \end{pmatrix}\right]\left[Y^T \cdot Y1 \cdot \begin{pmatrix} 0 \\ 1 \end{pmatrix}\right]$$

The computations are over. The Mathematical expression for SS39 shows us the absurdity of trying to express the final computations as a function of Mathematical equivalents. The final form of the 2-Variable Analysis of Covariance will be in terms of the computed SS values.

STEP 40: Subtract SS38 from SS29.

$$SS40 := SS29 - SS38$$

$$SS40 = 93$$

STEP 41: SS28 - Square SS40 and divide by SS15.

$$SS41 := SS28 - \frac{SS40^2}{SS15}$$

$SS41 = 28.481$

STEP 42: Add SS9 + SS15.

$SS42 := SS9 + SS15$

$SS42 = 2985.65$

STEP 43: Add SS22 + SS28.

$SS43 := SS22 + SS28$

$SS43 = 75.65$

STEP 44: Add SS40 + SS34.

$SS44 := SS40 + SS34$

$SS44 = -140.45$

STEP 45: Square SS44 and divide by SS42, then subtract this quotient and SS41 from SS43.

$$SS45 := SS43 - \frac{SS44^2}{SS42} - SS41$$

$SS45 = 40.562$

STEP 46: Add SS12 + SS15.

$SS46 := SS12 + SS15$

$SS46 = 2350.85$

STEP 47: Add SS28 + SS25.

SS47 := SS28 + SS25

SS47 = 33.65

STEP 48: Add SS37 + SS40.

SS48 := SS37 + SS40

SS48 = 98.75

STEP 49: Square SS48 and divide by SS46, then subtract this quotient and SS41 from SS46.

$$SS49 := SS47 - \frac{SS48^2}{SS46} - SS41$$

SS49 = 1.0209

STEP 50: Add SS14 + SS15.

SS50 := SS14 + SS15

SS50 = 1707.65

STEP 51: Add SS27 + SS28.

SS51 := SS27 + SS28

SS51 = 109.65

STEP 52: Add SS39 + SS40.

SS52 := SS39 + SS40

SS52 = 55.95

STEP 53: Square SS52 and divide by SS50, then add SS41TO this calculation from SS51.

$$SS53 := SS51 - \left(SS41 + \frac{SS52^2}{SS50} \right)$$

$SS53 = 79.3358$

MEAN SQUARES AND F RATIOS (STEPS 54-60)

STEP 54: Divide SS41 (which is the Adjusted Error Sum of Squares) by the Degrees of Free (df) for the Adjusted Error. df for the adjusted error = N - CL - 1 where N = total # of X (20 in this example), C is the # of experimental conditions on the first variable (2 in the example, Exp1 Vs Exp2), and L = # of experimental conditions on the other variable (2 in the example, Exp3 Vs no Exp3). So N-CL-1 = 20 - (2x2)-1 = **15.**

$N := 20 \quad C := 2 \quad L := 2 \quad N - C \cdot L - 1 = 15$

$$SS54 := \frac{SS41}{N - C \cdot L - 1}$$

$SS54 = 1.8987$

$SSems := SS54$

STEP 55: Divide SS45 (adjusted sum of squares for teaching methods) by df = C - 1 where C = the # of experimental conditions in this variable (2 in this example, Exp1 Vs Exp2).

$$SS55 := \frac{SS45}{C - 1} \qquad SSecond := SS55$$

$SS55 = 40.562$

STEP 56: This is the computation of the F-ratio. Divide SS55 by SS54.

$$SS56 := \frac{SS55}{SS54}$$

SS56 = 21.3626

SS57 = 1.0209

F := SS56

STEP 57: Divide SS49(the Adjusted Sum of Squares for the 3rd Experimental condition usage) by df = L-1 where L = # of Experimental Conditions in this Variable. L= 2 in this example (Exp 3 vs. no Exp 3).

$$SS57 := \frac{SS49}{L-1} \qquad\qquad S57 := msExp3$$

SS57 = 1.0209

Checking Appendix E, we find that for a df of 1/15, a F-ratio of .5377 is non-significant in the 10% range. It is concluded that having Experimental Condition 3 or not makes no difference.

STEP 58: Divide SS57/SS54 to get the F-ratio for Exp 3 usage:

$$df := \frac{L-1}{N - C \cdot L - 1} \qquad\qquad df := \frac{1}{15}$$

$$Fexp3 := \frac{SS57}{SS54} \qquad\qquad Fexp3 = 0.5377$$

From Appendix E. (Computational Handbook of Statistics, 2nd Ed.) with a df of 1/15, an F-ratio greater than 5.1 would be expected by chance alone one in a thousand times, therefore, this F is significant at the .001% level. It is concluded that the different Experimental methods were significantly different in their effectiveness.

STEP 59: Divide SS53 (the adjuasted Sum of Squares for the methods by Experimental Condition 3) by,

$$df := (C-1) \cdot (L-1)$$

$$SS59 := \frac{SS53}{(C-1)\cdot(L-1)} \qquad SS59 = 79.3358$$

This is the mean square for Experimental Condition 1 by Experimental Condition 3. ms1x3

STEP 60: Divide SS59/ms1x3 to yield the F-ratio for Exp 1 and Exp 2 by Exp 3 interaction.

$$F := \frac{SS59}{SS54}$$

$F = 41.7835$

Appendix E shows that this F is significant at the .001 level. This means at least one of the first 2 Experimental Factors (Exp1 or Exp2 interacted with Exp3). To check out the source of interaction, use the procedures found in Sect. 3.10.

SIMPLE CHI-SQUARE AND THE *PHI* COEFFICIENT

COMPUTATIONAL HANDBOOK OF STATISTICS,

2ND EDITION, SECTION 5.5, PG. 230-232.

When we have frequency data comparing the effects of two variables and there are two groups associated with each variable, we can compute the phi coefficient to find the degree of relationship between the two variables, while the chi-square test will determine whether the variables are related. **A significant chi-square is interpreted as showing a relationship between the two variables. The phi coefficient gives a numerical value (from 0 to 1) for that relationship.**

STATISTICAL FORMULA

$$\chi^2 = \frac{N(AD-BC)^2}{(A+B)\,(C+D)\,(A+C)\,(B+D)} \qquad PHI\ (\phi) = (\chi^2/N)^{1/2}$$

$$A := \begin{pmatrix} 85 & 60 \\ 30 & 40 \end{pmatrix} \qquad ONE2 := (1 \quad 1) \qquad TWO1 := \begin{pmatrix} 1 \\ 1 \end{pmatrix}$$

STEP 1: Organize the data into a contingency table, the values represent the number of people in each category.

	EXP COND 1	EXP COND 2
GROUP1	85	60
GROUP2	30	40

STEP 2: Add all the numbers in the Table. This is the Grand Sum.

$$[1]_{1,2}\,[A]_{2,2}\,[1]_{2,1} = GS$$

$ONE2 \cdot A \cdot TWO1 = 215$

The math looks like:

$$\textbf{GS} := (1 \quad 1) \cdot \begin{pmatrix} 85 & 60 \\ 30 & 40 \end{pmatrix} \cdot \begin{pmatrix} 1 \\ 1 \end{pmatrix}$$

$\textbf{GS} = 215$

STEP 3: Find the Row and Column sums.

$\textbf{ROW SUM} \quad = ([A]_{2,2}[1]_{2,1})^{T}$

$\textbf{COLUMN SUM} = ([1]_{1,2}[A]_{2,2}$

NOTE: We transpose the Row Sum to get it from a Column Matrix to a Row Matrix. Now to compute the 2 values.

$\textbf{SUMCOL} := \textbf{ONE2} \cdot \textbf{A}$

$$(1 \quad 1) \cdot \begin{pmatrix} 85 & 60 \\ 30 & 40 \end{pmatrix} = (115 \quad 100)$$

$\textbf{SUMROW} := (\textbf{A} \cdot \textbf{TWO1})^{T}$

$$\left[\begin{pmatrix} 85 & 60 \\ 30 & 40 \end{pmatrix} \cdot \begin{pmatrix} 1 \\ 1 \end{pmatrix} \right]^{T} = (145 \quad 70)$$

$\textbf{SUMCOL} = (115 \quad 100)$

$\textbf{SUMROW} = (145 \quad 70)$

STEP 4: Multiply the 4 Sums together computed in STEP 3. The way below seems to be the easiest way to multiply them all together.

$$\textbf{MULTIPLYALL} := 10^{\left(\overleftarrow{\log(\textbf{SUMCOL})} + \overrightarrow{\log(\textbf{SUMROW})} \right)} \cdot \textbf{TWO1}$$

$\textbf{MULTIPLYALL} = 116725000$

Mathematically, this looks like:

$$\mathbf{MULTIPLYALL} := 10^{\left(\overrightarrow{\log\left[(1\quad 1)\cdot\begin{pmatrix}85 & 60\\ 30 & 40\end{pmatrix}\right]}+\overrightarrow{\log\left(\left[\begin{pmatrix}85 & 60\\ 30 & 40\end{pmatrix}\cdot\begin{pmatrix}1\\ 1\end{pmatrix}\right]^{T}\right)}\right)\cdot\begin{pmatrix}1\\ 1\end{pmatrix}} \quad\blacksquare$$

We cannot compute this as it is, so we'll just compute the products of the Matrices.

$$\mathbf{MULTIPLYALL} := 10^{\left(\overrightarrow{\log((115\quad 100))}+\overrightarrow{\log((145\quad 70))}\right)\cdot\mathbf{TWO1}}$$

$$\mathbf{MULTIPLYALL} = 116725000$$

STEPS 5,6 AND 7: Compute the Determinant of A.

We can do this with the built in MathCad function for determining determinants.

$$|\mathbf{A}| = 1600$$

Or we can compute the determinant as follows:

$$\mathbf{A} = \begin{pmatrix}85 & 60\\ 30 & 40\end{pmatrix}$$

$$\mathbf{detA} := 85\cdot 40 - 30\cdot 60$$

$$\mathbf{detA} = 1600$$

STEP 8: Square the determinate of A and multiply by the Grandsum of the Matrix GS.

PROGRAM: $|\mathbf{A}|_{2,2}{}^2 \times [1]_{1,2}[\mathbf{A}]_{2,2}[1]_{2,1}$

$$\mathbf{SS8} := \left(|\mathbf{A}|\right)^2\cdot(\mathbf{ONE2}\cdot\mathbf{A}\cdot\mathbf{TWO1})$$

$$\mathbf{SS8} = 550400000$$

The Math looks like:

$$\mathbf{SS8} := \left[\left|\begin{pmatrix}85 & 60\\ 30 & 40\end{pmatrix}\right|\right]^2\cdot\left[(1\quad 1)\cdot\begin{pmatrix}85 & 60\\ 30 & 40\end{pmatrix}\cdot\begin{pmatrix}1\\ 1\end{pmatrix}\right]$$

$$\mathbf{SS8} = 550400000$$

STEP 9: Divide SS8 by STEP 4, this is the final computation for Chi-Square.

$$SS9 := \frac{SS8}{MULTIPLYALL}$$

$$SS9 = 4.715$$

To get to the Math, we proceed as follows. NOTE: Since both values are single element Matrices (1x1 Matrices) we do not have to multiply by the inverse of the Matrix to divide. NOTE: We have this in a single computation also.

PROGRAM:
$$\frac{(\,|A|\,)^2\,[1]_{1,2}[A]_{2,2}[1]_{2,1}}{(10EEX(LOG(ROWSUMS)^T + LOG(COLSUM)\,)\,[TWO1]\,)}$$

The equation for SS9 is

$$SS9 := \frac{(\,|A|\,)^2 \cdot (ONE2 \cdot A \cdot TWO1)}{10^{\overrightarrow{\log(SUMCOL)} + \overrightarrow{\log(SUMROW)}} \cdot TWO1}$$

$$SS9 = 4.715$$

Now lets look at the Math:

$$SS9 := \frac{\left[\left|\begin{pmatrix} 85 & 60 \\ 30 & 40 \end{pmatrix}\right|\right]^2 \cdot \left[(1 \quad 1) \cdot \begin{pmatrix} 85 & 60 \\ 30 & 40 \end{pmatrix} \cdot \begin{pmatrix} 1 \\ 1 \end{pmatrix}\right]}{10^{\overrightarrow{\log\left[(1 \quad 1) \cdot \begin{pmatrix} 85 & 60 \\ 30 & 40 \end{pmatrix}\right]} + \overrightarrow{\log\left(\left[\begin{pmatrix} 85 & 60 \\ 30 & 40 \end{pmatrix} \cdot \begin{pmatrix} 1 \\ 1 \end{pmatrix}\right]^T\right)}} \cdot \begin{pmatrix} 1 \\ 1 \end{pmatrix}}$$

This reduces to:

$$SS9 := \frac{2560000 \cdot 215}{10^{\overrightarrow{\log((115 \quad 100))} + \overrightarrow{\log((145 \quad 70))}} \cdot TWO1}$$

Which reduces to

$$SS9 := \frac{2560000 \cdot 215}{116725000}$$

SS9 = 4.715

The # of degrees of freedom for a 2x2 chi-square will always be one. From Appendix D, we find a chi-square value greater than 3.8 is significant at the .05 level. **Therefore the relationship of Experimental Condition 1 and Experimental Condition 2 between the 2 Groups are related.**

NOTE: If we can compute the denominator in its present form, we have written the Simple Chi-Square in a single equation.

STEP 10: Computation of the phi coefficient ϕ.

$$\phi := \left(\frac{\chi^2}{N} \right)^{\frac{1}{2}^{\blacksquare}} \qquad\qquad \phi := \left(\frac{SS9}{GS} \right)$$

$\phi = 0.022$

ϕ represents the degree of relationship between the 2 variables. (EXP1 and EXP2).

SUPPLEMENT

THE YATES CORRECTION FOR CONTINUITY

In a 2x2 contingency table, if the frequency of **any** cell is less than 10, it is recommended that the Yates correction for continuity be used to acount for the fact that x2 computed with very small N's will over-estimate the true x2 value. Yates correction is defined as:

$$\chi^2 = \frac{N((AD-BC) - N/2)^2}{(A+B)(C+D)(A+C)(B+D)} \qquad PHI\ (\phi) = (\ \chi^2/N\ )^{1/2}$$

$$GS = 215$$

Or using what we have already computed we have

$$YATESCORRECTIONCHISQ := \frac{GS \cdot \left(|A| - \frac{GS}{2} \right)^2}{\underset{10}{}\left(\overrightarrow{\log(SUMCOL)} + \overrightarrow{\log(SUMROW)} \right) \cdot TWO1}$$

YATESCORRECTIONCHISQ $= 4.103$

$$\dfrac{\mathbf{GS} \cdot \left(|\mathbf{A}| - \dfrac{\mathbf{GS}}{2} \right)^2}{10^{\overrightarrow{\left(\overrightarrow{\log\,(\mathbf{SUMCOL})} + \overrightarrow{\log\,(\mathbf{SUMROW})} \right) \cdot \mathbf{TWO1}}}} = 4.103$$

Chi-Square computed above =

$$\chi^2 := 4.715 \quad \blacksquare$$

We do not need the Yates Correction here as all elements are greater than 10.

COMPLEX CHI-SQUARE AND THE CONTINGENCY COEFFICIENT

Note: I've already worked this problem out in advance, and it's solution is really neat, for it uses everything mathematicians say is illegal. Chi-square complex is strikingly similar to Perturbation Theory in Quantum Chemistry, and I predict that this method is one of the connecting equations between Statistics and Physics.

When we have frequency data comparing the effects of two variables and there are more than two greoups associated with either of the two variables, the complex chi-square can be used to test the hyptheiss of no relationship between the variables. If we find a relationship, we can compute the contingency coefficicent to determine the degree of relationship.

Suppose we have 3 groups of people and 4 experimental conditions, each group is categorized according to the desired experimental condition and a frequency count is made for each category. We wish to find if the groups and Experimental Conditions are related.

STATISTICAL FORMULA

$$X^2 := \frac{\sum (O - E)^2 {}^{\blacksquare}}{E}$$

Where O is the observed frequency for a particular cell in the Contingency Table, and E = the expected frequency for a cell, based upon marginal totals (STEPS 2 - 6.

And the Contingency Coefficient is given by

$$C := \left(\frac{X^2}{X^2 + N} \right)^{\frac{1}{2}{}^{\blacksquare}}$$

where N is the Grand Sum of the Contingency Table.

VARIABLES:

$$A := \begin{pmatrix} 50 & 20 & 30 & 40 \\ 40 & 50 & 30 & 60 \\ 20 & 30 & 40 & 20 \end{pmatrix} \qquad \textbf{FOUR1} := \begin{pmatrix} 1 \\ 1 \\ 1 \\ 1 \end{pmatrix} \qquad \textbf{ONE3} := (1 \quad 1 \quad 1)$$

This is a chart of the Contingency Table used to obtain the Grand Sum.

G1	EXP1	EXP2	EXP3	EXP4
G1	50	20	30	40
G2	40	50	30	60
G3	20	30	40	20

PROGRAM: $\text{SUM1} = [1]_{1,3}[A]_{3,4}[1]_{4,1}$

GS := **ONE3** · **A** · **FOUR1**

GS = 430

The Math looks like:

$$\textbf{GS} := (1 \quad 1 \quad 1) \cdot \begin{pmatrix} 50 & 20 & 30 & 40 \\ 40 & 50 & 30 & 60 \\ 20 & 30 & 40 & 20 \end{pmatrix} \cdot \begin{pmatrix} 1 \\ 1 \\ 1 \\ 1 \end{pmatrix}$$

GS = 430

STEP 3: Add the numbers in each Row.

PROGRAM: $\text{ROWSUM} = [A]_{3,4}[1]_{4,1}$
$\qquad\qquad\quad \text{ROWSUM} := (A \cdot \text{FOUR1})^{\text{T}}$
$\qquad\qquad\quad \text{ROWSUM} = (140 \quad 180 \quad 110)$

The Math looks like:

$$\textbf{ROWSUM} := \begin{pmatrix} 50 & 20 & 30 & 40 \\ 40 & 50 & 30 & 60 \\ 20 & 30 & 40 & 20 \end{pmatrix} \cdot \begin{pmatrix} 1 \\ 1 \\ 1 \\ 1 \end{pmatrix}$$

Without transposing the solution gives a Column Matrix.

$$\mathbf{ROWSUM} = \begin{pmatrix} 140 \\ 180 \\ 110 \end{pmatrix}$$

We Transpose this solution to make future computations using logarithms simpler.

STEP 4: Add the numbers in each Column.

PROGRAM: COLUMNSUM $= [1]_{1,4}[A]_{3,4}$

COLUMNSUM = ONE3 A

COLUMNSUM $= (110 \quad 100 \quad 100 \quad 120)$

The Math looks like:

$$\mathbf{COLUMNSUM} := (1 \quad 1 \quad 1) \cdot \begin{pmatrix} 50 & 20 & 30 & 40 \\ 40 & 50 & 30 & 60 \\ 20 & 30 & 40 & 20 \end{pmatrix}$$

$$\mathbf{COLUMNSUM} = (110 \quad 100 \quad 100 \quad 120)$$

STEPS 5 & 6: Make a new Table, but unlike the book, we make the Table by multiplying the Sums by the Column Matrix FOUR1 and ONE3. NOTE: This is a more generalized form of the Perturbation relation $\Psi o \Psi$. We divide by the Grand Sum or the Matrix.

$$\mathbf{SS56} := \left(\frac{1}{\mathbf{GS}}\right) \cdot [\, (\mathbf{A} \cdot \mathbf{FOUR1}) \cdot (\mathbf{ONE3} \cdot \mathbf{A}) \,] \qquad\qquad \mathbf{GS} = 430$$

$$\mathbf{SS56} = \begin{pmatrix} 35.814 & 32.558 & 32.558 & 39.07 \\ 46.047 & 41.86 & 41.86 & 50.233 \\ 28.14 & 25.581 & 25.581 & 30.698 \end{pmatrix}$$

The Math looks like:

$$\mathbf{SS56} := \left(\frac{1}{430}\right) \cdot \left[\begin{pmatrix} 50 & 20 & 30 & 40 \\ 40 & 50 & 30 & 60 \\ 20 & 30 & 40 & 20 \end{pmatrix} \cdot \begin{pmatrix} 1 \\ 1 \\ 1 \\ 1 \end{pmatrix} \right] \cdot \left[(1 \quad 1 \quad 1) \cdot \begin{pmatrix} 50 & 20 & 30 & 40 \\ 40 & 50 & 30 & 60 \\ 20 & 30 & 40 & 20 \end{pmatrix} \right]$$

Which reduces to:

$$SS56 = \begin{pmatrix} 35.814 & 32.558 & 32.558 & 39.07 \\ 46.047 & 41.86 & 41.86 & 50.233 \\ 28.14 & 25.581 & 25.581 & 30.698 \end{pmatrix}$$

NOTE: This is just like the Perturbation Condition in the section on Quantum Chemistry. We now subtract the perturbed expression SS56 from the Contingency Table A.

DIFFERENCE := **A** − **SS56**

$$DIFFERENCE = \begin{pmatrix} 14.186 & -12.558 & -2.558 & 0.93 \\ -6.047 & 8.14 & -11.86 & 9.767 \\ -8.14 & 4.419 & 14.419 & -10.698 \end{pmatrix}$$

The Math looks like:

$$DIFFERENCE := A - \left[\left(\frac{1}{GS} \right) \cdot [\, (A \cdot FOUR1) \cdot (ONE3 \cdot A) \,] \right]$$

$$DIFFERENCE = \begin{pmatrix} 14.186 & -12.558 & -2.558 & 0.93 \\ -6.047 & 8.14 & -11.86 & 9.767 \\ -8.14 & 4.419 & 14.419 & -10.698 \end{pmatrix}$$

Now we invert each individual element in the Matrix SS56. NOTE: We do not take the inverse of the Matrix, this is an illegal transformation as Mathematics views Matrix Division today.

$$SS56inverse := \overrightarrow{SS56}^{-1}$$

$$\overrightarrow{SS56}^{-1} = \begin{pmatrix} 0.028 & 0.031 & 0.031 & 0.026 \\ 0.022 & 0.024 & 0.024 & 0.02 \\ 0.036 & 0.039 & 0.039 & 0.033 \end{pmatrix}$$

The Math looks like:

$$SS56inverse := \left(\frac{1}{GS} \right) \cdot \overrightarrow{[\, (A \cdot FOUR1) \cdot (ONE3 \cdot A) \,]}^{-1}$$

MathCad cannot do the Math, doubly in this case, so we will make the following double substitution.

$$\textbf{AxFOUR1xONE3xA} := (\textbf{A} \cdot \textbf{FOUR1}) \cdot (\textbf{ONE3} \cdot \textbf{A})$$

Now we can solve the equation.

$$\textbf{SS56inverse} := \left[\left(\frac{1}{\textbf{GS}} \right) \cdot \overrightarrow{\textbf{AxFOUR1xONE3xA}} \right]^{-1}$$

$$\textbf{SS56inverse} = \begin{pmatrix} 0.028 & 0.031 & 0.031 & 0.026 \\ 0.022 & 0.024 & 0.024 & 0.02 \\ 0.036 & 0.039 & 0.039 & 0.033 \end{pmatrix}$$

Now we Square each individual element in the DIFFERENCESQUARED Matrix. NOTE: We do not Square the Matrix, but we Square each individual element in the DIFFERENCE Matrix. This is also an illegal mathematical transformation as Mathematics views Matrices today, but it seems to work very well for Statistics.

$$\textbf{DIFFERENCESQUARED} := \overrightarrow{\textbf{DIFFERENCE}}^2$$

$$\textbf{DIFFERENCESQUARED} = \begin{pmatrix} 201.244 & 157.707 & 6.544 & 0.865 \\ 36.56 & 66.252 & 140.671 & 95.403 \\ 66.252 & 19.524 & 207.896 & 114.44 \end{pmatrix}$$

The Math looks like:

$$\textbf{DIFFERENCESQUARED} := \overrightarrow{\left[\begin{pmatrix} 14.186 & -12.558 & -2.558 & 0.93 \\ -6.047 & 8.14 & -11.86 & 9.767 \\ -8.14 & 4.419 & 14.419 & -10.698 \end{pmatrix} \cdot \begin{pmatrix} 14.186 & -12.558 & -2.558 & 0.93 \\ -6.047 & 8.14 & -11.86 & 9.767 \\ -8.14 & 4.419 & 14.419 & -10.698 \end{pmatrix} \right]}$$

$$\textbf{DIFFERENCESQUARED} = \begin{pmatrix} 201.243 & 157.703 & 6.543 & 0.865 \\ 36.566 & 66.26 & 140.66 & 95.394 \\ 66.26 & 19.528 & 207.908 & 114.447 \end{pmatrix}$$

STEP 7: We now multiply each individual element in the SS56inverse Matrix by its corresponding element in DIFFERENCESQUARED Matrix, we do not multiply the matrix column by row. This is also an illegal mathematical transformation as mathematics views matrix multiplication today.

$$SS7 := \left(\overrightarrow{(\text{DIFFERENCESQUARED} \cdot \text{SS56inverse})} \right)$$

$$SS7 = \begin{pmatrix} 5.619 & 4.844 & 0.201 & 0.022 \\ 0.794 & 1.583 & 3.36 & 1.899 \\ 2.355 & 0.763 & 8.127 & 3.728 \end{pmatrix}$$

The Math looks like: (we have to use the substitutions for AxFOUR1 AND ONE3xA derived above for this equation to compute).

$$\text{DIFFERENCESQUARED} := \left[\overrightarrow{A - \left(\frac{1}{GS} \right) \cdot \text{AxFOUR1xONE3xA}} \right]^2$$

$$\text{SS56inverse} := \left[\overrightarrow{\left(\frac{1}{GS} \right) \cdot \text{AxFOUR1xONE3xA}} \right]^{-1}$$

$$SS7 := \left[\left[\overrightarrow{A - \left(\frac{1}{GS} \right) \cdot \text{AxFOUR1xONE3xA}} \right]^2 \cdot \left[\overrightarrow{\left(\frac{1}{GS} \right) \cdot \text{AxFOUR1xONE3xA}} \right]^{-1} \right]$$

$$SS7 = \begin{pmatrix} 5.619 & 4.844 & 0.201 & 0.022 \\ 0.794 & 1.583 & 3.36 & 1.899 \\ 2.354 & 0.763 & 8.127 & 3.728 \end{pmatrix}$$

STEP 8: Add all the elements in SS7 to compute X^2.

$$X^2 := \text{ONE3} \cdot SS7 \cdot \text{FOUR1} \qquad\qquad \text{CHISQ} := \text{ONE3} \cdot SS7 \cdot \text{FOUR1}$$

$$\text{ONE3} \cdot SS7 \cdot \text{FOUR1} = 33.295$$

Expanding the equation by inserting SS7 above we get

$$\text{ONE3} \cdot \left(\left[\left[\overrightarrow{A - \left(\frac{1}{GS} \right) \cdot \text{AxFOUR1xONE3xA}} \right]^2 \cdot \left[\overrightarrow{\left(\frac{1}{GS} \right) \cdot \text{AxFOUR1xONE3xA}} \right]^{-1} \right] \right) \cdot \text{FOUR1} = 33.295$$

NOTE: We have computed Chi-Square in a single equation. The only outside computation we need is the Grand Sum of the Contingency Table.

STEP 9: Compute the degrees of freedom:

df = (# rows-1)(# columns-1) = (3-1)(4-1) = 2x3 = 6

A Chi-Square value larger than 12.6 with 6 degrees of freedom is significant at the .05 % level. (Appendix D).

STEP 10: Find the Contingency Coefficient by computing the following equation

$$C := \left(\frac{x^2}{x^2 + N} \right)^{\frac{1}{2}}$$

This translates to:

$$C := \left(\frac{CHISQ}{CHISQ + GS} \right)^{\frac{1}{2}}$$

Which becomes:

$$\left(\frac{33.295}{33.295 + 430} \right)^{\frac{1}{2}} = 0.268$$

and

C = 0.268

The Contingency Coefficient C is significantly different from zero, it is concluded that the Groups vs experimental conditions are related.

SUMMARY

All the math just done here in this problem is impossible according to the knowledge of today's mathematics; we cannot square the individual elements of a matrix; we cannot invert the individual elements of a matrix; we cannot multiply element by element with Matrices and have them mean everything. Therefore the answers we have just computed must be incorrect, (because of the method) even if the answers are right. But element by element operations work for statistics! They actually make the computations simple! But remember, we are not doing Mathematic here, we are doing simple arithmetic.

A number can be considered as a one by one Matrix (a Tensor of rank zero), but a row, column, square or rectangular Matrix can itself be considered a single number that follows the rules of single numbers. Each individual row or column can be added or subtracted from each other, element from corresponding element. Each individual element in the row or column can be multiplied by its corresponding element in the second Matrix, the solution being the same as doing the multiplication's one at a time on each element of the two Matrices. But we can do these all at the same time using the Half-Multiplier Operator. Each element can be multiplied by its corresponding element in another Matrix on a one-to-one correspondence without summing, or we can transpose the pre-multiplier, matrix multiply in Gaussian fashion and sum the products if we wish.

QUANTUM STATISTICS

Suppose we have completed the experimental data on an experiment and put into the matrix form $[A]_{MN}$. Now $[A]_{MN}$ may not be a square matrix, so we must make a square out of it before we can do any sort of quantum statistical analysis on it. Squaring the data changes the matrix $[A]$ from a statistical matrix to a probability matrix. To keep track of each of the sums of the squares for each column in the matrix, we must square it such that we end up with an NxN matrix.

$$[A]^{T}_{MN}[A]_{MN} = [A]_{NM}[A]_{MN} = [A]^{2}_{NN}$$

Then we do the following:

$$\frac{1/N[DB1]_{iN}[A]^2_{NN}[DB1]^T_{iN}}{[DB1]_{iN}[DB1]^T_{iN}} = \frac{1/N[DB1]_{iN}[A]^2_{NN}[DB1]_{Ni}}{[DB1]_{iN}[DB1]_{Ni}} = \frac{1/N[DB1]_{iN}[A]^2_{NN}[DB1]_{Ni}}{[DB1]^2_{ii}} =$$

$$\frac{1/N[C]^2_{ii}}{[DB1]^2_{ii}}$$

EXAMPLE: Since we've already done a series of different analysis on the problem in section 2.7, I will do the quantum statistics on that set of data. I will not try to interpret the results since I do not know what it means, but I can do the math. Professional statisticians will have to determine whether there is any significance in the answers or not. The data of the problem in section forms a 11x9 matrix, $[A]_{11,9}$.

$$A := \begin{bmatrix} 3 & 3 & 4 & 2 & 3 & 6 & 2 & 4 & 8 \\ 3 & 4 & 4 & 1 & 4 & 5 & 3 & 6 & 3 \\ 1 & 3 & 4 & 3 & 6 & 7 & 4 & 7 & 7 \\ 1 & 2 & 3 & 2 & 4 & 6 & 1 & 7 & 4 \\ 2 & 3 & 5 & 1 & 2 & 3 & 3 & 7 & 12 \\ 4 & 5 & 6 & 4 & 5 & 5 & 1 & 4 & 2 \\ 4 & 5 & 6 & 1 & 3 & 4 & 1 & 5 & 3 \\ 1 & 4 & 5 & 1 & 2 & 4 & 2 & 5 & 6 \\ 1 & 4 & 5 & 2 & 3 & 4 & 3 & 6 & 6 \\ 2 & 3 & 3 & 3 & 5 & 6 & 1 & 5 & 12 \\ 2 & 2 & 4 & 4 & 6 & 7 & 4 & 9 & 7 \end{bmatrix}$$

$$DB1 := \begin{bmatrix} 1 & 0 & 0 \\ 1 & 0 & 0 \\ 1 & 0 & 0 \\ 0 & 1 & 0 \\ 0 & 1 & 0 \\ 0 & 1 & 0 \\ 0 & 0 & 1 \\ 0 & 0 & 1 \\ 0 & 0 & 1 \end{bmatrix}$$

$$DB3 := \begin{bmatrix} 1 & 0 & 0 \\ 0 & 1 & 0 \\ 0 & 0 & 1 \\ 1 & 0 & 0 \\ 0 & 1 & 0 \\ 0 & 0 & 1 \\ 1 & 0 & 0 \\ 0 & 1 & 0 \\ 0 & 0 & 1 \end{bmatrix}$$

THIS IS THE MATHCAD +6 PROGRAM FOR THE QUANTUM STATISTIC FOR DB1

$$\Psi 1 := \left(\frac{1}{99}\right) \cdot \left(DB1^T \cdot A^T \cdot A \cdot DB1\right)$$

$$\Psi 2 := DB1^T \cdot DB1$$

$$\Psi := \Psi 1 \cdot \Psi 2^{-1}$$

$$\Psi = \begin{pmatrix} 4.037 & 4.101 & 5.158 \\ 4.101 & 5.158 & 6.148 \\ 5.158 & 6.148 & 8.552 \end{pmatrix}$$

THE FOLLOWING IS THE ABOVE PROGRAM ELUCIDATED STEP BY TEP:

$$A^T \cdot A = \begin{bmatrix} 66 & 90 & 113 & 53 & 94 & 122 & 49 & 133 & 138 \\ 90 & 142 & 178 & 80 & 144 & 189 & 82 & 214 & 224 \\ 113 & 178 & 229 & 105 & 186 & 245 & 110 & 284 & 298 \\ 53 & 80 & 105 & 66 & 108 & 135 & 56 & 145 & 153 \\ 94 & 144 & 186 & 108 & 189 & 239 & 102 & 262 & 269 \\ 122 & 189 & 245 & 135 & 239 & 313 & 133 & 343 & 363 \\ 49 & 82 & 110 & 56 & 102 & 133 & 71 & 160 & 168 \\ 133 & 214 & 284 & 145 & 262 & 343 & 160 & 407 & 423 \\ 138 & 224 & 298 & 153 & 269 & 363 & 168 & 423 & 560 \end{bmatrix}$$

$$DB1^T = \begin{pmatrix} 1 & 1 & 1 & 0 & 0 & 0 & 0 & 0 & 0 \\ 0 & 0 & 0 & 1 & 1 & 1 & 0 & 0 & 0 \\ 0 & 0 & 0 & 0 & 0 & 0 & 1 & 1 & 1 \end{pmatrix}$$

QUANTUM STATISTIC FOR DB1

THIS IS THE TERM IN THE DENOMINATOR:

$$\begin{pmatrix} 1 & 1 & 1 & 0 & 0 & 0 & 0 & 0 & 0 \\ 0 & 0 & 0 & 1 & 1 & 1 & 0 & 0 & 0 \\ 0 & 0 & 0 & 0 & 0 & 0 & 1 & 1 & 1 \end{pmatrix} \cdot \begin{bmatrix} 1 & 0 & 0 \\ 1 & 0 & 0 \\ 1 & 0 & 0 \\ 0 & 1 & 0 \\ 0 & 1 & 0 \\ 0 & 1 & 0 \\ 0 & 0 & 1 \\ 0 & 0 & 1 \\ 0 & 0 & 1 \end{bmatrix} = \begin{pmatrix} 3 & 0 & 0 \\ 0 & 3 & 0 \\ 0 & 0 & 3 \end{pmatrix}$$

$$\left(\frac{1}{99}\right) \cdot \left[\begin{pmatrix} 1 & 1 & 1 & 0 & 0 & 0 & 0 & 0 & 0 \\ 0 & 0 & 0 & 1 & 1 & 1 & 0 & 0 & 0 \\ 0 & 0 & 0 & 0 & 0 & 0 & 1 & 1 & 1 \end{pmatrix} \cdot \begin{bmatrix} 66 & 90 & 113 & 53 & 94 & 122 & 49 & 133 & 138 \\ 90 & 142 & 178 & 80 & 144 & 189 & 82 & 214 & 224 \\ 113 & 178 & 229 & 105 & 186 & 245 & 110 & 284 & 298 \\ 53 & 80 & 105 & 66 & 108 & 135 & 56 & 145 & 153 \\ 94 & 144 & 186 & 108 & 189 & 239 & 102 & 262 & 269 \\ 122 & 189 & 245 & 135 & 239 & 313 & 133 & 343 & 363 \\ 49 & 82 & 110 & 56 & 102 & 133 & 71 & 160 & 168 \\ 133 & 214 & 284 & 145 & 262 & 343 & 160 & 407 & 423 \\ 138 & 224 & 298 & 153 & 269 & 363 & 168 & 423 & 560 \end{bmatrix} \cdot \begin{bmatrix} 1 & 0 & 0 \\ 1 & 0 & 0 \\ 1 & 0 & 0 \\ 0 & 1 & 0 \\ 0 & 1 & 0 \\ 0 & 1 & 0 \\ 0 & 0 & 1 \\ 0 & 0 & 1 \\ 0 & 0 & 1 \end{bmatrix} \begin{pmatrix} 3 & 0 & 0 \\ 0 & 3 & 0 \\ 0 & 0 & 3 \end{pmatrix}^{-1} \right] = \begin{pmatrix} 4.037 & 4.101 & 5.158 \\ 4.101 & 5.158 & 6.148 \\ 5.158 & 6.148 & 8.552 \end{pmatrix}$$

THE QUANTUM STATISTIC FOR DB3

$$\begin{bmatrix} 1 & 0 & 0 \\ 0 & 1 & 0 \\ 0 & 0 & 1 \\ 1 & 0 & 0 \\ 0 & 1 & 0 \\ 0 & 0 & 1 \\ 1 & 0 & 0 \\ 0 & 1 & 0 \\ 0 & 0 & 1 \end{bmatrix}^T = \begin{pmatrix} 1 & 0 & 0 & 1 & 0 & 0 & 1 & 0 & 0 \\ 0 & 1 & 0 & 0 & 1 & 0 & 0 & 1 & 0 \\ 0 & 0 & 1 & 0 & 0 & 1 & 0 & 0 & 1 \end{pmatrix}$$

$[DB3]^T_{3,9} \ [DB3]_{9,3} = $ VALUE IN THE DENOMINATOR

$$\begin{pmatrix} 1 & 0 & 0 & 1 & 0 & 0 & 1 & 0 & 0 \\ 0 & 1 & 0 & 0 & 1 & 0 & 0 & 1 & 0 \\ 0 & 0 & 1 & 0 & 0 & 1 & 0 & 0 & 1 \end{pmatrix} \cdot \begin{bmatrix} 1 & 0 & 0 \\ 0 & 1 & 0 \\ 0 & 0 & 1 \\ 1 & 0 & 0 \\ 0 & 1 & 0 \\ 0 & 0 & 1 \\ 1 & 0 & 0 \\ 0 & 1 & 0 \\ 0 & 0 & 1 \end{bmatrix} = \begin{pmatrix} 3 & 0 & 0 \\ 0 & 3 & 0 \\ 0 & 0 & 3 \end{pmatrix}$$

$\Psi_{DB3} = (1/99)([DB3]^T[A]^T[A][DB3])([DB3]^T[DB3])^{-1}$

$$\left(\frac{1}{99}\right) \cdot \begin{pmatrix} 1 & 0 & 0 & 1 & 0 & 0 & 1 & 0 & 0 \\ 0 & 1 & 0 & 0 & 1 & 0 & 0 & 1 & 0 \\ 0 & 0 & 1 & 0 & 0 & 1 & 0 & 0 & 1 \end{pmatrix} \cdot \begin{bmatrix} 66 & 90 & 113 & 53 & 94 & 122 & 49 & 133 & 138 \\ 90 & 142 & 178 & 80 & 144 & 189 & 82 & 214 & 224 \\ 113 & 178 & 229 & 105 & 186 & 245 & 110 & 284 & 298 \\ 53 & 80 & 105 & 66 & 108 & 135 & 56 & 145 & 153 \\ 94 & 144 & 186 & 108 & 189 & 239 & 102 & 262 & 269 \\ 122 & 189 & 245 & 135 & 239 & 313 & 133 & 343 & 363 \\ 49 & 82 & 110 & 56 & 102 & 133 & 71 & 160 & 168 \\ 133 & 214 & 284 & 145 & 262 & 343 & 160 & 407 & 423 \\ 138 & 224 & 298 & 153 & 269 & 363 & 168 & 423 & 560 \end{bmatrix} \begin{bmatrix} 1 & 0 & 0 \\ 0 & 1 & 0 \\ 0 & 0 & 1 \\ 1 & 0 & 0 \\ 0 & 1 & 0 \\ 0 & 0 & 1 \\ 1 & 0 & 0 \\ 0 & 1 & 0 \\ 0 & 0 & 1 \end{bmatrix} \cdot \begin{pmatrix} 3 & 0 & 0 \\ 0 & 3 & 0 \\ 0 & 0 & 3 \end{pmatrix}^{-1} = \begin{pmatrix} 1.747 & 3.347 & 3.963 \\ 3.347 & 6.66 & 7.862 \\ 3.963 & 7.862 & 9.811 \end{pmatrix}$$

The point we are at is actually Quantum Accounting, to get the
Quantum Statistic we must sum the Matrix to a single value. i.e.

$$ONE3 \left[\left(\frac{1}{99} \right) \cdot \left[\left(DB3^T \right) \left(A^T \right) (DB3) \right] \cdot \left[\left(DB3^T \right) (DB3) \right]^{-1} \right] \cdot THREE$$

or

$$ONE3 \begin{pmatrix} 1.747 & 3.347 & 3.963 \\ 3.347 & 6.66 & 7.862 \\ 3.963 & 7.862 & 9.811 \end{pmatrix} \cdot THREE1 = 48.562$$

I do not know if these values have any meaning. I do not know how to compute the Correction Factor. We may just have to look at the diagonal for the statistical significance, or we may have to take the grand sum of the solution and check it. Or we may find that both the diagonal and grand sum are useful, or that both of them have no meaning at all. I leave this to the professionals to determine.

Just for the heck of it, (and because they do it in quantum mechanics) lets normalize the solutions. Going back to the chapter on quantum chemistry (I do not have my book paged yet so I can't tell you the page, it is all in separate files in my word processor waiting to be put together into final publishing form, so I have to say it is on the 40th page of that chapter), you'll find that the equation for normalizing the solution is:

$$\gamma_{jj} = \text{diag}(C^T_{jk} C_{jk})$$

then

$$\Psi_{\text{NORM}} = (\gamma_{jj}^{-1})^{1/2} \Psi$$

I'll solve the two problems side by side. The matrix solutions using DB1 and DB3 are:

$$\Psi DB1 := \begin{pmatrix} 4.037 & 4.101 & 5.158 \\ 4.101 & 5.158 & 6.148 \\ 5.158 & 6.148 & 8.552 \end{pmatrix} \qquad \Psi DB3 := \begin{pmatrix} 1.747 & 3.347 & 3.963 \\ 3.347 & 6.66 & 7.862 \\ 3.963 & 7.862 & 9.811 \end{pmatrix}$$

$$I3 := \begin{pmatrix} 1 & 0 & 0 \\ 0 & 1 & 0 \\ 0 & 0 & 1 \end{pmatrix}$$

$\gamma jj1 := \Psi DB1^T \cdot \Psi DB1$ $\gamma jj4 := \Psi DB3^T \cdot \Psi DB3$ Here I am squaring the DB matrices

$\gamma jj2 := \overrightarrow{(I3 \cdot \gamma jj1)}$ $\gamma jj5 := \overrightarrow{(I3 \cdot \gamma jj4)}$ Here I am making a matrix only of the diagonal

$\gamma jj3 := \left(\gamma jj2^{-1}\right)$ $\gamma jj6 := \left(\gamma jj5^{-1}\right)$ Here I'm taking the inverse of the matrix

$\gamma jj := \overrightarrow{\gamma jj3}^{\frac{1}{2}}$ $\gamma jjDB3 := \overrightarrow{\gamma jj6}^{\frac{1}{2}}$ Here I take the square root of each element

$$\gamma jj = \begin{pmatrix} 0.129 & 0 & 0 \\ 0 & 0.111 & 0 \\ 0 & 0 & 0.085 \end{pmatrix} \quad \gamma jjDB3 = \begin{pmatrix} 0.183 & 0 & 0 \\ 0 & 0.092 & 0 \\ 0 & 0 & 0.076 \end{pmatrix}$$ These are the normalizing multipliers

$\Psi NORMDB1 := \gamma jj \cdot \Psi DB1$ $\Psi NORMDB3 := \gamma jjDB3 \cdot \Psi DB3$

Here I multiply the normalizing factor times the solution Matrix.

$$\Psi NORMDB1 = \begin{pmatrix} 0.522 & 0.531 & 0.667 \\ 0.455 & 0.572 & 0.682 \\ 0.44 & 0.524 & 0.729 \end{pmatrix} \quad \Psi NORMDB3 = \begin{pmatrix} 0.319 & 0.611 & 0.724 \\ 0.309 & 0.615 & 0.726 \\ 0.301 & 0.596 & 0.744 \end{pmatrix}$$

Normalized solutions.

CHECK: FOR DB1 AND DB3, THE NORMALIZED SOLUTIONS SQUARED SHOULD EQUAL 1.

$CHK\Psi NORMDB1 := \Psi NORMDB1 \cdot \Psi NORMDB1^T$

$$CHK\Psi NORMDB1 = \begin{pmatrix} 1 & 0.997 & 0.995 \\ 0.997 & 1 & 0.998 \\ 0.995 & 0.998 & 1 \end{pmatrix}$$

$$\text{CHECK}\Psi\text{NORMDB1} := \begin{pmatrix} 0.522 & 0.531 & 0.667 \\ 0.455 & 0.572 & 0.682 \\ 0.44 & 0.524 & 0.729 \end{pmatrix} \cdot \begin{pmatrix} 0.522 & 0.531 & 0.667 \\ 0.455 & 0.572 & 0.682 \\ 0.44 & 0.524 & 0.729 \end{pmatrix}^{T}$$

$$\text{CHECK}\Psi\text{NORMDB1} = \begin{pmatrix} 0.999 & 0.996 & 0.994 \\ 0.996 & 0.999 & 0.997 \\ 0.994 & 0.997 & 1 \end{pmatrix}$$

$$\text{CHECK}\Psi\text{NORMDB3} := \Psi\text{NORMDB3} \cdot \Psi\text{NORMDB3}^{T}$$

$$\text{CHECK}\Psi\text{NORMDB3} = \begin{pmatrix} 1 & 1 & 1 \\ 1 & 1 & 1 \\ 1 & 1 & 1 \end{pmatrix}$$

LINEAR REGRESSION

Note: I did not develop this, it is how linear regression is computed on computers and calculators. I thought I'd not only show how it works, but also show how close Statisticians were to finding that Statistics is truly a matrix mathematics rather than a random variable mathematics.

No book on Statistics would be complete without a discussion of Linear Regression, or the best fit of a line or a curve through a set of experimental data points. In regular algebra, the equation for a line is given by $Y = mX + b$. The equation for a line in matrix form is given by $Y = MX$. Suppose we generate a set of data points in a laboratory experiment such that:

$$y_1 = ax_1 + b$$
$$y_2 = ax_2 + b$$

.   .   .

.   .   .

$$y_n = ax_n + b$$

The matrix equation looks like:

$$
\begin{matrix}
\mathbf{M} & \mathbf{X} & = & \mathbf{Y}
\end{matrix}
$$

$$
\begin{vmatrix}
x_1 & 1 \\
x_2 & 1 \\
x_3 & 1 \\
. & . \\
. & . \\
x_n & 1
\end{vmatrix}
\begin{vmatrix}
a \\
b
\end{vmatrix}
=
\begin{vmatrix}
y_1 \\
y_2 \\
y_3 \\
. \\
. \\
y_n
\end{vmatrix}
$$

Now take a general vector MV such that V = to the perpendicular distance from the x axis. We then multiply the matrix equation for a line by the general vector MV such that:

$$(MV)^T (MX-Y) = 0$$

expanding we get

$$V^T (M^T MX - M^T Y) = 0$$
$$V^T (M^T MX - M^T Y) = 0$$

We divide out the V^T term, and get

$$M^TMX - M^TY = 0$$
$$M^TMX = M^TY$$

and solving for X we get

$$X = (M^TM)^{-1}M^TY$$

First let's look at regular linear regression to see how to apply this equation to our experimental data.

REGULAR LINEAR REGRESSION

THE FORCE CONSTANT OF A SPRING

The equation for a force acting on a spring is given by $\mathbf{F} = k\mathbf{X} + b$. Suppose the data recorded in our experiment was as follows:

Y_{pounds}	$X_{spring\ stretch}$
0	6.1"
2	7.6"
4	8.7"
6	10.4"

$$Y := \begin{bmatrix} 0 \\ 2 \\ 4 \\ 6 \end{bmatrix} \qquad M := \begin{bmatrix} 6.1 & 1 \\ 7.6 & 1 \\ 8.7 & 1 \\ 10.4 & 1 \end{bmatrix} \qquad \left(M^T \cdot M\right)^{-1} \cdot M^T \cdot Y = \begin{pmatrix} 1.42 \\ -8.643 \end{pmatrix}$$

Then

$$M^T = \begin{pmatrix} 6.1 & 7.6 & 8.7 & 10.4 \\ 1 & 1 & 1 & 1 \end{pmatrix} = \begin{pmatrix} k \\ b \end{pmatrix}$$

or

$$\left[\begin{pmatrix} 6.1 & 7.6 & 8.7 & 10.4 \\ 1 & 1 & 1 & 1 \end{pmatrix} \cdot \begin{bmatrix} 6.1 & 1 \\ 7.6 & 1 \\ 8.7 & 1 \\ 10.4 & 1 \end{bmatrix} \right]^{-1} \cdot \begin{pmatrix} 6.1 & 7.6 & 8.7 & 10.4 \\ 1 & 1 & 1 & 1 \end{pmatrix} \cdot \begin{bmatrix} 0 \\ 2 \\ 4 \\ 6 \end{bmatrix} = \begin{pmatrix} 1.42 \\ -8.643 \end{pmatrix}$$

The first value is the value of the slope m, and the lower value is the value of the y intercept b. The equation F = kX + b becomes $\mathbf{F = 1.42X - 8.643}$.

SEMI-LOG LINEAR REGRESSION

Suppose you are a doctor or pharmacist. Your patient needs to have at least a .50 mg per 100 cc's concentration of Dilantin in their blood to keep them from having epileptic seizures. If the concentration falls below this level the patient may experience seizures. The patient is injected with a .5 gram per 70 Kg body weight sample of the drug. After 30 minutes a sample of blood is drawn and analyzed. For the next hour a sample is drawn every 30 minutes. Thereafter a sample is drawn every two hours and analyzed. Calculate the half-life of Dilantin in your patients body.

The semi-log equation is given by $Y = ae^{bx}$. Rewriting this as a logarithmic equation we get:

$$Log\ Y = Log\ A + bX$$

Let $x = X$; $Log\ Y = Y$ and $Log\ A = A$

Substituting these changes of variables into the logarithmic equation, we get:

$$Y = bX + A$$

The logarithmic equation is now linear.

Suppose the experimental values are as follows:

time	mg/100cc
0.5	1.12
1.0	.90
1.5	.80
2.0	.73
4.0	.50
6.0	.36
8.0	.20

The logarithmic equation looks like:

$$
\begin{vmatrix} x_1 & 1 \\ x_2 & 1 \\ x_3 & 1 \\ \cdot & \cdot \\ \cdot & \cdot \\ x_n & 1 \end{vmatrix} \quad \begin{vmatrix} b \\ Log\ a \end{vmatrix} = \begin{vmatrix} Log\ y_1 \\ Log\ y_2 \\ Log\ y_3 \\ \cdot \\ \cdot \\ Log\ y_n \end{vmatrix}
$$

$$
M := \begin{bmatrix} .5 & 1 \\ 1.0 & 1 \\ 1.5 & 1 \\ 2 & 1 \\ 4 & 1 \\ 6 & 1 \\ 8 & 1 \end{bmatrix} \quad M^T = \begin{pmatrix} 0.5 & 1 & 1.5 & 2 & 4 & 6 & 8 \\ 1 & 1 & 1 & 1 & 1 & 1 & 1 \end{pmatrix} \quad Y := \begin{bmatrix} log(1.12) \\ log(.90) \\ log(.80) \\ log(.73) \\ log(.50) \\ log(.36) \\ log(.20) \end{bmatrix} \quad Y = \begin{bmatrix} 0.0492 \\ -0.0458 \\ -0.0969 \\ -0.1367 \\ -0.301 \\ -0.4437 \\ -0.699 \end{bmatrix}
$$

$$
\left(M^T \cdot M\right)^{-1} \cdot M^T \cdot Y = \begin{pmatrix} -0.0918 \\ 0.0624 \end{pmatrix}
$$

$$
\left[\begin{pmatrix} 0.5 & 1 & 1.5 & 2 & 4 & 6 & 8 \\ 1 & 1 & 1 & 1 & 1 & 1 & 1 \end{pmatrix} \cdot \begin{bmatrix} .5 & 1 \\ 1.0 & 1 \\ 1.5 & 1 \\ 2 & 1 \\ 4 & 1 \\ 6 & 1 \\ 8 & 1 \end{bmatrix}\right]^{-1} \cdot \begin{pmatrix} 0.5 & 1 & 1.5 & 2 & 4 & 6 & 8 \\ 1 & 1 & 1 & 1 & 1 & 1 & 1 \end{pmatrix} \cdot \begin{bmatrix} 0.0492 \\ -0.0458 \\ -0.0969 \\ -0.1367 \\ -0.301 \\ -0.4437 \\ -0.699 \end{bmatrix} = \begin{pmatrix} -0.09177 \\ 0.06239 \end{pmatrix}
$$

Where $\begin{bmatrix} -.09018 \\ Log\ .0624 \end{bmatrix} = \begin{bmatrix} b \\ Log\ a \end{bmatrix}$

But $Y = ae^{bx}$; $Log\ Y = Log\ a + bx$; $a = .0624$; anti-log $a = 10^{.0624} = 1.1545$ and the equation for the best fit becomes:

$$Y = 1.1545e^{-.0918x}$$

To find the half-life, use the equation

$$Log\ .5000 = Log\ 1.1545 - .0918t$$

$$tHALFLIFE = \frac{\log(.5000) - \log(1.1545)}{-.0918}$$

tHALFLIFE = 3.9589 Hours

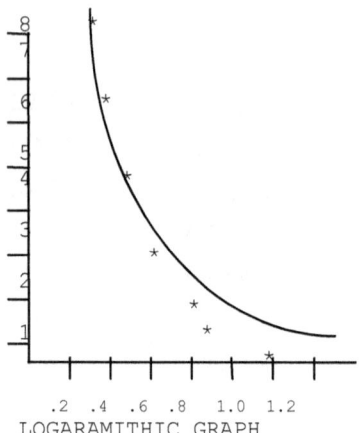

LOGARAMITHIC GRAPH

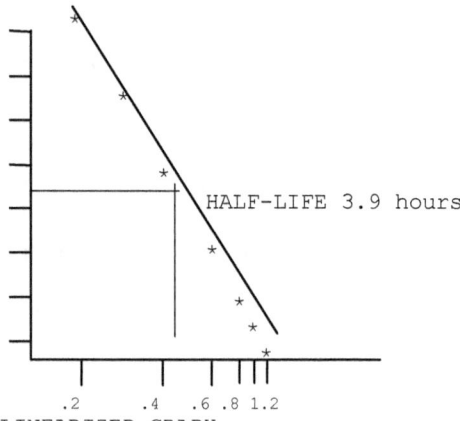

HALF-LIFE 3.9 hours

LINEARIZED GRAPH

LOG-LOG LINEAR REGRESSION

The equation for the Log-Log relationship is given by:

$$Y = kX^n$$

Changing this to a logarithmic equation we get:

$$\text{Log } Y = \text{Log } k + n\text{Log } X$$

Letting Log Y = **Y**; Log k = **K**; and Log X = **X** we get:

$$\mathbf{Y} = n\mathbf{X} + \mathbf{K}$$

The equation is now linear.

Suppose through experiment we obtain the following set of data:

X	1	2	3	4
Y	2.5	8.0	19.0	50.0

The matrix equation becomes:

$$
\begin{vmatrix}
\text{Log } x_1 & 1 \\
\text{Log } x_2 & 1 \\
 & \cdot \\
 & \cdot \\
\text{Log } x_n & 1
\end{vmatrix}
\begin{vmatrix}
n \\
\text{Log } K
\end{vmatrix}
=
\begin{vmatrix}
\text{Log } y_1 \\
\text{Log } y_2 \\
\cdot \\
\cdot \\
\text{Log } y_n
\end{vmatrix}
$$

$$
Y := \begin{bmatrix} \log(2.5) \\ \log(8.0) \\ \log(19.0) \\ \log(50.0) \end{bmatrix}
\qquad
M := \begin{bmatrix} \log(1) & 1 \\ \log(2) & 1 \\ \log(3) & 1 \\ \log(4) & 1 \end{bmatrix}
\qquad
M^T = \begin{pmatrix} 0 & 0.301 & 0.4771 & 0.6021 \\ 1 & 1 & 1 & 1 \end{pmatrix}
$$

Then $\left(M^T \cdot M\right)^{-1} \cdot M^T \cdot Y = \begin{pmatrix} 2.0952 \\ 0.3467 \end{pmatrix}$

Or expanding the above multiplication out we have:

$$\left[\begin{pmatrix} 0 & 0.301 & 0.4771 & 0.6021 \\ 1 & 1 & 1 & 1 \end{pmatrix} \cdot \begin{bmatrix} \log(1) & 1 \\ \log(2) & 1 \\ \log(3) & 1 \\ \log(4) & 1 \end{bmatrix}\right]^{-1} \cdot \begin{pmatrix} 0 & 0.301 & 0.4771 & 0.6021 \\ 1 & 1 & 1 & 1 \end{pmatrix} \cdot \begin{bmatrix} \log(2.5) \\ \log(8.0) \\ \log(19.0) \\ \log(50.0) \end{bmatrix} = \begin{pmatrix} 2.0953 \\ 0.3467 \end{pmatrix}$$

But

$$\begin{vmatrix} 2.0953 \\ \text{Log } .3467 \end{vmatrix} = \begin{vmatrix} n \\ \text{Log K} \end{vmatrix} \qquad \begin{array}{l} n = 2.0953 \\ k = \text{anti-log } K = 10^{.3467} = 2.2218 \end{array}$$

But $Y = kX^n$, so the best fit is **Y = 2**.

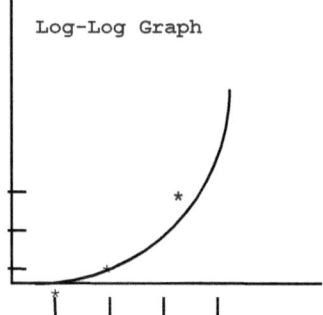

Log-Log Graph

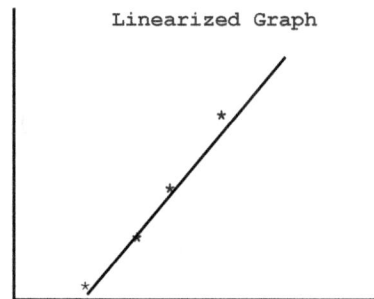

Linearized Graph

PARABOLIC FIT

Let's see how linear regression fits a parabolic equation $Y = a + bX + cX^2$.

X	=	-3	-2	0	3	4
Y	=	18	10	2	2	5

In the matrix X1, the first column consists of the values X^2, the second column consists of the values of X, and the third column consists of the value 1. The Y matrix is a column matrix consisting of the values of Y.

$$X1 := \begin{bmatrix} 9 & -3 & 1 \\ 4 & -2 & 1 \\ 0 & 0 & 1 \\ 9 & 3 & 1 \\ 16 & 4 & 1 \end{bmatrix} \qquad Y := \begin{bmatrix} 18 \\ 10 \\ 2 \\ 2 \\ 5 \end{bmatrix}$$

$$\left(X1^T \cdot X1\right)^{-1} \cdot X1^T \cdot Y = \begin{pmatrix} 0.8728 \\ -2.6455 \\ 1.8247 \end{pmatrix}$$

$Y = 1.8247 - 2.6455X + .8728X^2$

Let's look at this step by step:

$$X1^T \cdot X1 = \begin{pmatrix} 9 & 4 & 0 & 9 & 16 \\ -3 & -2 & 0 & 3 & 4 \\ 1 & 1 & 1 & 1 & 1 \end{pmatrix} \cdot \begin{bmatrix} 9 & -3 & 1 \\ 4 & -2 & 1 \\ 0 & 0 & 1 \\ 9 & 3 & 1 \\ 16 & 4 & 1 \end{bmatrix} = \begin{pmatrix} 434 & 56 & 38 \\ 56 & 38 & 2 \\ 38 & 2 & 5 \end{pmatrix}$$

$$\left(X1^T \cdot X1\right)^{-1} = \begin{pmatrix} 434 & 56 & 38 \\ 56 & 38 & 2 \\ 38 & 2 & 5 \end{pmatrix}^{-1} = \begin{bmatrix} 9.955 \cdot 10^{-3} & -0.0109 & -0.0713 \\ -0.0109 & 0.0389 & 0.0674 \\ -0.0713 & 0.0674 & 0.7148 \end{bmatrix}$$

$$\left(X1^T \cdot X1\right)^{-1} \cdot X1^T =$$

$$\begin{bmatrix} 9.955\ 10^{-3} & -0.0109 & -0.0713 \\ -0.0109 & 0.0389 & 0.0674 \\ -0.0713 & 0.0674 & 0.7148 \end{bmatrix} \cdot \begin{pmatrix} 9 & 4 & 0 & 9 & 16 \\ -3 & -2 & 0 & 3 & 4 \\ 1 & 1 & 1 & 1 & 1 \end{pmatrix} = \begin{bmatrix} 0.051 & -9.68 \cdot 10^{-3} & -0.0713 & -0.0144 & 0.0444 \\ -0.1474 & -0.054 & 0.0674 & 0.086 & 0.0486 \\ -0.1291 & 0.2948 & 0.7148 & 0.2753 & -0.1564 \end{bmatrix}$$

$$\left(X1^T \cdot X1\right)^{-1} \cdot X1^T \cdot Y = \begin{bmatrix} 0.051 & -9.68\ 10^{-3} & -0.0713 & -0.0144 & 0.0444 \\ -0.1474 & -0.054 & 0.0674 & 0.086 & 0.0486 \\ -0.1291 & 0.2948 & 0.7148 & 0.2753 & -0.1564 \end{bmatrix} \cdot \begin{bmatrix} 18 \\ 10 \\ 2 \\ 2 \\ 5 \end{bmatrix} = \begin{pmatrix} 0.8718 \\ -2.6434 \\ 1.8224 \end{pmatrix} = \begin{pmatrix} c \\ b \\ a \end{pmatrix}$$

$$Y = .8718X^2 - 2.6434X + 1.8224$$

CRYPTOGRAPHY

Just recently, I have found an example that follows the rules of quantum statistics. In the following example, though, the datastream matrix must be a square matrix. We cannot transpose and square because it is important and necessary that we do not add the matrix elements to each other. They must always keep their pristine identity. Since the matrix is square, we can multiply in two ways: This is not new code, but is well known in cryptographic circles.

[DB1][A][DB1] = [SOL1] or **[DB1]^T[A][DB1]= [SOL2]**. We do not normalize the resulting solution matrix. In the middle to last part of the following example, if [V] is defined as the Vinegeir transform, and [DB11] is defined as the pre-scrambler matrix, the above equations become (computing the values inside the parenthesis first):

[DB1] (([A] + [V])[DB11]) [DB1] =[SOL3] or **[DB1]^T(([A] + [V]) [DB11])[DB1] = [SOL4]**.

To reverse the above transformation and return to the original matrix, we put the solution back in the place of [A], subtract [V] and multiply by the transpose of the database matrices. i.e.

$$[DB1]^T[SOL1][DB1]^T = [A] \text{ or } [DB1][SOL2][DB1]^T = [A].$$

Or for the more complex forms:

$$[[DB1]^T[SOL3][DB1]^T)[DB11]^T = [SOL3B] = [SOL3A][DB11]^T$$

or

$$([[DB1][SOL4][DB1]^T)[DB11]^T) - [V] = [A] = [SOL4B] - [V]$$

PLAINTEXT

$$\begin{bmatrix} H & I & M & Y & N \\ A & M & E & I & S \\ C & L & I & N & T \\ W & H & A & T & S \\ Y & O & U & R & S \end{bmatrix}$$

CIPHERCODE

$$\begin{bmatrix} 0 & 0 & 1 & 0 & 0 \\ 0 & 0 & 0 & 1 & 0 \\ 1 & 0 & 0 & 0 & 0 \\ 0 & 0 & 0 & 0 & 1 \\ 0 & 1 & 0 & 0 & 0 \end{bmatrix}$$

CHANGING LETTERS TO CORRESPONDING NUMBER IN ALPHABET

$$
\begin{bmatrix}
H & I & M & Y & N \\
A & M & E & I & S \\
C & L & I & N & T \\
W & H & A & T & S \\
Y & O & U & R & S
\end{bmatrix}
=
\begin{bmatrix}
8 & 9 & 13 & 25 & 14 \\
1 & 13 & 5 & 9 & 19 \\
3 & 12 & 9 & 14 & 20 \\
23 & 8 & 1 & 20 & 19 \\
25 & 15 & 21 & 18 & 19
\end{bmatrix}
$$

We change the letters in the text where H is the 8th letter in the alphabet, I is the 9th letter in the alphabet, etc.

COLUMNS MIXED UP   **[A][DB1]**

$$
\begin{bmatrix}
8 & 9 & 13 & 25 & 14 \\
1 & 13 & 5 & 9 & 19 \\
3 & 12 & 9 & 14 & 20 \\
23 & 8 & 1 & 20 & 19 \\
25 & 15 & 21 & 18 & 19
\end{bmatrix}
\cdot
\begin{bmatrix}
0 & 0 & 1 & 0 & 0 \\
0 & 0 & 0 & 1 & 0 \\
1 & 0 & 0 & 0 & 0 \\
0 & 0 & 0 & 0 & 1 \\
0 & 1 & 0 & 0 & 0
\end{bmatrix}
=
\begin{bmatrix}
13 & 14 & 8 & 9 & 25 \\
5 & 19 & 1 & 13 & 9 \\
9 & 20 & 3 & 12 & 14 \\
1 & 19 & 23 & 8 & 20 \\
21 & 19 & 25 & 15 & 18
\end{bmatrix}
\begin{bmatrix}
M & N & H & I & Y \\
E & S & A & M & I \\
I & T & C & L & N \\
A & S & W & H & T \\
U & S & Y & O & R
\end{bmatrix}
$$

ROWS MIXED UP   **[DB1][A]**

$$
\begin{bmatrix}
0 & 0 & 1 & 0 & 0 \\
0 & 0 & 0 & 1 & 0 \\
1 & 0 & 0 & 0 & 0 \\
0 & 0 & 0 & 0 & 1 \\
0 & 1 & 0 & 0 & 0
\end{bmatrix}
\cdot
\begin{bmatrix}
8 & 9 & 13 & 25 & 14 \\
1 & 13 & 5 & 9 & 19 \\
3 & 12 & 9 & 14 & 20 \\
23 & 8 & 1 & 20 & 19 \\
25 & 15 & 21 & 18 & 19
\end{bmatrix}
=
\begin{bmatrix}
3 & 12 & 9 & 14 & 20 \\
23 & 8 & 1 & 20 & 19 \\
8 & 9 & 13 & 25 & 14 \\
25 & 15 & 21 & 18 & 19 \\
1 & 13 & 5 & 9 & 19
\end{bmatrix}
\begin{bmatrix}
C & L & I & N & T \\
W & H & A & T & S \\
H & I & M & Y & N \\
Y & O & U & R & S \\
A & M & E & I & S
\end{bmatrix}
$$

ROWS AND COLUMNS MIXED UP.   **[DB1][A][DB1] = [SOL1]**

$$
\begin{bmatrix}
0 & 0 & 1 & 0 & 0 \\
0 & 0 & 0 & 1 & 0 \\
1 & 0 & 0 & 0 & 0 \\
0 & 0 & 0 & 0 & 1 \\
0 & 1 & 0 & 0 & 0
\end{bmatrix}
\cdot
\begin{bmatrix}
8 & 9 & 13 & 25 & 14 \\
1 & 13 & 5 & 9 & 19 \\
3 & 12 & 9 & 14 & 20 \\
23 & 8 & 1 & 20 & 19 \\
25 & 15 & 21 & 18 & 19
\end{bmatrix}
\cdot
\begin{bmatrix}
0 & 0 & 1 & 0 & 0 \\
0 & 0 & 0 & 1 & 0 \\
1 & 0 & 0 & 0 & 0 \\
0 & 0 & 0 & 0 & 1 \\
0 & 1 & 0 & 0 & 0
\end{bmatrix}
=
\begin{bmatrix}
9 & 20 & 3 & 12 & 14 \\
1 & 19 & 23 & 8 & 20 \\
13 & 14 & 8 & 9 & 25 \\
21 & 19 & 25 & 15 & 18 \\
5 & 19 & 1 & 13 & 9
\end{bmatrix}
\begin{bmatrix}
I & T & C & L & N \\
A & S & W & H & T \\
M & N & H & I & Y \\
U & S & Y & O & R \\
E & S & A & M & I
\end{bmatrix}
$$

ROWS UN-MIXED

$$\begin{bmatrix} 0 & 0 & 1 & 0 & 0 \\ 0 & 0 & 0 & 1 & 0 \\ 1 & 0 & 0 & 0 & 0 \\ 0 & 0 & 0 & 0 & 1 \\ 0 & 1 & 0 & 0 & 0 \end{bmatrix}^{T} \cdot \begin{bmatrix} 3 & 12 & 9 & 14 & 20 \\ 23 & 8 & 1 & 20 & 19 \\ 8 & 9 & 13 & 25 & 14 \\ 25 & 15 & 21 & 18 & 19 \\ 1 & 13 & 5 & 9 & 19 \end{bmatrix} = \begin{bmatrix} 8 & 9 & 13 & 25 & 14 \\ 1 & 13 & 5 & 9 & 19 \\ 3 & 12 & 9 & 14 & 20 \\ 23 & 8 & 1 & 20 & 19 \\ 25 & 15 & 21 & 18 & 19 \end{bmatrix} \begin{bmatrix} H & I & M & Y & N \\ A & M & E & I & S \\ C & L & I & N & T \\ W & H & A & T & S \\ Y & O & U & R & S \end{bmatrix}$$

COLUMNS UN-MIXED

$$\begin{bmatrix} 13 & 14 & 8 & 9 & 25 \\ 5 & 19 & 1 & 13 & 9 \\ 9 & 20 & 3 & 12 & 14 \\ 1 & 19 & 23 & 8 & 20 \\ 21 & 19 & 25 & 15 & 18 \end{bmatrix} \cdot \begin{bmatrix} 0 & 0 & 1 & 0 & 0 \\ 0 & 0 & 0 & 1 & 0 \\ 1 & 0 & 0 & 0 & 0 \\ 0 & 0 & 0 & 0 & 1 \\ 0 & 1 & 0 & 0 & 0 \end{bmatrix}^{T} = \begin{bmatrix} 8 & 9 & 13 & 25 & 14 \\ 1 & 13 & 5 & 9 & 19 \\ 3 & 12 & 9 & 14 & 20 \\ 23 & 8 & 1 & 20 & 19 \\ 25 & 15 & 21 & 18 & 19 \end{bmatrix} \begin{bmatrix} H & I & M & Y & N \\ A & M & E & I & S \\ C & L & I & N & T \\ W & H & A & T & S \\ Y & O & U & R & S \end{bmatrix}$$

ROWS AND COLUMNS UM-MIXED $[DB1]^{T}[SOL1][DB1]^{T} = [A]$

$$\begin{bmatrix} 0 & 0 & 1 & 0 & 0 \\ 0 & 0 & 0 & 1 & 0 \\ 1 & 0 & 0 & 0 & 0 \\ 0 & 0 & 0 & 0 & 1 \\ 0 & 1 & 0 & 0 & 0 \end{bmatrix}^{T} \cdot \begin{bmatrix} 9 & 20 & 3 & 12 & 14 \\ 1 & 19 & 23 & 8 & 20 \\ 13 & 14 & 8 & 9 & 25 \\ 21 & 19 & 25 & 15 & 18 \\ 5 & 19 & 1 & 13 & 9 \end{bmatrix} \cdot \begin{bmatrix} 0 & 0 & 1 & 0 & 0 \\ 0 & 0 & 0 & 1 & 0 \\ 1 & 0 & 0 & 0 & 0 \\ 0 & 0 & 0 & 0 & 1 \\ 0 & 1 & 0 & 0 & 0 \end{bmatrix}^{T} = \begin{bmatrix} 8 & 9 & 13 & 25 & 14 \\ 1 & 13 & 5 & 9 & 19 \\ 3 & 12 & 9 & 14 & 20 \\ 23 & 8 & 1 & 20 & 19 \\ 25 & 15 & 21 & 18 & 19 \end{bmatrix} \begin{bmatrix} H & I & M & Y & N \\ A & M & E & I & S \\ C & L & I & N & T \\ W & H & A & T & S \\ Y & O & U & R & S \end{bmatrix}$$

NOTE: These letters are not coded, they are scrambled.

To code the message first, we will use a Vinegeir transform
letting C = 2 (A = C = 3;
B = D = 4; Or:

A B C D E F G H I J K L M N O P Q R S T U V W X Y Z
C D E F G H I J K L M N O P Q R S T U V W X Y Z A B
=

$$\begin{bmatrix} H & I & M & Y & N \\ A & M & E & I & S \\ C & L & I & N & T \\ W & H & A & T & S \\ Y & O & U & R & S \end{bmatrix} = \begin{bmatrix} 8 & 9 & 13 & 25 & 14 \\ 1 & 13 & 5 & 9 & 19 \\ 3 & 12 & 9 & 14 & 20 \\ 23 & 8 & 1 & 20 & 19 \\ 25 & 15 & 21 & 18 & 19 \end{bmatrix}$$

Applying the Vinegeir transform:

$$
\begin{bmatrix} 2 & 2 & 2 & 2 & 2 \\ 2 & 2 & 2 & 2 & 2 \\ 2 & 2 & 2 & 2 & 2 \\ 2 & 2 & 2 & 2 & 2 \\ 2 & 2 & 2 & 2 & 2 \end{bmatrix} + \begin{bmatrix} 8 & 9 & 13 & 25 & 14 \\ 1 & 13 & 5 & 9 & 19 \\ 3 & 12 & 9 & 14 & 20 \\ 23 & 8 & 1 & 20 & 19 \\ 25 & 15 & 21 & 18 & 19 \end{bmatrix} = \begin{bmatrix} 10 & 11 & 15 & 27 & 16 \\ 3 & 15 & 7 & 11 & 21 \\ 5 & 14 & 11 & 16 & 22 \\ 25 & 10 & 3 & 22 & 21 \\ 27 & 17 & 23 & 20 & 21 \end{bmatrix} \begin{bmatrix} J & K & O & A & P \\ C & O & G & K & U \\ E & N & K & P & V \\ Y & J & C & V & U \\ A & Q & W & T & U \end{bmatrix}
$$

Note the number 27, it is greater than 26, the number of letters in our alphabet. What we can do is replace the 27 with the remainder of 27/26 = 1, but leaving the numbers unchanged is vastly simpler. If we wish to keep all numbers from 1 - 26, we divide by 26 in all cases and replace the number by it's remainder.

This is the pre-coded message before dis-assembly.
[DB1] ([A] + [V]) [DB1] = [SOL3]

$$
\begin{bmatrix} 0 & 0 & 1 & 0 & 0 \\ 0 & 0 & 0 & 1 & 0 \\ 1 & 0 & 0 & 0 & 0 \\ 0 & 0 & 0 & 0 & 1 \\ 0 & 1 & 0 & 0 & 0 \end{bmatrix} \begin{bmatrix} 10 & 11 & 15 & 27 & 16 \\ 3 & 15 & 7 & 11 & 21 \\ 5 & 14 & 11 & 16 & 22 \\ 25 & 10 & 3 & 22 & 21 \\ 27 & 17 & 23 & 20 & 21 \end{bmatrix} \cdot \begin{bmatrix} 0 & 0 & 1 & 0 & 0 \\ 0 & 0 & 0 & 1 & 0 \\ 1 & 0 & 0 & 0 & 0 \\ 0 & 0 & 0 & 0 & 1 \\ 0 & 1 & 0 & 0 & 0 \end{bmatrix} = \begin{bmatrix} 11 & 22 & 5 & 14 & 16 \\ 3 & 21 & 25 & 10 & 22 \\ 15 & 16 & 10 & 11 & 27 \\ 23 & 21 & 27 & 17 & 20 \\ 7 & 21 & 3 & 15 & 11 \end{bmatrix} = \begin{bmatrix} K & V & E & N & P \\ C & U & Y & J & V \\ O & P & J & K & A \\ W & U & A & Q & T \\ G & U & C & O & K \end{bmatrix}
$$

To re-assemble, we proceed as follows:

$$
\begin{bmatrix} 0 & 0 & 1 & 0 & 0 \\ 0 & 0 & 0 & 1 & 0 \\ 1 & 0 & 0 & 0 & 0 \\ 0 & 0 & 0 & 0 & 1 \\ 0 & 1 & 0 & 0 & 0 \end{bmatrix}^T \begin{bmatrix} 11 & 22 & 5 & 14 & 16 \\ 3 & 21 & 25 & 10 & 22 \\ 15 & 16 & 10 & 11 & 27 \\ 23 & 21 & 27 & 17 & 20 \\ 7 & 21 & 3 & 15 & 11 \end{bmatrix} \cdot \begin{bmatrix} 0 & 0 & 1 & 0 & 0 \\ 0 & 0 & 0 & 1 & 0 \\ 1 & 0 & 0 & 0 & 0 \\ 0 & 0 & 0 & 0 & 1 \\ 0 & 1 & 0 & 0 & 0 \end{bmatrix}^T = \begin{bmatrix} 10 & 11 & 15 & 27 & 16 \\ 3 & 15 & 7 & 11 & 21 \\ 5 & 14 & 11 & 16 & 22 \\ 25 & 10 & 3 & 22 & 21 \\ 27 & 17 & 23 & 20 & 21 \end{bmatrix}
$$

$$
\begin{bmatrix} 10 & 11 & 15 & 27 & 16 \\ 3 & 15 & 7 & 11 & 21 \\ 5 & 14 & 11 & 16 & 22 \\ 25 & 10 & 3 & 22 & 21 \\ 27 & 17 & 23 & 20 & 21 \end{bmatrix} - \begin{bmatrix} 2 & 2 & 2 & 2 & 2 \\ 2 & 2 & 2 & 2 & 2 \\ 2 & 2 & 2 & 2 & 2 \\ 2 & 2 & 2 & 2 & 2 \\ 2 & 2 & 2 & 2 & 2 \end{bmatrix} = \begin{bmatrix} 8 & 9 & 13 & 25 & 14 \\ 1 & 13 & 5 & 9 & 19 \\ 3 & 12 & 9 & 14 & 20 \\ 23 & 8 & 1 & 20 & 19 \\ 25 & 15 & 21 & 18 & 19 \end{bmatrix}
$$

$$\begin{bmatrix} 0 & 1 & 0 & 0 & 0 \\ 1 & 0 & 0 & 0 & 0 \\ 0 & 0 & 0 & 1 & 0 \\ 0 & 0 & 0 & 0 & 1 \\ 0 & 0 & 1 & 0 & 0 \end{bmatrix} = \textbf{[DB11]}.$$

And I will define the complex Vinegeir transform [V] as:

$$\begin{bmatrix} 3 & 12 & 15 & 6 & 18 \\ 5 & 7 & 1 & 10 & 2 \\ 9 & 11 & 4 & 16 & 20 \\ 8 & 22 & 13 & 19 & 14 \\ 14 & 26 & 25 & 21 & 17 \end{bmatrix} = \textbf{[V]}$$

First we change the letters one for the other: **[A] + [V]**

$$\begin{bmatrix} 8 & 9 & 13 & 25 & 14 \\ 1 & 13 & 5 & 9 & 19 \\ 3 & 12 & 9 & 14 & 20 \\ 23 & 8 & 1 & 20 & 19 \\ 25 & 15 & 21 & 18 & 19 \end{bmatrix} + \begin{bmatrix} 3 & 12 & 15 & 6 & 18 \\ 5 & 7 & 1 & 10 & 2 \\ 9 & 11 & 4 & 16 & 20 \\ 8 & 22 & 13 & 19 & 14 \\ 14 & 26 & 25 & 21 & 17 \end{bmatrix} = \begin{bmatrix} 11 & 21 & 28 & 31 & 32 \\ 6 & 20 & 6 & 19 & 21 \\ 12 & 23 & 13 & 30 & 40 \\ 31 & 30 & 14 & 39 & 33 \\ 39 & 41 & 46 & 39 & 36 \end{bmatrix}$$

Now we pre-scramble the text:

$$\begin{bmatrix} 11 & 21 & 28 & 31 & 32 \\ 6 & 20 & 6 & 19 & 21 \\ 12 & 23 & 13 & 30 & 40 \\ 31 & 30 & 14 & 39 & 33 \\ 39 & 41 & 46 & 39 & 36 \end{bmatrix} \cdot \begin{bmatrix} 0 & 1 & 0 & 0 & 0 \\ 1 & 0 & 0 & 0 & 0 \\ 0 & 0 & 0 & 1 & 0 \\ 0 & 0 & 0 & 0 & 1 \\ 0 & 0 & 1 & 0 & 0 \end{bmatrix} = \begin{bmatrix} 21 & 11 & 32 & 28 & 31 \\ 20 & 6 & 21 & 6 & 19 \\ 23 & 12 & 40 & 13 & 30 \\ 30 & 31 & 33 & 14 & 39 \\ 41 & 39 & 36 & 46 & 39 \end{bmatrix}$$ **(([A] + [V])[DB11])**

Now we scramble according to the way discussed above:
[DB1] (([A] + [V])[DB11]) [DB1] = [SOL3]

$$\begin{bmatrix} 0 & 0 & 1 & 0 & 0 \\ 0 & 0 & 0 & 1 & 0 \\ 1 & 0 & 0 & 0 & 0 \\ 0 & 0 & 0 & 0 & 1 \\ 0 & 1 & 0 & 0 & 0 \end{bmatrix} \begin{bmatrix} 21 & 11 & 32 & 28 & 31 \\ 20 & 6 & 21 & 6 & 19 \\ 23 & 12 & 40 & 13 & 30 \\ 30 & 31 & 33 & 14 & 39 \\ 41 & 39 & 36 & 46 & 39 \end{bmatrix} \cdot \begin{bmatrix} 0 & 0 & 1 & 0 & 0 \\ 0 & 0 & 0 & 1 & 0 \\ 1 & 0 & 0 & 0 & 0 \\ 0 & 0 & 0 & 0 & 1 \\ 0 & 1 & 0 & 0 & 0 \end{bmatrix} = \begin{bmatrix} 40 & 30 & 23 & 12 & 13 \\ 33 & 39 & 30 & 31 & 14 \\ 32 & 31 & 21 & 11 & 28 \\ 36 & 39 & 41 & 39 & 46 \\ 21 & 19 & 20 & 6 & 6 \end{bmatrix}$$

Again, to de-code the cipher, we do the problem backwards:
[DB1]^T[SOL3][DB1]^T = [SOL3A]

$$\begin{bmatrix} 0 & 0 & 1 & 0 & 0 \\ 0 & 0 & 0 & 1 & 0 \\ 1 & 0 & 0 & 0 & 0 \\ 0 & 0 & 0 & 0 & 1 \\ 0 & 1 & 0 & 0 & 0 \end{bmatrix}^T \begin{bmatrix} 40 & 30 & 23 & 12 & 13 \\ 33 & 39 & 30 & 31 & 14 \\ 32 & 31 & 21 & 11 & 28 \\ 36 & 39 & 41 & 39 & 46 \\ 21 & 19 & 20 & 6 & 6 \end{bmatrix} \cdot \begin{bmatrix} 0 & 0 & 1 & 0 & 0 \\ 0 & 0 & 0 & 1 & 0 \\ 1 & 0 & 0 & 0 & 0 \\ 0 & 0 & 0 & 0 & 1 \\ 0 & 1 & 0 & 0 & 0 \end{bmatrix}^T = \begin{bmatrix} 21 & 11 & 32 & 28 & 31 \\ 20 & 6 & 21 & 6 & 19 \\ 23 & 12 & 40 & 13 & 30 \\ 30 & 31 & 33 & 14 & 39 \\ 41 & 39 & 36 & 46 & 39 \end{bmatrix}$$ **= [SOL3A]**

Even if they discovered the proper ciphercode above, the message is so garbled they would never know they found the key. Next we un-scramble the pre-scrambled message. **[[DB1]^T[SOL3][DB1]^T) [DB11]^T = [SOL3B] = [SOL3A][DB11]^T**

$$\begin{bmatrix} 21 & 11 & 32 & 28 & 31 \\ 20 & 6 & 21 & 6 & 19 \\ 23 & 12 & 40 & 13 & 30 \\ 30 & 31 & 33 & 14 & 39 \\ 41 & 39 & 36 & 46 & 39 \end{bmatrix} \begin{bmatrix} 0 & 1 & 0 & 0 & 0 \\ 1 & 0 & 0 & 0 & 0 \\ 0 & 0 & 0 & 1 & 0 \\ 0 & 0 & 0 & 0 & 1 \\ 0 & 0 & 1 & 0 & 0 \end{bmatrix}^T = \begin{bmatrix} 11 & 21 & 28 & 31 & 32 \\ 6 & 20 & 6 & 19 & 21 \\ 12 & 23 & 13 & 30 & 40 \\ 31 & 30 & 14 & 39 & 33 \\ 39 & 41 & 46 & 39 & 36 \end{bmatrix}$$ **= [SOL3B]**

Finally we revert from the Vinegeir transform back to our regular alphabet:

([[DB1]^T[SOL3][DB1]^T) [DB11]^T) - [V] = [A] = [SOL3B] - [V]

$$
\begin{bmatrix}
11 & 21 & 28 & 31 & 32 \\
6 & 20 & 6 & 19 & 21 \\
12 & 23 & 13 & 30 & 40 \\
31 & 30 & 14 & 39 & 33 \\
39 & 41 & 46 & 39 & 36
\end{bmatrix}
-
\begin{bmatrix}
3 & 12 & 15 & 6 & 18 \\
5 & 7 & 1 & 10 & 2 \\
9 & 11 & 4 & 16 & 20 \\
8 & 22 & 13 & 19 & 14 \\
14 & 26 & 25 & 21 & 17
\end{bmatrix}
=
\begin{bmatrix}
8 & 9 & 13 & 25 & 14 \\
1 & 13 & 5 & 9 & 19 \\
3 & 12 & 9 & 14 & 20 \\
23 & 8 & 1 & 20 & 19 \\
25 & 15 & 21 & 18 & 19
\end{bmatrix}
$$

Now I will solve the encryption using the equation for quantum statistics. ([DB]^T[A][DB])

[DB1]^T(([A] + [V])[DB11])[DB1] = [SOL4].

$$
\begin{bmatrix}
0 & 0 & 1 & 0 & 0 \\
0 & 0 & 0 & 1 & 0 \\
1 & 0 & 0 & 0 & 0 \\
0 & 0 & 0 & 0 & 1 \\
0 & 1 & 0 & 0 & 0
\end{bmatrix}^T
=
\begin{bmatrix}
0 & 0 & 1 & 0 & 0 \\
0 & 0 & 0 & 0 & 1 \\
1 & 0 & 0 & 0 & 0 \\
0 & 1 & 0 & 0 & 0 \\
0 & 0 & 0 & 1 & 0
\end{bmatrix}
$$

$$
\begin{bmatrix}
0 & 0 & 1 & 0 & 0 \\
0 & 0 & 0 & 0 & 1 \\
1 & 0 & 0 & 0 & 0 \\
0 & 1 & 0 & 0 & 0 \\
0 & 0 & 0 & 1 & 0
\end{bmatrix}
\begin{bmatrix}
8 & 9 & 13 & 25 & 14 \\
1 & 13 & 5 & 9 & 19 \\
3 & 12 & 9 & 14 & 20 \\
23 & 8 & 1 & 20 & 19 \\
25 & 15 & 21 & 18 & 19
\end{bmatrix}
\cdot
\begin{bmatrix}
0 & 0 & 1 & 0 & 0 \\
0 & 0 & 0 & 1 & 0 \\
1 & 0 & 0 & 0 & 0 \\
0 & 0 & 0 & 0 & 1 \\
0 & 1 & 0 & 0 & 0
\end{bmatrix}
=
\begin{bmatrix}
9 & 20 & 3 & 12 & 14 \\
21 & 19 & 25 & 15 & 18 \\
13 & 14 & 8 & 9 & 25 \\
5 & 19 & 1 & 13 & 9 \\
1 & 19 & 23 & 8 & 20
\end{bmatrix}
$$

For the simple case.

For the more complex case, I am going to create here a special Vinegeir transform, just to show how really difficult de-coding this message will be without the key.

$$
\begin{bmatrix}
3 & 2 & 24 & 12 & 23 \\
10 & 24 & 6 & 2 & 18 \\
8 & 25 & 2 & 23 & 17 \\
14 & 3 & 10 & 17 & 18 \\
12 & 22 & 16 & 19 & 18
\end{bmatrix}
= \textbf{[V]}
$$

[A] + [V] =

$$\begin{bmatrix} 8 & 9 & 13 & 25 & 14 \\ 1 & 13 & 5 & 9 & 19 \\ 3 & 12 & 9 & 14 & 20 \\ 23 & 8 & 1 & 20 & 19 \\ 25 & 15 & 21 & 18 & 19 \end{bmatrix} + \begin{bmatrix} 3 & 2 & 24 & 12 & 23 \\ 10 & 24 & 6 & 2 & 18 \\ 8 & 25 & 2 & 23 & 17 \\ 14 & 3 & 10 & 17 & 18 \\ 12 & 22 & 16 & 19 & 18 \end{bmatrix} = \begin{bmatrix} 11 & 11 & 37 & 37 & 37 \\ 11 & 37 & 11 & 11 & 37 \\ 11 & 37 & 11 & 37 & 37 \\ 37 & 11 & 11 & 37 & 37 \\ 37 & 37 & 37 & 37 & 37 \end{bmatrix} = \begin{bmatrix} K & K & K & K & K \\ K & K & K & K & K \\ K & K & K & K & K \\ K & K & K & K & K \\ K & K & K & K & K \end{bmatrix}$$

We do not have to scramble the letters, multiplying this by different databases returns the same solution. But you can see how difficult to de-code the cipher is, where do we start, how many E's are in the message? S's? A's? But let's see what happens when we try to scramble it:

The column pre-scrambling operation is:

$$\begin{bmatrix} 11 & 11 & 37 & 37 & 37 \\ 11 & 37 & 11 & 11 & 37 \\ 11 & 37 & 11 & 37 & 37 \\ 37 & 11 & 11 & 37 & 37 \\ 37 & 37 & 37 & 37 & 37 \end{bmatrix} \begin{bmatrix} 0 & 1 & 0 & 0 & 0 \\ 1 & 0 & 0 & 0 & 0 \\ 0 & 0 & 0 & 1 & 0 \\ 0 & 0 & 0 & 0 & 1 \\ 0 & 0 & 1 & 0 & 0 \end{bmatrix} = \begin{bmatrix} 11 & 11 & 37 & 37 & 37 \\ 37 & 11 & 37 & 11 & 11 \\ 37 & 11 & 37 & 11 & 37 \\ 11 & 37 & 37 & 11 & 37 \\ 37 & 37 & 37 & 37 & 37 \end{bmatrix}$$

and the final scramble becomes:

final ciphertext

$$\begin{bmatrix} 0 & 0 & 1 & 0 & 0 \\ 0 & 0 & 0 & 1 & 0 \\ 1 & 0 & 0 & 0 & 0 \\ 0 & 0 & 0 & 0 & 1 \\ 0 & 1 & 0 & 0 & 0 \end{bmatrix}^T \begin{bmatrix} 11 & 11 & 37 & 37 & 37 \\ 37 & 11 & 37 & 11 & 11 \\ 37 & 11 & 37 & 11 & 37 \\ 11 & 37 & 37 & 11 & 37 \\ 37 & 37 & 37 & 37 & 37 \end{bmatrix} \begin{bmatrix} 0 & 0 & 1 & 0 & 0 \\ 0 & 0 & 0 & 1 & 0 \\ 1 & 0 & 0 & 0 & 0 \\ 0 & 0 & 0 & 0 & 1 \\ 0 & 1 & 0 & 0 & 0 \end{bmatrix} = \begin{bmatrix} 37 & 37 & 37 & 11 & 11 \\ 37 & 37 & 37 & 37 & 37 \\ 37 & 37 & 11 & 11 & 37 \\ 37 & 11 & 37 & 11 & 11 \\ 37 & 37 & 11 & 37 & 11 \end{bmatrix} = \begin{bmatrix} K & K & K & K & K \\ K & K & K & K & K \\ K & K & K & K & K \\ K & K & K & K & K \\ K & K & K & K & K \end{bmatrix}$$

If you intercepted this code, you wouldn't even know if it was a code, or if it were, how to go about de-coding it. So let's de-scramble it.

$[DB1][SOL4][DB1]^T = [SOL4A]$

$$
\begin{bmatrix}
0 & 0 & 1 & 0 & 0 \\
0 & 0 & 0 & 0 & 1 \\
1 & 0 & 0 & 0 & 0 \\
0 & 1 & 0 & 0 & 0 \\
0 & 0 & 0 & 1 & 0
\end{bmatrix}
\begin{bmatrix}
37 & 37 & 37 & 11 & 11 \\
37 & 37 & 37 & 37 & 37 \\
37 & 37 & 11 & 11 & 37 \\
37 & 11 & 37 & 11 & 11 \\
37 & 37 & 11 & 37 & 11
\end{bmatrix}
\cdot
\begin{bmatrix}
0 & 0 & 1 & 0 & 0 \\
0 & 0 & 0 & 1 & 0 \\
1 & 0 & 0 & 0 & 0 \\
0 & 0 & 0 & 0 & 1 \\
0 & 1 & 0 & 0 & 0
\end{bmatrix}^T
=
\begin{bmatrix}
11 & 11 & 37 & 37 & 37 \\
11 & 37 & 37 & 11 & 37 \\
37 & 11 & 37 & 11 & 37 \\
37 & 37 & 37 & 37 & 37 \\
37 & 11 & 37 & 11 & 11
\end{bmatrix}
= [SOL4A]
$$

$([DB1][SOL4][DB1]^T)[DB11]^T = [SOL4B] = [SOL4A][DB11]^T$

$$
\begin{bmatrix}
11 & 11 & 37 & 37 & 37 \\
11 & 37 & 37 & 11 & 37 \\
37 & 11 & 37 & 11 & 37 \\
37 & 37 & 37 & 37 & 37 \\
37 & 11 & 37 & 11 & 11
\end{bmatrix}
\cdot
\begin{bmatrix}
0 & 1 & 0 & 0 & 0 \\
1 & 0 & 0 & 0 & 0 \\
0 & 0 & 0 & 1 & 0 \\
0 & 0 & 0 & 0 & 1 \\
0 & 0 & 1 & 0 & 0
\end{bmatrix}^T
=
\begin{bmatrix}
11 & 11 & 37 & 37 & 37 \\
37 & 11 & 11 & 37 & 37 \\
11 & 37 & 11 & 37 & 37 \\
37 & 37 & 37 & 37 & 37 \\
11 & 37 & 11 & 11 & 37
\end{bmatrix}
= [SOL4B]
$$

$(([DB1][SOL4][DB1]^T)[DB11]^T) - [V] = [A] = [SOL4B] - [V]$

$$
\begin{bmatrix}
11 & 11 & 37 & 37 & 37 \\
37 & 11 & 11 & 37 & 37 \\
11 & 37 & 11 & 37 & 37 \\
37 & 37 & 37 & 37 & 37 \\
11 & 37 & 11 & 11 & 37
\end{bmatrix}
-
\begin{bmatrix}
3 & 2 & 24 & 12 & 23 \\
10 & 24 & 6 & 2 & 18 \\
8 & 25 & 2 & 23 & 17 \\
14 & 3 & 10 & 17 & 18 \\
12 & 22 & 16 & 19 & 18
\end{bmatrix}
=
\begin{bmatrix}
8 & 9 & 13 & 25 & 14 \\
27 & -13 & 5 & 35 & 19 \\
3 & 12 & 9 & 14 & 20 \\
23 & 34 & 27 & 20 & 19 \\
-1 & 15 & -5 & -8 & 19
\end{bmatrix}
$$

Adding the negative numbers to 26 we get:

$$
\begin{bmatrix}
8 & 9 & 13 & 25 & 14 \\
1 & 13 & 5 & 9 & 19 \\
3 & 12 & 9 & 14 & 20 \\
23 & 8 & 1 & 20 & 19 \\
25 & 15 & 21 & 18 & 19
\end{bmatrix}
$$

Which is where we started.

Here I just wish to show that the eleven's matrix (or whatever same number matrix) is independent of all pre and post multiplier operations.

$$
\begin{bmatrix}
8 & 9 & 13 & 25 & 14 \\
1 & 13 & 5 & 9 & 19 \\
3 & 12 & 9 & 14 & 20 \\
23 & 8 & 1 & 20 & 19 \\
25 & 15 & 21 & 18 & 19
\end{bmatrix}
+
\begin{bmatrix}
3 & 2 & -2 & -14 & -3 \\
10 & -2 & 6 & 2 & -8 \\
8 & -1 & 2 & -3 & -9 \\
-12 & 3 & 10 & -9 & -8 \\
-14 & -4 & -10 & -7 & -8
\end{bmatrix}
=
\begin{bmatrix}
11 & 11 & 11 & 11 & 11 \\
11 & 11 & 11 & 11 & 11 \\
11 & 11 & 11 & 11 & 11 \\
11 & 11 & 11 & 11 & 11 \\
11 & 11 & 11 & 11 & 11
\end{bmatrix}
$$

$$
\begin{bmatrix}
11 & 11 & 11 & 11 & 11 \\
11 & 11 & 11 & 11 & 11 \\
11 & 11 & 11 & 11 & 11 \\
11 & 11 & 11 & 11 & 11 \\
11 & 11 & 11 & 11 & 11
\end{bmatrix}
\cdot
\begin{bmatrix}
0 & 1 & 0 & 0 & 0 \\
1 & 0 & 0 & 0 & 0 \\
0 & 0 & 0 & 1 & 0 \\
0 & 0 & 0 & 0 & 1 \\
0 & 0 & 1 & 0 & 0
\end{bmatrix}
=
\begin{bmatrix}
11 & 11 & 11 & 11 & 11 \\
11 & 11 & 11 & 11 & 11 \\
11 & 11 & 11 & 11 & 11 \\
11 & 11 & 11 & 11 & 11 \\
11 & 11 & 11 & 11 & 11
\end{bmatrix}
$$

$$
\begin{bmatrix}
0 & 0 & 1 & 0 & 0 \\
0 & 0 & 0 & 1 & 0 \\
1 & 0 & 0 & 0 & 0 \\
0 & 0 & 0 & 0 & 1 \\
0 & 1 & 0 & 0 & 0
\end{bmatrix}^{T}
\cdot
\begin{bmatrix}
11 & 11 & 11 & 11 & 11 \\
11 & 11 & 11 & 11 & 11 \\
11 & 11 & 11 & 11 & 11 \\
11 & 11 & 11 & 11 & 11 \\
11 & 11 & 11 & 11 & 11
\end{bmatrix}
\cdot
\begin{bmatrix}
0 & 0 & 1 & 0 & 0 \\
0 & 0 & 0 & 1 & 0 \\
1 & 0 & 0 & 0 & 0 \\
0 & 0 & 0 & 0 & 1 \\
0 & 1 & 0 & 0 & 0
\end{bmatrix}
=
\begin{bmatrix}
11 & 11 & 11 & 11 & 11 \\
11 & 11 & 11 & 11 & 11 \\
11 & 11 & 11 & 11 & 11 \\
11 & 11 & 11 & 11 & 11 \\
11 & 11 & 11 & 11 & 11
\end{bmatrix}
$$

$$
\begin{bmatrix}
0 & 0 & 1 & 0 & 0 \\
0 & 0 & 0 & 1 & 0 \\
1 & 0 & 0 & 0 & 0 \\
0 & 0 & 0 & 0 & 1 \\
0 & 1 & 0 & 0 & 0
\end{bmatrix}
\cdot
\begin{bmatrix}
11 & 11 & 11 & 11 & 11 \\
11 & 11 & 11 & 11 & 11 \\
11 & 11 & 11 & 11 & 11 \\
11 & 11 & 11 & 11 & 11 \\
11 & 11 & 11 & 11 & 11
\end{bmatrix}
\cdot
\begin{bmatrix}
0 & 0 & 1 & 0 & 0 \\
0 & 0 & 0 & 1 & 0 \\
1 & 0 & 0 & 0 & 0 \\
0 & 0 & 0 & 0 & 1 \\
0 & 1 & 0 & 0 & 0
\end{bmatrix}^{T}
=
\begin{bmatrix}
11 & 11 & 11 & 11 & 11 \\
11 & 11 & 11 & 11 & 11 \\
11 & 11 & 11 & 11 & 11 \\
11 & 11 & 11 & 11 & 11 \\
11 & 11 & 11 & 11 & 11
\end{bmatrix}
$$

$$
\begin{bmatrix}
11 & 11 & 11 & 11 & 11 \\
11 & 11 & 11 & 11 & 11 \\
11 & 11 & 11 & 11 & 11 \\
11 & 11 & 11 & 11 & 11 \\
11 & 11 & 11 & 11 & 11
\end{bmatrix}
\cdot
\begin{bmatrix}
0 & 1 & 0 & 0 & 0 \\
1 & 0 & 0 & 0 & 0 \\
0 & 0 & 0 & 1 & 0 \\
0 & 0 & 0 & 0 & 1 \\
0 & 0 & 1 & 0 & 0
\end{bmatrix}
=
\begin{bmatrix}
11 & 11 & 11 & 11 & 11 \\
11 & 11 & 11 & 11 & 11 \\
11 & 11 & 11 & 11 & 11 \\
11 & 11 & 11 & 11 & 11 \\
11 & 11 & 11 & 11 & 11
\end{bmatrix}
$$

$$
\begin{bmatrix}
11 & 11 & 11 & 11 & 11 \\
11 & 11 & 11 & 11 & 11 \\
11 & 11 & 11 & 11 & 11 \\
11 & 11 & 11 & 11 & 11 \\
11 & 11 & 11 & 11 & 11
\end{bmatrix}
-
\begin{bmatrix}
3 & 2 & -2 & -14 & -3 \\
10 & -2 & 6 & 2 & -8 \\
8 & -1 & 2 & -3 & -9 \\
-12 & 3 & 10 & -9 & -8 \\
-14 & -4 & -10 & -7 & -8
\end{bmatrix}
=
\begin{bmatrix}
8 & 9 & 13 & 25 & 14 \\
1 & 13 & 5 & 9 & 19 \\
3 & 12 & 9 & 14 & 20 \\
23 & 8 & 1 & 20 & 19 \\
25 & 15 & 21 & 18 & 19
\end{bmatrix}
$$

In this special case, we do not have to go through all the
de-scrambling steps, just take the K matrix and subtract the
Vinegeir transform to get the plain text message back. So it
is best to keep all the numbers different rather than the same
if we wish to make the de-coding of the message more difficult
for hackers.

Yeah, having a code with all the same letters is neat, but won't
it be a hassle to transform many pages of plaintext into code
this way? I mean, is there an easier way to encode your text
than to do it letter by letter and by hand? Yes! And it is very
simple to do. We go about it in this manner. In basic algebra,
we have the equation (for the letter S): $19 - x = 11$; Where x
is the number we wish to put in the key for our code. In this
case $x = 19 - 11 = 8$. In matrix form we will let A = whatever
letter we wish to transform and X be the number in the key.
Then we have $A - X = 11$; $X = A - 11$. Let's see how this works
for the example above. Changing each letter into it's numerical
equivalent we get:

$$
\begin{bmatrix}
H & I & M & Y & N \\
A & M & E & I & S \\
C & L & I & N & T \\
W & H & A & T & S \\
Y & O & U & R & S
\end{bmatrix}
=
\begin{bmatrix}
8 & 9 & 13 & 25 & 14 \\
1 & 13 & 5 & 9 & 19 \\
3 & 12 & 9 & 14 & 20 \\
23 & 8 & 1 & 20 & 19 \\
25 & 15 & 21 & 18 & 19
\end{bmatrix}
$$

Solving the equation A-11 = X we get:

$$
\begin{bmatrix}
8 & 9 & 13 & 25 & 14 \\
1 & 13 & 5 & 9 & 19 \\
3 & 12 & 9 & 14 & 20 \\
23 & 8 & 1 & 20 & 19 \\
25 & 15 & 21 & 18 & 19
\end{bmatrix}
-
\begin{bmatrix}
11 & 11 & 11 & 11 & 11 \\
11 & 11 & 11 & 11 & 11 \\
11 & 11 & 11 & 11 & 11 \\
11 & 11 & 11 & 11 & 11 \\
11 & 11 & 11 & 11 & 11
\end{bmatrix}
=
\begin{bmatrix}
-3 & -2 & 2 & 14 & 3 \\
-10 & 2 & -6 & -2 & 8 \\
-8 & 1 & -2 & 3 & 9 \\
12 & -3 & -10 & 9 & 8 \\
14 & 4 & 10 & 7 & 8
\end{bmatrix}
$$

The matrix to the right of the equal sign is the key.

$$
\text{KEY} =
\begin{bmatrix}
-3 & -2 & 2 & 14 & 3 \\
-10 & 2 & -6 & -2 & 8 \\
-8 & 1 & -2 & 3 & 9 \\
12 & -3 & -10 & 9 & 8 \\
14 & 4 & 10 & 7 & 8
\end{bmatrix}
\qquad
\text{PLAINTEXT} =
\begin{bmatrix}
8 & 9 & 13 & 25 & 14 \\
1 & 13 & 5 & 9 & 19 \\
3 & 12 & 9 & 14 & 20 \\
23 & 8 & 1 & 20 & 19 \\
25 & 15 & 21 & 18 & 19
\end{bmatrix}
$$

$$
\text{PLAINTEXT} - \text{KEY} =
\begin{bmatrix}
11 & 11 & 11 & 11 & 11 \\
11 & 11 & 11 & 11 & 11 \\
11 & 11 & 11 & 11 & 11 \\
11 & 11 & 11 & 11 & 11 \\
11 & 11 & 11 & 11 & 11
\end{bmatrix}
$$

or

$$
\begin{bmatrix}
8 & 9 & 13 & 25 & 14 \\
1 & 13 & 5 & 9 & 19 \\
3 & 12 & 9 & 14 & 20 \\
23 & 8 & 1 & 20 & 19 \\
25 & 15 & 21 & 18 & 19
\end{bmatrix}
-
\begin{bmatrix}
-3 & -2 & 2 & 14 & 3 \\
-10 & 2 & -6 & -2 & 8 \\
-8 & 1 & -2 & 3 & 9 \\
12 & -3 & -10 & 9 & 8 \\
14 & 4 & 10 & 7 & 8
\end{bmatrix}
=
\begin{bmatrix}
11 & 11 & 11 & 11 & 11 \\
11 & 11 & 11 & 11 & 11 \\
11 & 11 & 11 & 11 & 11 \\
11 & 11 & 11 & 11 & 11 \\
11 & 11 & 11 & 11 & 11
\end{bmatrix}
$$

Which is where we wish to be with this code.

Every message using this method will have it's unique key. A friend cannot have a copy in advance to use to decode your message, EXCEPT IN THE MANNER DESCRIBED BELOW. This is a code that is best used when you put information into a computer and do not wish for a hacker to download it. You transform the plaintext into ciphertext and copy the key onto a floppy disk or CD then erase it from your computer memory. If anyone hacks into your files, all they will get is a code of all the same letters, but the key is inaccessable to anyone but you because the key is not in computer memory where someone might discover it. For instance, any 25 letter message is a solution for the 5x5 eleven's matrix above. The solution "my dog had fleas and died" is a solution, but is incorrect. There are thousands of 25 letter messages that are solutions, but which one is your message? Without the key, deciphering the message is almost impossible. Recall the coded 5x5 eleven matrix message, recall the KEY, add them together and your message is returned.

$$KEY := \begin{bmatrix} -3 & -2 & 2 & 14 & 3 \\ -10 & 2 & -6 & -2 & 8 \\ -8 & 1 & -2 & 3 & 9 \\ 12 & -3 & -10 & 9 & 8 \\ 14 & 4 & 10 & 7 & 8 \end{bmatrix}$$

$$KEY1 := \begin{bmatrix} 0&0&0&1&0 \\ 0&1&0&0&0 \\ 1&0&0&0&0 \\ 0&0&0&0&1 \\ 0&0&1&0&0 \end{bmatrix} \quad KEY2 := \begin{bmatrix} 1&0&0&0&0 \\ 0&0&1&0&0 \\ 0&0&0&0&1 \\ 0&0&0&1&0 \\ 0&1&0&0&0 \end{bmatrix} \quad KEY3 := \begin{bmatrix} 1&0&0&0&0 \\ 0&0&0&1&0 \\ 0&0&1&0&0 \\ 0&0&0&0&1 \\ 0&1&0&0&0 \end{bmatrix}$$

Code1 means it needs KEY1 to decode it, CODE2 means it needs KEY2 to decode it, CODE3 means it needs KEY3 to decode it. In a 5x5 message, there are 120 possible KEYS (5!).

$$CODE1 := KEY1\,KEY\,KEY1 \quad CODE2 := KEY2\,KEY\,KEY2 \quad CODE3 := KEY3\,KEY\,KEY3$$

$$CODE1 = \begin{bmatrix} -10 & -3 & 8 & 12 & 9 \\ -6 & 2 & 8 & -10 & -2 \\ 2 & -2 & 3 & -3 & 14 \\ 10 & 4 & 8 & 14 & 7 \\ -2 & 1 & 9 & -8 & 3 \end{bmatrix} \quad CODE2 = \begin{bmatrix} -3 & 3 & -2 & 14 & 2 \\ -8 & 9 & 1 & 3 & -2 \\ 14 & 8 & 4 & 7 & 10 \\ 12 & 8 & -3 & 9 & -10 \\ -10 & 8 & 2 & -2 & -6 \end{bmatrix} \quad CODE3 = \begin{bmatrix} -3 & 3 & 2 & -2 & 14 \\ 12 & 8 & -10 & -3 & 9 \\ -8 & 9 & -2 & 1 & 3 \\ 14 & 8 & 10 & 4 & 7 \\ -10 & 8 & -6 & 2 & -2 \end{bmatrix}$$

$$\text{CODE1T} = \text{KEY1}^T \cdot \text{CODE1 KEY1}^T$$

$$\text{CODE1T} = \begin{bmatrix} -3 & -2 & 2 & 14 & 3 \\ -10 & 2 & -6 & -2 & 8 \\ -8 & 1 & -2 & 3 & 9 \\ 12 & -3 & -10 & 9 & 8 \\ 14 & 4 & 10 & 7 & 8 \end{bmatrix} \quad \text{KEY} = \begin{bmatrix} -3 & -2 & 2 & 14 & 3 \\ -10 & 2 & -6 & -2 & 8 \\ -8 & 1 & -2 & 3 & 9 \\ 12 & -3 & -10 & 9 & 8 \\ 14 & 4 & 10 & 7 & 8 \end{bmatrix}$$

Suppose there are 3 agencies, one the headquarters and it has all three keys, the second agency has keys 1 and 3, the third agency has keys 2 and three. If the message is for agency one only, it is sent encoded in KEY1, the third agency can't decode it. If the third agency is the only recipient it is sent in KEY2 and the second agency can't decode it. If it is meant for both agencies it is sent in KEY3. It is the same if the two sub-agencies are writing to headquarters, they can send it so it can be read by HQ only by sending it in KEY1, if the third agency needs to read it, it is sent in KEY3, etc.

$$\text{KEY} + \begin{bmatrix} 11 & 11 & 11 & 11 & 11 \\ 11 & 11 & 11 & 11 & 11 \\ 11 & 11 & 11 & 11 & 11 \\ 11 & 11 & 11 & 11 & 11 \\ 11 & 11 & 11 & 11 & 11 \end{bmatrix} = \begin{bmatrix} 8 & 9 & 13 & 25 & 14 \\ 1 & 13 & 5 & 9 & 19 \\ 3 & 12 & 9 & 14 & 20 \\ 23 & 8 & 1 & 20 & 19 \\ 25 & 15 & 21 & 18 & 19 \end{bmatrix} \begin{bmatrix} H & I & M & Y & N \\ A & M & E & I & S \\ C & L & I & N & T \\ W & H & A & T & S \\ Y & O & U & R & S \end{bmatrix}$$

$$\text{CODE1} + \begin{bmatrix} 11 & 11 & 11 & 11 & 11 \\ 11 & 11 & 11 & 11 & 11 \\ 11 & 11 & 11 & 11 & 11 \\ 11 & 11 & 11 & 11 & 11 \\ 11 & 11 & 11 & 11 & 11 \end{bmatrix} = \begin{bmatrix} 1 & 8 & 19 & 23 & 20 \\ 5 & 13 & 19 & 1 & 9 \\ 13 & 9 & 14 & 8 & 25 \\ 21 & 15 & 19 & 25 & 18 \\ 9 & 12 & 20 & 3 & 14 \end{bmatrix} \begin{bmatrix} A & H & S & W & T \\ E & M & S & A & I \\ M & I & N & H & Y \\ U & O & S & Y & R \\ I & L & T & C & N \end{bmatrix}$$

PHYSICS 101

EXAMPLE: A particle moving in one direction has a speed of v m/s at time t goven by the equation $v = at^2 + bt + c$. We define (for this case) that a = 3, b = 4 and c = 5.

Find the initial speed.

$$\mathbf{V}(0) = c = 5 \text{ m/s}$$

$$3\ 4\ 5] \ \begin{vmatrix} 0 \\ 0 \\ 1 \end{vmatrix} = 5 \text{ m/s}$$

Find the speed when 3 seconds have passed.

$$\mathbf{V}(3) = 3(3)^2 + 4(3) + 5 = 27 + 12 + 5 = 44 \text{m/s}$$

$$[3\ 4\ 5] \ \begin{vmatrix} 9 \\ 3 \\ 1 \end{vmatrix} = [27+12+5] = [44 \text{ m/s}]$$

What is the acceleration when 4 seconds have passed?

$$a(t) = d\mathbf{v}(t)/dt = 2at + b = 2(3)t + 4 = 6t + 4.$$
$$\text{At } t = 4 \text{ sec., } a(t) = 6(4) + 4 = 28 \text{ m/s}^2$$

To do the calculus derivative using matrices, shrink the matrix by one element and multiply the first number in the new matrix by its position number (in this case it is the number 2) and the second number in the new matrix by it's position number (in this case it is in the first position).

$$d[3\ 4\ 5]/dt = \begin{bmatrix} 2 \\ 1 \end{bmatrix} \circ \begin{bmatrix} 3 \\ 4 \end{bmatrix} = \begin{bmatrix} 6 \\ 4 \end{bmatrix} = 6t + 4$$

or $[3\ 4] \otimes [2\ 1] = [6\ 4] = 6t + 4.$

At t = 4s, we get

$$6\ 4] \begin{bmatrix} 4 \\ 1 \end{bmatrix} = [24 + 4] = 28 \text{ m/s}.$$

Find an equation for the displacement x(t) using the equation
v = 3t² + 4t +5.

$$dx/dt = at^2 + bt + c.$$

$$\int_{x_0}^{t} 1\,dx = \int_{0}^{t} at^2\,dt + \int_{0}^{t} bt\,dt + \int_{0}^{t} c\,dt$$

$$= x - x_0 = a\ t^3/3 + bt^2/2 + ct$$
$$x = a\ t^3/3 + bt^2/2 + ct + x_0 \quad \text{Substituting the given}$$
$$\text{values for a, b and c we get}$$
$$x = 3\ t^3/3 + 4t^2/2 + 5t + x_0 = \mathbf{t^3 + 2t^2 + 5t + x_0.}$$

Using matrices, to integrate, we expand the matrix by one element
and divide each element by it's corresponding position in the
new matrix (we ignore the position of the new constant). i.e.

$$\int_{0}^{t} (3\ 4\ 5)\,dt \begin{aligned} &= [3\ 4\ 5\ x_0\ ] \otimes [1/3\ 1/2\ 1\ 1] = [3/3\ 4/2\ 5/1\ x_0/1] \\ &= [1\ 2\ 5\ x_0] \end{aligned}$$

$$x = \mathbf{t^3 + 2t^2 + 5t + x_0.}$$

Let's look at the above problem using vectors.

Let **V** = 3t²**I** + 4t**j** +5**k**.

Using the displacement vector, find the displacement in 10 seconds.

$$[3i\ 4j\ 6k\ 0] \begin{vmatrix} 10^3 \\ 10^2 \\ 10 \\ 1 \end{vmatrix} = [3(10^3)i + 4(10^2j + 50k]$$

or

$$\begin{bmatrix} 3i \\ 4j \\ 5k \\ 0 \end{bmatrix} \circ \begin{bmatrix} 1000 \\ 100 \\ 10 \\ 1 \end{bmatrix} = \begin{bmatrix} 3000i \\ 400j \\ 50k \\ 0 \end{bmatrix}$$

Find the instantaneous and average velocity vectors.

$$d[3\mathbf{i}\ 4\mathbf{j}\ 5\mathbf{k}\ x_0]/dt = [3\mathbf{i}\ 4\mathbf{j}\ 5\mathbf{k}] \otimes \overset{t^2\ t\ 1}{[3\ 2\ 1]} = [9\mathbf{i}\ 8\mathbf{j}\ 5\mathbf{k}]$$ evaluating at 10 seconds:

$$[9\mathbf{i}\ 8\mathbf{j}\ 5\mathbf{k}] \begin{vmatrix} 10^2 \\ 10 \\ 1 \end{vmatrix} = [900\mathbf{i} + 80\mathbf{j} + 5\mathbf{k}]$$

The average velocity is given by:

$$(1/10)[3\mathbf{i}\ 4\mathbf{j}\ 5\mathbf{k}\ 0]\begin{bmatrix} 10^3 \\ 10^2 \\ 10 \\ 1 \end{bmatrix} = (1/10)[3000\mathbf{i} + 400\mathbf{j} + 50\mathbf{k} + 0] = [300\mathbf{i} + 40\mathbf{j} + 5\mathbf{k}]$$

Find the instantaneous acceleration.

$$\mathbf{a} = d\mathbf{v}/dt = 18t\mathbf{i} + 8\mathbf{j}\quad \text{or}$$

$$d[9\mathbf{i}\ 8\mathbf{j}\ 5\mathbf{k}]/dt = [9\mathbf{i}\ 8\mathbf{j}] \otimes [2\ 1] = [18\mathbf{i}\ 8\mathbf{j}]\quad \text{and at } t = 10s$$

$$[18\mathbf{i}\ 8\mathbf{j}]\begin{bmatrix} 10 \\ 1 \end{bmatrix} = [180\mathbf{i}\ 8\mathbf{j}]$$

EXAMPLE: A particle moves along the x-axis. At t = 0, it's position is -4 m. It's instantaneous velocity in meters/sec is given by v = 6t² - 5t + 4.

What is it's velocity at t = 2 seconds?

$$[6\ -5\ 4]\begin{bmatrix} 4 \\ 2 \\ 1 \end{bmatrix}\begin{matrix} t^2 \\ t \\ 1 \end{matrix} = [24 - 10 + 4] = \mathbf{18\ m/s}$$

What is it's instantaneous acceleration at t = 3 seconds?

$$d[6\ -5\ 4]dt = [6\ -5] \otimes [2\ 1] = [12\ -5]$$

$$\mathbf{a} = [12\ -5]\begin{bmatrix} 3 \\ 1 \end{bmatrix} = [36 - 5] = \mathbf{[31m/s^2]}$$

What is it's average acceleration for the time interval t = 2 to t = 4 seconds?

$$\overline{\mathbf{a}} = \frac{\mathbf{v}(4) - \mathbf{v}(2)}{2}$$

$$\mathbf{v}(2) = 18\ m/s;\ \mathbf{v}(4) = 6(4)^2 - 5(4) + 4 = 96 - 20 + 4 = 80m/s$$

$\bar{a} = (80 - 11)/2 = 34.5 \text{m/s}^2$

or

$(1/2) [1]_{1,3} ([\mathbf{v}]_{1,3} [\mathbf{\Delta a}]_{3,2} [\pm 1]_{2,1})$

$(1/2) [1\ 1\ 1]\ (6\ -5\ 4]\ \begin{bmatrix} 16 & 4 \\ 4 & 2 \\ 1 & 1 \end{bmatrix} \begin{bmatrix} 1 \\ -1 \end{bmatrix}) =$

$$\frac{1}{2} \cdot \left[(6\ -5\ 4) \cdot \begin{pmatrix} 16 & 4 \\ 4 & 2 \\ 1 & 1 \end{pmatrix} \cdot \begin{pmatrix} 1 \\ -1 \end{pmatrix} \right] = 31\ \text{m/s}^2$$

Where is the particle at t = 3 s ?

$$\int_0^3 (6\ -5\ 4)\, dt = [6\ -5\ 4\ 1] \otimes [1/3\ 1/2\ 1\ 1] = [2\ -5/2\ 4\ 1] \Big|_0^3$$

Then

$$\left(2\ -\frac{5}{2}\ 4\ 1 \right) \cdot \begin{bmatrix} 27 \\ 9 \\ 3 \\ -4 \end{bmatrix} = 39.5\ \text{m}$$

Where −4 is the initial displacement at −4 meters, $27 = 3^3$; $9 = 3^2$ and $3 = 3$ seconds.

Or we can solve this in the normal manner:

$$x(3) - x(0) = \int_0^3 v\, dt = \int_0^3 6t^2 - 5t + 4\, dt = x(3) - (-4) = \left(\frac{6t^3}{3} - \frac{5t^2}{2} + 4t \right) \Big|_0^3 =$$

$$= \frac{6}{3} \cdot (27) - \frac{5}{2} \cdot (9) + 4 \cdot (3) - 4 = 39.5\ \text{m}$$

EXAMPLE: The position of a particle is described by r = (4t³ − 12t +9)I + (6t + 4)j meters with t in seconds.

Calculate the instantaneous velocity at t = 0, 1, 2, and 3 seconds.

$\mathbf{r}$ = [4 0 -12 9]$\mathbf{i}$ + [6 4]$\mathbf{j}$
$\mathbf{v}$ = [4 0 -12]$\mathbf{I}$ ⊗ [3 2 1] + [6]$\mathbf{j}$ ⊗ [1] = **[12 0 -12]i + [6]j** = $(12\,t^2\,-12)$**i** +6**j**

$$\mathbf{v_0} = [12\ 0\ -12]\mathbf{i}\begin{bmatrix} 0 \\ 0 \\ 1 \end{bmatrix} + [0\ 0\ 6]\mathbf{j}\begin{bmatrix} 0 \\ 0 \\ 1 \end{bmatrix} = -12\mathbf{i} + 6\mathbf{j}$$

Or we can solve this using nested arrays, i.e.

$$\overset{\mathbf{i}}{[12\ 0\ -12]}\ \overset{\mathbf{j}}{[0\ 0\ 6]}\ ]\ \begin{bmatrix} 0 \\ 0 \\ 1 \end{bmatrix}$$

Transposing the nested array (so we can properly multiply them) we get

$$\begin{bmatrix} 12 & 0 & -12 \\ 0 & 0 & 6 \end{bmatrix}\begin{bmatrix} 0 \\ 0 \\ 1 \end{bmatrix} = \begin{bmatrix} 0 \\ 6 \end{bmatrix} = 0\mathbf{i} + 6\mathbf{j}$$

Re-transposing the solution, we get [[-12] [6]], or removing the matrices and re-formatting into equation form we get -12**i** +6**j**.

$$\mathbf{v_1} = \overset{\mathbf{v}}{\begin{bmatrix} 12 & 0 & -12 \\ 0 & 0 & 6 \end{bmatrix}}\overset{\mathbf{t}}{\begin{bmatrix} 1 \\ 1 \\ 1 \end{bmatrix}} \qquad \begin{bmatrix} 0 \\ 6 \end{bmatrix}$$

Or perhaps we don't need to transpose the nested array, but multiply each sub-matrix as if it were a single element, i.e.

$$[\ \overset{\mathbf{i}}{[12\ 0\ -12]}\ \overset{\mathbf{j}}{[0\ 0\ 6]}\ ]\ \begin{bmatrix} 1 \\ 1 \\ 1 \end{bmatrix} = [\ 0\quad 6]$$

$$\mathbf{v_1} = \begin{bmatrix} 12 & 0 & -12 \\ 0 & 0 & 6 \end{bmatrix}\begin{bmatrix} 4 \\ 2 \\ 1 \end{bmatrix} = \begin{bmatrix} 36 \\ 6 \end{bmatrix} = 36\mathbf{i} + 6\mathbf{j}\ .$$

$$\mathbf{v_1} = \begin{bmatrix} 12 & 0 & -12 \\ 0 & 0 & 6 \end{bmatrix}\begin{bmatrix} 9 \\ 3 \\ 1 \end{bmatrix} = \begin{bmatrix} 96 \\ 6 \end{bmatrix} = 96\mathbf{i} + 6\mathbf{j}\ .$$

Let's solve them all at once:

$$V := \begin{pmatrix} 12 & 0 & -12 \\ 0 & 0 & 6 \end{pmatrix} \qquad T := \begin{pmatrix} 0 & 1 & 4 & 9 \\ 0 & 1 & 2 & 3 \\ 1 & 1 & 1 & 1 \end{pmatrix} \qquad V \cdot T = \begin{pmatrix} -12 & 0 & 36 & 96 \\ 6 & 6 & 6 & 6 \end{pmatrix}$$

Let

$$VT := \begin{pmatrix} -12 & 0 & 36 & 96 \\ 6 & 6 & 6 & 6 \end{pmatrix}$$

$$VT^{<0>} = \begin{pmatrix} -12 \\ 6 \end{pmatrix} \quad VT^{<1>} = \begin{pmatrix} 0 \\ 6 \end{pmatrix} \quad VT^{<2>} = \begin{pmatrix} 36 \\ 6 \end{pmatrix} \quad VT^{<3>} = \begin{pmatrix} 96 \\ 6 \end{pmatrix}$$

All these values are in meters/second (velocity).

b) Compute the average velocity between t = 1 and t = 2 seconds.

$$\begin{vmatrix} 4 & 0 & -12 & 9 \\ 0 & 0 & 6 & 4 \end{vmatrix} \quad \begin{vmatrix} 8 & 1 \\ 4 & 1 \\ 2 & 1 \\ 1 & 1 \end{vmatrix}$$

$$T := \begin{bmatrix} 8 & 1 \\ 4 & 1 \\ 2 & 1 \\ 1 & 1 \end{bmatrix} \qquad r := \begin{pmatrix} 4 & 0 & -12 & 9 \\ 0 & 0 & 6 & 4 \end{pmatrix} \qquad r \cdot T = \begin{pmatrix} 17 & 1 \\ 16 & 10 \end{pmatrix}$$

To subtract r(1) from r(2), define

$$PLUSMIN1 := \begin{pmatrix} 1 \\ -1 \end{pmatrix}$$

$$\begin{pmatrix} 4 & 0 & -12 & 9 \\ 0 & 0 & 6 & 4 \end{pmatrix} \cdot \begin{bmatrix} 8 & 1 \\ 4 & 1 \\ 2 & 1 \\ 1 & 1 \end{bmatrix} \cdot \begin{pmatrix} 1 \\ -1 \end{pmatrix} = \begin{pmatrix} 16 \\ 6 \end{pmatrix} = 16\mathbf{i} + 6\mathbf{j}$$

c) Find the speed at t = 1 second.

$$v_1 = \begin{bmatrix} 12 & 0 & -12 \\ 0 & 0 & 6 \end{bmatrix} \begin{bmatrix} 1 \\ 1 \\ 1 \end{bmatrix} = \begin{bmatrix} 0 \\ 6 \end{bmatrix}$$

Speed $= \left| v(t) \right| = \left| 0i + 6j \right| = 6$ m/s

d) Find the instantaneous acceleration at t = 2 seconds.

$a = dv/dt = d[12\ 0\ -12]/dt + d[0\ 0\ 6]/dt = [12\ 0] \otimes [2\ 1] + [0\ 0] \otimes [2\ 1] = [24i + 0j]$. Evaluating at t=2 we get

$a = dv/dt = d[12\ 0\ -12]/dt + d[0\ 0\ 6]/dt = [12\ 0] \otimes [2\ 1] + [0\ 0] \otimes [2\ 1] =$

$[24i + 0j]$. Evaluating at t=2 we get

$[24\ 0] \begin{bmatrix} 2 \\ 1 \end{bmatrix} = 48i$ m/s^2

e) Find the average acceleration between t = 1 and 3 seconds.

$$a_{av} = \frac{v(3) - v(1)}{3 - 1} = \frac{(96i + 6j) - 6j}{2} = \frac{96i}{2} = 48i \text{ m/s}$$

$$V := \begin{pmatrix} 12 & 0 & -12 \\ 0 & 0 & 6 \end{pmatrix} \qquad T := \begin{pmatrix} 0 & 1 & 4 & 9 \\ 0 & 1 & 2 & 3 \\ 1 & 1 & 1 & 1 \end{pmatrix} \qquad V \cdot T = \begin{pmatrix} -12 & 0 & 36 & 96 \\ 6 & 6 & 6 & 6 \end{pmatrix}$$

$$\text{minplus40} := \begin{bmatrix} -1 \\ 0 \\ 0 \\ 1 \end{bmatrix} \qquad \text{minplus42} := \begin{bmatrix} 0 \\ -1 \\ 1 \\ 0 \end{bmatrix} \qquad \text{minplus41} := \begin{bmatrix} 0 \\ -1 \\ 0 \\ 1 \end{bmatrix}$$

$\text{aaverage1to3sec} := \left(\dfrac{1}{2}\right) \cdot V \cdot T \cdot \text{minplus41}$ $\qquad$ $\text{aaverage1to3sec} = \begin{pmatrix} 48 \\ 0 \end{pmatrix}$

$\text{aaverage0to3sec} := \left(\dfrac{1}{3}\right) \cdot V \cdot T \cdot \text{minplus40}$ $\qquad$ $\text{aaverage0to3sec} = \begin{pmatrix} 36 \\ 0 \end{pmatrix}$

$$\text{aaverage1to2sec} := \left(\frac{1}{2-1}\right)\cdot V\cdot T\cdot minplus42 \qquad\qquad \text{aaverage1to2sec} = \begin{pmatrix} 36 \\ 0 \end{pmatrix}$$

Let's solve all possible acceleration states for this example. The first column represents 0-1 seconds, the second column represents 0-2 seconds, the third column represents 0-3 seconds, the last column represents 2-3 seconds. We divide by the number of seconds difference between the two time values.

$$\text{deltaV} := \begin{bmatrix} -1 & -1 & -1 & 0 & 0 & 0 \\ 1 & 0 & 0 & -1 & -1 & 0 \\ 0 & 1 & 0 & 1 & 0 & -1 \\ 0 & 0 & 1 & 0 & 1 & 1 \end{bmatrix} \qquad\qquad \text{deltaT} := \begin{bmatrix} 1 \\ .5 \\ \frac{1}{3} \\ 1 \\ .5 \\ 1 \end{bmatrix}$$

$$\text{AV} := V\cdot T\cdot \text{deltaV}$$

$$\text{AV} = \begin{pmatrix} 12 & 48 & 108 & 36 & 96 & 60 \\ 0 & 0 & 0 & 0 & 0 & 0 \end{pmatrix}$$

$$\text{AVav} := V\cdot T\cdot \text{deltaV}\cdot \text{diag}(\text{deltaT})$$

$$\text{AVav} = \begin{pmatrix} 12 & 24 & 36 & 36 & 48 & 60 \\ 0 & 0 & 0 & 0 & 0 & 0 \end{pmatrix}$$

i.e., in 0-1 sec, the average acceleration is 12m/s², 0-2 = 24m/s², 0-3= 36m/s², 1-2sec = 36m/s², 1-3sec= 48m/s² and 2-3sec = 60 m/s².

EXAMPLE: Consider a weight W suspended from two ropes as shown. At what angle will the tensions in T₁ and T₂ be equal?

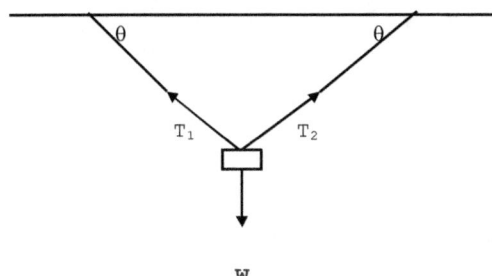

Computing the forces along the x-axis, we get

$$F_x = -T_1\cos\theta + T_2\cos\theta = 0$$

$$T_1 = T_2$$

Computing the forces acting along the y-axis we get:

$$F_y = T_1\sin\theta + T_2\sin\theta - W = 0$$

Putting the two equations in matrix form, letting the top row represent the forces on x ($\cos\theta$) and the bottom row represent the forces along y ($\sin\theta$)

$$\begin{pmatrix} -T_1 & T_2 & 0 \\ T_1 & T_2 & -W \end{pmatrix}$$

Using Gaussian Reduction and adding row 1 to row 2, we get

$$\begin{pmatrix} -T_1 & T_2 & 0 \\ 0 & 2T_2 & -W \end{pmatrix}$$

We want the tensions to be equal, so we will define $W = T2$; we now have

$$\begin{pmatrix} -T_1 & T_2 & 0 \\ 0 & 2T_2 & -T_2 \end{pmatrix}$$

Dividing out T2 in the second row, we get

$$\begin{pmatrix} -T_1 & T_2 & 0 \\ 0 & 2 & -1 \end{pmatrix}$$

Remembering the second row is the sin row, we get

$$\sin 2\theta - 1 = 0$$

$$\sin\theta = 1/2$$

$$\theta = 30°$$

When both the angles are equal to 30°, the tensions in the ropes will be equal.

b) **Find the angle θ which makes T_2 twice T_1. Find the tension in T_1.**

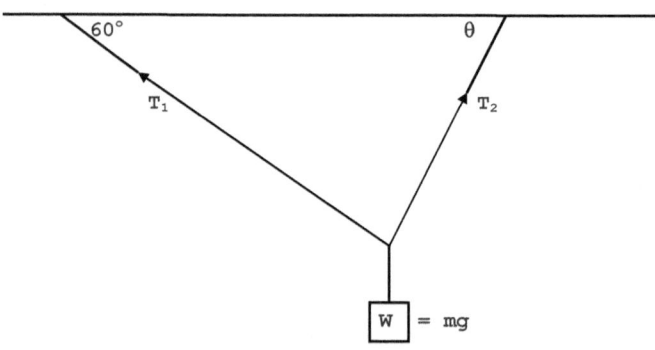

$$F_x = T_2\cos\theta - T_1\cos 60° = 0$$

$$F_y = T_2\sin\theta ° + T_1\sin 60 = mg$$

Row one represents the cos terms, row 2 represents the sin terms

$$\begin{matrix} \theta & 60 & \\ \begin{pmatrix} T_2 & -T_1 & 0 \\ T_2 & T_1 & -W \end{pmatrix} \end{matrix}$$

$T2 = 2T1,$

$$\begin{pmatrix} 2T_1 & -T_1 & 0 \\ 2T_1 & T_1 & -W \end{pmatrix}$$

Looking at row one we have

$2T_1 \cdot COS\theta = T_1 \cdot COS60$

$2 \cdot COS\theta = COS60$

$COS\theta := \dfrac{\cos(60)}{2}$

$COS\theta = -0.476$ RADIANS, OR 75.52 DEGREES

To find the tension in T_1, go to the second row of the matrix

$$\begin{pmatrix} 2T_1 & -T_1 & 0 \\ 2T_1 & T_1 & -W \end{pmatrix}$$

$2 \cdot T_1 \cdot \sin(75.52) + T_1 \cdot \sin(60) - 9.8 \cdot W = 0$

$T_1 \cdot (2 \cdot \sin(75.52) + \sin(60)) = 9.8 \cdot W$

$T_1 = 9.8 \cdot \dfrac{W}{(2 \cdot \sin(75.52) + \sin(60))}$

$T_1 = 9.8 \cdot \dfrac{W}{2.80249} = 3.49688 \, W$

EXAMPLE: A bar one meter long weighing 100 grams supports a
200g weight at one end and a 75 gram weight at the other end.
It is balanced on a fulcrum 20 cm from the 200g weight. Find
the center of gravity for this system.

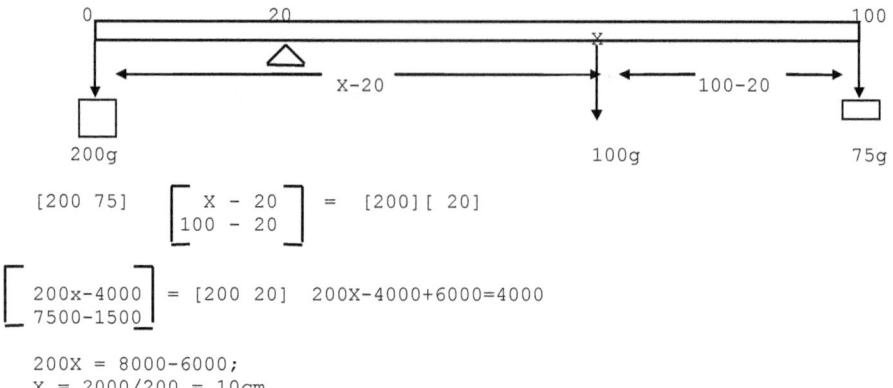

X = point of center of gravity

$$[200\ 75]\begin{bmatrix} X - 20 \\ 100 - 20 \end{bmatrix} = [200][\ 20]$$

$$\begin{bmatrix} 200x-4000 \\ 7500-1500 \end{bmatrix} = [200\ 20]\quad 200X-4000+6000=4000$$

$200X = 8000-6000;$
$X = 2000/200 = 10\text{cm}$

EXAMPLE:

a) Three particles have masses $m_1 = 3\text{kg}$, $m_2 = 2\text{kg}$ and $m_3 = 5\text{kg}$.
    They lie along a line such that $x_1 = -4\text{m}$, $x_2 = 3\text{m}$ and $x_3 =$
    8m. Find the position of the center of mass.

$m_1 = 3\text{kg}$ $m_2 = 2\text{kg}$ $m_3 = 5\text{kg}$

$x_1 = -4\text{m}$ $x_2 = 3\text{m}$ $x_3 = 8\text{m}$

Then:

$$\frac{[m_1\ m_2\ m_3]\begin{vmatrix} x_1 \\ x_2 \\ x_3 \end{vmatrix}}{[m_1\ m_2\ m_3]\begin{vmatrix} 1 \\ 1 \\ 1 \end{vmatrix}} = \quad [3\ 2\ 5]\begin{vmatrix} -4 \\ 3 \\ 8 \end{vmatrix}\quad [10]^{-1} = [34][10]^{-1} = 3.4\text{kg m/kg} = 3.4\text{m}$$

b) Two particles have mass $m_1 = .3$ kg and $m_2 = .4$ kg with
    velocities $v_1 = (3i +4j)$ and $v_2 = (5i -2j)$. Find the velocity
    of the center of mass of this two particle system.

$$\frac{[m_1\ m_2\ ]\begin{bmatrix} v_1 \\ v_2 \end{bmatrix}}{[m_1\ m_2\ ]\begin{bmatrix} 1 \\ 1 \end{bmatrix}} = \frac{[.3\ .4]\begin{bmatrix} 3 & 4 \\ 5 & -2 \end{bmatrix}}{[.3\ .4]\begin{bmatrix} 1 \\ 1 \end{bmatrix}} = [2.9\ .04][.7]^{-1} =$$

$$(.3 \quad .4) \cdot \begin{pmatrix} 3 & 4 \\ 5 & -2 \end{pmatrix} \cdot .7^{-1} = (4.143 \quad 0.571)$$

$\mathbf{v}_{cm} = (4.143\mathbf{i} + .571\mathbf{j})\,m/s$

EXAMPLE: Two forces F_1 and F_2 act on a mass of 1.5kg. F_1 = 12N and is directed 30° below the horizontal, F_2 = 5N and is directed 45° above the horizontal. Find the magnitude and direction of the resultant force and the magnitude and direction of it's acceleration.

COSθ   SINθ

$$\begin{bmatrix} (-30)^{\circ} & -30^{\circ} \\ 45^{\circ} & 45^{\circ} \end{bmatrix} = \begin{pmatrix} .8660 & -.500 \\ .7071 & .7071 \end{pmatrix}$$

$$(12 \quad 5) \cdot \begin{pmatrix} .8660 & -.500 \\ .7071 & .7071 \end{pmatrix} = (13.927 \quad -2.465)\ N$$

$F_x = 13.927N,\ F_y = -2.465N$

$$F = \sqrt{\left(F_X\right)^2 + \left(F_Y\right)^2}$$

$$\left[\left(\overrightarrow{(13.927 \ -2.465)^2}\right) \cdot \begin{pmatrix} 1 \\ 1 \end{pmatrix} \right]^{\frac{1}{2}}$$

$$\left(\overrightarrow{(13.927 \ -2.465)^2}\right) \cdot \begin{pmatrix} 1 \\ 1 \end{pmatrix} = 200.038$$

$$\sqrt{200.038} = 14.143\ N$$

$$TAN\theta = \frac{F_Y}{F_X} = -\frac{2.465}{13.927} = -0.177$$

$$\tan^{-1}\theta = (-.177)$$

$\theta = -10^{\circ}$ (10° below the horizontal)

b) **a** = **F**/m = 14N/1.5kg = 9.3m/s²

The acceleration is in the same direction as **F**.

COMPUTATION OF SEVERAL SETS
OF DATA ALL AT ONCE

For this set of data, in A2 I added 1 to each value in A1, in A3
I added 2 to each value, in A4 I subtracted 1 from each value
in A1 and in A5 I subtracted 2 from each value in A1.

$$A1 := \begin{bmatrix} 15 & 23 & 11 & 25 & 14 & 24 & 11 & 39 \\ 8 & 16 & 16 & 27 & 6 & 16 & 17 & 32 \\ 9 & 17 & 8 & 29 & 3 & 10 & 12 & 38 \\ 7 & 17 & 14 & 34 & 5 & 18 & 14 & 39 \\ 8 & 14 & 20 & 38 & 6 & 13 & 21 & 33 \\ 4 & 11 & 16 & 26 & 7 & 17 & 18 & 31 \end{bmatrix}$$

$$A2 := \begin{bmatrix} 16 & 24 & 12 & 26 & 15 & 25 & 12 & 30 \\ 9 & 17 & 17 & 28 & 7 & 17 & 18 & 33 \\ 10 & 18 & 9 & 30 & 4 & 11 & 13 & 39 \\ 8 & 18 & 15 & 35 & 6 & 19 & 15 & 40 \\ 9 & 15 & 21 & 39 & 7 & 14 & 22 & 34 \\ 5 & 12 & 17 & 27 & 8 & 18 & 19 & 32 \end{bmatrix}$$

$$A3 := \begin{bmatrix} 17 & 25 & 13 & 27 & 16 & 26 & 13 & 31 \\ 10 & 18 & 18 & 29 & 8 & 18 & 19 & 34 \\ 11 & 19 & 10 & 31 & 5 & 12 & 14 & 40 \\ 9 & 19 & 16 & 36 & 7 & 20 & 16 & 41 \\ 10 & 16 & 22 & 40 & 8 & 15 & 23 & 35 \\ 6 & 13 & 18 & 28 & 9 & 19 & 20 & 33 \end{bmatrix}$$

$$A4 := \begin{bmatrix} 14 & 22 & 10 & 24 & 13 & 23 & 10 & 38 \\ 7 & 15 & 15 & 26 & 5 & 15 & 16 & 31 \\ 8 & 16 & 7 & 28 & 2 & 9 & 11 & 37 \\ 6 & 16 & 13 & 33 & 4 & 17 & 13 & 38 \\ 7 & 13 & 19 & 37 & 5 & 12 & 20 & 32 \\ 3 & 10 & 15 & 25 & 6 & 16 & 17 & 30 \end{bmatrix}$$

$$A5 := \begin{bmatrix} 13 & 21 & 9 & 23 & 12 & 22 & 9 & 37 \\ 6 & 14 & 14 & 25 & 4 & 14 & 15 & 30 \\ 7 & 15 & 6 & 27 & 1 & 8 & 10 & 36 \\ 5 & 15 & 12 & 32 & 3 & 16 & 12 & 37 \\ 6 & 12 & 18 & 36 & 4 & 11 & 19 & 31 \\ 2 & 9 & 14 & 24 & 5 & 15 & 16 & 29 \end{bmatrix}$$

A6 := stack(A1, A2)
A7 := stack(A6, A3)
A8 := stack(A7, A4)
AALL := stack(A8, A5)

Here we have the nested array of all the statistical data taken
over the past 5 days.

$$
AALL =
\begin{bmatrix}
15 & 23 & 11 & 25 & 14 & 24 & 11 & 39 \\
8 & 16 & 16 & 27 & 6 & 16 & 17 & 32 \\
9 & 17 & 8 & 29 & 3 & 10 & 12 & 38 \\
7 & 17 & 14 & 34 & 5 & 18 & 14 & 39 \\
8 & 14 & 20 & 38 & 6 & 13 & 21 & 33 \\
4 & 11 & 16 & 26 & 7 & 17 & 18 & 31 \\
16 & 24 & 12 & 26 & 15 & 25 & 12 & 30 \\
9 & 17 & 17 & 28 & 7 & 17 & 18 & 33 \\
10 & 18 & 9 & 30 & 4 & 11 & 13 & 39 \\
8 & 18 & 15 & 35 & 6 & 19 & 15 & 40 \\
9 & 15 & 21 & 39 & 7 & 14 & 22 & 34 \\
5 & 12 & 17 & 27 & 8 & 18 & 19 & 32 \\
17 & 25 & 13 & 27 & 16 & 26 & 13 & 31 \\
10 & 18 & 18 & 29 & 8 & 18 & 19 & 34 \\
11 & 19 & 10 & 31 & 5 & 12 & 14 & 40 \\
9 & 19 & 16 & 36 & 7 & 20 & 16 & 41 \\
10 & 16 & 22 & 40 & 8 & 15 & 23 & 35 \\
6 & 13 & 18 & 28 & 9 & 19 & 20 & 33 \\
14 & 22 & 10 & 24 & 13 & 23 & 10 & 38 \\
7 & 15 & 15 & 26 & 5 & 15 & 16 & 31 \\
8 & 16 & 7 & 28 & 2 & 9 & 11 & 37 \\
6 & 16 & 13 & 33 & 4 & 17 & 13 & 38 \\
7 & 13 & 19 & 37 & 5 & 12 & 20 & 32 \\
3 & 10 & 15 & 25 & 6 & 16 & 17 & 30 \\
13 & 21 & 9 & 23 & 12 & 22 & 9 & 37 \\
6 & 14 & 14 & 25 & 4 & 14 & 15 & 30 \\
7 & 15 & 6 & 27 & 1 & 8 & 10 & 36 \\
5 & 15 & 12 & 32 & 3 & 16 & 12 & 37 \\
6 & 12 & 18 & 36 & 4 & 11 & 19 & 31 \\
2 & 9 & 14 & 24 & 5 & 15 & 16 & 29
\end{bmatrix}
$$

Now we construct a matrix so we can sum each individual nested array

$$
B1 :=
\begin{pmatrix}
1 & 1 & 1 & 1 & 1 & 1 & 0 & 0 & 0 & 0 & 0 & 0 & 0 & 0 & 0 & 0 & 0 & 0 & 0 & 0 & 0 & 0 & 0 & 0 & 0 & 0 & 0 & 0 & 0 & 0 \\
0 & 0 & 0 & 0 & 0 & 0 & 1 & 1 & 1 & 1 & 1 & 1 & 0 & 0 & 0 & 0 & 0 & 0 & 0 & 0 & 0 & 0 & 0 & 0 & 0 & 0 & 0 & 0 & 0 & 0 \\
0 & 0 & 0 & 0 & 0 & 0 & 0 & 0 & 0 & 0 & 0 & 0 & 1 & 1 & 1 & 1 & 1 & 1 & 0 & 0 & 0 & 0 & 0 & 0 & 0 & 0 & 0 & 0 & 0 & 0
\end{pmatrix}
$$

$$
B2 :=
\begin{pmatrix}
0 & 0 & 0 & 0 & 0 & 0 & 0 & 0 & 0 & 0 & 0 & 0 & 0 & 0 & 0 & 0 & 0 & 0 & 1 & 1 & 1 & 1 & 1 & 1 & 0 & 0 & 0 & 0 & 0 & 0 \\
0 & 0 & 0 & 0 & 0 & 0 & 0 & 0 & 0 & 0 & 0 & 0 & 0 & 0 & 0 & 0 & 0 & 0 & 0 & 0 & 0 & 0 & 0 & 0 & 1 & 1 & 1 & 1 & 1 & 1
\end{pmatrix}
$$

B := stack(B1 , B2)

$$B = \begin{bmatrix} 1 & 1 & 1 & 1 & 1 & 1 & 0 & 0 & 0 & 0 & 0 & 0 & 0 & 0 & 0 & 0 & 0 & 0 & 0 & 0 & 0 & 0 & 0 & 0 & 0 & 0 & 0 & 0 & 0 & 0 \\ 0 & 0 & 0 & 0 & 0 & 0 & 1 & 1 & 1 & 1 & 1 & 1 & 0 & 0 & 0 & 0 & 0 & 0 & 0 & 0 & 0 & 0 & 0 & 0 & 0 & 0 & 0 & 0 & 0 & 0 \\ 0 & 0 & 0 & 0 & 0 & 0 & 0 & 0 & 0 & 0 & 0 & 0 & 1 & 1 & 1 & 1 & 1 & 1 & 0 & 0 & 0 & 0 & 0 & 0 & 0 & 0 & 0 & 0 & 0 & 0 \\ 0 & 0 & 0 & 0 & 0 & 0 & 0 & 0 & 0 & 0 & 0 & 0 & 0 & 0 & 0 & 0 & 0 & 0 & 1 & 1 & 1 & 1 & 1 & 1 & 0 & 0 & 0 & 0 & 0 & 0 \\ 0 & 0 & 0 & 0 & 0 & 0 & 0 & 0 & 0 & 0 & 0 & 0 & 0 & 0 & 0 & 0 & 0 & 0 & 0 & 0 & 0 & 0 & 0 & 0 & 1 & 1 & 1 & 1 & 1 & 1 \end{bmatrix}$$

SUM15 := B·AALL

$$B \cdot AALL = \begin{bmatrix} 51 & 98 & 85 & 179 & 41 & 98 & 93 & 212 \\ 57 & 104 & 91 & 185 & 47 & 104 & 99 & 208 \\ 63 & 110 & 97 & 191 & 53 & 110 & 105 & 214 \\ 45 & 92 & 79 & 173 & 35 & 92 & 87 & 206 \\ 39 & 86 & 73 & 167 & 29 & 86 & 81 & 200 \end{bmatrix}$$

Now we will compute the Grand Sum of Squares for each data matrix:

$$ONE6 := (1 \ 1 \ 1 \ 1 \ 1 \ 1) \qquad EIGHT1 := \begin{bmatrix} 1 \\ 1 \\ 1 \\ 1 \\ 1 \\ 1 \\ 1 \\ 1 \end{bmatrix}$$

$ONE6 \cdot \overrightarrow{(A1 \cdot A1)} \cdot EIGHT1 = 20083$
$ONE6 \cdot \overrightarrow{(A2 \cdot A2)} \cdot EIGHT1 = 21145$
$ONE6 \cdot \overrightarrow{(A3 \cdot A3)} \cdot EIGHT1 = 22983$
$ONE6 \cdot \overrightarrow{(A4 \cdot A4)} \cdot EIGHT1 = 18417$
$ONE6 \cdot \overrightarrow{(A5 \cdot A5)} \cdot EIGHT1 = 16847$

And enter them into the SST1 Matrix by hand because the
computer cannot handle the math of multiplying a matrix inside
a matrix

$$SST1 := \begin{bmatrix} 0 & 20083 & 0 & 0 & 0 & 0 & 0 & 0 & 0 \\ 0 & 21145 & 0 & 0 & 0 & 0 & 0 & 0 & 0 \\ 0 & 22983 & 0 & 0 & 0 & 0 & 0 & 0 & 0 \\ 0 & 18417 & 0 & 0 & 0 & 0 & 0 & 0 & 0 \\ 0 & 16847 & 0 & 0 & 0 & 0 & 0 & 0 & 0 \end{bmatrix}$$

$$SUM15 \cdot DBALL = \begin{bmatrix} 857 & 0 & 413 & 444 & 288 & 569 & 270 & 587 & 149 & 264 & 139 & 305 & 136 & 277 & 134 & 310 & 92 & 196 & 178 & 391 & 51 & 98 & 85 & 179 & 41 & 98 & 93 & 212 \\ 895 & 0 & 437 & 458 & 312 & 583 & 294 & 601 & 161 & 276 & 151 & 307 & 148 & 289 & 146 & 312 & 104 & 208 & 190 & 393 & 57 & 104 & 91 & 185 & 47 & 104 & 99 & 208 \\ 943 & 0 & 461 & 482 & 336 & 607 & 318 & 625 & 173 & 288 & 163 & 319 & 160 & 301 & 158 & 324 & 116 & 220 & 202 & 405 & 63 & 110 & 97 & 191 & 53 & 110 & 105 & 214 \\ 809 & 0 & 389 & 420 & 264 & 545 & 246 & 563 & 137 & 252 & 127 & 293 & 124 & 265 & 122 & 298 & 80 & 184 & 166 & 379 & 45 & 92 & 79 & 173 & 35 & 92 & 87 & 206 \\ 761 & 0 & 365 & 396 & 240 & 521 & 222 & 539 & 125 & 240 & 115 & 281 & 112 & 253 & 110 & 286 & 68 & 172 & 154 & 367 & 39 & 86 & 73 & 167 & 29 & 86 & 81 & 200 \end{bmatrix}$$

Here I am going to cheat a little so I won't have to rename all the variables, I'm going to let

$SS15 := SUM15 \cdot DBALL$

Now we must create the N matrix, since we are dividing element by element, this matrix must be the same size as SS1

$N1 := stack(N, N)$

$N2 := stack(N1, N1)$

$N := stack(N2, N)$

$$N = \begin{bmatrix} 48 & 1 & 24 & 24 & 24 & 24 & 24 & 24 & 12 & 12 & 12 & 12 & 12 & 12 & 12 & 12 & 12 & 12 & 12 & 12 & 6 & 6 & 6 & 6 & 6 & 6 & 6 & 6 \\ 48 & 1 & 24 & 24 & 24 & 24 & 24 & 24 & 12 & 12 & 12 & 12 & 12 & 12 & 12 & 12 & 12 & 12 & 12 & 12 & 6 & 6 & 6 & 6 & 6 & 6 & 6 & 6 \\ 48 & 1 & 24 & 24 & 24 & 24 & 24 & 24 & 12 & 12 & 12 & 12 & 12 & 12 & 12 & 12 & 12 & 12 & 12 & 12 & 6 & 6 & 6 & 6 & 6 & 6 & 6 & 6 \\ 48 & 1 & 24 & 24 & 24 & 24 & 24 & 24 & 12 & 12 & 12 & 12 & 12 & 12 & 12 & 12 & 12 & 12 & 12 & 12 & 6 & 6 & 6 & 6 & 6 & 6 & 6 & 6 \\ 48 & 1 & 24 & 24 & 24 & 24 & 24 & 24 & 12 & 12 & 12 & 12 & 12 & 12 & 12 & 12 & 12 & 12 & 12 & 12 & 6 & 6 & 6 & 6 & 6 & 6 & 6 & 6 \end{bmatrix}$$

Computation of the uncorrected sum of Squares

$$\left[\left(\overrightarrow{SS15 \cdot SS15 \cdot N^{-1}}\right) \cdot DBSUM\right] = \begin{bmatrix} 15301.021 & 0 & 15321.042 & 16946.042 & 17394.542 & 17020.25 & 17440.083 & 19287.083 & 19391.5 \\ 16688.021 & 0 & 16697.208 & 18218.042 & 18651.542 & 18262.25 & 18673.75 & 20385.75 & 20443.5 \\ 18526.021 & 0 & 18535.208 & 20056.042 & 20489.542 & 20100.25 & 20511.75 & 22223.75 & 22281.5 \\ 13635.021 & 0 & 13655.042 & 15280.042 & 15728.542 & 15354.25 & 15774.083 & 17621.083 & 17725.5 \\ 12065.021 & 0 & 12085.042 & 13710.042 & 14158.542 & 13784.25 & 14204.083 & 16051.083 & 16155.5 \end{bmatrix}$$

Computation of the effects of the 3rd factor (col. 4); 2nd factor effects col. 3; 1st factor effects, col. 2 after adding in Grand Sum of Squares

$$\left[\left[\left[\left(\overrightarrow{SS15 \cdot SS15 \cdot N^{-1}}\right) \cdot DBSUM\right] + SST1\right] \cdot DBSSTLR\right] = \begin{bmatrix} 4781.979 & 20.021 & 1645.021 & 2093.521 & 1719.229 & 2139.063 & 3986.063 & 4090.479 & 0 \\ 4456.979 & 9.188 & 1530.021 & 1963.521 & 1574.229 & 1985.729 & 3697.729 & 3755.479 & 0 \\ 4456.979 & 9.188 & 1530.021 & 1963.521 & 1574.229 & 1985.729 & 3697.729 & 3755.479 & 0 \\ 4781.979 & 20.021 & 1645.021 & 2093.521 & 1719.229 & 2139.063 & 3986.063 & 4090.479 & 0 \\ 4781.979 & 20.021 & 1645.021 & 2093.521 & 1719.229 & 2139.063 & 3986.063 & 4090.479 & 0 \end{bmatrix}$$

The Database Matrices are:

$$DB11 = \begin{bmatrix}
1 & 0 & 0 & 0 & 0 & 0 & 0 & 0 & 0 \\
0 & 1 & 0 & 0 & 0 & 0 & 0 & 0 & 0 \\
0 & 0 & 1 & 0 & 0 & 0 & 0 & 0 & 0 \\
0 & 0 & 0 & 1 & 0 & 0 & 0 & 0 & 0 \\
0 & 0 & 0 & 0 & 0 & 0 & 0 & 0 & 0 \\
0 & 0 & 0 & 0 & 0 & 0 & 0 & 0 & 0 \\
0 & 0 & 0 & 0 & 0 & 0 & 0 & 0 & 0 \\
0 & 0 & 0 & 0 & 0 & 0 & 0 & 1 & 0 \\
0 & 0 & 0 & 0 & 0 & 0 & 0 & 0 & 0
\end{bmatrix}
\qquad
DB12 = \begin{bmatrix}
0 & 0 & 0 & 0 & 0 & 0 & 0 & 0 & 0 \\
0 & 0 & 0 & 0 & -1 & -1 & 0 & 0 & 0 \\
0 & 0 & 0 & 0 & -1 & 0 & -1 & 0 & 0 \\
0 & 0 & 0 & 0 & 0 & -1 & -1 & 0 & 0 \\
0 & 0 & 0 & 0 & 1 & 0 & 0 & 0 & 0 \\
0 & 0 & 0 & 0 & 0 & 1 & 0 & 0 & 0 \\
0 & 0 & 0 & 0 & 0 & 0 & 1 & 0 & 0 \\
0 & 0 & 0 & 0 & 0 & 0 & 0 & 0 & 0 \\
0 & 0 & 0 & 0 & 0 & 0 & 0 & 0 & 0
\end{bmatrix}$$

$$DB13 = \begin{bmatrix}
1 & 0 & 0 & 0 & 0 & 0 & 0 & 0 & 0 \\
0 & 1 & 0 & 0 & 0 & 0 & 0 & -1 & 0 \\
0 & 0 & 1 & 0 & 0 & 0 & 0 & -1 & 0 \\
0 & 0 & 0 & 1 & 0 & 0 & 0 & -1 & 0 \\
0 & 0 & 0 & 0 & 1 & 0 & 0 & -1 & 0 \\
0 & 0 & 0 & 0 & 0 & 1 & 0 & -1 & 0 \\
0 & 0 & 0 & 0 & 0 & 0 & 1 & -1 & 0 \\
0 & 0 & 0 & 0 & 0 & 0 & 0 & 1 & 0 \\
0 & 0 & 0 & 0 & 0 & 0 & 0 & 0 & 0
\end{bmatrix}
\qquad
DB14 = \begin{bmatrix}
1 & 0 & 0 & 0 & 0 & 0 & 0 & 0 & 1 \\
0 & 1 & 0 & 0 & 0 & 0 & 0 & 0 & -1 \\
0 & 0 & 1 & 0 & 0 & 0 & 0 & 0 & -1 \\
0 & 0 & 0 & 1 & 0 & 0 & 0 & 0 & -1 \\
0 & 0 & 0 & 0 & 1 & 0 & 0 & 0 & -1 \\
0 & 0 & 0 & 0 & 0 & 1 & 0 & 0 & -1 \\
0 & 0 & 0 & 0 & 0 & 0 & 1 & 0 & -1 \\
0 & 0 & 0 & 0 & 0 & 0 & 0 & 1 & -1 \\
0 & 0 & 0 & 0 & 0 & 0 & 0 & 0 & 0
\end{bmatrix}$$

$$\left[\left[\left[\left[\left(\overline{(SS15 \cdot SS15 \cdot N^{-1})}\right) \cdot DBSUM\right] + SST1\right] \cdot DBSSTLR\right] \cdot ((DB11) + DB12)\right] = \begin{bmatrix}
4781.979 & 20.021 & 1645.021 & 2093.521 & 54.188 & 25.521 & 247.521 & 4090.479 & 0 \\
4456.979 & 9.188 & 1530.021 & 1963.521 & 35.021 & 13.021 & 204.188 & 3755.479 & 0 \\
4456.979 & 9.188 & 1530.021 & 1963.521 & 35.021 & 13.021 & 204.188 & 3755.479 & 0 \\
4781.979 & 20.021 & 1645.021 & 2093.521 & 54.188 & 25.521 & 247.521 & 4090.479 & 0 \\
4781.979 & 20.021 & 1645.021 & 2093.521 & 54.188 & 25.521 & 247.521 & 4090.479 & 0
\end{bmatrix}$$

Since this solution won't fit on a single page, we'll complete
it below.

$$\begin{pmatrix}
4781.979 & 20.021 & 1645.021 & 2093.521 & 54.188 & 25.521 & 247.521 & 4090.479 & 0 \\
4781.979 & 20.021 & 1645.021 & 2093.521 & 54.188 & 25.521 & 247.521 & 4090.479 & 0 \\
4781.979 & 20.021 & 1645.021 & 2093.521 & 54.188 & 25.521 & 247.521 & 4090.479 & 0 \\
4781.979 & 20.021 & 1645.021 & 2093.521 & 54.188 & 25.521 & 247.521 & 4090.479 & 0 \\
4781.979 & 20.021 & 1645.021 & 2093.521 & 54.188 & 25.521 & 247.521 & 4090.479 & 0
\end{pmatrix}$$

Computation of the interactive effects of 1st and 2nd, 1st and
3rd and 2nd and 3rd factors (col. 5, 6 and 7)

$$\left[\left[\left[\left[\left[\overrightarrow{\left(SS15 \cdot SS15 \cdot N^{-1}\right)}\right) \cdot DBSUM\right] + SST1\right] \cdot DBSSTLR\right] \cdot ((DB11) + DB12)\right] \cdot DB13\right] = \begin{bmatrix} 4781.979 & 20.021 & 1645.021 & 2093.521 & 54.188 & 25.521 & 247.521 & 4.687 & 0 \\ 4456.979 & 9.188 & 1530.021 & 1963.521 & 35.021 & 13.021 & 204.188 & 0.521 & 0 \\ 4456.979 & 9.188 & 1530.021 & 1963.521 & 35.021 & 13.021 & 204.188 & 0.521 & 0 \\ 4781.979 & 20.021 & 1645.021 & 2093.521 & 54.188 & 25.521 & 247.521 & 4.688 & 0 \\ 4781.979 & 20.021 & 1645.021 & 2093.521 & 54.188 & 25.521 & 247.521 & 4.688 & 0 \end{bmatrix}$$

The complete Matrix looks like,

$$\begin{pmatrix} 4781.979 & 20.021 & 1645.021 & 2093.521 & 54.188 & 25.521 & 247.521 & 4.687 & 0 \\ 4781.979 & 20.021 & 1645.021 & 2093.521 & 54.188 & 25.521 & 247.521 & 4.687 & 0 \\ 4781.979 & 20.021 & 1645.021 & 2093.521 & 54.188 & 25.521 & 247.521 & 4.688 & 0 \\ 4781.979 & 20.021 & 1645.021 & 2093.521 & 54.188 & 25.521 & 247.521 & 4.688 & 0 \\ 4781.979 & 20.021 & 1645.021 & 2093.521 & 54.188 & 25.521 & 247.521 & 4.688 & 0 \end{pmatrix}$$

Computation of the interactive effects of the 1st, 2nd and 3rd factors (col.8) and error term sum of squares (col. 9)

$$\left[\left[\left[\left[\left[\overrightarrow{\left(SS15 \cdot SS15 \cdot N^{-1}\right)}\right) \cdot DBSUM\right] + SST1\right] \cdot DBSSTLR\right] \cdot ((DB11) + DB12)\right] \cdot DB13\right] \cdot DB14 = \begin{bmatrix} 4781.979 & 20.021 & 1645.021 & 2093.521 & 54.188 & 25.521 & 247.521 & 4.687 & 691.5 \\ 4456.979 & 9.188 & 1530.021 & 1963.521 & 35.021 & 13.021 & 204.188 & 0.521 & 701.5 \\ 4456.979 & 9.188 & 1530.021 & 1963.521 & 35.021 & 13.021 & 204.188 & 0.521 & 701.5 \\ 4781.979 & 20.021 & 1645.021 & 2093.521 & 54.188 & 25.521 & 247.521 & 4.688 & 691.5 \\ 4781.979 & 20.021 & 1645.021 & 2093.521 & 54.188 & 25.521 & 247.521 & 4.688 & 691.5 \end{bmatrix}$$

The complete Matrix looks like,

$$\begin{pmatrix} 4781.979 & 20.021 & 1645.021 & 2093.521 & 54.188 & 25.521 & 247.521 & 4.687 & 691.5 \\ 4781.979 & 20.021 & 1645.021 & 2093.521 & 54.188 & 25.521 & 247.521 & 4.687 & 691.5 \\ 4781.979 & 20.021 & 1645.021 & 2093.521 & 54.188 & 25.521 & 247.521 & 4.688 & 691.5 \\ 4781.979 & 20.021 & 1645.021 & 2093.521 & 54.188 & 25.521 & 247.521 & 4.688 & 691.5 \\ 4781.979 & 20.021 & 1645.021 & 2093.521 & 54.188 & 25.521 & 247.521 & 4.688 & 691.5 \end{pmatrix}$$

$$SS := \left[\left[\left[\left[\left[\overrightarrow{\left(SS15 \cdot SS15 \cdot N^{-1}\right)}\right) \cdot DBSUM\right] + SST1\right] \cdot DBSSTLR\right] \cdot ((DB11) + DB12)\right] \cdot DB13\right] \cdot DB14$$

To find the F values, I think this is the easiest way, just make a matrix of identical rows and half-multiply across.

$df1 := (47 \ 1 \ 1 \ 1 \ 1 \ 1 \ 1 \ 1 \ 40)$

$df2 := stack(df1, df1)$

$df3 := stack(df2, df2)$

$df := stack(df3, df1)$

Here we will set up the ERROR Matrix, remember, the error term
might not be the same for all values, check the ninth column
in the SS Matrix.

$$\text{ERROR1} := \left(\frac{40}{692} \quad \frac{40}{692} \quad .058 \quad .058 \quad .058 \quad .058 \quad .058 \quad .058 \quad .058 \right)$$

$$\text{ERROR2} := \left(\frac{40}{701.5} \quad \frac{40}{701.5} \quad 0.057 \quad 0.057 \quad 0.057 \quad 0.057 \quad 0.057 \quad 0.057 \quad 0.057 \right)$$

$$\text{ERROR3} := \left(\frac{40}{701.5} \quad \frac{40}{701.5} \quad 0.057 \quad 0.057 \quad 0.057 \quad 0.057 \quad 0.057 \quad 0.057 \quad 0.057 \right)$$

$$\text{ERROR4} := \left(\frac{40}{692} \quad \frac{40}{692} \quad .058 \quad .058 \quad .058 \quad .058 \quad .058 \quad .058 \quad .058 \right)$$

$$\text{ERROR5} := \left(\frac{40}{692} \quad \frac{40}{692} \quad .058 \quad .058 \quad .058 \quad .058 \quad .058 \quad .058 \quad .058 \right)$$

ERRORA := stack(ERROR1 , ERROR2)
ERRORB := stack(ERRORA , ERROR3)
ERRORC := stack(ERRORB , ERROR4)
ERROR := stack(ERRORC , ERROR5)

$$\text{ERROR} = \begin{bmatrix} 0.058 & 0.058 & 0.058 & 0.058 & 0.058 & 0.058 & 0.058 & 0.058 & 0.058 \\ 0.057 & 0.057 & 0.057 & 0.057 & 0.057 & 0.057 & 0.057 & 0.057 & 0.057 \\ 0.057 & 0.057 & 0.057 & 0.057 & 0.057 & 0.057 & 0.057 & 0.057 & 0.057 \\ 0.058 & 0.058 & 0.058 & 0.058 & 0.058 & 0.058 & 0.058 & 0.058 & 0.058 \\ 0.058 & 0.058 & 0.058 & 0.058 & 0.058 & 0.058 & 0.058 & 0.058 & 0.058 \end{bmatrix}$$

Here we find the degrees of freedom, then find their inverses

$$\text{df} = \begin{bmatrix} 47 & 1 & 1 & 1 & 1 & 1 & 1 & 1 & 40 \\ 47 & 1 & 1 & 1 & 1 & 1 & 1 & 1 & 40 \\ 47 & 1 & 1 & 1 & 1 & 1 & 1 & 1 & 40 \\ 47 & 1 & 1 & 1 & 1 & 1 & 1 & 1 & 40 \\ 47 & 1 & 1 & 1 & 1 & 1 & 1 & 1 & 40 \end{bmatrix}$$

$$\overrightarrow{df}^{-1} = \begin{bmatrix} 0.021 & 1 & 1 & 1 & 1 & 1 & 1 & 1 & 0.025 \\ 0.021 & 1 & 1 & 1 & 1 & 1 & 1 & 1 & 0.025 \\ 0.021 & 1 & 1 & 1 & 1 & 1 & 1 & 1 & 0.025 \\ 0.021 & 1 & 1 & 1 & 1 & 1 & 1 & 1 & 0.025 \\ 0.021 & 1 & 1 & 1 & 1 & 1 & 1 & 1 & 0.025 \end{bmatrix}$$

Here I've already computed ERROR as it's inverse

$$\text{ERROR} = \begin{bmatrix} 0.058 & 0.058 & 0.058 & 0.058 & 0.058 & 0.058 & 0.058 & 0.058 & 0.058 \\ 0.057 & 0.057 & 0.057 & 0.057 & 0.057 & 0.057 & 0.057 & 0.057 & 0.057 \\ 0.057 & 0.057 & 0.057 & 0.057 & 0.057 & 0.057 & 0.057 & 0.057 & 0.057 \\ 0.058 & 0.058 & 0.058 & 0.058 & 0.058 & 0.058 & 0.058 & 0.058 & 0.058 \\ 0.058 & 0.058 & 0.058 & 0.058 & 0.058 & 0.058 & 0.058 & 0.058 & 0.058 \end{bmatrix}$$

It's easier just to multiply one on one to find the F values.

$$F := \overrightarrow{\left(SS \cdot \overrightarrow{df}^{-1} \cdot ERROR \right)}$$

$$F = \begin{bmatrix} 5.881 & 1.157 & 95.411 & 121.424 & 3.143 & 1.48 & 14.356 & 0.272 & 1.003 \\ 5.407 & 0.524 & 87.211 & 111.921 & 1.996 & 0.742 & 11.639 & 0.03 & 1 \\ 5.407 & 0.524 & 87.211 & 111.921 & 1.996 & 0.742 & 11.639 & 0.03 & 1 \\ 5.881 & 1.157 & 95.411 & 121.424 & 3.143 & 1.48 & 14.356 & 0.272 & 1.003 \\ 5.881 & 1.157 & 95.411 & 121.424 & 3.143 & 1.48 & 14.356 & 0.272 & 1.003 \end{bmatrix}$$

We have computed the F values of all five sets of data in one operation.

A ∘ B ∘ C

And finally, lets explore what might happen when we multiply A ∘ B ∘ C.

First of all, A43 ∘ B43 = 3 4x3 Nested Arrays.

$$
\begin{pmatrix} A_{11} & A_{21} & A_{31} \\ A_{12} & A_{22} & A_{32} \\ A_{13} & A_{23} & A_{33} \\ A_{14} & A_{24} & A_{34} \end{pmatrix} \cdot\circ\cdot \begin{pmatrix} B_{11} & B_{21} & B_{31} \\ B_{12} & B_{22} & B_{32} \\ B_{13} & B_{23} & B_{33} \\ B_{14} & B_{24} & B_{34} \end{pmatrix} := \left[\begin{pmatrix} A_{11}B_{11} & A_{11}B_{21} & A_{11}B_{31} \\ A_{12}B_{12} & A_{12}B_{22} & A_{12}B_{32} \\ A_{13}B_{13} & A_{13}B_{23} & A_{13}B_{33} \\ A_{14}B_{14} & A_{14}B_{24} & A_{14}B_{34} \end{pmatrix} \begin{pmatrix} A_{21}B_{11} & A_{21}B_{21} & A_{21}B_{31} \\ A_{22}B_{12} & A_{22}B_{22} & A_{22}B_{32} \\ A_{23}B_{13} & A_{23}B_{23} & A_{23}B_{33} \\ A_{24}B_{14} & A_{24}B_{24} & A_{24}B_{34} \end{pmatrix} \begin{pmatrix} A_{31}B_{11} & A_{31}B_{21} & A_{31}B_{31} \\ A_{32}B_{12} & A_{32}B_{22} & A_{32}B_{32} \\ A_{33}B_{13} & A_{33}B_{23} & A_{33}B_{33} \\ A_{34}B_{14} & A_{34}B_{24} & A_{34}B_{34} \end{pmatrix} \right]
$$

A43 ∘ B43 ∘ C43 = 9 4X3 Nested Arrays. I will transpose the Nesrted Arrays so they will fit on a single page.

$$
\left[\begin{pmatrix} A_{11}B_{11} & A_{11}B_{21} & A_{11}B_{31} \\ A_{12}B_{12} & A_{12}B_{22} & A_{12}B_{32} \\ A_{13}B_{13} & A_{13}B_{23} & A_{13}B_{33} \\ A_{14}B_{14} & A_{14}B_{24} & A_{14}B_{34} \end{pmatrix} \begin{pmatrix} A_{21}B_{11} & A_{21}B_{21} & A_{21}B_{31} \\ A_{22}B_{12} & A_{22}B_{22} & A_{22}B_{32} \\ A_{23}B_{13} & A_{23}B_{23} & A_{23}B_{33} \\ A_{24}B_{14} & A_{24}B_{24} & A_{24}B_{34} \end{pmatrix} \begin{pmatrix} A_{31}B_{11} & A_{31}B_{21} & A_{31}B_{31} \\ A_{32}B_{12} & A_{32}B_{22} & A_{32}B_{32} \\ A_{33}B_{13} & A_{33}B_{23} & A_{33}B_{33} \\ A_{34}B_{14} & A_{34}B_{24} & A_{34}B_{34} \end{pmatrix} \right] \cdot\circ\cdot \begin{pmatrix} C_{11} & C_{21} & C_{31} \\ C_{12} & C_{22} & C_{32} \\ C_{13} & C_{23} & C_{33} \\ C_{14} & C_{24} & C_{34} \end{pmatrix} =
$$

$$
\left[\begin{pmatrix} A_{11}B_{11}C_{11} & A_{11}B_{21}C_{11} & A_{11}B_{31}C_{11} \\ A_{12}B_{12}C_{12} & A_{12}B_{22}C_{12} & A_{12}B_{32}C_{12} \\ A_{13}B_{13}C_{13} & A_{13}B_{23}C_{13} & A_{13}B_{33}C_{13} \\ A_{14}B_{14}C_{14} & A_{14}B_{24}C_{14} & A_{14}B_{34}C_{14} \end{pmatrix} \begin{pmatrix} A_{21}B_{11}C_{11} & A_{21}B_{21}C_{11} & A_{21}B_{31}C_{11} \\ A_{22}B_{12}C_{12} & A_{22}B_{22}C_{12} & A_{22}B_{32}C_{12} \\ A_{23}B_{13}C_{13} & A_{23}B_{23}C_{13} & A_{23}B_{33}C_{13} \\ A_{24}B_{14}C_{14} & A_{24}B_{24}C_{14} & A_{24}B_{34}C_{14} \end{pmatrix} \begin{pmatrix} A_{31}B_{11}C_{11} & A_{31}B_{21}C_{11} & A_{31}B_{31}C_{11} \\ A_{32}B_{12}C_{12} & A_{32}B_{22}C_{12} & A_{32}B_{32}C_{12} \\ A_{33}B_{13}C_{13} & A_{33}B_{23}C_{13} & A_{33}B_{33}C_{13} \\ A_{34}B_{14}C_{14} & A_{34}B_{24}C_{14} & A_{34}B_{34}C_{14} \end{pmatrix} \right.
$$

$$
\begin{pmatrix} A_{21}B_{11}C_{21} & A_{21}B_{21}C_{21} & A_{21}B_{31}C_{21} \\ A_{22}B_{12}C_{22} & A_{22}B_{22}C_{22} & A_{22}B_{32}C_{22} \\ A_{23}B_{13}C_{23} & A_{23}B_{23}C_{23} & A_{23}B_{33}C_{23} \\ A_{24}B_{14}C_{24} & A_{24}B_{24}C_{24} & A_{24}B_{34}C_{24} \end{pmatrix} \begin{pmatrix} A_{21}B_{11}C_{21} & A_{21}B_{21}C_{21} & A_{21}B_{31}C_{21} \\ A_{22}B_{12}C_{22} & A_{22}B_{22}C_{22} & A_{22}B_{32}C_{22} \\ A_{23}B_{13}C_{23} & A_{23}B_{23}C_{23} & A_{23}B_{33}C_{23} \\ A_{24}B_{14}C_{24} & A_{24}B_{24}C_{24} & A_{24}B_{34}C_{24} \end{pmatrix} \begin{pmatrix} A_{31}B_{11}C_{21} & A_{31}B_{21}C_{21} & A_{31}B_{31}C_{21} \\ A_{32}B_{12}C_{22} & A_{32}B_{22}C_{22} & A_{32}B_{32}C_{22} \\ A_{33}B_{13}C_{23} & A_{33}B_{23}C_{23} & A_{33}B_{33}C_{23} \\ A_{34}B_{14}C_{24} & A_{34}B_{24}C_{24} & A_{34}B_{34}C_{24} \end{pmatrix}
$$

$$
\left. \begin{pmatrix} A_{21}B_{11}C_{31} & A_{21}B_{21}C_{31} & A_{21}B_{31}C_{31} \\ A_{22}B_{12}C_{32} & A_{22}B_{22}C_{32} & A_{22}B_{32}C_{32} \\ A_{23}B_{13}C_{33} & A_{23}B_{23}C_{33} & A_{23}B_{33}C_{33} \\ A_{24}B_{14}C_{34} & A_{24}B_{24}C_{34} & A_{24}B_{34}C_{34} \end{pmatrix} \begin{pmatrix} A_{21}B_{11}C_{31} & A_{21}B_{21}C_{31} & A_{21}B_{31}C_{31} \\ A_{22}B_{12}C_{32} & A_{22}B_{22}C_{32} & A_{22}B_{32}C_{32} \\ A_{23}B_{13}C_{33} & A_{23}B_{23}C_{33} & A_{23}B_{33}C_{33} \\ A_{24}B_{14}C_{34} & A_{24}B_{24}C_{34} & A_{24}B_{34}C_{34} \end{pmatrix} \begin{pmatrix} A_{31}B_{11}C_{31} & A_{31}B_{21}C_{31} & A_{31}B_{31}C_{31} \\ A_{32}B_{12}C_{32} & A_{32}B_{22}C_{32} & A_{32}B_{32}C_{32} \\ A_{33}B_{13}C_{33} & A_{33}B_{23}C_{33} & A_{33}B_{33}C_{33} \\ A_{34}B_{14}C_{34} & A_{34}B_{24}C_{34} & A_{34}B_{34}C_{34} \end{pmatrix} \right]
$$

Lets try a mind experiment multiplying A43 ∘ B43 ∘ C43 ∘ D43. The solution could be Nested Arrays stacked on top of each other like a package of crackers in 3 dimensions. The solution to A ∘ B ∘ C ∘ D would be A 3 X 3 X 3 Nested Array (the above NA but stacked 3 deep including the D term).

With such a stacked NA, it could be possible to:

Inscribe a sphere, tilt it 23 1/2 degrees, give it a virtual spin and use the spherical matrix to help solve ballistic problems of objects travelling across different lines of latitude and longitude at angles.

Inscribe s, p, d and f orbitals of atoms.

Inscribe complete 3-D 'pictures' of molecules, instead of flat pictures as in the section of Quantum Chemistry in this book.

Make possible 3-D television.

Now let's try multiplying A o B o C o D o E. We would stack over the 3-D Matrix creating a tesseract in 4-D. Suppose we multiply A o B o C o D o . . . o U. This is 20 dimensions. Could we inscribe a superstring within this Nested Array? This is for better Mathematicians to ponder and verify or deny.

NOTE: When we multiply Nested Arrays, it is necessary (at this time) to partition the Matrix to allow the computations. It is also possible to take a single Matrix, not a Nested Array, and treat it as if it's elements were Nested Arrays. In other words, change a regular Matrix into a Nested Array.

And finally, a bit of humor. This last piece was written while taking a class in Applied Mathematics under Dr. Charles Votaw at Fort Hays State University. Hint: Rite is mis-spelled as ryte, Knott is trying to get Wright to change the spelling.

Y-NAUGHT RYTE

Knott: Write ryte rite, 'right Wright?

Wright: Right! . . . . Write ryte right?

Knott: Right, Wright, write rite right!

Wright: Write ryte right?

Knott: Right!

Wright: Ryte's right Knott, right?

Knott: Not right Wright, rite not ryte!

Wright: Ryte's not right Knott, right?

Knott: Wright, why write rite ryte? Ryte's not right Wright, rite's right. Why not write ryte right, 'right?

Wright: Why, Knott?

Knott: Why not?

Wright: Write Y-Naught ryte right? Y-Naught's right, right Knott?

Knott: Y-Naught rite's right Wright. Wright . . . why write Y-Naught rite, why not write Yo rite. (Yo pronounced Y-Naught).

Wright: o's not naught, Knott . . . . 's'not???? . . . . Yo not Y-Naught, Knott! Why not, Knott!

Knott: Why why not write Yo rite, not write Yo ryte? R-I-T-E, Wright, not R-Y-T-E! Write rite not ryte!!!!

Wright: R-I-T-E? Rite?! RIGHT! Yo rite not Yo ryte, Knott! Write ryte rite not write rite ryte, Knott!

Knott: Right, Wright, write ryte rite! Why not write Y-naught
         ryte Yo rite? Why not, Wright? Why not write rite
         right, 'right?

Wright: Right!

Yo RITE